Frustrated Spin Systems

3rd Edition

Frustrated Spin Systems

3rd Edition

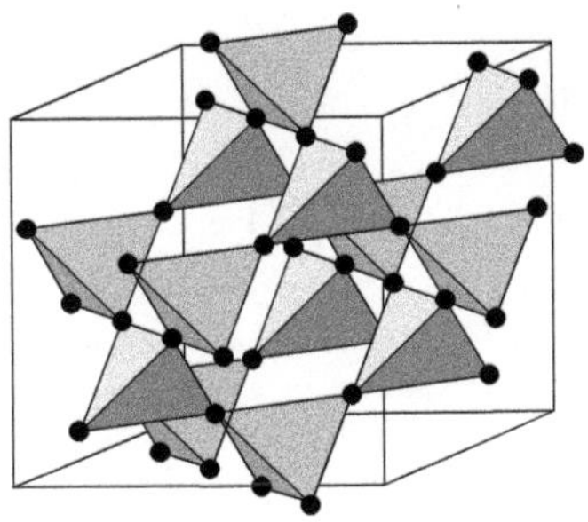

H T Diep

University of Cergy-Pontoise, France

Editor

World Scientific

NEW JERSEY • LONDON • SINGAPORE • BEIJING • SHANGHAI • HONG KONG • TAIPEI • CHENNAI • TOKYO

Published by

World Scientific Publishing Co. Pte. Ltd.
5 Toh Tuck Link, Singapore 596224
USA office: 27 Warren Street, Suite 401-402, Hackensack, NJ 07601
UK office: 57 Shelton Street, Covent Garden, London WC2H 9HE

British Library Cataloguing-in-Publication Data
A catalogue record for this book is available from the British Library.

FRUSTRATED SPIN SYSTEMS
Third Edition

ISBN 978-981-121-413-4 (hardcover)
ISBN 978-981-121-414-1 (ebook for institutions)
ISBN 978-981-121-415-8 (ebook for individuals)

For any available supplementary material, please visit
https://www.worldscientific.com/worldscibooks/10.1142/11660#t=suppl

Desk Editor: Nur Syarfeena Binte Mohd Fauzi

Typeset by Stallion Press
Email: enquiries@stallionpress.com

PREFACE OF THE FIRST EDITION

This book covers the recent theoretical, numerical and experimental developments in the field of frustrated spin systems. The first edition of the book appeared in 1994 under the title "Magnetic Systems with Competing Interactions" (World Scientific) where most of the works until that year have been reviewed.

This book contains nine chapters, of which seven are new. Two chapters of the first edition have been revised with new references and comments added: these are Chapter 1, Frustration — Exactly Solved Frustrated Models, and Chapter 2, Properties and Phase Transition in Frustrated Ising Spin Systems.

Frustrated spin systems were first investigated five decades ago. Well-known examples include the Ising model on the antiferromagnetic triangular lattice studied by G. H. Wannier in 1950 and the Heisenberg helical structure discovered independently by A. Yoshimori, J. Villain and T. A. Kaplan in 1959. However, extensive investigations on frustrated spin systems have undoubtedly started with the concept of frustration introduced at the same time by G. Toulouse and J. Villain in 1977 in the context of spin glasses. The frustration is generated by the competition of different kinds of interactions and/or by lattice geometry. As a result, in the ground state, all bonds are not fully satisfied. In frustrated Ising spin systems, a number of spins behave as free spins. In frustrated vector spin systems, the ground-state configuration is usually non-collinear. The ground state of frustrated spin systems is therefore highly degenerate, and new induced symmetries give rise to spectacular and often unexpected behaviors at finite temperatures.

Many properties of frustrated systems are still not well understood at present. Recent studies shown in this book reveal that established theories, numerical simulations as well as experimental techniques have encountered many difficulties in dealing with frustrated systems. In some sense, frustrated systems provide an excellent testing ground for approximations and theories.

The chapters of this book are written by researchers who have actively contributed to the field. Many results are from recent works of the authors.

The book is intended for post-graduate students as well as researchers in statistical physics, magnetism, materials science and various domains where real systems can be described with the spin language. Explicit demonstrations of formulae and full arguments leading to important results are given where possible.

The book is organized as follows. The first two chapters deal with properties and phase transition in frustrated Ising spin systems. The two following chapters deal with the nature of the phase transition in frustrated vector spin systems. Chapters 5 and 6 treat low-dimensional frustrated quantum spin models. Chapter 7 studies the spin ice, and Chapter 8 shows recent experimental results on pyrochlores which are real frustrated materials. Chapter 9 deals with the classical and quantum spin glasses where both disorder and frustration act simultaneously.

I summarize in the following the contents of each chapter.

Chapter 1 shows the frustration effects in exactly solved two-dimensional Ising models. The systems considered in this chapter are periodically defined (without bond disorder). The frustration due to competing interactions will itself induce disorder in the spin orientations. After a detailed presentation of 16- and 32-vertex models, applications are made to some selected systems which possess most of the spectacular features due to the frustration such as high ground-state degeneracy, reentrance, successive phase transitions and disorder solutions. In some simple models, up to five transitions separated by two reentrant paramagnetic phases are found. A conjecture is made on the origin of the paramagnetic reentrant phase. The nature of ordering as well as the relation between the considered systems and the random-field Ising model are discussed. The relevance of disorder solutions for the reentrance phenomena is also pointed out. Evidence of the existence of partial disorder and of reentrance in complicated, non-exactly solved systems is shown and discussed.

Chapter 2 deals mainly with the Ising model on the antiferromagnetic triangular and stacked triangular lattices. Ground-state properties and the nature of the phase transition are studied by various methods, as functions of the spin magnitude S and nearest- and next-nearest-neighbor interactions. It is shown in this chapter that the symmetry of spin ordering strongly depends on S. Furthermore, due to the frustration, there exist "free" spins or "free" linear chains, on which internal fields are cancelled out. These free spins and free linear chains play an important role as for spin orderings. Another characteristic feature of frustrated Ising spin systems is the existence of various metastable states, which is closely related to the degeneracy of ground state and also the excited states. These metastable states may give rise to a first-order phase transition as found in some models. The effects of the far-neighbor interactions on the antiferromagnetic triangular and stacked triangular lattices are clarified.

Chapter 3 is devoted to the recent advances in the renormalization group (RG) approaches to the physics of frustrated classical vector spin systems in three dimensions. The main features of the field theoretical approach to these systems, including considerations on symmetries, symmetry breaking schemes, continuum limits, topological contents, etc., are recalled here. An overview is provided on the phenomenological situation with emphasis laid on the most striking aspect of the physics of frustrated magnets: the existence of *non-universal scaling* behaviors. A review is then provided on the various perturbative and non-perturbative RG approaches that have been used to investigate frustrated magnets. Finally, a large part of this chapter is devoted to a recent non-perturbative approach that has clarified the intricate physical situation of frustrated magnets.

Chapter 4 is devoted to a review on recent numerical studies dealing with frustrated vector spin systems between two and four dimensions. It is shown that various breakdowns of symmetry can occur, contrary to the case of ferromagnetic systems. The author shows that in three dimensions the transition is always of first order in the thermodynamic limit. However, for "small" sizes in numerical simulations or for temperatures not "too close" to the transition temperatures in experiments, the system could display an "almost universality class" for an $O(N)/O(N-2)$ breakdown of symmetry. Many compounds studied experimentally are in this class. In two dimensions, the situation is much less clear. Indeed, the topological defects can play a fundamental role, and their couplings with a discrete symmetry (Ising or Potts models) are not well known. In contrast to the two-dimensional case, our understanding of the three-dimensional case has increased considerably in the last decade.

Chapter 5 is devoted to a review on some theoretical advances in the field of quantum magnetism in two-dimensional systems. It is known that the spin-1/2 nearest-neighbor 2D Heisenberg models on Bravais lattices (square, triangular) are Néel ordered at $T = 0$. Frustration, small coordination number and competition between interactions can lead to specific quantum phases without magnetic long-ranged order. This long-standing subject is revived by the discovery of high-T_c superconductivity in the doped cuprates and fuelled by numerous experimental studies of 2D antiferromagnetic insulators. The authors of this chapter show results on several models. They also show general properties of valence bond crystals (VBCs) and related states, as well as large-N generalizations of the Heisenberg model. Some results of quantum dimer models (QDM) are presented. They provide useful insights on the phenomenology of VBC and other systems. The authors also review some results concerning models with multiple-spin exchange and the Heisenberg model on the Kagomé lattice (and related models). Despite an important activity on this subject, the understanding of the low-energy physics of the spin-1/2 Kagomé antiferromagnet remains a challenging problem.

Chapter 6 is devoted to an overview of some of the zero-temperature quantum spin liquid phases with unbroken SU(2) spin symmetry that have been found in one dimension. The main characteristics of these phases are discussed by means of the bosonization approach. A special emphasis is laid on the interplay between frustration and quantum fluctuations in one dimension. The author presents the different spin liquid phases that occur in spin chains and spin ladders. The main effects of frustration in one-dimensional spin liquids are described. In particular, it is observed that frustration plays its trick by allowing deconfined spinons (carrying fractional $S = 1/2$ quantum number) as elementary excitations and it provides a non-trivial source of incommensurability.

Chapter 7 is devoted to the theoretical and experimental study of the so-called "spin ice", the magnetic equivalent of Pauling's model of hydrogen disorder in water ice. This represents the prototypical frustrated system, with a macroscopically degenerate ground state and extensive zero-point entropy. Pauling's concept was extended by Wannier, Anderson and others to include magnetic systems and was later developed to include statistical "vertex models", some of which are exactly soluble (see Chapter 1). Spin Ice is a 16-vertex model of "ferromagnetic frustration" that, remarkably, is found to apply to certain rare earth oxide materials. The search for a detailed microscopic understanding of these spin ice materials has cast light on many aspects of magnetic frustration and uncovered several new features of magnetic interactions (particularly the dipole interaction). For these reasons, spin ice represents an ideal laboratory in which to develop our understanding of frustrated spin systems. This chapter is a comprehensive review of the physics of spin ice including both theoretical and experimental aspects. The authors start with the concept of spin ice and its relation to the historic problem of water ice and to other frustrated systems. The following sections review the current understanding of the zero-field spin ice state, the numerous field-induced states and the magnetic dynamics of the spin ice materials. Some materials related to spin ice are briefly described.

Chapter 8 describes experimental results on geometrically frustrated magnetic systems. Much recent experimental progress has been made in the study of magnetic materials made up of antiferromagnetically coupled magnetic moments residing on networks of corner-sharing tetrahedra. They are found in nature in a variety of cubic pyrochlore, spinel and Laves phase materials, with the magnetic moments arising from either rare earth or transition metal electrons. This review focusses on experimental progress in this area from the last 10 years primarily due to neutron scattering studies. Even within this subset of three frustrated materials, quite different exotic ground states are found: the enigmatic cooperative paramagnetic ground state of $Tb_2Ti_2O_7$, the spin glass ground state of $Y_2Mo_2O_7$ and the composite spin degrees of freedom and distorted pyrochlore lattice in $ZnCr_2O_4$. It is very rewarding to appreciate the richness of the complex ground states that geometrical frustration has enabled in these real materials. There seems little doubt that this richness will continue as a theme in the elucidation of the physical properties of geometrically frustrated magnets in the near future.

Chapter 9 is devoted to a review on recent progress in spin glasses. Both the equilibrium properties and the dynamic properties are covered. The authors focus on progress in theoretical, in particular numerical, studies, while its relationship to real magnetic materials is also mentioned. A brief overview of two well-known paradigms on spin glasses is given, and a summary of the predictions derived from them is presented. Then, equilibrium properties of Ising spin glass models are shown and discussed. In particular, a recent active debate, as to which paradigm is appropriate for realistic short-range spin glass models in three dimensions, is presented. Then, the dynamical properties are examined using these paradigms on a different ground, with an emphasis on aging phenomena. Some other non-equilibrium properties are also discussed. Models with continuous degrees of freedom as well as the Potts spin glass models are mentioned. The effects of weak disorder are discussed and compared to those of strong disorder. Several exact relations can be found and play an important role in shedding light on the issue under debate. Finally, results from the interplay between quantum fluctuations and randomness in spin glasses are also reviewed.

Since several problems treated in this book are currently being investigated, I alert the reader that the authors of each chapter have taken the liberty to express their viewpoint on each unsettled issue.

H. T. Diep
University of Cergy-Pontoise
Cergy-Pontoise, France
2004

PREFACE OF THE SECOND EDITION

In the first edition, the chapters have been written in a pedagogical manner: an introduction and methods used to investigate subjects treated in each chapter have been presented in detail to facilitate the reader to enter the field before showing relevant research results published by 2004 in the literature. In this second edition, those parts of background remain unchanged. New important results since 2005 have been added and commented where this was necessary in this edition. The importance of the subject treated in each chapter, which was described in the preface of the first edition, remains true at the present time. Thus, I just outline hereafter changes which have been made in the second edition.

Chapter 1 was devoted to an introduction to frustrated spin systems and exactly solved frustrated spin models. In this field, there are no other models which have been solved since the first edition of this book. This chapter therefore remains unchanged.

Chapter 2 concerned other frustrated two- and three-dimensional Ising systems. This chapter was not updated because no significant progress has been noticed.

Chapter 3 was devoted to a review on theoretical investigations, mainly by renormalization group. A new section has been added to present and comment new interesting results which have been published since the first edition of this book. New figures and references have been added.

Chapter 4 was devoted to numerical studies of frustrated spin systems. A new section has been added to show new works which allowed to settle a long-lasting controversial subject on the phase transition in the stacked triangular antiferromagnet and in some other frustrated systems. New references have been added.

Chapter 5 treating the problem of 2D frustrated quantum magnets has been updated. The authors have rewritten Sections 5.7 and 5.8, taking into account new results since 2004.

Chapter 6 dealing with 1D quantum magnets has also been updated with an added section signaling recent breakthroughs on the subject.

Subjects on spin ice treated in Chapter 7 have greatly evolved during these 10 years. The authors have added at the end of their chapter a section describing the current tendency and recommended some recent reviews on the new subjects.

Chapters 8 and 9 remain unchanged. The materials provided in these chapters are always of fundamental interest.

The contents of the book cover several important aspects of frustrated spin systems. This second edition allows the reader to keep himself informed on the rapid development of this field during the last 20 years.

H. T. Diep
University of Cergy-Pontoise
Cergy-Pontoise, France
2013

PREFACE OF THE THIRD EDITION

In the previous editions, each chapter has been written in a pedagogical manner with an introduction and a presentation of basic methods to help the reader enter the field with a self-contained background.

In this third edition, those parts of background remain unchanged. New important results since the second edition have been added and commented.

The importance of the subject treated in each chapter, which was described in the preface of the first edition in 2004, remains relevant at the present time. I invite the reader to read that first-edition preface. Thus, I just outline hereafter changes which have been made in the third edition.

Chapter 1 was devoted to an introduction to frustrated spin systems and exactly solved frustrated spin models. In this very specialized field, there are only a small number of models which have been exactly solved since the first edition of this book. These works are mentioned in this third edition.

Chapter 2 concerned other frustrated 2D and 3D Ising systems. This chapter has been throughout reviewed and updated with modifications and new references.

Chapter 3 was devoted to a review on theoretical investigations, mainly by renormalization group. A new section has been added to present and comment new interesting results which have been published since the first edition of this book. New figures and references have been added. Results presented in this chapter are still timely. No significant new results are found since the second edition.

Chapter 4 was devoted to numerical studies of frustrated spin systems. A new section has been added in the second edition to show new works which allowed to settle a long-lasting controversial subject on the phase transition in the stacked triangular antiferromagnet and in some other frustrated systems. New references have been added. No other cases have been found since the second edition.

Chapter 5 treating the problem of 2D frustrated quantum magnets has been updated. The authors have rewritten for the second edition Sections 7 and 8, taking into account new results since 2004 up to 2013.

Chapter 6 dealing with 1D quantum magnets has also been updated in the second edition with an added section signaling recent breakthroughs on the subject. This chapter is still timely.

Subjects on spin ice treated in Chapter 7 have greatly evolved. In the second edition, the authors have added at the end of their chapter a section describing the current tendency and recommended some recent reviews on the new subjects. This chapter remains unchanged in the third edition.

Chapters 8 and 9 remain unchanged. The materials provided in these chapters are always of fundamental interest.

In the third edition, a new chapter (Chapter 10) has been added. This chapter treats the frustration in magnetic thin films, a subject with a rapid development due to new applications in nanotechnology and spintronics. The surface effect combined with the frustration gives rise to interesting phenomena such as surface transition and surface ordering, which are distinct from the bulk ones. In this chapter, we present the simple Green's function method, which allows to study spin systems with non-collinear spin configurations due the the frustration. Some concrete cases such as helimagnetic thin films are shown in detail. Monte Carlo results are also presented. The Dzyaloshinskii–Moriya interaction is next introduced in thin films. Spin waves are calculated in zero field. Under an applied magnetic field, the system shows the formation of a skyrmion crystal in a region of the phase space. The stability and the phase transition of the skyrmion crystal are shown. The last part of this chapter is devoted to the effect of elasticity on the skyrmions. The effect of uniaxial strains on the skyrmion structure is shown.

The contents of the book cover several important aspects of frustrated spin systems. This third edition allows the reader to keep himself informed on the progress of this field during the last 20 years.

H. T. Diep
University of Cergy-Pontoise
Cergy-Pontoise, France
2019

CONTENTS

CHAPTER 1

FRUSTRATION — EXACTLY SOLVED FRUSTRATED MODELS

H. T. Diep*,‡ and H. Giacomini†,§

**Laboratoire de Physique Théorique et Modélisation*
Université de Cergy-Pontoise, CNRS UMR 8089
2, Avenue Alphonse Chauvin, 95302 Cergy-Pontoise Cedex, France
†Laboratoire de Mathématiques et Physique Théorique,
Université de Tours, CNRS UMR 6083
Parc de Grandmont, 37200 Tours, France
‡diep@u-cergy.fr
§giacomini@univ-tours.fr

After a short introduction on frustrated spin systems, we study in this chapter several 2D frustrated Ising spin systems that can be exactly solved by using vertex models. We show that these systems contain most of the spectacular effects due to the frustration: high ground-state (GS) degeneracy, existence of several phases in the GS phase diagram, multiple phase transitions with increasing temperature, reentrance, disorder lines, and partial disorder at equilibrium. Evidences of such effects in non-solvable models are also shown and discussed.

1.1. Frustration: An Introduction

The study of order–disorder phenomena is a fundamental task of equilibrium statistical mechanics. Great efforts have been made to understand the basic mechanisms responsible for spontaneous ordering as well as the nature of the phase transition in many kinds of systems. In particular, during the last 25 years, much attention has been paid to frustrated models.[1] The word "frustration" has been introduced[2,3] to describe the situation where a spin (or a number of spins) in the system cannot find an orientation to *fully* satisfy all the interactions with its neighboring spins (see below). This definition can be applied to Ising spins, Potts models, and vector spins. In general, frustration is caused either by competing interactions (such as the Villain model[3]) or by lattice structure as in the triangular, face-centered cubic (fcc), and hexagonal-close-packed (hcp) lattices, with antiferromagnetic nearest-neighbor (NN) interaction. The effects of frustration are rich and often unexpected. Many of them are not understood yet at present (see the other chapters of this book).

In addition to the fact that real magnetic materials are often frustrated due to several kinds of interactions (see Chapter 8), frustrated spin systems have their own interest in statistical mechanics. Recent studies show that many established statistical methods and theories have encountered many difficulties in dealing with frustrated systems. In some sense, frustrated systems are excellent candidates to test approximations and improve theories. Since the mechanisms of many phenomena are not understood in real systems (disordered systems, systems with long-range interaction, 3D systems, etc.), it is worth searching for the origins of those phenomena in exactly solved systems. These exact results will help to qualitatively understand the behavior of real systems which are in general much more complicated.

1.1.1. *Definition*

Let us give here some basic definitions to help readers unfamiliar with these subjects to read the remaining chapters of this book.

Consider two spins $\mathbf{S}_i$ and $\mathbf{S}_j$ with an interaction J. The interaction energy is $E = -J\,(\mathbf{S}_i \cdot \mathbf{S}_j)$. If J is positive (ferromagnetic interaction) then the minimum of E is $-J$ corresponding to the configuration in which $\mathbf{S}_i$ is parallel to $\mathbf{S}_j$. If J is negative (antiferromagnetic interaction), the minimum of E corresponds to the configuration where $\mathbf{S}_i$ is antiparallel to $\mathbf{S}_j$. It is easy to see that in a spin system with NN ferromagnetic interaction, the ground state (GS) of the system corresponds to the spin configuration where all spins are parallel: the interaction of every pair of spins is fully satisfied. This is true for any lattice structure. If J is antiferromagnetic, the spin configuration of the GS depends on the lattice structure: (i) For lattices containing no elementary triangles, i.e. bipartite lattices (such as square lattice, simple cubic lattices), the GS is the configuration in which each spin is antiparallel to its neighbors, i.e. every bond is fully satisfied. (ii) For lattices containing elementary triangles such as the triangular lattice,

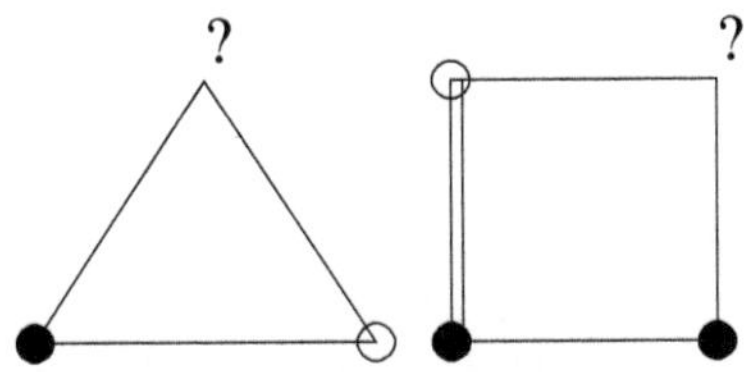

Fig. 1.1. Examples of frustrated plaquettes: ferro- and antiferromagnetic interactions, J and $-J$, are shown by single and double lines, $\uparrow$ and $\downarrow$ Ising spins by filled and void circles, respectively. Choosing any orientation for the spin marked by the question mark will leave one of its bonds unsatisfied (frustrated bond).

the fcc lattice, and the hcp lattice, one cannot construct a GS where all bonds are fully satisfied (see Fig. 1.1). The GS does not correspond to the minimum of the interaction of every spin pair. In this case, one says that the system is frustrated.

We consider another situation where the spin system can be frustrated: this is the case with different kinds of conflicting interactions and the GS does not correspond to the minimum of each kind of interaction. For example, consider a chain of spins where the NN interaction J_1 is ferromagnetic while the next NN (NNN) interaction J_2 is antiferromagnetic. As long as $|J_2| \ll J_1$, the GS is ferromagnetic: every NN bond is then satisfied but the NNN ones are not. Of course, when $|J_2|$ exceeds a critical value, the ferromagnetic GS is no longer valid (see example below): both the NN and NNN bonds are not fully satisfied.

In a general manner, we can say that a spin system is frustrated when one cannot find a configuration of spins to fully satisfy the interaction (bond) between every pair of spins. In other words, the minimum of the total energy does not correspond to the minimum of each bond. This situation arises when there is a competition between different kinds of interactions acting on a spin by its neighbors or when the lattice geometry does not allow to satisfy all the bonds simultaneously. With this definition, the chain with NN ferromagnetic and NNN antiferromagnetic interactions discussed above is frustrated even in the case where the ferromagnetic spin configuration is its GS ($|J_2| \ll J_1$).

The first frustrated system which was studied in 1950 is the triangular lattice with Ising spins interacting with each other via an NN antiferromagnetic interaction.[4] For vector spins, non-collinear spin configurations due to competing interactions were first discovered in 1959 independently by Yoshimori,[5] Villain,[6] and Kaplan.[7]

Consider an elementary cell of the lattice. This cell is a polygon formed by faces hereafter called "plaquettes". For example, the elementary cell of the simple cubic lattice is a cube with six square plaquettes, the elementary cell of the fcc lattice is a tetrahedron formed by four triangular plaquettes. Let $J_{i,j}$ be the interaction between two NN spins of the plaquette. According to the definition of Toulouse,[2] the plaquette is frustrated if the parameter P defined below is negative:

$$P = \prod_{\langle i,j \rangle} \text{sign}(J_{i,j}), \tag{1.1}$$

where the product is performed over all $J_{i,j}$ around the plaquette. Two examples of frustrated plaquettes are shown in Fig. 1.1: a triangle with three antiferromagnetic bonds and a square with three ferromagnetic bonds and one antiferromagnetic bond. P is negative in both cases. One sees that if one tries to put Ising spins on those plaquettes, at least one of the bonds around the plaquette will not be satisfied. For vector spins, we show below that in the lowest energy state, each bond is only partially satisfied.

One sees that for the triangular plaquette, the degeneracy is three, and for the square plaquette it is four, in addition to the degeneracy associated with returning all spins. Therefore, the degeneracy of an infinite lattice composed of such plaquettes is infinite, in contrast to the unfrustrated case.

At this stage, we note that although in the above discussion we have taken the interaction between two spins to be of the form $E = -J\,(\mathbf{S}_i \cdot \mathbf{S}_j)$, the concept of frustration can be applied to other types of interactions such as the Dzyaloshinski–Moriya interaction $E = -J\,|(\mathbf{S}_i \wedge \mathbf{S}_j)|$: a spin system is frustrated whenever the minimum of the system energy does not correspond to the minimum of all local interactions, whatever the form of interaction. We note however that this definition of frustration is more general than the one using Eq. (1.1).

In this chapter, we will determine the GS of various frustrated Ising spin systems and discuss their properties as well. In Section 1.1.2, we analyze the GS of XY and Heisenberg spins.

1.1.2. *Non-collinear spin configurations*

Let us return to the plaquettes shown in Fig. 1.1. In the case of XY spins, one can calculate the GS configuration by minimizing the energy of the plaquette E while keeping the spin modulus constant. In the case of the triangular plaquette, suppose that spin $\mathbf{S}_i$ $(i = 1, 2, 3)$ of amplitude S makes an angle θ_i with the **Ox** axis. Writing E and minimizing it with respect to the angles θ_i, one has

$$\begin{aligned} E &= J(\mathbf{S}_1 \cdot \mathbf{S}_2 + \mathbf{S}_2 \cdot \mathbf{S}_3 + \mathbf{S}_3 \cdot \mathbf{S}_1) \\ &= JS^2 \left[\cos(\theta_1 - \theta_2) + \cos(\theta_2 - \theta_3) + \cos(\theta_3 - \theta_1)\right], \\ \frac{\partial E}{\partial \theta_1} &= -JS^2 \left[\sin(\theta_1 - \theta_2) - \sin(\theta_3 - \theta_1)\right] = 0, \\ \frac{\partial E}{\partial \theta_2} &= -JS^2 \left[\sin(\theta_2 - \theta_3) - \sin(\theta_1 - \theta_2)\right] = 0, \\ \frac{\partial E}{\partial \theta_3} &= -JS^2 \left[\sin(\theta_3 - \theta_1) - \sin(\theta_2 - \theta_3)\right] = 0. \end{aligned}$$

A solution of the last three equations is $\theta_1 - \theta_2 = \theta_2 - \theta_3 = \theta_3 - \theta_1 = 2\pi/3$. One can also write

$$E = J(\mathbf{S}_1 \cdot \mathbf{S}_2 + \mathbf{S}_2 \cdot \mathbf{S}_3 + \mathbf{S}_3 \cdot \mathbf{S}_1) = -\frac{3}{2}JS^2 + \frac{J}{2}(\mathbf{S}_1 + \mathbf{S}_2 + \mathbf{S}_3)^2.$$

The minimum here evidently corresponds to $\mathbf{S}_1 + \mathbf{S}_2 + \mathbf{S}_3 = 0$, which yields the 120° structure. This is true for Heisenberg spins also.

We can do the same calculation for the case of the frustrated square plaquette. Suppose that the antiferromagnetic bond connects the spins $\mathbf{S}_1$ and $\mathbf{S}_2$. We find

$$\theta_2 - \theta_1 = \theta_3 - \theta_2 = \theta_4 - \theta_3 = \frac{\pi}{4} \quad \text{and} \quad \theta_1 - \theta_4 = \frac{3\pi}{4} \tag{1.2}$$

If the antiferromagnetic bond is equal to $-\eta J$, the solution for the angles is[8]

$$\cos\theta_{32} = \cos\theta_{43} = \cos\theta_{14} \equiv \theta = \frac{1}{2}\left[\frac{\eta + 1}{\eta}\right]^{1/2} \tag{1.3}$$

and $|\theta_{21}| = 3|\theta|$, where $\cos\theta_{ij} \equiv \cos\theta_i - \cos\theta_j$.

This solution exists if $|\cos\theta| \leq 1$, namely $\eta > \eta_c = 1/3$. One can check that when $\eta = 1$, one has $\theta = \pi/4$, $\theta_{21} = 3\pi/4$.

We show the frustrated triangular and square lattices in Fig. 1.2 with XY spins ($N = 2$).

One observes that there is a two-fold degeneracy resulting from the symmetry by mirror reflecting with respect to an axis, for example the y axis in Fig. 1.2. Therefore, the symmetry of these plaquettes is of Ising type O(1), in addition to the symmetry SO(2) due to the invariance by global

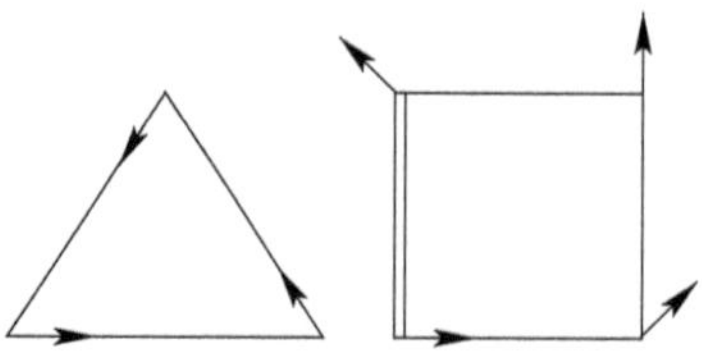

Fig. 1.2. Non-collinear spin configuration of frustrated triangular and square plaquettes with XY spins: ferro- and antiferromagnetic interactions J and $-J$ are indicated by thin and double lines, respectively.

rotation of the spins in the plane. The lattices formed by these plaquettes will be called "antiferromagnetic triangular lattice" and "Villain lattice", respectively.

It is expected from the GS symmetry of these systems that the transitions due to the breaking of the O(1) and SO(2) symmetries, if they occur at different temperatures, belong, respectively, to the 2D Ising universality class and to the Kosterlitz–Thouless universality class. The question of whether the two phase transitions would occur at the same temperature and the nature of their universality remains at present an open question. For more discussion on this, see Chapter 4.

The reader can find in Refs. [9, 10] the derivation of the non-trivial classical GS configuration of the fully frustrated simple cubic lattice formed by stacking the 2D Villain lattices, in the case of Heisenberg and XY spins.

Another example is the case of a chain of Heisenberg spins with ferromagnetic interaction $J_1(>0)$ between NN and antiferromagnetic interaction $J_2(<0)$ between NNN. When $\varepsilon = |J_2|/J_1$ is larger than a critical value ε_c, the spin configuration of the GS becomes non-collinear. One shows that the helical configuration displayed in Fig. 1.3 is obtained by minimizing the interaction energy:

$$\begin{aligned}
E &= -J_1 \sum_i \mathbf{S}_i \cdot \mathbf{S}_{i+1} + |J_2| \sum_i \mathbf{S}_i \cdot \mathbf{S}_{i+2} \\
&= S^2 \left[-J_1 \cos\theta + |J_2| \cos(2\theta)\right] \sum_i 1 \\
\frac{\partial E}{\partial \theta} &= S^2 \left[J_1 \sin\theta - 2|J_2| \sin(2\theta)\right] \sum_i 1 = 0 \\
&= S^2 \left[J_1 \sin\theta - 4|J_2| \sin\theta \cos\theta\right] \sum_i 1 = 0,
\end{aligned} \tag{1.4}$$

where one has supposed that the angle between NN spins is θ.

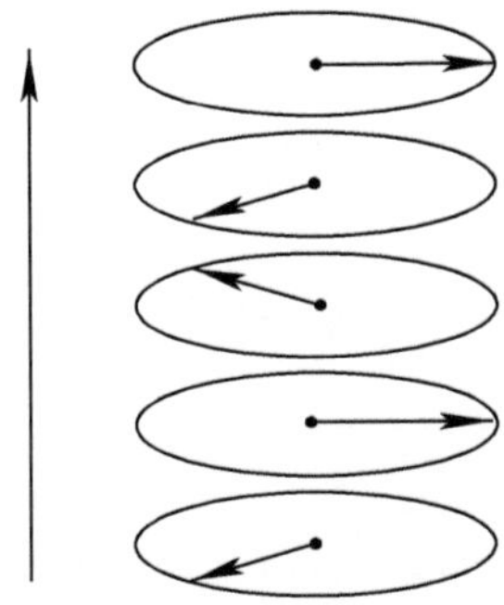

Fig. 1.3. Helical configuration when $\varepsilon = |J_2|/J_1 > \varepsilon_c = 1/4$ ($J_1 > 0$, $J_2 < 0$).

The two solutions are

$$\sin\theta = 0 \longrightarrow \theta = 0 \quad \text{(ferromagnetic solution)}$$

and

$$\cos\theta = \frac{J_1}{4|J_2|} \longrightarrow \theta = \pm\arccos\left(\frac{J_1}{4|J_2|}\right). \tag{1.5}$$

The last solution is possible if $-1 \leq \cos\theta \leq 1$, i.e. $J_1/(4|J_2|) \leq 1$ or $|J_2|/J_1 \geq 1/4 \equiv \varepsilon_c$.

Again in this example, there are two degenerate configurations: clockwise and counter-clockwise.

One defines in the following a chiral-order parameter for each plaquette. For example, in the case of a triangular plaquette, the chiral parameter is given by

$$\boldsymbol{\kappa} = \frac{2}{3\sqrt{3}}\left[\mathbf{S}_1 \wedge \mathbf{S}_2 + \mathbf{S}_2 \wedge \mathbf{S}_3 + \mathbf{S}_3 \wedge \mathbf{S}_1\right], \tag{1.6}$$

where coefficient $2/\left(3\sqrt{3}\right)$ was introduced so that the $\pm 2\pi/3$ degeneracy corresponds to $\kappa_i = \pm 1$.

We can form a triangular lattice using plaquettes as shown in Fig. 1.4. The GS corresponds to the state where all plaquettes of the same orientation have the same chirality: $\triangle$ plaquettes have positive chirality ($\kappa = 1$) and $\bigtriangledown$ plaquettes have negative chirality ($\kappa = -1$). In terms of Ising spins, we have a perfect antiferromagnetic order. This order is broken at a phase transition temperature where κ vanishes.

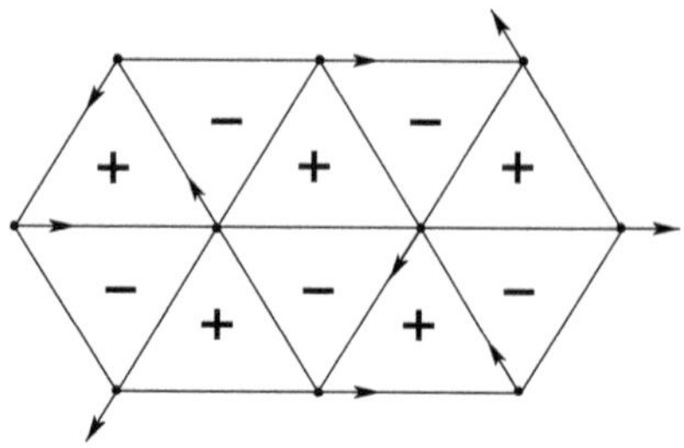

Fig. 1.4. Antiferromagnetic triangular lattice with XY spins. The positive and negative chiralities are indicated by $+$ and $-$.

Let us enumerate two frequently encountered frustrated spin systems where the NN interaction is antiferromagnetic: the fcc lattice and the hcp lattice. These two lattices are formed by stacking tetrahedra with four triangular faces. The frustration due to the lattice structure such as in these cases is called "geometry frustration".

1.2. Frustrated Ising Spin Systems

We are interested here in frustrated Ising spin systems without disorder. A review of early works (up to about 1985) on frustrated Ising systems with periodic interactions, i.e. no bond disorder, has been given by Liebmann.[1] These systems have their own interest in statistical mechanics because they are periodically defined and thus subject to exact treatment. To date, very few systems are exactly solvable. They are limited to 1D and 2D.[11] A few well-known systems showing remarkable properties include the centered square lattice[12] and its generalized versions,[13,14] the Kagomé lattice,[15–17] an anisotropic centered honeycomb lattice,[18] and several periodically dilute centered square lattices.[19] Complicated cluster models[20] and a particular 3D case have also been solved.[21] The phase diagrams in frustrated models show a rich behavior. Let us mention a few remarkable consequences of the frustration which are in connection with what will be shown in this chapter. The degeneracy of the GS is very high, often infinite. At finite temperatures, in some systems the degeneracy is reduced by thermal fluctuations which select a number of states with largest entropy. This has been called "order by disorder",[22] in the Ising case. Quantum fluctuations and/or thermal fluctuations can also select particular spin configurations in the case of vector spins.[23,24] Another striking phenomenon is the coexistence of order and disorder at equilibrium: a number of spins in the system are disordered at all temperatures even in an ordered phase.[16] The frustration is also at the origin of the reentrance phenomenon. A reentrant phase can be defined as a phase with no long-range order, or no order at all, occurring in a region below an ordered phase on the temperature scale. In addition, the frustration can also give rise to disordered lines in the phase diagram of many systems as will be shown below.

In this chapter, we confine ourselves to exactly solved Ising spin systems that show remarkable features in the phase diagram such as the reentrance, successive transitions, disorder lines, and partial disorder. Other Ising systems are treated in Chapter 2. Also, the reentrance in disordered systems such as spin glasses is discussed in Chapter 9.

The systems we consider in this chapter are periodically defined (without bond disorder). The frustration due to competing interactions will itself induce disorder in the spin orientations. The results obtained can be applied to physical systems that can be mapped into a spin language. The chapter is organized as follows. In Section 1.3, we outline the method which allows to calculate the partition function and the critical varieties of 2D Ising models without crossing interactions. In particular, we show in detail the mapping of these models onto the 16- and 32-vertex models. We also explain a decimation method for finding disorder solutions. The purpose of this section is to give the reader enough mathematical details so that, if he wishes, he can apply these techniques to 2D Ising models with non-crossing interactions. In Section 1.4, we shall apply the results of Section 1.3 in some systems that present remarkable physical properties. The systems studied in Section 1.4 contain most of the interesting features of the frustration: high GS degeneracy, reentrance, partial disorder, disorder lines, successive phase transitions, and some aspects of the random-field Ising model. In Section 1.4, we show some evidences of reentrance and partial disorder found in 3D systems and in systems with spins other than the Ising model (Potts model, classical vector spins, quantum spins). A discussion on the origin of the reentrance phenomenon and concluding remarks are given in Section 1.5.

1.3. Mapping between Ising Models and Vertex Models

The 2D Ising model with non-crossing interactions is exactly soluble. The problem of finding the partition function can be transformed in a free-fermion model. If the lattice is a complicated one, the mathematical problem to solve is very cumbersome.

For numerous 2D Ising models with non-crossing interactions, there exists another method, by far easier, to find the exact partition function. This method consists in mapping the model on a 16-vertex model or a

32-vertex model. If the Ising model does not have crossing interactions, the resulting vertex model will be exactly soluble. We will apply this method for finding the exact solution of several Ising models in 2D lattices with non-crossing interactions.

Let us at first introduce the 16-vertex model and the 32-vertex model and the cases for which these models satisfy the free-fermion condition.

1.3.1. *The 16-vertex model*

The 16-vertex model which we will consider is a square lattice of N points, connected by edges between neighboring sites. These edges can assume two states, symbolized by right- and left- or up- and down-pointing arrows, respectively. The allowed configurations of the system are characterized by specifying the arrangement of arrows around each lattice point. In characterizing these so-called vertex configurations, we follow the enumeration of Baxter[11] (see Fig. 1.5).

To each vertex we assign an energy ϵ_k ($k = 1, 2, \ldots, 16$) and a corresponding vertex weight (Boltzmann factor) $\omega_k = e^{\beta\epsilon_k}$, where $\beta = (1)/(k_B T)$, with T being the temperature and k_B the Boltzmann constant. Then the partition function is

$$Z = \sum_C e^{-\beta(n_1\epsilon_1 + \cdots + n_{16}\epsilon_{16})}, \tag{1.7}$$

where the sum is over all allowed configurations C of arrows on the lattice, n_j is the number of vertex arrangements of type j in configuration C. It is

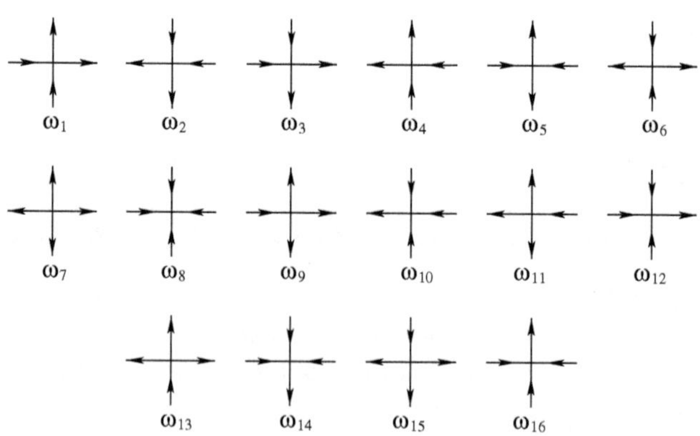

Fig. 1.5. Arrow configurations and vertex weights of the 16-vertex model.

clear from Eq. (1.7) that Z is a function of the eight Boltzmann weights ω_k $(k = 1, 2, \ldots, 16)$:

$$Z = Z(\omega_1, \ldots, \omega_{16}). \tag{1.8}$$

So far, exact results have only been obtained for three subclasses of the general 16-vertex model, i.e. the 6-vertex (or ferroelectric) model, the symmetric 8-vertex model and the free-fermion model.[11,25] Here, we will consider only the case where the free-fermion condition is satisfied, because in these cases the 16-vertex model can be related to 2D Ising models without crossing interactions. Generally, a vertex model is soluble if the vertex weights satisfy certain conditions so that the partition function is reducible to the S matrix of a many-fermion system.[25] In the present problem, these constraints are the following:

$$\begin{aligned}
\omega_1 &= \omega_2,\ \omega_3 = \omega_4,\\
\omega_5 &= \omega_6,\ \omega_7 = \omega_8,\\
\omega_9 &= \omega_{10} = \omega_{11} = \omega_{12},\\
\omega_{13} &= \omega_{14} = \omega_{15} = \omega_{16},\\
\omega_1\omega_3 + \omega_5\omega_7 &- \omega_9\omega_{11} - \omega_{13}\omega_{15} = 0.
\end{aligned} \tag{1.9}$$

If these conditions are satisfied, the free energy of the model can be expressed, in the thermodynamical limit, as follows:

$$f = -\frac{1}{4\pi\beta}\int_0^{2\pi} d\phi \log\{A(\phi) + [Q(\phi)]^{1/2}\}, \tag{1.10}$$

where

$$\begin{aligned}
A(\phi) &= a + c\cos(\phi)\\
Q(\phi) &= y^2 + z^2 - x^2 - 2yz\cos(\phi) + x^2\cos^2(\phi)\\
a &= \frac{1}{2}(\omega_1^2 + \omega_3^2 + 2\omega_1\omega_3 + \omega_5^2 + \omega_7^2 + 2\omega_5\omega_7) + 2(\omega_9^2 + \omega_{13}^2)\\
c &= 2[\omega_9(\omega_1 + \omega_3) - \omega_{13}(\omega_5 + \omega_7)]\\
y &= 2[\omega_9(\omega_1 + \omega_3) + \omega_{13}(\omega_5 + \omega_7)]\\
z &= \frac{1}{2}[(\omega_1 + \omega_3)^2 - (\omega_5 + \omega_7)^2] + 2(\omega_9^2 - \omega_{13}^2)\\
x^2 &= z^2 - \frac{1}{4}[(\omega_1 - \omega_3)^2 - (\omega_5 - \omega_7)^2]^2.
\end{aligned} \tag{1.11}$$

Phase transitions occur when one or more pairs of zeros of the expression $Q(\phi)$ close in on the real ϕ axis and "pinch" the path of integration in the expression on the right-hand side of Eq. (1.10). This happens when $y^2 = z^2$, i.e. when

$$\omega_1 + \omega_3 + \omega_5 + \omega_7 + 2\omega_9 + 2\omega_{13} = 2\max\{\omega_1 + \omega_3, \omega_5 + \omega_7, 2\omega_9, 2\omega_{13}\} \tag{1.12}$$

The type of singularity in the specific heat depends on whether

$$(\omega_1 - \omega_3)^2 - (\omega_5 - \omega_7)^2 \neq 0 \quad \text{(logarithmic singularity)}$$

or

$$(\omega_1 - \omega_3)^2 - (\omega_5 - \omega_7)^2 = 0 \quad \text{(inverse square-root singularity).} \tag{1.13}$$

1.3.2. *The 32-vertex model*

The 32-vertex model is defined by a triangular lattice of N points, connected by edges between neighboring sites. These edges can assume two states, symbolized by an arrow pointing in or pointing out of a site. In the general case, there are 64 allowed vertex configurations. If only an odd number of arrows pointing into a site are allowed, we have 32 possible vertex configurations. This is the constraint that characterizes the 32-vertex model. To each allowed vertex configuration we assign an energy ϵ_k $(k = 1, 2, \ldots, 32)$ and a corresponding vertex weight, defined as shown in Fig. 1.6, where $\omega = e^{-\beta\epsilon_1}$, $\overline{\omega} = e^{-\beta\epsilon_2}$, $\omega_{56} = e^{-\beta\epsilon_3}$, $\overline{\omega}_{56} = e^{-\beta\epsilon_4}$, etc.

This notation for the Boltzmann vertex weights was introduced by Sacco and Wu[26] and was also used by Baxter.[11] This model is not exactly soluble in the general case, but there are several particular cases that are soluble.[26] Here, we will consider one such case, when the model satisfies the free-fermion condition:

$$\begin{aligned} \omega\overline{\omega} &= \omega_{12}\overline{\omega}_{12} - \omega_{13}\overline{\omega}_{13} + \omega_{14}\overline{\omega}_{14} - \omega_{15}\overline{\omega}_{15} + \omega_{16}\overline{\omega}_{16}, \\ \omega\omega_{mn} &= \omega_{ij}\omega_{kl} - \omega_{ik}\omega_{jl} + \omega_{il}\omega_{jk} \end{aligned} \tag{1.14}$$

for all permutations i, j, k, l, m, n of $1, 2, \ldots, 6$ such that $m < n$ and $i < j < k < l$. There are 15 such permutations (corresponding to the 15 choices of m and n), and hence a total of 16 conditions.

The rather complicated notation for the Boltzmann weights is justified by the condensed form of the free-fermion conditions Eq. (1.14).

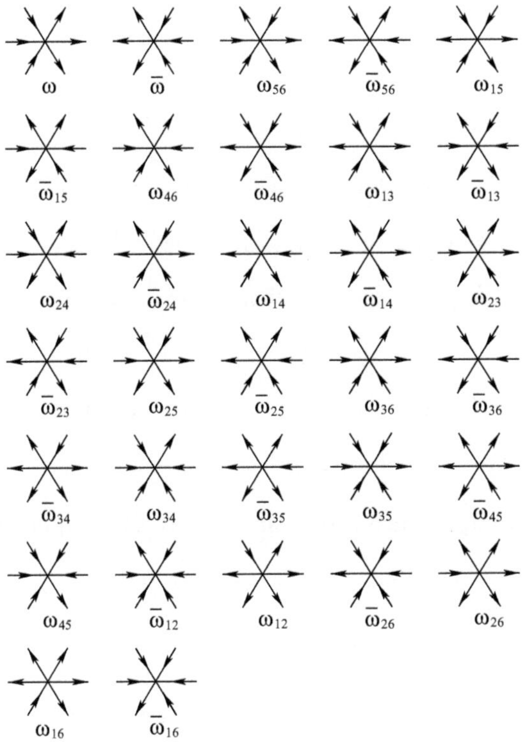

Fig. 1.6. Arrow configurations and vertex weights of the 32-vertex model.

When these conditions are satisfied, the free energy in the thermodynamical limit can be expressed as

$$f = -\frac{1}{8\pi^2\beta}\int_0^{2\pi} d\theta \int_0^{2\pi} d\phi \log[\omega^2 D(\theta,\phi)], \tag{1.15}$$

where

$$\begin{aligned}\omega^2 D(\theta,\phi) = {} & \Omega_1^2+\Omega_2^2+\Omega_3^2+\Omega_4^2-2(\Omega_1\Omega_3-\Omega_2\Omega_4)\cos(\theta)\\ & -2(\Omega_1\Omega_4-\Omega_2\Omega_3)\cos(\phi)+2(\Omega_3\Omega_4-\Omega_5\Omega_6)\cos(\theta+\phi)\\ & +2(\Omega_5\Omega_6-\Omega_1\Omega_2)\cos(\theta-\phi)-4a\sin(\phi)\sin(\theta+\phi)\\ & -4b\sin(\theta)\sin(\theta+\phi)-2c\sin^2(\theta+\phi)-2d\sin^2(\theta)-2e\sin^2(\phi)\end{aligned} \tag{1.16}$$

with

$$\begin{aligned}
\Omega_1 &= \omega + \overline{\omega}, \quad \Omega_2 = \omega_{25} + \overline{\omega}_{25}, \\
\Omega_3 &= \omega_{14} + \overline{\omega}_{14}, \quad \Omega_4 = \omega_{36} + \overline{\omega}_{36}, \\
\Omega_5\Omega_6 &= \omega_{15}\omega_{24} + \overline{\omega}_{15}\overline{\omega}_{24} + \omega_{14}\overline{\omega}_{25} + \omega_{25}\overline{\omega}_{14}, \\
a &= \omega_{12}\omega_{45} + \overline{\omega}_{12}\overline{\omega}_{45} - \omega\overline{\omega}_{36} - \overline{\omega}\omega_{36}, \\
b &= \omega_{23}\omega_{56} + \overline{\omega}_{23}\overline{\omega}_{56} - \omega\overline{\omega}_{14} - \overline{\omega}\omega_{14}, \\
c &= \omega\overline{\omega} + \omega_{13}\overline{\omega}_{13} - \omega_{12}\overline{\omega}_{12} - \omega_{23}\overline{\omega}_{23}, \\
d &= \omega\overline{\omega} + \omega_{26}\overline{\omega}_{26} - \omega_{16}\overline{\omega}_{16} - \omega_{12}\overline{\omega}_{12}, \\
e &= \omega\overline{\omega} + \omega_{15}\overline{\omega}_{15} - \omega_{56}\overline{\omega}_{56} - \omega_{16}\overline{\omega}_{16}.
\end{aligned} \tag{1.17}$$

The critical temperature is determined from the equation

$$\Omega_1 + \Omega_2 + \Omega_3 + \Omega_4 = 2\max(\Omega_1, \Omega_2, \Omega_3, \Omega_4) \tag{1.18}$$

We will show now how different 2D Ising models without crossing interactions can be mapped onto the 16-vertex model or the 32-vertex model, with the free-fermion condition automatically satisfied in such cases.

Let us consider at first an Ising model defined on a Kagomé lattice, with two-spin interactions between nearest neighbors (NN) and between next-nearest neighbors (NNN), J_1 and J_2, respectively, as shown in Fig. 1.7.

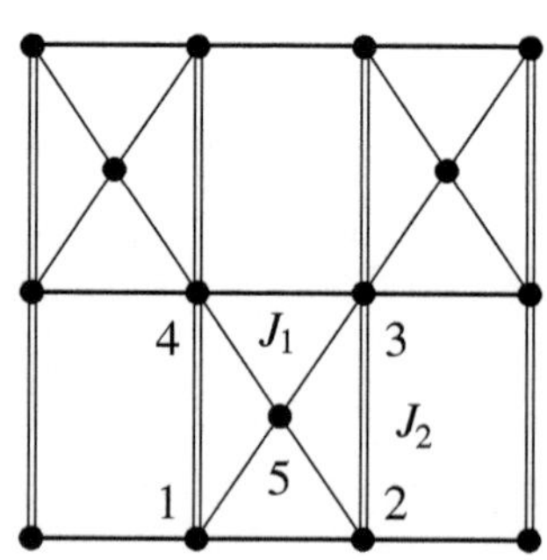

Fig. 1.7. Kagomé lattice. Interactions between nearest neighbors and between next-nearest neighbors, J_1 and J_2, are shown by single and double bonds, respectively. The lattice sites in a cell are numbered for decimation demonstration.

The Hamiltonian is written as

$$H = -J_1 \sum_{(ij)} \sigma_i \sigma_j - J_2 \sum_{(ij)} \sigma_i \sigma_j, \tag{1.19}$$

where and the first and second sums run over the spin pairs connected by single and double bonds, respectively.

The partition function is written as

$$Z = \sum_{\sigma} \prod_{c} \exp[K_1(\sigma_1\sigma_5 + \sigma_2\sigma_5 + \sigma_3\sigma_5 + \sigma_4\sigma_5 + \sigma_1\sigma_2 + \sigma_3\sigma_4) + K_2(\sigma_1\sigma_4 + \sigma_3\sigma_2)], \tag{1.20}$$

where $K_{1,2} = J_{1,2}/k_B T$ and where the sum is performed over all spin configurations and the product is taken over all elementary cells of the lattice.

Since there are no crossing bond interactions, the system can be transformed into an exactly solvable free-fermion model. We decimate the central spin of each elementary cell of the lattice. In doing so, we obtain a checkerboard Ising model with multispin interactions (see Fig. 1.8).

The Boltzmann weight associated with each shaded square is given by

$$W(\sigma_1, \sigma_2, \sigma_3, \sigma_4) = 2\cosh(K_1(\sigma_1 + \sigma_2 + \sigma_3 + \sigma_4)) \exp[K_2(\sigma_1\sigma_4 + \sigma_2\sigma_3) + K_1(\sigma_1\sigma_2 + \sigma_3\sigma_4)]. \tag{1.21}$$

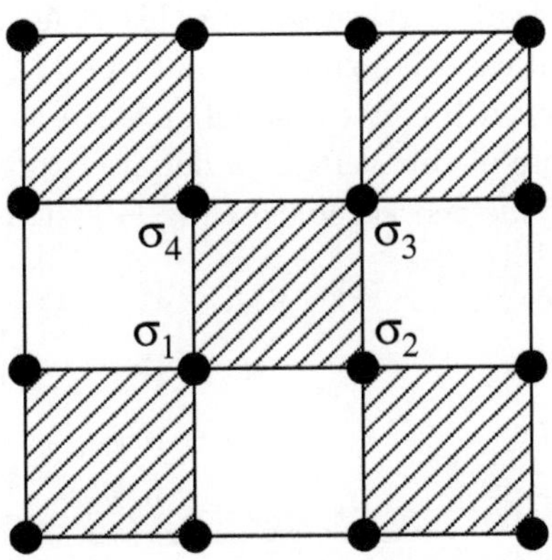

Fig. 1.8. The checkerboard lattice. Each shaded square is associated with the Boltzmann weight $W(\sigma_1, \sigma_2, \sigma_3, \sigma_4)$, given in the text.

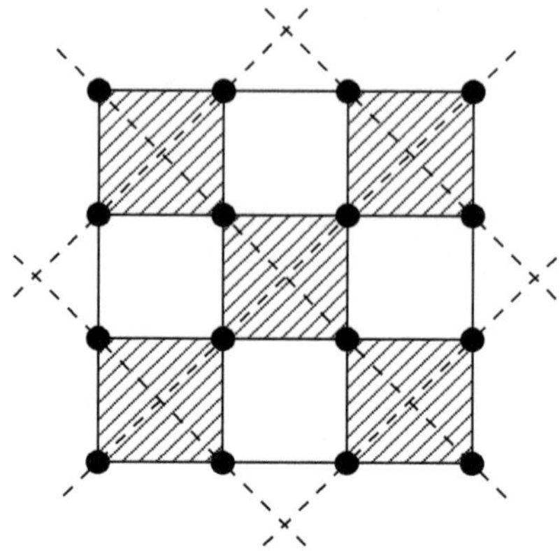

Fig. 1.9. The checkerboard lattice and the associated square lattice with their bonds indicated by dashed lines.

The partition function of this checkerboard Ising model is given by

$$Z = \sum_{\sigma} \prod W(\sigma_1, \sigma_2, \sigma_3, \sigma_4), \tag{1.22}$$

where the sum is performed over all spin configurations and the product is taken over all the shaded squares of the lattice.

In order to map this model onto the 16-vertex model, let us introduce another square lattice where each site is placed at the center of each shaded square of the checkerboard lattice, as shown in Fig. 1.9.

At each bond of this lattice we associate an arrow pointing out of the site if the Ising spin that is traversed by this bond is equal to $+1$, and pointing into the site if the Ising spin is equal to -1, as shown in Fig. 1.10.

In this way, we have a 16-vertex model on the associated square lattice. The Boltzmann weights of this vertex model are expressed in terms of the Boltzmann weights of the checkerboard Ising model, as follows:

$$\begin{aligned}
\omega_1 &= W(-,-,+,+) & \omega_5 &= W(-,+,-,+), \\
\omega_2 &= W(+,+,-,-) & \omega_6 &= W(+,-,+,-), \\
\omega_3 &= W(-,+,+,-) & \omega_7 &= W(+,+,+,+), \\
\omega_4 &= W(+,-,-,+) & \omega_8 &= W(-,-,-,-), \\
\omega_9 &= W(-,+,+,+) & \omega_{13} &= W(+,-,+,+), \\
\omega_{10} &= W(+,-,-,-) & \omega_{14} &= W(-,+,-,-),
\end{aligned}$$

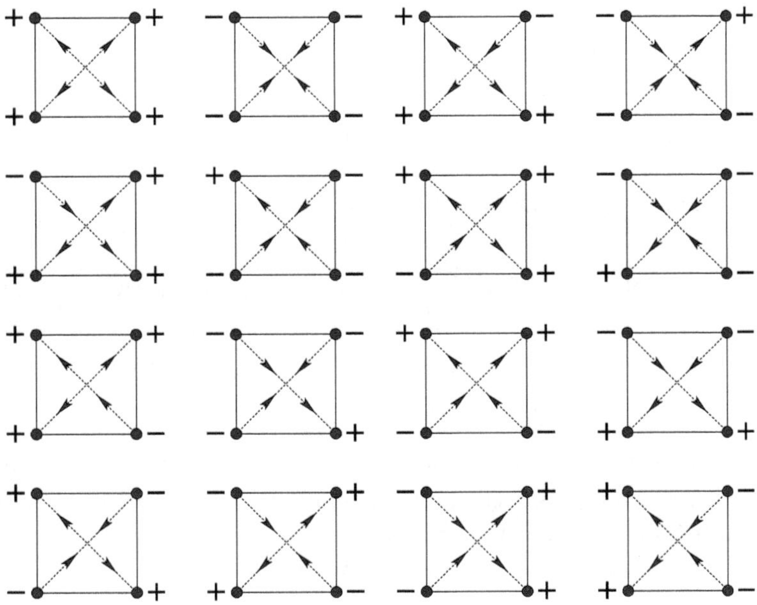

Fig. 1.10. The relation between spin configurations and arrow configurations of the associated vertex model.

$$\omega_{11} = W(+,+,-,+) \qquad \omega_{15} = W(+,+,+,-),$$
$$\omega_{12} = W(-,-,+,-) \qquad \omega_{16} = W(-,-,-,+). \tag{1.23}$$

Taking Eq. (1.21) into account, we obtain

$$\begin{aligned}
&\omega_1 = \omega_2 = 2e^{-2K_2+2K_1},\\
&\omega_3 = \omega_4 = 2e^{2K_2-2K_1},\\
&\omega_5 = \omega_6 = 2e^{-2K_2-2K_1},\\
&\omega_7 = \omega_8 = 2e^{2K_2+2K_1}\cosh(4K_1),\\
&\omega_9 = \omega_{10} = \omega_{11} = \omega_{12} = \omega_{13} = \omega_{14} = \omega_{15} = \omega_{16} = 2\cosh(2K_1).
\end{aligned} \tag{1.24}$$

As can be verified easily, the free-fermion conditions Eq. (1.9) are identically satisfied by the Boltzmann weights Eq. (1.24), for arbitrary values of K_1 and K_2. If we replace Eq. (1.24) in Eqs. (1.10) and (1.11), we can obtain the explicit expression of the free energy of the model. Moreover, by replacing Eq. (1.24) in Eq. (1.12), we obtain the critical condition for this system:

$$\begin{aligned}&\frac{1}{2}\,[\exp(2K_1+2K_2)\cosh(4K_1)+\exp(-2K_1-2K_2)]\\&\quad+\cosh(2K_1-2K_2)+2\cosh(2K_1)\\&=2\max\left\{\frac{1}{2}\,[\exp(2K_1+2K_2)\cosh(4K_1)\right.\\&\quad\left.+\exp(-2K_1-2K_2)];\ \cosh(2K_2-2K_1);\ \cosh(2K_1)\right\},\end{aligned} \tag{1.25}$$

which is decomposed into four critical lines depending on the values of J_1 and J_2.

The singularity of the free energy is everywhere logarithmic.

Now, we will consider another 2D Ising model with two-spin interactions and without crossing bonds. This model is defined on a centered honeycomb lattice, as shown in Fig. 1.11.

The Hamiltonian of this model is given as follows:

$$H=-J_1\sum_{(ij)}\sigma_i\sigma_j-J_2\sum_{(ij)}\sigma_i\sigma_j-J_3\sum_{(ij)}\sigma_i\sigma_j, \tag{1.26}$$

where $\sigma_i=\pm1$ is an Ising spin occupying the lattice site i, and the first, second, and third sums run over the spin pairs connected by heavy, light, and doubly light bonds, respectively (see Fig. 1.11). When $J_2=J_3=0$, one recovers the honeycomb lattice, and when $J_1=J_2=J_3$, one has the triangular lattice.

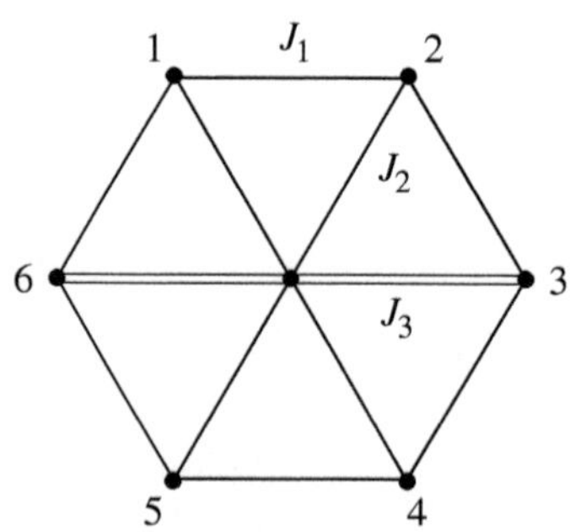

Fig. 1.11. Unit cell of the centered honeycomb lattice: heavy, light, and double-light bonds denote the interactions J_1, J_2, and J_3, respectively. The sites on the honeycomb are numbered from 1 to 6 for decimation demonstration (see text).

Let us denote the central spin in a lattice cell, shown in Fig. 1.11, by σ and number the other spins from σ_1 to σ_6. The Boltzmann weight associated with the elementary cell is given by

$$W = \exp[K_1(\sigma_1\sigma_2 + \sigma_2\sigma_3 + \sigma_3\sigma_4 + \sigma_4\sigma_5 + \sigma_5\sigma_6 + \sigma_6\sigma_1) + K_2\sigma(\sigma_1 + \sigma_2 + \sigma_4 + \sigma_5) + K_3\sigma(\sigma_3 + \sigma_6)]. \tag{1.27}$$

The partition function of the model is written as

$$Z = \sum_{\sigma}\prod_{c} W, \tag{1.28}$$

where the sum is performed over all spin configurations and the product is taken over all elementary cells of the lattice. Periodic boundary conditions are imposed. Since there is no crossing-bond interaction, the model is exactly soluble. To obtain the exact solution, we decimate the central spin of each elementary cell of the lattice. In doing so, we obtain a honeycomb Ising model with multispin interactions.

After decimation of each central spin, the Boltzmann factor associated with an elementary cell is given by

$$W' = 2\exp[K_1(\sigma_1\sigma_2 + \sigma_2\sigma_3 + \sigma_3\sigma_4 + \sigma_4\sigma_5 + \sigma_5\sigma_6 + \sigma_6\sigma_1)] \times \cosh[K_2(\sigma_1 + \sigma_2 + \sigma_4 + \sigma_5) + K_3(\sigma_3 + \sigma_6)]. \tag{1.29}$$

We will show in the following that this model is equivalent to a special case of the 32-vertex model on the triangular lattice that satisfies the free-fermion condition.

Let us consider the dual lattice of the honeycomb lattice, i.e. the triangular lattice.[11] The sites of the dual lattice are placed at the center of each elementary cell and their bonds are perpendicular to the bonds of the honeycomb lattice, as shown in Fig. 1.12.

Each site of the triangular lattice is surrounded by 6 sites of the honeycomb lattice. At each bond of the triangular lattice we associate an arrow. We take the arrow configuration shown in Fig. 1.13 as the standard one. We can establish a two-to-one correspondence between spin configurations of the honeycomb lattice and arrow configurations in the triangular lattice. This can be done in the following way: if the spins on either side of a bond of the triangular lattice are equal (different), place an arrow on the bond pointing in the same (opposite) direction as the standard. If we do this for all bonds, then at each site of the triangular lattice, there must be an even

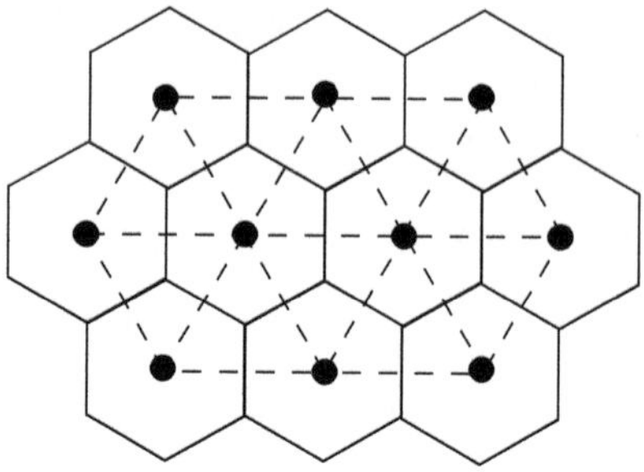

Fig. 1.12. The honeycomb lattice and the dual triangular lattice, with their bonds indicated by dashed lines.

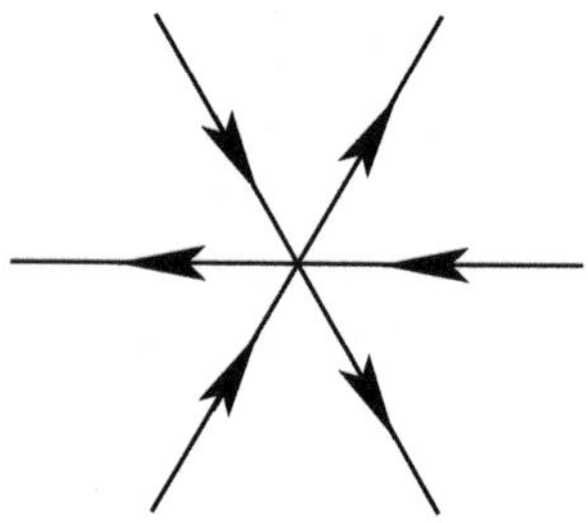

Fig. 1.13. The standard arrow configuration for the triangular lattice.

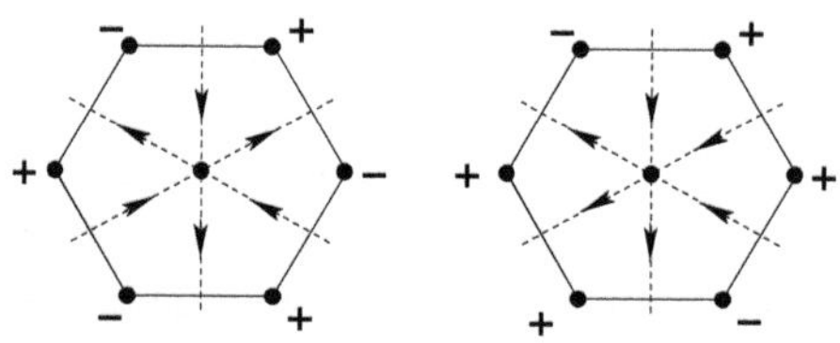

Fig. 1.14. Two cases of the correspondence between arrow configurations and spin configurations.

number of non-standard arrows on the six incident bonds, and hence an odd number of incoming (and outgoing) arrows. This is the property that characterizes the 32-vertex model on the triangular lattice.

In Fig. 1.14, we show two cases of the relation between arrow configurations on the triangular lattice and spin configurations on the honeycomb lattice.

In consequence, the Boltzmann weights of the 32-vertex model will be a function of the Boltzmann weights $W'(\sigma_1, \sigma_2, \sigma_3, \sigma_4, \sigma_5, \sigma_6)$, associated with a face of the honeycomb lattice. By using the relation between vertex and spin configurations described above and expression Eq. (1.29), we find

$$
\begin{aligned}
\omega &= W'(+,-,-,-,+,+) = 2e^{2K_1}\\
\overline{\omega} &= W'(+,+,-,+,+,-) = 2e^{-2K_1}\cosh(4K_2 - 2K_3)\\
\omega_{56} &= W'(+,-,+,-,+,+) = 2e^{-2K_1}\cosh(2K_3)\\
\overline{\omega}_{56} &= W'(+,+,+,+,+,-) = 2e^{2K_1}\cosh(4K_2)\\
\omega_{15} &= W'(+,+,+,-,+,+) = 2e^{2K_1}\cosh(2K_2 + 2K_3)\\
\overline{\omega}_{15} &= W'(+,-,+,+,+,-) = 2e^{-2K_1}\cosh(2K_2)\\
\omega_{46} &= W'(+,-,+,+,+,+) = 2e^{2K_1}\cosh(2K_2 + 2K_3)\\
\overline{\omega}_{46} &= W'(+,+,+,-,+,-) = 2e^{-2K_1}\cosh(2K_2)\\
\omega_{13} &= W'(+,+,+,+,-,+) = 2e^{2K_1}\cosh(2K_2 + 2K_3)\\
\overline{\omega}_{13} &= W'(+,-,+,-,-,-) = 2e^{-2K_1}\cosh(2K_2)\\
\omega_{24} &= W'(+,-,-,-,-,-) = 2e^{2K_1}\cosh(2K_2 + 2K_3)\\
\overline{\omega}_{24} &= W'(+,+,-,+,-,+) = 2e^{-2K_1}\cosh(2K_2)\\
\omega_{14} &= W'(+,+,+,+,+,+) = 2e^{6K_1}\cosh(4K_2 + 2K_3)\\
\overline{\omega}_{14} &= W'(+,-,+,-,+,-) = 2e^{-6K_1}\\
\omega_{23} &= W'(+,-,-,-,+,-) = 2e^{-2K_1}\cosh(2K_3)\\
\overline{\omega}_{23} &= W'(+,+,-,+,+,+) = 2e^{2K_1}\cosh(4K_2)\\
\omega_{25} &= W'(+,-,-,+,-,-) = 2e^{-2K_1}\cosh(2K_3)\\
\overline{\omega}_{25} &= W'(+,+,-,-,-,+) = 2e^{2K_1}\\
\omega_{36} &= W'(+,-,+,+,-,+) = 2e^{-2K_1}\cosh(2K_3)\\
\overline{\omega}_{36} &= W'(+,+,+,-,-,-) = 2e^{2K_1}\\
\overline{\omega}_{34} &= W'(+,+,-,+,-,-) = 2e^{-2K_1}\cosh(2K_2 - 2K_3)\\
\omega_{34} &= W'(+,-,-,-,-,+) = 2e^{2K_1}\cosh(2K_2)\\
\overline{\omega}_{35} &= W'(+,+,-,-,-,-) = 2e^{2K_1}\cosh(2K_3)
\end{aligned}
$$

$$
\begin{aligned}
\omega_{35} &= W'(+,-,-,+,-,+) = 2e^{-2K_1} \\
\overline{\omega}_{45} &= W'(+,+,-,-,+,-) = 2e^{-2K_1}\cosh(2K_2 - 2K_3) \\
\omega_{45} &= W'(+,-,-,+,+,+) = 2e^{2K_1}\cosh(2K_2) \\
\overline{\omega}_{12} &= W'(+,-,+,-,-,+) = 2e^{-2K_1}\cosh(-2K_2 + 2K_3) \\
\omega_{12} &= W'(+,+,+,+,-,-) = 2e^{2K_1}\cosh(2K_2) \\
\overline{\omega}_{26} &= W'(+,+,+,-,-,+) = 2e^{2K_1}\cosh(2K_3) \\
\omega_{26} &= W'(+,-,+,+,-,-) = 2e^{-2K_1} \\
\omega_{16} &= W'(+,+,-,-,+,+) = 2e^{2K_1}\cosh(2K_2) \\
\overline{\omega}_{16} &= W'(+,-,-,+,+,-) = 2e^{-2K_1}\cosh(2K_2 - 2K_3). \qquad (1.30)
\end{aligned}
$$

Using the above expressions in Eqs. (1.15)–(1.17), we can obtain the expression of the free energy of the centered honeycomb lattice Ising model.

Taking into account Eqs. (1.30), (1.17), and (1.18), the critical temperature of the model is determined from the below equation:

$$
\begin{aligned}
&e^{2K_1} + e^{-2K_1}\cosh(4K_2 - 2K_3) + 2e^{-2K_1}\cosh(2K_3) + 2e^{2K_1} \\
&\quad + e^{6K_1}\cosh(4K_2 + 2K_3) + e^{-6K_1} \\
&= 2\max\{e^{2K_1} + e^{-2K_1}\cosh(4K_2 - 2K_3);\ e^{2K_1} + e^{-2K_1}\cosh(2K_3); \\
&\quad e^{6K_1}\cosh(4K_1 + 2K_3) + e^{-6K_1}\}. \qquad (1.31)
\end{aligned}
$$

The solutions of this equation are analyzed in Section 1.3.3.

We think that with the two cases studied above, the reader will be able to apply this procedure to other 2D Ising models without crossing bonds as, for instance, the Ising model on the centered square lattice. After decimation of the central spin in each square, this model can be mapped into a special case of the 16-vertex model, by following the same procedure that we have employed for the honeycomb lattice model.

1.3.3. *Disorder solutions for 2D Ising models*

Disorder solutions are very useful for clarifying the phase diagrams of anisotropic models and also imply constraints on the analytical behavior of the partition function of these models.

A great variety of anisotropic models (with different coupling constants in the different directions of the lattice) are known to posses remarkable submanifolds in the space of parameters, where the partition function is computable and takes a very simple form. These are the disorder solutions.

All the methods applied for obtaining these solutions rely on the same mechanism: a certain local decoupling of the degrees of freedom of the model, which results in an effective reduction of dimensionality for the lattice system. Such a property is provided by a simple local condition imposed on the Boltzmann weights of the elementary cell generating the lattice.[27]

Some completely integrable models present disorder solutions, e.g. the triangular Ising model and the symmetric 8-vertex model. But very important models that are not integrable, also present this type of solutions, e.g. the triangular Ising model with a field, the triangular q-state Potts model, and the general 8-vertex model. Here, we will consider only 2D Ising models.

In order to introduce the method, we will analyze, at the first place, the simplest case, i.e. the anisotropic Ising model on the triangular lattice (see Fig. 1.15).

The Boltzmann weight of the elementary cell is

$$W(\sigma_1, \sigma_2, \sigma_3) = \exp\left[\frac{1}{2}(K_1\sigma_1\sigma_3 + K_2\sigma_2\sigma_3 + K_3\sigma_1\sigma_2)\right]. \tag{1.32}$$

In every case, the local criterion will be defined by the following condition: after summation over some of its spins (to be defined in each case), the Boltzmann weight associated with the elementary cell of the model must not depend on the remaining spins any longer. For instance, for the

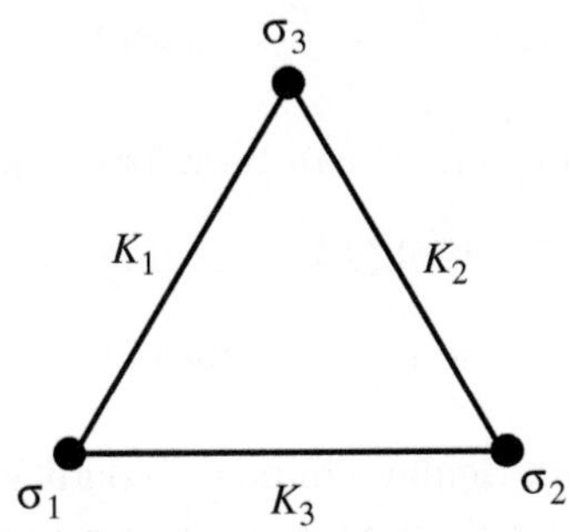

Fig. 1.15. The elementary cell of the triangular lattice, with three interactions K_1, K_2, and K_3.

triangular lattice, we will require

$$\sum_{\sigma_3} W(\sigma_1, \sigma_2, \sigma_3) = \lambda(K_1, K_2, K_3), \tag{1.33}$$

where λ is a function only of K_1, K_2, and K_3 (it is independent of σ_1 and σ_2). By using Eq. (1.32) we find

$$\sum_{\sigma_3} W(\sigma_1, \sigma_2, \sigma_3) = \exp\left(\frac{1}{2}K_3\sigma_1\sigma_2\right)\cosh\left[\frac{1}{2}(K_1\sigma_1 + K_2\sigma_2)\right]. \tag{1.34}$$

But, as it is well known, we can write

$$\cosh\left[\frac{1}{2}(K_1\sigma_1 + K_2\sigma_2)\right] = A\exp(K\sigma_1\sigma_2) \tag{1.35}$$

with

$$A = \left[\cosh\left(\frac{K_1 + K_2}{2}\right)\cosh\left(\frac{K_1 - K_2}{2}\right)\right]^{\frac{1}{2}}, \tag{1.36}$$

$$K = \frac{1}{2}\log\left[\frac{\cosh\left(\frac{K_1+K_2}{2}\right)}{\cosh\left(\frac{K_1-K_2}{2}\right)}\right]. \tag{1.37}$$

In order that $\sum_{\sigma_3} W(\sigma_1, \sigma_2, \sigma_3)$ be independent of σ_1 and σ_2, we must impose the condition $K = -\frac{1}{2}K_3$. From this condition we find

$$e^{K_3}\cosh\left(\frac{K_1 + K_2}{2}\right) = \cosh\left(\frac{K_1 - K_2}{2}\right) \tag{1.38}$$

from which we can determine the expression of λ

$$\lambda(K_1, K_2, K_3) = \left[\cosh\left(\frac{K_1 + K_2}{2}\right)\cosh\left(\frac{K_1 - K_2}{2}\right)\right]^{\frac{1}{2}}. \tag{1.39}$$

It is easy to verify that Eq. (1.38) can be written as

$$\tanh(K_1)\tanh(K_2) + \tanh(K_3) = 0. \tag{1.40}$$

This 2D subvariety in the space of parameters is called the disorder variety of the model.

Let us now impose particular boundary conditions for the lattice (see Fig. 1.16): on the upper layer, all interactions are missing, so that the spins of the upper layer only interact with those of the lower layer. It immediately follows that if one sums over all the spins of the upper layer and if one

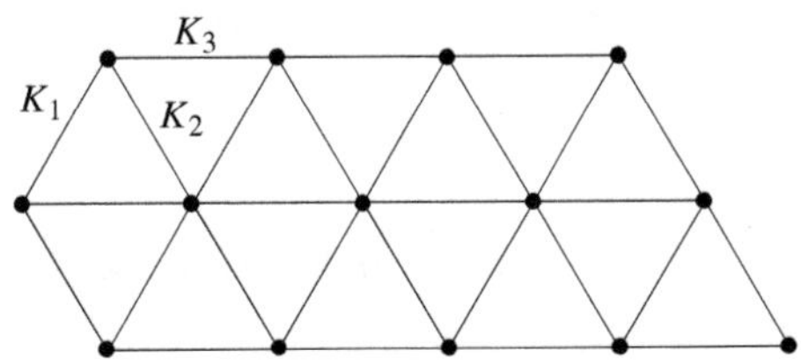

Fig. 1.16. Two layers of the triangular lattice.

requires the disorder condition Eq. (1.40), the same boundary conditions reappear for the next layer.

Iterating the procedure leads one to an exact expression for the partition function, when restricted to subvariety Eq. (1.40)

$$Z = \lambda(K_1, K_2, K_3)^N, \tag{1.41}$$

where N is the number of sites of the lattice. The free energy in the thermodynamic limit is given by

$$f = -\frac{1}{2\beta} \log \left[\cosh\left(\frac{K_1 + K_2}{2}\right) \cosh\left(\frac{K_1 - K_2}{2}\right) \right]. \tag{1.42}$$

The partition function Eq. (1.41) corresponds to lattices with unusual boundary conditions. In the physical domain, where the coupling constants are real, these do not affect the partition function per site (or the free energy per site) in the thermodynamic limit, and the expression Eq. (1.42) also corresponds to the free energy per site with standard periodic boundary conditions. On the contrary, in the non-physical domain (complex coupling constants), the boundary conditions are known to play an important role, even after taking the thermodynamic limit.

Let us consider now the Kagomé lattice Ising model with two-spin interactions between NN and NNN, studied in Section 1.4. If we apply the same procedure as that used for the triangular lattice Ising model, we obtain for the disorder variety:

$$e^{4K_2} = \frac{2(e^{4K_1} + 1)}{e^{8K_1} + 3}. \tag{1.43}$$

This disorder variety does not have an intersection with the critical variety of the model.

Following the method that we have exposed for the Ising model on the triangular lattice, the reader will be able to find the disorder varieties for other 2D Ising models with anisotropic interactions.

1.4. Reentrance in Exactly Solved Frustrated Ising Spin Systems

In this section, we show and discuss the phase diagrams of several selected 2D frustrated Ising systems that have been recently solved. For general exact methods, the reader is referred to the book by Baxter[11] and to the preceding section. In what follows, we consider only frustrated systems that exhibit the *reentrance phenomenon.* A reentrant phase can be defined as a phase with no long-range order, or no order at all, occurring in a region below an ordered phase on the temperature (T) scale. A well-known example is the reentrant phase in spin-glasses (see the review by Binder and Young[28]). The origin of the reentrance in spin-glasses is not well understood. It is believed that it is due to a combination of frustration and bond disorder. In order to see the role of the frustration alone, we show here the exact results on a number of periodically frustrated Ising systems. The idea behind the works shown in this section is to search for the ingredients responsible for the occurrence of the reentrant phase. Let us review in the following a few models showing a reentrant phase. Discussion on the origin of the reentrance will be given in the conclusion.

1.4.1. *Centered square lattice*

Even before the concept of frustration was introduced,[2] systems with competing interactions were found to possess rich critical behavior and nontrivial ordered states. Among these models, the centered square lattice Ising model (see Fig. 1.17), introduced by Vaks *et al.*,[12] with NN and NNN interactions, J_1 and J_2, respectively, is to our knowledge the first exactly soluble model which exhibits successive phase transitions with a reentrant paramagnetic phase at low T. Exact expression for the free energy, some correlation functions, and the magnetization of one sublattice were given in the original work of Vaks *et al.*

We distinguish two sublattices 1 and 2. Sublattice 1 contains the spins at the square centers, and sublattice 2 generates a square lattice with interaction J_2 in both the horizontal and vertical directions. Spins of sublattice 1 interact only with spins of sublattice 2 via diagonal interactions J_1. The GS properties of this model are as follows: for $a = J_2/ \mid J_1 \mid > -1$, spins of

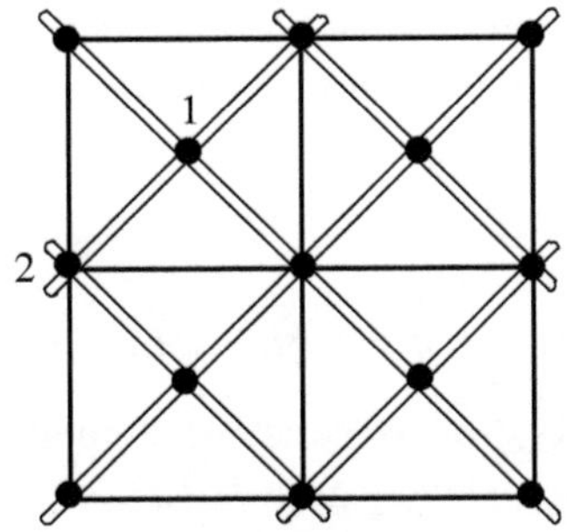

Fig. 1.17. Centered square lattice. Interactions between NN and NNN, J_1 and J_2, are denoted by white and black bonds, respectively. The two sublattices are numbered 1 and 2.

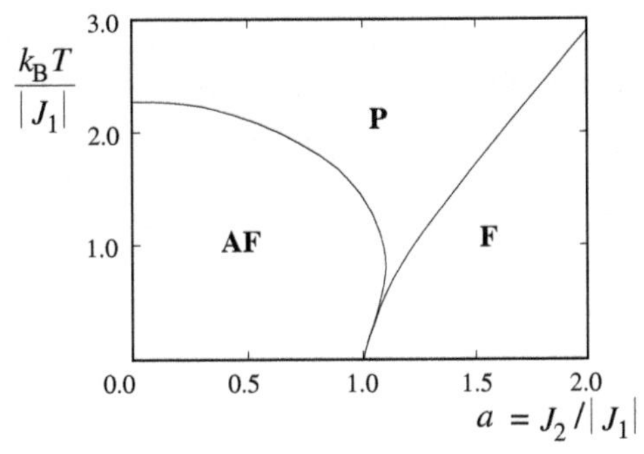

Fig. 1.18. Phase diagram of the centered square lattice.[12]

sublattice 2 order ferromagnetically and the spins of sublattice 1 are parallel (antiparallel) to the spins of sublattice 2 if $J_1 > 0$ (< 0); for $a < -1$, spins of sublattice 2 order antiferromagnetically, leaving the centered spins free to flip.

1.4.1.1. *Phase diagram*

The phase diagram of this model is given by Vaks *et al.*[12] Except for $a = -1$, there is always a finite critical temperature.

When J_2 is antiferromagnetic (>0) and J_2/J_1 is in a small region near 1, the system is successively in the paramagnetic state, an ordered state, the *reentrant* paramagnetic state, and another ordered state, with decreasing temperature (see Fig. 1.18).

The centered square Ising lattice has been generalized to include several kinds of interactions.[13,14] For example, when the vertical interaction J_2 is different from the horizontal one, say J_3, the phase diagram becomes more complicated.

1.4.1.2. *Nature of ordering and disorder solutions*

For the sake of simplicity, let us consider hereafter the case of NN and NNN interactions only, namely J_1 and J_2 ($J_3 = J_2$). Note that though an exact critical line was obtained,[12] the order parameter was not calculated, though the magnetization of one sublattice was given in the original work of Vaks *et al.*[12] Later, Choy and Baxter[29] obtained the total magnetization for this model in the ferromagnetic region. However, the ordering in the antiferromagnetic (frustrated) region has not been exactly calculated, despite the fact that it may provide an interesting ground for understanding the reentrance phenomenon. We have studied this aspect by means of Monte Carlo (MC) simulations.[30] The question which naturally arises is whether or not the disorder of sublattice 1 at $T = 0$ remains at finite T. If the spins of sublattice 1 remain disordered at finite T in the antiferromagnetic region, we have a remarkable kind of ordered state: namely the coexistence between order and disorder. This behavior has been observed in 3D Ising spin models[31,32] and in an exactly soluble model (the Kagomé lattice).[16] In the latter system, which is similar to the present model (discussed in Section 1.4.2), it was shown that the coexistence of order and disorder at finite T sheds some light on the reentrance phenomena. To verify the coexistence between order and disorder in the centered square lattice, we have performed MC simulations. The results for the Edwards–Anderson sublattice order parameters q_i and the staggered susceptibility of sublattice 2, as functions of T, are shown in Fig. 1.19 in the case $a = -2$.

As is seen, sublattice 2 is ordered up to the transition at T_c while sublattice 1 stays disordered at all T. This result shows a new example where order and disorder coexists in an equilibrium state. This result supports the conjecture formulated by Azaria *et al.*,[16] namely the coexistence of order and disorder is a necessary condition for the reentrant behavior to occur. The partial disorder just compensates the loss of entropy due to the partial ordering of the high-T phase. In a previous paper,[16] the importance of the disorder line in understanding the reentrance phenomenon has been emphasized. It has been suggested that this type of line may be necessary for the change of ordering from the high-T ordered phase to the low-T one.

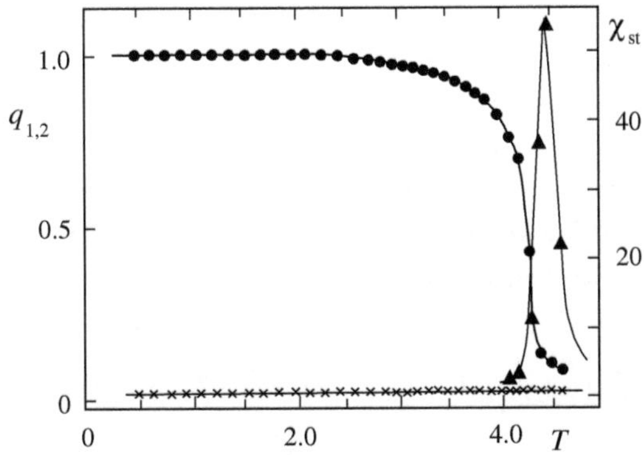

Fig. 1.19. Temperature dependence of sublattice Edwards–Anderson order parameters, q_1 and q_2 (crosses and black circles, respectively) in the case $a = J_2/ \mid J_1 \mid = -2$, by MC simulation. Susceptibility calculated by fluctuations of magnetization of sublattice 2 is also shown. The lattice used contains $N = 2 \times 60 \times 60$ spins with periodic boundary conditions.[30]

In the narrow reentrant paramagnetic region, preordering fluctuations with different symmetries exist near each critical line. Therefore, the correlation functions change their behavior as the temperature is varied in the reentrant paramagnetic region. As a consequence of the change of symmetries, there exist spins for which the two-point correlation function (between NN spins) has different signs, near the two critical lines, in the reentrant paramagnetic region. Hence, it is reasonable to expect that it has to vanish at a disorder temperature T_D. This point can be considered as a non-critical transition point that separates two different paramagnetic phases. The two-point correlation function defined above may be thought of as a non-local "disorder parameter". This particular point is just the one which has been called a disorder point by Stephenson[33] in analyzing the behavior of correlation functions for systems with competing interactions. For the centered square lattice Ising model considered here, the Stephenson disorder line is[30]

$$\cosh(4J_1/k_BT_D) = \exp(-4J_2/k_BT_D). \tag{1.44}$$

The two-point correlation function at T_D between spins of sublattice 2 separated by a distance r is zero for odd r and decay like $r^{-1/2}[\tanh(J_2/k_BT_D)]^r$ for r even.[33] However,there is *no dimensional reduction* on the Stephenson line given above. Usually, one defines the disorder point as the temperature where there is an effective reduction of dimensionality in such a way that physical quantities become simplified spectacularly.[27] In general, these two types of disorder lines are equivalent, as for example, in the case of the Kagomé lattice Ising model (see below). This is not the case here. In order to calculate this disorder line for the centered square lattice, we recall that this model is equivalent to an 8-vertex model that verifies the free-fermion condition.[34] The disorder line corresponding to dimensional reduction was given for the general 8-vertex model by Giacomini.[35] When this result is applied to the centered square lattice, one finds that the disorder variety is given by

$$\exp(4J_2/k_BT) = (1 - i\sinh(4J_1/k_BT))^{-1}, \qquad (1.45)$$

where $i^2 = -1$. This disorder line lies on the unphysical (complex) region of the parameter space of this system. When calculated on the line Eq. (1.45), the magnetization of this model, evaluated recently by Choy and Baxter[29] in the ferromagnetic region, becomes singular, as it is usually the case for the disorder solutions with dimensional reduction.[36] Since in the centered square lattice, the two kinds of disorder lines are not equivalent, we conclude, according to the arguments presented above, that the Stephenson disorder line Eq. (1.44) is the relevant one for the reentrance phenomenon.

Disorder solutions have recently found interesting applications, as for example in the problem of cellular automata (for a review, see Ref. [37]). Moreover, they also serve to build a new kind of series expansion for lattice spin systems.[27]

1.4.2. *Kagomé lattice*

1.4.2.1. *Model with NN and NNN interactions*

Another model of interest is the Kagomé lattice shown in Fig. 1.7. The Kagomé Ising lattice with NN interaction J_1 has been solved a long time ago[15] showing no phase transition at finite T when J_1 is antiferromagnetic. Taking into account the NNN interaction J_2 (see Fig. 1.7 and Eq. (1.19)], we have solved[16] this model by transforming it into a 16-vertex model that satisfies the free-fermion condition.

The critical condition is given by Eq. (1.25). For the whole phase diagram, the reader is referred to the paper by Azaria *et al.*[16] We show in Fig. 1.20 only the small region of J_2/J_1 in the phase diagram which has the reentrant paramagnetic phase and a disorder line.

Phase X indicates a partially ordered phase where the central spins are free (the nature of ordering was determined by MC simulations).[16] Here again, the reentrant phase takes place between a low-T ordered phase

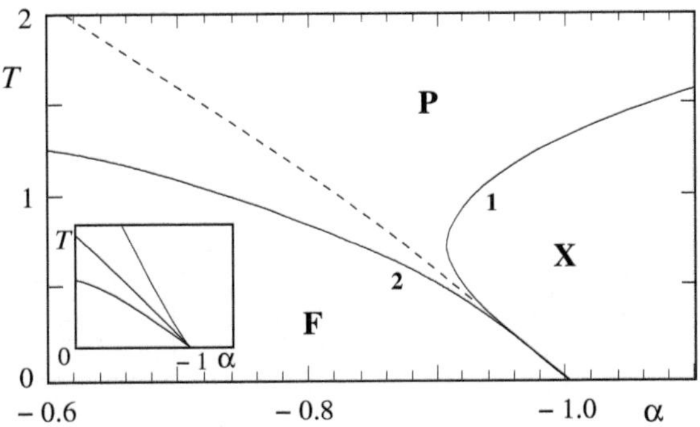

Fig. 1.20. Phase diagram of the Kagomé lattice with NNN interaction in the region $J_1 > 0$ of the space ($\alpha = J_2/J_1, T$). T is measured in the unit of J_1/k_B. Solid lines are critical lines, and dashed line is the disorder line. P, F, and X stand for paramagnetic, ferromagnetic, and partially disordered phases, respectively. The inset shows a schematically enlarged region of the endpoint.

and a partially disordered phase. This suggests that a partial disorder at the high-T phase is necessary to ensure that the entropy is larger than that of the reentrant phase.

1.4.2.2. *Generalized Kagomé lattice*

When all the interactions are different in the model shown in Fig. 1.7, i.e. the horizontal bonds J_3, the vertical bonds J_2, and the diagonal ones are not equal (see Fig. 1.21), the phase diagram becomes complicated with new features[17]: in particular, we show that reentrance can occur in an *infinite region* of the phase space. In addition, there may be *several reentrant phases* occurring for a given set of interactions when T varies.

The Hamiltonian is written as

$$H = -J_1 \sum_{(ij)} \sigma_i \sigma_j - J_2 \sum_{(ij)} \sigma_i \sigma_j - J_3 \sum_{(ij)} \sigma_i \sigma_j, \tag{1.46}$$

where $\sigma_i = \pm 1$ is an Ising spin occupying the lattice site i, and the first, second, and third sums run over the spin pairs connected by diagonal, vertical, and horizontal bonds, respectively. When $J_2 = 0$ and $J_1 = J_3$, one recovers the original NN Kagomé lattice.[15] The effect of J_2 in the case $J_1 = J_3$ has been shown above.

The phase diagram at temperature $T = 0$ is shown in Fig. 1.21 in the space ($\alpha = J_2/J_1$, $\beta = J_3/J_1$) for positive J_1. The GS spin configurations are also displayed. The hatched regions indicate the three partially

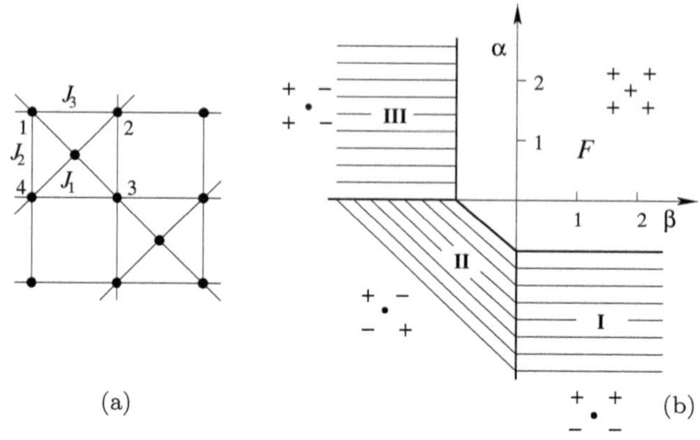

Fig. 1.21. (a) Generalized Kagomé lattice: diagonal, vertical and horizontal bonds denote the interactions J_1, J_2 and J_3, respectively. (b) Phase diagram of the GS shown in the plane ($\alpha = J_2/J_1, \beta = J_3/J_1$). Heavy lines separate different phases and spin configuration of each phase is indicated (up, down and free spins are denoted by +, − and o, respectively). The three kinds of partially disordered phases and the ferromagnetic phase are denoted by I, II, III and F, respectively.

disordered phases (I, II, and III) where the central spins are free. Note that the phase diagram is mirror-symmetric with respect to the change of the sign of J_1. With negative J_1, it suffices to reverse the central spin in the spin configuration shown in Fig. 1.21. Furthermore, the interchange of J_2 and J_3 leaves the system invariant, since it is equivalent to a $\pi/2$ rotation of the lattice. Let us consider the effect of the temperature on the phase diagram shown in Fig. 1.21. Partial disorder in the GS often gives rise to the reentrance phenomenon as in the systems shown above. Therefore, similar effects are to be expected in the present system. As it will be shown below, we find a new and richer behavior of the phase diagram: in particular, the reentrance region is found to be extended to infinity, unlike systems previously studied, and for some given set of interactions, there exist *two disorder lines* that divide the paramagnetic phase into regions of different kinds of fluctuations with a reentrant behavior.

Following the method discussed in Section 1.3, one obtains a checkerboard Ising model with multispin interactions. This resulting model is equivalent to a symmetric 16-vertex model that satisfies the free-fermion condition.[25,38,39] The critical temperature of the model is given by

$$\cosh(4K_1)\exp(2K_2 + 2K_3) + \exp(-2K_2 - 2K_3)$$
$$= 2\cosh(2K_3 - 2K_2) \pm 4\cosh(2K_1). \quad (1.47)$$

Note that Eq. (1.47) is invariant when changing $K_1 \to -K_1$ and interchanging K_2 and K_3 as stated earlier. The phase diagram in the 3D space (K_1, K_2, K_3) is rather complicated to show. Instead, we show in the following the phase diagram in the plane $(\beta = J_3/J_1, T)$ for typical values of $\alpha = J_2/J_1$. To describe each case and to follow the evolution of the phase diagram, let us move along the direction of decreasing α:

A. $\alpha > 0$.

This case is shown in Fig. 1.22. Two critical lines are found with the paramagnetic reentrance having a usual shape (Fig. 1.22(a)) between the partially disordered (PD) phase of type III (see Fig. 1.21) and the ferromagnetic (F) phase with an endpoint at $\beta = -1$. The width of the reentrance region $[-1, \beta_1]$ decreases with decreasing α, from $\beta_1 = 0$ for α at infinity to $\beta_1 = -1$ for $\alpha = 0$ (zero width). Note that as α decreases, the PD phase III is depressed and disappears at $\alpha = 0$, leaving only the F phase (one critical line, see Fig. 1.22(b)). The absence of order at zero α for β smaller than -1 results from the fact that in the GS, this region of parameters corresponds to a superdegenerate line separating the two PD phases II and III (see Fig. 1.21). So, along this line, the disorder contaminates the system for all T. As for disorder solutions, for positive α we find in the reentrant paramagnetic region a disorder line with dimension reduction[33] given by

$$\exp(4K_3) = 2\cosh(2K_2)/[\cosh(4K_1)\exp(2K_2) + \exp(-2K_2)]. \quad (1.48)$$

This is shown by the dotted lines in Fig. 1.22.

B. $0 > \alpha > -1$.

In this range of α, there are three critical lines. The critical line separating the F and P phases and the one separating the PD phase I from the P phase have a common horizontal asymptote as β tends to infinity. They form a reentrant paramagnetic phase between the F phase and the PD phase I for positive b between a value β_2 and infinite β (Fig. 1.23). An infinite region of reentrance like this has never been found before this model. As α decreases, β_2 tends to zero and the F phase is contracted. For $\alpha < -1$, the F phase disappears together with the reentrance.

In the interval $0 > \alpha > -1$, the phase diagram possesses two disorder lines, the first being given by Eq. (1.48) and the second being given by

$$\exp(4K_3) = 2\sinh(2K_2)/[-\cosh(4K_1)\exp(2K_2) + \exp(-2K_2)]. \quad (1.49)$$

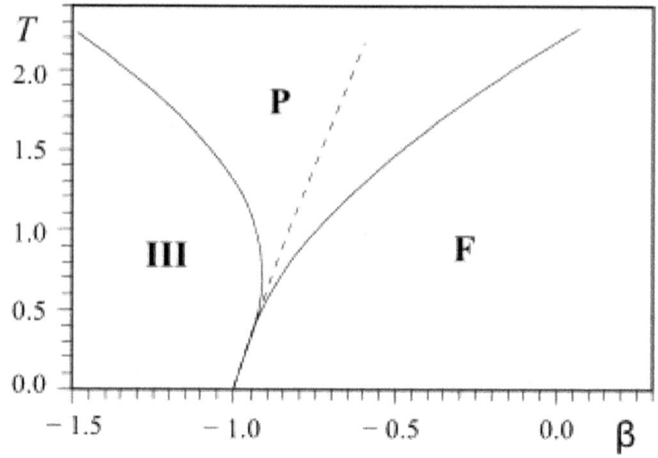

(a)

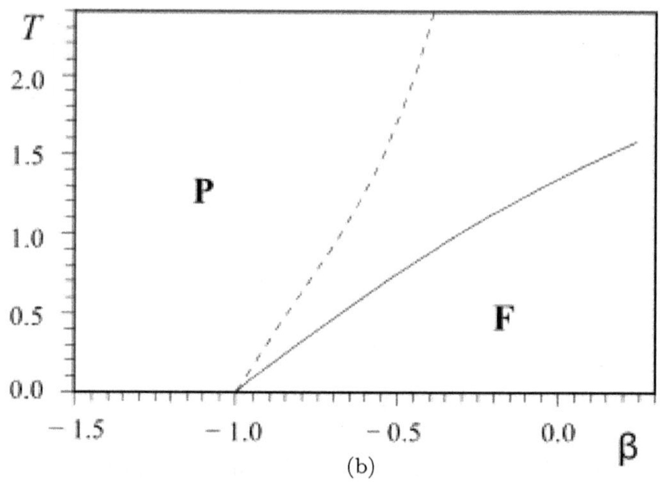

(b)

Fig. 1.22. Phase diagram in the plane ($\beta = J_3/J_1, T$) for positive values of $\alpha = J_2/J_1$: (a) $\alpha = 1$, (b) $\alpha = 0$. Solid lines are critical lines that separate different phases: paramagnetic (P), ferromagnetic (F), and partially disordered phase of type III (III). Dotted line shows the disorder line.

These two disorder lines are issued from a point near $\beta = -1$ for small negative α; this point tends to zero as α tends to -1. The disorder line given by Eq. (1.49) enters the reentrant region which separates the F phase and the PD phase I (Fig. 1.23(a)), and the one given by Eq. (1.48) tends to infinity with the asymptote $\beta = 0$ as $T \to \infty$. The most striking feature is the behavior of these two disorder lines at low T: they cross each other in the

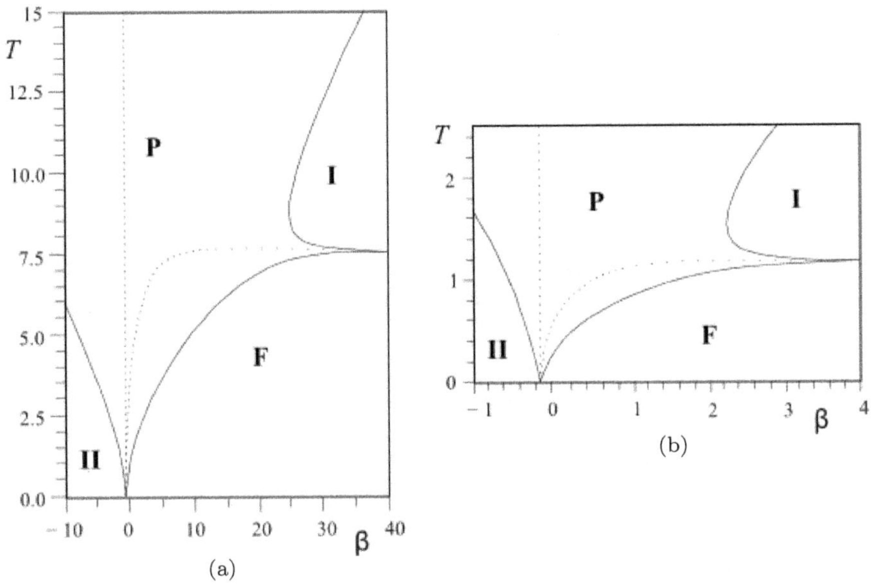

Fig. 1.23. Phase diagram in the plane ($\beta = J_3/J_1, T$) for negative values of $\alpha = J_2/J_1$. (a) $\alpha = -0.25$, (b) $\alpha = -0.8$. Solid lines are critical lines that separate the different phases: paramagnetic (P), ferromagnetic (F), and partially disordered phases of type I and III. Dotted lines show the disorder lines.

P phase for $0 > \alpha > -0.5$, forming regions of fluctuations of different nature (Fig. 1.24(a)). For $-0.5 > \alpha > -1$, the two disorder lines no longer cross each other (see Fig. 1.24(b)). The one given by Eq. (1.48) has a reentrant aspect: in a small region with negative values of β, one crosses three times this line in the P phase with decreasing T.

C. $\alpha \leq -1$.

For α smaller than -1, there are two critical lines and no reentrance (Fig. 1.25). Only the disorder line given by Eq. (1.48) survives with a reentrant aspect: in a small region of negative values of β, one crosses twice this line in the P phase with decreasing T. This behavior, being undistiguishable in the scale of Fig. 1.25, is schematically enlarged in Fig. 1.24(c). The multicritical point where the P, I, and II phases meet is found at $\beta = 0$ and $T = 0$.

At this stage, it is interesting to note that while reentrance and disorder lines occur along the horizontal axis $\alpha = -1$ and along the vertical axis $\beta = -1$ of Fig. 1.21 when the temperature is switched on, the most frustrated region ($\alpha < 0$ and $\beta < 0$) of the GS does not show successive

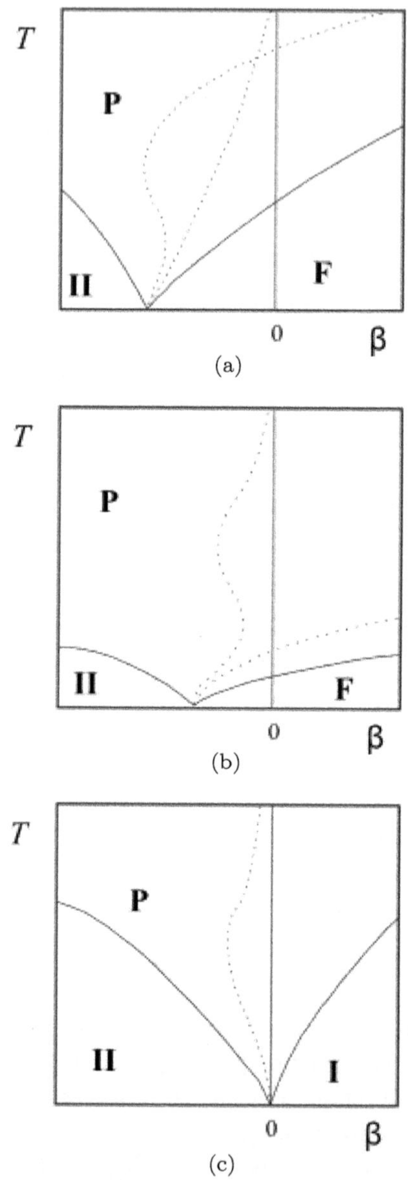

Fig. 1.24. The behavior of the disorder lines (dotted) is schematically enlarged in the cases (a) $\alpha = -0.25$, (b) $\alpha = -0.8$, and (c) $\alpha = -1.5$.

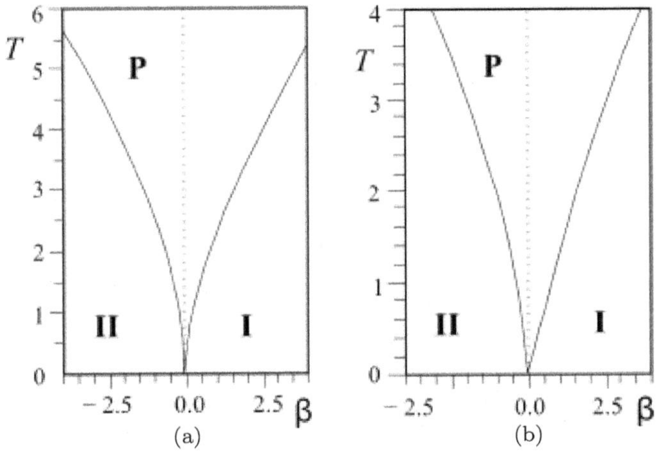

Fig. 1.25. The same caption as that of Fig. 1.23 with (a) $\alpha = -1$, (b) $\alpha = -1.5$.

phase transitions (see Fig. 1.25, for example). Therefore, the existence of a reentrance may require a sufficient frustration, but not overfrustration. Otherwise, the system may have either a PD phase (Fig. 1.25) or no order at all (Fig. 1.22(b)).

The origin of the reentrance phenomenon will be discussed again in the conclusion.

1.4.3. *Centered honeycomb lattice*

In order to find common aspects of the reentrance phenomenon, we have constructed a few other models that possess a partially disordered phase next to an ordered phase in the GS. Let us mention here the anisotropic centered honeycomb lattice shown in Fig. 1.11.[18] The Hamiltonian is given by Eq. (1.25), with three kinds of interactions J_1, J_2, and J_3 denoting the interactions between the spin pairs connected by heavy, light, and double-light bonds, respectively. We recall that when $J_2 = J_3 = 0$, one recovers the honeycomb lattice, and when $J_1 = J_2 = J_3$, one has the triangular lattice.

Figure 1.26 shows the phase diagram at temperature $T = 0$ for three cases $(J_1 \neq J_2 = J_3)$, $(J_1 \neq J_3, J_2 = 0)$, and $(J_1 \neq J_2, J_3 = 0)$. The GS spin configurations are also indicated.

The phase diagram is symmetric with respect to the horizontal axis: the transformation $(J_2, \sigma) \to (-J_2, -\sigma)$ or $(J_3, \sigma) \to (-J_3, -\sigma)$, leaves the system invariant. In each case, there is a phase where the central spins are free to flip ("partially disordered phase"). In view of this common feature

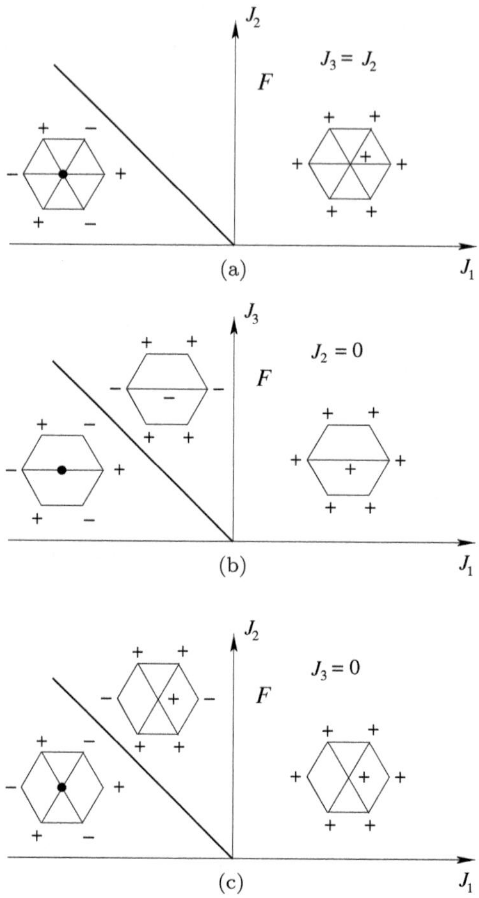

Fig. 1.26. Phase diagram of the GS shown in the space: (a) $(J_1, J_2 = J_3)$; (b) (J_1, J_3) with $J_2 = 0$; (c) (J_1, J_2) with $J_3 = 0$. Heavy lines separate different phases, and the spin configuration of each phase is indicated (up, down, and free spins are denoted by $+$, $-$, and o, respectively).

with other models studied so far, one expects a reentrant phase occurring between the partially disordered phase and its neighboring phase at finite T. As it will be shown below, though a partial disorder exists in the GS, it does not in all cases studied here yield a reentrant phase at finite T. Only the case $(J_1 \neq J_2, J_3 = 0)$ shows a reentrance.

To obtain the exact solution of our model, we decimate the central spin of each elementary cell of the lattice as outlined in Section 1.3. The resulting model is equivalent to a special case of the 32-vertex model[26] on a triangular lattice that satisfies the free-fermion condition. The explicit expression of the free energy as a function of interaction parameters K_1, K_2, and K_3 is very complicated, as seen by replacing Eq. (1.29) in Eqs. (1.15)–(1.17). The critical temperature is given by Eq. (1.30).

We have analyzed, in particular, the three cases ($K_1 \neq K_2 = K_3$), ($K_1 \neq K_3, K_2 = 0$), and ($K_1 \neq K_2, K_3 = 0$).

When $K_2 = K_3$, the critical line obtained from Eq. (1.30) is

$$\exp(3K_1)\cosh(6K_2) + \exp(-3K_1) = 3[\exp(K_1) + \exp(-K_1)\cosh(2K_2)]. \tag{1.50}$$

In the case $K_2 = 0$, the critical line is given by

$$\exp(3K_1)\cosh(2K_3) + \exp(-3K_1) = 3[\exp(K_1) + \exp(-K_1)\cosh(2K_3)]. \tag{1.51}$$

Note that these equations are invariant with respect to the transformation $K_2 \to -K_2$ (see Eq. (1.50)) and $K_3 \to -K_3$ (see Eq. (1.51)).

The phase diagrams obtained from Eqs. (1.44) and (1.45) are shown in Figs. 1.27(a) and 1.27(b), respectively.

These two cases do not present the reentrance phenomenon though a partially disordered phase exists next to an ordered phase in the GS (this is seen by plotting a line from the origin, i.e. from infinite T: this line never crosses twice a critical line whatever its slope, i.e. the ratio $K_{2,3}/K_1$, is). In the ordered phase II, the partial disorder, which exists in the GS, remains so up to the phase transition. This has been verified by examining the Edwards–Anderson order parameter associated with the central spins in MC simulations.[18] Note that when $K_2 = K_3 = 0$, one recovers the transition at finite temperature found for the honeycomb lattice[40] and when $K_2 = K_3 = K_1 = -1$ one recovers the antiferromagnetic triangular lattice which has no phase transition at finite temperature.[4] The case $K_2 = 0$ (Fig. 1.27(b)) does not have a phase transition at finite T in the range $-\infty < K_3/K_1 < -1$, and phase II has a partial disorder similar to that in Fig. 1.22(a).

On the other hand the case $K_3 = 0$ shows a reentrant phase. The critical lines are determined from the equations

$$\cosh(4K_2) = \frac{\exp(4K_1) + 2\exp(2K_1) + 1}{[1 - \exp(4K_1)]\exp(2K_1)}, \tag{1.52}$$

$$\cosh(4K_2) = \frac{3\exp(4K_1) + 2\exp(2K_1) - 1}{[\exp(4K_1) - 1]\exp(2K_1)}. \tag{1.53}$$

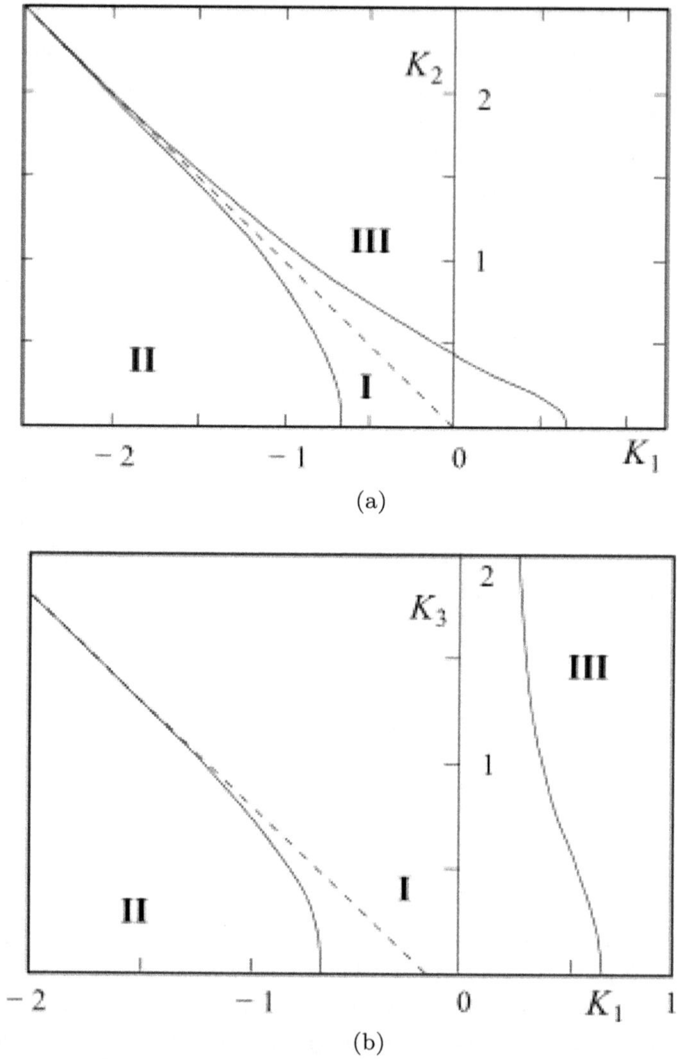

Fig. 1.27. Phase diagram shown in the space (a) $(K_1, K_2 = K_3)$; (b) $(K_1, K_3, K_2 = 0)$. Solid lines are critical lines that separate paramagnetic (I), partially disordered (II), and ordered (III) phases. Discontinued lines of slope -1 are the asymptotes.

Figure 1.28 shows the phase diagram obtained from Eqs. (1.52) and (1.53). The reentrant paramagnetic phase goes down to zero temperature with an end point at $\alpha = -0.5$ (see Fig. 1.28(b)).

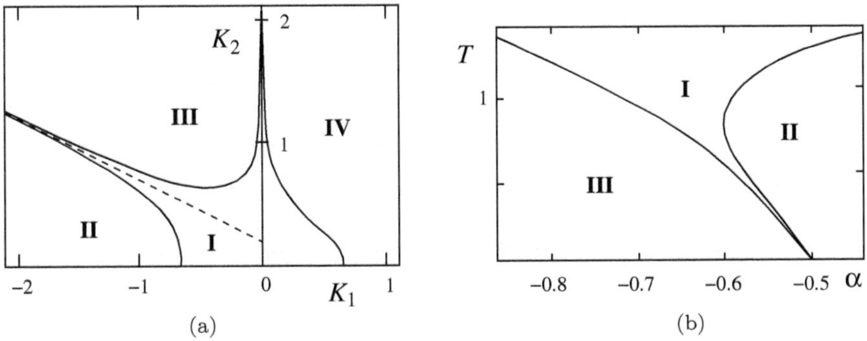

Fig. 1.28. Phase diagram of the centered honeycomb lattice with reentrance in the space (K_1, K_2) (a) and in the space $(T, \alpha = K_2/K_1)$ (b). I, II, III phases are paramagnetic, partially disordered, and ordered phases, respectively. Discontinued line is the asymptote.

Note that the honeycomb model that we have studied here does not present a disorder solution with a dimensional reduction.

1.4.4. *Periodically dilute centered square lattices*

In this section, we show the exact results on several periodically dilute centered square Ising lattices by transforming them into 8-vertex models of *different vertex statistical weights* that satisfy the free-fermion condition. The dilution is introduced by taking away a number of centered spins in a periodic manner. For a given set of interactions, there may be five transitions with decreasing temperature with two reentrant paramagnetic phases. These two phases extend to infinity in the space of interaction parameters. Moreover, two additional reentrant phases are found, each in a limited region of phase space.[19]

Let us consider several *periodically dilute centered square lattices* defined from the centered square lattice shown in Fig. 1.29.

The Hamiltonian of these models is given by

$$H = -J_1 \sum_{(ij)} \sigma_i \sigma_j - J_2 \sum_{(ij)} \sigma_i \sigma_j - J_3 \sum_{(ij)} \sigma_i \sigma_j, \tag{1.54}$$

where $\sigma_i = \pm 1$ is an Ising spin occupying the lattice site i, and the first, second, and third sums run over the spin pairs connected by diagonal, vertical, and horizontal bonds, respectively. All these models have at least one partially disordered phase in the GS, caused by the competing interactions.

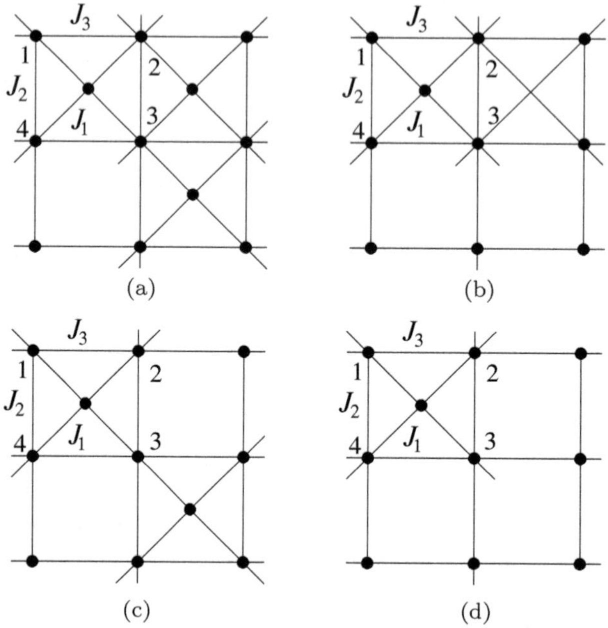

Fig. 1.29. Elementary cells of periodically dilute centered square lattice: (a) three-center case, (b) two-adjacent-center case, (c) two-diagonal-center case, (d) one-center case. Interactions along diagonal, vertical, and horizontal bonds are J_1, J_2, and J_3, respectively.

The model shown in Fig. 1.29(c) is in fact the generalized Kagomé lattice[17] shown above. The other models are less symmetric and require different vertex weights as will be seen below.

Let us show in Fig. 1.30 the phase diagrams, at $T = 0$, of the models shown in Figs. 1.25(a), 1.25(b) and 1.25(d), in the space (a, b) where $a = J_2/J_1$ and $b = J_3/J_1$. The spin configurations in different phases are also displayed. The three-center case (Fig. 1.30(a)) has six phases (numbered from I to VI), five of which (I, II, IV, V, and VI) are partially disordered (with, at least, one centered spin being free), while the two-center case (Fig. 1.30(b)) has five phases, three of which (I, IV, and V) are partially disordered. Finally, the one-center case has seven phases with three partially disordered ones (I, VI, and VII).

As will be shown later, in each model, the reentrance occurs along most of the critical lines when the temperature is switched on. This is a

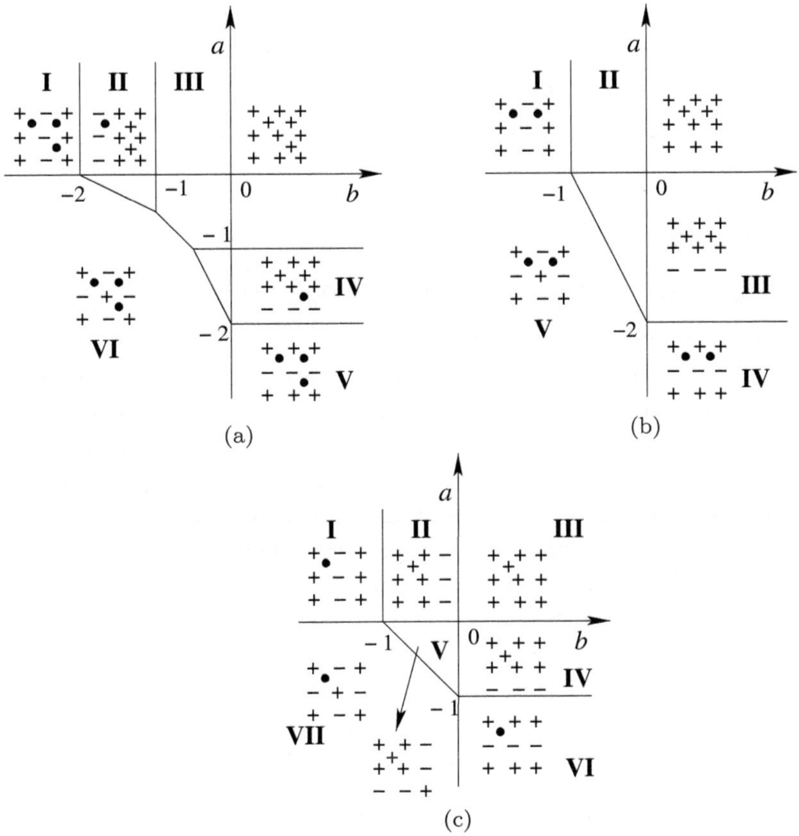

Fig. 1.30. Phase diagrams in the plane ($a = J_2/J_1$, $b = J_3/J_1$) at $T = 0$ are shown for the three-center case (a), two-adjacent center case (b), and one-center case (c). Critical lines are drawn by heavy lines. Each phase is numbered and the spin configuration is indicated (+, −, and o are up, down, and free spins, respectively). Degenerate configurations are obtained by reversing all spins.

very special feature of the models shown in Fig. 1.29 and has not been found in other models.

The partition function is written as

$$Z = \prod_j \sum_\sigma W_j, \tag{1.55}$$

where the sum is performed over all spin configurations and the product is performed over all the elementary squares. W_j is the statistical weight of the jth square. Let us denote the centered spin (when it exists) by σ and the spins at the square corners by σ_1, σ_2, σ_3, and σ_4. If the centered site exists, the statistical weight W_j of the square is

$$W_j = \exp[K_1(\sigma_1\sigma_2 + \sigma_3\sigma_4) + K_2(\sigma_1\sigma_4 + \sigma_2\sigma_3) + K_3\sigma(\sigma_1 + \sigma_2 + \sigma_3 + \sigma_4)]. \tag{1.56}$$

Otherwise, it is given by

$$W_j = \exp[K_1(\sigma_1\sigma_2 + \sigma_3\sigma_4) + K_2(\sigma_1\sigma_4 + \sigma_2\sigma_3)], \tag{1.57}$$

where $K_i = J_i/k_BT$ $(i = 1, 2, 3)$.

In order to obtain the exact solution of these models, we decimate the central spins of the centered squares. The resulting system is equivalent to an 8-vertex model on a square lattice, but with *different vertex weights*. For example, when one center is missing (Fig. 1.29(d)), three squares over four have the same weight W_i, and the fourth has a weight $W_i' \neq W_i$. So, we have to define four different sublattices with different statistical weights. The problem has been studied by Hsue, Lin, and Wu for two different sublattices[41] and by Lin and Wang[42] for four sublattices. They showed that an exact solution can be obtained provided that all different statistical weights satisfy the free-fermion condition.[11,25,41,42] This is indeed our case, and we get the exact partition function in terms of interaction parameters. The critical surfaces of our models are obtained by

$$\Omega_1 + \Omega_2 + \Omega_3 + \Omega_4 = 2\max(\Omega_1, \Omega_2, \Omega_3, \Omega_4), \tag{1.58}$$

where W_i are functions of K_1, K_2, and K_3. We explicit this equation and we obtain a second-order equation for X, which is a function of K_2 only:

$$A(K_1, K_3)X^2 + B(K_1, K_3)X + C(K_1, K_3) = 0 \tag{1.59}$$

with *a priori* four possible values of A, B, and C for each model.

For given values of K_1 and K_3, the critical surface is determined by the value of K_2 which satisfies Eq. (1.59) through X. X must be real positive. We show in the following the expressions of A, B, and C for which this condition is fulfilled for each model. Equation (1.58) may have as much as five solutions for the critical temperature,[42] and the system may, for some given values of interaction parameters, exhibit up to five phase transitions. This happens for the model with three centers, when one of the interaction is large positive and the other slightly negative, the diagonal one being taken as unit. In general, we obtain one or three solutions for T_c.

1.4.4.1. *Model with three centers*

(Fig. 1.29(a))

The quantities that satisfy Eq. (1.59) are given by

$$
\begin{aligned}
X &= \exp(4K_2),\\
A &= \exp(4K_1)\cosh^3(4K_3) + \exp(-4K_1) - \cosh^2(4K_3) - \cosh(4K_3),\\
B &= \pm\{1 + 3\cosh(4K_3) + 8\cosh^3(2K_3)\\
&\quad + [\cosh(4K_3) + \cosh^2(4K_3)]\exp(4K_1) + 2\exp(-4K_1)\},\\
C &= [\exp(2K_1) - \exp(-2K_1)]^2,\\
A &= \exp(4K_1)\cosh^3(4K_3) + \exp(-4K_1) + \cosh^2(4K_3) + \cosh(4K_3),\\
B &= [1 + 3\cosh(4K_3) + 8\cosh^3(2K_3)\\
&\quad - (\cosh(4K_3) + \cosh^2(4K_3))\exp(4K_1) - 2\exp(-4K_1)],\\
C &= [\exp(2K_1) + \exp(-2K_1)]^2.
\end{aligned}
\tag{1.60}
$$

Let us now describe in detail the phase diagram of the three-center model (Fig. 1.29(a)).

For clarity, we show in Fig. 1.31 the phase diagram in the space ($a = J_2/J_1$, T) for typical values of $b = J_3/J_1$.

For $b < -1$, there are two reentrances. Figure 1.31(a) shows the case of $b = -1.25$ where the nature of the ordering in each phase is indicated using the same numbers of corresponding GS configurations (see Fig. 1.30). Note that all phases (I, II, and VI) are partially disordered: the centered spins that are disordered at $T = 0$ (Fig. 1.30(a)) remain so at all T. As seen, one paramagnetic reentrance is found in a small region of negative a (schematically enlarged in the inset of Fig. 1.31(a)), and the other on the positive a extending to infinity. The two critical lines in this region have a common horizontal asymptote.

For $-1 < b < -0.5$, there are three reentrant paramagnetic regions as shown in Fig. 1.31(b): the reentrant region on the negative a is very narrow (inset), and the two on the positive a become so narrower while a goes to infinity that they cannot be seen on the scale of Fig. 1.31. Note that the critical lines in these regions have horizontal asymptotes. For a large value of a, one has five transitions with decreasing T: paramagnetic state — partially disordered phase I; reentrant paramagnetic phase — partially disordered phase II; and reentrant paramagnetic phase — ferromagnetic phase (see

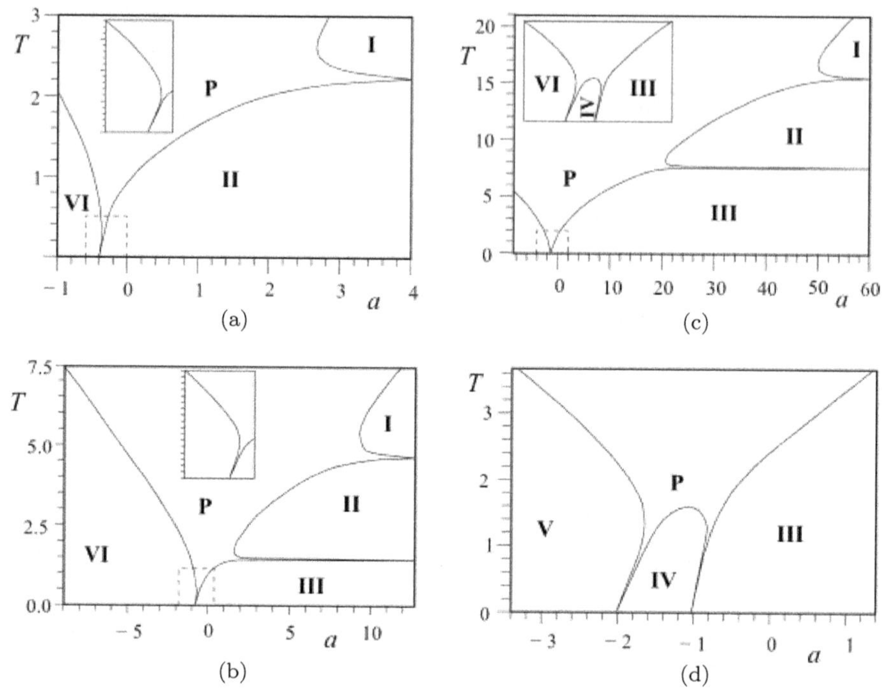

Fig. 1.31. Phase diagrams in the plane $(T, a = J_2/J_1)$ for several values of $b = J_3/J_1$: (a) $b = -1.25$, (b) $b = -0.75$, (c) $b = -0.25$, and (d) $b = 0.75$. Reentrant regions on negative sides of a (limited by discontinued lines) are schematically enlarged in the insets. The nature of ordering in each phase is indicated by a number which is referred to the corresponding spin configuration in Fig. 1.30. P is paramagnetic phase.

Fig. 1.31(b)). So far, this is the first model that exhibits such successive phase transitions with two reentrances.

For $-0.5 < b < 0$, there is an additional reentrance for $a < -1$: this is shown in the inset of Fig. 1.31(c). As b increases from negative values, the ferromagnetic region (III) in the phase diagram "pushes" the two partially disordered phases (I and II) toward higher T. At $b = 0$, these two phases disappear at infinite T, leaving only the ferromagnetic phase. For positive b, there are thus only two reentrances remaining on a negative region of a, with endpoints at $a = -2$ and $a = -1$, at $T = 0$ (see Fig. 1.31(d)).

1.4.4.2. *Model with two adjacent centers*

(Fig. 1.29(b))

The quantities that satisfy Eq. (1.59) are given by

$$
\begin{aligned}
X &= \exp(2K_2),\\
A &= \exp(2K_1)\cosh(4K_3) + \exp(-2K_1),\\
B &= 2[\exp(2K_1)\cosh^2(2K_3) - \exp(-2K_1)],\\
C &= \exp(2K_1) + \exp(-2K_1),\\
A &= \exp(2K_1)\cosh(4K_3) - \exp(-2K_1),\\
B &= \pm 2[\exp(2K_1)\cosh^2(2K_3) + \exp(-2K_1)],\\
C &= \exp(2K_1) - \exp(-2K_1).
\end{aligned}
\tag{1.61}
$$

The phase diagram is shown in Fig. 1.32.

For $b < -1$, this model shows only one transition at a finite T for a given value of a, except when $a = 0$ where the paramagnetic state goes down to $T = 0$ (see Fig. 1.32(a)).

However, for $-1 < b < 0$, two reentrances appear, the first one separating phases I and II goes to infinity with increasing a, and the second one exists in a small region of negative a with an endpoint at ($a = -2 - 2b$, $T = 0$). The slope of the critical lines at $a = 0$ is vertical (see inset of Fig. 1.32(b)).

As b becomes positive, the reentrance on the positive side of a disappears (Fig. 1.32(c)), leaving only phase III (ferromagnetic).

1.4.4.3. *Model with one center*

(Fig. 1.29(d))

For this case, the quantities that satisfy Eq. (1.59) are given by

$$
\begin{aligned}
X &= \exp(4K_2),\\
A &= \exp(4K_1)\cosh(4K_3) + \exp(-4K_1) - 2\cosh^2(2K_3),\\
B &= \pm 2\{[\cosh(2K_3) + 1]^2 + [\exp(2K_1)\cosh(2K_3) + \exp(-2K_1)]^2\},
\end{aligned}
$$

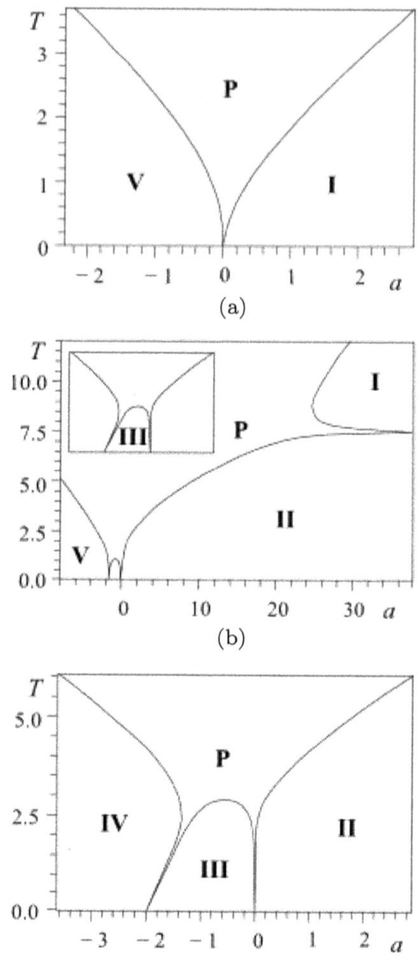

Fig. 1.32. Two-center model: the same caption as that of Fig. 1.31 with (a) $b = -1.25$, (b) $b = -0.25$, and (c) $b = 2$.

$$
\begin{aligned}
C &= [\exp(2K_1) - \exp(-2K_1)]^2, \\
A &= \exp(4K_1)\cosh(4K_3) + \exp(-4K_1) + 2\cosh^2(2K_3), \\
B &= \pm 2\{[\cosh(2K_3) + 1]^2 - [\exp(2K_1)\cosh(2K_3) - \exp(-2K_1)]^2\}, \\
C &= [\exp(2K_1) + \exp(-2K_1)]^2. \qquad (1.62)
\end{aligned}
$$

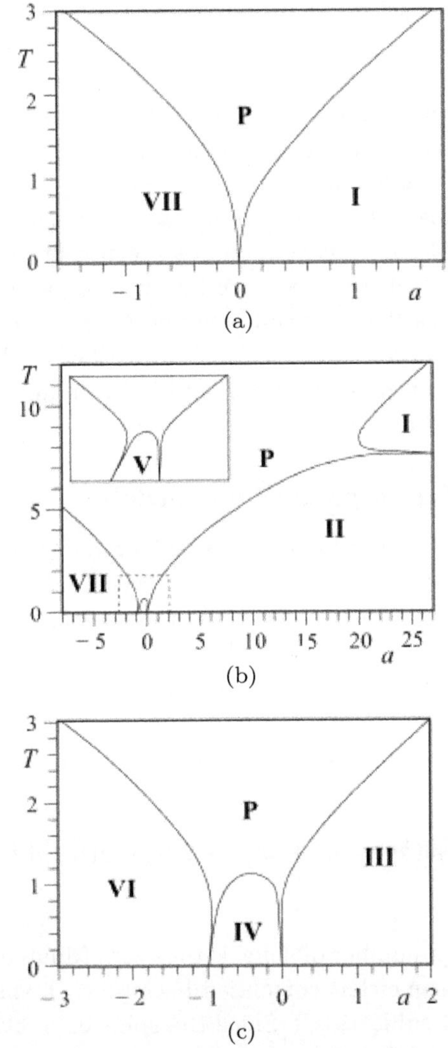

Fig. 1.33. One-center case: the same caption as that of Fig. 1.31 with (a) $b = -1.25$, (b) $b = -0.25$, and (c) $b = 0.5$.

The phase diagrams of this model shown in Fig. 1.33 for $b < -1$, $-1 < b < 0$ and $b > 0$ are very similar to those of the two-center model shown in Fig. 1.32. This is not unexpected if one examines the GS phase diagrams of the two cases (Figs. 1.25(b) and 1.25(c)): their common point is the existence of a partially disordered phase next to an ordered phase. The difference between the one- and two-center cases and the three-center case shown above is that the latter has, in addition, two boundaries, each of which separates two partially disordered phases (see Fig. 1.30(a)). It is along these boundaries that the two additional reentrances take place at finite T in the three-center case.

In concluding this section, we summarize that in simple models such as those shown in Fig. 1.29, we have found two reentrant phases occurring on the temperature scale at a given set of interaction parameters. A striking feature is the existence of a reentrant phase between *two partially disordered phases* which has not been found so far in any other model (we recall that in other models, a reentrant phase is found between an ordered phase and a partially disordered phase).

1.4.5. *Random-field aspects of the models*

Let us touch upon the random-field aspect of the model. A connection between the Ising models presented above and the random-field problem can be established. Consider for instance the centered square lattice (Fig. 1.17). In the region of antiferromagnetic ordering of sublattice 2 ($a = J_2/|J_1| < -1$), the spins on sublattice 1 (centered spins) are free to flip. They act on their neighboring spins (sublattice 2) as an annealed random field h. The probability distribution of this random field at a site of sublattice 2 is given by

$$P(h) = \frac{1}{16}[6\delta(h) + 4\delta(h+2J_1) + 4\delta(h-2J_1) + \delta(h+4J_1) + \delta(h-4J_1)]. \tag{1.63}$$

The random field at a number of spins is thus zero (diluted). Moreover, this field distribution is somewhat correlated because each spin of sublattice 1 acts on four spins of sublattice 2. Since the spins on sublattice 1 are completely disordered at all T, it is reasonable to consider this effective random-field distribution as quenched. In addition to possible local annealed effects, in this model the phase transition at a finite T may be a consequence of the above-mentioned dilution and correlations of the field distribution, because it is known that in a 2D random-field Ising model (without dilution) there is no such a transition.[43]

The same argument is applied to other models studied above.

1.5. Evidence of Partial Disorder and Reentrance in Other Frustrated Systems

The partial disorder and the reentrance that occur in exactly solved Ising systems shown above are also expected to occur in models other than the Ising one as well as in some 3D systems. Unfortunately, these systems cannot be exactly solved. One has to use approximations or numerical simulations to study them. This renders difficult the interpretation of the results. Nevertheless, in the light of what has been found in exactly solved systems, we can introduce the necessary ingredients into the model under study if we expect the same phenomenon to occur.

As seen above, the most important ingredient for a partial disorder and a reentrance to occur at low T in the Ising model is the existence of a number of free spins in the GS.

In three dimensions, apart from a particular exactly solved case[21] showing a reentrance, a few Ising systems such as the fully frustrated simple cubic lattice,[31,32] a stacked triangular Ising antiferromagnet (see Chapter 2),[44] and a body-centered cubic (bcc) crystal[45] exhibit a partially disordered phase in the GS. We believe that reentrance should also exist in the phase space of such systems though evidence is found numerically only for the bcc case.[45]

In two dimensions, a few non-Ising models also show evidence of a reentrance. For the q-state Potts model, evidence of a reentrance is found in a recent study of the 2D frustrated Villain lattice (the so-called piled-up domino model) by a numerical transfer matrix calculation.[46,47] It is noted that the reentrance occurs near the fully frustrated situation, i.e. $\alpha_c = J_{AF}/J_F = -1$ (equal antiferromagnetic and ferromagnetic bond strengths), for q between $\simeq$1.0 and $\simeq$4. Note that there is no reentrance in the case $q = 2$. Below (above) this q value, the reentrance occurs above (below) the fully frustrated point α_c as shown in Figs. 1.34 and 1.35. For q larger than $\simeq$4, the reentrance disappears.[47]

A frustrated checkerboard lattice with XY spins also shows evidence of a paramagnetic reentrance.[48]

In vector spin models such as the Heisenberg and XY models, the frustration is shared by all bonds so that no free spins exist in the GS. However, one can argue that if there are several kinds of local fields in the GS due to several kinds of interaction, then there is a possibility that a subsystem

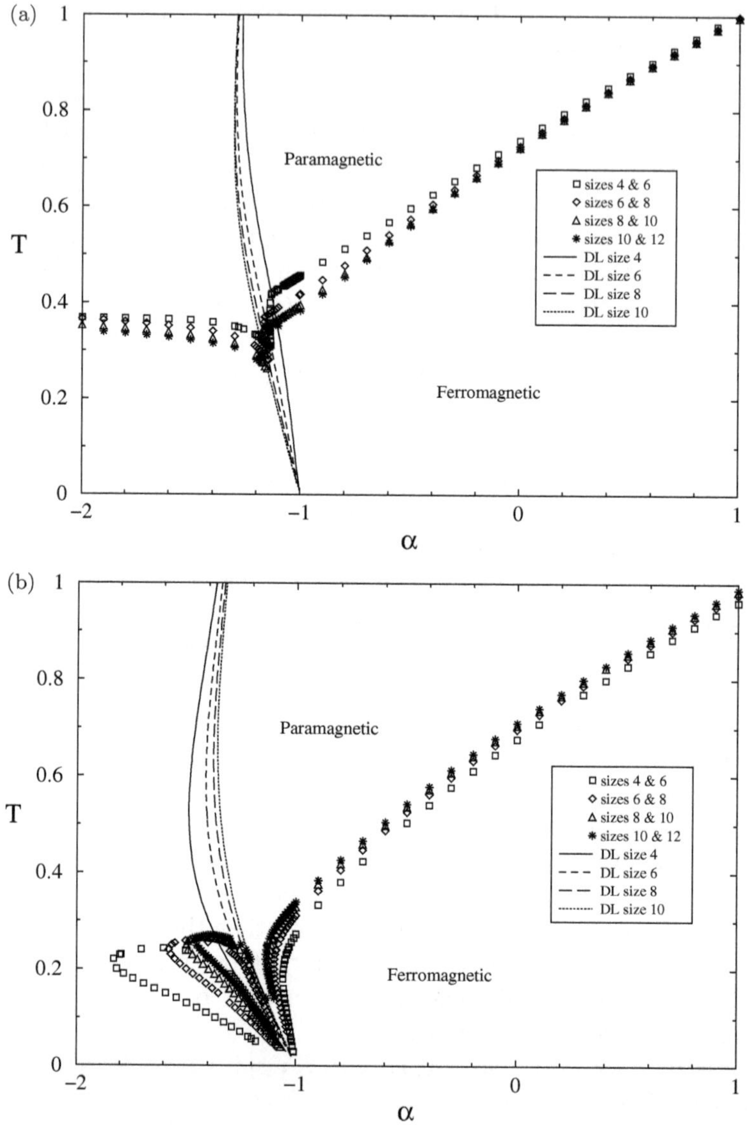

Fig. 1.34. Phase diagram for the Potts piled-up-domino model with $q = 3$: periodic boundary conditions (a) and free boundary conditions (b). The disorder lines are shown as lines and the phase boundaries as symbols. The numerical uncertainty is smaller than the size of the symbols.[47]

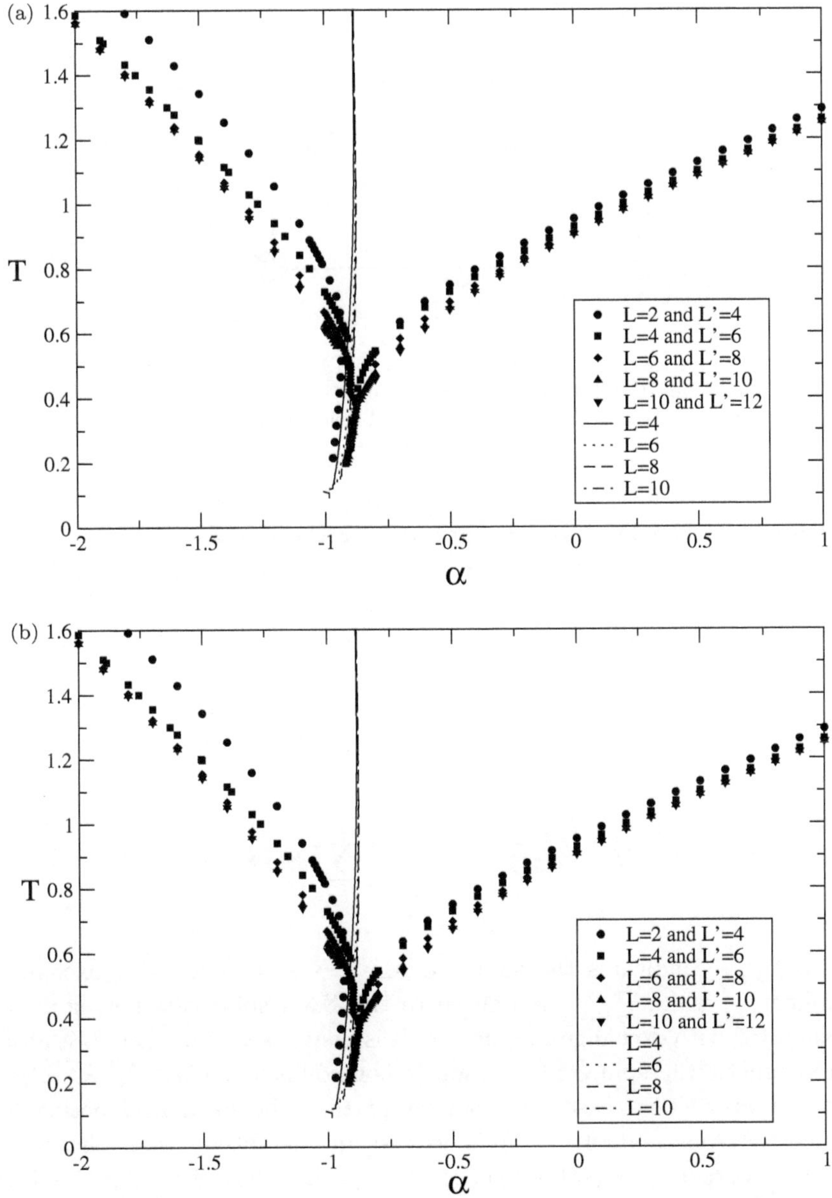

Fig. 1.35. The phase diagram for $q = 1.5$ found using transfer matrices and the phenomenological renormalization group with periodic (a) and free boundary conditions (b). The points correspond to finite-size estimates for T_c, while the lines correspond to the estimates for the disorder line ($\alpha = J_2/J_1$).[47]

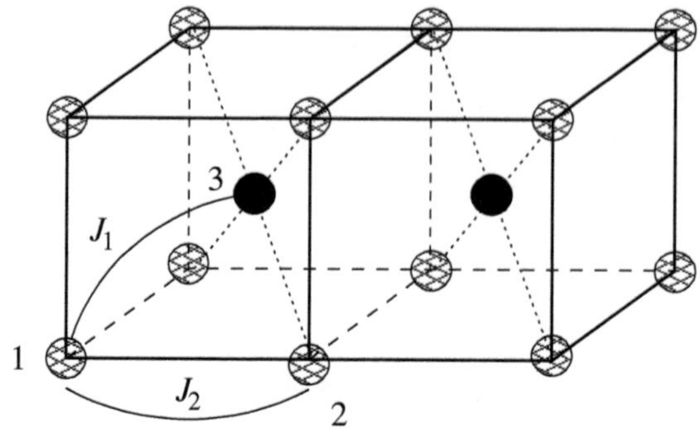

Fig. 1.36. bcc lattice. Spins are shown by gray and black circles. Interaction between NN (spins numbered 1 and 3) is denoted by J_1 and that between NNN (spins 1 and 2) by J_2. Note that there is no interaction between black spins.

with weak local field is disordered at low T while those of stronger local field stay ordered up to higher temperatures. This conjecture has been verified in a number of recent works on classical[49] and quantum spins.[50, 51] Consider, for example, Heisenberg spins $\mathbf{S_i}$ on a bcc lattice with a unit cell shown in Fig. 1.36.[49]

For convenience, let us call sublattice 1 the sublattice containing the sites at the cube centers and sublattice 2 the other sublattice. The Hamiltonian reads

$$H = -\frac{1}{2}\sum_{\langle i,j\rangle_1} \mathbf{S_i}.\mathbf{S_j} - \frac{1}{2}\sum_{\langle i,j\rangle_2} \mathbf{S_i}\cdot\mathbf{S_j},$$

where $\sum_{\langle i,j\rangle_1}$ indicates the sum over the NN spin pairs with exchange coupling J_1, while $\sum_{\langle i,j\rangle_2}$ is limited to the NNN spin pairs belonging to sublattice 2 with exchange coupling J_2. It is easy to see that when J_2 is antiferromagnetic the spin configuration is non-collinear for $J_2/|J_1| < -2/3$. In the non-collinear case, one can verify that the local field acting at the center spin (sublattice 1) is weaker in magnitude than that acting on a corner spin (sublattice 2). The partial disorder is observed in Fig. 1.37: the sublattice of black spins (sublattice 1) is disordered at a low T.

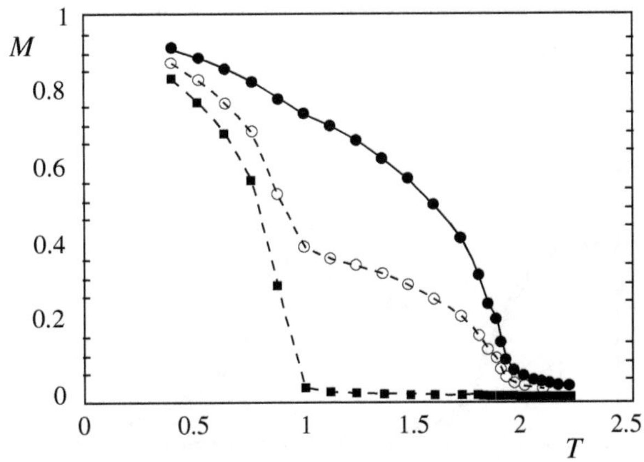

Fig. 1.37. MC results for sublattice magnetizations vs. T in the case $J_1 = -1$, $J_2 = -1.4$: black squares and black circles are for sublattices 1 and 2, respectively. Void circles indicate the total magnetization.

The same argument is applied for quantum spins.[51] Consider the bcc crystal as shown in Fig. 1.36, but the sublattices are now supposed to have different spin magnitudes, for example $S_A = 1/2$ (sublattice 1) and $S_B = 1$ (sublattice 2). In addition, one can include NNN interactions in both sublattices, namely J_{2A} and J_{2B}. The Green function technique is then applied for this quantum system.[51] We show in Fig. 1.38 the partial disorder observed in two cases: sublattice 1 is disordered (Fig. 1.38(a)) or sublattice 2 is disordered (Fig. 1.38(b)). In each case, one can verify, using the corresponding parameters, that in the GS the spin of the disordered sublattice has an energy lower than a spin in the other sublattice.

We show in Fig. 1.39 the specific heat vs. T for the parameters used in Fig. 1.38. One observes the two peaks corresponding to the two phase transitions associated with the loss of sublattice magnetizations.

The necessary condition for the occurrence of a partial disorder at finite T is thus the existence of several kinds of sites with different energies in the GS. This has been so far verified in a number of systems as shown above.

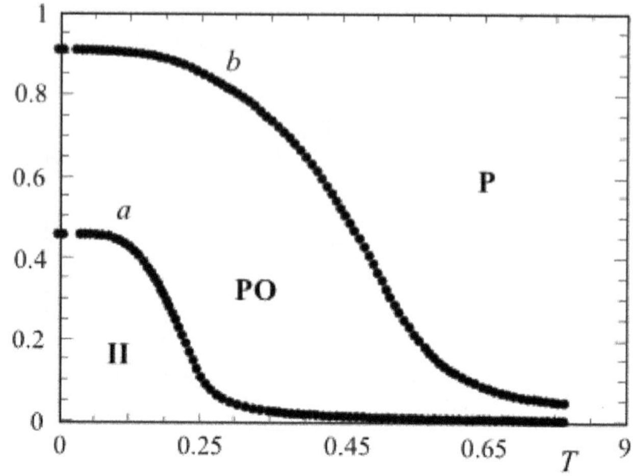

(a)

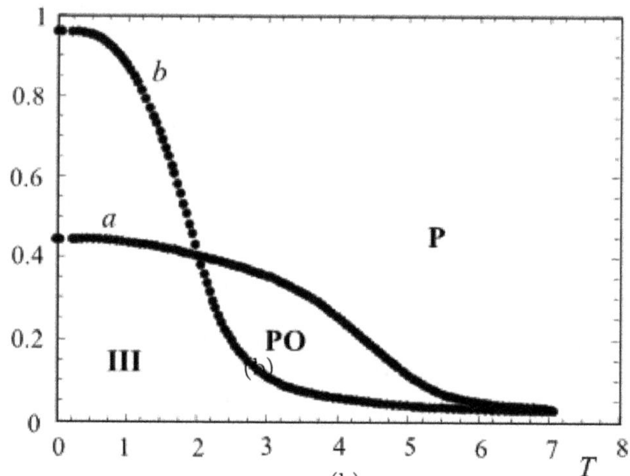

(b)

Fig. 1.38. Sublattice magnetizations vs. T in the case $S_A = 1/2$, $S_B = 1$: (a) curve a (b) is the sublattice-1(2) magnetization in the case $J_{2A}/|J_1| = 0.2$ and $J_{2B}/|J_1| = 0.9$ (b) curve a (b) is the sublattice-1(2) magnetization in the case $J_{2A}/|J_1| = 2.2$ and $J_{2B}/|J_1| = 0.1$. P is the paramagnetic phase, PO the partial order phase (only one sublattice is ordered), and II and III are non-collinear spin configuration phases. See the text for comments.

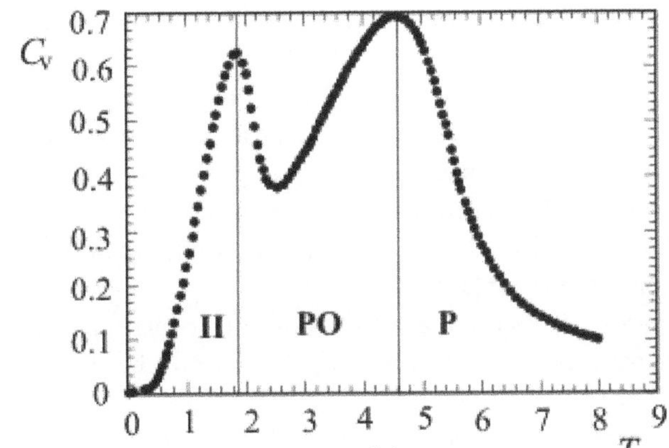

(a)

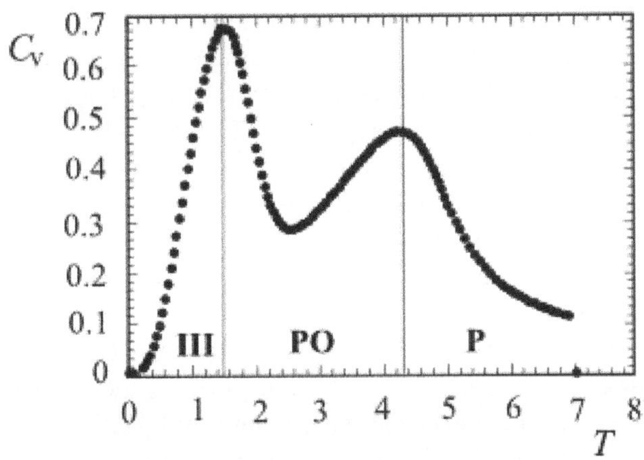

(b)

Fig. 1.39. Specific heat vs. T of the parameters used in Fig. 1.38, $S_A = 1/2$, $S_B = 1$: (a) $J_{2A}/|J_1| = 0.2$ and $J_{2B}/|J_1| = 0.9$ (b) $J_{2A}/|J_1| = 2.2$ and $J_{2B}/|J_1| = 0.1$. See the caption of Fig. 1.38 for the meaning of P, PO, II, and III. See the text for comments.

1.6. Conclusion

In this chapter, we have discussed some properties of periodically frustrated Ising systems. We have limited the discussion to exactly solved models which possess at least a reentrant phase. Other Ising systems that involve approximations are discussed in Chapter 2 and in the book by Liebmann.[1]

Let us emphasize that simple models having no bond disorder like those presented in this chapter can possess complicated phase diagrams due to the frustration generated by competing interactions. Many interesting physical phenomena such as successive phase transitions, disorder lines, and reentrance are found. In particular, a reentrant phase can occur in an infinite region of parameters. For a given set of interaction parameters in this region, successive phase transitions take place on the temperature scale, with one or two paramagnetic reentrant phases.

The relevance of disorder solutions for the reentrance phenomena has also been pointed out. An interesting finding is the occurrence of two disorder lines that divide the paramagnetic phase into regions of different kinds of fluctuations (Section 1.4.2). Therefore, care should be taken in analyzing experimental data such as correlation functions, susceptibility, etc., in the paramagnetic phase of frustrated systems.

Although the reentrance is found in the models shown above by exact calculations, there is no theoretical explanation why such a phase can occur. In other words, what is the necessary and sufficient condition for the occurrence of a reentrance? We have conjectured[16, 19] that the necessary condition for a reentrance to take place is the existence of at least a partially disordered phase next to an ordered phase or another partially disordered phase in the GS. The partial disorder is due to the competition between different interactions.

The existence of a partial disorder yields the occurrence of a reentrance in most known cases,[12–14, 16, 17, 19] except in some particular regions of interaction parameters in the centered honeycomb lattice (Section 1.4.3): the partial disorder alone is not sufficient to make a reentrance as shown in Figs. 1.27(a) and 1.27(b), the finite zero-point entropy due to the partial disorder of the GS is the same for the three cases considered in Fig. 1.26, i.e. $S_0 = \log(2)/3$ per spin, but only one case yields a reentrance. Therefore, the existence of a partial disorder is a necessary, but not sufficient, condition for the occurrence of a reentrance.

The anisotropic character of the interactions can also favor the occurrence of the reentrance. For example, the reentrant region is enlarged by anisotropic interactions as in the centered square lattice[14] and becomes infinite in the generalized Kagomé model (Section 1.4.2). But again, this alone cannot cause a reentrance as seen by comparing the anisotropic cases shown in Figs. 1.26(b) and 1.26(c): only in the latter case does a reentrance occur. The presence of a reentrance may also require a coordination number at a disordered site large enough to influence the neighboring ordered sites.

When it is too small such as in the case shown in Fig. 1.26(b) (equal to two), it cannot induce a reentrance. However, it may have an upper limit to avoid the disorder contamination of the whole system such as in the case shown in Fig. 1.27(a) where the coordination number is equal to six. So far, the "right" number is four in the known reentrant systems shown above. Systematic investigations of all possible ingredients are therefore desirable to obtain a sufficient condition for the existence of a reentrance.

Finally, let us emphasize that when a phase transition occurs between states of different symmetries that have no special group–subgroup relation, it is generally accepted that the transition is of first order. However, the reentrance phenomenon is a symmetry-breaking alternative that allows one ordered phase to change into another incompatible ordered phase by going through an intermediate reentrant phase. A question which naturally arises is under which circumstances does a system prefer an intermediate reentrant phase to a first-order transition. In order to analyze this aspect, we have generalized the centered square lattice Ising model into three dimensions.[45] This is a special bcc lattice. We have found that at low T the reentrant region observed in the centered square lattice shrinks into a first-order transition line which is ended at a multicritical point from which two second-order lines emerge forming a narrow reentrant region.[45]

As a final remark, let us mention that although the exactly solved systems shown in this chapter are models in statistical physics, we believe that the results obtained in this work have qualitative bearing on real frustrated magnetic systems. In view of the simplicity of these models, we believe that the results found here will have several applications in various areas of physics.

1.7. Note Added for the Third Edition

A number of papers dealing with exactly solved frustrated models have been published by J. Strečka and coworkers since 2006. These models are essentially decorated Ising models in one or two dimensions.

In Ref. [52], GS and finite-temperature properties of the mixed spin-1/2 and spin-S Ising–Heisenberg diamond chains are examined within an exact analytical approach based on the generalized decoration–iteration map. A particular emphasis is laid on the investigation of the effect of geometric frustration, which is generated by the competition between Heisenberg- and Ising-type exchange interactions. It is found that an interplay between the geometric frustration and quantum effects gives rise to several quantum

GSs with entangled spin states in addition to some semi-classically ordered ones. Among the most interesting results to emerge from our study one could mention a rigorous evidence for quantized plateaux in magnetization curves, an appearance of the round minimum in the thermal dependence of susceptibility times temperature data, double-peak zero-field specific heat curves, or an enhanced magneto-caloric effect when the frustration comes into play. The triple-peak specific heat curve is also detected when applying small external field to the system driven by the frustration into the disordered state.

In Ref. [53], the geometric frustration of the spin-1/2 Ising–Heisenberg model on the triangulated Kagomé "triangles-in-triangles" lattice is investigated within the framework of an exact analytical method based on the generalized star-triangle mapping transformation. GS and finite-temperature phase diagrams are obtained along with other exact results for the partition function, Helmholtz free energy, internal energy, entropy, and specific heat, by establishing a precise mapping relationship to the corresponding spin-1/2 Ising model on the Kagomé lattice. It is shown that the residual entropy of the disordered spin liquid phase for the quantum Ising–Heisenberg model is significantly lower than that for its semiclassical Ising limit ($S_0/NTk_B = 0.2806$ and 0.4752, respectively), which implies that quantum fluctuations partially lift a macroscopic degeneracy of the GS manifold in the frustrated regime.

In Ref. [54], a spin-1/2 Ising model with a spin–phonon coupling on decorated planar lattices partially amenable to lattice vibrations is examined using the decoration–iteration transformation and harmonic approximation. It is shown that the magneto-elastic coupling gives rise to an effective antiferromagnetic next-nearest-neighbor interaction, which competes with the nearest-neighbor interaction and is responsible for a frustration of decorating spins. A strong enough spin–phonon coupling consequently leads to an appearance of striking partially ordered and partially disordered phase, where a perfect antiferromagnetic alignment of nodal spins is accompanied by a complete disorder of decorating spins.

In the above works, the decorations are local couplings of an Ising spin to decorated Heisenberg spins or to phonons which can be summed up to renormalize the interactions between Ising spins. This leaves the systems solvable by star-triangle transformations, vertex models, or other methods. Their results are interesting. We find again in these models striking features shown above in this chapter for 2D solvable frustrated Ising models. In particular, partially disordered systems, multiple phase transitions, and the reentrant phase due to the frustration are shown to exist in the phase diagrams.

For details, the reader is referred to Refs. [52–54].

Acknowledgments

The authors thank Patrick Azaria and Marianne Debauche for their close collaboration in the work presented in this chapter. They are grateful to Ngo Van Thanh, Sylvain Reynal, and Karim Drouiche for technical help.

References

1. R. Liebmann, *Statistical Mechanics of Periodic Frustrated Ising Systems*, Lecture Notes in Physics, vol. 251 (Springer-Verlag, Berlin, 1986).
2. G. Toulouse, *Commun. Phys.* **2**, 115 (1977).
3. J. Villain, *J. Phys. C* **10**, 1717 (1977).
4. G. H. Wannier, *Phys. Rev.* **79**, 357 (1950); *Phys. Rev. B* **7**, 5017 (E) (1973).
5. A. Yoshimori, *J. Phys. Soc. Jpn.* **14**, 807 (1959).
6. J. Villain, *Phys. Chem. Solids* **11**, 303 (1959).
7. T. A. Kaplan, *Phys. Rev.* **116**, 888 (1959).
8. B. Berge, H. T. Diep, A. Ghazali, and P. Lallemand, *Phys. Rev. B* **34**, 3177 (1986).
9. P. Lallemand, H. T. Diep, A. Ghazali, and G. Toulouse, *Physique Lett.* **46**, L-1087 (1985).
10. H. T. Diep, A. Ghazali, and P. Lallemand, *J. Phys. C* **18**, 5881 (1985).
11. R. J. Baxter, *Exactly Solved Models in Statistical Mechanics* (Academic, New York, 1982).
12. V. Vaks, A. Larkin, and Y. Ovchinnikov, *Sov. Phys. JEPT* **22**, 820 (1966).
13. T. Morita, *J. Phys. A* **19**, 1701 (1987).
14. T. Chikyu and M. Suzuki, *Prog. Theor. Phys.* **78**, 1242 (1987).
15. K. Kano and S. Naya, *Prog. Theor. Phys.* **10**, 158 (1953).
16. P. Azaria, H. T. Diep, and H. Giacomini, *Phys. Rev. Lett.* **59**, 1629 (1987).
17. M. Debauche, H. T. Diep, P. Azaria, and H. Giacomini, *Phys. Rev. B* **44**, 2369 (1991).
18. H. T. Diep, M. Debauche, and H. Giacomini, *Phys. Rev. B* **43**, 8759 (1991).
19. M. Debauche and H. T. Diep, *Phys. Rev. B* **46**, 8214 (1992); H. T. Diep, M. Debauche, and H. Giacomini, *J. Mag. Mag. Mater.* **104**, 184 (1992).
20. H. Kitatani, S. Miyashita, and M. Suzuki, *J. Phys. Soc. Jpn.* **55** (1986) 865; *Phys. Lett. A* **158**, 45 (1985).
21. T. Horiguchi, *Physica A* **146**, 613 (1987).
22. J. Villain, R. Bidaux, J. P. Carton, and R. Conte, *J. Physique* **41**, 1263 (1980).
23. T. Oguchi, H. Nishimori, and Y. Taguchi, *J. Phys. Jpn.* **54**, 4494 (1985).
24. C. Henley, *Phys. Rev. Lett.* **62**, 2056 (1989).
25. A. Gaff and J. Hijmann, *Physica A* **80**, 149 (1975).
26. J. E. Sacco and F. Y. Wu, *J. Phys. A* **8**, 1780 (1975).

27. J. Maillard, *Second Conference on Statistical Mechanics*, California Davies (1986), unpublished.
28. K. Binder and A. P. Young, *Rev. Mod. Phys.* **58**, 801 (1986).
29. T. Choy and R. Baxter, *Phys. Lett. A* **125**, 365 (1987).
30. P. Azaria, H. T. Diep, and H. Giacomini, *Phys. Rev. B* **39**, 740 (1989).
31. D. Blankschtein, M. Ma, and A. Berker, *Phys. Rev. B* **30**, 1362 (1984).
32. H. T. Diep, P. Lallemand, and O. Nagai, *J. Phys. C* **18**, 1067 (1985).
33. J. Stephenson, *J. Math. Phys.* **11**, 420 (1970); *Can. J. Phys.* **48**, 2118 (1970); *Phys. Rev. B* **1**, 4405 (1970).
34. F. Y. Wu and K. Y. Lin, *J. Phys. A* **20**, 5737 (1987).
35. H. Giacomini, *J. Phys. A* **19**, L335 (1986).
36. R. Baxter, *Proc. R. Soc. A* **404**, 1 (1986).
37. P. Rujan, *J. Stat. Phys.* **49**, 139 (1987).
38. M. Suzuki and M. Fisher, *J. Math. Phys.* **12**, 235 (1971).
39. F. Y. Wu, *Solid Stat. Comm.* **10**, 115 (1972).
40. L. Onsager, *Phys. Rev.* **65**, 117 (1944).
41. C. S. Hsue, K. Y. Lin, and F. Y. Wu, *Phys. Rev. B* **12**, 429 (1975).
42. K. Y. Lin and I. P. Wang, *J. Phys. A* **10**, 813 (1977).
43. J. Imbrie, *Phys. Rev. Lett.* **53**, 1747 (1984).
44. D. Blankschtein, M. Ma, A. Nihat Berker, G. S. Grest, and C. M. Soukoulis, *Phys. Rev. B* **29**, 5250 (1984).
45. P. Azaria, H. T. Diep, and H. Giacomini, *Europhys. Lett.* **9**, 755 (1989).
46. D. P. Foster, C. Gérard, and I. Puha, *J. Phys. A: Math. Gen.* **34**, 5183 (2001).
47. D. P. Foster and C. Gérard, *Phys. Rev. B* to appear (2004).
48. E. H. Boubcheur, R. Quartu, H. T. Diep, and O. Nagai, *Phys. Rev. B* **58**, 400 (1998).
49. C. Santamaria and H. T. Diep, *J. Appl. Phys.* **81**, 5276 (1997).
50. C. Santamaria, R. Quartu, and H. T. Diep, *J. Appl. Phys.* **84**, 1953 (1998).
51. R. Quartu and H. T. Diep, *Phys. Rev. B* **55**, 2975 (1997).
52. L. Čanová, J. Strečka, and M. Jaščur, Geometric frustration in the class of exactly solvable Ising-Heisenberg diamond chains, *J. Phys.: Condens. Matter* **18**, 4967 (2006).
53. J. Strečka, L. Čanová, M. Jaščur, and M. Hagiwara, Exact solution of the geometrically frustrated spin-1/2 Ising-Heisenberg model on the triangulated (triangles-in-triangles) lattice, *Phys. Rev. B* **78**, 024427 (2008).
54. J. Strečka, O. Rojas, and S. M. de Souza, Spin–phonon coupling induced frustration in the exactly solved spin-1/2 Ising model on a decorated planar lattice, *Phys. Lett. A* **376**, 197 (2012).

CHAPTER 2

PROPERTIES AND PHASE TRANSITIONS IN FRUSTRATED ISING SYSTEMS

Ojiro Nagai*, Tsuyoshi Horiguchi† and Seiji Miyashita‡

**Department of Physics, Faculty of Science, Kobe University, Kobe 657-0013, Japan*
†Department of Computer and Mathematical Sciences, GSIS, Tohoku University, Sendai 980-0845, Japan
‡Department of Physics, Graduate School of Science, The University of Tokyo, Tokyo 113-8656, Japan

2.1. Introduction

Frustration is one of the most interesting concepts in condensed matter physics. Historically, a striking affair concerning frustration was the discovery of the helical spin structure in 1959.[1–3] Until that year, nobody had identified this structure, despite the fact that many substances were noticed later as helical magnets and also despite the fact that the mathematical procedure to derive the helical spin structure is quite easy to understand. The concept of frustration has a very important role on spin systems, including Ising spin systems, classical vector spin systems such as XY model, classical Heisenberg model and so on, and also quantum spin systems. Spin glasses are one of the most interesting problems in spin systems, in which frustration plays an important role. Experimentally, in the Ising spin systems, there exist various Ising magnets with hexagonal symmetry including the ordered problem of adsorbates on a crystal plane. Note that the binary

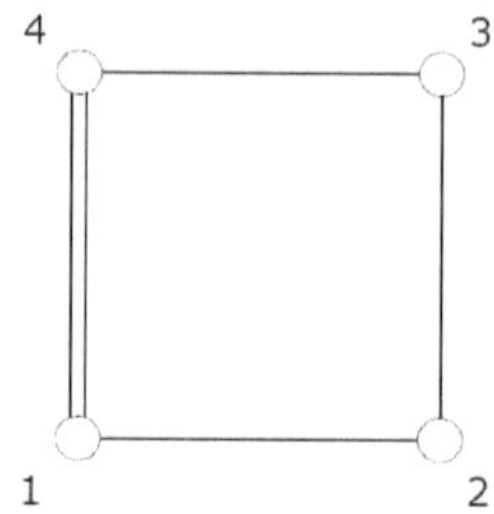

Fig. 2.1. A frustrated plaquette. The single bond and the double bond denote a ferromagnetic interaction and an antiferromagnetic interaction, respectively.

alloy problem is equivalent to Ising models. According to the recent findings, neuron systems are greatly related to the Ising spin systems with frustration.[4]

The word "frustration" was first introduced by Toulouse[5,6] in 1977 for a plaquette shown in Fig. 2.1. Here the plaquette is composed of three ferromagnetic bonds ($J_{12} = J_{23} = J_{34} = J > 0$) and one antiferromagnetic bond ($J_{41} = -J$). An Ising spin, $\sigma_j = \pm 1$, is located at the jth vertex (or site). The Hamiltonian of this four-spin system is written as

$$H = -J_{12}\sigma_1\sigma_2 - J_{23}\sigma_2\sigma_3 - J_{34}\sigma_3\sigma_4 - J_{41}\sigma_4\sigma_1. \tag{2.1}$$

The degeneracy of ground state (GS) for this system with Hamiltonian H is eight. A frustration parameter G defined by the sign of the product of interaction constants $\{J_{ij}\}$, namely $G = \text{sign}(J_{12}J_{23}J_{34}J_{41})$, is negative.

If G is negative (or positive), the plaquette is frustrated (or non-frustrated). A spin model on a frustrated lattice is also called a frustrated spin model. To discuss the properties of frustrated lattice, one may use the concept of local gauge invariant. When we change the way of expressing the variable at a site i, in order to maintain the same energy structure as the Hamiltonian, one has to change the expression of the interactions including the site i. If both changes are done consistently, then the Hamiltonian is kept invariant. That is, we replace a spin σ_i by $-\sigma_i$, and simultaneously replace the coupling constant J_{ij} by $-J_{ij}$ for all j. Using the local gauge transformation, we can change the distribution of ferro- and antiferromagnetic bonds, but cannot change the distribution of the frustrated plaquettes. Therefore, the important characteristic of the lattice is the distribution of the frustrated plaquettes but not ferro- and antiferro-magnetic bonds.[7]

In this chapter, we are concerned with Ising and Ising-like (uniaxial anisotropy) models on the lattices with frustration, e.g., the antiferromagnetic triangular (AFT) lattice and the Villain lattice, and also the 3D models with those lattices stacked. We explicitly demonstrate important formulae and develop full arguments leading to important results for these systems. A comprehensive review of frustrated Ising spin systems has been presented by Liebmann.[8] The articles and monographs published until 1984 may be found in that textbook.

Frustrated Ising spin systems are characterized by a large degeneracy of the GS. Thus, frustrated systems exhibit various interesting ordering processes with additional perturbations. We will study the effects of dimensionality in stacked AFT lattices and also the effects of the next-nearest-neighbor (NNN) interaction, which enhances the sublattice correlations. In particular, the nature of intermediate temperature phase is discussed in detail.

Moreover, since the number of degeneracy is strongly related to the spin ordering, the internal degree of freedom of spin, or the magnitude S of spin (i.e. $M = -S, -S+1, \ldots, S$), is an important physical parameter in frustrated Ising spin systems. We show in this chapter that the spin ordering is strongly dependent on the magnitude S of spin. There exist "free" spins or "free" linear chains, on which internal fields are canceled out, in some frustrated Ising spin systems. These free spins and free linear chains play an important role as for spin orderings.

Another characteristic feature of frustrated Ising spin systems is that various metastable states exist in these systems. Existence of metastable states is closely related to the degeneracy of GS and also to the excited states. Furthermore, the metastable states may give rise to a first-order phase transition. Actually, some models introduced in this chapter show a first-order phase transition. From this viewpoint, we try to understand the physics of frustrated Ising spin systems in this chapter.

2.2. Frustrated Lattices

2.2.1. *Antiferromagnetic triangular lattice (2D)*

The simplest fully frustrated lattice is the antiferromagnetic triangular lattice. We abbreviate antiferromagnetic triangular lattice as the AFT in the following. The Ising model is considered on the AFT lattice shown in Fig. 2.2; the lattice is divided into A-, B-, and C-sublattices. We will call

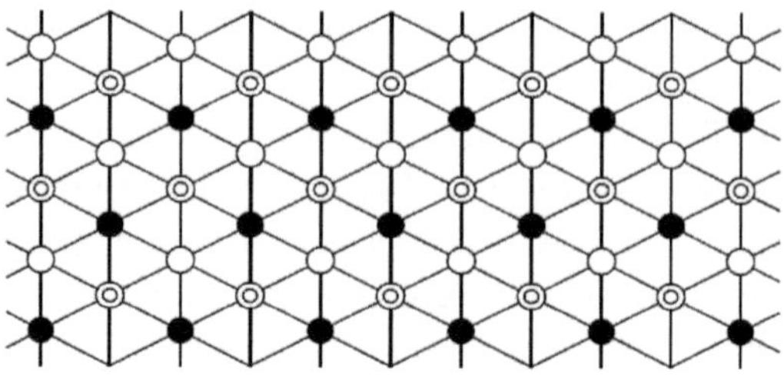

Fig. 2.2. The AFT lattice. The solid circles, the open circles and the double circles denote the lattice sites belonging to A-, B-, and C-sublattice, respectively.

spins on the A-sublattice as A spins, those on the B-sublattice as B spins, and those on the C-sublattice as C spins hereafter. We consider the nearest-neighbor (NN) interaction, and the Hamiltonian is written as

$$H_0 = J \sum_{(ij)} \sigma_i \sigma_j, \tag{2.2}$$

where the summation with respect to (ij) means that the sum is taken over all the NN pairs of lattice sites. The frustrated plaquettes are formed on all the triangles with antiferromagnetic interactions, and the frustration parameter G is negative for all plaquettes.

The Ising model on the antiferromagnetic triangular lattice was exactly solved by Wannier,[9] Houtappel,[10] Husimi and Syoji,[11,12] and no long-range order exists at any finite temperature. There are many degenerate states in the GS, and the number of degeneracy is $\exp(0.323N)$, where N denotes the total number of the lattice sites or spins. Thus, zero-point entropy of this system is of the order of N. Because of this macroscopic degeneracy, if a very small perturbational energy is added to Eq. (2.2), an ordered spin state may be chosen due to the perturbation.

Although no long-range order exists in this system even at zero temperature, it was found by Stephenson[13,14] that the sublattice order develops at low temperatures, and at $T = 0$, the correlation function in the sublattice decays by a power law with the exponent $\eta = 1/2$ at $T = 0$[13,14]

$$\langle \sigma_i \sigma_j \rangle \propto r_{ij}^{-\eta}, \tag{2.3}$$

where $\eta = 1/2$ and r_{ij} denotes the distance between the sites i and j. Moreover, critical properties at $T = 0$ have been clarified.[15]

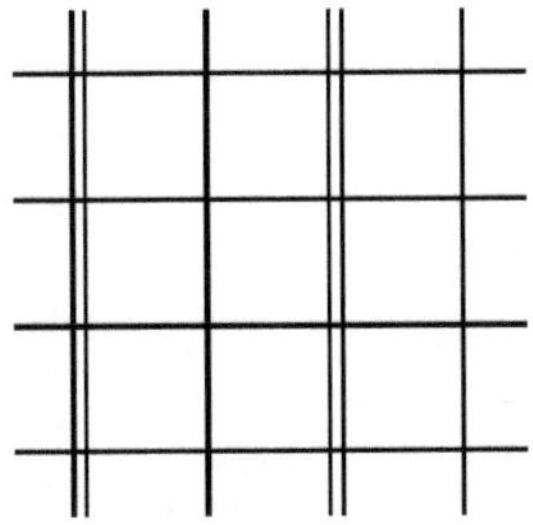

Fig. 2.3. The odd model or Villain lattice. The lattice is fully frustrated.

This semi-long-range order at $T = 0$ suggests that the sublattice order tends to develop. If we add the next-nearest-neighbor (NNN) ferromagnetic interaction to enhance the sublattice order,[16] the model exhibits various phase transitions, which will be introduced later.

2.2.2. *Villain lattice (2D)*

When all the plaquettes of the square lattice are frustrated with the plaquette depicted in Fig. 2.1, it is called the odd model or Villain lattice (Fig. 2.3).[17–19] The partition function of the Ising model on the Villain lattice has also been exactly calculated.[17] Zero-point entropy is of the order of N and no long-range order of spins exists in this system at any finite temperature as well as for the triangular lattice. As well as 2DAFTI, $T = 0$ is the critical point with $\eta = 1/2$ and the ferromagnetic interactions among the sublattice provide an interesting nature of the ordering.[18,19]

2.3. Stacked Frustrated Lattices (3D)

The critical temperature of the Ising model on the Villain lattice is zero.[17] However, the lattice structure of the stacked Villain lattice is shown in Fig. 2.4(a) and the stacked triangular lattice is shown in Fig. 2.4(b). The Hamiltonian of the Ising model on the stacked AFT lattice is written as

$$H = J \sum_{(i,j)}^{xy} \sigma_i \sigma_j - J_z \sum_{(i,j)}^{z} \sigma_i \sigma_j, \tag{2.4}$$

where J denotes the nearest-neighbor (NN) interaction constant in the xy-plane and J_z is the NN interaction constant along the z axis. Here, J and J_z are both positive. The sum runs over the NN pairs either in the

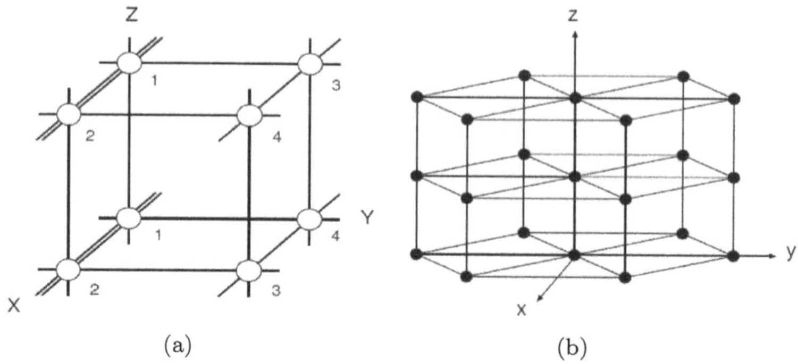

Fig. 2.4. (a) The stacked Villain lattice. (b) The stacked AFT lattice.

xy-plane or along the z direction. In these lattices, the interaction in the xy-plane is frustrated, but none of planes including the z-axis are frustrated. Because of this fact, the critical temperatures of the Ising models on the stacked lattices are not zero.[20,21] The detailed analysis will be given later.

2.4. Ising Models on Antiferromagnetic Triangular Lattice: Effect of the Ferromagnetic Next-Nearest-Neighbor Interactions

2.4.1. *Model S3*

Now, we study the effect of ferromagnetic next-nearest-neighbor interaction. First, we choose the spin states by adding a small ferromagnetic interaction, J_2, between next-nearest-neighbor (NNN) pairs. We see in Fig. 2.2 that an NNN pair corresponds to an NN pair within a sublattice. The Hamiltonian H_3 for the system is written as

$$H_3 = H_0 - J_2 \left(\sum_{(i,i')}^{A} \sigma_i \sigma_{i'} + \sum_{(j,j')}^{B} \sigma_j \sigma_{j'} + \sum_{(k,k')}^{C} \sigma_k \sigma_{k'} \right), \qquad (2.5)$$

where $\sum^A$ ($\sum^B$ and $\sum^C$) denotes the sum over NN pairs within the A sublattice (B and C sublattices). We use notations such as these throughout this chapter. We call this system with the NNN interaction in all the three sublattices S3. The NNN interaction enhances a spin ordering within the sublattice.

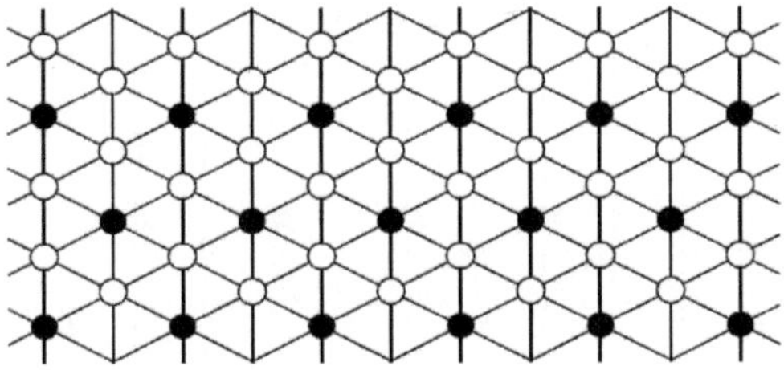

Fig. 2.5. The ferrimagnetic (FR) structure $(1, -1, -1)$. The solid and open circles show up- and down-spins, respectively.

In order to study the properties of the spin ordering, we investigate the set of sublattice magnetizations, $\boldsymbol{M}$:

$$\boldsymbol{M} = \frac{3}{N}\left(\sum_{i\in A}\sigma_i, \sum_{j\in B}\sigma_j, \sum_{k\in C}\sigma_k\right). \tag{2.6}$$

The spin state for this system at low temperatures is of the so-called ferrimagnetic (FR) structure. The FR structure, which is shown in Fig. 2.5, is one of the spin states in the GS occurring in the system with $J_2 = 0$. Temperature dependence of the case S3 will be studied later in detail.

2.4.2. *Model S2*

Next, we consider the system in which the ferromagnetic interaction, J_2, exists between NN pairs within the A-sublattice and also between those within the B-sublattice. We call this system S2. At low temperatures, we assume that all the A spins take $+1$ and all the B spins take -1. Then the internal field for each of the C spins vanishes. This spin state corresponds to the partial disorder (PD) structure $(1, -1, 0)$. We show an example of the PD structure in Fig. 2.6. In other words, all the C spins are free. Thus, the number of degeneracy of this PD structure is $2^{N/3}$. The MC data on the sublattice magnetizations for the case $J_2 = 0.1J$ are shown in Fig. 2.7(a). The temperature dependence of specific heat is shown in Fig. 2.7(b). Here we have only one phase transition.

2.4.3. *Model S1*

Finally, let us consider the system where the ferromagnetic interaction, J_2, exists only between NN pairs within the A-sublattice. We call this system S1. The spin state in the GS for this system is represented by

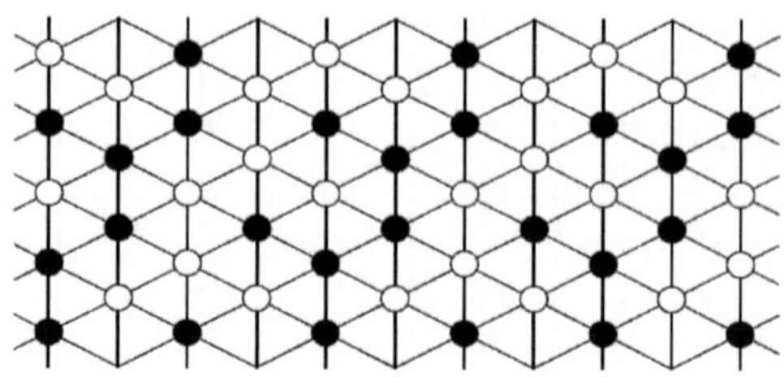

Fig. 2.6. One of the PD structures, $(1, -1, 0)$.

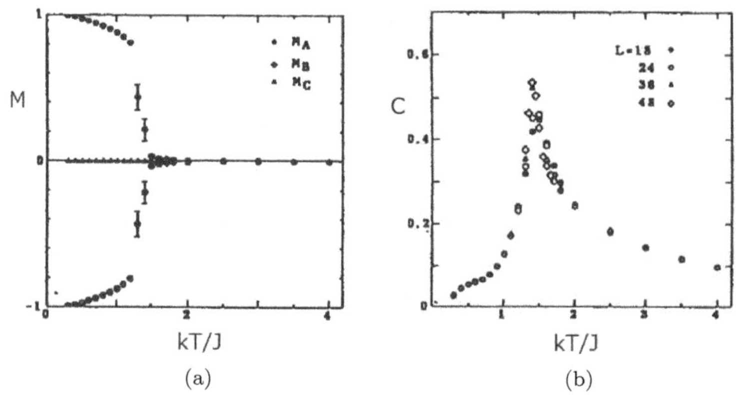

Fig. 2.7. MC data of (a) sublattice magnetizations (M_1, M_2, M_3), and (b) the specific heat C for S2.

$(1, -1/2, -1/2)$. This spin state was proposed by Blankschtein *et al.*[20] as a low-temperature phase for the Ising model on the stacked AFT lattice (see Fig. 2.4(b)). We call this spin state the BL structure. The reason why the system takes the BL structure at low temperatures will be explained as follows. Let us assume all the A spins take $+1$. The B spins or the C spins may be free. If one of the B spins is reversed, then the directions of three C spins surrounding the B spin are fixed. The MC results for the sublattice magnetizations for the case $J_2 = 0.1J$ are shown in Fig. 2.8. The specific heat shows a temperature dependence similar to that in Fig. 2.7(b). The spin state of the BL structure is included in the Wannier state for the system with $J_2 = 0$. This BL structure is realized as the GS, when a uniform magnetic field is applied to the spins in one of sublattices, as can be easily supposed.

We have seen that the FR, PD, and BL structures, respectively, appear for the S3, S2, and S1 at $T = 0$. If a perturbational interaction between

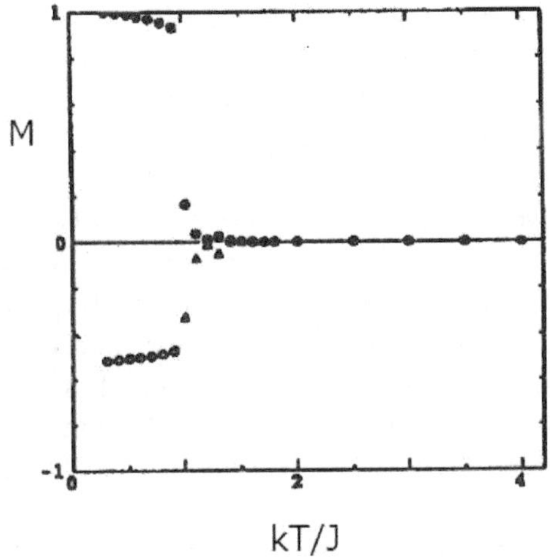

Fig. 2.8. MC data of sublattice magnetizations (M_1, M_2, M_3) for S1.

far-neighbor spins is assumed, there may appear a multi-sublattice spin structure as the GS.

2.5. Thermal Properties of Ising Models on Stacked Antiferromagnetic Triangular Lattice

As was mentioned, in the stacked Villain model and AFT model, the plaquettes including the z-axis are not frustrated. In the Villain model, as shown in Fig. 2.4(a), each ferromagnetic yz-plane is connected by the ferro- and antiferromagnetic bonds with its nearest yz-planes. Since the number of ferromagnetic bonds and antiferromagnetic bonds are the same between two successive yz-planes, it is easily supposed that the critical temperature of this system is non-zero and close to that of the Ising model on the square lattice. Actually, according to the Monte Carlo (MC) simulations, the critical temperature T_c of the Ising model on the stacked Villain lattice is given by $kT_c/J \cong 2.8$.[21] Here, we note that the critical temperature of the Ising model on the square lattice is given by $kT_c/J \cong 2.27$.[22] These two values are in fact close.

In the stacked AFT lattice shown in Fig. 2.4(b), the nearest-neighbor interaction between spins on each xy-plane is $-J(J > 0)$ and that between a spin on an xy-plane and one on its nearest xy-plane is $J_z > 0$. Similar to the stacked Villain lattice in this system, none of planes including the

z axis are frustrated. The square lattices in the zx-plane or yz-plane are composed of the antiferromagnetic bond J and the ferromagnetic bond J_z. However, the effect of the J bonds cancels out each other on the ordering problem of those xy-planes. Therefore, the critical temperature of the Ising model on the stacked AFT lattice is supposed to be close to that of the Ising model on the square lattice composed of J and J_z. It was found that $kT/J \cong 2.9$ when $J/J_z = 1$.[23,24] The long-range order of the type of the PD structure $(M, 0, -M)$ appears below a critical temperature.

At $T = 0$, since the spins along the z axis are parallel to each other, the spin state in the xy-plane should be the same as that of the AFT lattice. That is, no long-range order exists in the xy-plane at $T = 0$. Hence, there occurs some crossover from the PD structure to the Wannier state at a low temperature.

In the MC simulations, two peaks are observed for the specific heat at $T \cong 2.9J/k(= T_1)$ and at $T \cong 0.9J/k(= T_2)$ in the system with $J = J_z$. In Fig. 2.9, the MC data of the specific heat are shown as a function of kT/J_z.

The peak at T_1 is due to the phase transition. There is a small peak or bump in the specific heat. The position, the height, and the shape of the bump at T_2, are not affected by the value of the J/J_z. The reason that the second peak at T_2 appears is as follows. We know that there exist many free spins in the AFT lattice at $T = 0$. Corresponding to these free spins, there exist many free linear chains in the stacked AFT lattice. The free linear chains in the system are parallel to the z axis. It is known that the specific heat of Ising model on the linear chain shows a Schottky-type peak at $kT/J \cong 0.9$. By using the MC simulations, we calculate the specific heat, c, for the Ising model on the stacked AFT lattice for various values of J/J_z;

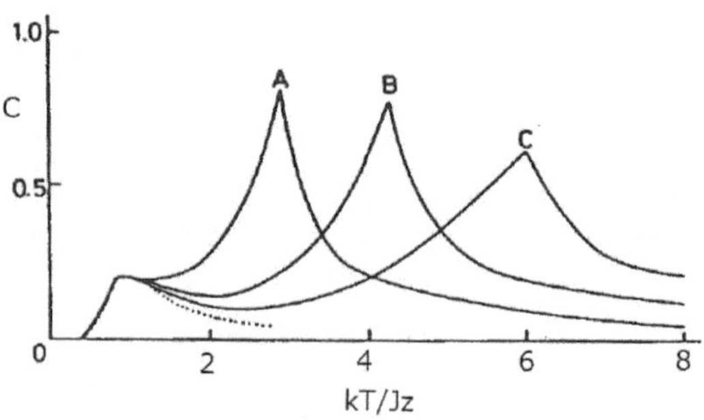

Fig. 2.9. MC data of the specific heat for various values of J/J_z: $J/J_z = 1$ (A), 2.0 (B), and 3.0 (C). The dotted line is explained in the text. Here J_0 is used in place of J_z.

we use c given by

$$c = \frac{1}{N}\left(\frac{J_z}{kT}\right)^2\left[\left\langle\left(\frac{H}{J_z}\right)^2\right\rangle - \left\langle\frac{H}{J_z}\right\rangle^2\right]. \tag{2.7}$$

In the Ising model on 2D fully frustrated lattices such as the Villain lattice and the AFT lattice, there appear many free spins.[25] Corresponding to the free spins, there appear many free linear chains in the Ising model on the stacked frustrated lattices. Hence, it is conjectured that this bump is due to the linear-chain-like excitations in the system, where the interaction constant of the "linear chain" is given by J_z.[26]

It should be noted that to study such systems with strongly interacting chains, the conventional single-spin-flip MC simulation method may fail for this system at low temperatures. Let us consider the MC simulations for the Ising model on the linear chain. As is well-known, the internal energy, U, of the Ising model on the linear chain is given by

$$U = -LJ_z \tanh\frac{J_z}{kT} = -LJ_z\left[1 - 2\exp\left(-\frac{2J_z}{kT}\right) + \cdots\right], \tag{2.8}$$

where L denotes the total number of spins in the linear chain. This expression indicates that the first excited energy is equal to $2J_z$ and also that the first excited state includes only one kink. Thus, it is needed to adopt the cluster-flip MC method, where the cluster in the system is a free linear chain. We schematically show the GS and the first excited state in the system in Figs. 2.10(a) and 2.10(b). We take up the temperature dependence of long-range order parameters in the whole temperature range. Let us define the order parameter X and the sublattice order parameter Y by the following equations:

$$X = \frac{1}{N^2}\left\langle\left(\sum_i \sigma_i\right)^2\right\rangle, \tag{2.9}$$

$$Y = \left(\frac{3}{N}\right)^2\left\langle\left(\sum_{i\in A}\sigma_i\right)^2 + \left(\sum_{j\in B}\sigma_j\right)^2 + \left(\sum_{k\in C}\sigma_k\right)^2\right\rangle. \tag{2.10}$$

It is easily seen that $X = 0$ and $Y = 2$ for the PD structure $(1, -1, 0)$, and $X = 1$ and $Y = 3$ for the FR structure $(1, -1, -1)$. The temperature dependencies of X and Y in the MC simulations are shown in Fig. 2.11. We see that $X = 0$ in the whole temperature range and the values of Y

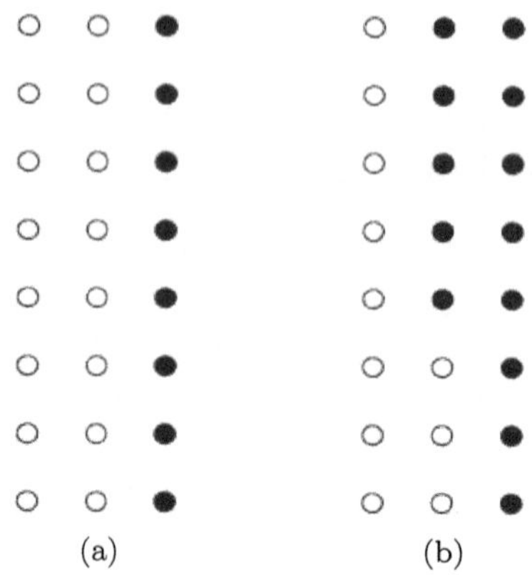

Fig. 2.10. (a) The GS of a free linear chain in the system. The solid and open circles denote the up- and down-spins, respectively. (b) The first exited state. The excitation energy is $2J_z$.

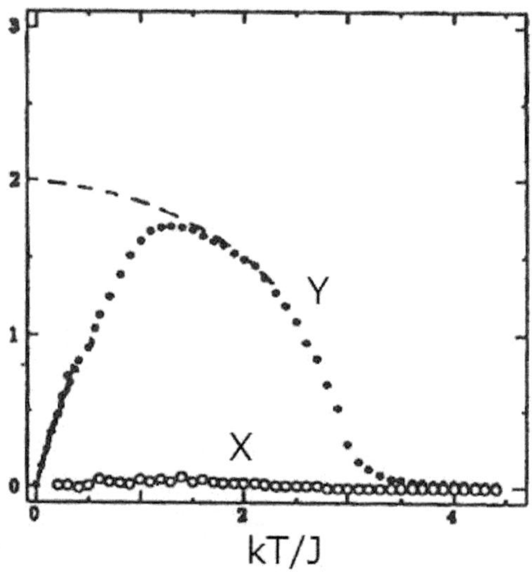

Fig. 2.11. MC data of X and Y for the Ising model on the stacked AFT lattice.

agree with those for the PD structure (dashed line) above $kT/J \cong 1.2$. Here, these spin pairs are belonging to the A-sublattice or B-sublattice. Remember system S2 in the Section 2.3. The dashed line in Fig. 2.11 shows an extrapolation to the state with the PD structure. We may assume that

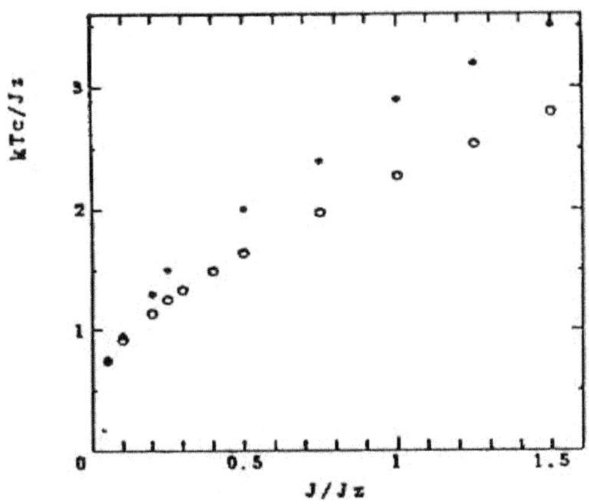

Fig. 2.12. J/J_z dependence of kT_c/J_z. The open circles denote the critical temperature for the Ising model on the square lattice and the solid circles denote the critical temperature for the Ising model on the stacked AFT lattice.

a temperature region between $kT/J = 0$ and 2.0 is the crossover region between a 3D ordered state and the Wannier state. The number of free linear chains, which is denoted by $\gamma N^{2/3}$ at low temperatures, may be estimated from the MC data of the specific heat. The dotted line for the specific heat in Fig. 2.9 is obtained by assuming $\gamma = 0.376$.[27]

We calculate kT_c/J_z as a function of J/J_z. The critical temperature for the Ising model on the square lattice with J and J_z is calculated from the following equation[22]:

$$\sinh(2K_c)\sinh(2K_c^z) = 1, \tag{2.11}$$

where $K_c = J/kT_c$ and $K_c^z = J_z/kT_c$. The critical temperature of the Ising model on the stacked AFT lattice is computed by the MC simulations. These two critical temperatures are compared in Fig. 2.12. We see that J/J_z dependencies of kT_c/J_z for these two systems show a similar behavior. This behavior supports our physical interpretation.

2.6. Ising Model on Stacked Antiferromagnetic Triangular Lattice with NNN Interaction

As mentioned earlier, the AFT-Ising ($S = 1/2$) model with only nearest-neighbor (NN) interaction does not have an ordered phase at finite temperatures,[9–11] but it shows a quasi-long-range order (power law decay) of the sublattice magnetization.[13,14]

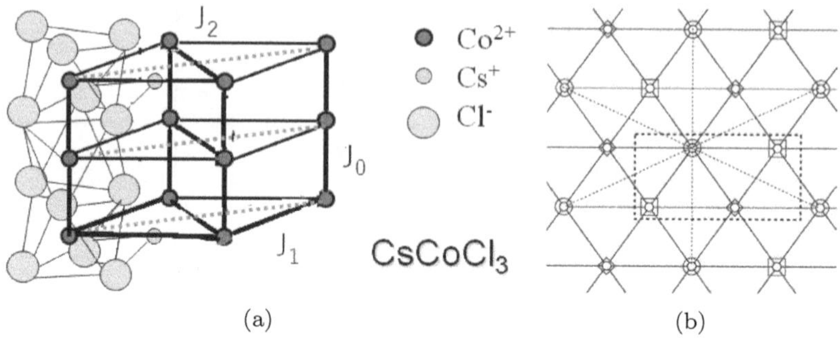

Fig. 2.13. (a) Lattice structure of an AFT-Ising material ($CsCoCl_3$) and its magnetic interactions: intra-layer NN interaction (J_1), intra-layer NNN interaction (J_2: dotted line) and inter-layer NN interaction (J_0). (b) View from the top: Only Co-ions are shown. In (a), only the part including two triangles enclosed within a dotted square is shown. The three sublattices are depicted by circles, squares, and diamonds.

$CsCoCl_3$, which has the ABX_3 structure (layered triangular lattice), was studied as a related material. The structure of $CsCoCl_3$ is depicted in Fig. 2.13(a). The magnetic moment is given by the Co^{2+}, which is expressed by an Ising spin $\sigma_i = \pm 1$ at the site i. The lattice forms a triangular lattice in the xy plane, which consists of three sublattices denoted by circles, squares, and diamonds as depicted in Fig. 2.13(b). Polarized neutron scattering shows Bragg peaks corresponding to the sublattice ordering, which indicates the existence of several patterns of sublattice magnetizations.[28]

To analyze the result, a model with the NNN ferromagnetic interaction (Model S3) which enhances the sublattice ordering was studied by Mekata[16] by a mean-field method.

As was explained, the GS configuration of the model S3 is the FR state. At very high temperatures, the system is in a disordered (D) state. In between, an intermediate temperature phase of the PD state was found. This appearance of the PD phase in an intermediate phase is a novel characteristic and many related works has been done, and finally the existence of PD phase as an intermediate phase phase in a short-range interacting system was confirmed. We perform an overview of the works here.

In the ABX_3 type materials, the inter-layer interaction J_0 is much larger than the intra-layer interaction, and the system is often regarded as an assembly of 1D chains. This 1D structure is important for the PD structure. Moreover, in the 3D case, the system exhibits phase transition even without the NNN interaction. However, here we study the case of $J_2 > 0$ and focus on the ordering in the layer (triangular lattice).

2.6.1. *Mean-field analysis*

Now, we review the Mekata's analysis.[16] Here, we treat the chain connecting the layers as a single spin and do not refer J_0. Let us denote the sublattice circle, square, and diamond by A, B, and C, respectively, and let the magnetization of the sublattices be m_A, m_B, and m_C. The expectation value of a spin in the A-sublattice is given by

$$m_A = \langle \sigma_A \rangle = \frac{\sum_{\sigma_A=\pm 1} \sigma_A e^{-\beta\sigma_A(J_1(3m_B+3m_C)-6J_2m_A)}}{\sum_{\sigma_A=\pm 1} e^{-\beta\sigma_A(J_1(3m_B+3m_C)-6J_2m_A)}}, \tag{2.12}$$

and similarly for m_B and m_C. Thus, the self-consistent equations for these are given by

$$\begin{cases} m_A = \tanh\left(-\beta\left(J_1(3m_B + 3m_C) + 6J_2m_A\right)\right) \\ m_B = \tanh\left(-\beta\left(J_1(3m_C + 3m_A) + 6J_2m_B\right)\right) \\ m_C = \tanh\left(-\beta\left(J_1(3m_A + 3m_B) + 6J_2m_C\right)\right). \end{cases} \tag{2.13}$$

For small NNN interactions, $J_2/J_1 < 0.8$, the solution of the self-consistent equations which minimizes the free energy shows the successive phase transitions depicted in Fig. 2.14.

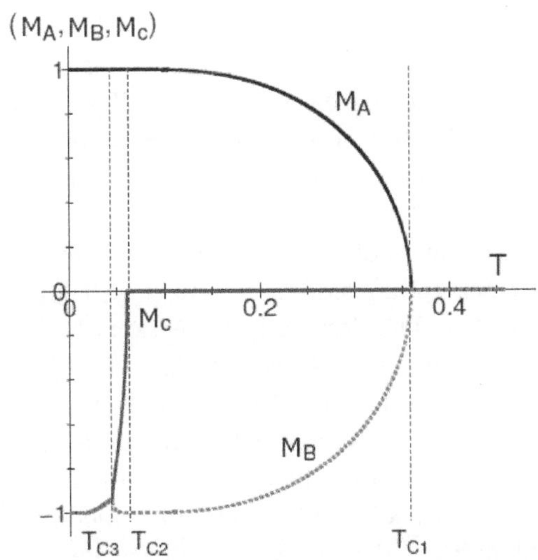

Fig. 2.14. Temperature dependences of the sublattice magnetizations depicted by a bold solid line ($m_A(T)$), a dashed line ($m_B(T)$), and a thin solid line ($m_C(T)$) for $\alpha = 0.2$.

At high temperatures, the trivial paramagnetic solution ($m_A = m_B = m_C = 0$) gives the thermal equilibrium, that is, the disordered (D) (or paramagnetic) phase. At $T = 0$, each sublattice is ordered perfectly and GS is given by the FR state, and at low temperatures $T < T_{C3}$, the FR phase $(m_A, m_B, m_C) = (m_1, -m_2, -m_2) \quad |m_1| > |m_2|$ appears.

Here, we should pay attention to the appearance of phases in the intermediate temperature region $T_{C3} < T < T_{C1}$. This phase appears as a consequence of the fact that the unit structure consists of three spins due to the frustration.

From the D phase, as the temperature is reduced, the PD phase appears.

The configurations of PD phase and FR phase do not match. In the present analysis, the magnetization of the disordered sublattice (C-sublattice in the figure) shows a symmetry breaking at T_{C2}, and the PD structure moves to the FR phase. The intermediate phase has a configuration $m_A = m_1 > 0$, $m_B = m_2 < 0$, and $m_C = m_3 < 0$. In general, the transition between phases with disconnected configurations tends to be of the first order. This model also exhibits the first-order phase transition when we adopt better mean-field methods taking into account the local fluctuation.[29]

2.6.2. *Monte Carlo analysis in two dimensions*

The MC results of the specific heat, $c = C/Nk$, vs. temperature, kT/J, are shown in Fig. 2.15 for the case $J_2 = 0.1$. Two peaks are observed in the specific heat. The higher transition temperature is estimated to be $kT_1/J \cong 1.5$ and the lower transition temperature is estimated to be $kT_2/J \cong 0.4$.

The ordering mechanism of the model S3 has been studied in relation with the six-state clock (6SC) model.[30–34] It is well known that the 6SC model has two Kosterlitz–Thouless (KT) transitions.[35,36] The two peaks in Fig. 2.15 correspond to the two KT transitions.

The nature of the intermediate temperature was checked first in 2D systems. As mentioned in the Section 2.4.1, the model shows two peaks of the specific heat, which indicates the occurrence of some successive phase transitions. But, the intermediate phase was identified to be the KT phase in which all the sublattice order decay via a power law.

The successive phase transition was studied by making use of a mapping of the present model to the 6SC model (Fig. 2.16).

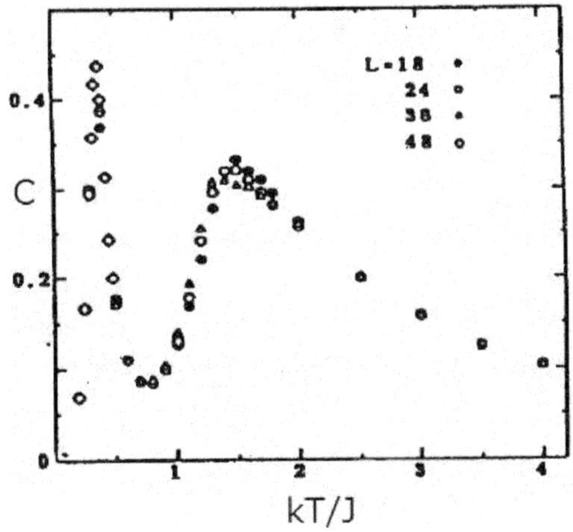

Fig. 2.15. Specific heat for S3 with $J_2 = 0.1J$.

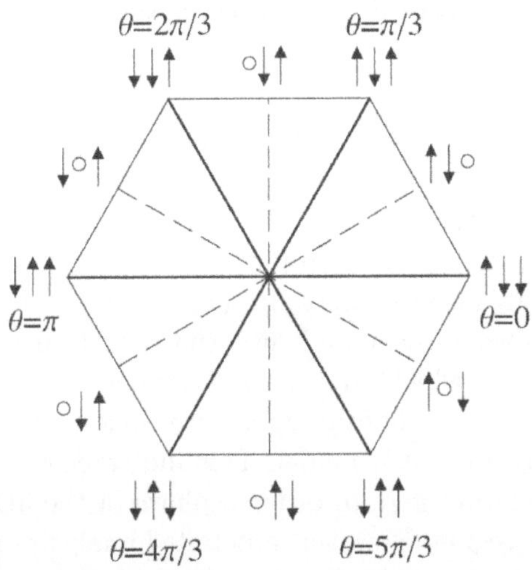

Fig. 2.16. Correspondence between the configuration of the AFTI model and the 6SC model.

Similar to the GS spin configuration on a triangle, there are six possible combinations which are energetically degenerate, i.e., $(1,-1,-1)$, $(1,-1,1)$, $(-1,-1,1)$, $(-1,1,1)$, $(-1,1,-1)$, and $(1,1,-1)$. We can map these six states with the 6SC model as depicted in Fig. 2.16.

The Hamiltonian of a simple 6SC model is given by

$$\mathcal{H}_{6\mathrm{SC}} = -J\sum_{\langle ij\rangle}\cos(\theta_i-\theta_j), \quad \theta_i = \frac{2\pi}{6}k, \quad k=0,1,\ldots,5 \tag{2.14}$$

As we will study later, the energy structure within the spin state is not unique and the nature of phase transition largely depends on it. It has been well understood that the 6SC model has double (dual) KT phase transitions and the intermediate phase is the KT phase. Thus, it was concluded that the Mekata model in two dimensions is of the same type as that of 6SC model and the model does not exhibit the PD phase.

2.6.3. *Monte Carlo analysis in three dimensions*

Next, phase transition of Mekata model was studied in three dimensions. But, because of large fluctuations, the intermediate phase was hardly characterized.[24] Then, the 6SC model was studied in three dimensions as a reference.[37,38] Temperature dependencies of the specific heat C and the squared magnetization

$$m^2 = \frac{\langle(\sum_i^N \cos\theta_i)^2\rangle + \langle(\sum_i^N \sin\theta_i)^2\rangle}{N^2}$$

are depicted in Fig. 2.17.

In Fig. 2.17(a), we find a sharp peak at a high temperature ($T_1 \sim 3J$) and a broad peak at a low temperature ($T_2 \sim 1.5J$). At low temperatures ($T < 0.5J$), the specific heat and also the deviation of the m^2 from 1 decrease exponentially, which reflects the discreteness of the model. On the other hand, in the intermediate temperature region ($0.5J < T < 3J$) the quantities behave like the XY model. This indicates that the discreteness of the model is irrelevant and an ordered phase of the 3D XY type exists in the temperature region. However, a detailed analysis on the size dependence of the discreteness helped in figuring out that the discreteness is always relevant below T_1. This means that the $3DXY$-like nature of the intermediate temperature region is only apparent, and the model has a single low-temperature phase. This picture was also supported by a renormalization group analysis.[39] From this point of view, one may consider that there is only one phase below the temperature where the sublatttice order develops, and the quantitative changes below this temperature are just due

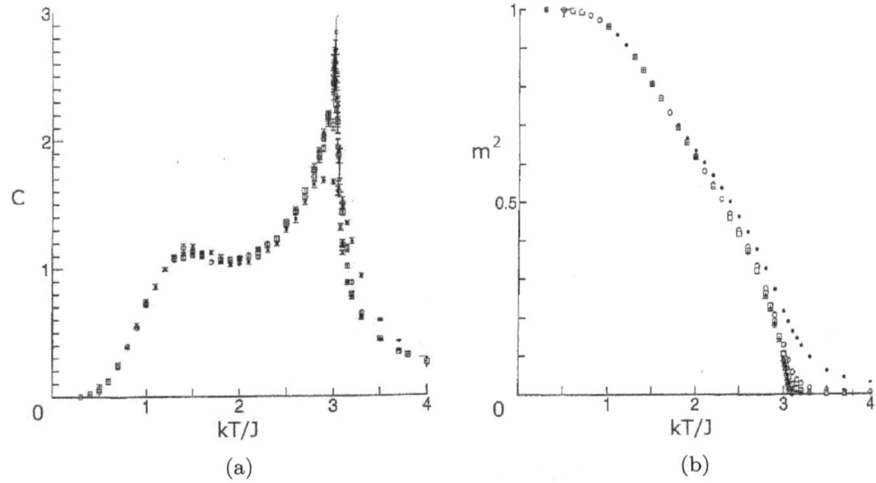

Fig. 2.17. Temperature dependencies of (a) the specific heat C and (b) the squared magnetization.

Source: Modified from Ref. [38].

to the crossover. This observation indicated that the large fluctuation of the Mekata model in the intermediate temperature region reflects the above-mentioned large fluctuation and there is no intermediate phase in three dimensions, which denied the existence of the PD phase.

2.6.4. *New type of intermediate phase in the generalized six-state clock model*

However, the nature was not so simple. The mapping of the six states of the FR phase to a Z_6 model is correct, but the Hamiltonian of the simple 6SC model (2.14) is found to be too simple. In the general idea of the universality of phase transitions, the details of the structure are irrelevant. However, it was found that the generalized Z_6 (G6SC) model exhibits various patterns of ordering processes depending on the parameters of the Hamiltonian, although the Z_6 symmetry holds in all the cases. The Hamiltonian of G6SC model is given by

$$\mathcal{H}_{GZ6} = \sum_{\langle ij \rangle} (\varepsilon_0 \delta_{S_i,S_j} + \varepsilon_1 \delta_{S_i,S_j \pm 1} + \varepsilon_2 \delta_{S_i,S_j \pm 2} + \varepsilon_3 \delta_{S_i,S_j \pm 3}), \quad \text{mod}(6), \tag{2.15}$$

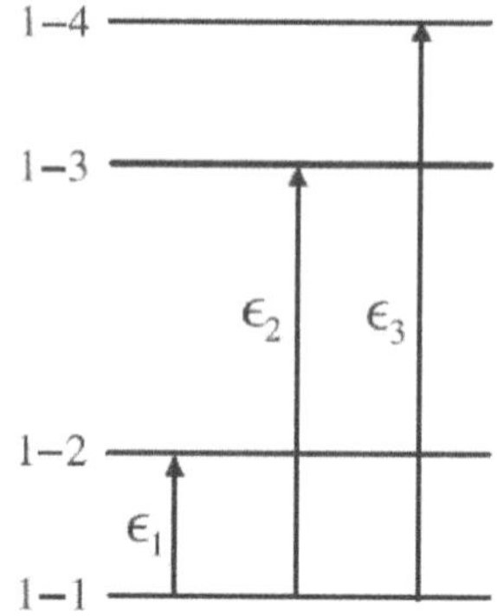

Fig. 2.18. Energy structure of the generalized Z_6 model.

where the six states are named as

$$S_i = 0, 2, 3, \ldots, 5. \tag{2.16}$$

The energy structure is depicted in Fig. 2.18. In the simple 6SC model, the interaction energy between the same states is $\varepsilon_0 = -J\cos(0) = -J$ and that for the states which differ by 1 it is $\varepsilon_1 = -J\cos(\pi/3) = -J/2$, that for the states which differ by 2 it is $\varepsilon_2 = -J\cos(2\pi/3) = J/2$, and that for the states which differ by 3 it is $\varepsilon_2 = -J\cos(2\pi/3) = J$.

It was pointed out that when $\varepsilon_1 - \varepsilon_0$ is much smaller than $\varepsilon_2 - \varepsilon_0$ and $\varepsilon_3 - \varepsilon_0$, there exists a phase in which the two states are mixed.[40] To realize such a phase, we have to choose two neighboring states. This is a kind of symmetry breaking. This state is indicated by the dashed lines in Fig. 2.16. As we see there, the mixed state of $(\uparrow, \uparrow, \downarrow)$ and $(\uparrow, \downarrow, \downarrow)$ gives the state $(\uparrow, 0, \downarrow)$, which corresponds to the PD phase.

At lower temperatures, the state must be the one for the FR phase, and an additional phase transition takes place. This phase transition was found to be a first-order phase transition.[40]

Once we know that such a state exists, the nature of the Mekata model was investigated carefully with the 3D structure with strong spin chain ($J_1/J_0 = 3 \times 10^{-2}$, $J_2/J_0 = 10^{-4}$).[29,41] The temperature dependence of specific heat and sublattice magnetizations are depicted in Figs. 2.19(a) and 2.19(b), respectively. The system exhibits a sharp second-order phase transition at T_{C1}, which is of the $3DXY$ universality class.

Besides the transition at T_{C1}, we find that the sublattice magnetizations exhibit a discontinuous jump at T_{C2} between the PD type and the FR type,

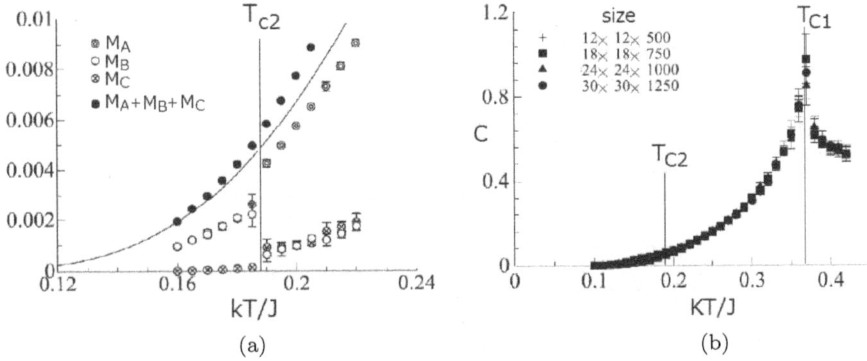

Fig. 2.19. The temperature dependence of (a) sublattice magnetizations and (b) the specific heat. The dotted line indicates the position of T_{C2}, where the first-order phase transition takes place.

Source: Modified from Ref. [29].

indicating the first-order phase transition. However, there seems no jump in the specific heat. This is due to the fact that a spin in the plane represents a spin chain. Thus, $\langle\sigma\rangle = 0$ means that the total magnetization of the chain is zero. Such a state can be realized with a domain wall in a chain. Thus, the energy difference per site between $\langle\sigma\rangle = 0$ and $\langle\sigma\rangle = 1$ is small. Thus, the latent heat becomes very small.[29]

Finally, let us look at the feature of the phase transitions from the view point of renormalization group (RG) treatment.[23,29,39] The discreteness is expressed by the parameter for the six-fold anisotropyλ:

$$\mathcal{H}_{\text{eff}} = -J\sum_{\langle ij\rangle}\cos(\theta_i - \theta_j) + \lambda\sum_i \cos(6\theta_i). \tag{2.17}$$

Here, we choose the sign of λ so that $\lambda > 0$ denotes the PD phase and $\lambda < 0$ denotes the FR phase.

The schematic flow diagram of the parameters (temperature and the six-fold anisotropy λ) is depicted in Fig. 2.20. The bold line denotes the bare parameters of the model. If the model is in the high-temperature side of the line denoted by an arrow to the point of $3DXY$ which is the fixed point of $3DXY$ model at $\lambda = 0$, the flow takes the system to the right, i.e., to the high temperatures. This means the system is in disorder phase. If the point is in the low-temperature side with positive λ, the flow takes the point to the small λ for a while, which causes the apparent XY phase in

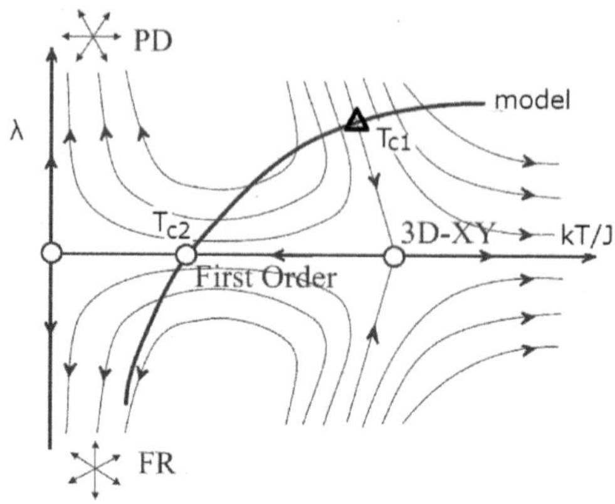

Fig. 2.20. The RG Flow diagram.

Source: Modified from Ref. [29].

the intermediate temperatures as was observed in the simple 6SC model. But, finally, the point moves up to the large λ region, which means that the state is essentially in PD phase. The point denoted by the bold triangle denotes the transition between the two regions, where the phase transition of the $3DXY$ universality takes place. On the other hand, after the line crosses the line of $\lambda = 0$ as shown in the figure, the point with negative λ moves to large negative λ which indicates that the system is in the FR phase. This is the case when the first-order phase transition occurs.

In this way, we understand the successive phase transitions from D, PD, and FR phases, and finally we find that the PD phase exists in a system with short-range interaction.

2.7. Ising Model with Large S on Antiferromagnetic Triangular Lattice

We know that there is no long-range order in the GS of the Ising model with spin-1/2 on the AFT lattice. This property may change if the magnitude S of spin changes. For the Ising model with infinite-spin, there exists a long-range order, as will be shown in this section, and hence, there is an evidence of a phase transition at a finite temperature. Frustrated systems such as the Ising model with spin-1/2 on the AFT lattice are characterized by a large degeneracy of the GS. The effect of magnitude S of spin on the ordering properties of frustrated Ising spin systems is an interesting problem.[42,43]

It is interesting to show in this section that the nature of the spin ordering for the Ising model with large S is quite different from that with spin-1/2 on the AFT lattice.[44–48]

The Hamiltonian of the Ising model with spin S on the AFT lattice is given by

$$H = J \sum_{(i,j)} \sigma_i \sigma_j, \tag{2.18}$$

where $\sigma_i = S_i/S$. S_i denotes the Ising spin of magnitude S on the ith lattice site and takes one of the $2S+1$ values, $S, S-1, \ldots, -S$. The spin with the limit of S infinity is also called the continuous Ising spin since σ_i takes on $[-1, 1]$; we assume that the limit of S infinity is taken before the thermodynamic limit. The set of sublattice magnetizations at $T = 0$ is given by Eq. (2.6). At free spins, the internal fields are canceled out. For the model considered here, each free spin can take a value out of $2S+1$ values. Therefore, n free spins cause $(2S+1)^n$-fold degeneracy of the GS. The spin structure with maximum number of the free spins is the PD structure (Fig. 2.21), where the spins in the A-sublattice takes $+S$, all the spins in the B-sublattice take $-S$, and all the spins in the C-sublattice are free. In this case, the set of the sublattice magnetizations, $\boldsymbol{M}$, is $(1, -1, 0)$ and the degeneracy of the spin states is equal to $(2S+1)^{N/3}$.

We consider a spin on the A-sublattice (A spin) in the partially disordered structure of $(1, -1, 0)$ type. In order to make the spin free, the neighboring three C spins must be fixed to be $+S$. Thus, the number of free spins is reduced by 2 as depicted in Fig. 2.22. We call the cluster of those four spins "a misfit cluster" of $(0, -1, 1)$ type. The boundary of the misfit cluster is drawn by the closed dashed lines in Fig. 2.22. We may also

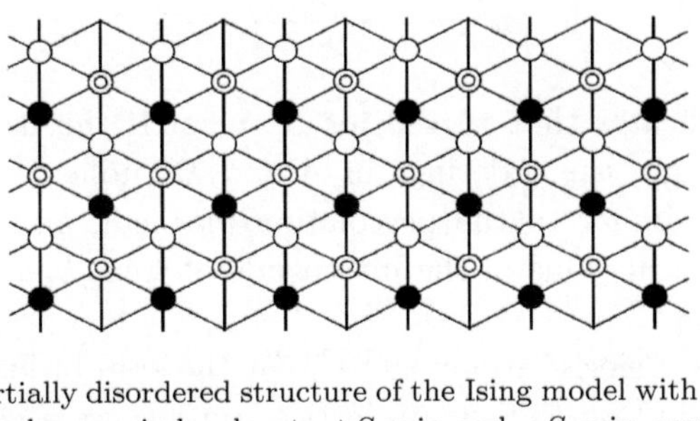

Fig. 2.21. The partially disordered structure of the Ising model with spin S on the AFT lattice. The solid and open circles denote $+S$ spin and $-S$ spin, respectively. Free spin is represented by the double circles.

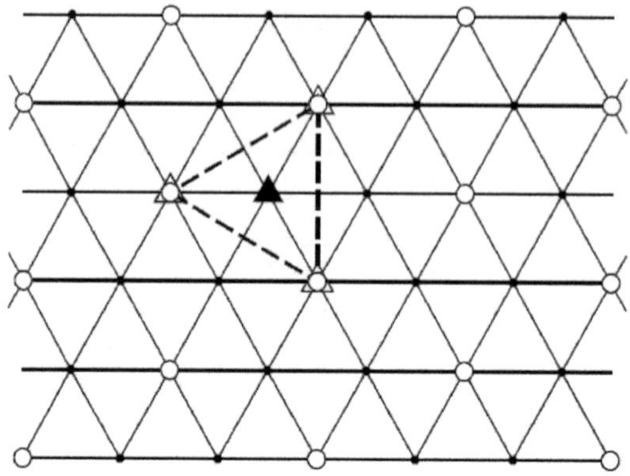

Fig. 2.22. A misfit cluster in the PD structure of $(1, -1, 0)$. The triangles represent the spins in the misfit cluster. The circles denote the free spins. The boundary of the misfit cluster is shown by the dashed lines.

consider a misfit cluster in a B spin free. In order to study the stability of the PD structure, we study the degeneracy due to the misfit clusters. When n independent misfit clusters exist, the GS degeneracy, W_n, is

$$W_n \cong (2S+1)^{\frac{N}{3}-2n} \binom{\frac{2N}{3}}{n}. \tag{2.19}$$

The condition for the maximum degeneracy is

$$\frac{\partial}{\partial n} \ln W_n = 0, \tag{2.20}$$

which yields

$$n = \frac{2N}{3} \frac{1}{(2S+1)^2}. \tag{2.21}$$

This relation indicates that $n = 0$ for $S = \infty$. If the density of misfits, n/N, is low enough, the orderings in A and B sublattice would remain. This problem is similar to the percolation problem, and we expect that there exists a critical value of the magnitude of spin, S_c for the sublattice long-range order.

Following the Peierls' argument[49, 50] for the spin ordering of the Ising model with spin-1/2 on the square lattice, we discuss a condition for the existence of long-range order in the Ising model with spin S on the AFT

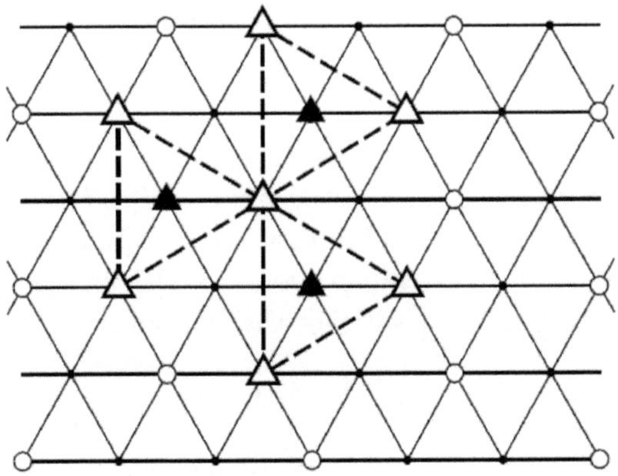

Fig. 2.23. A misfit cluster of large size.

lattice.[47] We consider spin configurations with a boundary condition which induces the order of $(1, -1, 0)$ type. There are various kinds of misfit clusters of larger size; an example is given in Fig. 2.23. As depicted in Figs. 2.22 and 2.23, we define the length of boundary, f, measuring in unit of $\sqrt{3} \times$ bond length. For example, $f = 3$ in Fig. 2.22 and $f = 9$ in Fig. 2.23. Let d denote the reduction in the number of free spins due to the appearance of misfit cluster, e.g., $d = 2$ in the case of Fig. 2.22 and $d = 4$ in the case of Fig. 2.23. Generally, the relation

$$f \leq 3d \tag{2.22}$$

is satisfied. Consider a spin on the A-sublattice. Let p_+ be the probability that the spin is found in the cluster of $(1, -1, 0)$ type or $(1, 0, -1)$ type. Similarly, let p_- be the probability that the spin is found in the cluster of $(-1, 1, 0)$ type or $(-1, 0, 1)$ type and let p_0 be the probability that the spin is found in the cluster of $(0, 1, -1)$ type or $(0, -1, 1)$ type. Since $p_+ + p_- + p_0 = 1$, we have

$$\langle \sigma_A \rangle = \frac{3}{NS} \left\langle \sum_{i \in A} S_i \right\rangle = p_+ - p_- = 1 - p_0 - 2p_-, \tag{2.23}$$

where $\langle \cdots \rangle$ denotes the average over all the spin states in the GS instead of the canonical weight in the case of Peierls' argument. As in the usual Peierls' argument, we compare the number of degeneracies of the cases with and without the boundary of p_0 and we have the inequality

$$p_0 < \sum_f \sum_k m_k (2S+1)^{-d}, \tag{2.24}$$

where $k = k(f)$ denotes the kth cluster and $d = d(k, f)$ denotes the reduction in the number of free spins due to the appearance of the cluster. m_k is the number of sites of the A-sublattice in the kth cluster. The number of clusters C_f with the boundary satisfies the relation

$$C_f \le 2(z-1)^f, \tag{2.25}$$

where $z(=6)$ is the number of the nearest-neighbor sites. The factor 2 is due to the fact that there are two kinds of misfit clusters with free A spins, namely the clusters of $(0, 1, -1)$ type and those of $(0, -1, 1)$ type. Using Eq. (2.25) and the relation, $m_k < f^2$, we have

$$\sum_k m_k (2S+1)^{-d} < 2f^2 \left(\frac{5^3}{2S+1} \right)^{f/3}. \tag{2.26}$$

Therefore, the series Eq. (2.24) converges if $2S+1 > 5^3$. p_- is estimated in the same way. Thus, the long-range sublattice order remains finite for large values of S.

The sublattice spin pair correlation function decays by a power law

$$\langle \sigma_i \sigma_{i+r} \rangle \sim r^{-\eta}, \quad \eta = 1/2 \tag{2.27}$$

for $S = 1/2$. For the sublattice long-range order, η is zero. It is an interesting problem to figure out how η reduces as S increases. In order to study η, we calculate the quantity A, which is defined by

$$A = \frac{1}{N} \left\langle \left(\sum_{i \in A} \sigma_i \right)^2 + \left(\sum_{i \in B} \sigma_i \right)^2 + \left(\sum_{i \in C} \sigma_i \right)^2 \right\rangle \tag{2.28}$$

by the MC simulations. The values of η are obtained from the size dependence of A

$$A \sim L^{2-\eta}. \tag{2.29}$$

The S dependence of η is shown in Fig. 2.24. The threshold S_c has been estimated to be 3.[51–54]

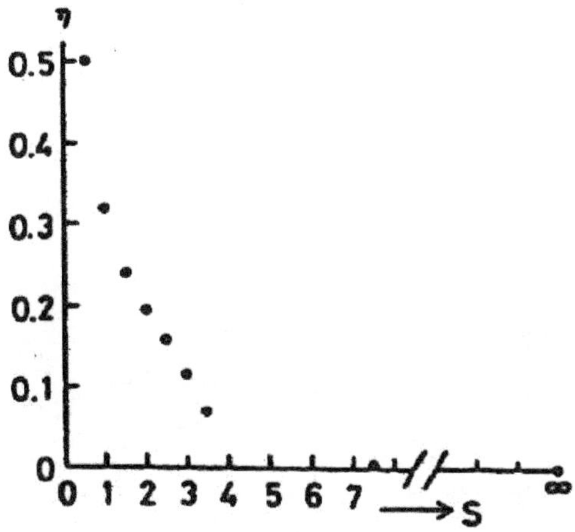

Fig. 2.24. S dependence of η; η is the exponent for the correlation function.

2.8. Ising Model with Infinite Spin on Antiferromagnetic Triangular Lattice

It is interesting to study the magnetic properties of the Ising model with infinite spin on the AFT lattice. We show the MC data of the specific heat for various values of S in Fig. 2.25. The shape around the peak of specific heat, c, for the Ising model with infinite spin is rather sharper than that for the Ising model with finite S, as can be seen in Fig. 2.25. However, any size dependence was not observed in the peak of c for the Ising model with infinite spin. The sublattice magnetizations of PD structure $(M, 0, -M)$ were apparently observed up to the temperature $T = 0.1J/k$ for a fairly large lattice size ($L = 240$) and a long MC average time ($t_{\text{MC}} = 10^6$ MC steps per spin).[46] However, the temperature dependence of $2 - \eta$ decreases with respect to T and reaches 1.75 at $T \sim 0.41J/k(= T_1)$ and suddenly decreases to zero as shown in Fig. 2.26. Here, η denotes the exponent for correlation function and the value of η is 0.25 at the KT phase transition temperature. Since η should be zero in the spin-ordered region, no long-range order exists at any finite temperatures except at very low temperatures for the Ising model of infinite spin on the AFT lattice. It is desired to be investigated in detail the behavior of η near $T = 0$ and whether the long-range order exists or not at very low temperatures. On the other hand, η should be equal to 2 in the paramagnetic region. Thus, the MC data of η seem to indicate that the KT-type phase transition occurs at T_1.

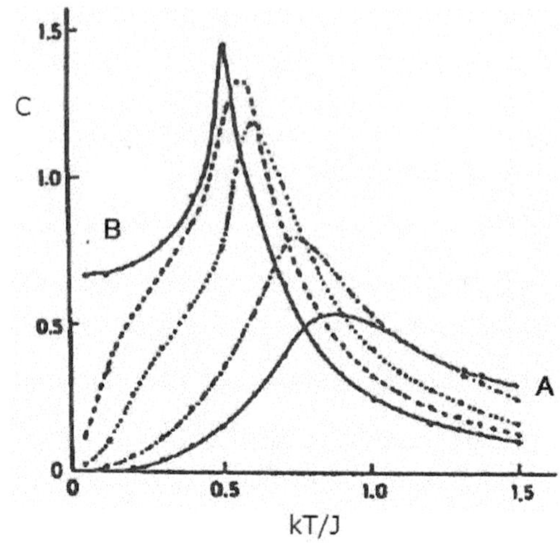

Fig. 2.25. MC data of specific heat for the Ising model with spin S on the AFT lattice for various values of S. The solid line (A) is for $S = 1/2$, the dash-dot line for $S = 3/2$, the dotted line for $S = 7/2$, the dashed line for $S = 15/2$, and the solid line (B) for $S = \infty$.

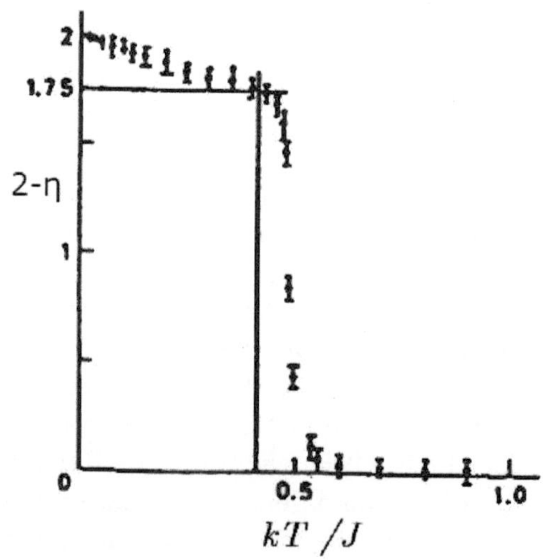

Fig. 2.26. MC data of $2 - \eta$ as a function of the temperature for the Ising model with infinite spin. The KT value, 1.75, is found at $T = 0.41J/k$.

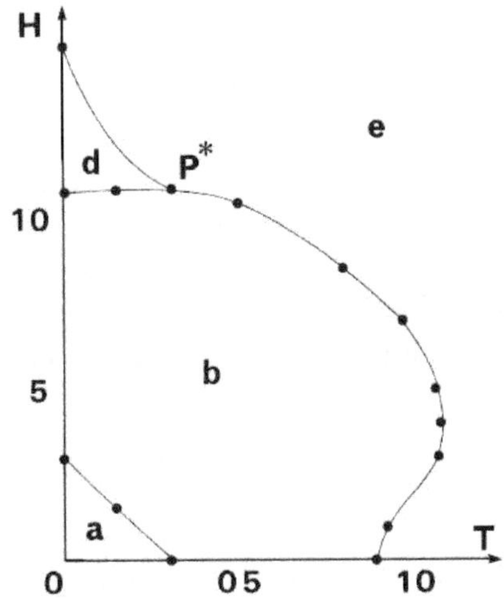

Fig. 2.27. The phase diagram in the (T, H) plane of the Ising-like Heisenberg model on an antiferromagnetic triangular lattice.

2.9. Ising-like Heisenberg Model on Antiferromagnetic Triangular Lattice

It has been pointed out that the transverse interaction, i.e., interaction of the xy components, causes successive phase transitions at T_1 and T_2. Below the higher critical temperature T_1, the z component of the spin is ordered in the FR type configuration, while the xy components are still disordered. Below the lower critical temperature, the xy components are also ordered to form a non-collinear structure.[55–57] It has been pointed out that the GS of this model has a non-trivial continuous degeneracy which cannot be resolved even if we take the entropy effect into account. Because of many nearly degenerate structures, this model has a rich structure of phases in the magnetic field.[58] The phase diagram in the (T, H) plane is depicted in Fig. 2.27, which has been found in $GdPd_2AL_3$.[59] Quantum effect on the magnetization process has been also studied.[60]

2.10. Ising Model with Infinite Spin on Stacked Antiferromagnetic Triangular Lattice

In the GS, the Ising model with spin-1/2 on the AFT lattice does not show any long-range order, while the Ising model with infinite spin on the AFT lattice shows a PD structure which is expressed as $(1, 0, -1)$, for example. Similarly, the spin structure for the Ising model with infinite spin on the stacked AFT lattice is quite different from that for the Ising model with spin-1/2. The Ising model with infinite spin shows a phase transition at a finite temperature. The spin structure in an intermediate temperature region is the PD structure, but it is the ferrimagnetic (FR) structure $(M, -m', -m')$ in a low-temperature region. This FR structure is quite stable up to $T = 0$. This can be seen by the use of low-temperature series expansion (LTSE).[44,45,61,62]

The Hamiltonian of the present model is given by

$$H = J \sum_{(i,j)}^{xy} x_i x_j - J_z \sum_{(i,j)}^{z} x_i x_j, \tag{2.30}$$

where x_i takes on the value $[-1, 1]$. We assume $J = J_z$ in the present section. The MC data for the sublattice magnetizations and the specific heat are shown in Figs. 2.28(a), and 2.28(b), respectively. As can be seen in Fig. 2.28, the system takes the PD structure in a temperature region, $T_1 < T < T_c$, where $T_1 \cong 0.8J/k$ and $T_c \cong 1.20J/k$. The FR structure is observed in a temperature region, $0 < T < T_1$. A sharp peak in the specific heat is observed at the temperature T_c. Although we think that the phase

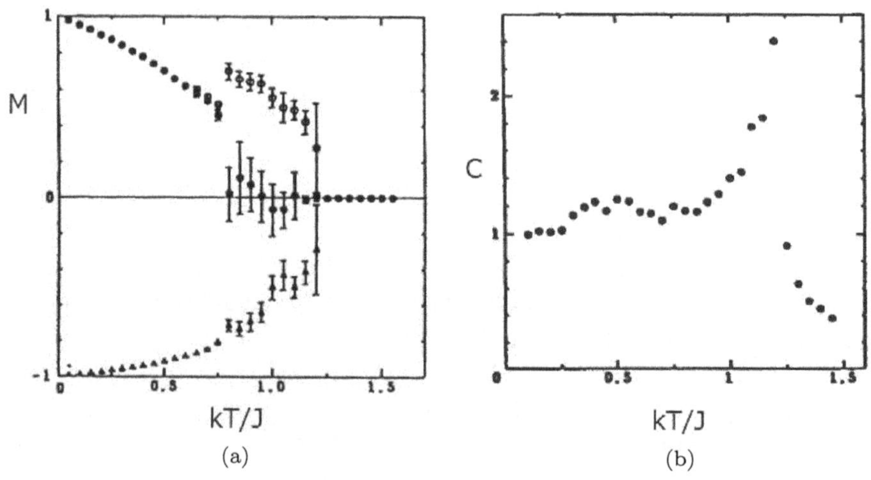

Fig. 2.28. (a) MC data of sublattice magnetizations. The spin structure is the FR structure for $T < T_1$ and it is the PD structure for $T_1 < T < T_c$. (b) MC data of specific heat.

transition at $T = T_1$ is of the first order, the singularity of the specific heat is very weak.

Let us study theoretically the stability of the FR structure at low temperatures observed by the MC simulations. The partition function Z is written as

$$Z = \int_{-1}^{1} dx_1 \cdots \int_{-1}^{1} dx_N \exp(-\beta H). \tag{2.31}$$

We introduce new variables u and v for up- and down-spins, respectively: $x_i = 1 - u_i$ and $x_j = -1 + v_j$. Then Z is written as

$$Z = \int_0^2 du_1 \cdots \int_0^2 du_{N/3} \int_0^2 dv_1 \cdots \int_0^2 dv_{2N/3} \exp(-\beta H), \tag{2.32}$$

where

$$H = -N(J + J_z) + (6J + 2J_z) \sum_i^A u_i + 2J_z \left(\sum_j^B v_j + \sum_k^C v_k \right) + H'. \tag{2.33}$$

Here, H' contains quadratic terms of u or v. A precise expression of H' will be omitted in this text. At low temperatures, we use the following replacement for the integrations:

$$\int_0^2 du \to \int_0^\infty du \quad \text{and} \quad \int_0^2 dv \to \int_0^\infty dv. \tag{2.34}$$

Then, the free energy is obtained by the LTSE as

$$\begin{aligned} F = &-N(J + J_z) - \frac{NkT}{3} \ln \left[\frac{kT}{6J + 2J_z} \right] \\ &- \frac{2NkT}{3} \ln \frac{kT}{2J} - \frac{N}{4} \left(\frac{kT}{J_z} \right)^2 Q + \cdots , \end{aligned} \tag{2.35}$$

where

$$Q = J(a^3 + 5a^2 + 6a - 9)/(a + 3)^2, \tag{2.36}$$

and $a = J_z/J$. The internal energy U is given by

$$U = -N(J + J_z) + NkT + \frac{N}{4} \left(\frac{kT}{J_z} \right)^2 Q + \cdots . \tag{2.37}$$

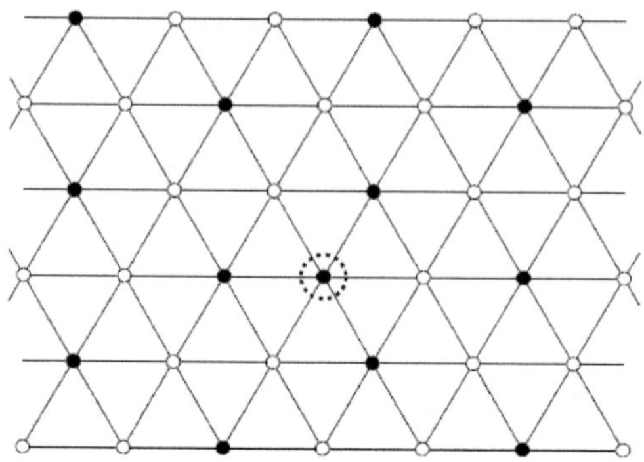

Fig. 2.29. The FR structure with a single misfit. The misfit is marked by a dashed circle.

The above expansions are not correct when $J_z < kT$, since free spins appear in the system and the above LTSE becomes invalid. When $J_z = J$, Eq. (2.37) agrees well with the MC data of U for $kT/J < 0.75$. Let us consider an FR structure with a single misfit which is defined in Fig. 2.29. This structure is constructed from the FR structure in Fig. 2.5 by replacing an open circle with a solid circle. Therefore, there exist $L(2L^2/3 - 1)$ down-spins and $L(L^2/3 + 1)$ up-spins in the spin state shown in Fig. 2.29. Let us consider the difference between the free energy of FR structure and that of ferrimagnetic-with-misfit (FRm) structure. The difference is given as

$$\begin{aligned}\Delta F &= F(\mathrm{FR}) - F(\mathrm{FRm}) \\ &= 3NkT\left\{\ln\left[\frac{2}{kT}(2J+J_z)\right] + \ln\left[\frac{2}{kT}(J+J_z)\right]\right. \\ &\quad \left. - \ln\left[\frac{2}{kT}(3J+J_z)\right] - \ln\frac{2J_z}{kT} + \cdots\right\}. \end{aligned} \tag{2.38}$$

The above equation is rewritten as

$$\Delta F = -3NkT\ln\left[\frac{(a+1)(a+2)}{a(a+3)}\right] + \cdots. \tag{2.39}$$

ΔF is negative for any positive value of a. Hence, the FR structure is more stable than the FRm structure. Furthermore, we notice that there are many

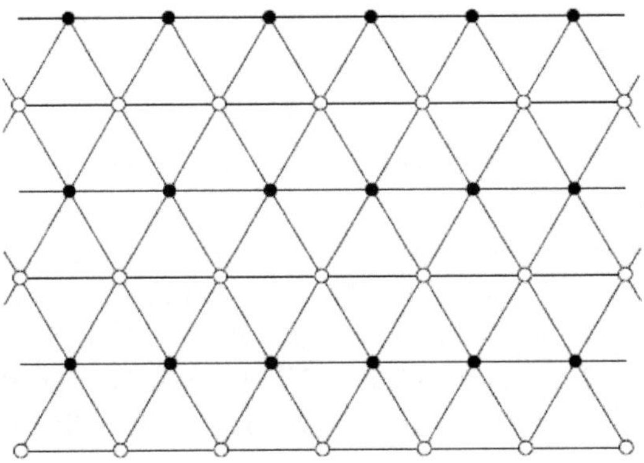

Fig. 2.30. A 2×1 structure.

metastable states. One of the possible metastable states will be the so-called (2×1) structure, which is shown in Fig. 2.30. The free energy of this state is calculated by LTSE as

$$F(2 \times 1) = -N(J + J_z) + NkT \ln \left[\frac{2}{kT}(J + J_z) \right] + \cdots . \tag{2.40}$$

The difference between $F(\text{FR})$ and $F(2 \times 1)$ is as follows:

$$F(\text{FR}) - F(2 \times 1) = -\frac{NkT}{3} \ln \frac{(a+1)^3}{a^2(a+3)} + \cdots . \tag{2.41}$$

We see from Eq. (2.41) that $F(\text{FR}) - F(2 \times 1)$ is negative for any positive value of a. Thus, the FR structure is more stable than the 2×1 structure at low temperatures. In this way, it turns out that the FRm structure and the (2×1) structure are both metastable.[63] In a similar way, it turns out that the PD structure is also metastable. Furthermore, the spin states in a "stacked" Wannier state are all metastable. Here, a "stacked" Wannier state is defined by a set of spin states each of which belongs to one of the Wannier states for the GS of the Ising model on the AFT lattice in any xy-plane and is stacked in such way that all the spins on a column along the z axis have the same direction. Hence, the number of metastable states in the system is of the order of $N^{2/3}$.

2.11. Phase Diagram in Spin Magnitude vs. Temperature for Ising Models with Spin S on Stacked Antiferromagnetic Triangular Lattice

The magnetic property of the Ising model with infinite spin on the stacked AFT lattice is well understood. That is, the system takes the PD structure in an intermediate temperature region and the FR structure in a low-temperature region. Although the Wannier state is realized in the GS, the FR structure is achieved in the limit of $T = 0$. On the other hand, the Ising model with spin-1/2 on the stacked AFT lattice takes the PD structure in an intermediate temperature region, but the system falls into the Wannier state in the limit of $T = 0$. Because, the spins on a column along the z axis are each parallel to others at low temperatures and then the column may be regarded as a single Ising spin. Therefore, there is a cross-over region in the temperature axis from the state of the 3D PD structure to the 2D Wannier state. Then, it is interesting to study the phase transition for the Ising model with general spin S.

We restrict our interest to the systems with $J = J_z$ in this section. The energy for the first excited state in the Ising model with spin S is of the order of J/S. Hence, if the temperature is lower than J/kS, the spins on a column along the z axis are each parallel to others and then the system will show the behavior of the Ising model on the AFT lattice. Hence, the boundary of the temperature for the cross-over region will depend on the value of J/S.[64]

In order to see the phase transitions in the systems, we calculate the sublattice magnetizations, X and Y defined by Eqs. (2.9) and (2.10), respectively, and also the specific heat. We show the MC data for the Ising model with spin-15/2. We have $X = 0$ and $Y = 2$ for the PD structure $(1, 0, -1)$, $X = 1$ and $Y = 3$ for the FR structure $(1, -1, -1)$, and $Y = 0$ and $X = 0$ for the Wannier state. The sublattice magnetizations are shown as functions of the temperature in Fig. 2.31(a), and the specific heat is shown in Fig. 2.31(b).

The temperature dependencies of Y and X are shown in Fig. 2.32(a). T_1 denotes the critical temperature at which the sublattice magnetizations appear, Y becomes non-zero, and the peak of the specific heat is observed. T_2 denotes a transition temperature at which the uniform magnetization X becomes non-zero and the curvature of Y changes. Note that any appreciable anomaly is not observed in the specific heat at T_2. There is no crossover region from the 3D ordered-state to the 2D Wannier-state for $kT/J > 0.1$

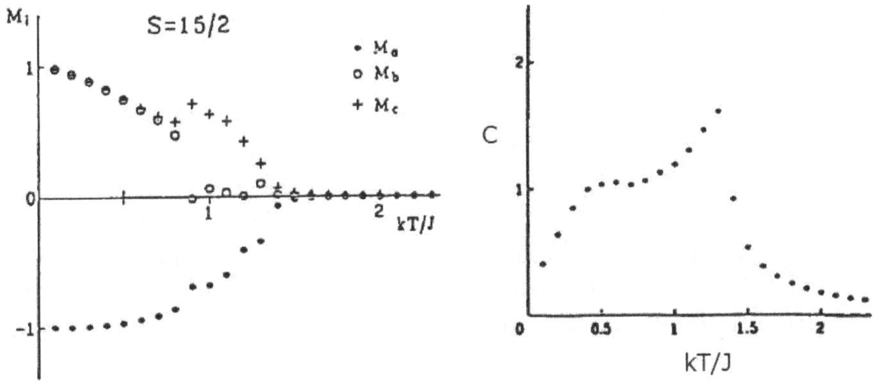

(a) (b)

Fig. 2.31. MC data of (a) the sublattice magnetizations and (b) the specific heat for the Ising model with spin-15/2. The magnetic structure is the PD structure in the intermediate temperature region and the FR structure in the low-temperature region.

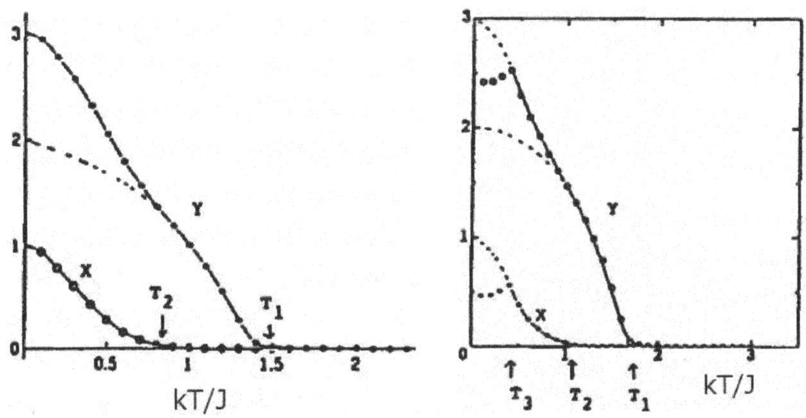

(a) (b)

Fig. 2.32. (a) MC data of Y and X for the Ising model with spin $S = 15/2$. The dashed line denotes the extrapolation to the zero temperature. The arrows indicate the transition points. (b) For $S = 5/2$. See the text for the explanation of T_1, T_2, and T_3.

in this case. Next, we show the MC data for the Ising model with spin 5/2. Since the statistical errors in the MC data of the sublattice magnetizations are very large, we show only the data of X and Y. The temperature dependencies of Y and X are shown in Fig. 2.32(b). In this figure, T_1 denotes the critical temperature of the phase transition from the paramagnetic state to the state with the PD structure. T_2 is a transition temperature of the first order from the state with the PD structure to that with the FR structure. T_3 denotes a turning point where the FR structure becomes unstable. The curve of Y between T_1 and T_2 tends toward the value 2 at $T = 0$, but the curve between T_2 and T_3 tends toward the value 3 at $T = 0$. However, the data of Y suddenly becomes small at T_3 and seem to tend toward zero at $T = 0$. Similarly, the MC data of X between T_2 and T_3 tend toward the value 1 at $T = 0$ but suddenly become small at T_3 and tend toward zero at $T = 0$. We think that the temperature region between T_3 and zero is a cross-over region for the system from the state with the FR structure to the Wannier state as the temperature decreases.

Thus, in this spin system there are two transition temperatures T_1 and T_2, and one turning point T_3 below which the system is in the cross-over region. Similar analyses are made for the Ising models with various spin S on the stacked AFT lattice and an obtained S–T phase diagram is shown in Fig. 2.33. In Fig. 2.33, the dashed line denotes the S dependence of T_3, which is an upper bound of the cross-over region obtained by the MC simulations. As stated above, the dashed line behaves as a linear function

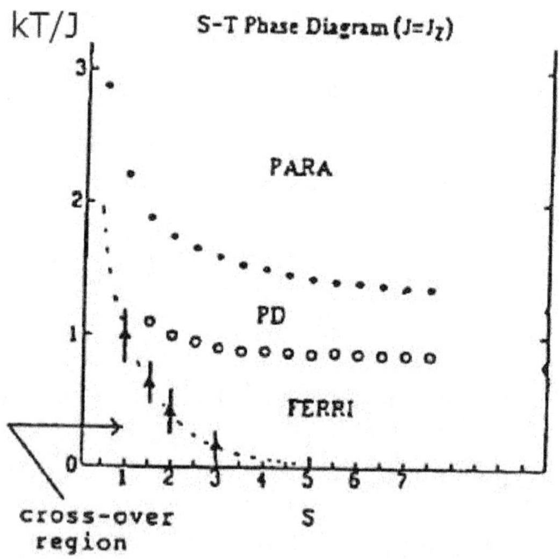

Fig. 2.33. S–T phase diagram for the Ising model with spin S on the stacked AFT lattice. The dashed line shows an upper bound of the cross-over region obtained by the MC simulations.

of J/S. The phase transition at T_2 seems to be of first order as we see from the discontinuous change in Fig. 2.31, which is similar to that of stacked $S = 1/2$ Ising AFT with NNN interaction.[29]

2.12. Effect of Antiferromagnetic Interaction between Next-Nearest-Neighbor Spins in xy-Plane

In this section, we discuss the effect of antiferromagnetic interaction J_2 in the Ising models with spin S and with infinite spin on the AFT lattice and on the stacked AFT lattice. The Hamiltonian for both systems is given as follows:

$$H = J\sum_{(i,j)}^{xy} \sigma_i\sigma_j - J_z\sum_{(i,j)}^{z} \sigma_i\sigma_j + J_2{\sum_{(i,j)}^{xy}}' \sigma_i\sigma_j. \tag{2.42}$$

Here, $\sum'$ denotes the sum over the NNN pairs in the xy-plane. The effect of J_2 on the Ising model with spin-1/2 on the AFT lattice was previously studied by Saito.[65] However, it has not been noticed that there exist long-range orders in the Ising models with spin S, including the Ising model with infinite spin, on the stacked AFT lattice.

The GS shows the 2×1 structure given in Fig. 2.30 if the antiferromagnetic interactions J_2 exist between the NNN pairs in the xy-plane. However, this 2×1 structure is metastable when $J_2 = 0$. In order to see this, we make the MC simulations for the case of $J_2 = 0$ by assuming the 2×1 structure as an initial spin configuration. We define the order parameter, Y_4, assuming a four-sublattice structure for a lattice in the xy-plane as follows:

$$Y_4 = \left(\frac{4}{N}\right)^2 \left\langle \left(\sum_i^A \sigma_i\right)^2 + \left(\sum_j^B \sigma_j\right)^2 + \left(\sum_k^C \sigma_k\right)^2 + \left(\sum_l^D \sigma_l\right)^2 \right\rangle. \tag{2.43}$$

We show the temperature dependence of Y_4 for the Ising model with spin-1/2 on the stacked AFT lattice in Fig. 2.34(a). The 2×1 structure is stable up to $T \cong 1.5J/k$, as can be seen in Fig. 2.34(a). The temperature dependence of Y_4 for the Ising model with infinite spin on the stacked AFT lattice is shown in Fig. 2.34(b) when $J_2 = 0$. We see the 2×1 structure is stable up to $T \cong 0.65J/k$. The existence of "stable" metastable states suggests the appearance of a first-order phase transition in the systems. Actually, we find a first-order phase transition in the systems.[63]

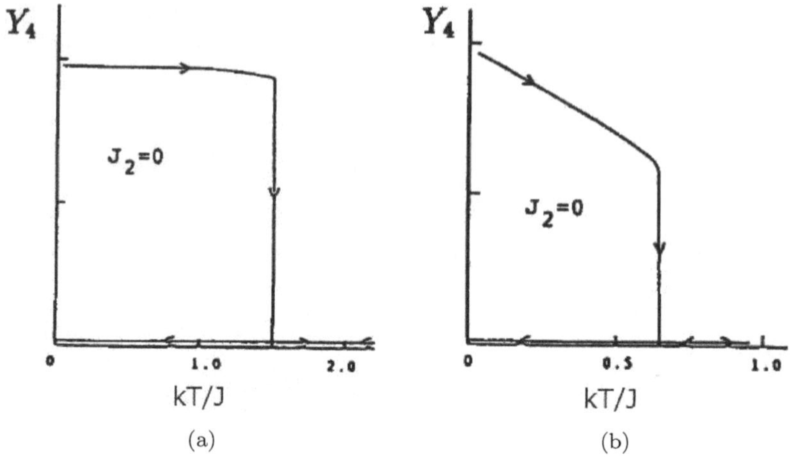

Fig. 2.34. MC data of Y_4 for the Ising model with (a) spin-1/2 and (b) infinite spin on the stacked AFT lattice with $J_z = J$ and $J_2 = 0$.

The energy for the 2×1 structure is given by

$$E_0 = -N(J + J_z + J_2). \tag{2.44}$$

When the GS is the 2×1 structure in the AFT lattice, the 2×1 structure also appears in each sublattice which is one of three sublattices in Section 2.4. Hence, we may suppose that the phase diagram of T vs. J is similar to that of T vs. J_2. We found this similarity by the MC simulations as seen below. Hence, we discuss only the case of $0 \le J_2 \le J$.

In Fig. 2.35, we show the phase diagram obtained by the MC simulations for the Ising model with spin-1/2 on the AFT lattice. In this system, there appear the paramagnetic state and the ordered state with the 2×1 structure. The paramagnetic state is denoted by P and the state with the 2×1 structure by 2×1 in Fig. 2.35 and also in the following figures.

In Fig. 2.36(a), we show the phase diagram obtained by the MC simulations for the Ising model with spin-1/2 on the stacked AFT lattice. In the system, the spin structure is not clarified yet in an intermediate temperature region for small J_2 and even in the case of $J_2 = 0$ at low temperatures. In fact, the spin state observed in the MC simulations shows some complex structure in this region and is very sensitive to the values of parameters, J_z, J_2, and T. Hence, we use the symbol C in Fig. 2.36(a) to denote this complex spin state. However, the phase boundaries are easily obtained by the MC simulations through Y_4 and other quantities. The spin structure at

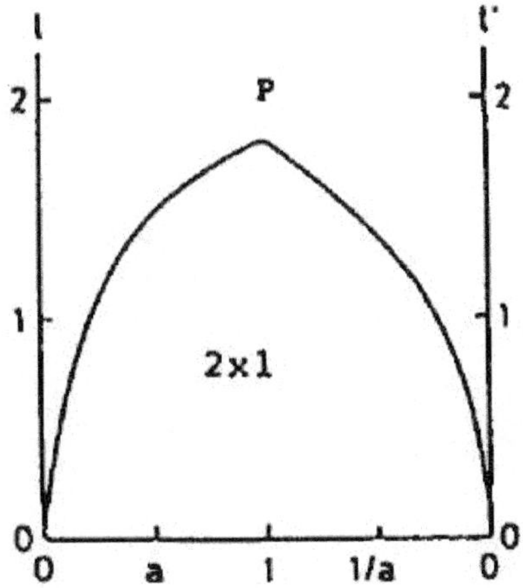

Fig. 2.35. Phase diagram for the Ising model with spin-1/2 on the AFT lattice. a, t, and t' denote $J_2/J, kT/J$, and kT/J_2, respectively.

low temperatures is the 2×1 structure. We calculate the free energies in the LTSE for various spin states. Since there are no free linear chains in the system for the case of $J_2 \neq 0$, it is sufficient to take single-spin-flip-type excitations into account. The free energy for the 2×1 structure, $F(2 \times 1)$, is given as

$$F(2 \times 1) = -N(J + J_z + J_2) - NkT \exp\left[-\frac{4}{kT}(J + J_z + J_2)\right] + \cdots . \tag{2.45}$$

The free energies for the FR structure and the PD structure are given as follows:

$$F(\mathrm{FR}) = -N(J + J_z - 3J_2) - \frac{2NkT}{3} \exp\left[-\frac{4}{kT}(J_z - 3J_2)\right] + \cdots , \tag{2.46}$$

$$F(\mathrm{PD}) = -N\left(J + J_z - \frac{5}{3}J_2\right) - \frac{NkT}{3} \exp\left[-\frac{4}{kT}(J_z + J_2)\right] + \cdots . \tag{2.47}$$

According to these equations, we have at $T = 0$

$$F(2 \times 1) < F(\mathrm{PD}) < F(\mathrm{FR}), \tag{2.48}$$

while the PD or FR structure becomes stable at $T \neq 0$.

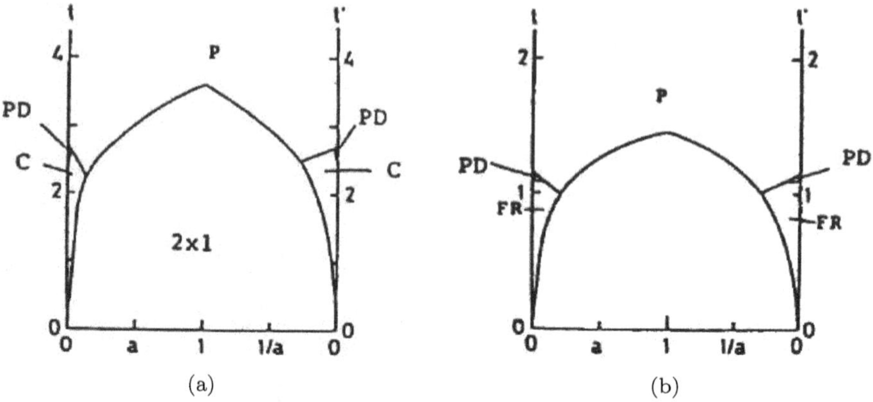

Fig. 2.36. Phase diagram for the Ising model with (a) spin-1/2 and (b) infinite spin on the stacked AFT lattice with $J = J_z$. a, t, and t' denote J_2/J, kT/J, and kT/J_2, respectively.

In Fig. 2.36(b), we show the phase diagram obtained by the MC simulations for the Ising model with infinite spin on the stacked AFT lattice. The spin structures of the ordered states in this system are well defined. For a very small value of J_2, we find that the spin structure is the 2×1 structure in a low-temperature region, the FR structure in an intermediate low-temperature region and PD structure in an intermediate high-temperature region, as seen in Fig. 2.36(b). In the LTSE, the free energy for the 2×1 structure is given as

$$F(2 \times 1) = -N(J + J_z + J_2) + NkT \ln \left[\frac{2}{kT}(J + J_z + J_2) \right] + \cdots . \quad (2.49)$$

The free energy for the FR structure is given as

$$F(\mathrm{FR}) = -N(J + J_z - 3J_2) + \frac{NkT}{3} \ln \left[\left(\frac{1}{kT} \right)^3 (6J + 2J_z - 6J_2)(2J_z - J_2)^2 \right] + \cdots . \quad (2.50)$$

The free energies for the 2×1 and the FR structure for the system with $J_z = 0.3J$ and $J_2 = 0.05J$ are shown in Fig. 2.37(a). In Fig. 2.37(b), we show the results by the LTSE for the internal energy $u = U/NJ$ together with the results by the MC simulations for the system with $J_z = 0.3J$ and $J_2 = 0.05J$. A first-order phase transition occurs at $T = 0.165J/k$.

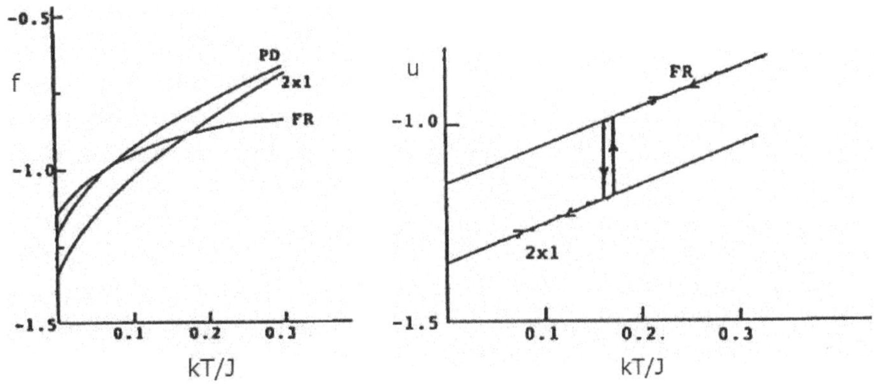

(a) (b)

Fig. 2.37. (a) The free energy by the low-temperature expansion. (b) The internal energy. The dots show the MC results and the solid lines indicate the theoretical results. A first-order phase transition is observed.

A small hysteresis, $\Delta(kT/J) = 0.01$, is observed in this case, as shown in Fig. 2.37(b).

2.13. Three-Dimensional Ising Paramagnet

Various frustrated Ising spin systems have been designed and studied to understand the effect of frustration in connection with the spin-glass problem.[66–68] It is well known that the frustration in general disturbs the formation of long-range order in spin systems. In spin systems on 3D lattices, however, some long-range orders driven by thermal fluctuations can be found easily even in fully frustrated Ising models. No long-range order appears in the Ising model on the AFT lattice as a typical example, but the Ising model on the antiferromagnetic face-centered cubic lattice shows the phase transition from the paramagnetic state to a periodic long-range order at a finite temperature.[8] The Ising model on the stacked Villain lattice, being different from the Ising model on the Villain lattice, also comes under this category. Then we have the following question: "Is there a 3d-Ising spin system which shows the paramagnetic behavior at all finite temperatures?" We take up this problem in this section. It is possible in principle because we may construct such a 3d-Ising model from the Ising model on the AFT lattice or the Villain lattice by using a glue-Hamiltonian.[69] But we propose a realistic candidate for the 3d-Ising paramagnet and demonstrate some magnetic properties of the candidate by the MC simulations in order to understand the paramagnetic nature of the system.[70,71]

An Ising model proposed for the 3d-paramagnet is composed in the following way. Consider a cube with lattice spacing a where an Ising spin of spin-1/2 is located at each corner of the cube and the neighboring spins are coupled by a ferromagnetic interaction J (see Figs. 2.38 and 2.39). This cube is named a J-cube. Imagine a simple cubic (sc) lattice with lattice spacing $2a$. Place the center of a J-cube on each lattice site of the sc lattice and let each edge of the J-cube be parallel to the sc lattice. Let an Ising spin on a corner of the J-cube interact with its nearest Ising spins on the other J-cubes through $+J_1$ or $-J_1$ interaction constant as shown in Fig. 2.38. When $J = J_1$, the xy-, yz-, or zx-plane in the resultant lattice corresponds to the so-called chessboard (CB) plane where the frustrated and non-frustrated plaquettes are alternately arranged, as shown in Fig. 2.39.

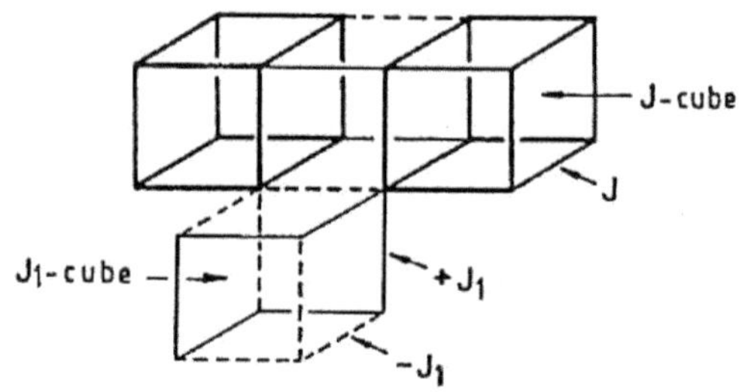

Fig. 2.38. A local structure of the CB3 system.

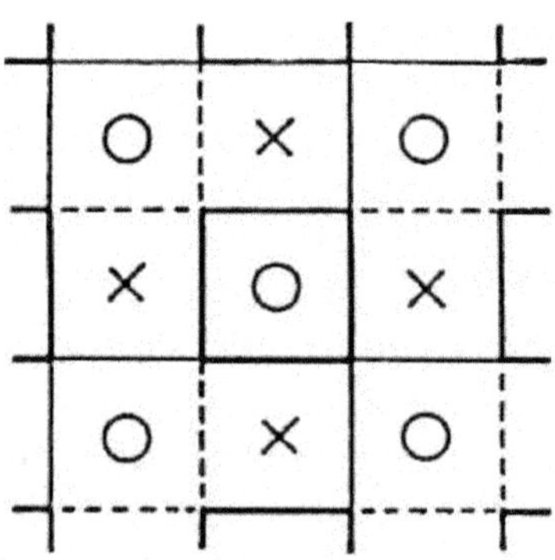

Fig. 2.39. A chessboard plane of alternating frustrated (marked by crosses) and non-frustrated (marked by circles) plaquettes.

We call this resultant lattice the CB3 lattice since all the three planes are of the CB type.

The CB3 lattice contains other kinds of cubes which are made up of $+J_1$ and $-J_1$ bonds. Each of these cubes is named a J_1-cube. Note that the J_1-cube is not frustrated. For the Ising model on the CB3 lattice with N sites, called the CB3 system hereafter, the number of both J-cubes and J_1-cubes amounts to $N/8$. The Hamiltonian of the CB3 system is written as

$$H = -\sum_{(i,j)} J_{ij}\sigma_i\sigma_j, \tag{2.51}$$

where J_{ij} stands for the interaction constant between σ_i and σ_j and $\sum_{(ij)}$ indicates the sum over all the NN spin pairs. It is easy to see that the GS energy of CB3 system is $E_0 = -(1/2)NJ$ and the zero-point entropy is $S_0 = (N/8)\ln 2$ if $J = J_1$. Similarly, if $J < J_1$, we have $E_0 = -(1/2)NJ_1$ and $S_0 = (N/8)\ln 2$. For the case $J = J_1$, the ferromagnetic state (F-state) is one of the spin states in the GS, since $E_0 = -(1/2)NJ$. Because there are at least $N/4$ free spins in this F-state, we have $S_0 \geq (N/4)\ln 2$ for the CB3 system when $J = J_1$. Let us make a thought experiment for the CB3 system according to the concept of the MC simulations. When $J_1 = 0$, the CB3 system is obviously paramagnetic at any finite temperatures.

However, if the conventional single-spin-flip (SSF) MC method is applied to the CB3 system, we are confronted with a crucial contradiction that a spin-glass order parameter tends to unity in the limit of temperature $T = 0$; the spin-glass order parameter, q, is defined by

$$q = \frac{1}{N}\sum_i |\langle\sigma_i\rangle|, \tag{2.52}$$

where the brackets denote the MC average. This discrepancy can be removed if we use the multi-spin-flip (MSF) MC method, which updates all the spin configurations in a spin cluster at the same time.

We now take a J-cube as a spin cluster in the MC calculation for the CB3 system. 2^8 states of a spin cluster are chosen as the achieved state of the cluster at each MC trial, according to the heat-bath importance sampling. We take advantage of the MSF as well as SSF methods for the present system. We calculate the specific heat, the susceptibility, and the spin-glass order-parameter. It is found that the spin-glass order parameter, q, vanishes at all temperatures for any J_1 value. In Fig. 2.40, the MC results of the specific heat, $c = C/Nk$, is shown for $J_1 = J$ (solid circles) and $J_1 = J/2$ (open circles). The exact result for $J_1 = 0$ is also plotted by the solid curve. No size dependence is observed in the specific heat data. Also, the peak position at $T \cong 1.9J/k$ is nearly fixed for different values of J_1. In the case of $J_1 = 0$, the peak of specific heat is undoubtedly caused by the energy fluctuation of the paramagnetic J-cubes. We suppose that the peak, even for the case of $J = J_1$, is due to nothing more than the energy fluctuations of J-cubes or J_1-cubes. Since the J-cubes (or J_1-cubes) are connected by $\pm J_1$ bonds (or J bonds), the above fluctuation is now suppressed with no shift of the peak position in the case of $J_1 \neq J$. The MC results in Fig. 2.40 are consistent with this tendency as the strength of J_1 increases.

In the limit of J infinity, the J-cube is regarded as large "block spins" whose value is $+8$ or -8. Since these $N/8$ block spins interact with each other via $\pm J_1$ in the CB3 system, the critical temperature should be zero in this limit.

A similar conjecture holds for the case of $J > J_1$ at very low temperatures, because of the strong correlation between spins on the J-cube. Then, the susceptibility of the system must be written as

$$\chi = \frac{8NJ}{kT}. \tag{2.53}$$

Actually, the MC results of susceptibility show this paramagnetic behavior at low temperatures as seen in Fig. 2.41. On the other hand, the low-temperature behavior of χ in the MC results behaves as the Curie law of $\chi = NJ/kT$ for the case $J = J_1$. This does not indicate that there exist

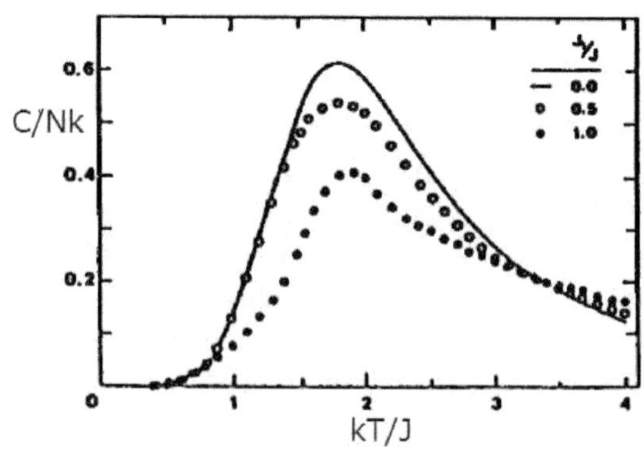

Fig. 2.40. MC data of the specific heat for the CB3 system.

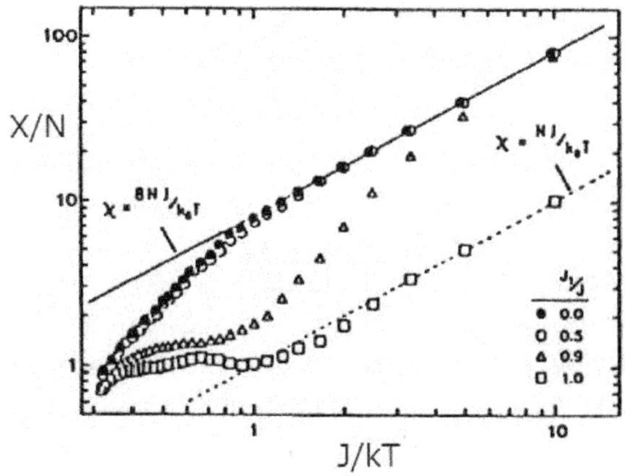

Fig. 2.41. MC data of the susceptibility for the CB3 system.

N free spins in the system, but it should be read as a suggestion that the system may include a large number of not only free block spins but also free spins, induced by the frustration effect. The MC data of χ at low temperatures give another reason for us to support why we have to use the MSF MC method for this system: the excitations within the J-cube are more dominant in the system.

It is interesting that the correlation functions of the Ising model on the 2D chessboard lattice[72] decay exponentially at all temperatures, in contrast to those on the 2D purely frustrated lattice where a power law decaying correlation appears in the GS. In view of its paramagnetic behaviors, the CB3 system is expected to have the correlation functions of the exponential decay at all temperatures.

2.14. Antiferromagnets on Corner-Sharing Lattices

Spin ordering in the so-called corner-sharing tetrahedron lattices, such as ferrite, pyrochlore, and the Laves phase, attracts interest.[73,74] They are extensions of the Kagomé lattice in two dimensions. The effect of frustration is very strong and an interesting ordering process reflecting spin configuration on a tetrahedron has been reported. In some cases, such as $Dy_2Ti_2O_7$, a strong uniaxial single-spin anisotropy exists and the spins can take only two states, i.e., inward or outward to the tetrahedron. Therefore, this system has another interesting ordering process. In these systems, the residual entropy plays an important role and they are called a "spin-ice" system.[75–78]

As to the Kagomé lattice antiferromagnet, it is known that any Ising($S = 1/2$), XY, and Heisenberg model do not have phase transition at all temperatures including $T = 0$.[79,80] Therefore, even the continuous Ising model does not have phase transition. However, it has been pointed out that the interaction of the transverse components causes a phase transition at a finite temperature.[81] The low-temperature phase has a uniform magnetization, and the phase transition belongs to the universality class of the ferromagnetic Ising model. This phase transition is attributed to the existence of uncompensated magnetic magnetization of the three spins on a triangle due to the non-collinear spin structure. The phase transition disappears in the Heisenberg limit where no uniaxial anisotropy exists and also in the limit of the continuous Ising model where the anisotropy is infinite. Thus, this phase transition occurs only in finitely uniaxial anisotropic systems.

2.15. Concluding Remarks

We discussed in this chapter several interesting features of the Ising spin systems with frustration. For the Ising model on the 2D frustrated lattices such as the antiferromagnetic triangular (AFT) lattice and the Villain lattice, exact solutions were known. Based on those results, we focused our attention mainly to the following problems: spin orderings, effects of interactions between next-nearest-neighbor spin pairs, and effects of magnitude S of spin in Ising spin systems on the AFT lattice and on the stacked AFT lattice.

One of characteristic features of the Ising spin systems on frustrated lattices is the existence of "free" spins. There also appear "free" spin-clusters such as "free" linear chains and "free" cubes. In those frustrated systems, cluster flip-type excitations exist in the low-lying energy states. Consequently, a cluster flip-type MC method is needed in the MC simulations for the those frustrated systems.

The effects of the far-neighbor interactions in the Ising model on the AFT lattice and the stacked AFT lattice were clarified in this chapter. The GS of the Ising model with only nearest-neighbor interactions on the AFT lattice is infinitely degenerate and is called the Wannier state. The Wannier state includes many states. Then if a perturbational interaction exists between spins on far-neighbor sites, the large degeneracy of the GS is lifted and there appears a multi-sublattice spin structure corresponding to the symmetry of the far-neighbor interaction. It is intriguing to find the phase diagrams for those systems, and the Ising models with the perturbational interaction between spins on the next-nearest-neighbor sites were studied. Recently, the effect of long-range ferromagnetic interaction on this model was studied and a new type of phase transition was found.[82]

As an alternate perturbation, effects of the interaction between the xy component were also briefly discussed in both classical and quantum cases. Moreover, in $S = 1/2$ systems, the interaction of the xy components causes the quantum effect, which would bring a new quantum state, the so-called resonating valence-bond (RVB).[83,84] Furthermore, the effect of the transverse field has been studied, where the GS degeneracy is resolved and the system has an extra peak of the specific heat at a low temperature.[85] As well as the case in the triangular lattice, the quantum effect on the Kagomé lattice also provides various peculiar properties such as the double peak in the specific heat and existence of many low-lying singlet states.[86–88] This is another of example of how the ordering process is sensitive to the different types of perturbations.

It has been noticed that the magnitude S of spin can be an important physical parameter in both classical spin systems and quantum spin systems. In the quantum spin systems, the Haldane problem is one of controversial problems. In frustrated Ising spin systems, the universality class may depend strongly on the magnitude S of spin, although no such S dependence is found in the Ising models on the non-frustrated lattices. We discussed S-dependencies of spin ordering for the Ising spin systems on the frustrated lattices. We hope that more active investigations will be done in order to clarify the S-dependent properties of the Ising spin systems on the frustrated lattices.

Acknowledgments

We thank M. Kang, Y. Yamada, M. Tanaka, and N. Todoroki for their help in numerical computations and in preparing this chapter.

References

1. J. Villain, *Phys. Chem. Solids* **11**, 303 (1959).
2. A. Yoshimori, *J. Phys. Soc. Jpn.* **14**, 807 (1959).
3. T. A. Kaplan, *Phys. Rev.* **116**, 888 (1959).

4. D. J. Amit, *Modeling Brain Function: The World of Attractor Neural Networks* (Cambridge University Press, Cambridge, 1989).
5. G. Toulouse, *Commun. Phys.* **2**, 115 (1977).
6. J. Vannimenus and G. Toulouse, *J. Phys.* **C10**, L537 (1977).
7. D. C. Mattis, *The Theory of Magnetism II* (Springer-Verlag, Berlin, 1985).
8. R. Liebmann, *Statistical Mechanics of Periodic Frustrated Ising Systems* (Springer-Verlag, Berlin, 1986).
9. G. H. Wannier, *Phys. Rev.* **79**, 357 (1950).
10. M. Houtappel, *Physica* **16**, 425 (1950).
11. K. Husimi and I. Syoji, *Prog. Theor. Phys.* **5**, 177 (1950).
12. I. Syoji, *Prog. Theor. Phys.* **5**, 341 (1950).
13. J. Stephenson, *Can. J. Phys.* **47**, 2621 (1969).
14. J. Stephenson, *J. Math. Phys.* **11**, 413 (1970).
15. T. Horiguchi, K. Tanaka, and T. Morita, *J. Phys. Soc. Jpn.* **61**, 64 (1992).
16. M. Mekata, *J. Phys. Soc. Jpn.* **42**, 76 (1977).
17. J. Villain, *J. Phys. C* **10**, 1717 (1977).
18. G. Forgacs, *Phys. Rev. B* **22**, 4478 (1980).
19. S. Miyashita, *J. Phys. Soc. Jpn.* **52**, 780 (1983).
20. D. Blankschtein, M. Ma, A. N. Berker, G. S. Grest, and C. M. Soukoulis, *Phys. Rev. B* **29**, 5250 (1984).
21. O. Nagai, Y. Yamada, and H. T. Diep, *Phys. Rev. B* **32**, 480 (1985).
22. L. Onsager, *Phys. Rev.* **65**, 117 (1944).
23. A. N. Berker, G. S. Grest, C. M. Soukoulis, D. Blankschtein, and M. Ma, *J. Appl. Phys.* **55**, 2416 (1984).
24. F. Matsubara and S. Inawashiro, *J. Phys. Soc. Jpn.* **56**, 2666 (1987).
25. O. Nagai, K. Nishino, J. J. Kim, and Y. Yamada, *Phys. Rev. B* **37**, 5448 (1988).
26. J. J. Kim, Y. Yamada, and O. Nagai, *Phys. Rev. B* **41**, 4760 (1990).
27. S. N. Coppersmith, *Phys. Rev. B* **32**, 1584 (1985).
28. M. Mekata and K. Adachi, *J. Phys. Soc. Jpn.* **44**, 806 (1978).
29. N. Todoroki and S. Miyashita, *J. Phys. Soc. Jpn.* **73**, 412 (2004).
30. H. Takayama, K. Matsumoto, H. Kawahara, and K. Wada, *J. Phys. Soc. Jpn.* **52**, 2888 (1983).
31. S. Fujiki, K. Shutoh, Y. Abe, and S. Katsura, *J. Phys. Soc. Jpn.* **52**, 1531 (1983).
32. L. P. Landau, *Phys. Rev. B* **27**, 5604 (1983).
33. S. Miyashita, H. Kitatani, and Y. Kanada, *J. Phys. Soc. Jpn.* **60**, 523 (1991).
34. O. Nagai, M. Kang, and S. Miyashita, *Phys. Lett. A* **196**, 101 (1994).
35. S. Elitzur, P. B. Pearson, and J. Shigemitsu, *Phys. Rev. D* **19**, 3698 (1979).
36. J. L. Cardy, *J. Phys. A* **26**, 6201 (1982).
37. J. R. Banavar, G. G. Grestand, and D. Jasnow, *Phys. Rev. B* **25**, 4639 (1982).
38. S. Miyashita, *J. Phys. Soc. Jpn.* **66**, 3411 (1997).
39. M. Oshikawa, *Phys. Rev. B* **61**, 3430 (2000).
40. N. Todoroki, Y. Ueno, and S. Miyashita, *Phys. Rev. B* **66**, 214405 (2002).
41. O. Koseki and F. Matsubara, *J. Phys. Soc. Jpn.* **69**, 1202 (2000).

42. D. C. Mattis, *Phys. Rev. Lett.* **42**, 1503 (1979).
43. S. Miyashita, *Prog. Theor. Phys. Suppl.* **87**, 112 (1986).
44. T. Horiguchi, O. Nagai, and S. Miyashita, *J. Phys. Soc. Jpn.* **60**, 1513 (1991).
45. T. Horiguchi, O. Nagai, and S. Miyashita, *J. Phys. Soc. Jpn.* **61**, 308 (1992).
46. T. Horiguchi, O. Nagai, S. Miyashita, Y. Miyatake, and Y. Seo, *J. Phys. Soc. Jpn.* **61**, 3114 (1992).
47. O. Nagai, S. Miyashita, and T. Horiguchi, *Phys. Rev. B* **47**, 202 (1993).
48. O. Nagai, T. Horiguchi, and S. Miyashita, *Phys. Lett. A* **186**, 186 (1994).
49. R. Peierls, *Proc. Cambridge Phil. Soc.* **32**, 477 (1936).
50. R. B. Griffiths, in *Phase Transitions and Critical Phenomena*, Vol. 3, ed. C. Domb and M. S. Green (Academic Press, London, 1972), p. 1.
51. A. Lipowski, T. Horiguchi, and D. Lipowska, *Phys. Rev. Lett.* **74**, 3888 (1995).
52. A. Lipowski and T. Horiguchi, *J. Phys. A: Math. Gen.* **28**, 3371 (1995).
53. Y. Honda, A. Lipowski, and T. Horiguchi, *Phys. Rev. B* **52**, 13429 (1995).
54. A. Lipowski, Y. Honda, T. Horiguchi, and Y. Hondai, *Physica A* **237**, 297 (1997).
55. S. Miyashita and H. Kawamura, *J. Phys. Soc. Jpn.* **54**, 3385 (1985).
56. S. Watarai and S. Miyashita, *J. Phys. Soc. Jpn.* **63**, 1916 (1994).
57. S. Watarai and S. Miyashita, *J. Phys. Soc. Jpn.* **67**, 816 (1998).
58. S. Miyashita, *J. Phys. Soc. Jpn.* **55**, 3605 (1986).
59. H. Kitazawa, H. Suzuki, H. Abe, J. Tang, and G. Kido, *Physica B* **259–261**, 890 (1999).
60. H. Nishimori and S. Miyashita, *J. Phys. Soc. Jpn.* **55**, 4448 (1986).
61. C. Domb, in *Phase Transitions and Critical Phenomena*, Vol. 3, ed. C. Domb and M. S. Green (Academic Press, London, 1974), p. 1.
62. T. Horiguchi, *J. Phys. Soc. Jpn.* **59**, 3142 (1990).
63. O. Nagai, M. Kang, T. Horiguchi, and H. T. Diep, *Phys. Lett. A* **186**, 186 (1994).
64. O. Nagai, M. Kang, Y. Yamada, and T. Horiguchi, *Phys. Lett. A* **195**, 195 (1994).
65. Y. Saito, *Phys. Rev. B* **24**, 6652 (1981).
66. K. Binder and A. P. Young, *Rev. Mod. Phys.* **58**, 801 (1986).
67. S. Kirkpatrick, in *Disordered Systems and Localization*, eds. C. Castellani, C. Di Castro and L. Pelti (Springer-Verlag, Berlin, 1981), p. 280.
68. A. P. Young, *Spin Glasses and Random Fields*, ed. A. P. Young (World Scientific, Singapore, 1998).
69. T. Horiguchi, *Physica* **146A**, 613 (1987).
70. H. T. Diep, P. Lallemand, and O. Nagai, *J. Phys. C* **18**, 1067 (1985).
71. J. J. Kim, K. Nishino, Y. Yamada, and O. Nagai, in *Proceeding of Fourth Asia Pacific Physics Conference*, eds. S. H. Ahn, S. H. Choh, H. T. Cheon and C. Lee (World Scientific, Singapore, 1990).
72. G. André, R. Bidaux, J.-P. Carton, R. Conte, and L. de Seze, *J. Physique* **40**, 479 (1979).
73. P. W. Anderson, *Phys. Rev.* **102**, 1008 (1996).
74. Y. Kobayashi, T. Takagi, and M. Mekata, *J. Phys. Soc. Jpn.* **67**, 3906 (1998).

75. M. J. Harris *et al. Phys. Rev. Lett.* **79**, 2554 (1997).
76. R. Siddharthan *et al. Phys. Rev. Lett.* **83**, 1854 (1999).
77. M. Udagawa, M. Ogata, and Z. Hiroi, *J. Phys. Soc. Jpn.* **71**, 2365 (2002).
78. T. Sakakibara, T. Tayama, Z. Hiroi, K. Matsuhira, and S. Takagi, *Phys. Rev. Lett.* **90**, 207205-1-4 (2003).
79. H. Barry, H. Khatun, and T. Tanaka, *Phys. Rev. B* **37**, 5193 (1998).
80. D. A. Huseand and A. D. Rutenberg, *Phys. Rev. B* **45**, 7536 (1992).
81. A. Kuroda and S. Miyashita, *J. Phys. Soc. Jpn.* **64**, 4509 (1995).
82. M. Nishino and S. Miyashita, *Phys. Rev. B* **92**, 184404 (2015).
83. P. W. Anderson, *Mater. Res. Bull* **8**, 153 (1973).
84. P. Fazekas and P. W. Anderson, *Philos. Mag.* **30**, 423 (1974).
85. M. V. Mostovoy, D. I. Khomskii, J. Knoester, and N. V. Prokof'ev, *Phys. Rev. Lett.* **90**, 14723 (2003).
86. V. Elser, *Phys. Rev. Lett.* **62**, 2405 (1989).
87. T. Nakamura and S. Miyashita, *Phys. Rev. B* **52**, 9174 (1995).
88. F. Mila, *Phys. Rev. Lett.* **81**, 2356 (1998).

CHAPTER 3

RENORMALIZATION GROUP APPROACHES TO FRUSTRATED MAGNETS IN $d = 3$

B. Delamotte[*,‡], D. Mouhanna[*,§] and M. Tissier[†,¶]

[*] *Laboratoire de Physique Théorique et Hautes Energies, CNRS UMR 7589*
Universités Paris VI-Pierre et Marie Curie, Paris VII-Denis Diderot
2 Place Jussieu, 75251 Paris Cedex 05, France
[†] *Laboratoire de Physique Théorique des Liquides, CNRS UMR 7600*
Université Paris VI-Pierre et Marie Curie
4 Place Jussieu, 75252 Paris Cedex 05, France
[‡] *delamotte@lpthe.jussieu.fr*
[§] *mouhanna@lpthe.jussieu.fr*
[¶] *tissier@lptl.jussieu.fr*

This chapter is devoted to the recent advances in the renormalization group (RG) approaches to the physics of classical frustrated magnets in three dimensions. We first recall the main features of the field theoretical approach to these systems including considerations on symmetries, symmetry breaking schemes, continuum limits, topological contents, etc. We then provide an overview of the phenomenological situation, stressing on the most striking aspect of the physics of frustrated magnets: the existence of *non-universal scaling* behaviors. We then review the various perturbative and non-perturbative RG approaches that have been used to investigate frustrated magnets. Finally, a large part of this chapter is devoted to a recent non-perturbative approach that has clarified the intricate physical situation of frustrated magnets.

3.1. Introduction

Magnetic systems with competing interactions — or *frustrated* magnets — like helimagnets or geometrically frustrated magnets represent one of the most celebrated class of systems whose critical physics, although studied for more than 25 years, is still strongly debated. The specificity of frustrated magnets is that the competition between the interactions generates non-trivial ground states (GSs) together with original symmetry breaking schemes. For continuous spins, for instance, the GS is no longer made of aligned or anti-aligned spins. This is well known from the case of the stacked triangular antiferromagnets (STA) whose GS gives rise to the famous 120° structure. Also, in the case of Heisenberg spins, the SO(3) rotational symmetry group is *fully* broken at low temperatures. This contrasts with the ferromagnetic case for which the corresponding symmetry breaking scheme is given by SO(3) $\rightarrow$ SO(2). These non-trivial GS and symmetry breaking schemes imply that both the spin wave physics and the topological content differ from those of non-frustrated systems. As a consequence of these features, the situation appears to be much more intricate for frustrated magnets than for non-frustrated ones, as we now describe.

First, experimentally and numerically, there is no clear, unambiguous, picture of the 3D physics for both XY and Heisenberg systems. On the contrary, the phenomenology of frustrated magnets displays quite unconventional features: almost all experiments exhibit scaling laws around the transition temperature but with *non-universal* critical exponents, i.e. with critical exponents that vary from one compound to the other. Moreover, other strange behaviors are encountered: scaling laws are violated, the anomalous dimension is negative at the transition, etc. This strongly contrasts with the situation for non-frustrated systems and makes very doubtful the proposition that frustrated magnets would belong to a new universality class, the so-called "chiral universality class".[1–6]

Second, the theoretical situation in these systems is also far from being clear: (i) independently of the phenomenological context, the results obtained within the usual perturbative approaches — in dimensions $d = 2+\epsilon$, in $d = 4-\epsilon$ or directly in $d = 3$ — conflict. Indeed, the nonlinear sigma (NLσ) model approach performed around two dimensions predicts a second-order phase transition of the standard $O(4)/O(3)$ universality class.[7,8] On the other hand, the weak coupling expansion performed on the suitable Ginzburg–Landau–Wilson (GLW) model in the vicinity of $d = 4$ leads to a first-order phase transition due to the lack of a stable fixed point.[1,2,9] Finally, high-order perturbative approaches performed directly in $d = 3$ predict a second-order phase transition different from the $O(4)/O(3)$ universality class. (ii) Neither the low-temperature expansion around $d = 2$

nor the high-order weak-coupling calculations performed around $d = 4$ or directly in $d = 3$ — at least, when interpreted in a conventional way — succeed in reproducing the whole phenomenology of frustrated magnets.

During the last few years, there have been considerable progress in the understanding of both the theoretical and the phenomenological features of the physics of frustrated magnets. These progress have been the consequence of several facts. First, the acknowledgment of the non-universal character of the phase transitions in these systems: it is now more and more accepted that frustrated magnets do not constitute a new universality class. Second, and greatly motivated by the first point, the recognition of the inadequacy of the conventional theoretical approaches. This has led to the use of new or improved methods — high-order perturbative approaches,[10–13] non-perturbative renormalization group (NPRG) methods[14–19] — and concepts — pseudo-fixed point,[14,15,17,19] continuously varying exponents[11,12,18,19] — suitable for the description of new kinds of critical behaviors.

It is the aim of this chapter to review these major advances in the RG analysis of frustrated systems that have led to an almost complete understanding of the physics of these systems in three dimensions, with a particular emphasis on the NPRG approach.

3.2. The STA Model and Generalization

3.2.1. *The lattice model, its continuum limit, and its symmetries*

We now describe the archetype of frustrated spin systems, the STA. This system is composed of 2D triangular lattices that are piled-up in the third direction. At each lattice site, there is a magnetic ion whose spin is described by a classical vector. The interaction between the spins is given by the usual lattice Hamiltonian:

$$H = \sum_{\langle ij \rangle} J_{ij} \vec{S}_i . \vec{S}_j, \tag{3.1}$$

where, depending on the anisotropies, the $\vec{S}_i$ are two- or three-component vectors and the sum runs over all pairs of nearest-neighbor spins. The coupling constants J_{ij} equal $J_{\parallel}$ for a pair of sites inside a plane and $J_{\perp}$ between planes. The interactions between nearest-neighbor spins within a plane are

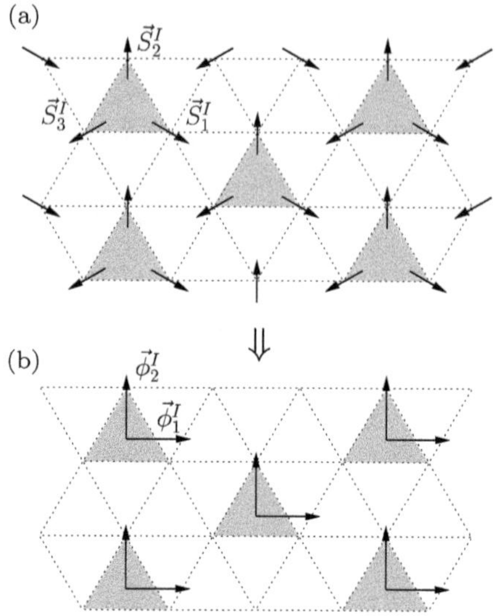

Fig. 3.1. The GS configurations (a) of the spins on the triangular lattice and (b) of the order parameter made of two orthonormal vectors. The plaquettes, which constitute the magnetic cell, are indexed by I and are shaded.

antiferromagnetic, i.e. $J_{\|} > 0$. This induces frustration in the system and, in the GS, gives rise to the famous 120° structure of the spins, see Fig. 3.1(a).

This non-trivial magnetic structure is invariant under translations of length $\sqrt{3}$ times the initial lattice spacing. The magnetic cell, indexed by I, which is replicated all over the system, is a plaquette of three spins $\vec{S}_1^I$, $\vec{S}_2^I$, and $\vec{S}_3^I$, see Fig. 3.1(a). Note that the nearest-neighbor out-of-plane interaction $J_\perp$ is, depending on the compounds, ferro- or antiferromagnetic, but the two cases can be treated simultaneously since no extra frustration appears through this interaction. Finally, interactions between more distant spins (next-to-nearest neighbors, etc.) also exist but are neglected in the following since they are supposed to be irrelevant.

There have been numerous derivations of the long-distance effective field theory supposed to describe the critical physics of this system.[1,2,6,20] We now sketch the derivation which is most appropriate for our purpose.

The Hamiltonian (3.1) has the usual rotational symmetry acting on the spin components: $O(2)$ or $O(3)$ for XY or Heisenberg spins, respectively. To identify the order parameter, it is also necessary to consider the symmetry of the magnetic cell. For the triangular lattice, this is the C_{3v} group that interchanges the spins inside a plaquette.

The identification of the order parameter is close in spirit to what is done in the non-frustrated case, e.g. for the antiferromagnets on a square lattice. At zero temperature, the sum of the three spins for a given plaquette I:

$$\vec{\Sigma}^I = \vec{S}_1^I + \vec{S}_2^I + \vec{S}_3^I \tag{3.2}$$

is vanishing.[a] In average, this is also the case at any finite temperature so that the thermal average of

$$\vec{\Sigma} = \sum_I \vec{\Sigma}^I, \tag{3.3}$$

where the sum runs on all plaquettes, cannot be an order parameter: the associated modes are never critical. We, therefore, replace $\vec{\Sigma}^I$ by its average value

$$\vec{\Sigma}^I \to \langle \vec{\Sigma}^I \rangle = \vec{0}, \tag{3.4}$$

which is equivalent to freezing the fluctuations of the spins inside each plaquette. The constraint $\vec{\Sigma}^I = \vec{0}$ is called the "local rigidity constraint". Having eliminated $\vec{\Sigma}$, we keep only two vectors per plaquette $(\vec{\phi}_1^I, \vec{\phi}_2^I)$ which represent the local order parameter. For $\vec{\phi}_2^I$, we choose one of the spins of the plaquette, see Figs. 3.1(a) and 3.1(b). For the other, $\vec{\phi}_1^I$, we choose the linear combination of the spins which is orthogonal to $\vec{\phi}_2^I$ and of unit norm, see Fig. 3.1(b). The local order parameter thus obeys on each plaquette:

$$\vec{\phi}_i^I . \vec{\phi}_j^I = \delta_{ij} \quad \text{with } i, j \in \{1, 2\}. \tag{3.5}$$

The dihedral $(\vec{\phi}_1^I, \vec{\phi}_2^I)$ plays a role analogous to the staggered magnetization in the non-frustrated case.

As usual, once the model is reformulated in terms of its order parameter, the effective interaction — from plaquette to plaquette — becomes ferromagnetic, see Fig. 3.1(b). By taking the dihedral $(\vec{\phi}_1^I, \vec{\phi}_2^I)$ on the center of the plaquette I, we indeed find that it interacts ferromagnetically with the dihedral $(\vec{\phi}_1^J, \vec{\phi}_2^J)$ defined on the center of the plaquette J — the plaquettes I and J being nearest neighbors — such that $\vec{\phi}_1^I$ interacts only with $\vec{\phi}_1^J$ and $\vec{\phi}_2^I$ only with $\vec{\phi}_2^J$. A more detailed analysis shows that the two

[a] $\vec{\Sigma}^I$ is analogous to the local magnetization of non-frustrated antiferromagnets — $\vec{\Sigma}^I = \vec{S}_1^I + \vec{S}_2^I$ in this last case — that also vanishes in the GS.

vectors $\vec{\phi}_1^I$ and $\vec{\phi}_2^I$ play symmetric roles.[2] As a consequence, the effective Hamiltonian reads

$$H = -J \sum_{\langle I,J \rangle} (\vec{\phi}_1^I . \vec{\phi}_1^J + \vec{\phi}_2^I . \vec{\phi}_2^J) \tag{3.6}$$

with the same coupling constant $J > 0$ for the $\vec{\phi}_1^I$ and $\vec{\phi}_2^I$ vectors. Moreover, since the anisotropies resulting from the stacked structure of the lattice are supposed to be irrelevant, we take the same coupling constant for the interactions inside a plane and between the planes. The continuum limit is now trivial and proceeds as in the usual ferromagnetic case. The effective Hamiltonian in the continuum thus writes, up to constants

$$H = -\int d^d\mathbf{x}((\partial\vec{\phi}_1(\mathbf{x}))^2 + (\partial\vec{\phi}_2(\mathbf{x}))^2) \tag{3.7}$$

with the constraint that $\vec{\phi}_1$ and $\vec{\phi}_2$ are orthonormal. This model is called the Stiefel $\mathrm{V}_{N,2}$ model with $N = 2$ in the XY case and $N = 3$ in the Heisenberg case. In $V_{N,2}$, the index 2 means that we are considering *two* orthonormal vectors $\vec{\phi}_1$ and $\vec{\phi}_2$.

It is convenient to gather the vectors $\vec{\phi}_1$ and $\vec{\phi}_2$ into a rectangular matrix

$$\Phi = (\vec{\phi}_1, \vec{\phi}_2) \tag{3.8}$$

and to rewrite H as

$$H = -\int d^d\mathbf{x}\,\mathrm{Tr}(\partial^t\Phi(\mathbf{x}).\partial\Phi(\mathbf{x})), \tag{3.9}$$

where $({}^t\Phi)_{ij} = \Phi_{ji}$.

In Sections 3.2.2 and 3.2.3, we consider successively the case of Heisenberg and XY spins.

3.2.2. *The Heisenberg case*

In this case, H is invariant under the usual left $O(3)$ rotation and inversion group acting on the spins

$$\Phi' = R\Phi, \quad R \in O(3). \tag{3.10}$$

It is also invariant under a right $O(2)$

$$\Phi' = \Phi U, \quad U \in O(2). \tag{3.11}$$

This last symmetry encodes the fact that $\vec{\phi}_1$ and $\vec{\phi}_2$ play the same role which, itself, is reminiscent of the C_{3v} symmetry of the triangular plaquette. The system is thus symmetric under $G = O(3) \times O(2)$. In the low-temperature phase, a typical GS configuration is given by (see Fig. 3.1(b))

$$\Phi_0 \propto \begin{pmatrix} 1 & 0 \\ 0 & 1 \\ 0 & 0 \end{pmatrix}. \tag{3.12}$$

It is symmetric under the diagonal group — $O(2)_{\text{diag}}$ — built from the right $O(2)$ and from a particular left $O(2)$ in $O(3)$:

$$\Phi_0 = \begin{pmatrix} \epsilon\cos\theta & -\sin\theta & 0 \\ \epsilon\sin\theta & \cos\theta & 0 \\ 0 & 0 & 1 \end{pmatrix} \Phi_0 \begin{pmatrix} \epsilon\cos\theta & \epsilon\sin\theta \\ -\sin\theta & \cos\theta \end{pmatrix}, \tag{3.13}$$

where $\epsilon = \pm 1$ encodes the $\mathbb{Z}_2$ part of $O(2)_{\text{diag}}$. Apart from the previous $\mathbb{Z}_2$ contained in the $O(2)_{\text{diag}}$, another $\mathbb{Z}_2$ is also left unbroken. It is the combination of a $\mathbb{Z}_2$ included into the right $O(2)$ of G, Eq. (3.11), and of a rotation of π around the x-axis contained in the rotation group $SO(3)$ of G. Thus, G is spontaneously broken down to $H = \mathbb{Z}_2 \times O(2)_{\text{diag}}$. As a consequence, the symmetry breaking scheme reads

$$G = O(3) \times O(2) \to H = \mathbb{Z}_2 \times O(2)_{\text{diag}}, \tag{3.14}$$

which is often referred to as the $SO(3) \times SO(2)/SO(2)$ model, when all the discrete $\mathbb{Z}_2$ groups have been cancelled.

Here appears the main feature of frustrated magnets: the $SO(3)$ group is *fully* broken in the low-temperature phase, whereas it is only broken down to $SO(2)$ in non-frustrated magnets. This has two important consequences that are at the very origin of the non-trivial critical behavior encountered in frustrated magnets.

First, there are three Goldstone modes in the broken phase instead of two in the non-frustrated case. This implies a spin-wave physics different from that of $O(3)/O(2)$ model. Second, in the order parameter space $SO(3)$ having a non-trivial first homotopy group[21]:

$$\pi_1(SO(3)) = \mathbb{Z}_2, \tag{3.15}$$

there exist stable non-trivial topological configurations called vortices. Because of the $\mathbb{Z}_2$ homotopy group, only one kind of vortex exists, in contrast to the well-known case of XY ferromagnets where there are infinitely many different kinds of vortices, each one being indexed by an integer, the winding number.

It has been established first by Kawamura and Miyashita[22] that the existence of vortices is important in two dimensions. This has been largely confirmed by subsequent works studying the temperature dependence of thermodynamical quantities such as the correlation length, the spin stiffness, etc.[23–28] Actually, although this has not been directly established, they certainly also play an important role for the critical physics of the STA in three dimensions. A simple argument allows to argue to that end: let us go back on the lattice and introduce, on each plaquette I, together with $\vec{\phi}_1^I$ and $\vec{\phi}_2^I$, a third vector $\vec{\phi}_3^I$ defined by

$$\vec{\phi}_3^I = \vec{\phi}_1^I \wedge \vec{\phi}_2^I. \tag{3.16}$$

Let us then gather them into a 3×3 matrix

$$\Phi^I = (\vec{\phi}_1^I, \vec{\phi}_2^I, \vec{\phi}_3^I). \tag{3.17}$$

Since $(\vec{\phi}_1^I, \vec{\phi}_2^I, \vec{\phi}_3^I)$ are three orthonormal vectors, one has ${}^t\Phi^I.\Phi^I = \mathbb{1}$, and therefore, Φ^I is an $SO(3)$ matrix. This allows to rewrite the Hamiltonian (3.6) on the lattice as

$$H = -\sum_{\langle I,J\rangle} \mathrm{Tr}(\mathcal{P}\,{}^t\Phi^I.\Phi^J), \tag{3.18}$$

where $\mathcal{P}$ is a diagonal matrix of coupling constants that characterizes the interaction between the $\vec{\phi}_1^I$ vectors, between the $\vec{\phi}_2^I$ vectors, and between the $\vec{\phi}_3^I$ vectors. One has, from the microscopic derivation, $\mathcal{P} = \mathrm{diag}(J, J, 0)$, i.e the interaction is the same between the $\vec{\phi}_1^I$ vectors and between the $\vec{\phi}_2^I$ vectors and there is no interaction between the $\vec{\phi}_3^I$ vectors. However, for the present purpose, we consider, without loss of generality, the case where the interaction is non-vanishing and identical between all vectors. One thus has $\mathcal{P} = J\mathbb{1}$. Now, we use the decomposition of a rotation matrix Φ^I of $SO(3)$ in terms of a four-component *unit* vector $\widetilde{S^I} = (S_0^I, S_1^I, S_2^I, S_3^I)$:

$$\Phi_{kl}^I = 2\left(S_k^I S_l^I - \frac{1}{4}\delta_{kl}\right) + 2\epsilon_{klm} S_0^I S_m^I + 2\left({S_0^I}^2 - \frac{1}{4}\right)\delta_{kl}. \tag{3.19}$$

In terms of the vector $\widetilde{S^I}$, the Hamiltonian (3.18) writes

$$H = -4J \sum_{\langle I,J\rangle} (\widetilde{S^I}.\widetilde{S^J})^2 \tag{3.20}$$

which is the Hamiltonian for *four*-component non-frustrated spins with the peculiarity that each vector $\widetilde{S^I}$ appears quadratically. Therefore, the Hamiltonian (3.20) is invariant under a global $O(4)$ group and under a *local* — gauge — $\mathbb{Z}_2$ group that changes $\widetilde{S^I}$ to $-\widetilde{S^I}$. It corresponds to the $RP^3 = SO(4)/(SO(3) \times \mathbb{Z}_2)$ model.[b] For three-component spins, an analogous Hamiltonian, the RP^2 model, had been introduced by Maier and Saupe[29] and by Lebwohl and Lasher[30] to investigate the isotropic–nematic transition in liquid crystals. An extensive study of the RP^2 model, as well as a detailed investigation of the role of vortices in this transition, has been performed by Lammert *et al.*[31,32] These authors have shown, in particular, that these non-trivial topological configurations favor the first-order character of the transition. In the case of four-component spins, no such detailed analysis has been performed. However, the $RP^N = SO(N)/(SO(N-1) \times \mathbb{Z}_2)$ models that generalize the Hamiltonian (3.20) to N-component spins with $2 \leq N \leq 4$ have been numerically studied by Kohring and Shrock.[33] These systems have been shown to undergo a first-order phase transition. Since the only difference between the RP^N and the $SO(N)/SO(N-1)$ models lies in their topological properties, one is naturally led to attribute the origin of the first-order character of the phase transition in the RP^N models to the $\mathbb{Z}_2$ vortices. Finally, since the Hamiltonian (3.18), relevant to STA, can be mapped onto the Hamiltonian (3.20) — up to the $O(4)$-breaking terms — one can expect that the topological configurations also favor first-order phase transitions in frustrated magnets in three dimensions.

3.2.3. *The XY case*

In the XY case, the Hamiltonian (3.9) is still invariant under the right $O(2)$ group, see Eq. (3.11), while the left symmetry group becomes $O(2)$. In the low-temperature phase, the rotational symmetry is broken and, since the spins are constrained to be in a plane, the permutation symmetry between $\vec{\phi}_1$ and $\vec{\phi}_2$ is also broken. As a consequence, the symmetry breaking scheme is

$$G = O(2) \times O(2) \to H = O(2)_{\text{diag}}. \tag{3.21}$$

[b]Note that, had we kept the microscopical content, with $\mathcal{P} = \text{diag}(J, J, 0)$, the Hamiltonian (3.20) would be supplemented by terms breaking the $SO(4)$ global symmetry, leaving the $\mathbb{Z}_2$ local symmetry untouched which is the important point for our purpose.

This symmetry breaking scheme is usually referred to as $SO(2) \times \mathbb{Z}_2 \to \mathbb{1}$. The $\mathbb{Z}_2$ degrees of freedom are known as chirality variables.[6,22,34,35]

In this case, there also exist topological defects since

$$\pi_1(SO(2)) = \mathbb{Z}. \tag{3.22}$$

These defects are identical to those of the ferromagnetic XY model that drive the famous Berezinskii–Kosterlitz–Thouless transition in two dimensions.[36,37] However, in the frustrated case, they very likely interact non-trivially with the chirality degrees of freedom which are critical in $d = 3$ at the same temperature as the spin-wave degrees of freedom. This is apparent from the fact that one observes a unique phase transition and not two distinct Ising-like and XY-like transitions.[38] As a consequence, one can expect, in the frustrated case, a physics different and probably more complicated than in the non-frustrated $O(2)$ model that undergoes a standard second-order phase transition in three dimensions.

3.2.4. *Generalization*

For reasons that will become clear in the following, we consider the generalization of Hamiltonian (3.1) to N-component spins. It is straightforward to extend the previous considerations to this case. One finds the following symmetry breaking scheme:

$$G = O(N) \times O(2) \to H = O(N-2) \times O(2)_{\text{diag}}. \tag{3.23}$$

In the following, we shall drop the "diag" index for simplicity. Note that the previous Heisenberg and XY cases are recovered trivially provided that we identify $O(0)$ with the trivial group $\mathbb{1}$ and $O(1)$ with $\mathbb{Z}_2$.

We now give an overview of the experimental and numerical situations in XY and Heisenberg cases.

3.3. Experimental and Numerical Situations

As said in Section 3.1, due to their specific symmetry breaking scheme, frustrated spin systems have been proposed to belong to a new universality class.[1–6] However, in view of the experimental and numerical results relevant to the physics of frustrated magnets, the situation is far from being so simple. In effect, as emphasized by the present authors,[18,19] the phenomenology of frustrated magnets shows that these systems display scaling *without* universality, a behavior not compatible with a usual second-order phase transition. Moreover, there also exist clear indications of weak first-order phase transitions.

3.3.1. *The XY systems*

Let us first discuss the XY case since the experimental situation is richer than in the Heisenberg case. Also, the symptoms of the existence of a problem in the interpretation of the results are clearer than in the latter case.

3.3.1.1. *The experimental situation*

Two classes of materials are supposed to be described by Hamiltonian (3.9). The first one is made of ABX_3 hexagonal perovskites — where A is an alkali metal, B a transition metal, and X is a halogen atom — which are physical realizations of XY STA. The most studied ones are $CsMnBr_3$, $CsCuCl_3$, $CsNiCl_3$, and $CsMnI_3$. (See Collins and Petrenko[39] for a review and Weber *et al.*[40] for $RbMnBr_3$. We have excluded this material since the measurement of its specific heat presents a shoulder like anomaly near T_c which renders the determination of α and β doubtful.) The second one is made of rare earth helimagnets: Ho, Dy, and Tb. For most materials, the transitions are found to be continuous but *not* with the same critical exponents. For $CsCuCl_3$, the transition is found to be a weak first-order transition, i.e. with small discontinuities. The results are summarized in Table 3.1.

We highlight four striking characteristics[18,19] of these data.[c] Their consequences for the physics of frustrated magnets will be then discussed.

(i) *There are two groups of incompatible exponents.* In the present discussion, we mainly use the exponent β to analyze the results since it is by far the most precisely measured exponent. Clearly, there are two groups of materials, each of which are being characterized by a set of exponents, β in particular.

In the first one — that we call group 1 — made up of

$$\text{group 1: } \mathrm{CsMnBr_3}, \mathrm{CsNiCl_3}, \mathrm{CsMnI_3}, \mathrm{Tb}, \tag{3.24}$$

one has

$$\beta \sim 0.237(4), \tag{3.25}$$

[c]The details concerning the computations of the average values of the critical exponents as well as their error bars have been given by the present authors.[18,19]

Table 3.1. The critical exponents of XY frustrated materials.

Materials	Critical exponents
$CsMnBr_3$	$\alpha = 0.39(9)$,[41] $0.40(5)$,[42] $0.44(5)$;[42] $\beta = 0.21(1)$,[38] $0.21(2)$,[43] $0.22(2)$,[44] $0.24(2)$,[45] $0.25(1)$;[46] $\gamma = 1.01(8)$,[43] $1.10(5)$;[47] $\nu = 0.54(3)$,[43] $0.57(3)$[47]
$CsNiCl_3$	$\alpha = 0.342(5)$,[48] $0.37(6)$,[49] $0.37(8)$;[49] $\beta = 0.243(5)$[50]
$CsMnI_3$	$\alpha = 0.34(6)$[49]
$CsCuCl_3$	$\alpha = 0.35(5)$;[40] $\beta = 0.23(2)$,[51] $0.24(2)$,[51] $0.25(2)$;[51] 1st order[40]
Tb	$\alpha = 0.20(3)$;[52] $\beta = 0.21(2)$,[53] $0.23(4)$;[54] $\nu = 0.53$[55]
Ho	$\alpha = 0.27(2)$,[56] 0.1–0.22;[41] $\beta = 0.30(10)$,[57] $0.37(10)$,[57] $0.38(1)$,[58] $0.39(3)$,[59] $0.39(2)$,[60] $0.39(4)$,[61] $0.39(4)$,[62] $0.41(4)$;[62] $\gamma = 1.14(10)$,[63] $1.24(15)$;[57] $\nu = 0.54(4)$,[57] $0.57(4)$;[63] 1st order[64]
Dy	$\alpha = 0.24(2)$;[65] $\beta = 0.335(10)$,[66] $0.38(2)$,[60] $0.39(1)$,[59] $0.39^{+0.04}_{-0.02}$;[67] $\gamma = 1.05(7)$;[63] $\nu = 0.57(5)$[63]

Note: For $CsCuCl_3$, the transition has been found to be of first-order and the exponents mentioned here hold only for a reduced temperature larger that 5.10^{-3} (see Ref. [40]).

while in the second — group 2 — made up of

$$\text{group 2: Ho, Dy,} \tag{3.26}$$

one has

$$\beta \sim 0.389(7). \tag{3.27}$$

Note that, as far as we know, there is no determination of the exponent β for $CsMnI_3$ that, being given its composition, has been included in the group 1 of materials. Anyway, our conclusions are not affected by this fact.

The two values of β given by Eqs. (3.25) and (3.27) are completely incompatible. Actually, we find for the average exponents for the materials of group 1 (except Tb for which the results are not fully under control)

$$\beta = 0.237(4), \quad \nu = 0.555(21), \quad \alpha = 0.344(5), \quad \gamma = 1.075(42). \quad (3.28)$$

For materials of group 2 (Ho and Dy) we find

$$\beta = 0.389(7), \quad \nu = 0.558(25), \quad \gamma = 1.10(5). \quad (3.29)$$

We do not give a value for α which is poorly determined.

Let us also indicate that the exponents vary much from compound to compound in group 1. Although less accurately determined than β, α is only marginally compatible between $CsNiCl_3$ and $CsMnBr_3$. Note moreover that, even for the same material, the data are not fully compatible among themselves: β in $CsMnBr_3$ shows a large dispersion.

(ii) *The anomalous dimension η is negative for group 1 while it cannot be so for a second-order phase transition.*[d]

If one assumes that the transition is of second order for group 1, one can use the scaling relations to compute η. In particular, the precise determination of β allows to use $\eta = 2\beta/\nu - 1$ to determine η rather accurately. The exponent ν itself can be obtained directly from the experiments or deduced using the scaling relation

$$\nu = (2 - \alpha)/3. \quad (3.30)$$

The large number of experiments devoted to the determination of α allows a precise determination of ν. By using the scaling relation Eq. (3.30), we find $\nu = 0.528(11)$ if we consider the experimental results for $CsMnBr_3$ alone and $\nu = 0.552(2)$ if we consider $CsMnBr_3$, $CsNiCl_3$, and $CsMnI_3$. By using the relation $\eta = 2\beta/\nu - 1$ together with Eq. (3.30) or the relation $\eta = 6\beta/(2 - \alpha) - 1$ and by considering the data of $CsMnBr_3$ alone or the data of the materials of group 1 (except Tb for which it is not sure that the data are reliable), we can obtain four determinations of η. In the four cases, we find η to be negative by at least 4.1 standard deviations and the probability to find it is positive and always less than 10^{-5}. In fact, the most precise determination is obtained by combining all the data of group 1, Eq. (3.28), and by using the relation $\eta = 6\beta/(2 - \alpha) - 1$. In this case,

[d] η is *always* positive if the underlying field theory is a Ginzburg–Landau–Wilson φ^4-like theory,[138] as it is the case here.[2]

we obtain $\eta = -0.141(14)$ and thus a (almost) vanishing probability to find it positive. Note also that, although β and ν are less accurately known in Tb — for which experiments are anyway delicate — η is also found negative.

(iii) *For group 2, the scaling relation* $\gamma + 2\beta - 3\nu = 0$ *is violated.* From Eq. (3.29), it is possible to check the scaling relations. We find $\gamma+2\beta-3\nu = 0.202(92)$ and thus a violation by 2.2 standard deviations.

(iv) $CsCuCl_3$ *undergoes a weak first-order phase transition.* Until recently, $CsCuCl_3$ was believed to undergo a second-order phase transition with exponents compatible with those of group 1, see Table 3.1. It has been finally found to display a weak first-order phase transition.[40]

3.3.1.2. *The numerical situation*

Monte Carlo (MC) simulations have been performed on five different kinds of XY systems. The first one is the STA itself.[4,5,68–72] The second model is the STAR (where R stands for rigidity) which consists in an STA for which the local rigidity constraint — Eq. (3.4) — has been imposed on each plaquette at all temperatures.[73] The third model is the Stiefel $V_{2,2}$ model whose Hamiltonian is given by Eq. (3.6).[73,74] This is a hard spin, discretized version of the NLσ model relevant to frustrated magnets. Also a soft spin, discretized version of the GLW model has recently been studied by Itakura[72] who also restudied the STA model for large sizes. Finally, a helimagnetic system defined on a body-centered tetragonal (BCT) lattice — the "BCT model" — has been investigated.[75]

Here, we emphasize that the local rigidity constraint — Eq. (3.4) — as well as the manipulations that lead to the STAR, Stiefel $V_{2,2}$, GLW, and BCT models only affect the *massive* — non-critical — modes. Thus, the STA, STAR, Stiefel $V_{2,2}$, GLW, and BCT models have the same *critical* modes, the same symmetries, and the same order parameter. Therefore, one could expect a common critical behavior for all these systems.

Let us comment the results of the simulations given in Table 3.2. Note that, due to the its novel character, we shall comment on the recent work of Itakura[72] separately.

(i) *For STA, scaling laws are found with exponents compatible with those of group 1.* Let us, however, note that, similar to what happens for the materials of group 1, there exists, in the numerical simulations of STA, a rather large dispersion of the results. For instance, the two extreme values of ν differ by 2.1 standard deviations.

Table 3.2. The critical exponents of the XY frustrated systems.

Systems	Critical exponents
STA	$\alpha = 0.34(6)$,[4,68] $0.46(10)$,[70] $0.43(10)$;[71]
	$\beta = 0.24(2)$,[70] $0.253(10)$;[4,68]
	$\gamma = 1.03(4)$,[70] $1.13(5)$;[4,68]
	$\nu = 0.48(2)$,[71] $0.50(1)$,[70] $0.54(2)$;[4,68]
	$\eta = -0.09(8)$,[4,68] $-0.06(4)$;[70]
	1st order[72]
STAR, $V_{2,2}$, BCT, GLW	1st order[72,73,75]

Let us make two other remarks. First, the good agreement between the numerical results for STA and the experimental ones for materials of group 1 has been repeatedly interpreted in the literature as a proof of the existence of a second-order transition and even as an evidence of the existence of the chiral fixed point of the GLW model.[6] We emphasize here that the fact that an MC simulation reproduces the experimental results only means that the Hamiltonian of the simulated system is a good approximation of the microscopic Hamiltonian describing the physics of real materials. However, this neither explains nor proves anything else — and *certainly does not prove* the existence of a second-order phase transition — since MC simulations suffer from problems analogous to those encountered in experiments: a weak first-order phase transition is very difficult to identify and to distinguish from a second-order one.

Let us now come to our second remark. In a beautiful experiment, Plakhty *et al.*[38] have measured the so-called chiral critical exponents β_c and γ_c in $CsMnBr_3$. They have found values compatible with those found numerically in STA.[68] Let us emphasize, again, that this agreement simply means that the parameters characterizing the numerical simulations are not too far from those associated to the experiments. By no means it implies — or gives a new indication of the existence of — a second-order transition. Let us notice that β_c has also been measured in Ho.[58] The value found completely disagrees with the result found in STA and in $CsMnBr_3$.

(ii) *The anomalous dimension η is negative for STA.* As shown by Loison and Schotte,[73] η is found to be negative using the two scaling relations $\eta = 2\beta/\nu - 1$ and $\eta = 2 - \gamma/\nu$ for the two simulations where these calculations can be performed.

(iii) *The simulations performed on STAR, $V_{2,2}$, and BCT models give first-order transitions.* Therefore, the modifications in the microscopic details which change STA into STAR, $V_{2,2}$, and BCT affect drastically the scaling behavior.

(iv) *In a remarkable work, Itakura has recently performed MC and MC RG approaches of the STA and its GLW model version that has led to a clear first-order behavior.*[72] Itakura has performed standard MC simulations of the STA involving sizes up to $126 \times 144 \times 126$ leading to clear first-order transitions. In particular, for these lattice sizes, the double-peak of the probability distribution of the energy at the transition is clearly identified. Itakura has also used an improved MC RG simulation of the STA and its GLW model version. One advantage of this approach compared with previous RG MC studies is that it allows to reach the asymptotic critical behavior using systems of moderately large lattice sizes. Within this approach, Itakura has found several evidences for a first-order behavior with, notably, a runaway behavior of the RG flow and the absence of any non-trivial fixed point.

3.3.1.3. *Summary*

Let us now summarize the XY case. The experimental and numerical data reveal some striking features:

(1) Scaling laws are found in STA and helimagnetic materials on a rather wide range of temperatures. This is also the case within all — but an important one[72] — numerical simulations of the STA.
(2) There are two groups of systems that differ by their exponents. The first one includes the group 1 materials and the numerical STA model. The second one corresponds to the group 2 materials. One also observes variations of critical exponents inside a given group of exponents.
(3) The anomalous dimension η is negative for the materials of group 1 and for the numerical STA model. This is very significant from the experimental results, less from the numerical ones.
(4) For group 2, the scaling relations are violated by 2.2 standard deviations.
(5) $CsCuCl_3$ is found to undergo a weak first-order transition.
(6) STAR, $V_{2,2}$, and BCT models undergo strong first-order transitions.
(7) Recent MC and MC RG approaches of STA and the soft spin discretized version of the GLW model give clear indications of first-order behaviors.

We now review the experimental and numerical results obtained for the Heisenberg systems.

3.3.2. *The Heisenberg systems*

3.3.2.1. *The experimental situation*

In contrast to the XY case, there is no Heisenberg helimagnets (see however Garel and Pfeuty[1]). Therefore, there remain, *a priori*, only the Heisenberg STA materials. In fact, the A/B phase transition of He_3 can be described by the same GLW Hamiltonian as the Heisenberg STA.[9,76] It is thus a candidate. Unfortunately, the narrowness of the critical region of this transition does not allow a reliable study of the critical behavior of this system and there are no available data concerning the transition.

Three classes of Heisenberg STA materials have been studied. The first class is made up of systems like VCl_2, VBr_2, $Cu(HCOO)_2 2CO(ND_2)_2$ $2D_2O$, and $Fe[S_2CN(C_2H_5)_2]_2Cl$ which are generically quasi-XY except in a particular range of temperature where their anisotropies are irrelevant. The second class is made up of systems which become isotropic thanks to a magnetic field that exactly counterbalances the anisotropies. This is the case of $CsNiCl_3$ and $CsMnI_3$ at their multi-critical point. Finally, the third class is made up of systems which become isotropic because they have been prepared in a fine-tuned stoichiometry such that the Ising-like and XY-like anisotropies cancel each other to form an isotropic material. This is the case of $CsMn(Br_{0.19}I_{0.81})_3$.

Let us comment on the experimental results summarized in Table 3.3.

(i) *As in the XY case, the Heisenberg materials fall into two groups.* The group 1, made up of

$$\begin{aligned} \text{group 1:}\ & Cu(HCOO)_2 2CO(ND_2)_2 2D_2O, \\ & Fe[S_2CN(C_2H_5)_2]_2Cl, \\ & VCl_2, VBr_2 \end{aligned} \tag{3.31}$$

is characterized by

$$\beta = 0.230(8), \tag{3.32}$$

while group 2, made up of

$$CsNiCl_3, CsMnI_3, CsMn(Br_{0.19}, I_{0.81})_3 \tag{3.33}$$

Table 3.3. The critical exponents of the Heisenberg frustrated materials.

Materials	Critical exponents
VCl_2	$\beta = 0.20(2);^{77}$ $\gamma = 1.05(3);;^{77}$ $\nu = 0.62(5)^{77}$
VBr_2	$\alpha = 0.30(5)^{78}$
A	$\beta = 0.22(2)^{79}$
B	$\alpha = 0.244(5);^{80}$ $\beta = 0.24(1);^{81-83}$ $\gamma = 1.16(3)^{81-83}$
$CsNiCl_3$	$\alpha = 0.25(8),^{49,84}$ $0.23(4);^{48}$ $\beta = 0.28(3)^{50}$
$CsMnI_3$	$\alpha = 0.28(6)^{49}$
C	$\alpha = 0.23(7);^{85}$ $\beta = 0.29(1);^{86}$ $0.28(2);^{87}$ $[\gamma = 0.75(4)],^{86}$ $[\nu = 0.42(3)]^{86}$

Notes: The abbreviations A, B, and C stand for $Cu(HCOO)_2 2CO(ND_2)_2$ $2D_2O$, $Fe[S_2CN(C_2H_5)_2]_2Cl$, and $CsMn(Br_{0.19}I_{0.81})_3$, respectively. The data in brackets are suspected to be incorrect. They are given for completeness.

one finds

$$\beta = 0.287(8), \tag{3.34}$$

Note that, strictly speaking, the values of β for VBr_2 and for $CsMnI_3$ are not known and, thus, our classification is somewhat improper. It seems however logical to suppose that VBr_2 is close to VCl_2 and $CsMnI_3$ close to $CsNiCl_3$. Anyway, it will be clear in the following that our analysis is almost insensitive to this point.

For group 1, the average values of the critical exponents are given by

$$\beta = 0.230(8), \quad \alpha = 0.272(35), \quad \nu = 0.62(5), \quad \gamma = 1.105(21). \tag{3.35}$$

A very severe difficulty in the study of the materials of group 1 is their 2D character and Ising-like anisotropies. The temperature range where the systems behave effectively as 3D Heisenberg systems is narrow. This is the case of VCl_2 where this range is less than two decades and where, closer to the critical temperature, the system becomes Ising-like. For this group of materials, the exponent β is very small and it has been noticed that such small values have also been found in materials where dimensional cross-over is suspected.[77] Thus, it is not clear whether the whole set of results really corresponds to a 3D Heisenberg STA.

For group 2, the experimental situation seems to be better under control. The average values of the critical exponents are given by

$$\beta = 0.287(9), \quad \alpha = 0.243(3), \quad \nu = 0.585(9), \quad \gamma = 1.181(33), \tag{3.36}$$

where the scaling relations have been used to compute ν and γ. Note that the values of ν and γ thus obtained differ significantly from those of $CsMn(Br_{0.19}I_{0.81})_3$ whose critical behavior has been claimed to be perturbed by disorder.

(ii) *For group 1, the anomalous dimension η is significantly negative.* Using the two exponents that have been measured at least twice in group 1 — β and γ — we can compute the anomalous dimension from $\eta = (4\beta - \gamma)/(2\beta+\gamma)$. We find $\eta = -0.118(25)$ which is thus negative by 4.8 standard deviations.

(iii) *For group 2, the anomalous dimension η is marginally negative.* Using the critical exponents given in Eq. (3.36), one obtains, for the anomalous dimension $\eta = -0.018(33)$. Thus, η is found to be negative but not significantly, in contrast to what happens in group 1.

(iv) *For group 1, the scaling relations $\gamma + 2\beta - 2 + \alpha = 0 = 2\beta + \gamma - 3\nu$ are violated.* Indeed, $\gamma + 2\beta - 2 + \alpha = -0.135(56)$ and $2\beta + \gamma - 3\nu = -0.29(15)$. Of course, none of these violations is completely significant in itself because of the lack of experimental data. However, since they are both independently violated, it remains only a very small probability that the scaling relations are actually valid.

3.3.2.2. *The numerical situation*

In the Heisenberg case, as in the XY case, five different kinds of systems: STA, STAR, Stiefel ($V_{3,2}$ in this case), BCT, and GLW models have been studied. The results of the simulations are given in Table 3.4.

Let us comment on them. Again, we put aside the work of Itakura.[72]

(i) *For the STA, scaling laws are found with an exponent β close to that of group 2.* The average values for the exponents of STA are

$$\beta = 0.288(6), \quad \gamma = 1.185(3), \quad \nu = 0.587(5). \tag{3.37}$$

β is thus extremely close to the experimental value of group 2 and ν and γ are extremely close to the experimental values deduced from the scaling

Table 3.4. The critical exponents of the Heisenberg frustrated systems.

Systems	Critical exponents
STA	$\alpha = 0.240(80)$,[3, 68] 0.242(24),[88] 0.245(27),[89] 0.230(30);[90]
	$\beta = 0.300(20)$,[3, 68] 0.285(11),[88] 0.289(15),[89] 0.280(15);[90]
	$\gamma = 1.170(70)$,[3, 68] 1.185(3),[88] 1.176(26);[89]
	$\nu = 0.590(20)$,[3, 68] 0.586(8),[88] 0.585(9),[89] 0.590(10),[90] 0.589(7);[91]
	$\eta = 0.020(180)$,[3, 68] $-0.033(19)$,[88] $-0.011(14)$,[89] 0.000(40);[90]
	1st order[72]
STAR	$\alpha = 0.488(30)$;[92] $\beta = 0.221(9)$;[92] $\gamma = 1.074(29)$;[92]
	$\nu = 0.504(10)$;[92] $\eta = -0.131(13)$[92]
$V_{3,2}$	$\alpha = 0.479(24)$,[92] 0.460(30);[74] $\beta = 0.193(4)$;[92]
	$\gamma = 1.136(23)$,[92] 1.100(100);[74] $\nu = 0.507(8)$,[92] 0.515(10);[74]
	$\eta = -0.240(10)$,[92] $-0.100(50)$;[74]
	1st order[72]
BCT	$\alpha = 0.287(30)$;[93] $\beta = 0.247(10)$;[93] $\gamma = 1.217(32)$;[93]
	$\nu = 0.571(10)$;[93] $\eta = -0.131(18)$[93]
GLW	1st order[72]

Note: η is computed by $\gamma/\nu = 2 - \eta$ and, apart in the work of Kawamura[68] and Kunz and Zumbach,[74] α is computed by $3\nu = 2 - \alpha$.

relations, Eq. (3.36). The scaling relation $\gamma + 2\beta - 3\nu = 0$ is very well verified since $\gamma + 2\beta - 3\nu = 10^{-4} \pm 6.10^{-2}$.

(ii) *For the STA, η is negative.* Using the values of β/ν and γ/ν obtained directly in the simulations, one can compute the average value of η: $-0.0182(89)$. The probability of it being positive is 0.02, and it is thus small although not vanishing.

(iii) *For the STAR, $V_{3,2}$, and BCT models, the values of β are all incompatible with those of STA (three standard deviations at least) and are all incompatible among each others.* This has been interpreted as an indication of very weak first-order phase transitions.[92] This is to be compared with the XY case, where the transitions for STAR and the $V_{2,2}$ model are strongly of first-order.

(iv) *For the BCT, STAR, and $V_{3,2}$ models, η is always found to be significantly negative.* See Table 3.4, where η has been calculated from γ/ν.

(v) *The MC and MC RG approaches of the STA, $V_{3,2}$, and GLW model performed by Itakura has led to clear first-order behaviors.*[72] For Heisenberg

STA, in contrast to the XY case, even for the largest lattice sizes — $84 \times 96 \times 84$ — the double peak of the probability distribution of the energy is not observed. However, the $V_{3,2}$ model displays a clear double peak. Moreover, for the STA and the $V_{3,2}$ model, the RG flow clearly does not exhibit any fixed-point. Instead, a runaway of the RG flow toward the region of instability is found which indicates first-order transitions. The transitions are thus — weak — of first-order. The transition is also weaker of first-order for Heisenberg than for XY spins.

3.3.2.3. *Summary*

We now summarize the experimental and numerical situations in the Heisenberg case. In this case, the experimental situation is much poorer than in the XY case and is still unclear on many aspects. On the contrary, the numerical results are numerous and more precise than in the XY case.

(1) Scaling laws are found in STA materials on a rather wide range of temperatures as well as in all MC simulations apart from the one based on MC RG.
(2) There are two groups of materials that do not have the same exponents. The exponent β of the numerical STA model agrees very well with that of group 2.
(3) The anomalous dimension is manifestly negative for group 1 and very marginally for group 2. For the numerical STA model, η is found to be negative although not completely significantly. For STAR, $V_{3,2}$, and BCT, η is found to be significantly negative.
(4) For group 1, the scaling relations are violated.
(5) STAR, $V_{3,2}$, and BCT exhibit scaling behaviors without universality. Also, the results are incompatible with those of the numerical STA model.
(6) An MC RG approach of the STA, $V_{3,2}$, and GLW models has led to clear first-order behaviors.

3.3.3. *The $N = 6$ STA*

Before concluding, let us quote a simulation of the STA with six-component spins that has been performed by Loison *et al.*[94] The results are given in Table 3.5. Six-component spins were chosen since it was expected that the transition was of second order. Loison *et al.* have clearly identified scaling laws at the transition with a positive anomalous dimension. Let us

Table 3.5. MC critical exponents for six-component spins in the STA system.[94]

System	α	β	γ	ν	η
$N = 6$ STA	$-0.100(33)$	0.359(14)	1.383(36)	0.700(11)	0.025(20)

emphasize that, even if the transition is actually of first-order, as suggested by the recent results of Calabrese *et al.*[12] it should be an extremely weak first-order transition — see the following. Thus, scaling laws should hold for all temperatures but those very close to T_c. In this respect, the exponents for $N = 6$ are therefore very trustable so that reproducing them is a challenge for the theoretical approaches.

3.3.4. *Summary*

Let us now draw some conclusions about the experimental and numerical situations. In the XY and in the Heisenberg cases, one observes the following phenomena: the different materials split into two groups, the anomalous dimension is found to be negative in many materials and in most numerical simulations, the scaling relations are violated in some materials and there is no universality in the exponents found in the simulations since changes of microscopic details induce changes of scaling behaviors. Therefore, it appears that an explanation of the critical physics of frustrated systems based on the existence of a standard and *unique* universality class is excluded. The theoretical challenge is to explain this situation and to reproduce the whole set of experimental and numerical data.

3.4. A Brief Chronological Survey of the Theoretical Approaches

Let us briefly review the most important theoretical developments concerning this subject.

The first microscopic derivation and RG study — at one- and two-loop order in $d = 4 - \epsilon$ — of the effective GLW model relevant for STA — see below — was performed for He_3 by Jones *et al.*[76] in 1976 and by Bailin *et al.*[9] in 1977. The model was rederived and restudied in the context of helimagnets (for general N) by several groups including Bak *et al.*[95] (1976), Garel and Pfeuty[1] (1976), and Barak and Walker[96] (1982). It was established at that time that, around $d = 4$, the transitions for Heisenberg spin systems had to be of first-order. More precisely, these authors found that there exists a critical value $N_c(d)$ of the number N of spin components, above which the transition is of second order and below which it is of first-order. They found[9,76]

$$N_c(d = 4 - \epsilon) = 4(3 + \sqrt{6}) - 4\left(3 + \frac{7}{\sqrt{6}}\right)\epsilon + O(\epsilon^2)$$
$$\sim 21.8 - 23.4\epsilon + O(\epsilon^2) \quad (3.38)$$

with $\epsilon = 4 - d$. A first large-N expansion was also studied, in particular for $d = 3$, by Bailin *et al.*[9] in 1977.

A group theoretical derivation of the GLW model relevant to the XY STA was performed by Yosefin and Domany[2] in 1985. They found the same Hamiltonian as for helimagnets. Between 1985 and 1988, Kawamura[3–6] has performed this analysis for an N-component STA. He has shown that the Hamiltonian is the same as for He_3 or helimagnets, the RG analysis giving obviously the same results. This author has also extrapolated the two-loop result for $N_c(d)$ of Eq. (3.38) in $d = 3$ and found $N_c(d = 3) < 2$. This led him to conjecture the existence of a new universality class for frustrated magnets. However, as well known, this direct extrapolation cannot be reliable since it is notorious that the perturbative series must be resummed.

In 1988, Dombre and Read[20] derived, in the quantum case, the NLσ model relevant to frustrated magnets. In 1990, Azaria *et al.*[7] studied the classical version of the NLσ model around $d = 2$. They found a fixed point of the RG flow in a two-loop calculation for any $N \geq 3$. For $N = 3$, they found the phenomenon of enlarged symmetry: at the fixed point the symmetry becomes $SO(3) \times SO(3) \sim SO(4)$ instead of $SO(3) \times SO(2)$. Thus, their conclusion was that, if the transition is of second order, it is characterized by $O(4)/O(3)$ critical exponents — at least for ν. Another possibility proposed by these authors was that the transition could be also mean-field tricritical or of first-order. However, *none* of the experimental or numerical results are compatible with the $O(4)/O(3)$ or mean-field tricritical exponents. Note finally that these authors supposed that, if tricritical, the behavior at the transition should be mean-field tricritical in $d = 3$, something which is mandatory only for $O(N)/O(N-1)$ models, but not for more complex models.

The first NPRG approach to frustrated magnets was performed by Zumbach[14,15,74] in 1993. He wrote down the NPRG equations for the GLW models suited to the description of frustrated systems. He studied them within the local potential approximation (LPA) of the Wilson–Polchinski equation — analogous to the Wegner–Houghton approximation[97] — and found $N_c(d = 3) \sim 4.7$. Since he found no fixed point for $N = 2$ and $N = 3$, he claimed that the transition is of first-order in these cases. In the case $N = 3$, he has shown that there is a minimum in the RG flow, a pseudo-fixed point, that fakes a true fixed point (see below for details). The transition was thus conjectured to be a *weak* first-order transition with pseudo-scaling characterized by pseudo-critical exponents. Note that, within the LPA, all derivative terms in the Hamiltonian are neglected so that the anomalous dimension is zero. This has two important consequences. First, the pseudo-critical exponents found by Zumbach were not very reliable and thus difficult to compare with the experimental and numerical results. Second, this approach neglects terms — the so-called current-term (see below) — that are fundamental within the perturbative approach of the NLσ model performed around two dimensions. Thus, within Zumbach's approach, it was *not* possible to match with these results. Finally note that, in the $N = 2$ case, *no* minimum in the RG flow was found and, thus, no pseudo-critical exponents were obtained, in contradiction with the scaling behaviors observed in the experimental and numerical contexts.

Then, three-loop calculations have been performed by Antonenko *et al.*[98,99] in 1994 and 1995 on the GLW model. In $d = 3$ this has led, after Padé-Borel resummation, to $N_c(d = 3) = 3.91$.[98] In $d = 4 - \epsilon$, they have determined the three-loop contribution — $7.1\,\epsilon^2$ — to $N_c(d = 4 - \epsilon)$, see Eq. (3.38). This has led to $N_c(d = 3) = 3.39$.[99] The authors have mentioned that, in contrast to the $O(N)$ models, their three-loop results were not well converged.

In 1996, Jolicœur and David[100] studied a generalization of the Stiefel model that involved N vectors with N components. They have shown within a mean-field approximation and a one-loop calculation performed in $d = 2 + \epsilon$ that a first-order line should appear in a non-trivial dimension above two. It should isolate the chiral fixed point in the metastability region in such a way that this point should no longer play any role. Above this dimension, the transition should therefore be of first-order.

In 2000–2002, using the technique of the effective average action, including derivative terms, the present authors performed a non-perturbative study of frustrated magnets for any dimension between two and four.[16–19] They recovered *all* known perturbative results at one loop in two and four dimensions as well as for $N \to \infty$. They determined $N_c(d)$ for all d and found $N_c(d = 3) = 5.1$. Accordingly, for $N = 6$, they found a second-order phase transition. Their exponents were in very good agreement with those found numerically. For $N = 3$, they recovered[17] Zumbach's result — presence of a minimum in the RG flow — and improved his approach: they found pseudo-critical exponents in good agreement with *some* experimental realizations of frustrated magnets. However, regarding the spreading of the experimental and numerical data, the recourse to a minimum, leading to a *unique* set of pseudo-critical exponents, was clearly not the end of the story. During the study of the $N = 2$ case,[18] the present authors realized that the property of pseudo-scaling and even more, generic pseudo-scaling, does not rely on the existence of a minimum of the flow. Pseudo-scaling appears as a consequence of the existence of a *whole region* in the flow diagram in which the flow is slow. This allowed them to account for the *non-universal scaling* that occurs in XY as well as in Heisenberg frustrated magnets.

In 2001, Pelissetto *et al.*[10] derived the six-loop series for the GLW model. They used sophisticated resummation methods in order to find the fixed points and to determine the critical exponents of the model. For $N \gtrsim 7$, they found a fixed point of the same nature as that obtained at large-N and in the $4 - \epsilon$ expansion. Thus, a second-order phase transition is expected in this case. For $5 \lesssim N \lesssim 7$, they considered that their resummed series were not well converged, the number of fixed points depending strongly on the number of loops considered. This led them to interpret this result as an indication that $N_c(d = 3) \sim 6$. Finally and surprisingly, for $N \lesssim 5$ and, in particular, for the physically relevant cases $N = 2$ and $N = 3$, they found stable fixed points. Thus, a second-order phase transition was also predicted in these cases. However, the critical exponents found were far from all experimental and numerical data (see the following). Moreover, regarding again the spreading of these data, an interpretation in terms of a unique set of exponents was clearly insufficient.

In another work,[101] assuming that $N_c(d = 2) = 2$, Pelissetto *et al.* have reformulated the three-loop version of the series of Eq. (3.38) (see below) to make it compatible with this last guess. The series seemed to have better convergence properties (see, however, below) and allowed Pelissetto *et al.* to compute $N_c(d)$. They found $N_c(d = 3) = 5.3$, in good agreement with the value — $N_c(d = 3) = 5.1$ — obtained from the NPRG approach.[17,19]

Recent reinvestigations of the five- and six-loop perturbative series[11,12] have led Calabrese *et al.* to conjecture that the fixed point found by Pelissetto *et al.* — that correspond to a *focus* fixed point — could explain the existence of the spreading of critical exponents encountered in frustrated magnets. Indeed, they observed that, due to the specific structure of the fixed point, the critical exponents display strong variations along the RG trajectories that could explain the lack of universality observed experimentally and numerically. They have also given estimates of the critical number of spin components for which there is a change of the order of the phase transition. They have found that there is a first-order phase transition in the whole domain $5.7(3) < N < 6.4(4)$ and a second-order phase transition for the other values of N and, in particular, for $N = 2$ and $N = 3$.

Finally, a very recent computation of the five-loop β function of the GLW model in a $4 - \epsilon$ expansion has lead to a novel estimation of $N_c(d)$. Calabrese and Parruccini[102] have found the value $N_c(d = 3) = 6.1(6)$, which is compatible with the value $N_c(d = 3) = 6.4(4)$ found within the six-loop computation performed in three dimensions.[11,12]

Since several aspects of the recent perturbative and non-perturbative approaches differ, in particular in their interpretations of the origin of the non-universal scaling found in frustrated magnets, we postpone the detailed discussion of these last developments of both methods to the following sections.

3.5. The Perturbative Situation

There are essentially two different methods to analyze the critical behavior of the system described by the Hamiltonian (3.9). They correspond to two different methods to deal with the constraints obeyed by the microscopic degrees of freedom, Eq. (3.5). They lead to the NLσ and GLW models that have been both perturbatively analyzed around their respective critical dimension, as well as directly in $d = 3$ for the latter model. Let us review the results of these approaches.

3.5.1. *The Nonlinear Sigma (NLσ) model approach*

The idea underlying the construction of this model is to consider the system in its low-temperature — symmetry broken — phase and to take into account small fluctuations of the fields around the direction of the order parameter. The corresponding treatment is thus, by construction, a low-temperature expansion. Its actual validity is in fact less stringent than that: it is enough that the system is *locally* ordered and that the temperature is small. This explains why this approach is valid even in two dimensions for systems obeying the Mermin–Wagner theorem. Note that this approach applies — *a priori* (see Section 3.2.2 and the discussion at the end of this section) — only for $N \geq 3$. Indeed, in the $N = 2$ case, the low-temperature expansion of the NLσ model leads to a trivial result, i.e. the theory is perturbatively free. This result is, however, not reliable since there exist topological as well as Ising-like degrees of freedom in the XY frustrated case (see Section 3.2). These degrees of freedom, that are completely missed within the low-temperature perturbative approach, drastically affect the physics at finite temperature as in the famous Berezinskii–Kosterlitz–Thouless phase transition.[36,37]

Within the NLσ model approach, the partition function of the $SO(3) \times SO(2)$-symmetric model follows from the Hamiltonian Eq. (3.7) together with the constraints of Eq. (3.5):[20]

$$\mathcal{Z} = \int \mathcal{D}\vec{\phi}_1 \mathcal{D}\vec{\phi}_2 \prod_{i \leq j} \delta(\vec{\phi}_i.\vec{\phi}_j - \delta_{ij}) \exp\left(-\frac{1}{2T}\int d^d\mathbf{x}((\partial\vec{\phi}_1)^2 + (\partial\vec{\phi}_2)^2)\right). \tag{3.39}$$

The delta-functionals allow the integration of the three massive modes among the six degrees of freedom of $\vec{\phi}_1$ and $\vec{\phi}_2$. Therefore, only the three — Goldstone — modes $\vec{\pi}$ remain, in terms of which the partition function is written as:[7,8,103]

$$Z = \int_{|\vec{\pi}| \leq 1} D\vec{\pi} \exp\left(-\frac{1}{2T}\int d^d\mathbf{x}\; g_{ij}(\pi)\partial\pi^i\partial\pi^j\right). \tag{3.40}$$

Equation (3.40), where $g_{ij}(\pi)$ embodies the interaction, is the suitable expression for a low-temperature expansion of the $SO(3) \times SO(2)/SO(2)$ NLσ model.

The low-temperature expansion of such NLσ models has been studied in general but rather abstract terms by Friedan.[103] The specific study of the $SO(3) \times SO(2)/SO(2)$ model and its generalization to N-component spins — the $O(N) \times O(2)/(O(N-2) \times O(2))$ model — has been performed by Azaria *et al.*[7,8] (see also Ref. [101]). The RG analysis requires to consider the most general Hamiltonian invariant under $O(N) \times O(2)$ and renormalizable around $d = 2$. This Hamiltonian involves not only the usual kinetic terms for $\vec{\phi}_1$ and $\vec{\phi}_2$, Eq. (3.39), but also a non-trivial derivative term, called the "current-term", which reads as follows:

$$\int d^d\mathbf{x}(\vec{\phi}_1.\partial\vec{\phi}_2 - \vec{\phi}_2.\partial\vec{\phi}_1)^2. \tag{3.41}$$

This term must be included in the model since it has the right symmetry, is power-counting renormalizable around $d = 2$, and is thus generated during the RG flow. The correct NLσ model — in the sense of stability under RG transformations — is given by (for any $N \geq 3$):[7]

$$H = \int d^d\mathbf{x} \left(\frac{\eta_1}{2} \left(\left(\partial\vec{\phi}_1\right)^2 + \left(\partial\vec{\phi}_2\right)^2 \right) + \left(\frac{\eta_2}{8} - \frac{\eta_1}{4}\right) \left(\vec{\phi}_1.\partial\vec{\phi}_2 - \vec{\phi}_2.\partial\vec{\phi}_1\right)^2 \right), \tag{3.42}$$

where we have chosen to reparametrize the coupling constants in a way convenient for what follows. Now, the Hamiltonian of the naïve continuum limit Eq. (3.39) is just the initial condition of the RG flow corresponding to $\eta_1 = \eta_2/2 = 1/T$.[e]

For the special case $N = 3$, it is convenient to rewrite the model differently. We define, as in Eq. (3.16), a third vector $\vec{\phi}_3$ by

$$\vec{\phi}_3 = \vec{\phi}_1 \wedge \vec{\phi}_2. \tag{3.43}$$

With this expression, it is easy to verify that the current term, Eq. (3.41), is nothing but a linear combination of the kinetic terms of $\vec{\phi}_1$, $\vec{\phi}_2$, and $\vec{\phi}_3$:

$$\int d^d\mathbf{x}(\vec{\phi}_1.\partial\vec{\phi}_2 - \vec{\phi}_2.\partial\vec{\phi}_1)^2 = 2\int d^d\mathbf{x}\left((\partial\vec{\phi}_1)^2 + (\partial\vec{\phi}_2)^2 - (\partial\vec{\phi}_3)^2\right). \tag{3.44}$$

One can then gather the three vectors $\vec{\phi}_1$, $\vec{\phi}_2$, and $\vec{\phi}_3$ into a 3×3 matrix

$$\Phi = (\vec{\phi}_1, \vec{\phi}_2, \vec{\phi}_3). \tag{3.45}$$

Since $(\vec{\phi}_1, \vec{\phi}_2, \vec{\phi}_3)$ are three orthonormal vectors, one has ${}^t\Phi\,\Phi = \mathbb{1}$ and Φ is therefore an $SO(3)$ matrix.

The partition function thus reads

$$\mathcal{Z} = \int \mathcal{D}\Phi\delta({}^t\Phi\,\Phi - \mathbb{1})\, e^{-\int d^d\mathbf{x}\,\mathrm{Tr}(\mathcal{P}\partial\, {}^t\Phi\partial\Phi)}, \tag{3.46}$$

where $\mathcal{P}$ is a diagonal matrix of coupling constants: $\mathcal{P} = \mathrm{diag}(p_1 = p_2 = \eta_2/4, p_3 = \eta_1/2 - \eta_2/4)$.

It is easy to check on Eq. (3.46) that the model is invariant under the right transformation

$$\Phi \to \Phi.V, \tag{3.47}$$

with V being the subset of $SO(3)$ matrices that commute with $\mathcal{P}$. When $p_3 \neq p_1$, i.e. $\eta_1 \neq \eta_2$, V is isomorphic to $SO(2)$. When $\mathcal{P}$ is proportional to

[e]We have included the temperature in the coupling constants.

the identity, V is isomorphic to the whole $SO(3)$ group. In this last case, the high-temperature symmetry group is $G = SO(3) \times SO(3) \sim SO(4)$. Note that this identity has to be understood at the level of the Lie algebras since $SO(3) \times SO(3)$ and $SO(4)$ are locally isomorphic but differ globally and have different topological properties.

The RG equations for the $O(N) \times O(2)/(O(N-2) \times O(2))$ model have been computed at two-loop order in $d = 2+\epsilon$.[7,8] We recall here the one-loop result

$$\begin{cases} \beta_{\eta_1} = -(d-2)\eta_1 + N - 2 - \dfrac{\eta_2}{2\eta_1}, \\ \beta_{\eta_2} = -(d-2)\eta_2 + \dfrac{N-2}{2}\left(\dfrac{\eta_2}{\eta_1}\right)^2. \end{cases} \tag{3.48}$$

A fixed point is found for any $N \geq 3$. For $N = 3$, it corresponds to $p_1^\star = p_3^\star$, i.e. $\eta_1^\star = \eta_2^\star$, and thus to an enlarged symmetry: $SO(3) \times SO(3)/SO(3) \sim SO(4)/SO(3)$. This fixed point has only one direction of instability — the direction of the temperature — and thus corresponds to a second-order phase transition. Surprisingly, the critical behavior is thus predicted to be governed by the usual ferromagnetic Wilson–Fisher fixed point with the subtlety that it corresponds to *four-component* spins. This situation corresponds precisely to the particular case considered in Section 3.2.2 (see in particular the discussion following Eq. (3.20)). Another subtlety is that since, here, the order parameter is a matrix instead of a vector — it is a $SO(4)$ tensor — the anomalous dimension is different from the usual anomalous dimension of the four-component vector model. Only the exponent ν is independent of the nature of the order parameter and is thus identical to the usual value of ν of the Wilson–Fisher $N = 4$ universality class.[7,8]

In fact, it is easy to convince oneself that this fixed point exists at all order of perturbation theory. Actually, the crucial fact is that, in $d = 2+\epsilon$, the *perturbative* β functions of an NLσ model associated with the symmetry breaking scheme $G \to H$ only depend on the *local geometrical* structure of the manifold G/H which is itself determined by the Lie algebras of G and H.[7,8] Since the Lie algebras of $SO(3) \times SO(3)$ and of $SO(4)$ are identical, the perturbative β function for the — remaining — coupling constant of the model with $p_1 = p_3$ is identical at all orders to the perturbative β function of the usual $SO(4)/SO(3)$ NLσ model. The existence of a fixed point for the $SO(3) \times SO(3)/SO(3)$ NLσ model at all orders of perturbation theory follows from the fact that its existence makes no doubt for the $SO(4)/SO(3)$ NLσ model. At the time of the first investigation of the $O(N) \times O(2)/(O(N-2) \times O(2))$ model, the most natural position was to extend this equivalence *beyond* perturbation theory and to assume that the $SO(3) \times SO(3)/SO(3)$ fixed point exists everywhere between two and four dimensions, as is the case for the $SO(4)/SO(3)$ fixed point. This was, in particular, the position advocated by Azaria *et al.*[7,8]

The outstanding fact is that although the $SO(4)$ behavior has indeed been seen numerically in $d = 2$,[24,28] it actually does *not exist* far from two dimensions. This is clear since *no* such fixed point is found in $d = 4 - \epsilon$ and since, as already emphasized, the $SO(4)$ behavior is not seen in any numerical or experimental data in $d = 3$. It is thus extremely probable that either the NLσ model fixed point disappears in a non-trivial dimension smaller than 3 or it survives in $d = 3$ while being no longer the usual $N = 4$ fixed point. Note that, in the first case, its $SO(4)$ nature can also change before it disappears. Anyway, this fixed point must disappear below $d = 4$. The situation is thus more involved than in the "usual" $SO(4)/SO(3)$ model. There must exist non-perturbative reasons explaining the disappearance of the fixed point and/or the loss of its $SO(4)$ character.

Actually, it is clear that the perturbative low-temperature expansion performed on the NLσ model misses several non-analytic terms in T — typically terms that behave as $\exp(-1/T)$ — that could be responsible for the disappearance of the fixed point and/or its change of nature. There are, at least, two origins for such terms.

(1) The first one consists in the non-trivial topological configurations — see the discussion in Section 3.2.2 following Eq. (3.15) — that are neglected in the low-temperature expansion of the NLσ models. Indeed, this expansion relies, by construction, on the local geometrical properties of the manifold G/H and is insensitive to its global — topological — structure. Thus, it ignores vortex-like configurations that likely play an important role in three dimensions.
(2) The second origin of non-analytic corrections to the low-temperature β function is more technical. The low-temperature expansion is performed in terms of the Goldstone — or pseudo-Goldstone in $d = 2$ — modes that are represented by fields constrained to have a modulus less than one, see Eq. (3.40). This inequality cannot be taken into account in the perturbative treatment[104] and is thus relaxed, leading to neglect terms of order $\exp(-1/T)$. All these terms are negligible for the critical behavior when the critical temperature is very small, which is the case

near $d = 2$. However, they become important when $T_c \sim 1$, which is typically the case in $d = 3$.

3.5.2. *The Ginzburg–Landau–Wilson model approach*

The GLW model for the $O(N) \times O(2)/(O(N-2) \times O(2))$ model can be deduced from a generalization of Eq. (3.39) to N-component vectors, by replacing the functional delta-constraint by the most general potential that favors the field configurations obeying the initial constraint. For convenience, we choose to parametrize it by

$$\prod_{i \leq j} \delta(\vec{\phi}_i.\vec{\phi}_j - \delta_{ij}) \to e^{-U} \tag{3.49}$$

with

$$U = \int d^d\mathbf{x} \left(\frac{r}{2}(\vec{\phi}_1^2 + \vec{\phi}_2^2) + \frac{\lambda + \mu}{16}(\vec{\phi}_1^2 + \vec{\phi}_2^2)^2 - \frac{\mu}{4}(\vec{\phi}_1^2 \vec{\phi}_2^2 - (\vec{\phi}_1.\vec{\phi}_2)^2) \right), \tag{3.50}$$

where, as usual, r is proportional to the reduced temperature while λ and μ are ϕ^4-like coupling constants.

All field-dependent terms in Eq. (3.50) can be rewritten in terms of the rectangular matrix Φ defined in Eq. (3.8). The corresponding Hamiltonian then reads

$$H = \int d^d\mathbf{x} \left(\frac{1}{2}\,\mathrm{Tr}(\partial^t \Phi \partial \Phi) + \frac{r}{2}\rho + \frac{\lambda}{16}\rho^2 + \frac{\mu}{4}\tau \right), \tag{3.51}$$

with $\rho = \mathrm{Tr}({}^t\Phi\Phi)$ and $\tau = \frac{1}{2}\mathrm{Tr}({}^t\Phi\Phi - \mathbb{1}\,\rho/2)^2$ being the only $O(N) \times O(2)$ independent invariants that can be built in the fields. Note that minimizing the term before of μ corresponds to imposing ${}^t\Phi\Phi \propto \mathbb{1}$, i.e. to imposing that $\vec{\phi}_1$ and $\vec{\phi}_2$ are orthogonal and of the same norm, in agreement with the characteristics of the GS of frustrated magnets — see Fig. 3.1(b).

3.5.2.1. *The RG flow*

The RG equations for the coupling constants entering in Hamiltonian (3.50) have been computed in the $\epsilon = 4 - d$-expansion up to five-loop order[102] and in a weak-coupling expansion in $d = 3$ up to six-loop order.[10] We recall

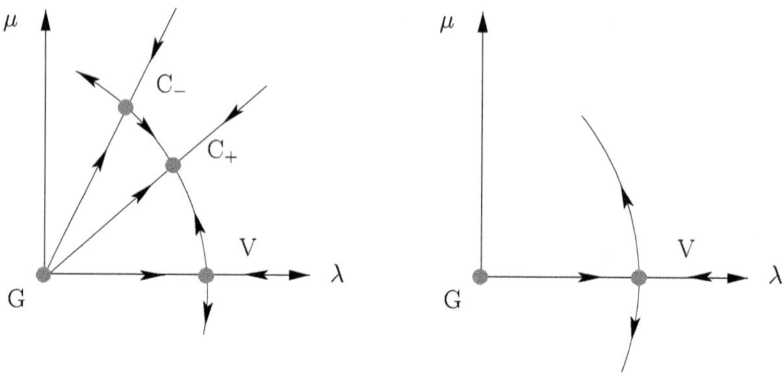

(a) $N > N_c(d)$ (b) $N_c'(d) \le N \le N_c(d)$

Fig. 3.2. Flow diagram for (a) N above $N_c(d)$ and (b) N below $N_c(d)$. The fixed points C_+ and C_- that exist above $N_c(d)$ coalesce at $N_c(d)$ and then disappear. G and V are the Gaussian and vector $O(2N)$ fixed points.

here only the one-loop result of the ϵ-expansion to discuss qualitatively the flow diagram

$$\begin{cases} \beta_\lambda = -\epsilon\lambda + \dfrac{1}{16\pi^2}(4\lambda\mu + 4\mu^2 + \lambda^2(N+4)), \\ \\ \beta_\mu = -\epsilon\mu + \dfrac{1}{16\pi^2}(6\lambda\mu + N\mu^2). \end{cases} \tag{3.52}$$

As is well known, for any $N > N_c(d = 4 - \epsilon) = 21.8 + O(\epsilon)$, there exist four fixed points: the Gaussian — G — the vector $O(2N)$ — V — and two others called the chiral — C_+ — and anti-chiral — C_- — fixed points. Among these fixed points, one C_+ is stable and governs the critical properties of the system and the others are unstable (see Fig. 3.2(a)). When, at a given dimension d close to four, N is decreased, C_+ and C_- move closer together, coalesce at $N_c(d)$, and then disappear (see Fig. 3.2(b)). More precisely, for $N < N_c(d)$, the roots of the β functions acquire an imaginary part. Since no stable fixed point exists below $N_c(d)$ and since the flow drives the system in a region of instability, it is believed that the transition is of first-order. Note that for $N < N_c'(d = 4 - \epsilon) = 2.2 + O(\epsilon)$, C_+ and C_- reappear but not in the physically relevant region to frustrated magnets.

For completeness, we give the exponent ν at one loop:

$$\nu = \frac{1}{2} + \frac{(N-3)(N+4)\sqrt{48-24N+N^2}+N(48+N+N^2)}{8(144-24N+4N^2+N^3)}\epsilon \qquad (3.53)$$

and recall that the anomalous dimension vanishes at this order. Note that the square root becomes complex for $2.2 < N < 21.8$, which is reminiscent of the critical values $N_c(d)$ and $N_c'(d)$ of the number of spin components, see above.

3.5.2.2. *The three- and five-loop results in $d = 4 - \epsilon$*

In $4 - \epsilon$ dimensions, the critical value $N_c(d)$ has been computed at the three-loop order[99] and, very recently, at the five-loop order:[102]

$$N_c(d = 4 - \epsilon) = 21.80 - 23.43\epsilon + 7.09\epsilon^2 - 0.03\epsilon^3 + 4.26\epsilon^4 + O(\epsilon^5). \qquad (3.54)$$

In fact, as it is often the case within this kind of expansion, the series are not well behaved and it is difficult to obtain reliable results even after resummation.[94,99,102] We, however, indicate the value found at three-loop order:[99] $N_c(d = 3) = 3.39$ and at the five-loop order:[102] $N_c(d = 3) = 5.45$.

3.5.2.3. *The improved three- and five-loop results*

It has been conjectured by Pelissetto *et al.*[101] that $N_c(d = 2) = 2$, a result which is however somewhat controversial.[102] It is possible to use this non-perturbative information to reformulate the series obtained within the $4-\epsilon$ expansion. Imposing the constraint $N_c(d = 2) = 2$ to the three-loop series, Pelissetto *et al.* have obtained.[101]

$$N_c(d = 4 - \epsilon) = 2 + (2 - \epsilon)(9.90 - 6.77\epsilon + 0.16\epsilon^2) + O(\epsilon^3). \qquad (3.55)$$

Reformulated in this way, the coefficients of the series decrease rapidly. It is thus reasonable to use this expression to estimate $N_c(d = 3)$. Pelissetto *et al.* have thus obtained:[101] $N_c(d = 3) = 5.3(2)$ where the error bar indicates how $N_c(d = 3)$ varies from two to three loops. However, Calabrese and Parruccini have shown that, extended to five loops, the same series behaves badly:[102]

$$
\begin{aligned}
&N_c(d = 4 - \epsilon)\\
&\quad = 2 + (2 - \epsilon)(9.90 - 6.77\epsilon + 0.16\epsilon^2 + 0.06\epsilon^3 + 2.16\epsilon^4) + O(\epsilon^5). \quad (3.56)
\end{aligned}
$$

Using different kinds of tricks, notably the inverse of the series, they have obtained, from the five-loop series, the value $N_c(d = 3) = 6.1(6)$.

3.5.2.4. *The three-loop results in $d = 3$*

A weak-coupling analysis has been performed by Antonenko and Sokolov[98] directly in $d = 3$ at three-loop order. This leads to $N_c(d = 3) = 3.91$. However, as already emphasized, this result is not well converged.

3.5.2.5. *The large-N results*

The large-N expansion was first performed by Bailin *et al.*[9] It was then re-examined by Kawamura[6] and Pelissetto *et al.*[101] A fixed point is found within this expansion in all dimensions between 2 and 4. The exponents ν and η have been computed up to order $1/N^2$ in $d = 3$:[101]

$$
\begin{cases}
\nu = 1 - \dfrac{16}{\pi^2}\dfrac{1}{N} - \left(\dfrac{56}{\pi^2} - \dfrac{640}{3\pi^4}\right)\dfrac{1}{N^2} + O(1/N^3),\\[2ex]
\eta = \dfrac{4}{\pi^2}\dfrac{1}{N} - \dfrac{64}{3\pi^4}\dfrac{1}{N^2} + O(1/N^3).
\end{cases}
\tag{3.57}
$$

Around $d = 4$ and $d = 2$, the perturbative results of, respectively, the GLW and NLσ models are recovered once the limit $N \to \infty$ has been performed. This suggests that, at least for sufficiently large N, the two models belong indeed to the same universality class in all dimensions. However, within this approach, no $N_c(d)$ line is found (see, however, Ref. [101]). It is thus impossible to extrapolate to finite N the results obtained within this approach.

3.5.2.6. *The six-loop results in $d = 3$*

In three dimensions, a six-loop computation has been performed by Pelissetto *et al.*[10] and re-examined by Calabrese *et al.*,[11,12] see below. The results are as listed below:

(1) For N sufficiently large — $N > 6.4(4)$ — there exist four fixed points, one stable and three unstable, in agreement with the usual picture given above, see Fig. 3.2. The transition is thus of second order.

Table 3.6. The six-loop perturbative results in $d = 3$ by Pelissetto *et al.*[10]

System	α	β	γ	ν
XY	0.29(9)	0.31(2)	1.10(4)	0.57(3)
Heisenberg	0.35(9)	0.30(2)	1.06(5)	0.55(3)

(2) For $5.7(3) < N < 6.4(4)$, there is no non-trivial fixed point and the transition is expected to be of first-order.
(3) For $N < 5.7(3)$ and, in particular, for $N = 2$ and $N = 3$ a *stable fixed point* is found and a second-order phase transition is expected.

According to Pelissetto *et al.*,[10,101] the fixed points found for $N = 2$ and $N = 3$ should be non-analytically connected with those found in the $1/N$ and $4 - \epsilon$ approaches. Therefore, it should be impossible to obtain them by following smoothly those obtained at large-N or close to $d = 4$.

The critical exponents obtained by Pelissetto *et al.* are given in Table 3.6. Let us discuss them.

The XY case. First, one should indicate that the exponents γ and ν computed from the six-loop approach compare reasonably well with the data of group 1. However, as already mentioned, the value of η found by the scaling relations must be positive when there exists a fixed point. One finds, with the data of Table 3.6, $\eta \sim 0.08$, which is significantly positive. Let us recall that this is *not* the case for the experiments performed on the materials of group 1 and for the numerical simulations performed on STA. Note, moreover, that the value of β found within the six-loop calculation is very far — around four theoretical error bars — from the average experimental ones, which are $\beta = 0.228(6)$ for $CsMnBr_3$ alone and $\beta = 0.237(4)$ for the whole group 1. It is also far from the numerical values obtained for STA: $\beta = 0.24$–0.25. Thus, in contrast to what is asserted by Pelissetto *et al.*,[10] it seems extremely improbable that the exponents found at six-loop order could fit with those of group 1 and with those of the numerical STA model. Actually, this is also the case for the materials of group 2 for which the average β is $\beta = 0.389(7)$.

The Heisenberg case. First, one notes that the agreement between the γ and ν exponents obtained from the six-loop approach and from the experimental or numerical data is not as good as it is in the XY case. Concerning η, one finds, with the data of Table 3.6, $\eta \sim 0.08$. This has to be compared with the value of η obtained (i) for the materials of group 1, which is significantly negative — $\eta = -0.118(25)$ — (ii) for materials of group 2, which is marginally negative — $\eta = -0.018(33)$ — and (iii) in the simulations of the STA which is also negative although not completely significantly: $\eta = -0.0182(89)$. The negativity of η is a first indication of a mismatch between the six-loop result and the data for the Heisenberg systems even if it cannot be used as a definitive argument against a second-order phase transition. The exponent γ obtained from the numerical simulations of the Heisenberg STA model provides a further information. Its average value — $\gamma = 1.185(3)$ — is rather far — 2.5 theoretical error bars — from the six-loop result.[f]

From the previous analysis, one can conclude that, as such, the fixed point obtained within the six-loop approach turns out to be *not* directly relevant to the phenomenology of XY and Heisenberg materials or simulated systems.

3.5.3. *The six-loop results in* $d = 3$ *re-examined*

In order to cope with the discrepancy between the six-loop results obtained by Pelissetto *et al.* and the experimental and numerical data, Calabrese *et al.* have reconsidered the resummed six-loop series.[11–13] They claim that they can account for the unusual properties of the critical exponents for XY and Heisenberg frustrated systems in $d = 3$ — negative anomalous dimension and weak universality — by the fact that the RG trajectories around the stable — focus — fixed point found at six-loop by Pelissetto *et al.* are spiral-like, see Fig. 3.3.

In practice, Calabrese *et al.* have integrated the resummed β functions for the two coupling constants $(\bar{u}, \bar{v})$ of the GLW model and computed the effective exponents $\eta(\bar{u}, \bar{v})$ and $\nu(\bar{u}, \bar{v})$ along the RG trajectories. They have found that these exponents display large variations in a transient regime, see Fig. 3.4. These authors then argue that the scaling properties of the system are governed over several decades of temperature by the preasymptotic regime so that the effective exponents observed experimentally can differ significantly from their asymptotic values, i.e. those defined at the fixed point.

Let us underline here several drawbacks concerning this scenario.

[f]This result would be almost unchanged if the numerical error bar — and especially the one quoted in Mailhot *et al.*[88] — was largely underestimated since it would surely be much less than the theoretical one.

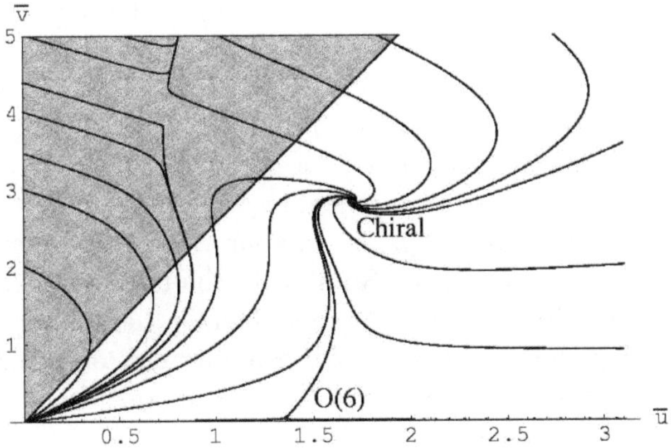

Fig. 3.3. RG flow for $N = 3$.

Source: From Ref. [11].

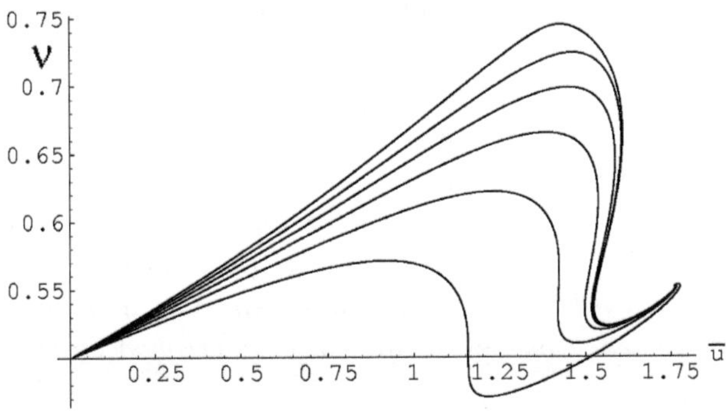

Fig. 3.4. Effective exponent ν for $N = 3$ along RG trajectories as a function of the effective coupling constant $\bar{u}$ for different initial conditions.

Source: From Ref. [11].

First, it is based on the existence of stable fixed points that are not related to any already known fixed point. In particular, the fixed points found for $N = 2$ and $N = 3$ within this computation in $d = 3$ are, according to Pelissetto *et al.* and Calabrese *et al.*, *non-analytically* related to those found in the large-N as well as in the $4-\epsilon$ expansions. This means that there is no way to check their existence using these perturbative methods. This is specifically problematic in the context of frustrated magnets where the properties of the fixed points appear to be very unusual: (i) the existence of the stable fixed points strongly depends on the order of perturbation — they are not present at three-loop order and only show up, as far as we know, at five-loop order, (ii) the location of the fixed points, as N and d are varied, seems to have, in the (N, d) plane, a very particular structure since, in three dimensions, they only exist when N is below *another* critical value of N — which is found to be equal to 5.7(3).

Second, it is very difficult, in the computation of Calabrese *et al.*, to relate the initial conditions of the RG flow to the microscopic characteristics of real systems. This would require to handle the infinity of coupling constants entering into the microscopic Hamiltonian obtained from the Hubbard–Stratonovitch transformation. This is impossible, at least within the usual perturbation theory.

Third, it is very difficult to account, in this framework, for the first-order behavior deduced from the numerical simulations of Itakura.[72] We have also already noted that XY-systems have a stronger tendency to undergo first-order transitions than Heisenberg systems. However, there is no natural explanation for this phenomenon in the scenario of Calabrese *et al.*

Fourth, in this scenario, it is also very difficult to explain why there is no physical system characterized by the asymptotic critical exponents, i.e. those corresponding the fixed point found by Pelissetto *et al.* This seems to require very unnatural experimental circumstances such that the initial conditions of the flow corresponding to the physical realizations of frustrated magnets are very close to the Gaussian fixed point in such a way that their long-distance properties are *never* controlled by the non-trivial fixed point.

Finally, there is no possible explanation of the breakdown of the NLσ model predictions.

3.5.3.1. *Summary*

XY and Heisenberg frustrated magnets represent systems for which the perturbative results obtained within a low-temperature expansion around two dimensions, within a weak-coupling expansion around four dimensions, or within a large-N expansion fail to describe their critical physics in three dimensions. Moreover, these different perturbative predictions are in contradiction with each other. In contrast to the $O(N)$ non-frustrated case, there is no possible smooth interpolation of these results between two and four dimensions and, at fixed dimension, between $N = \infty$ and $N = 2, 3$. More surprisingly, and again in contradiction with what happens in the $O(N)$ non-frustrated case, high-order calculations performed directly in $d = 3$ also fail to reproduce the phenomenology when they are interpreted in the usual way. This situation reveals the difficulties of the conventional approaches to tackle with the physics of frustrated magnets. Only new interpretations or methods can allow to shed light on the problems encountered here. We have presented the solution proposed by Calabrese *et al.* on the basis of a high-order pertubative calculation and underlined its difficulties. We now present the non-perturbative method which has been used by the present authors to explain the unusual behavior of frustrated magnets.

3.6. The Effective Average Action Method

3.6.1. *The effective average action equation*

Recently, the present authors, following the pioneering work of Zumbach,[14,15,74] have used a non-perturbative method to investigate the physics of frustrated magnets. They have used the effective average action method that has allowed them to smoothly interpolate between the $d = 2$ and $d = 4$ results and, therefore, to understand the mismatch between the NLσ and the GLW model approaches. This method has also allowed them to improve the determination of pseudo-critical exponents found by Zumbach in three dimensions for Heisenberg system. But the main feature of this approach is that it has allowed to show that the property of pseudo-scaling encountered in frustrated magnets does not, in general, rely on the concept of minimum flow or pseudo-fixed point introduced by Zumbach. In fact, pseudo-scaling appears as a consequence of the existence of a *whole region* in the flow diagram in which the flow is slow. This has allowed the present authors to account for and to reproduce the *non-universal scaling* that occurs in XY as well as in Heisenberg frustrated magnets.

The effective average action method,[105–109] as well as many other NPRG techniques, is based on the well-known concept of "block spin"[110–112]: when dealing with any strongly correlated system, it is fruitful to integrate out the fluctuations step by step and, more precisely, scale by scale. In practice, one first gathers the initial — microscopic — degrees of freedom into small "blocks". It is then possible, at least formally, to integrate out, in the partition function of the system, the internal fluctuations of the blocks. This "decimation" is followed by a rescaling of length scales, coupling constants, and fields. In this way, starting from a "bare"

GLW Hamiltonian, one gets an effective Hamiltonian for the block degrees of freedom, i.e. for the low-energy modes. By iterating this procedure, one generates a sequence of — scale-dependent — effective Hamiltoninans, parametrized by a running scale k, that all share the same long-distance physics. This sequence defines a Wilsonian RG flow.[112] At a fixed point of this flow, the system displays scale invariance. This allows to obtain the critical quantities through an analysis of the neighborhood of the fixed point in the flow of effective Hamiltonians.[112]

Although it has been originally formulated in terms of Hamiltonians, the most recent and successful implementation of the block spin idea in the continuum involves the effective (average) action.[108, 113, 114] In the same way as in the original Wilson's approach, one builds a *running* Gibbs free energy Γ_k, also called effective average action,[108] that only includes high-energy fluctuations with momenta $q^2 > k^2$. This implies that, on the one hand, at the underlying microscopic scale $k = \Lambda$, Γ_k coincides with the classical Hamiltonian H since no fluctuation has yet been taken into account. On the other hand, when the running scale is lowered to $k = 0$, i.e. when *all* fluctuations have been integrated out, the standard Gibbs free energy Γ is recovered. To summarize, Γ_k continuously interpolates between the microscopic Hamiltonian H and the free energy Γ:

$$\begin{cases} \Gamma_{k=\Lambda} = H \\ \Gamma_{k=0} = \Gamma. \end{cases} \tag{3.58}$$

Let us consider, for simplicity, the case of a system described, *at a typical microscopic scale* Λ, by a scalar field $\zeta(\mathbf{x})$. The construction of the effective average action proceeds in two steps. First, one should decouple the low-energy modes — with momenta $\boldsymbol{q}^2 > k^2$ — in the partition function in order to get a theory involving only the high-energy ones that will be summed over. Second, in this modified theory, one builds the Gibbs free energy by a Legendre transform. This gives Γ_k. Let us now study how this is implemented in practice.

The first step is conveniently implemented by changing the partition function $\mathcal{Z}$ into $\mathcal{Z}_k$ for which a k-dependent term, quadratic in the fields and thus analogous to a mass-term is added to the microscopic Hamiltonian.[108, 109] With this "mass-term", the partition function in the presence of a source J is written as

$$\mathcal{Z}_k[J] = \int \mathcal{D}\zeta \exp(-H[\zeta] - \Delta H_k[\zeta] + J.\zeta) \tag{3.59}$$

with $J.\zeta = \int d^d\mathbf{q} J(\mathbf{q})\zeta(-\mathbf{q})$ and

$$\Delta H_k[\zeta] = \frac{1}{2}\int \frac{d^d\mathbf{q}d^d\mathbf{q}'}{(2\pi)^{2d}} \mathcal{R}_k(\mathbf{q},\mathbf{q}')\zeta(\mathbf{q})\zeta(\mathbf{q}') \tag{3.60}$$

$$= \frac{1}{2}\int \frac{d^d\mathbf{q}}{(2\pi)^d} R_k(\mathbf{q}^2)\zeta(\mathbf{q})\zeta(-\mathbf{q}) \tag{3.61}$$

with $\mathcal{R}_k(\mathbf{q},\mathbf{q}') = (2\pi)^d\ \delta(\mathbf{q}+\mathbf{q}')R_k(\mathbf{q}^2)$. In Eq. (3.61), $R_k(\mathbf{q}^2)$ is the cut-off function that controls the separation between the low- and high-energy modes. To decouple the low-energy modes, it must act as a large-mass term for small $\mathbf{q}$, whereas it must vanish for large $\mathbf{q}$ to keep the high-energy sector of the theory unchanged. Thus,

$$R_k(\mathbf{q}^2) \sim k^2 \quad \text{for } \mathbf{q}^2 \ll k^2 \tag{3.62}$$

and

$$R_k(\mathbf{q}^2) \to 0 \quad \text{when } \mathbf{q}^2 \gg k^2. \tag{3.63}$$

The first constraint means that, for momenta lower than k, $R_k(\mathbf{q}^2)$ essentially acts as a mass — i.e. an IR cut-off — which prevents the propagation of the low-energy modes. The second ensures that the high-energy modes are fully taken into account in $\mathcal{Z}_k[J]$ and thus in the effective average action. Moreover, since we want to recover the original theory when $k \to 0$, i.e. when all fluctuations have been integrated out, $R_k(\mathbf{q}^2)$ must vanish in this limit. Thus, one requires

$$R_k(\mathbf{q}^2) \to 0 \quad \text{identically when } k \to 0, \tag{3.64}$$

which ensures that $\mathcal{Z}_{k=0}[J] = \mathcal{Z}[J]$. On the other hand, when $k \to \Lambda$, i.e. when no fluctuation has been integrated out, Γ_k should coincide with the microscopic Hamiltonian. This is achieved by requiring (see below for the proof)

$$R_k(\mathbf{q}^2) \to \infty \quad \text{identically when } k \to \Lambda. \tag{3.65}$$

Note that, since we shall not be interested in the precise relation between the microscopic characteristics — defined at scale Λ — of a given system and its critical or pseudo-critical properties, we set $\Lambda = \infty$ in the following.

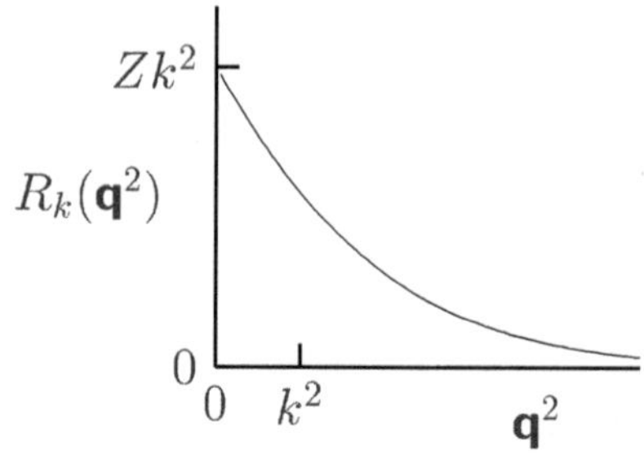

Fig. 3.5. A typical realization of the separation of high- and low-momentum modes provided by the cut-off function $R_k(\mathbf{q}^2)$. At low momentum, $R_k(\mathbf{q}^2)$ acts as an effective mass of order Zk^2 while the high-momentum behavior is not modified.

A widely used cut-off function is provided by[115]

$$R_k(\mathbf{q}^2) = \frac{Z\mathbf{q}^2}{e^{\mathbf{q}^2/k^2} - 1}, \tag{3.66}$$

where Z is the field renormalization. Including it in R_k allows to suppress the explicit Z dependence in the final RG equations. The cut-off function $R_k(\mathbf{q}^2)$ corresponding to Eq. (3.66) is plotted in Fig. 3.5. Another useful cut-off function has been proposed by Litim.[116] It writes

$$R_k(\mathbf{q}^2) = Z(k^2 - \mathbf{q}^2)\Theta(k^2 - \mathbf{q}^2), \tag{3.67}$$

where Θ is the usual step function.

The second step consists in defining the effective average action. The free energy at scale k is given — up to a factor $-k_BT$ — by

$$W_k[J] = \ln \mathcal{Z}_k[J]. \tag{3.68}$$

From (3.68), one defines the order parameter $\phi_k(\mathbf{q})$ at scale k as the average value of the microscopic field $\zeta(\mathbf{q})$ in the modified theory

$$\phi_k(\mathbf{q}) = \langle\zeta(\mathbf{q})\rangle = \frac{\delta W_k[J]}{\delta J(-\mathbf{q})}. \tag{3.69}$$

Thanks to the properties of $R_k(\mathbf{q}^2)$, the contribution to the average value in Eq. (3.69) coming from modes with momenta $\mathbf{q}^2 \ll k^2$ is strongly suppressed. Also, $\phi_k(\mathbf{q})$ identifies with the true order parameter in the limit

$k \to 0$. Note that, for simplicity, we omit, in the following, the index k to ϕ_k.

The effective average action is defined by[108]

$$\Gamma_k[\phi] = -W_k[J] + J.\phi - \Delta H_k[\phi], \tag{3.70}$$

where $J = J[\phi]$, see Eq. (3.69). Thus, $\Gamma_k[\phi]$ essentially corresponds to a Legendre transform of $W_k[J]$ for the macroscopic field ϕ — up to the mass-like term ΔH_k. The relation (3.70) implies several unconventional relations. First, taking its derivative with respect to $\phi(\mathbf{q})$ provides the relation between the source and $\Gamma_k[\phi]$

$$J(-\mathbf{q}) = \frac{\delta\Gamma_k}{\delta\phi(\mathbf{q})} + \int \frac{d^d\mathbf{q}'}{(2\pi)^{2d}} \mathcal{R}_k(\mathbf{q},\mathbf{q}')\phi(\mathbf{q}'). \tag{3.71}$$

Taking the derivative of this relation with respect to $\phi(\mathbf{q}')$ implies a second important relation

$$\Gamma_k^{(2)}(\mathbf{q},\mathbf{q}') + \frac{\mathcal{R}_k(\mathbf{q},\mathbf{q}')}{(2\pi)^{2d}} = \frac{\delta J(-\mathbf{q})}{\delta\phi(\mathbf{q}')} = (2\pi)^{-2d}\left(\frac{\delta^2 W_k}{\delta J(\mathbf{q})\delta J(\mathbf{q}')}\right)^{-1}, \tag{3.72}$$

where $\Gamma_k^{(2)}(\mathbf{q},\mathbf{q}') = \delta^2\Gamma_k/\delta\phi(\mathbf{q})\delta\phi(\mathbf{q}')$.

Let us now show that the definition of Γ_k, Eq. (3.70), ensures that it indeed interpolates between the microscopic Hamiltonian H for $k = \infty$ and the (true) effective action Γ for $k \to 0$, see Eq. (3.58). This last property follows directly from Eq. (3.70) and the fact that for $k = 0$ the IR cut-off $R_k(\mathbf{q}^2)$ identically vanishes. The fact that Γ_k identifies with H when $k \to \infty$ can be shown in the following way. One has from Eqs. (3.59), (3.68), (3.70), and (3.71) the functional identity

$$e^{-\Gamma_k[\phi]} = \int \mathcal{D}\zeta \exp\left(-H[\zeta] + \frac{\delta\Gamma_k[\phi]}{\delta\phi}.(\zeta - \phi) - \Delta H_k[\zeta - \phi]\right). \tag{3.73}$$

In the limit $k \to \infty$, $R_k(\mathbf{q}^2)$ goes to infinity. In this limit, the mass-term $\exp(-\Delta H_k[\zeta - \phi])$ acts as a hard constraint on the functional integration — $\exp(-\Delta H_k[\zeta - \phi]) \simeq \delta[\zeta - \phi]$ — so that $\Gamma_{k=\infty}[\phi] = H[\phi]$. With these properties, $\Gamma_k[\phi]$ has the meaning of a coarse-grained Gibbs free energy at scale k^{-1}: lowering k corresponds to including more and more fluctuations.

The effective average action Γ_k follows an exact equation which controls its evolution when the running scale k is lowered, i.e. when more and more (low-energy) fluctuations are integrated out[108,109,114,117]:

$$\partial_t \Gamma_k[\phi] = \frac{1}{2} \int \frac{d^d\mathbf{q}}{(2\pi)^d} \dot{R}_k(\mathbf{q}^2)\, \mathrm{Tr}((2\pi)^{2d}\Gamma_k^{(2)}[\phi] + \mathcal{R}_k)^{-1}(\mathbf{q}, -\mathbf{q}), \qquad (3.74)$$

with $t = \ln k/\Lambda$ and $\dot{R}_k = \partial_t R_k$. In Eq. (3.74), Tr must be understood as a trace on internal indices — vectorial or tensorial — if ζ spans a non-trivial representation of a group.

Equation (3.74) defines an RG flow for the effective action Γ_k analogous to the RG flow of effective Hamiltonians provided by the Wilson–Kadanoff block spin procedure.

3.6.2. *Properties*

We now give some important properties of Eq. (3.74). The reader interested in more details can consult the review of Berges *et al.*[115]

(1) Equation (3.74) is *exact*. It thus contains all perturbative[118,119] and nonperturbative features of the underlying theory: weak- or strong-coupling behaviors, tunneling between different minima,[120] bound states,[121,122] topological excitations,[123] etc.

(2) While it is written here in the case of a one-component scalar field theory, Eq. (3.74) obviously holds for any number of components and, more generally, for any kind of order parameter. The extension to fermions is also trivial (see the review of Berges *et al.*,[115] for instance).

(3) With a cut-off function $R_k(\mathbf{q}^2)$ which meets the condition (3.62) or, more generally, with a finite limit when $\mathbf{q}^2 \to 0$, the integral in Eq. (3.74) is infrared (IR) *finite* for any $k > 0$. This IR finiteness is ensured by the presence of the mass-term R_k which makes the quantity $\Gamma_k^{(2)}[\phi] + R_k$ positive for $k > 0$ even *at* the critical temperature. This allows to explore the low-temperature phase even in the presence of massless — Goldstone — modes. From the UV side, the finiteness of the integral in Eq. (3.74) is ensured by a requirement of the fast decaying behavior of $\dot{R}_k(\mathbf{q}^2)$.

(4) One can give a graphical representation of Eq. (3.74), see Fig. 3.6. It displays a *one-loop* structure. Obviously, this one-loop structure must not be mistaken for that of a weak-coupling expansion. Actually, the loop involves here the *full* — i.e. field-dependent — inverse propagator $\Gamma_k^{(2)}[\phi]$ so that the graphical representation of Fig. 3.6 implicitly contains *all* powers of the coupling constants entering the model. Note also that this one-loop structure automatically ensures that all integrals over the internal momenta

$$\partial_t \Gamma_k = \frac{1}{2}$$ 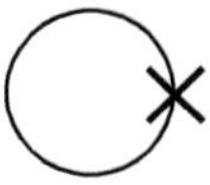

Fig. 3.6. A graphical representation of Eq. (3.74). The cross represents the function $\dot{R}_k$ and the line the propagator $(\Gamma_k^{(2)}(\phi) + R_k)^{-1}$.

involved in this formalism have a one-loop structure and are thus 1D. Thus, they can be easily evaluated numerically and, when some particular cut-off is used, analytically. This radically differs from a weak-coupling expansion which leads to multiple loop diagrams and thus, multiple integrals. Another important feature of Eq. (3.74) is that very simple ansätze on Γ_k allow to recover in a unique framework the one-loop perturbative results obtained by standard perturbative calculations around two and four dimensions as well as in a large-N expansion.

3.6.3. *Truncations*

Unfortunately, Eq. (3.74) is too complicated to be solved exactly. As a consequence, the functional Γ_k has to be truncated. Different kinds of truncations have been considered[109]:

(1) Field expansion where Γ_k is expanded in powers of the order parameter ϕ. For a scalar field theory, one has

$$\Gamma_k[\phi] = \sum_{n=0}^{\infty} \frac{1}{n!} \int \prod_{i=0}^{n} d^d\mathbf{x_i}\phi(\mathbf{x_1}) \cdots \phi(\mathbf{x_n}) \Gamma_k^{(n)}(\mathbf{x_1}, \ldots, \mathbf{x_n}), \tag{3.75}$$

where $\Gamma_k^{(n)}(\mathbf{x_1}, \ldots, \mathbf{x_n})$ denotes the nth functional derivative of Γ_k.

(2) Derivative expansion where Γ_k is expanded in powers of the derivatives of the order parameter. Still for a scalar field theory, one has

$$\Gamma_k[\phi] = \int d^d\mathbf{x} \left(U_k(\phi) + \frac{1}{2} Z_k(\phi)(\partial\phi)^2 + O(\partial^4) \right). \tag{3.76}$$

$U_k(\phi)$ being the potential — i.e. derivative-independent — part of Γ_k while $Z_k(\phi)$ corresponds to the kinetic term. The physical motivation for such an expansion in which one neglects high-order derivative terms is that since the anomalous dimension η is small, these terms should not drastically affect the physics.

(3) Combined derivative and field expansions where one further expands, in Eq. (3.76), the functions $U_k(\phi)$ and $Z_k(\phi)$ in powers of ϕ around a given field configuration ϕ_0. Technically, this kind of approximation allows to transform the functional equation (3.74) into a set of ordinary coupled differential equations for the coefficients of the expansion. In practice, it is interesting to consider an expansion around (one of) the field configuration ϕ_0 that minimizes the potential U_k. For the simplest — Ising — model, this expansion is written as follows:

$$\Gamma_k[\phi] = \int d^d\mathbf{x} \left(\frac{1}{2} U_k''(\tilde{\rho}_0)(\rho - \tilde{\rho}_0)^2 + \frac{1}{3!} U_k'''(\tilde{\rho}_0)(\rho - \tilde{\rho}_0)^3 + \cdots + \right.$$
$$\left. + \frac{1}{2} Z_k(\tilde{\rho}_0)(\partial\phi)^2 + \frac{1}{2} Z_k'(\tilde{\rho}_0)(\rho - \tilde{\rho}_0)(\partial\phi)^2 + \cdots \right), \qquad (3.77)$$

where $\tilde{\rho} = \frac{1}{2}\phi^2$ and $\tilde{\rho}_0 = \frac{1}{2}{\phi_0}^2$. In Eq. (3.77), the quantity ϕ_0 has the meaning of a magnetization at scale k. The rational behind the choice (3.77) is that the minimum of U_k is physically the location that one wants to describe the best since thermodynamical quantities at vanishing external field are determined from the minimum of U_k at $k = 0$. The relevance of such a parametrization is confirmed by many works showing that the convergence of the critical quantities, when more and more powers of the field ϕ are added in the truncation, is improved when compared with the same calculation performed with an expansion of $U_k(\rho)$ and $Z_k(\rho)$ around the $\phi = 0$ configuration.[124, 125]

The choice of a good truncation is a complex problem. One has to choose a manageable truncation that however encodes the relevant physics. In practice, it appears that, surprisingly, even at low orders in powers of derivatives and fields, Eq. (3.74) provides correct qualitative features of the RG flow. However, the precise determination of the critical quantities requires to push the expansion to rather large orders in the field and involves a heavy algebra.[17, 116, 126–128]

3.6.4. *Principle of the calculation*

We do not enter in the details of the computation of the RG flow that have been given in several articles[105, 106, 109, 115] and thus only indicate the four steps of the procedure.

(i) First, one defines the running coupling constants entering in Γ_k from functional derivatives of the ansätze of Γ_k. The coupling constants are defined as (combinations of) functional derivatives of Γ_k — the "vertex functions" — taken in a specific field configuration of the model, for instance the minimum of the potential ϕ_0.

(ii) One then applies the operator ∂_t on these definitions. This is implemented by the use of the evolution equation (3.74). The flow equations for the coupling constants are then expressed as traces of products of vertex functions that are evaluated from the ansätze for Γ_k, which is given, in the case of a scalar theory, by Eq. (3.77).

(iii) The flow equations involve integrals over the internal momentum. It is convenient to express these integrals as dimensionless functions, known as threshold functions[109] that embody the non-perturbative content of the model.

(iv) Also, as usual, one introduces dimensionless renormalized quantities to study the scale-invariant solutions of the RG equations.

3.6.5. *The $O(N) \times O(2)$ model*

We now consider frustrated magnets in this framework. As already said, since the NPRG equation (3.74) cannot be solved exactly, a truncation for Γ_k is needed. The present authors[18,19] have used truncations involving terms having at most two derivatives. At this order, the most general form of the $O(N) \times O(2)$ effective average action is written as[19]:

$$\begin{aligned}\Gamma_k[\vec{\phi}_1, \vec{\phi}_2] = \int d^d\mathbf{x} \Big(& U_k(\rho,\tau) + \frac{1}{2} Z_k(\rho,\tau)((\partial\vec{\phi}_1)^2 + (\partial\vec{\phi}_2)^2) \\ & + \frac{1}{4} Y_k^{(1)}(\rho,\tau)(\vec{\phi}_1 \cdot \partial\vec{\phi}_2 - \vec{\phi}_2 \cdot \partial\vec{\phi}_1)^2 \\ & + \frac{1}{4} Y_k^{(2)}(\rho,\tau)(\vec{\phi}_1 \cdot \partial\vec{\phi}_1 + \vec{\phi}_2 \cdot \partial\vec{\phi}_2)^2 \\ & + \frac{1}{4} Y_k^{(3)}(\rho,\tau)((\vec{\phi}_1 \cdot \partial\vec{\phi}_1 - \vec{\phi}_2 \cdot \partial\vec{\phi}_2)^2 + (\vec{\phi}_1 \cdot \partial\vec{\phi}_2 + \vec{\phi}_2 \cdot \partial\vec{\phi}_1)^2)\Big).\end{aligned} \tag{3.78}$$

We recall that $\vec{\phi}_1$ and $\vec{\phi}_2$ are the two N-component vectors that constitute the order parameter $\Phi = (\vec{\phi}_1, \vec{\phi}_2)$ while $\rho = \mathrm{Tr}({}^t\Phi.\Phi)$ and $\tau = \frac{1}{2}\mathrm{Tr}({}^t\Phi.\Phi - \mathbb{1}\,\rho/2)^2$ are the two independent $O(N) \times O(2)$ invariants. Here $U_k(\rho,\tau)$ is the potential part of the effective average action, while $Z_k(\rho,\tau)$ and $Y_k^{(i)}(\rho,\tau)$ are kinetic functions. The ansätze (3.78) can been further simplified by expanding all the functions in powers of the fields around a nonvanishing field configuration which minimizes U_k. This last constraint is satisfied when $\vec{\phi}_1$ and $\vec{\phi}_2$ are orthogonal, with the same norm. One can choose a particular configuration:

$$\Phi^{\rm Min}(\mathbf{x}) = \sqrt{\tilde{\kappa}} \begin{pmatrix} 1 & 0 \\ 0 & 1 \\ 0 & 0 \\ \vdots & \vdots \\ 0 & 0 \end{pmatrix} \tag{3.79}$$

the physical results being, of course, independent of this particular choice. The quantity $\sqrt{\tilde{\kappa}}$ entering in (3.79) is analogous to the amplitude of the magnetization ϕ_0 in the Ising case — see Eq. (3.77). While studying the critical properties of frustrated systems, the present authors have considered various truncations combining derivative and field expansions and differing by the number of monomials in ρ and τ included in the field expansion. In order to keep our presentation concise here, we present only the simple truncation[17, 19]:

$$\Gamma_k[\vec{\phi}_1, \vec{\phi}_2] = \int d^d\mathbf{x} \left(\frac{Z}{2}((\partial\vec{\phi}_1)^2 + (\partial\vec{\phi}_2)^2) + \frac{\tilde{\omega}}{4}(\vec{\phi}_1 \cdot \partial\vec{\phi}_2 - \vec{\phi}_2 \cdot \partial\vec{\phi}_1)^2 \right.$$
$$\left. + \frac{\tilde{\lambda}}{4}\left(\frac{\rho}{2} - \tilde{\kappa}\right)^2 + \frac{\tilde{\mu}}{4}\tau \right). \tag{3.80}$$

The set $\{\tilde{\kappa}, \tilde{\lambda}, \tilde{\mu}, Z, \tilde{\omega}\}$ denotes the scale-dependent coupling constants that parametrize the model at this order of the truncation. The first quantity in Eq. (3.80) corresponds to the standard kinetic term while the third and fourth correspond to the potential part. Actually, apart from the second term — called the current term — Γ_k in Eq. (3.80) looks very much like the usual GLW Hamiltonian used to study perturbatively the critical physics of the $O(N) \times O(2)$ model, up to trivial reparametrizations, see Eq. (3.50). There is however a fundamental difference since, here, Γ_k is *not* studied within a weak-coupling perturbative approach. This allows the presence of the current term which corresponds to a non-standard kinetic term. This term is irrelevant by power counting around four dimensions since it is quartic in the fields and quadratic in derivatives. However, its presence is *necessary* around two dimensions to recover the results of the low-temperature approach of the NLσ model since it contributes to the field renormalization of the Goldstone modes. Being not constrained by the usual power counting, one includes this term in the ansätze.

3.6.5.1. *The flow equations*

To write the flow equations, one has to introduce the dimensionless renormalized quantities defined as

$$\begin{cases} \kappa = Zk^{2-d}\tilde{\kappa} \\ \lambda = Z^{-2}k^{d-4}\tilde{\lambda} \\ \mu = Z^{-2}k^{d-4}\tilde{\mu} \\ \omega = Z^{-2}k^{d-2}\tilde{\omega}. \end{cases} \tag{3.81}$$

We do not provide the details of the computation. The general technique is given in several publications and its implementation on the specific $O(N)\times O(2)$ model has been given by the present authors.[19] The β functions for the different coupling constants entering in (3.80) are given by[17, 19]

$$\beta_\kappa = -(d-2+\eta)\kappa + 4v_d\left[\frac{1}{2}l_{01}^d(0,0,\kappa\omega) + (N-2)l_{10}^d(0,0,0)\right.$$
$$\left.+\frac{3}{2}l_{10}^d(\kappa\lambda,0,0) + \left(1+2\frac{\mu}{\lambda}\right)l_{10}^d(\kappa\mu,0,0) + \frac{\omega}{\lambda}l_{01}^{2+d}(0,0,\kappa\omega)\right],$$

$$\beta_\lambda = (d-4+2\eta)\lambda + v_d\left[2\lambda^2(N-2)l_{20}^d(0,0,0) + \lambda^2 l_{02}^d(0,0,\kappa\omega)\right.$$
$$+9\lambda^2 l_{20}^d(\kappa\lambda,0,0) + 2(\lambda+2\mu)^2 l_{20}^d(\kappa\mu,0,0)$$
$$\left.+4\lambda\omega l_{02}^{2+d}(0,0,\kappa\omega) + 4\omega^2 l_{02}^{4+d}(0,0,\kappa\omega)\right],$$

$$\beta_\mu = (d-4+2\eta)\mu - 2v_d\mu\left[-\frac{2}{\kappa}l_{01}^d(0,0,\kappa\omega) + \frac{3(2\lambda+\mu)}{\kappa(\mu-\lambda)}l_{10}^d(\kappa\lambda,0,0)\right.$$
$$\left.+\frac{8\lambda+\mu}{\kappa(\lambda-\mu)}l_{10}^d(\kappa\mu,0,0) + \mu l_{11}^d(\kappa\mu,0,\kappa\omega) + \mu(N-2)l_{20}^d(0,0,0)\right],$$

$$\eta = -\frac{d\ \ln Z}{dt} = 2\frac{v_d}{d\kappa}\left[(4-d)\kappa\omega l_{01}^d(0,0,\kappa\omega) + 2\kappa^2\omega^2 l_{02}^{2+d}(0,0,\kappa\omega)\right.$$
$$+2m_{02}^d(0,0,\kappa\omega) - 4m_{11}^d(0,0,\kappa\omega) + 2(-2+d)\kappa\omega l_{10}^d(0,0,0)$$
$$+2m_{20}^d(0,0,\kappa\omega) + 2\kappa^2\lambda^2 m_{2,2}^d(\kappa\lambda,0,0) + 4\kappa^2\mu^2 m_{2,2}^d(\kappa\mu,0,0)$$
$$\left.+4\kappa\omega n_{02}^d(0,0,\kappa\omega) - 8\kappa\omega n_{11}^d(0,0,\kappa\omega) + 4\kappa\omega n_{20}^d(0,0,\kappa\omega)\right],$$

$$\beta_\omega = (d-2+2\eta)\omega + \frac{4v_d}{d\kappa^2}\Bigg[\kappa\omega\Bigg\{\frac{(4-d)}{2}l_{01}^d(0,0,\kappa\omega) + \frac{(d-16)}{2}l_{01}^d(\kappa\lambda,0,\kappa\omega)$$
$$+\kappa\omega l_{02}^{2+d}(0,0,\kappa\omega) - 3\kappa\omega l_{02}^{2+d}(\kappa\lambda,0,\kappa\omega) + (d-2)l_{10}^d(0,0,0)$$
$$-(d-8)l_{10}^d(\kappa\lambda,0,0) + 8\kappa\lambda l_{11}^d(\kappa\lambda,0,\kappa\omega) + 2\kappa\omega l_{20}^{2+d}(\kappa\mu,0,0)$$
$$+2\kappa\omega(N-2)l_{20}^{2+d}(0,0,0)\Bigg\} + m_{02}^d(0,0,\kappa\omega) - m_{02}^d(\kappa\lambda,0,\kappa\omega)$$
$$-2m_{11}^d(0,0,\kappa\omega) + 2m_{11}^d(\kappa\lambda,0,\kappa\omega) + m_{20}^d(0,0,\kappa\omega) - m_{20}^d(\kappa\lambda,0,\kappa\omega)$$
$$+\kappa^2\lambda^2 m_{22}^d(\kappa\lambda,0,0) + 2\kappa^2\mu^2 m_{22}^d(\kappa\mu,0,0) + 2\kappa\omega n_{02}^d(0,0,\kappa\omega)$$
$$-4\kappa\omega n_{02}^d(\kappa\lambda,0,\kappa\omega) - 4\kappa\omega n_{11}^d(0,0,\kappa\omega) + 8\kappa\omega n_{11}^d(\kappa\lambda,0,\kappa\omega)$$
$$+2\kappa\omega n_{20}^d(0,0,\kappa\omega) - 4\kappa\omega n_{20}^d(\kappa\lambda,0,\kappa\omega)\Bigg]. \tag{3.82}$$

In the previous set of equations appear the dimensionless functions $l_{n1,n2}^d(w_1,w_2,a), m_{n1,n2}^d(w_1,w_2,a), n_{n1,n2}^d(w_1,w_2,a)$, called threshold functions. They encode the non-perturbative content of the flow equations (3.82). They are complicated integrals over momenta and are given by[17,19]

$$l_{n_1,n_2}^d(w_1,w_2,a)$$
$$= -\frac{1}{2}\int_0^\infty dy\, y^{d/2}\frac{\eta r + 2yr'}{(P_1+w_1)^{n_1}(P_2+w_2)^{n_2}}\left(\frac{n_1}{P_1+w_1} + \frac{n_2}{P_2+w_2}\right),$$
$$n_{n_1,n_2}^d(w_1,w_2,a)$$
$$= -\frac{1}{2}\int_0^\infty dy\, y^{d/2}\frac{1}{(P_1+w_1)^{n_1}(P_2+w_2)^{n_2}}\Bigg(y(1+a+r+yr')(\eta r + 2yr')$$
$$\cdot\left(\frac{n_1}{P_1+w_1} + \frac{n_2}{P_2+w_2}\right) - \eta(r+yr') - 2y(2r'+yr'')\Bigg),$$
$$m_{n_1,n_2}^d(w_1,w_2,a)$$
$$= -\frac{1}{2}\int_0^\infty dy\, y^{d/2}\frac{1+a+r+yr'}{(P_1+w_1)^{n_1}(P_2+w_2)^{n_2}}\Bigg(y(1+a+r+yr')(\eta r + 2yr')$$
$$\cdot\left(\frac{n_1}{P_1+w_1} + \frac{n_2}{P_2+w_2}\right) - 2\eta(r+yr') - 4y(2r'+yr'')\Bigg), \tag{3.83}$$

where one has introduced

$$P_1 = P_1(y,a) = y(1+r(y)+a), \tag{3.84}$$

$$P_2 = P_2(y, a) = y(1 + r(y)), \tag{3.85}$$

$$r(y) = \frac{R_k(yk^2)}{Zk^2y}. \tag{3.86}$$

3.6.6. *Tests of the method and first results*

This section is devoted to all possible tests of the NPRG method as well as new results obtained with it.[17,19]

(1) Around $d = 4$, one easily checks that, in the limit of small coupling constants λ and μ, the set of equations (3.82) degenerate from those obtained from the weak-coupling expansion at one loop

$$\begin{cases} \beta_\lambda = -\epsilon\lambda + \dfrac{1}{16\pi^2}\left(4\lambda\mu + 4\mu^2 + \lambda^2(N+4)\right) \\ \beta_\mu = -\epsilon\mu + \dfrac{1}{16\pi^2}\left(6\lambda\mu + N\mu^2\right) \\ \beta_\kappa = -(2-\epsilon)\kappa + \dfrac{l_1^4(0)}{8\pi^2}\left(N + 1 + \dfrac{2\mu}{\lambda}\right) - \dfrac{3\kappa\lambda}{16\pi^2} - \dfrac{\kappa\mu}{8\pi^2}\left(1 + \dfrac{2\mu}{\lambda}\right) \end{cases} \tag{3.87}$$

from which one can deduce, for instance, the expression of ν at order $\epsilon = 4 - d$. It, of course, coincides with the one-loop result of Eq. (3.53).

(2) Also, around $d = 2$, performing a low-temperature expansion of the Eq. (3.82)

$$\begin{cases} \beta_\kappa = -(d - 2 + \eta)\kappa + \dfrac{N-2}{2\pi} + \dfrac{1}{4\pi(1+\kappa\omega)} \\ \beta_\omega = (-2 + d + 2\eta)\omega + \dfrac{1 + \kappa\omega + (N-1)\kappa^2\omega^2 + (N-2)\kappa^3\omega^3}{2\pi\kappa^2(1+\kappa\omega)} \\ \eta = \dfrac{3 + 4\kappa\omega + 2\kappa^2\omega^2}{4\pi\kappa(1+\kappa\omega)} \end{cases} \tag{3.88}$$

and making the change of variables

$$\begin{cases} \eta_1 = 2\pi\kappa \\ \eta_2 = 4\pi\kappa(1+\kappa\omega), \end{cases} \tag{3.89}$$

one recovers the NLσ model β functions at one-loop order, Eq. (3.48).

(3) The non-perturbative equations (3.82) allow to recover the critical exponents at leading order in $1/N$. The present authors have computed

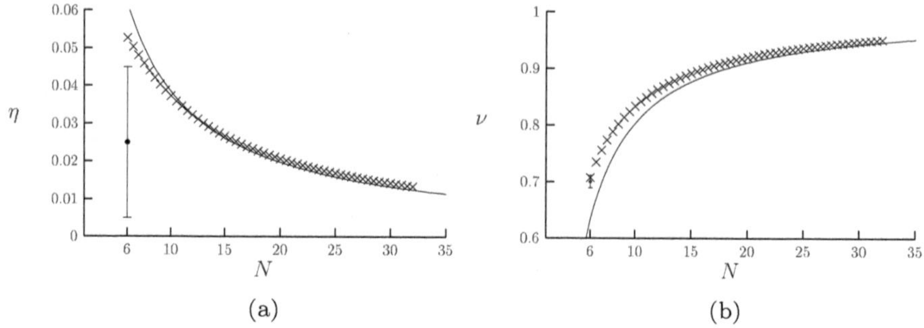

Fig. 3.7. The exponents η and ν as functions of N in $d = 3$. The crosses represent the NPRG results and the full line represents the values obtained from the $1/N$ expansion, Eq. (3.57). The circles and error bars are the MC results obtained for $N = 6$.[94] (a) η as a function of N, (b) ν as a function of N.

Table 3.7. Exponents ν and η computed from the $1/N$ expansion, by the NPRG method and from the six-loop calculation.

N	Method	ν	η
16	1/N[101]	0.885	0.0245
	NPRG[19]	0.898	0.0252
	six-loop	0.858(4),[10] 0.863(4)[12]	0.0246(2)[12]
32	1/N[101]	0.946	0.0125
	NPRG[19]	0.950	0.0134
	six-loop	0.936(2),[10] 0.936(1)[12]	0.01357(1)[12]

η and ν for a large range of values of N and have compared their results with those calculated perturbatively at order $1/N^2$, Eq. (3.57). They have found an excellent agreement — better than 1% — for ν, *for all* $N > 10$, see Fig. 3.7, a domain of values of N where one expects the $1/N^2$ results to be very close to the exact values. We also quote, in Table 3.7, the NPRG results and those obtained by the six-loop calculation for the cases $N = 16$ and $N = 32$.

(4) Let us now discuss the curve $N_c(d)$ that separates the regions of first and second order. The present authors have computed $N_c(d)$ with their best truncation — involving 10 powers of the fields — and with the cut-off function (3.67). In Fig. 3.8, we give the result obtained from the NPRG[19] together with the improved three-loop result given by Eq. (3.56)

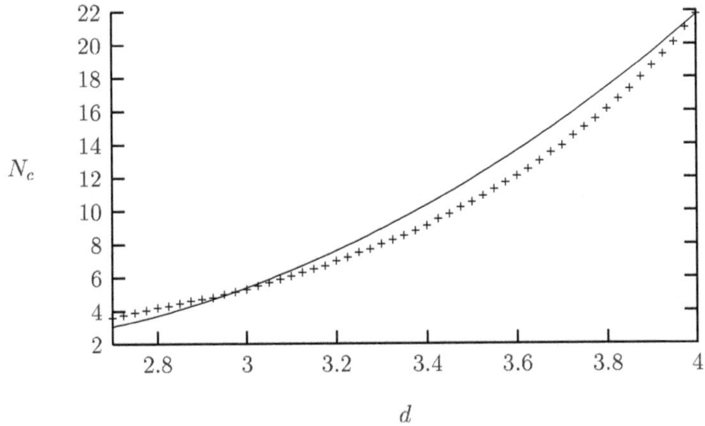

Fig. 3.8. The full line represents the curve $N_c(d)$ obtained by the three-loop result improved by the constraint $N_c(d = 2) = 2$, Eq. (3.56). The crosses represent the NPRG calculation.[19]

for comparison.[g] The two curves agree quite well. Note that it is probably a coincidence that the curves cross very close to $d = 3$. In this dimension, the NPRG method leads to $N_c(d = 3) \simeq 5.1$ and the improved three-loop result: $N_c(d = 3) \simeq 5.3(2)$. Let us emphasize that, within the NPRG method, the quantity $N_c(d)$ is very sensitive to the order of the truncation,[129] much more than the exponents. This means that one probably should not consider the previous results as very reliable. In this respect, we recall the result obtained by means of the six-loop calculation of Calabrese *et al.*[12]: $N_c(d = 3) = 6.4(4)$ and by means of the $4 - \epsilon$ expansion at five-loop order[102]: $N_c(d = 3) \simeq 6.1(6)$.

(5) Let us now consider the consequences of the NPRG approach for the NLσ model predictions. The very probable existence of a line $N_c(d)$ going from $d = 2$ to $d = 4$ confirms what has been already anticipated in Section 3.5: the predictions based on this approach are incorrect as for the physics in $d = 3$. Indeed, the shape of this line implies that the $O(4)$ fixed point found in the NLσ model approach disappears very probably between two and three dimensions. Actually, following this fixed point,

[g]Note that, due to artifact of the field expansion performed at order ϕ^{10}, it is not possible to follow the chiral fixed point C_+ in dimensions typically between $d = 2.5$ and $d = 2.1$ and thus to determine reliably the curve $N_c(d)$ in these dimensions. This artifact could be overcome by keeping the full field dependence of the effective potential $U(\rho, \tau)$.

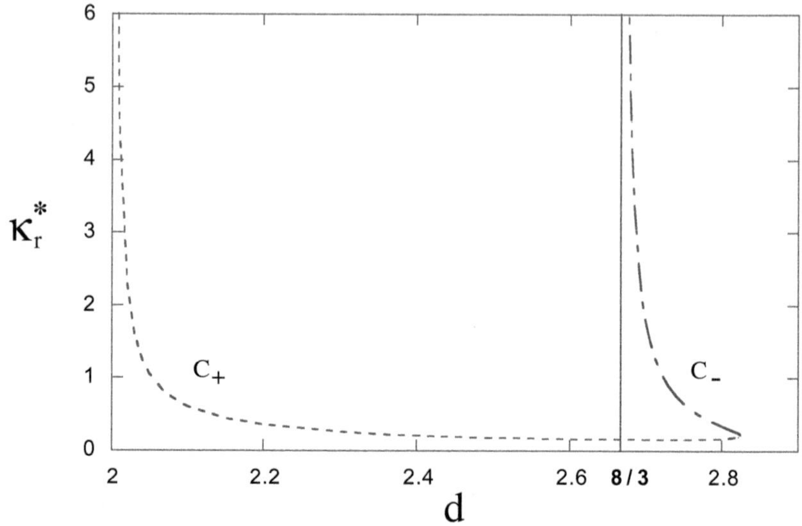

Fig. 3.9. The stable — C_+ — and unstable — C_- — fixed points as functions of the dimension d. The fixed points are parametrized by the quantity κ_r^* which is proportional to the inverse temperature of the NLσ model. The fixed point C_- appears in a dimension 8/3 in the simplest truncation and collapses with the stable fixed point C_+ in $d \simeq 2.83$.

called C_+ for an obvious reason, from $d = 2$ to $d = 4$ with the simplest ϕ^4-like truncation, the present authors have found several interesting features. First, infinitesimally close to $d = 2$, one finds that C_+ is characterized by an exponent ν of the $O(4)$ universality class. Second, as d is increased, the exponent ν associated with C_+ becomes more and more different from that characterizing an $O(4)$ transition. Third, one finds that an unstable fixed point, called C_-, shows up in a dimension $d > 2$. As the dimension is further increased, the fixed points C_+ and C_- get closer together and eventually coalesce in a dimension less than three. This phenomenon is illustrated in Fig. 3.9 in the case of the $O(3) \times O(3)$ model at the lowest order in the field expansion.[16]

The collapse of the fixed points for different values of N generates the curve $N_c(d)$. This curve is well known from the perturbative expansion performed around four dimensions. Within the NPRG approach, this curve can be followed when the dimension is lowered down to $d = 2$. There, for a given — low — value of N, the curve $N_c(d)$ provides the value of $d_c(N)$ for which the stable fixed point obtained within the NLσ model approach collapses with another — unstable — fixed point. Since this unstable fixed

Table 3.8. The exponents for $N = 6$ obtained from the NPRG — first line — and from the MC simulation — second line.

Method	α	β	γ	ν	η
NPRG[18, 19]	−0.121	0.372	1.377	0.707	0.053
MC[94]	−0.100(33)	0.359(14)	1.383(36)	0.700(11)	0.025(20)

point is *not* found in the low-temperature perturbative expansion, one therefore obtains a non-perturbative solution to the breakdown of the NLσ model approach. For $N = 3$, one gets $d_c \simeq 2.8$ which fits well with the results of Pelissetto *et al.*,[101] see Fig. 3.8.

(6) As already said, for $N = 6$,[94] the transition is either a second-order transition or an extremely weak first-order transition. In both cases, scaling should exist on a large domain of temperature. For the best truncation used by the present authors, the data are given in Table 3.8. Note that ν and η are computed directly while γ, β, and α are computed using the scaling relations. The NPRG results agree very well with the numerical ones. In particular, the error on ν, which is as usual the best determined exponent, is only 1%. This constitutes a success of the NPRG approach from the methodological point of view.

(7) Finally, let us discuss the search, within the NPRG method, of fixed points in $d = 3$ and for $N < N_c(d = 3) \simeq 5.1$. We recall that, for this critical value of N, the two fixed points C_+ and C_- — see Fig. 3.2 — coalesce. This means that these fixed points — that can be followed smoothly in the (d, N) plane from the gaussian in $d = 4$ — cease to be real below this value of N. However, this does not imply the absence of other real fixed points. One has to test the existence of fixed points non-trivially connected with C_+ and C_-, as advocated by Pelissetto *et al.*[10] The present authors have thus looked for such fixed points both by directly looking for zeroes of the β-functions and by integrating numerically the RG flow — see below. After an intensive search, they have found *no* such fixed point. This result will be discussed in the following.

We are now in a position to discuss the physics of frustrated magnets in three dimensions.

3.6.7. *The physics in $d = 3$ according to the NPRG approach*

We now discuss the physics in three dimensions. Let us start by studying the physics corresponding to values of N just below $N_c(d)$.

3.6.7.1. *The physics in $d = 3$ just below $N_c(d)$: Scaling with a pseudo-fixed point and minimum of the flow*

In a fixed dimension d, the disappearance of the non-trivial fixed points C_+ and C_-, when N crosses $N_c(d)$, could seem to be an abrupt process: the two fixed points collapse and disappear. Actually, when extended to the space of complex coupling constants, this process is continuous since the only change is that, when going from $N > N_c(d)$ to $N < N_c(d)$, the fixed points acquire a small complex part.[14] This continuous character manifests itself as smooth changes of the RG flow that can be explained thanks to continuity arguments.

To understand the evolution of the RG flow as N is decreased, one needs to consider the space of *all* coupling constants, i.e. the space such that to each point corresponds a microscopic Hamiltonian of a frustrated system. In this space, we focus on the subspace $\mathcal{E}$ containing the representative points, at $T = T_c$, of STA, STAR, $V_{N,2}$, and BCT, of all real materials studied experimentally and, more generally, of all systems of physical interest. Let us now describe qualitatively the change of the RG trajectories as N crosses $N_c(d)$.

(i) When N is larger than $N_c(d)$, there exists a true stable fixed point of the RG flow so that all trajectories emerging from $\mathcal{E}$ eventually end on this fixed point, see Fig. 3.10(a). All systems exhibit scaling around the critical temperature and universality holds.

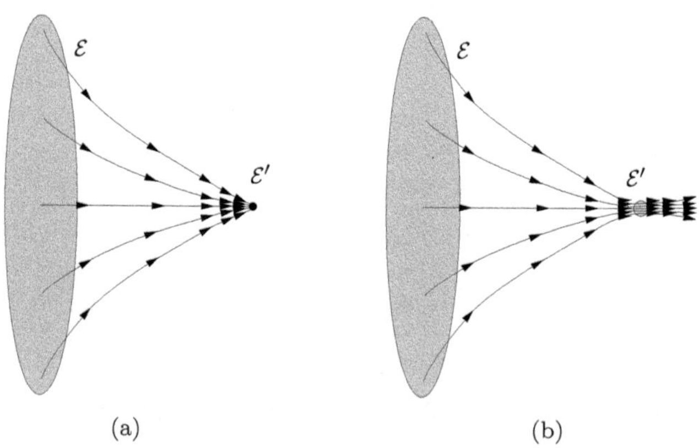

Fig. 3.10. Schematic representation of the flow for N above and just below $N_c(d)$. For the sake of clarity, we have represented $\mathcal{E}'$ outside $\mathcal{E}$ while it can be included in it. (a) $N > N_c(d)$, (b) N just below $N_c(d)$.

(ii) As already stated, when N is decreased slightly below $N_c(d)$, the fixed point C_+ gets complex coordinates and looses its direct physical meaning. In particular, the flow no longer stops at a point, see Fig. 3.10(b). Consequently, the correlation lengths of systems in $\mathcal{E}$ do not diverge at T_c. Strictly speaking, all systems undergo first-order phase transitions. However, two facts must be noted. *First*, all the trajectories emerging from $\mathcal{E}$ are attracted toward a small region in coupling constant space, denoted by $\mathcal{E}'$ in Fig. 3.10(b). *Second*, the flow in $\mathcal{E}'$ is very slow.

From the second observation, one deduces that for all systems in $\mathcal{E}$ the correlation lengths at the transition are very large — although finite — since they typically behave as the exponential of the RG time spent in $\mathcal{E}'$, time which is large. Therefore, the transitions are all extremely weak first-order transition for systems in $\mathcal{E}$. One thus expects scaling behaviors with pseudo-critical exponents for all physical quantities, with the subtlety that this scaling aborts very close to T_c where the true first-order nature of the transitions shows up.

As for the first observation — i.e. all trajectories are attracted toward a small region $\mathcal{E}'$ — it allows to conclude that all phase transitions are governed by a small region in coupling constant space and that, therefore, universality almost holds. In particular, the pseudo-critical exponents should be roughly the same for all systems whose microscopic Hamiltonian corresponds to a point in $\mathcal{E}$.

Let us study in greater detail the case where N is just below $N_c(d = 3)$. For such values of N, it is reasonable to approximate $\mathcal{E}'$ by a point. The best approximation is clearly to choose the point in $\mathcal{E}'$ that mimics best a fixed point, i.e. the point where the flow is the slowest: the minimum of the flow.[14,15] To determine this so-called pseudo-fixed point, Zumbach[14,15] has proposed to introduce a norm for the flow and to determine the point where the norm is minimum. He has performed this approach in the context of an NPRG equation (LPA of the Polchinski equation) where he has built the needed mathematical structures. He has shown that, when a minimum exists, pseudo-critical exponents characterizing pseudo-scaling can be associated with the pseudo-fixed point, in the same way that true exponents are associated with a true fixed point.

A natural assumption to explain the pseudo-scaling behaviors observed in real systems is that the minimum of the RG flow mimics a true fixed point even for values of N not very close to $N_c(d = 3)$. For the Heisenberg

Table 3.9. The critical exponents and pseudo-critical exponents for $N = 3$. α, β, and γ have been computed assuming that the scaling relations hold.

Method	α	β	γ	ν	η
NPRG[17]	0.38	0.29	1.04	0.54	0.072
LPA[15]	0.11	0.31	1.26	0.63	0.0
6-loop[10]	0.35(9)	0.30(2)	1.06(5)	0.55(3)	0.08

Notes: The first line corresponds to the NPRG approach of the present authors, the second to Zumbach's work. In the third line, we have recalled the six-loop results of Pelissetto *et al.* for comparison.

systems, this position has been advocated by Zumbach[15] and by the present authors.[17]

Within the NPRG approach the present authors have confirmed the existence of a minimum of the flow, for values of N just below $N_c(d = 3)$, leading to pseudo-scaling and pseudo-universality.[17] By following this minimum, they have confirmed that it persists down to $N = 3$ and have computed the associated pseudo-critical exponents, see Table 3.9. We also give in this table the exponents found by Zumbach within the LPA of the Polchinski equation for the same model[15] and recall those found within the six-loop approach of Pelissetto *et al.*[10]

The values of the critical exponents obtained within the NPRG are not too far from — some of — those found experimentally for group 2 of materials, see (3.36), as well as those found numerically for the STA, see Table 3.4. As usual, the truncation used overestimates η and thus, at fixed β, underestimates ν. It is remarkable that the values of the pseudo-critical exponents found at the minimum are in good agreement with those obtained within the six-loop approach. This strongly suggests that there is a common origin to these two sets of critical exponents. We shall come back on this point in the conclusion.

3.6.7.2. *Scaling with or without pseudo-fixed point: The Heisenberg and XY cases*

Let us now argue that the preceding analysis, based solely on the notion of minimum, is too naïve to give an explanation of the pseudo-critical behaviors in the physically interesting cases. Let us also give a qualitative picture that supplements the concept of minimum.

The present authors have found that, when N is lowered below $N = 3$, the minimum of the flow is less and less pronounced and that, for some value of N between 2 and 3, it completely disappears. Since several XY systems exhibit pseudo-scaling in experiments or in numerical simulations, this means that the concept of minimum of the flow does not constitute the definitive explanation of scaling in the absence of a fixed point. One encounters here the limit of the concept of minimum of the flow. First, it darkens the important fact that the notion relevant to scaling is not the existence of a minimum of the flow but that of a whole region in coupling constant space in which the flow is slow, i.e. the β functions are small. Put it differently, the existence of a minimum does not guarantee that the flow is sufficiently slow to produce large correlation lengths. Reciprocally, one can encounter situations where the RG flow is slow, the correlation length being large so that scaling occurs even in the absence of a minimum. The existence of a minimum is thus neither necessary nor sufficient to explain pseudo-scaling. Second, even when the minimum exists, reducing the region $\mathcal{E}'$ to a point rules out the possibility of testing the violation of universality. For instance, one knows that for $N = 3$ universality is violated, see Table 3.4, while a minimum of the RG flow is found. This feature cannot be reproduced by the unique set of exponents computed at the minimum. The opposite assumption, done first by Zumbach[15] and by the present authors,[17] was thus unjustified.

Thus, even for very weak first-order transitions, the beautiful simplicity of second-order transitions is lost and the finite extent of the attractive region $\mathcal{E}'$ has to be taken into account. To be precise, one needs to define two subsets of $\mathcal{E}$ and $\mathcal{E}'$: $\mathcal{D}$, which is the region in $\mathcal{E}$ leading to pseudo-scaling, and $\mathcal{R}$, the subset of $\mathcal{E}'$ which is the image of $\mathcal{D}$ in the RG flow, see Figs. 3.11(a) and 3.11(b). Let us now consider the characteristics of the flow when N is varied.

Since for $N > N_c(d = 3)$ all the systems in $\mathcal{E}$ undergo a second-order phase transition, one expects — thanks to continuity arguments — that for N slightly below $N_c(d = 3)$, all systems in $\mathcal{E}$ exhibit pseudo-scaling and thus that $\mathcal{D} = \mathcal{E}$. At the same time, $\mathcal{E}'$, the image of $\mathcal{E}$, is almost point-like — see Fig. 3.10(b) — and universality holds.

As N is decreased below $N_c(d)$, two phenomena occur.

(i) While $\mathcal{D}$ remains equal to $\mathcal{E}$, the domain $\mathcal{E}'$, which is initially point-like, grows, see Fig. 3.11(a). This means that while pseudo-scaling should be generically observed, universality starts to be significantly violated:

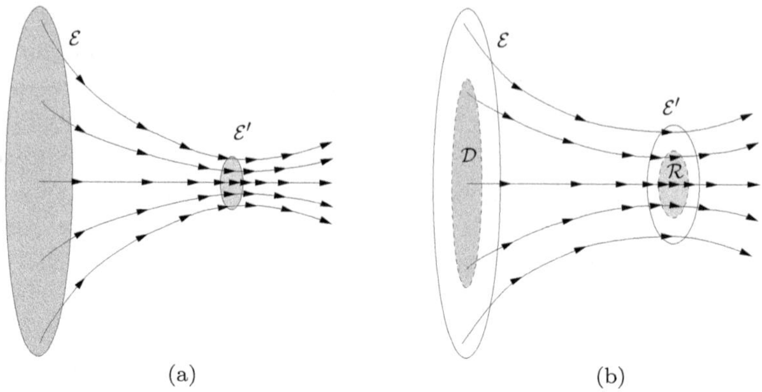

Fig. 3.11. Schematic representation of the flow (a) for $N \simeq 3$ and (b) for $N \simeq 2$. For the sake of clarity, we have represented $\mathcal{E}'$ outside $\mathcal{E}$ while it can be included in it. $\mathcal{D}$ and $\mathcal{R}$ are represented in gray. In (a), $\mathcal{D} = \mathcal{E}$ and $\mathcal{R} = \mathcal{E}'$. (a) N below $N_c(d)$, (b) N well below $N_c(d)$.

a whole spectrum of exponents should be observed, the size of $\mathcal{E}'$ providing a measure of this violation of universality.

(ii) For low values of N, the region $\mathcal{D}$ leading to pseudo-scaling can become smaller than $\mathcal{E}$, see Fig. 3.11(b). For systems defined by initial conditions in $\mathcal{D}$, the correlation lengths are still relatively large but the pseudo-critical exponents can vary from system to system according to the size of $\mathcal{R}$. For systems defined by initial conditions in $\mathcal{E}$ but not in $\mathcal{D}$, the RG flow is always fast, producing small correlation lengths at T_c. The corresponding systems undergo strong first-order phase transitions. Moreover, as N decreases, the flow in $\mathcal{E}'$ should become more and more rapid so that, for systems in $\mathcal{E}$, the correlation lengths at the phase transitions should decrease. The transitions are thus expected to become more strongly of first-order for lower N.

The precise values of N for which these changes of behaviors occur as well as the shapes and extents of $\mathcal{D}$, $\mathcal{R}$, and $\mathcal{E}'$ can only be obtained from a detailed analysis of both the microscopic Hamiltonian and of the RG flow. However, independent of the details of the model under study, of the precise value of $N_c(d)$, etc., one expects the following behavior: as N is decreased, a system that undergoes at large N a second-order transition undergoes, for N just smaller than $N_c(d)$, a very weak first-order transition governed by the minimum. Then, it should undergo a weak first-order transition where the notion of minimum is irrelevant and for which universality does not hold anymore. Finally, it should undergo a strong first-order phase transition. In the spectrum of models studied numerically, it is easy to see that the STAR, $V_{N,2}$, and BCT models with XY and Heisenberg spins nicely obey this prediction. For $N = 3$, they all show scaling and the phase transitions should be very weak first-order transitions. However, their exponents are clearly incompatible with those of STA and with those associated with the minimum, see Tables 3.4 and 3.9. The RG trajectories associated with these systems are thus expected to pass through $\mathcal{R}$, but far from the minimum. One thus naturally expects that, when N is decreased down to $N = 2$, no scaling behavior is observed for these systems. This is indeed what is found in numerical simulations, see Table 3.2. This strongly suggests that $\mathcal{D}$ has shrunk between $N = 3$ and $N = 2$ and that $N = 3$ corresponds to Fig. 3.11(a) and $N = 2$ to Fig. 3.11(b).

3.6.7.3. *The integration of the RG flow*

In the previous section, it has been shown that the notion of minimum — or pseudo-fixed point — in the RG flow is neither sufficient nor necessary to explain the existence of scaling without a fixed point. For this reason, one has to resort to another method to study the physics of XY and Heisenberg frustrated magnets. In practice, one integrates numerically the RG flow around the transition temperature T_c and determines the behavior of the physical quantities such as the correlation length, the susceptibility, and the "magnetization" — defined as $\sqrt{\bar{\kappa}}$, see Eq. (3.79) — as functions of the reduced temperature $t_r = (T - T_c)/T_c$. In principle, in the absence of universality, one should study each system independently of the others. Thus, to correctly specify the initial conditions of the RG flow, one should also keep all the microscopic information relevant to the description of a given material. This program remains, in the most general case, a theoretical challenge since this would consist in keeping track of the lattice structure as well as of the infinite number of coupling constants involved in the microscopic Hamiltonian. However it is, in principle, possible. Actually, this has been done — with much success — for certain classes of magnetic systems and fluids described by $O(N)$ models,[130] and this is mostly within the LPA.[131,132] The truncations used by the present authors — even the best one — are too restricted approximations to reach this goal since this would at least require to keep the *full* field dependence of the potential $U_k(\rho, \tau)$. They have thus used their flow equations to explain the generic occurrence of pseudo-scaling in frustrated systems without trying to describe the behavior of a specific system. In practice, they have computed the correlation length, magnetization, and susceptibility using a simplified version of their truncation keeping only the potential part expanded up to order eight in the fields, a field-independent field renormalization, and discarding all the current terms involving four fields and two derivatives. They have checked that this ansätze leads to stable results with respect to the addition of higher powers of the fields and inclusion of current terms.

A related difficulty encountered in the integration of the flow is that, in the absence of universality, the temperature dependence of the physical quantities relies on the precise temperature dependence of the microscopic coupling constants. The present authors have used several ansätze for the temperature dependence of the coupling constants and have observed that, although it could be important for the details of the results, it does not affect much the general conclusions. Thus, we illustrate the results with the simplest ansätze consisting in fixing all the couplings to temperature-independent values and by taking a linear temperature dependence for κ at the lattice scale

$$\kappa_{k=\Lambda} = a + bT. \tag{3.90}$$

For each temperature, the present authors have integrated the flow equations and have deduced the t_r-dependence of the physical quantities, such as the "magnetization", the correlation length, etc., around T_c. The different coupling constants parametrizing the initial condition of the flow have been varied to test the robustness of their conclusions. This has allowed the present authors[19] to establish the following facts.

3.6.7.4. *The Heisenberg case*

In the Heisenberg case, the following results have been obtained:

- For $N = 3$, there exist initial conditions of the flow such that for a wide range of reduced temperatures — up to four decades — the physical quantities behave as power laws. From an experimental viewpoint, this is all what is needed since scaling has been found on temperature ranges that are even smaller. The kind of pseudo-critical behaviors found is illustrated in Fig. 3.12.
- Varying the initial conditions of the flow, one observes that this phenomenon happens in a wide domain of the coupling constant space. This corresponds to the domain $\mathcal{D}$ previously defined, see Fig. 3.11(a).

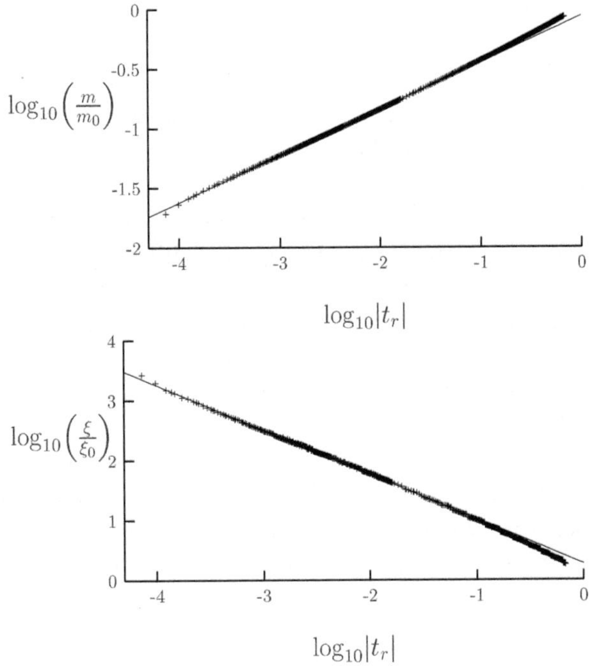

Fig. 3.12. Log–log plot of the magnetization m and of the correlation length ξ for $N = 3$ as functions of the reduced temperature t_r. The straight lines correspond to the best power law fit of the data.

- Within $\mathcal{D}$, the pseudo-critical exponents vary smoothly: β varies typically between 0.27 and 0.42 and ν between 0.56 and 0.71. These are only typical values since it has been impossible to explore the whole space of coupling constants. Since for $\beta \simeq 0.27$, one can find $\nu \simeq 0.56$, the exponents of group 2 are satisfactorily reproduced, see Tables 3.3 and 3.4. This shows in particular that there exists, in $\mathcal{D}$, a set of "microscopic" coupling constants that lead to the behavior observed in group 2.
- It is easy to find initial conditions leading to pseudo-critical exponents in good agreement with those obtained in the six-loop calculation, see Table 3.6. Actually, a whole set of initial conditions lead to exactly the same (pseudo-) exponents as those found at six-loop: $\beta = 0.30(2)$, $\nu = 0.55(3)$. This corresponds to the region of the minimum of the flow, see Table 3.9.
- In contrast, the present authors have not found initial conditions of the RG flow reproducing correctly the critical exponents of group 1, of STAR,

$V_{3,2}$, and BCT as well as negative values for η. This can originate (i) in the overestimation of η produced by the truncation of Γ_k in powers of the derivatives at order ∂^2, Eq. (3.78), (ii) in the impossibility to sample the whole coupling constant space, (iii) in the too simple temperature dependence of κ_Λ that have been considered, see Eq. (3.90).

- For a given value of one exponent, it is possible to find several values for the other exponents. Thus, one expects to find systems sharing, for instance, almost the same β but having quite different values for ν and γ.
- At the border of $\mathcal{D}$, the temperature ranges over which power laws become smaller and smaller. In a log–log plot, the t_r-dependence becomes less and less linear and the pseudo-critical exponents more and more sensitive to the choice of T_c made for the fit. Finally, outside $\mathcal{D}$, no more power-law behavior is observed.
- When going from $N = 3$ to $N = 4$, one observes, as expected, that $\mathcal{D}$ becomes far wider and that the power laws hold generically on larger temperature ranges. This is consistent with the discussion in Section 3.6.7.2. Reciprocally, and as also expected, when going from $N = 3$ to $N = 2$, $\mathcal{D}$ becomes much smaller and the power laws hold generically on smaller temperature ranges. Let us discuss this point in greater detail now.

3.6.7.5. *The XY case*

In the XY case, one founds the following results:

- For $N = 2$, one observes qualitatively the same type of behaviors as for $N = 3$. However, as predicted above, $\mathcal{D}$ is smaller and the power laws hold at best only on two decades of reduced temperature, which is consistent with what is observed experimentally. This is illustrated in Fig. 3.13 where we have represented log–log plots of the magnetization and correlation length as functions of the reduced temperature.
- Within $\mathcal{D}$ the exponents vary on the intervals $0.25 < \beta < 0.38$ and $0.47 < \nu < 0.58$.
- One finds initial conditions leading to exponents close to those of group 2 (for Ho and Dy, see Table 3.1): $\beta = 0.38$, $\nu = 0.58$, $\gamma = 1.13$. These results are quite stable with respect to changes of microscopic parameters. This is in agreement with the stability of β in group 2. Interestingly, these initial conditions correspond to small $\tilde{\mu}$ in the truncation Eq. (3.80), i.e. to initial conditions close the $O(4)$-invariant line: $\tilde{\mu} = 0$, see Fig. 3.2

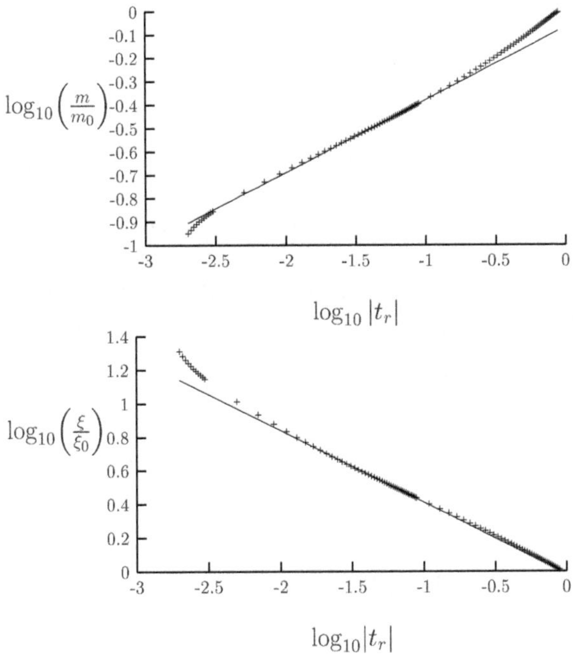

Fig. 3.13. Log–log plot of the of the magnetization m and of the correlation length ξ for $N = 2$ as functions of the reduced temperature t_r. The straight lines correspond to the best power law fit of the data. The power-law behavior observed far from the critical temperature breaks down for small t_r. The behavior of the correlation length at small t_r is an artifact of the truncation.

where the $O(4)$ fixed point is denoted by V.[h] Thus, during a large part of the flow, the trajectory remains close to the $O(4)$ fixed point before bifurcating away from this point. This is perhaps the reason why the value of β of materials of group 2 is close to that associated with an $O(4)$ behavior — $\beta_{O(4)} = 0.382$ — a fact that has been repeatedly noticed by experimentalists. Note however that the other exponents are not close to the $O(4)$ values: $\nu_{O(4)} = 0.738, \gamma_{O(4)} = 1.449$.

- One also easily finds initial conditions leading to $\beta = 0.25$, corresponding to group 1, essentially composed of STA systems. The power laws then hold on smaller ranges of temperatures and the critical exponent β is more sensitive to the determination of T_c and to the initial conditions.

[h] Let us emphasize that this $O(4)$ fixed point has nothing to do with that of the frustrated Heisenberg system found around $d = 2$ in the NLσ model approach, see Section 3.5.1.

For such values of β, one finds that ν varies between 0.47 and 0.49, which is somewhat below the value found for $CsMnBr_3$, see Table 3.1.

- The two previous points suggest that both helimagnets — such as Ho or Dy — and STA — such as $CsMnBr_3$ — can be described by the same field theory but with exponents at the two ends of the spectrum. It is actually also possible that helimagnets display a different kind of physics because of the presence of long-range interactions or because of the presence of surface effects.[133]
- As in the $N = 3$ case, one can easily find initial conditions leading to pseudo-critical exponents close to those found in the six-loop calculation, Table 3.6. For instance, for initial conditions leading to $\beta = 0.33$, one finds typically: $\nu = 0.56$ and $\gamma = 1.07$.
- As in the Heisenberg case, it has not been possible to find initial conditions of the RG flow leading to negative values of η.

3.6.8. *Summary*

The main feature of the physics of Heisenberg and XY frustrated magnets — scaling behaviors *without* universality — is reproduced, at least qualitatively and, to some extent, quantitatively. This behavior finds a *natural* explanation: there exists a whole domain $\mathcal{D}$ in the space of coupling constants such that the RG trajectories starting in $\mathcal{D}$ are "attracted" toward a region $\mathcal{R}$ where the RG flow is slow so that there is pseudo-scaling. Since $\mathcal{R}$ is *not* reduced to a point, there exists a whole spectrum of exponents and not a *unique* set. The occurrence of strong first-order phase transitions, which are observed in some materials and simulated systems, is explained by the RG trajectories starting out of $\mathcal{D}$.

Let us now stress that since universality is lost, the determination of the precise pseudo-critical exponents associated with a given material or system is obviously more difficult than the determination of the usual — universal — critical exponents characterizing a second-order phase transition. As already said, computing them would indeed require to know precisely the microscopic structure of the materials or system studied — providing the initial conditions of the RG flow — and to take into account the full field dependence of the potential $U_k(\rho, \tau)$.

3.7. Conclusion and Prospects

On the basis of their specific symmetry breaking scheme, it has been proposed[1–6] that the critical physics of XY and Heisenberg frustrated systems in three dimensions could be characterized by critical exponents associated with a *new* universality class. From this point of view, the study of frustrated magnets has been rather disappointing, the experimental and numerical contexts excluding such an hypothesis. At the same time, the phenomenology of frustrated magnets has displayed a novel kind of critical behavior — the existence of *generic* scaling *without* universality[18, 19] — requiring the use of new theoretical approaches.

Within the framework of an NPRG approach, this generic and non-universal scaling find a natural explanation in terms of slowness and "geometry" of the flow. This method has also explained the mismatch between the different perturbative approaches by means of a mechanism of annihilation of fixed points in a dimension between two and three that invalidates the low-temperature perturbative approach performed from the NLσ model for low values of N. In this respect, it would be satisfying to understand the very origin of this failure. The influence of non-trivial topological configurations on the phase transition in three dimensions has been invoked (see Sections 3.2.2 and 3.5.1). It remains, however, to confirm that these configurations indeed play such a crucial role and to understand if they are really responsible for the first-order character of the transitions in three dimensions. This is a delicate question. Indeed, the perturbative and non-perturbative RG approaches performed on the GLW model are very likely sensitive to such non-trivial topological configurations.[i] However, within these approaches, disentangling vortices and spin waves remain a theoretical challenge.

High-order perturbative calculations in $d = 3$[10–12] provide another explanation of the lack of universality in frustrated magnets: the focus character of the fixed point found within this approach induces spiral-like RG trajectories from which, according to Calabrese *et al.*,[12] follows varying effective critical exponents.[11, 12]

Even if it could appear as a rather formal question, it would be also satisfying to clarify the discrepancy between the NPRG and the six-loop perturbative approaches. In the Heisenberg case, the matching between the set of exponents found within the NPRG approach at the minimum and that found by Pelissetto *et al.* suggests a common origin to these two set of exponents. One can formulate some hypothesis to reconcile the two approaches.

[i]Both approaches correctly reproduce the physics of 3D XY non-frustrated spin systems which is probably driven by vortices at the transition.[139, 140]

First, one can invoke the specificities of the perturbative series in the case of frustrated magnets: the series display bad convergence properties[10–13] and the critical properties strongly depend on the order of the series. As already said, there is no non-trivial fixed point for $N = 2$ and $N = 3$ up to three loops; the fixed points only appear at five-loop order. Due to these specific features one can suspect that the real fixed point found by Pelissetto *et al.* is, actually, a complex one.

Second, reciprocally, one can suspect that, within the NPRG method, the lack of fixed point in the XY and Heisenberg cases is due to artifacts of the truncation in fields and/or derivatives. Only the recourse to other kinds of expansions of the effective action Γ_k — involving either the *full* function $U_k(\rho, \tau)$ or the *full* momentum dependence — could lead to unambiguous statements.

Finally, the major characteristics of XY and Heisenberg frustrated magnets, i.e. the existence of scaling laws with continuously varying exponents, are probably encountered in other physical contexts, generically systems with a critical value $N_c(d)$ of the number of components of the order parameter, separating a true second-order behavior and a naïvely first-order one: Abelian Higgs model,[134–136] $O(p)$ gauge theory coupled to N scalar fields,[137] etc.

3.8. Note Added in the Second Edition

Several theoretical and numerical studies have recently clarified some of the main issues discussed in this chapter.

First, from the theoretical point of view, in a series of articles[141–143] a careful reanalysis of the resummation process of the perturbative series obtained at fixed dimension — which includes both the $\overline{\text{MS}}$ scheme without ϵ-expansion[144] and the zero momentum massive scheme in $d = 3$ and in $d = 2$[10–12, 98] — has been performed. The aim of this analysis was to understand the discrepancy between these approaches, that lead to predict a second-order phase transition in $d = 3$, see Fig. 3.14, and the perturbative approach performed in the $\overline{\text{MS}}$ scheme with the ϵ-expansion together with the NPRG approach, that both lead to predict a (possibly weak) first-order phase transition in this dimension, see Fig. 3.14.

The main outcome of this reanalysis is the conclusion that the perturbative series obtained at fixed dimension lead very likely to *unphysical* predictions concerning the critical behavior of frustrated magnets in $d = 3$. This means in particular that the dotted curve N_c^{FD} of Fig. 3.14 is very

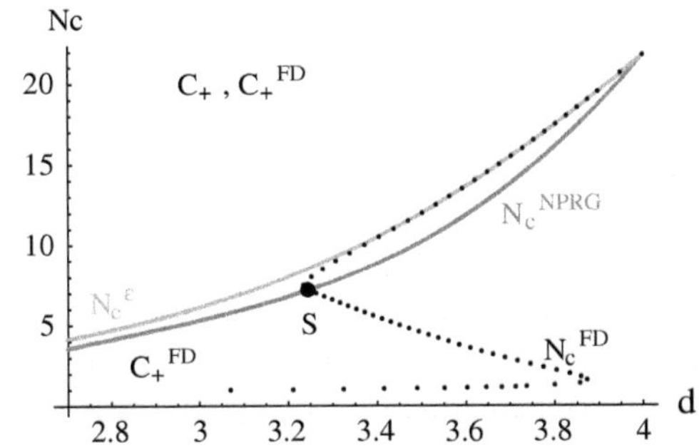

Fig. 3.14. Curves $N_c(d)$ obtained within the $\overline{\text{MS}}$ scheme with ϵ-expansion (N_c^ϵ), within the $\overline{\text{MS}}$ scheme without ϵ-expansion (N_c^{FD}) and within the NPRG approach (N_c^{NPRG}). These curves separate a region of first-order behaviors for $N < N_c$ and a region of second-order behavior for $N > N_c$. The point C_+ correspond to the stable fixed point obtained through the NPRG and the $\overline{\text{MS}}$ scheme with ϵ-expansion. The point C_+^{FD} correspond to the stable fixed point obtained through the $\overline{\text{MS}}$ scheme without ϵ-expansion, i.e. at fixed dimension. As a consequence, the transition in $d = 3$ for all values of $N > 1$ is expected to be of second order from the $\overline{\text{MS}}$ scheme without ϵ-expansion while it is expected to be of second order only for $N > N_c(d = 3)$ from the other approaches.

likely an artifact of the fixed dimension perturbative expansion. As a consequence, the behavior of frustrated magnets in $d = 3$ should not be of second order, as predicted by this approach, but of first order as predicted by the NPRG and ϵ-expansion approaches.

At the heart of these works is the remark that, within the fixed dimension approach, generically the (non-resummed) series obtained at L loops for the β-functions of the two couplings u_1 and u_2 are polynomials of order $L+1$ in the coupling constant, say u (we consider, in order to simplify, a theory with one coupling constant). Thus, the fixed point equation $\beta(u^*) = 0$ admits $L + 1$ roots u^* that are either real or complex. Contrary to what occurs within the ϵ-expansion, where the coupling constant is by definition of order ϵ, in the fixed dimension approach no real root can be *a priori* selected or, reciprocally, discarded. As a result, the generic situation is that the number of fixed points as well as their stability vary with the order L: at a given order, there can exist several real and stable fixed points or none instead of a single one. This artifact of the fixed dimension approach is already known and was first noticed in the massive scheme in $d = 3$.[145] The way to cope with it is also known: resumming the perturbative expansion of $\beta(u)$ is supposed both to restore the non-trivial Wilson–Fisher fixed point and to suppress the non-physical or "spurious" roots. This is indeed what occurs for the $O(N)$ model for which the fixed point analysis performed on the resummed β-function of fixed dimension approaches enables to discriminate between physical and "spurious" fixed points. However this ability of the resummation procedures to eradicate spurious solutions of the fixed dimension approach was not questioned in the past and, *a fortiori*, evaluated in the context of more complex models and, in particular, for models with several coupling constants.

It has been convincingly argued by Delamotte *et al.*[141] that, indeed, the situation for frustrated magnets and probably for several other models is very different from what occurs in the $O(N)$ case. Considering the β-functions derived at five loops in the MS scheme and using standard resummation procedures, it has been shown that the fixed point found in $d = 3$ *without expanding in* ϵ, although it persists after resummation, is in fact spurious. These conclusions are based on several facts:

(i) The critical exponents computed at the fixed point supposed to control the second-order behavior of frustrated magnets display bad convergence properties with the order of computation in the controversial cases of XY and Heisenberg spin systems. Indeed while for systems for which the critical properties the physics is well established, like the $O(N)$ model, the error bars on the critical exponents are typically of a few percents, for XY or Heisenberg frustrated magnets these error bars are of order 30%.

(ii) When analyzed with the same fixed dimension approach, the field theory relevant to ferromagnetic systems with cubic anisotropy displays a similar fixed point "having no counterpart within the ϵ-expansion" in contradiction with its well-established critical physics. The critical exponents computed at this supernumerary fixed point display the same bad convergence properties with the loop order as in the case of XY and Heisenberg frustrated magnets.

(iii) The coordinates (u_1^*, u_2^*) of the attractive fixed point found in the fixed dimension approach of frustrated magnets are multivalued functions of (d, N) — N being the number of spin components — because of the existence of a topological singularity in the mapping between (d, N) and the fixed point coordinates $(u_1^*(d, N), u_2^*(d, N))$, see Fig. 3.15. This singularity provides strong indications of the existence of pathologies in the RG equations obtained at fixed dimensions.

(iv) Finally, strong arguments have been provided showing that the supernumerary fixed points found in the frustrated and cubic models survive

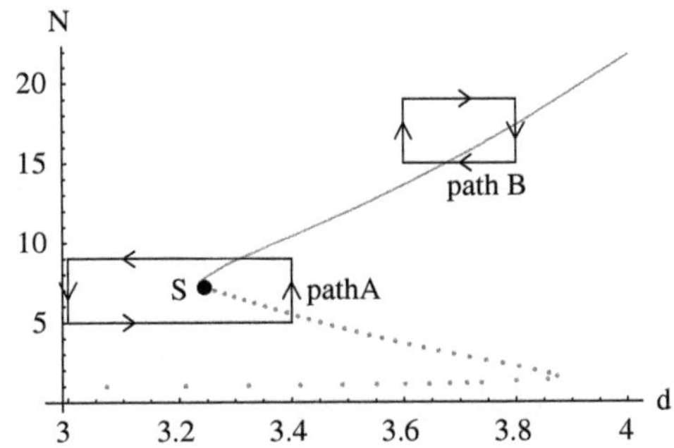

Fig. 3.15. The point S of the curve $N_c^{\rm FD}$ as a singularity of the (d, N) plane: two different paths in the (N, d) plane. Path A encloses S and is such that the coordinates of the fixed point $C_+^{\rm FD}$ do not go back to their original values after a round trip along this path. This is at variance with the paths that do not enclose S, as path B for instance.

in the upper critical dimension $d = 4$ where they are found to be non-Gaussian, a behavior deeply connected with the existence of the above mentioned topological singularity. Being given the state of knowledge of ϕ^4-like theories that are very likely trivial in $d = 4$, this fact confirms the serious doubts on the actual existence of these supernumerary fixed points.

The previous considerations have been extended[142, 143] to the case where the fixed dimension perturbative expansion is obtained within the zero momentum massive scheme in $d = 3$ at six loops[10] and in $d = 2$ at five loops.[146] It has then be confirmed that the fixed points found in $d = 3$ for $N = 2$ and $N = 3$ are most likely spurious. In $d = 2$, the situation is more delicate. Indeed we recall that, already for the (non-frustrated) $O(N)$ models, the critical exponents found from the $(\boldsymbol{\phi}^2)^2$ field theory are quantitatively wrong, although apparently converged for all N. This striking phenomenon, already mentioned in the past,[147] relies on the presence of non-analytic contributions to the β-function at the fixed point.[148] The same kind of problem has been mentioned in the case of frustrated magnets,[11, 146] but it was assumed to leave unaffected the qualitative predictions, in particular, the existence of a non-trivial fixed point in the Heisenberg case. On the contrary, it was shown by Delamotte *et al.*[142] that the phenomenon of apparent convergence toward erroneous values, together with the presence of instabilities of some critical exponents with respect to the resummation parameters, leads to seriously question the conclusions drawn in the past as for the critical behavior of these systems. This study has shown clear evidences that the fixed dimension perturbative approaches to 3D frustrated magnets are not reliable, at least at the orders studied and that there is no convincing evidence of a genuine phase transition induced by vortices in 2D Heisenberg spin systems. Finally, for the same reasons as in the Heisenberg case, it has been shown that the behavior of XY spins in $d = 2$ cannot be elucidated from the five-loop perturbative data.

On the numerical side, MC simulations have been recently performed on frustrated systems and have provided further evidence of a weak first-order phase transition of frustrated magnets in $d = 3$. An approach based on the short-time critical dynamics analysis of a weak first-order phase transition has been used to identify in an unambiguous way the weak first-order character of the transition for XY magnets.[149] This work has then been extended in the Heisenberg case[150] with the same conclusion. Finally, still in the Heisenberg case, using the Wang–Landau flat-histogram MC method for very large lattice sizes, $N \times N \times N$ where $N = 72, 84, 90, 96, 108, 120,$ and 150, Ngo and Diep[151] have reached the same conclusion: the system undergoes a clear first-order transition.

From all these recent works, one is led to the conclusion that frustrated magnets undergo a (possibly weak) first-order transition in $d = 3$ both in the XY and Heisenberg cases. Further investigations are needed to understand the situation in $d = 2$. From this point of view, the NPRG will be surely helpful.

3.9. Note Added by the Editor for the Third Edition

There is a very recent paper by Kompaniets, A. Kudlisa and A. I. Sokolov[152] using the Landau-Wilson field theory with $O(n) \times O(m)$ symmetry which describes the critical thermodynamics of frustrated spin systems with noncollinear and noncoplanar ordering analyzed in $4-\epsilon$ dimensions within the minimal subtraction scheme in the six-loop approximation. The ϵ expansions for marginal dimensionalities of the order parameter $n^H(m, 4-\epsilon)$, $n^-(m, 4-\epsilon)$, $n^+(m, 4-\epsilon)$ separating different regimes of critical behavior are extended up to ϵ^5 terms. Concrete series with coefficients in decimals are presented for $m = [2, \ldots, 6]$. The diagram of stability of nontrivial fixed points, including the chiral one, in (m, n) plane is constructed by means of summing up of corresponding ϵ expansions using various resummation techniques. Numerical estimates of the chiral critical exponents for several couples $[m, n]$ are also found. Comparative analysis of the results with their counterparts obtained earlier within the lower-order approximations and by means of alternative approaches is performed. It is confirmed, in particular, that in the physically interesting cases $(n = 2, m = 2)$ and $(n = 2, m = 3)$, phase transitions into chiral phases should be first-order. The reader is referred to this paper for technical details and for recent references on the subject.

Acknowledgments

We thank D. Loison for useful remarks and J. Vidal for a careful reading of the manuscript and helpful remarks.

References

1. T. Garel and P. Pfeuty, *J. Phys. C: Solid St. Phys.* **9**, L245 (1976).
2. M. Yosefin and E. Domany, *Phys. Rev. B* **32**, 1778 (1985).
3. H. Kawamura, *J. Phys. Soc. Jpn.* **54**, 3220 (1985).
4. H. Kawamura, *J. Phys. Soc. Jpn.* **55**, 2095 (1986).
5. H. Kawamura, *J. Phys. Soc. Jpn.* **56**, 474 (1987).
6. H. Kawamura, *Phys. Rev. B* **38**, 4916 (1988).
7. P. Azaria, B. Delamotte, and T. Jolicœur, *Phys. Rev. Lett.* **64**, 3175 (1990).
8. P. Azaria, B. Delamotte, F. Delduc, and T. Jolicœur, *Nucl. Phys. B [FS]* **408**, 485 (1993).
9. D. Bailin, A. Love, and M. A. Moore, *J. Phys. C: Solid State Phys.* **10**, 1159 (1977).
10. A. Pelissetto, P. Rossi, and E. Vicari, *Phys. Rev. B* **63**, 140414 (2001).
11. P. Calabrese, P. Parruccini, and A. I. Sokolov, *Phys. Rev. B* **66**, 180403 (2002).
12. P. Calabrese, P. Parruccini, and A. I. Sokolov, *Phys. Rev. B* **68**, 094415 (2003).
13. P. Parruccini, cond-mat/0305287.
14. G. Zumbach, *Phys. Rev. Lett.* **71**, 2421 (1993).
15. G. Zumbach, *Nucl. Phys. B* **413**, 771 (1994).
16. M. Tissier, D. Mouhanna, and B. Delamotte, *Phys. Rev. B* **61**, 15327 (2000).
17. M. Tissier, B. Delamotte, and D. Mouhanna, *Phys. Rev. Lett.* **84**, 5208 (2000).
18. M. Tissier, B. Delamotte, and D. Mouhanna, *Phys. Rev. B* **67**, 134422 (2003).
19. B. Delamotte, D. Mouhanna, and M. Tissier, cond-mat/0309101.
20. T. Dombre and N. Read, *Phys. Rev. B* **38**, 7181 (1988).
21. N. D. Mermin, *Rev. Mod. Phys.* **51**, 591 (1979).
22. H. Kawamura and S. Miyashita, *J. Phys. Soc. Jpn.* **53**, 4138 (1984).
23. W. Apel, M. Wintel, and H. Everts, *Z. Phys. B* **86**, 139 (1992).
24. B. W. Southern and A. P. Young, *Phys. Rev. B* **48**, 13170 (1993).

25. M. Wintel, H. U. Everts, and W. Apel, *Europhys. Lett.* **25**, 711 (1994).
26. B. W. Southern and H.-J. Xu, *Phys. Rev. B* **52**, 3836 (1995).
27. W. Stephan and B. W. Southern, *Phys. Rev. B* **61**, 11514 (2000).
28. M. Caffarel, P. Azaria, B. Delamotte, and D. Mouhanna, *Phys. Rev. B* **64**, 014412 (2001).
29. W. Maier and A. Saupe, *Z. Naturforsch. A* **14**, 882 (1959).
30. P. A. Lebwohl and G. Lasher, *Phys. Rev. A* **6**, 426 (1972).
31. P. E. Lammert, D. S. Rokhsar, and J. Toner, *Phys. Rev. Lett.* **70**, 1650 (1993).
32. P. E. Lammert, D. S. Rokhsar, and J. Toner, *Phys. Rev. E* **52**, 1778 (1995).
33. G. Kohring and R. E. Shrock, *Nucl. Phys. B* **285**, 504 (1987).
34. J. Villain, *J. Phys. C: Solid St. Phys.* **10**, 4793 (1977).
35. S. Miyashita and H. Shiba, *J. Phys. Soc. Jpn.* **53**, 1145 (1985).
36. V. L. Berezinskii, *Sov. Phys. JETP* **32**, 493 (1970).
37. J. M. Kosterlitz and D. J. Thouless, *J. Phys. C* **6**, 1181 (1973).
38. V. P. Plakhty, J. Kulda, D. Visser, E. V. Moskvin, and J. Wosnitza, *Phys. Rev. Lett.* **85**, 3942 (2000).
39. M. F. Collins and O. A. Petrenko, *Can. J. Phys.* **75**, 605 (1997).
40. H. B. Weber, T. Werner, J. Wosnitza, H. v. Löhneysen, and U. Schotte, *Phys. Rev. B* **54**, 15924 (1996).
41. J. Wang, D. P. Belanger, and B. D. Gaulin, *Phys. Rev. Lett.* **66**, 3195 (1991).
42. R. Deutschmann, H. von Löhneysen, J. Wosnitza, R. K. Kremer, and D. Visser, *Europhys. Lett.* **17**, 637 (1992).
43. T. E. Mason, B. D. Gaulin, and M. F. Collins, *Phys. Rev. B* **39**, 586 (1989).
44. T. E. Mason, M. F. Collins, and B. D. Gaulin, *J. Phys. C: Solid St. Phys.* **20**, L945 (1987).
45. B. D. Gaulin, T. E. Mason, M. F. Collins, and J. Z. Larese, *Phys. Rev. Lett.* **62**, 1380 (1989).
46. Y. Ajiro, T. Nakashima, Y. Unno, H. Kadowaki, M. Mekata, and N. Achiwa, *J. Phys. Soc. Jpn.* **57**, 2648 (1988).
47. H. Kadowaki, S. M. Shapiro, T. Inami, and Y. Ajiro, *J. Phys. Soc. Jpn.* **57**, 2640 (1988).
48. M. Enderle, G. Fortuna, and M. Steiner, *J. Phys.: Condens. Matter* **6**, L385 (1994).
49. H. Weber, D. Beckmann, J. Wosnitza, H. von Löhneysen, and D. Visser, *Int. J. Mod. Phys. B* **9**, 1387 (1995).
50. M. Enderle, R. Schneider, Y. Matsuoka, and K. Kakurai, *Physica B* **234–236**, 554 (1997).
51. U. Schotte, N. Stusser, K. D. Schotte, H. Weinfurter, H. M. Mayer, and M. Winkelmann, *J. Phys.: Condens. Matter* **6**, 10105 (1994).
52. K. D. Jayasuriya, A. M. Stewart, S. J. Campbell, and E. S. R. Gopal, *J. Phys. F* **14**, 1725 (1984).
53. C. C. Tang, W. G. Stirling, D. L. Jones, C. C. Wilson, P. W. Haycock, A. J. Rollason, A. H. Thomas, and D. Fort, *J. Magn. Magn. Mater.* **103**, 86 (1992).

54. C. C. Tang, P. W. Haycock, W. G. Stirling, C. C. Wilson, D. Keen, and D. Fort, *Physica B* **205**, 105 (1995).
55. K. Hirota, Y. Nakazawa, and M. Ishikawa, *J. Phys.: Condens. Matter* **3**, 4721 (1991).
56. K. D. Jayasuriya, S. J. Campbell, and A. M. Stewart, *J. Phys. F: Met. Phys.* **15**, 225 (1985).
57. T. R. Thurston, G. Helgesen, J. P. Hill, D. Gibbs, B. D. Gaulin, and P. J. Simpson, *Phys. Rev. B* **49**, 15730 (1994).
58. V. P. Plakhty, W. Schweika, T. Brückel, J. Kulda, S. V. Gavrilov, L.-P. Regnault, and D. Visser, *Phys. Rev. B* **64**, 100402 (2001).
59. G. H. F. Brits and P. de V. du Plessis, *J. Phys. F* **18**, 2659 (1988).
60. P. de V. Du Plessis, A. M. Venter, and G. H. F. Brits, *J. Phys.: Condens. Matter* **7**, 9863 (1995).
61. J. Eckert and G. Shirane, *Solid State Commun.* **19**, 911 (1976).
62. G. Helgesen, J. P. Hill, T. R. Thurston, D. Gibbs, J. Kwo, and M. Hong, *Phys. Rev. B* **50**, 2990 (1994).
63. B. D. Gaulin, M. Hagen, and H. R. Child, *J. Phys. (Paris) Colloq.* **49**, C8 (1988).
64. D. A. Tindall, M. O. Steinitz, and M. L. Plumer, *J. Phys. F* **7** (1977).
65. K. D. Jayasuriya, S. J. Campbell, and A. M. Stewart, *Phys. Rev. B* **31**, 6032 (1985).
66. E. Loh, C. L. Chien, and J. C. Walker, *Phys. Lett. A* **49**, 357 (1974).
67. P. de V. Du Plessis, C. F. van Doorn, and D. C. van Delden, *J. Magn. Magn. Mater.* **40**, 91 (1983).
68. H. Kawamura, *J. Phys. Soc. Jpn.* **61**, 1299 (1992).
69. H. Kawamura, *J. Phys. Soc. Jpn.* **58**, 584 (1989).
70. M. L. Plumer and A. Mailhot, *Phys. Rev. B* **50**, 16113 (1994).
71. E. H. Boubcheur, D. Loison, and H. T. Diep, *Phys. Rev. B* **54**, 4165 (1996).
72. M. Itakura, *J. Phys. Soc. Jpn.* **72**, 74 (2003).
73. D. Loison and K. D. Schotte, *Euro. Phys. J. B* **5**, 735 (1998).
74. H. Kunz and G. Zumbach, *J. Phys. A: Math. Gen.* **26**, 3121 (1993).
75. H. Diep, *Phys. Rev. B* **39**, 397 (1989).
76. D. R. T. Jones, A. Love, and M. A. Moore, *J. Phys. C: Solid St. Phys.* **9**, 743 (1976).
77. H. Kadowaki, K. Ubukoshi, K. Hirakawa, J. L. Martinez, and G. Shirane, *J. Phys. Soc. Jpn.* **56**, 4027 (1987).
78. J. Wosnitza, R. Deutschmann, H. von Löhneysen, and R. K. Kremer, *J. Phys.: Condens. Matter* **6**, 8045 (1994).
79. K. Koyama and M. Matsuura, *J. Phys. Soc. Jpn.* **54**, 4085 (1985).
80. G. C. DeFotis, M. L. Laccheo, and H. A. Katori, *Phys. Rev. B* **65**, 94403 (2002).
81. G. C. DeFotis, S. A. Palacio, and R. L. Carlin, *Physica B+C* **95**, 380 (1978).
82. G. C. DeFotis and S. A. Pugh, *Phys. Rev. B* **24**, 6497 (1981).
83. G. C. DeFotis and J. R. Laughlin, *J. Magn. Magn. Matter.* **54–57**, 713 (1986).

84. D. Beckmann, J. Wosnitza, H. v. Löhneysen, and D. Visser, *Phys. Rev. Lett.* **71**, 2829 (1993).
85. R. Bügel, J. Wosnitza, H. v. Löhneysen, T. Ono, and H. Tanaka, *Phys. Rev. B* **64**, 094406 (2001).
86. T. Ono, H. Tanaka, T. Kato, K. Iio, K. Nakajima, and K. Kakurai, *J. Magn. Magn. Mater.* **177–181**, 735 (1998).
87. T. Ono, H. Tanaka, T. Kato, K. Nakajima, and K. Kakurai, *J. Phys.: Condens. Matter* **11**, 4427 (1999).
88. A. Mailhot, M. L. Plumer, and A. Caillé, *Phys. Rev. B* **50**, 6854 (1994).
89. T. Bhattacharya, A. Billoire, R. Lacaze, and T. Jolicœur, *J. Phys. I (Paris)* **4**, 181 (1994).
90. D. Loison and H. T. Diep, *Phys. Rev. B* **50**, 16453 (1994).
91. A. Peles and B. W. Southern, *Phys. Rev. B* **67**, 184407 (2003).
92. D. Loison and K. D. Schotte, *Euro. Phys. J. B* **14**, 125 (2000).
93. D. Loison, *Physica A* **275**, 207 (1999).
94. D. Loison, A. I. Sokolov, B. Delamotte, S. A. Antonenko, K. D. Schotte, and H. T. Diep, *JETP Lett.* **72**, 337 (2000).
95. P. Bak, S. Krinsky, and D. Mukamel, *Phys. Rev. Lett.* **36**, 52 (1976).
96. Z. Barak and M. B. Walker, *Phys. Rev. B* **25**, 1969 (1982).
97. F. J. Wegner and A. Houghton, *Phys. Rev. A* **8**, 401 (1973).
98. S. A. Antonenko and A. I. Sokolov, *Phys. Rev. B* **49**, 15901 (1994).
99. S. A. Antonenko, A. I. Sokolov, and V. B. Varnashev, *Phys. Lett. A* **208**, 161 (1995).
100. T. Jolicœur and F. David, *Phys. Rev. Lett.* **76**, 3148 (1996).
101. A. Pelissetto, P. Rossi, and E. Vicari, *Nucl. Phys. B [FS]* **607**, 605 (2001).
102. P. Calabrese and P. Parruccini, *Nucl.Phys. B* **679**, 568 (2004).
103. D. H. Friedan, *Ann. Phys. (N.Y.)* **163**, 318 (1985).
104. J. Zinn-Justin, *Quantum Field Theory and Critical Phenomena*, 3rd edn. (Oxford University Press, New York, 1989).
105. C. Wetterich, *Nucl. Phys. B* **352**, 529 (1991).
106. C. Wetterich, *Z. Phys. C* **57**, 451 (1993).
107. C. Wetterich, *Z. Phys. C* **60**, 461 (1993).
108. C. Wetterich, *Phys. Lett. B* **301**, 90 (1993).
109. N. Tetradis and C. Wetterich, *Nucl. Phys. B [FS]* **422**, 541 (1994).
110. L. P. Kadanoff, *Physics* **2**, 263 (1966).
111. L. P. Kadanoff, W. Götze, D. Hamblen, R. Hecht, E. A. S. Lewis, V. V. Palciauskas, M. Rayl, J. Swift, D. Aspnes, and J. Kane, *Rev. Mod. Phys.* **39**, 395 (1967).
112. K. G. Wilson and J. Kogut, *Phys. Rep. C* **12**, 75 (1974).
113. U. Ellwanger, *Z. Phys. C* **62**, 503 (1994).
114. T. R. Morris, *Int. J. Mod. Phys. A* **9**, 2411 (1994).
115. J. Berges, N. Tetradis, and C. Wetterich, *Phys. Rep.* **363**, 223 (2002).
116. D. F. Litim, *Nucl. Phys. B* **631**, 128 (2002).
117. U. Ellwanger, *Z. Phys. C* **58**, 619 (1993).

118. M. Bonini, M. D'Attanasio, and G. Marchesini, *Nucl. Phys. B* **409**, 441 (1993).
119. T. R. Morris and J. F. Tighe, *JHEP* **08**, 007 (1999).
120. K. I. Aoki, A. Horikoshi, M. Taniguchi, and H. Terao, *Prog. Theor. Phys.* **108**, 571 (2002).
121. U. Ellwanger, M. Hirsch, and A. Weber, *Eur. Phys. J.* p. 563 (1998).
122. H. Gies and C. Wetterich, *Phys. Rev. D* **65**, 065001 (2002).
123. G. v. Gersdorff and C. Wetterich, *Phys. Rev. B* **64**, 054513 (2001).
124. T. R. Morris, *Phys. Lett. B* **334**, 355 (1994).
125. K. I. Aoki, K. Morikawa, W. Souma, J. I. Sumi, and H. Terao, *Prog. Theor. Phys.* **95**, 409 (1996).
126. K. I. Aoki, K. Morikawa, W. Souma, J. I. Sumi, and H. Terao, *Prog. Theor. Phys.* **99**, 451 (1998).
127. L. Canet, B. Delamotte, D. Mouhanna, and J. Vidal, *Phys. Rev. D* **67**, 065004 (2003).
128. L. Canet, B. Delamotte, D. Mouhanna, and J. Vidal, *Phys. Rev. B* **68**, 064421 (2003).
129. M.Tissier, D. Mouhanna, J. Vidal, and B. Delamotte, *Phys. Rev. B* **65**, 140402 (2002).
130. S. Seide and C. Wetterich, *Nucl. Phys. B* **562**, 524 (1999).
131. A. Parola and L. Reatto, *Phys. Rev. A* **31**, 3309 (1985).
132. A. Parola and L. Reatto, *Adv. Phys.* **44**, 211 (1995).
133. T. R. Thurston, G. Helgesen, D. Gibbs, J. P. Hill, B. D. Gaulin, and G. Shirane, *Phys. Rev. Lett.* **70**, 3151 (1993).
134. T. C. Lubensky, B. I. Halperin, and S. K. Ma, *Phys. Rev. Lett.* **32**, 292 (1974).
135. I. D. Lawrie, *Nucl. Phys. B [FS]* **200**, 1 (1982).
136. P. Arnold and L. G. Yaffe, *Phys. Rev. D* **49**, 3003 (1994), err. *ibid* **55** 1114 (1997).
137. P. Ginsparg, *Nucl. Phys. B [FS]* **170**, 388 (1980).
138. J. Zinn-Justin, *Quantum Field Theory and Critical Phenomena*, 3rd edn. (Oxford University Press, New York, 1989), pp. 151, 578.
139. B. I. Halperin, *Physics of Defects* (Les Houches XXXV NATO ASI, Eds. Balian, Kleman and Poirier, North Holland, 1981).
140. N. D. Antunes, L. M. A. Bettencourt, and A. Yates, *Phys. Rev. D* **64**, 065020 (2001).
141. B. Delamotte, Yu. Holovatch, D. Ivaneyko, D. Mouhanna, and M. Tissier, *J. Stat. Mech.* P03014 (2008).
142. B. Delamotte, M. Dudka, Yu. Holovatch, and D. Mouhanna, *Phys. Rev. B* **82**, 104432 (2010).
143. B. Delamotte, M. Dudka, Yu. Holovatch, and D. Mouhanna, *Condens. Matter Phys.* **13**, 43703 (2010).
144. P. Calabrese, P. Parruccini, A. Pelissetto, and E. Vicari, *Phys. Rev. B* **70**, 174439 (2004).
145. G. Parisi, *J. Stat. Phys.* **23**, 49 (1980).

146. P. Calabrese, E. V. Orlov, P. Parruccini, and A. I. Sokolov, *Phys. Rev. B* **67**, 024413 (2003).
147. E. V. Orlov, and A. I. Sokolov, *Phys. Sol. State* **42**, 2151 (2000).
148. P. Calabrese, M. Caselle, A. Celli, A. Pelissetto, and E. Vicari, *J. Phys. A* **33**, 8155 (2000).
149. S. Bekhechi, B. W. Southern, A. Peles, and D. Mouhanna, *Phys. Rev. E* **74**, 016109 (2006).
150. M. Zelli, K. Boese, and B. W. Southern, *Phys. Rev. B* **76**, 224407 (2007).
151. V. Thanh Ngo, and H. T. Diep, *Phys. Rev. E* **78**, 031119 (2008).
152. M. V. Kompaniets, A. Kudlisa, A. I. Sokolov, Six-loop ϵ expansion study of three-dimensional $O(n) \times O(m)$ spin models, *Nuclear Physics B* **950**, 114874 (2020); https://doi.org/10.1016/j.nuclphysb.2019.114874; arXiv:1911.01091v3 [cond-mat.stat-mech] 6 Dec 2019.

CHAPTER 4

PHASE TRANSITIONS IN FRUSTRATED VECTOR SPIN SYSTEMS: NUMERICAL STUDIES

Damien Loison

Institut für Theoretische Physik, Freie Universität Berlin,
Arnimallee 14, 14195 Berlin, Germany
Damien.Loison@physik.fu-berlin.de

4.1. Introduction

We present in this chapter a review on recent numerical studies dealing with frustrated vector spin systems in two and three dimensions. A system of spins is frustrated when all interactions between spin pairs cannot have simultaneously their optimal values. In other words, a system is frustrated when the global order is incompatible with the local one, a definition applicable in a broader sense and not restricted to spins. For spin systems, frustration has the consequence that the ordered state is different from the collinear order found for common unfrustrated antiferromagnets or ferromagnets. To study and to classify the phase transitions between the ordered and less ordered states, the symmetries of these states play an important role. The critical behavior of second-order transitions is principally governed by the change of symmetries. Therefore, even very different systems like superfluid helium and XY spin systems can have the same critical properties. This fact is called universality, and our objective is to analyze the corresponding universality classes for frustrated spin systems. We will mainly review magnetic models since they are the easiest to analyze numerically and theoretically. Nevertheless, the results should be valid for any system belonging to the same universality class.

During the last decade, important progress has been made in the understanding of the physics of frustrated spin systems. For example, there is now convincing evidence on the genuine first-order nature of the phase transition for XY and Heisenberg spins for the 3D stacked triangular antiferromagnet (STA). This still contradicts the latest renormalization group (RG) expansion based on resummation. We think that studying phase transitions of vector spins theoretically, numerically, and experimentally should have implications beyond this special field, that is for the understanding and the theory of phase transition in general.

We will concentrate our attention to the phase transition of the physical XY and Heisenberg spin systems in two and three dimensions. However, to understand these systems, we have to analyze the phase transitions of frustrated spins of N components as well, where N takes all integer values from 1 to ∞ and not only 2 and 3 for XY and Heisenberg spins. We will also present studies for dimension d varying between two and four. Then we will review the particular case of strictly two dimensions where topological defects have a dominant role.

Frustrated Ising models also are reviewed in Chapters 1 and 2.

Since most of the numerical simulations presented here use the Monte Carlo (MC) method, a short appendix (Section 4.7) at the end is devoted to this technique. In addition, since the RG is fundamental to the understanding of phase transitions, a small appendix (Section 4.7) is added to discuss the methods used here for frustrated spin systems.

4.2. Breakdown of Symmetry

We first briefly review the fundamental concept of the reduction or the breakdown of symmetry (BS) in the transition from the high- to the low-temperature phases. The classification of phase transitions in universality classes is based on this concept.

4.2.1. *Symmetry in the high-temperature region*

We will consider the Hamiltonian

$$H = -J_1 \sum_{(ij)} \mathbf{S}_i.\mathbf{S}_j, \tag{4.1}$$

where $\mathbf{S}_i$ are N component classical vectors of unit length and the sum is usually restricted to the nearest-neighbors (NNs) or at least to short-range interactions. The symmetry of this Hamiltonian is $O(N)$ or equivalently

$Z_2 \otimes SO(N)$, where $O(N)$ is the orthogonal transformation group of N-dimensional Euclidean space. For $SO(N)$, the determinant is unity, and the Ising symmetry Z_2 corresponds to the mirror operation. If we add other terms, like dipolar interactions or anisotropies, the symmetry group will be reduced. Some other changes, like a cubic term $(\mathbf{S}_i.\mathbf{S}_j),^3$ will not reduce the symmetry. In many experiments, anisotropies will reduce Heisenberg ($N = 3$) symmetry to XY ($N = 2$) or Ising ($N = 1$) symmetry. Long-range interactions could also be present, and the interpretation of experimental results could be problematic (see the cross-over Section in 4.7).

We will also encounter the Potts symmetry Z_q. The Potts model has the Hamiltonian

$$H = -J_1 \sum_{(ij)} \delta_{q_i q_j}, \tag{4.2}$$

$\delta_{q_i q_j}$ refers to the q states Potts spin with $\delta_{q_i q_j} = 0$ when $q_i \neq q_j$ and $\delta_{q_i q_j} = 1$ when $q_i = q_j$.

In addition, we have to consider the symmetry of the lattice which is the sum of all symmetry elements which let the lattice to be invariant. For example, the triangular lattice has C_{3v} symmetry and the square lattice C_4 symmetry.[1] Therefore, we have to take into account the total BS composed of the $O(N)$ spin rotations and the symmetries of the lattices.

4.2.2. *Breakdown of symmetry for ferromagnetic systems*

If J_1 in (4.1) is taken positive, the ground state (GS), for any lattice, is ferromagnetic with collinear spins pointing in the same direction. If J_1 is negative (antiferromagnetic) but there is no frustration as for the square lattice, the GS is collinear with alternating direction. Then the BS will be identical to the ferromagnetic case. The symmetry at low temperature is $O(N-1)$ for the spins and the lattice symmetries are preserved. Therefore, the BS is $O(N) \longrightarrow O(N-1)$, where $O(N)$ is the symmetry of the phase at high temperatures and $O(N-1)$ is the symmetry at low temperatures. We will denote this BS $O(N)/O(N-1)$. Keeping in mind that $O(0) \equiv 1$, $O(1) \equiv Z_2$, $SO(1) \equiv 1$, and $O(N) \equiv Z_2 \otimes SO(N)$ for $N \geq 2$, the BS $O(N)/O(N-1)$ can be written as Z_2 for $N = 1$, $SO(2)$ for $N = 2$, and $SO(N)/SO(N-1)$ for $N > 2$.

All the ferromagnetic systems (cubic, stacked triangular, etc.) with a Hamiltonian of type (4.1) have an identical BS and consequently belong to the same universality class. The only relevant variables are the dimension

Table 4.1. Critical exponents for the ferromagnetic systems in three dimensions calculated by RG.

BS	α	β	γ	ν	η
$O(1)/O(0) \equiv Z_2$	0.107	0.327	1.239	0.631	0.038
$O(2)/O(1) \equiv SO(2)$	−0.010	0.348	1.315	0.670	0.039
$O(3)/O(2) \equiv SO(3)/SO(2)$	−0.117	0.366	1.386	0.706	0.038
$O(4)/O(3) \equiv SO(4)/SO(3)$	−0.213	0.382	1.449	0.738	0.036
$O(6)/O(5) \equiv SO(6)/SO(5)$	−0.370	0.407	1.556	0.790	0.031
1*st* order[a] (Z_q if $q > 3$)	1	0	1	1/3	−1[b]

Note: [a]We cannot define exponents in a first-order transition; however, in the case of a weak first-order transition, the exponents found with MC and in experiments must tend to these values. [b]Calculated by $\gamma/\nu = 2 - \eta$.

of the spin space, i.e. the number of components of the spin N, and the dimension of the real-space d.

For Potts ferromagnetic system with a Hamiltonian of the type (4.2) and $J_1 > 0$, the Potts symmetry is broken in the low-temperatures phase and the BS is simply Z_q. Again, this does not depend on details of the system. This model has a second-order phase transition for $q \leq q_c$ and a first-order phase transition for $q > q_c$. We know that $q_c = 4$ in two dimensions and $q_c < 3$ in three dimensions. Therefore, in three dimensions, for any $q \geq 3$, the transition will be of first order.

In Table 4.1, we give the value of the critical exponents calculated by the RG[2] depending on N and q in three dimensions ($d = 3$). Even if $N > 3$ does not correspond to physical systems, we will get very useful information for frustrated systems.

4.2.3. *Breakdown of symmetry for frustrated systems*

The BS in frustrated systems is more complicated. We will present examples where the $O(N)$ symmetry is reduced to $O(N-1), O(N-2), \ldots, O(O)$ (completely broken in the last case). In addition, it is possible that the lattice symmetry is also reduced, usually giving an additional Z_q (mostly $q = 2$ and $q = 3$) broken symmetry. Therefore, the possible BS are $S_{\text{lattice}} \otimes O(N)/O(N-P)$ with $S_{\text{lattice}} = 1$, Z_2 or Z_3 and P varies from 1 to N.

4.2.3.1. *Stacked triangular antiferromagnetic lattices*

We consider here the STA lattice with NN and next-nearest-neighbor (NNN) J_2 antiferromagnetic interactions, J_1 and J_2, respectively. There is no frustration along the z-axis (the direction of stacking), and therefore, the interaction can be ferromagnetic or antiferromagnetic along z. We will explain this case in detail since it shows many phenomena appearing in frustrated systems.

- $\mathbf{J_2 = 0}$: Consider the case without second neighbor interaction ($J_2 = 0$). It is not possible for all three spins at the corners of a triangle to have the optimal antiparallel orientation which would minimize the energy of individual pair interactions. The resulting compromise in the case of vector spins is the so-called 120° structure, as shown in Fig. 4.1 (see Chapter 1). In frustrated systems, local minimization of the energy is not compatible with the global energy minimum (or minima). A formal definition proposed by Toulouse[3] (and also Villain[4]) in the study of spin glasses states that a geometry is frustrated if the sign of the product of exchange interactions J_i around a plaquette C

$$\Phi_C = \text{sign}\left[\prod_{i \epsilon C} J_i\right] \tag{4.3}$$

is negative (where $J_i < 0$ implies antiferromagnetic interactions). An antiferromagnetic triangular plaquette is thus frustrated as it involves a product of three $J_i < 0$. The triangular lattice is *fully frustrated* since all plaquettes satisfy this rule. The principal effect of the frustration here is that it gives rise to a non-collinear magnetic order. This spin-order GS (I) is planar and stable as long as the NNN interaction is small, that is $0 \leq J_2/J_1 \leq 0.125$.

The BS between the disordered phase at high temperatures and the phase at low temperatures is $O(N)/O(N-2)$, which is equivalent to

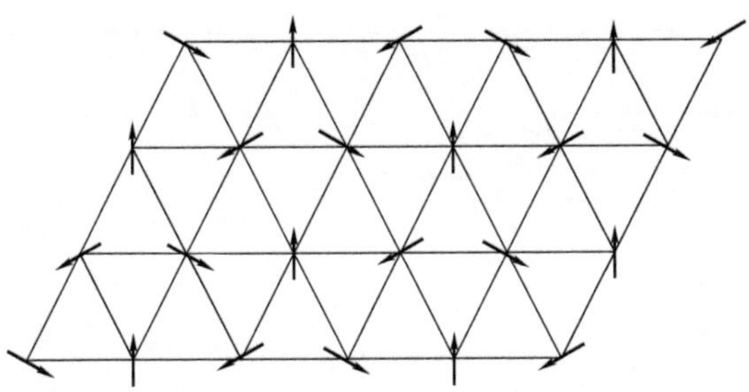

Fig. 4.1. GS I of the antiferromagnetic triangular lattice (first neighbors only). The GS has a 120° spin structure.

$Z_2 \otimes SO(2)$ for $N = 2$ and $SO(N)/SO(N-2)$ for $N > 2$. We note that the exact value of the angle between the spin directions plays no role for the phase transition.

This system has been extensively studied in three dimensions for XY spins[5–9] for Heisenberg spin,[5, 8, 10–12] and for the unphysical systems $N = 6$ in Ref. [13] and $N = 8$ in Ref. [8]. The 2D systems have also been the subject of several studies for XY spins[106–110] and also for Heisenberg spins.[145]

- $\mathbf{0.125 \leq J_2/J_1 \leq 1}$**:** Consider the presence of an antiferromagnetic second-neighbor interactions (J_2) in the xy plane. The GS can be determined by minimizing the energy after a Fourier transform.[14] For $0 \leq J_2/J_1 \leq 0.125$, the GS is still the 120° structure. For $0.125 \leq J_2/J_1 \leq 1$, the GS is degenerate with θ taking any value between 0 and 2π (see Fig. 4.2). However, the degeneracy will be lifted by thermal fluctuations (spin waves) and only a collinear GS (II) will be chosen.[15] This phenomenon, called "order-by-disorder" following Villain,[16] is general in frustrated systems and we will see other examples (fcc, hcp) later. There are three ways to choose the parallel spins and the BS of the lattice symmetry C_{3v} is a three-state Potts symmetry Z_3. The total BS is $Z_3 \otimes O(N)/O(N-1)$. For XY spins ($N = 2$), it is equivalent to $Z_3 \otimes SO(2)$ and to $Z_3 \otimes SO(N)/SO(N-1)$ for $N > 2$.

  Numerical studies have been done for the 3D case by Loison *et al.*[7, 11]
- $\boldsymbol{J_2/J_1 > 1}$**:** Now consider that $J_2/J_1 > 1$. The GS is also degenerate, but this degeneracy is lifted by thermal fluctuations. The GS (III) is no more collinear, but still planar (see Fig. 4.3) and there are still three ways to choose the parallel spins. α is determined by $\cos(\alpha) = -0.5 \cdot (1 + J_1/J_2)$ and can be incommensurable, i.e. does not correspond to a rational value.

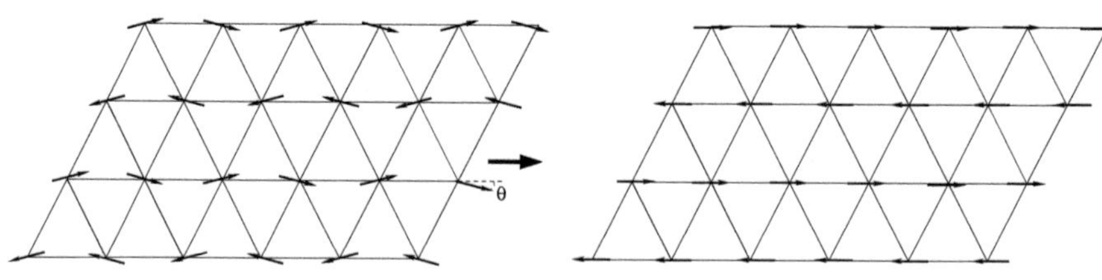

(a) (b)

Fig. 4.2. GS (II) of the antiferromagnetic triangular lattice with first- (J_1) and second (J_2)-neighbor antiferromagnetic interactions. $0.125 \leq J_2/J_1 \leq 1$. (a) θ can take any value between 0 and 2π. (b) At $T > 0$ only the "most collinear" GS is chosen.

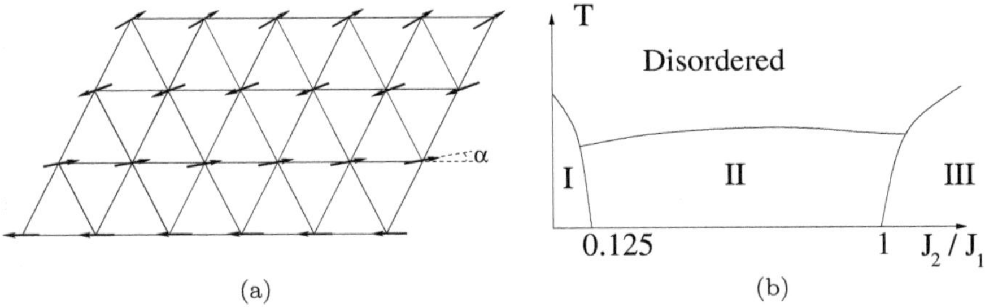

Fig. 4.3. (a) GS III of the antiferromagnetic triangular lattice with first (J_1) and second (J_2)-neighbor antiferromagnetic interactions. $1 < J_2/J_1$. (b) Phase diagram in the $(J_2/J_1, T)$ space.

Furthermore, α varies slowly as a function of the temperature. The BS is $Z_3 \otimes O(N)/O(N-2)$, which is equivalent to $Z_3 \otimes Z_2 \otimes SO(2)$ for $N = 2$ (XY spins), to $Z_3 \otimes SO(3)$ for $N = 3$ (Heisenberg spins), and to $Z_3 \otimes SO(N)/SO(N-2)$ for $N \geq 4$. Numerically, the incommensurable angle is problematic. Indeed, we have to impose periodic boundary conditions and since α varies with the temperature, the size chosen compatible with $\alpha(T = 0)$ is no more compatible with $\alpha(T > 0) < \alpha(T = 0)$ for higher temperatures. Therefore, the use of Finite Size Scaling technique (FSS, see Section 4.7) to calculate the critical exponents will be problematic.[17] In addition, the GS III may not be stable under the new boundary constraint. Then the Potts symmetry could not be broken and the BS could be just $O(N)/O(N-2)$ and does not belong to the same universality class as seen previously. We will have an akin problem for helimagnets (see later) and for triangular lattices with two distinct NN interactions.[17,18]

This model has been studied in three dimensions by Loison *et al.*[7,11]

- **Other BS:** We have seen the three BS which appear between the disordered phase and the GS. In addition, two other transitions appear for a small range of J_2/J_1 near 0.125 and 1: between the GS I and II for $0.120 < J_2/J_1 \leq 0.125$, and between the GS II and III for $1 \leq J_2/J_1 < 1.05$.[7,11] This is a general scheme: Due to thermal fluctuation, the most collinear state is favored when the temperature increases. Looking at the symmetries, we can get the order and even the universality class of these transitions.

The GS III and II are compatible in the sense that at the transition α will go smoothly to zero and the transition could be of second order.

The BS between III and II is $O(N-1)/O(N-2)$ and consequently the transition should be $N-1$ ferromagnetic type.

The GS I and II are incompatible because there is no way to go smoothly from one to the other and the transition should be of first order.

4.2.3.2. *BCT Helimagnets*

We consider the body-centered-tetragonal (bct) helimagnets. The competition between the J_1 and J_2 interactions gives rise to a helical ordering along the z axis (see Fig. 4.4). This GS is characterized by the turn angle α between spins belonging to two adjacent planes perpendicular to the z axis. α is given by the formula $\cos(\alpha) = -J_2/J_1$ and α decreases slowly as a function of the temperature. There is no breakdown of the symmetry of the lattice, the GS is non-collinear but planar and the BS is $O(N)/O(N-2)$, i.e. identical as the STA with first NN interaction only.

Numerically, we have a similar problem as for the STA with large NNN interaction: $\alpha(T)$ varies as a function of the temperature, but with periodic boundary conditions a constraint is present.

Nevertheless, the effect of this constraint should not be too strong. Indeed, a small constraint will not break a symmetry of the lattice, and for the XY, it will not change the BS of the spin rotation because all the

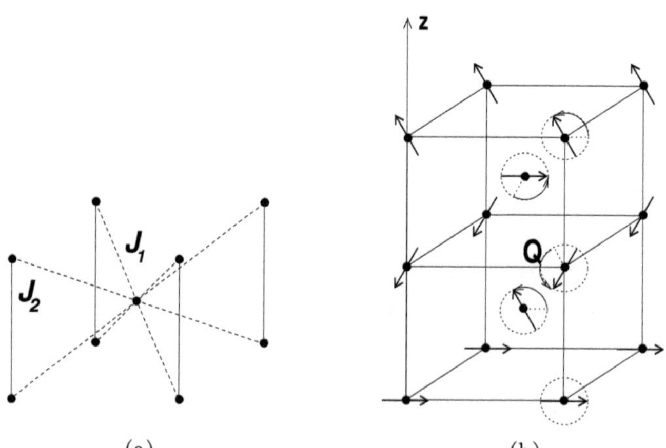

Fig. 4.4. (a) Interactions for the bct. (b) GS for the bct for $J_2/J_1 = 0.5$.

symmetries of the rotation group are already broken. For Heisenberg spins, the GS could become non coplanar, but this is unlikely. Therefore, we should find the same universality as that of the STA. The only difference will be a new correction to the scaling laws for a second-order phase transition. But for a first-order one, the boundary conditions would not matter.

This model has been studied by Diep and Loison.[19,20] A quantum version has also been considered by Quartu and Diep.[21]

4.2.3.3. *Stacked J_1–J_2 square lattices*

The J_1–J_2 simple cubic lattice is made by stacking along the z-axis the square lattices with antiferromagnetic (or ferromagnetic) first NN interaction (J_1) and antiferromagnetic second NN interaction (J_2) in the xy plane. For $J_2/J_1 > 0.5$, the GS is degenerate, but by "order-by-disorder", only collinear configurations appear (see Fig. 4.5). There are two ways to place the parallel spins (following the x or y axis) and a Z_2 Ising symmetry of the lattice symmetry is broken. The BS will be $Z_2 * O(N)/O(N-1)$ or equivalently $Z_2 \otimes SO(2)$ for $N = 2$ and $Z_2 \otimes SO(N)/SO(N-1)$ for $N > 2$. For XY spins, there exists an identical BS as for the STA with first NN interaction only. Therefore, the two systems for $N = 2$ should belong to the same universality class.

No numerical studies have been done on this model for the 3D case. The 2D case has been studied by Loison and Simon.[22]

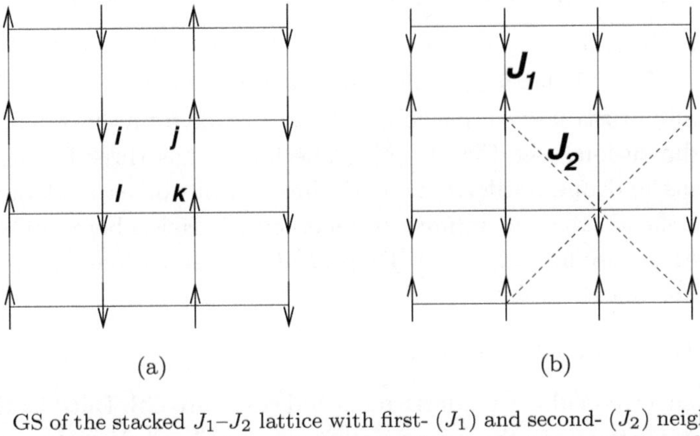

Fig. 4.5. GS of the stacked J_1–J_2 lattice with first- (J_1) and second- (J_2) neighbor antiferromagnetic interactions. $J_2/J_1 > 0.5$. (a) and (b) show the two possible configurations at non-zero temperatures.

4.2.3.4. *The simple cubic J_1–J_2 lattice*

The simple cubic J_1–J_2 lattice is similar to the stacked J_1–J_2 square lattices shown above, but the second-neighbor interactions are also present in the xz and yz planes. Similarly, by "order-by-disorder", the spin configuration is collinear, but now there are three ways to choose the direction of parallel spins (the z axis in addition to the x and y axes). Therefore, the BS is $Z_3 \otimes O(N)/O(N-1) \equiv Z_3 \otimes SO(N)/SO(N-1)$. It is equivalent to the BS of the STA with intermediate NNN interaction.

This model has been studied for Heisenberg spin by Alonso *et al.*[25] and Pinettes and Diep.[26]

4.2.3.5. *J_1–J_2–J_3 lattice*

The addition of a third-neighbor antiferromagnetic interaction to the previous model leads to new GS and therefore to a new BS. For some value of the interactions, a non-planar GS state appears.[27] In addition, a Z_2 symmetry is broken from the lattice group and the BS is hence $Z_2 \otimes O(N)/O(N-2)$ between the ordered phase and the disordered phase. For XY spins, the BS is equivalently $Z_2 \otimes Z_2 \otimes SO(2)$, and for Heisenberg spins $Z_2 \otimes SO(3)$. There exist also various transitions between various phases, but following the discussion at the end of Section 4.2.3.1 concerning STA with NNN interaction, it is not difficult to find that the transition will be of first order or of the ferromagnetic type $SO(N-1)/SO(N-2)$.

No numerical studies have been done so far on this model.

4.2.3.6. *Villain lattice and fully frustrated simple cubic lattice*

The fully frustrated square lattice, called Villain lattice, is shown in Fig. 4.6(a). This 2D lattice has been extensively studied.[4, 106, 111–118, 122]

The fully frustrated simple cubic lattice is made from the Villain lattice in three dimensions (see Fig. 4.6(b)). The lattice has three ferromagnetic interactions and one antiferromagnetic interaction for each plaquette. In this case, the GS has an infinite degeneracy for Heisenberg spins and is 12-fold degenerate for XY spins. Diep *et al.*[23, 24] and Alonso *et al.*[25] have studied this model.

4.2.3.7. *Face-centered cubic lattice*

The face-centered cubic (fcc) lattice has a degenerate GS. But like the STA with NNN interaction, the "most collinear phase" is selected by thermal fluctuations. If we consider the smaller unit tetrahedral cell (see Fig. 4.7), the GS consists of two parallel and two antiparallel spin pairs. There are

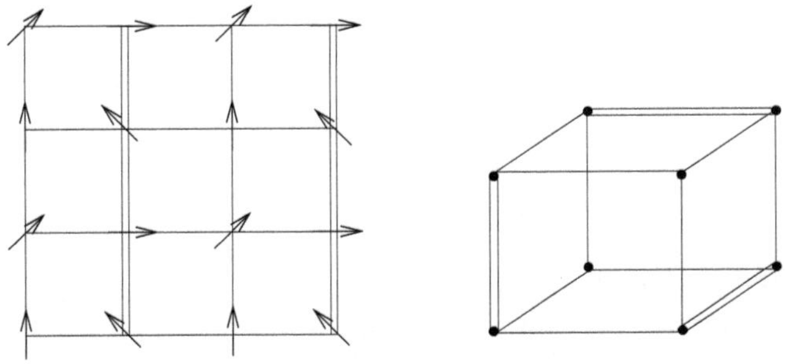

(a) (b)

Fig. 4.6. (a) GS of the stacked Villain lattice. The double lines are antiferromagnetic interactions. (b) The fully frustrated cubic lattice.

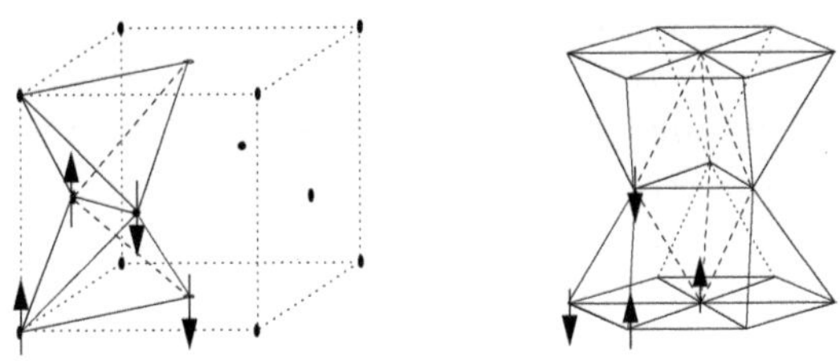

(a) (b)

Fig. 4.7. (a) GS of the fcc lattice. (b) GS of the hcp lattice.

three ways to construct. This is equivalent to a Potts Z_3 symmetry. Therefore, since the GS is collinear, the BS will be $Z_3 \otimes O(N)/O(N-1)$ or, equivalently, $Z_3 \otimes SO(N)/SO(N-1)$. This BS is identical to the STA with intermediate NNN interaction.

This model has been studied by Diep and Kawamura[28] and Alonso *et al.*[25]

4.2.3.8. *Hexagonal-close-packed lattice (hcp)*

The hexagonal-close-packed (hcp) lattice has many common features with the fcc. It is constructed by stacking tetrahedra (see Fig. 4.7), and an "order-by-disorder" mechanism lifts one part of the degeneracy of the GS resulting in a collinear GS composed with two parallel and two antiparallel spins. Equivalent to the fcc case, the BS will be $Z_3 \otimes O(N)/O(N-1)$ or equivalently $Z_3 \otimes SO(N)/SO(N-1)$. This BS is identical to the STA with intermediate NNN interaction.

This model has been studied by Diep.[29]

4.2.3.9. *Pyrochlores*

The pyrochlore lattice can be made by stacking Kagomé lattices along (111) direction and is composed of an arrangement of corner-sharing tetrahedra (see Chapters 7 and 8). In the presence of the first and third NN interactions Reimers[30] has proved that the GS is collinear. Similar to the fcc and hcp cases, there are three ways to place the spins on each tetrahedron. The BS should therefore be $Z_3 \otimes O(N)/O(N-1)$ or equivalently $Z_3 \otimes SO(N)/SO(N-1)$. This BS is identical to the STA with intermediate NNN interaction.

Reimers *et al.*[30] have studied this model.

4.2.3.10. *Other lattices*

We could stack a zig-zag model[31] and various phases and phase transitions should appear. With an equivalent analysis described at the end of Section 4.2.3.1 concerning STA with NNN interaction, it is not difficult to find the nature of the transitions. Still, other lattices can be considered; however, as will be shown later, they will have equivalent BS.

4.2.3.11. *STAR lattices*

In addition, one can construct spin systems having an identical breakdown of symmetry even if they do not correspond to a real system.

The first example is derived from the STA. Following Chapter 3, certain modes are irrelevant near the critical point. In particular, we can construct cells of three spins which are always in the GS 120° configuration. One gets a system of cells in interaction but with a rigidity imposed inside the cells (see Fig. 4.8). We call this system stacked triangular antiferromagnetic with rigidity (STAR). The BS is then $O(N)/O(N-2)$, identical to STA with NN interaction.

This model has been studied in three dimensions by Dobry and Diep[32] and by Loison and Schotte.[33–35]

4.2.3.12. *Dihedral lattices $V_{N,2}$*

The second example is derived from the STAR. It is composed of two vector spins $\mathbf{e}_1$ and $\mathbf{e}_2$ constrained to be orthogonal to each other at each lattice site. The interactions are set to be ferromagnetic and the spin $\mathbf{e}_1$ ($\mathbf{e}_2$) interacts only with the other spins $\mathbf{e}_1$ ($\mathbf{e}_2$) at other sites (see Fig. 4.8). This model is referred to as the dihedral model $V_{N,2}$. We note that for XY spins $N = 2$, the model can be right-handed or left-handed.

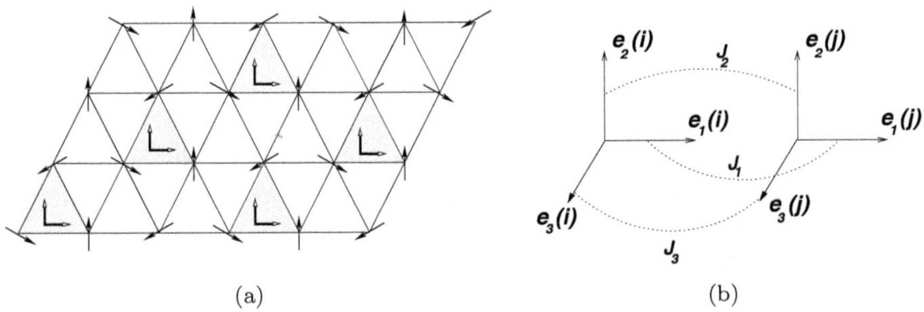

Fig. 4.8. (a) STAR model and dihedral model. (b) Trihedral model and $V_{3,P}$ model. $P = 1$ for $J_2 = J_3 = 0$, $P = 2$ for $J_3 = 0$ and $J_1 = J_2 = 1$, and $P = 3$ for $J_1 = J_2 = J_3 = 1$.

The Hamiltonian is defined by

$$H = J \sum_{\langle ij \rangle} \sum_{k=1}^{P} \langle \mathbf{e}_k(i) \cdot \mathbf{e}_k(j) \rangle, \tag{4.4}$$

where $P = 2$.

At high temperatures, the symmetry is $O(N)$ for the first vector and $O(N-1)$ for the second one, since the two vectors must be orthogonal. At low temperatures, the symmetry is $O(N-1)$ for the first vector and $O(N-2)$ for the second one for the same reason. Therefore, the BS is $O(N)/O(N-2)$, identical to the STA with NN interaction.

This model has been studied in three dimensions by Kunz and Zumbach[36] and by Loison and Schotte,[33,34] and in two dimensions for XY spins by Nightingale *et al.*[120,121]

4.2.3.13. *Right-handed trihedral lattices* $V_{3,3}$

For Heisenberg spins, one can construct another model from the dihedral $V_{3,2}$ with an identical BS. By adding a third vector $\mathbf{e}_3 = \mathbf{e}_1 \times \mathbf{e}_2$ to the dihedral model, no degree of freedom is added. Therefore, the BS is unchanged and identical to the dihedral $V_{3,2}$ and to the STA with NN interaction, $O(3)/O(1) \equiv SO(3)$.

This model has been studied in three dimensions by Loison and Diep.[37]

4.2.3.14. *P-hedral lattices* $V_{N,P}$

We can generalize the dihedral $V_{N,2}$ model to $V_{N,P}$ with P orthogonal vectors at each site. For $P = N$, the N vectors can be right- or left-handed. The BS of symmetry is then $O(N)/O(N-P)$ with $P \leq N$.

For $P = 1$, we have the ferromagnetic case, i.e. a collinear GS. For $P = 2$, we have a coplanar GS but no more collinear like in the STA with NN interaction. For $P = 3$, the GS is no more coplanar but restricted to a space in three dimensions, and so on. The case $N = P = 3$ could correspond to some experimental systems, and spin glasses should also have this kind of breakdown of symmetry but in the presence of disorder. For $P = N$, the BS is $O(N)/O(0) \equiv Z_2 \otimes SO(N)$.

This model has been studied by Loison.[38]

4.2.3.15. *Ising and Potts-$V_{N,1}$ model*

We define the following Hamiltonian:

$$H = -J_1 \sum_{(ij)} \mathbf{S}_i.\mathbf{S}_j.\delta_{q_i q_j}, \tag{4.5}$$

where $\mathbf{S}_i$ are N-component classical vectors of unit length, $\delta_{q_i q_j}$ are the q-state Potts spin ($q = 2$ for Ising spin) with $\delta_{q_i q_j} = 0$ when $q_i \neq q_j$, and the interaction constant J_1 is taken positive (ferromagnetic interaction). The sum runs over all NNs. For $N = 2$, this model is exactly equivalent to the dihedral $V_{2,2}$ model introduced previously.

Obviously, the BS is $Z_q \otimes O(N)/O(N-1)$ or equivalently $Z_q \otimes SO(N)/SO(N-1)$, which is identical to the stacked J_1–J_2 model if $q = 2$ and to the STA with NNN interaction, fcc, hcp if $q = 3$.

This model has been studied in three dimensions by Loison.[39] and in two dimensions for XY spins by Nightingale *et al.*[120,121]

4.2.3.16. *Ising and Potts-$V_{N,2}$ model*

We can also define a dihedral model with the Hamiltonian (4.4) coupled with a Potts model in the same way as in (4.5). In this case, the BS is $Z_q \otimes O(N)/O(N-2)$, which is identical to the BS for the stacked J_1–J_2–J_3 model if $q = 2$ and to the STA with a large NNN for $q = 3$.

This model has been studied by Loison.[39]

4.2.3.17. *Landau–Ginzburg model*

The Landau–Ginzburg Hamiltonian is constructed from the dihedral $V_{N,2}$ model (see Chapter 3). We release the constraints on the orthogonality of the spins and on the norm unity, and replacing them by a potential, we arrive at the following Hamiltonian:

$$H = K\sum_{\langle ij\rangle}(\vec{\phi}(i)-\vec{\phi}(j))^2 + r\sum_i \vec{\phi}(i)^2 + u\sum_i \left(\vec{\phi}(i)^2\right)^2 + v\sum_i[(\vec{\phi}_a(i)\cdot\vec{\phi}_b(i))^2 - \vec{\phi}_a(i)^2\vec{\phi}_b(i)^2], \tag{4.6}$$

where $\vec{\phi}(i) = \left(\vec{\phi}_a(i), \vec{\phi}_b(i)\right)$ is an $N+N$–component vector defined on the lattice site i, and the summation $\sum_{\langle ij\rangle}$ runs over all NN pairs of the lattice. The first term represents the interactions between the sites and the last term the constraint that the spins $\phi_a(i)$ and $\phi_b(i)$ are orthogonal. If $v=0$, the Hamiltonian is reduced to a standard $O(2N)$ ferromagnetic problem. This is our third model that we call the chiral ϕ^4 model. For $v>0$, the BS is $O(N)/O(N-2)$, identical to STA with NN interaction.

This model has been studied numerically by Itakura.[8]

4.2.3.18. *Cubic term in Hamiltonian*

One can introduce a model where the frustration is not geometrical but included on each link resulting in a non-collinear GS. Consider the Hamiltonian for two spins

$$H = J_1\mathbf{S}_i.\mathbf{S}_j + J_3(\mathbf{S}_i.\mathbf{S}_j)^3, \tag{4.7}$$

where $\mathbf{S}_i$ is an N component classical vector of unit length, J_1 is a ferromagnetic coupling constant with $J_1<0$, and J_3 is an antiferromagnetic $J_3>0$. This cubic term has the same symmetries as the linear term, and for a ferromagnetic system ($J_3<0$), the universality class will be identical.[40] For $1/3 \le -J_3/J_1 \le 4/3$, the minimum of H occurs when the two spins are canted with an angle $\cos(\alpha) = \frac{1}{\sqrt{3J_3/J_1}}$. On a cubic lattice, the GS is planar; on a triangular lattice, it is in three dimensions (if $N\ge 3$); on an fcc lattice, it is in four dimensions (if $N\ge 4$); etc. This phase has, at finite temperature, a transition to a ferromagnetic collinear phase. Therefore, the BS will be $O(N-1)/O(N-P)$ with $P=2$ for the cubic lattice (planar GS), $P=3$ for the stacked triangular, $P=4$ for the fcc lattice, etc. The case $P=2$ has been tested for Heisenberg spins, and it indeed gives an $O(2)/O(1)\equiv SO(2)$ transition, i.e. a ferromagnetic XY transition.[40] The case $P=3$ has an identical BS as the STA, but it has not been tested.

This model has been studied numerically by Loison.[40]

4.2.3.19. *Summary*

The number of models that can be constructed by adding interactions is unlimited. However, the number of BS can only be finite. As has been shown in the previous sections, the symmetry breaking is limited to $Z_q \otimes O(N)/O(N-P)$. For the physically relevant cases, it is restricted to $q=1$ (Identity), $q=2$ (Ising), or $q=3$ (Potts), $N=2$ (XY spins), $N=3$ (Heisenberg spins), and $P=1$, 2, or 3. In total for Heisenberg spins, eight cases will exist (two are identical: $(q=1, P=3) \equiv (q=2, P=2)$) and for XY spins 5 cases will exist.

Tables 4.2 and 4.3 summarize the probable BS for physically frustrated systems with XY or Heisenberg spins.

Table 4.2. Most probable BS for frustrated XY spin systems with the corresponding lattices or models.

BS	Lattice model
$SO(2)$	Ferromagnetic, small frustration
$Z_2 \otimes SO(2)$	STA, STAR, $V_{2,2} \equiv$ Ising-$V_{2,1}$, bct, Stacked J_1–J_2, Stacked Villain, Fully Frustrated cubic, Stacked Zig-Zag, chiral ϕ^4 model, $(S_i.S_j)^3$ term
$Z_3 \otimes SO(2)$	STA+NNN, cubic J_1–J_2, fcc, hcp, pyrochlore, Potts-$V_{2,1}$
$Z_2 \otimes Z_2 \otimes SO(2)$	Stacked J_1–J_2–J_3, Ising-$V_{2,2}$
$Z_3 \otimes Z_2 \otimes SO(2)$	STA+NNN, Potts-$V_{2,2}$

Table 4.3. Most probable BS for frustrated Heisenberg spin systems with the corresponding lattices or models.

BS	Lattice model
$SO(3)/SO(2)$	Ferromagnetic, small frustration
$SO(3)$	STA, STAR, $V_{3,2}$, right-handed trihedral, bct, Stacked Villain, Stacked Zig–Zag, Fully frustrated cubic(?), chiral ϕ^4 model, $(S_i.S_j)^3$ term
$Z_2 \otimes SO(3)/SO(2)$	Stacked J_1–J_2, Ising-$V_{3,1}$
$Z_2 \otimes SO(3)$	Stacked J_1–J_2–J_3, $V_{3,3} \equiv$ Ising-$V_{3,2}$, $(S_i.S_j)^3$ term, Fully frustrated cubic(?)
$Z_2 \otimes Z_2 \otimes SO(3)$	Ising-$V_{3,3}$
$Z_3 \otimes SO(3)/SO(2)$	STA+NNN, cubic J_1–J_2, fcc, hcp, pyrochlore, Potts-$V_{3,1}$
$Z_3 \otimes SO(3)$	STA+NNN, Potts-$V_{3,2}$
$Z_3 \otimes Z_2 \otimes SO(3)$	Potts-$V_{3,3}$

4.3. Phase Transitions between Two and Four Dimensions: $2 < d \leq 4$

In this section, we will concentrate on the nature of the various transitions mentioned in the previous section. Especially the transition for the $O(N)/O(N-2)$ BS is considered in detail since it appears in numerous systems and was extensively debated. Then we will discuss the other BS which are less problematic.

4.3.1. $O(N)/O(N-2)$ *breakdown of symmetry*

4.3.1.1. *Fixed points*

Since many models (STA, STAR, dihedral, chiral ϕ^4 model, etc., see Tables 4.2 and 4.3) have an identical BS, they should have equivalent critical behavior, i.e. they belong to the same universality class. However, the situation is more complicated and even with an identical BS two systems could show different behavior. To understand this fact, we have plotted in Fig. 4.9 the fixed points in the critical plan for this model. Since there are two fields (Eq. (4.6)) to allow coplanar non-collinear GS, we will have four possible fixed points. There are two more than the ferromagnetic case which has a collinear GS and consequently only one field.

The fixed points are as follows:

1. The Gaussian fixed G point at $u^* = v^* = 0$ with mean-field critical exponents.
2. The $O(2N)$ fixed point H at $v^* = 0$ and $u^* = u_H \neq 0$ with $O(2N)$ exponents (see Table 4.1).

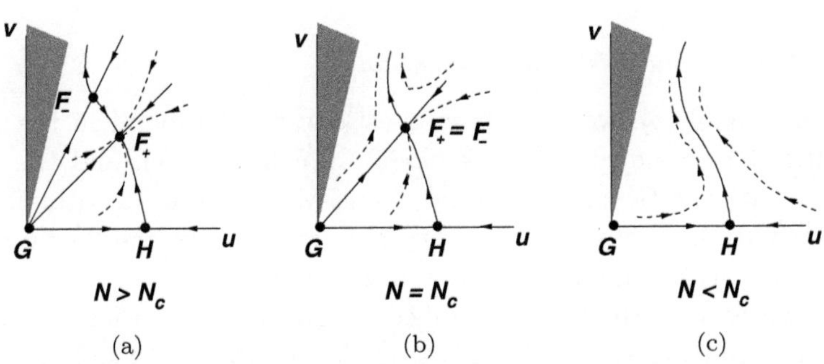

Fig. 4.9. (a)–(c) Hamiltonian flow induced by RG transformations. The arrows indicate the direction of flow under iterations.

3. Two fixed points F_+ and F_- at location (u_{F_+}, v_{F_+}) and (u_{F_-}, v_{F_-}) different from zero. These are the fixed points associated with a new universality class.

The existence and stability of the fixed points depend on the number of components N:

(a) $\boldsymbol{N > N_c}$: Four fixed points are present, but three are unstable (G, H, F_-) and one is stable (F_+). Therefore, the transition belongs to a new universality class different from the standard $SO(N)/SO(N-1)$ class. If the initial point for the RG flow is to the left of the line (G, F_-), see Fig. 4.9(a), the flow is unstable and the transition will be of first order. Therefore, two systems with the same Hamiltonian and the same breakdown of symmetry could have different critical behaviors.
(b) $\boldsymbol{N = N_c}$: The fixed points F_- and F_+ coalesce to a marginally stable fixed point. One would think that the transition is "tricritical" but the exponents are different and not given by the tricritical mean-field values contrary to common belief. The reason is that there are two non-zero quartic coupling constants, see Fig. 4.9(b), in contrast to the "standard" tricritical point where the quartic term disappears and a sextic term takes over.
(c) $\boldsymbol{N < N_c}$: F_- and F_+ move into the complex parameter space, see Fig. 4.9(c). The absence of stable fixed points is interpreted as a signature of a first-order transition.

There are at least three questions: the value of N_c, the location of the initial point in the RG flow, and the nature of the transition.

4.3.1.2. *MCRG and first-order transition*

The most reliable answers to these questions can be found in the paper by Itakura.[8] He studied the STA model, dihedral model, and the chiral ϕ^4 model with the Hamiltonian (4.6), using Monte Carlo Renormalization Group (MCRG). This method is very powerful to give the flow diagram but not the critical exponents with a great precision.

He showed that the STA and dihedral models have an initial point in the RG flow under the line GF_- in Fig. 4.9. Therefore, they should belong to the same universality class provided the fixed point F_+ exists, i.e. $N > N_c$. He found that N_c is between 3 and 8 in three dimensions which means that the real physical systems XY ($N = 2$) and Heisenberg spins ($N = 3$) have a first-order transition.

In addition, he did a standard canonical MC simulation for XY spins on the STA lattice for very large sizes of 96^3 and 126^3. He found a first-order transition in agreement with MCRG study, contrary to smaller size systems which seem to have a second-order transition.[5–7] For Heisenberg spins, he could not find a first-order transition, but using the results of the MCRG, he concludes that the first-order transition could only be seen for a size larger than 800^3 which is not accessible for actual computer resources. For the dihedral model with a canonical, MC, he found a clear first-order transition for large sizes for Heisenberg spins. For XY spins, Loison and Schotte have already shown that the transition is of first order. Diep[19] in 1989 was the first to find a first-order transition in helimagnetics bct lattice with XY spins. This result was considered not conclusive because of problems of periodic boundary conditions in numerical simulation. However, as noted above, this is not relevant for the XY spins (contrary to Heisenberg spins) and hence this conclusion is indeed correct for this BS.

We have now to address the problem why phase transitions appear continuous for small sizes but show the true first-order nature only for larger sizes. This phenomenon is not restricted to frustrated spin systems such as STA but appears also in the weakly first-order transition of the 2D Potts model with $q = 5$ components.[41–43] More generally, it appears when two fixed points collapse and disappears following one variable (like F_- and F_+ in Fig. 4.9). The Potts model is discussed latex in the chapter.

4.3.1.3. *Complex fixed point or minimum in the flow*

The change from continuous to discontinuous transition can be understood using the concept of "complex fixed point" or "minimum in the flow" first introduced by Zumbach[44] and then developed by Loison and Schotte.[33,34] In Fig. 4.9(c), for $N < N_c$ the fixed points F_- and F_+ have collapsed and no solution exists in real parameter space. Instead, there exists an imaginary solution plotted in Fig. 4.10(a). These solutions should have an influence on the flow in the real plane as shown in Fig. 4.10(b). Zumbach showed that there exists a basin of attraction due to the complex fixed points where the RG flow is very slow.

If the RG flow of a system goes through this region, the number of RG iterations $L \to L/b$ to get outside will be large. Therefore, the size L of the numerical system considered should be made large enough (see Section 4.7).

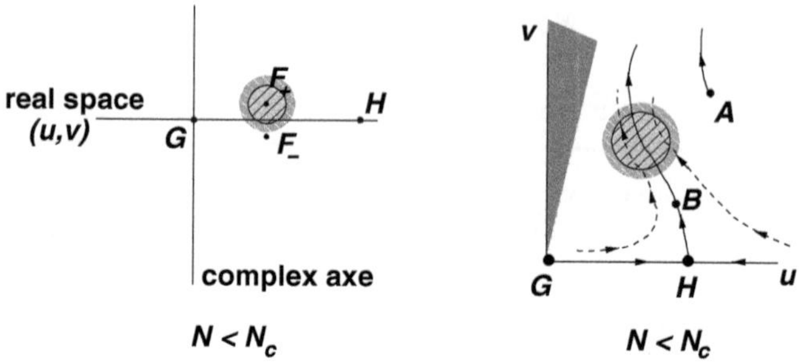

(a) (b)

Fig. 4.10. (a) For $N < N_c$, the fixed points F_+ and F_- become complex. (b) Hypothesis on the Hamiltonian flow induced by RG transformations for $N < N_c$ (see Fig. 4.9(c)). The arrows give the direction of flow. The two circles correspond to a low velocity region. The inner one corresponds to a minimum and hence an "almost" second-order transition. The outer circle corresponds to a slow cross-over where the critical exponents will vary as a function of the system size.

If the size is "too small", the flow is trapped inside of this low-velocity region and the transition seems to be continuous. We immediately see that the size L for which the true first-order transition is visible, that is when the flow is outside of the basin of attraction, will depend on the starting point. If it is located outside of the domain of low velocity and if the flow does not approach this region (point A in Fig. 4.10), then the first-order nature of transition can be seen already for small sizes. In the other case, if the initial position is such that the flow has to go through the entire region of low velocity (point B in Fig. 4.10), then the necessary size to reach the true first-order region will be very large. This explains why STA, STAR, dihedral, and right-handed trihedral (for $N = 3$) models having an identical breakdown of symmetry behave differently. Similarly, a single system can show different behaviors for different sizes. For the Heisenberg case, for example, the right-handed trihedral model shows a strong first-order behavior even for small sizes. The dihedral model with the same BS for $N = 3$ has a weaker first-order behavior visible only for sizes bigger than 80^3 spins while the STA or bct helimagnets show a second-order transition for similar sizes.

Furthermore, the size of this region of low velocity will vary as a function of the distance between the complex fixed point F_+ and the real plane shown in Fig. 4.10(a). It has been shown that N_c, where the two fixed points F_- and F_+ collapse and become complex, is bigger than 3 and the distance of F_+ to the real plane will be larger for $N = 2$ than for $N = 3$. Therefore, the size of the domain of low velocity will be bigger for $N = 3$ than for $N = 2$, and the true first-order behavior will be seen for smaller sizes in the XY case ($N = 2$) than in the Heisenberg case ($N = 3$). Indeed, the first-order transition in the dihedral, STAR, and STA can be seen from the size 12, 18, and 96, respectively, for XY spin systems (see Table 4.4). For Heisenberg spins, this has been seen for the dihedral model with a size $L = 80$ (see Table 4.6) while the estimate size for the STA should be larger than 800.[8] We note that a first-order transition has been found in quasi-1D STA[9]: the initial point in the flow diagram is outside the domain of low velocity.

For the XY case, Delamotte *et al.*[47] using a non-perturbative RG approach found that there is no minimum in the flow but there is always a low-velocity region (outer circle of Fig. 4.10(b)). Following the initial point in the RG flow, the systems will show different sets of exponents, i.e. the system is in a cross-over region. Numerical simulations tend to support this interpretation. Indeed, for XY spins, the STAR and the STA for small sizes ($L < 40$) display a second-order transition with different critical exponents (see Tables 4.4 and 4.5), and for large sizes, the STA shows a first-order transition as discussed above.[8]

In conclusion, the transition for Heisenberg spins is of first order but an "almost second-order transition" could exist for a range of "finite" sizes. This "almost second-order transition" will have a set of exponents different from the ferromagnetic ones because the breakdown of symmetry is

Table 4.4. Critical exponents associated to the $SO(2)$ symmetry by MC for XY spins ($N = 2$) and a BS $Z_2 \otimes SO(2)$.

System	Ref.	$L_{\max}$	α	β	γ	ν	η
STA	[5]	60	0.34(6)	0.253(10)	1.13(5)	0.54(2)	−0.09(8)[a]
STA	[6]	33	0.46(10)	0.24(2)	1.03(4)	0.50(1)	−0.06(4)[a]
STA	[7]	36	0.43(10)			0.48(2)	
STA	[8]	126			First order		
STA	[9]	35			First order		
bct	[19]	24			First order		
STAR	[33]	36			First order		
$V_{2,2}$	[33]	36			First order		

Notes: [a]Calculated by $\gamma/\nu = 2 - \eta$. The first result[5] comes from a study at high and low temperatures and uses the FSS. The second[6] uses the Binder parameter to find T_c and uses the FSS, and the third[7] uses the maxima in FSS region.

Table 4.5. Critical exponents associated to the Z_2 symmetry (chirality κ) by Monte Carlo for XY spins ($N = 2$) and a BS $Z_2 \otimes SO(2)$.

System	Ref.	$L_{\max}$	α	β_κ	γ_κ	ν_κ	η_κ
STA	[5]	60	0.34(6)	0.55(4)	0.72(8)	0.60(3)	0.80(19)^a
STA	[6]	33	0.46(10)	0.38(1)	0.90(2)	0.55(1)	0.28(3)^a
STA	[8]	126			First order		
STA	[9]	35			First order		
bct	[19]	24			First order		
STAR	[33]	36			First order		
$V_{2,2}$	[33]	36			First order		

Notes: ^{a}Calculated by $\gamma/\nu = 2 - \eta$. The first result[5] comes from a study at high and low temperature and uses FSS. The second[6] uses the Binder parameter to find T_c and uses the FSS.

Table 4.6. Critical exponents by Monte Carlo for Heisenberg spins ($N = 3$) and a BS $SO(3)$.

System	Ref.	$L_{\max}$	α	β	γ	ν	η
STA	[5]	60	0.240(80)	0.300(20)	1.170(70)	0.590(20)	+0.020(180)^a
STA	[10]	36	0.242(24)^b	0.285(11)	1.185(3)	0.586(8)	−0.033(19)^a
STA	[12]	48	0.245(27)^b	0.289(15)	1.176(26)	0.585(9)	−0.011(14)^a
STA	[11]	36	0.230(30)^b	0.280(15)		0.590(10)	0.000(40)^c
bct	[20]	42	0.287(30)^b	0.247(10)	1.217(32)	0.571(10)	−0.131(18)^a
STAR	[34]	42	0.488(30)^b	0.221(9)	1.074(29)	0.504(10)	−0.131(13)^a
$V_{3,2}$	[34]	40	0.479(24)^b	0.193(4)	1.136(23)	0.507(8)	−0.240(10)^a
$V_{3,2}$	[8]	80			First order		

Notes: Calculated by $^a\gamma/\nu = 2 - \eta$, $^b d\nu = 2 - \alpha$, $^c 2\beta/\nu = d - 2 + \eta$.

different: $O(N)/O(N-2)$ in comparison to $O(N)/O(N-1)$ for ferromagnetic systems. This "new" universality class has been called chiral class.[5] For the XY case, there is a cross-over region with a slow velocity and exponents will vary between those of the chiral class and of weak first-order transitions ($\nu = 1/3$, see Table 4.1). The critical exponents of this chiral class are given in Tables 4.4–4.8 for N varying from 2 to 6 for the STA, STAR, and dihedral models. With numerical simulations, exponents γ/ν β/ν and ν are usually calculated using FSS while other exponents are calculated using the scaling relation $d\nu = 2 - \alpha$ and $\gamma/\nu = 2 - \eta$ (see Section 4.7).

There is another indication given by Loison and Schotte[33] which suggests that the phase transition is indeed due of a "complex fixed point". Using the scaling formula $\gamma/\nu = 2 - \eta$, one gets a negative η for $N = 2$ and

Table 4.7. Critical exponents by Monte Carlo for spins with four components ($N = 4$) and a BS $O(4)/O(2) \equiv SO(4)/SO(2)$.

System	Ref.	L_{max}	α	β	γ	ν	η
STAR	[35]	42	0.287(27)^b	0.291(11)	1.133(28)	0.571(9)	+0.015(18)^a
$V_{4,2}$	[35]	40	0.278(30)^b	0.290(12)	1.142(34)	0.574(10)	+0.011(25)^a

Notes: Calculated by $^a\gamma/\nu = 2 - \eta$, $^b d\nu = 2 - \alpha$.

Table 4.8. Critical exponents by MC for spins with six components ($N = 6$) and a BS $O(6)/O(4) \equiv SO(6)/SO(4)$.

System	Ref.	L_{max}	α	β	γ	ν	η
STA	[13]	36	−0.100(33)^b	0.359(14)	1.383(36)	0.700(11)	+0.025(20)^a

Notes: Calculated by $^a\gamma/\nu = 2 - \eta$, $^b d\nu = 2 - \alpha$.

$N = 3$ for the STA, STAR, or dihedral model (see Tables 4.4–4.8) which contradicts the theorem that η must always be positive.[45,46] Negative η can also be calculated using experimental results.[47] Therefore, the fixed point cannot be real. Using this indication, we can conclude that the continuous transition found in fully frustrated XY lattice[23] is indeed of first order.

4.3.1.4. *Experiment*

A resembling situation occurs in the analysis of experiments. There the correlation length ξ plays the role of the system size in numerical simulations. For a second-order transition, one has $\xi \sim (T - T_c)^{-\nu}$. Therefore, we should observe a cross-over between the region of low velocity ("almost second-order transition" in Zumbach's words) to the true first-order behavior for temperatures "closer" to the critical temperature. Unfortunately, the situation is even more complicated in experimental systems with the omnipresence of planar or axial anisotropies and the 1D or 2D characters of the compounds. Then a succession of crossovers (see Section 4.7 for more details about cross-overs) from 2D to 3D and from Heisenberg to Ising or XY behavior could lead to difficulties in the interpretation. Furthermore, other (small) interactions could also dominate the behavior near the critical temperature and change the universality class.

Experiments for XY spins: All experimental results can be found in Ref. [33]. Several AXB_3 compounds have the STA structure. The experiments on $CsMnBr_3$,[48–58] $RbMnBr_3$,[59,60] and $CsVBr_3$[61] give critical exponents compatible with those of MC simulation on STA and a second-order transition. We can interpret this result by the fact that the systems are under the influence of a complex fixed point, and $t \propto T - T_c$ is too small to observe a first-order transition.

The case $CsCuCl_3$[62] is different since the authors observe a cross-over from a second-order region with exponents compatible with MC results on STA for $10^{-3} < t < 5.10^{-2}$ to a region of first-order transition for $5.10^{-5} < t < 5.10^{-3}$. For $t < t_0 \approx 10^{-3}$, one seems to observe the true first-order region which corroborates the scenario introduced previously.

Three kinds of helimagnetic structures have been studied: Holmium, Dysprosium, and Terbium. Essentially, three types of results exist: those compatible with MC ones of the STA, those with a large β incompatible with STA, and those showing a weak first-order transition.

The results compatible with those of MC on STA for Ho[64–66] Dy,[65,72,73] and Tb[77–81] can be interpreted as before: the systems are under the influence of F_+. The first-order transition for Ho[63] and Dy[70,71] is due to the fact that the measurements were done in the first-order region near the transition temperature. The values of the exponent $\beta \sim 0.39$ in the case of Ho[66–69] and Dy[67,74–76] are not compatible with those found by MC ($\beta \sim 0.25$). This fact can be explained by the presence of a second length scale in the critical fluctuations near T_c related to random strain fields which are localized at or near the sample surface.[66] Thus, the critical exponent β measured depends on this second length.

Experiments for Heisenberg spins: All experimental results can be found in Ref. [33]. As explained previously, before the first-order region is reached, the cross-over from Heisenberg to Ising or XY behavior prevents a first-order transition of Heisenberg type. Nevertheless, the second-order transition can be studied for VCl_2,[82] VBr_2,[83] $Cu(HCOO)_2 2CO(ND_2)_2 2D_2O$,[84] and $Fe[S_2CN(C_2H_5)_2]_2Cl$.[85] For the last two examples the observed exponents might be influenced by the cross-over from 2D to 3D Ising behavior. The experimental results agree quite well with the MC simulations.

4.3.1.5. *Value of N_c*

Tables 4.4–4.8 allow us to get an idea of the value of N_c. The MCRG gives an estimate $3 < N_c < 8$, but the values of η can give a better estimate following Ref. [33]. As can be seen for the XY and Heisenberg systems, negative values of η appear for the STA, STAR, or dihedral model. But η must always be positive.[45,46] This is due to the use of the scaling relation $\gamma/\nu = 2 - \eta$. Indeed, Zumbach[44] has shown that for an "almost second-order transition", i.e. when the solution becomes complex, this relation has to be modified to $\gamma/\nu = 2 - \eta + c$, c being a constant different from zero. We can use this relation as a criterion for real or complex fixed points. In three dimensions, η is usually small and is almost independent of N for the ferromagnetic case, that is ~ 0.03 (see Table 4.1). Our hypothesis is that it is also true for the frustrated case. Indeed, for $N = 6$ we found $\eta \sim 0.03$ (see Table 4.8). Accepting this value, c becomes zero around $N_c \sim 4.5$. Obviously, if a bigger value of η is chosen, N_c will increase.

4.3.1.6. *Phase diagram (N,d)*

In Fig. 4.11, we have plotted a phase diagram where the abscissa is the spin dimension N and the ordinate the space dimension d.

There is a line which divides a region of first-order transition from a region of second-order transition. At, and near, $d = 4$ we can use the RG $d = 4 - \epsilon$ expansion to find $N_c = 21.8 - 23.4\,\epsilon$, and for $d = 3$ we have seen that $N_c \sim 4.5$. Furthermore, there is a region near this line where the transition is "almost second order", i.e. under the influence of the complex fixed point for "small" sizes. The system will display a second-order transition with stable critical exponents. Besides, there is a slow cross-over region where critical exponents vary with the system size. The sizes of

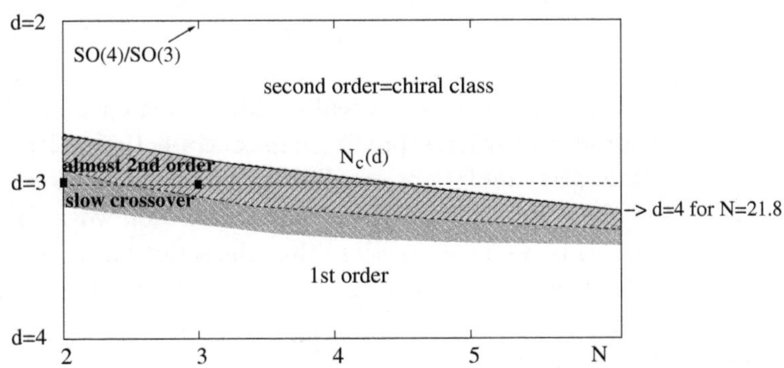

Fig. 4.11. The critical curve $N_c(d)$ separating the first-order region for small N and large d from the second-order region for large N and low d. The squares represent the physical systems of interest ($N = 2, 3$ in three dimensions).

these regions depend on the initial point in the flow diagram and therefore on the model.

Near two dimensions and for Heisenberg spins, the transition is of the type $SO(4)/SO(3)$, i.e. ferromagnetic for spins with four components (see the following section and Chapter 3 for more details).

4.3.1.7. *Renormalization group expansions*

We present here a review of the RG expansions ("non-perturbative", $d = 4 - \epsilon$, in fixed dimension $d = 3$, and $d = 2 + \epsilon$). For more details see Section 4.7 and Chapter 3.

We discuss now the $4 - \epsilon$ expansion. Using a continuous limit of the Landau–Ginzburg Hamiltonian (4.6), it involves a perturbative extension (u, v) around the Gaussian solution with the dimension d and the number of spin components N as parameters. To get the result for $d = 3$, one puts $\epsilon = 1$. Of course, the series are at best asymptotic and must be resummed, even for the ferromagnetic case.[86] The extension until ϵ^2 in Ref. [87] has been "resummed" (only three terms) and the value of $N_c(d = 3) \sim 3.39$ is compatible with the numerical result $N_c^{\text{numerical}}(d = 3) \sim 4.5$. The calculations have been extended to ϵ^4 in Ref. [88], but unfortunately, no resummation has been done on this series yet.

The expansion for fixed dimension $d = 3$ is very similar to the $d = 4 - \epsilon$ one. The series and the resummation will differ slightly. At three loops, the result is $N_c \sim 3.91$[89] which is again not far from $N_c^{\text{numerical}}(d = 3) \sim 4.5$. A very interesting picture emerges from the six-loop calculations.[90,91] First, for $N \sim 5$, the resummation does not converge. Second, for $N = 3$, the resummation gives rise to a very large variation of the critical exponents following the flow chosen to arrive at a single critical exponent at the critical point (see Fig. 5 of Ref. [92]), which is a really "new" phenomenon. Third, for $N = 2$ a second-order transition is predicted in contradiction to Itakura's results[8] and the numerical simulations (see Tables 4.4 and 4.5). Manifestly, we could conclude that the resummation chosen does not work for this series. To understand this, we have to stress that the series has many more terms because of the double expansions in u and v compared to a single one for the ferromagnetic case. Hence, it is much more difficult to find a "good" resummation scheme.

On the other hand, Delamotte *et al.*[92] found that the transition must be of $SO(4)/SO(3)$ type, i.e. equivalent to a ferromagnetic $O(4)$ transition, near two dimension for Heisenberg spins. This study was done using the

$d = 2 + \epsilon$ expansion or equivalently the nonlinear σ model and the result is valid only near two dimensions. This prediction was checked by Southern and Young[93] by MC in two dimensions. But we can give arguments[34] using $1/N$-expansion and the value of the critical exponents to rule out this kind of transition in three dimensions. These arguments hold as long as the relevant operators are identical in $1/N$ expansion and in the nonlinear σ model. Indeed for the $1/N$ expansion, and equivalently for the $4 - \epsilon$ expansion and in fixed dimension, we keep in the Hamiltonian only the terms which are renormalizable near $d = 4$ or infinite N and discard the others. Nevertheless, it seems that some non-renormalizable terms (see Section 4.7) become relevant and important for low N and d and, whatever the number of loops, we cannot find the correct behavior using these expansions.

Because of the problems encountered by the standard expansions, a "non-perturbative" approach could be very useful. This was introduced in this model by Tissier *et al.*[47] We quote "non-perturbative" because it is not an expansion resembling the other methods where the extension parameters (u, v, ϵ) are usually not small. In this non-perturbative method, even if we introduce only a few terms in the action, results will be very good and no resummation is necessary. We notice that the simplest action was studied by Zumbach,[44] which allowed him to introduce the "almost second-order" transition. Adding more terms, Tissier *et al.* found that they are able to retrieve all the previous expansions with additional information. Their value of $N_c \sim 5.1$ is comparable to $N_c^{\text{numerical}} \sim 4.5$. In three dimensions, they found critical exponents very close to those calculated by MC for $N = 6$ (see Table 4.8), a minimum in the flow for $N = 3$, and a slow cross-over for $N = 2$ as explained in the previous section. They also found that the critical behavior is indeed that of a ferromagnetic $O(4)$ transition near two dimensions for Heisenberg spins. Some non-renormalizable operators excluded in the $4 - \epsilon$ and $1/N$ expansions are included in this "non-perturbative" method. These operators are always relevant between two and four dimensions whatever N is, but have an influence on the values of the critical exponents only for small N and near two dimensions. This explains the discrepancy between the $d = 2 + \epsilon$ and $d = 4 - \epsilon$ expansions.

4.3.1.8. *Short historical review*

In this section, we give a short historical review of studies on frustrated systems. Indeed, the history was not straight if we look back at the last 25 years. The development began with an RG $4 - \epsilon$ expansion by Jones *et al.*[94] in 1978. It is only in 1984 and in the following years that Kawamura[5] started with the first numerical simulations and, in combination with the results of several experiments, proposed a "new" universality class.

Then Azaria *et al.*[92] found that the transition near two dimensions for Heisenberg spins should belong to the $O(4)$ ferromagnetic class. Several groups (see Tables 4.4–4.8) have done numerical simulations and experiments which favored either the new universality class, a first-order transition, an $O(4)$ transition, and even a mean-field tricritical transition. During those years, Sokolov *et al.*[87,89] have extended the RG expansion to three loops and found a first-order behavior for XY and Heisenberg spins.

Surely, one of the most important articles to find the key to understand the physics was written by Zumbach,[44] using a non-perturbative approach. He predicts an "almost second-order transition" for XY and Heisenberg spins. Loison and Schotte,[33,34] using this concept, were able to get a clear picture for both the numerical and experimental studies. Then Tissier *et al.*,[47] by extending the work of Zumbach, were able to understand the whole phase diagram for the dimension between two and four. To conclude, Itakura[8] confirmed the picture given by Zumbach *et al.*, which we think is the definitive answer from a numerical point of view.

4.3.1.9. *Relations with the Potts model*

We would like now to stress the similarities between the Potts model and the frustrated case just studied.[43] In Fig. 4.12, we have plotted the RG flow diagram[95] as a function of the first- and second-neighbor ferromagnetic interactions (J_1 and J_2) and the chemical potential Δ (corresponding to the site vacancy).

F_+ is the standard stable ferromagnetic fixed point and F_- is an unstable fixed point. D is a first-order fixed point. O is the initial point of the system. This figure resembles the one of the frustrated case (Fig. 4.9). For $q < q_c$, the flow goes to the fixed point F_+ and has a second-order transition. At $q = q_c = 4$ (in two dimensions), the two points F_+ and F_- collapse and a second-order phase transition with logarithm correction results. For $q > q_c$, there is no longer a solution and the flow goes directly to the fixed point D, i.e. the system has a first-order transition. Our hypothesis is that even if the solution of F_+ is complex, it has an influence on the real space and there is a region of low velocity in the flow diagram. It can be a true minimum for $q \gtrsim q_c$ with an "almost" second-order transition. For q bigger, no minimum

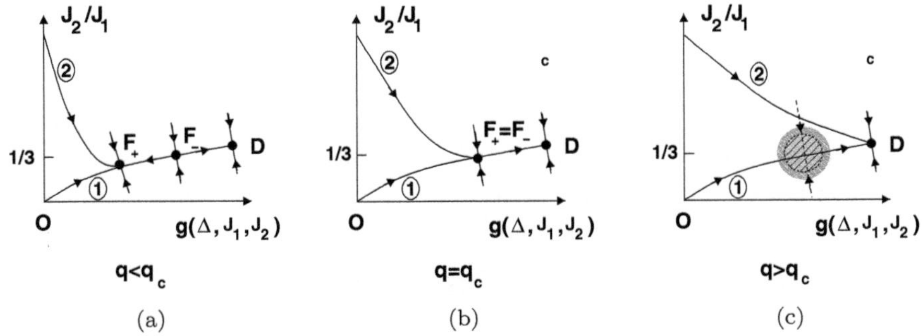

Fig. 4.12. Hypothesis of the renormalization flow in the presence of second-neighbor interaction. For $q \leq q_c$, the initial point of the flow is irrelevant to the critical behavior. For $q > q_c$, the behavior for accessible lattice size will be different for curve 1 (going through the basin of attraction of the fixed points F_+) and curve 2 (going directly to the first-order fixed point D).

should exist but, following the sizes studied, a variation of the critical exponents could occur. Indeed, for $q = 5$ a set of critical exponents was found,[42] corresponding to our hypothesis of "almost second-order" transition. The cases $q = 6$ and $q = 7$ have not been studied, but possibly one could observe a variation of critical exponents corresponding to a slow cross-over to the first-order point.

We remark that depending on q we can observe clearly whether the transition is of first order or second order. In Fig. 4.12, we have plotted the flow (4.2) in the presence of the second-neighbor interaction. We observe that this flow is less influenced by the low-velocity region than the flow (1) if $q > q_c$ and it reaches F_+ if $q \leq q_c$. Adding the third-NN interaction, we are able to find numerically that $q_c = 4$ for the cross-over from the first-order to second-order transitions. Contrary to the frustrated case, we know roughly the coordinate of the fixed point F_+ $(J_2/J_1 \sim 0.3)$[96] and $\Delta \neq 0$.[95] By consequence, we know which kind of interaction must be added.

4.3.2. $O(N)/O(N-P)$ breakdown of symmetry for $d = 3$

In mean-field theory, for $N > P$ the model shows a usual second-order type, but for $N = P$ the transition is a special one.[44] The BS in this case is $Z_2 \otimes SO(N)$ and the coupling between the two symmetries leads to a special behavior which, for $N = P = 2$ in two dimensions, has been extensively debated (see later in this chapter).

These generalized chiral models have been studied by applying the RG technique ($d = 4 - \epsilon$ expansion).[44,97,98] The picture is very similar for all $N \geq P \geq 2$. At the lowest order in ϵ, there are up to four fixed points, depending on the values of N and P. Among them are the trivial Gaussian fixed point and the standard isotropic $O(NP)$ Heisenberg fixed point. These two fixed points are unstable. In addition, a pair of new fixed points, one stable and the other unstable, appear when $N \geq N_c(d)$ with

$$N_c(d) = 5P + 2 + 2\sqrt{6(P+2)(P-1)} - \left[5P + 2 + \frac{25P^2 + 22P - 32}{2\sqrt{6(P+2)(P-1)}}\right]\epsilon\,. \tag{4.8}$$

For $P = 2$, we find the standard result $N_c = 21.8 - 23.4\,\epsilon$. In contrast, for $P = 3$ we obtain $N_c = 32.5 - 33.7\,\epsilon$ and for $P = 4$ we obtain $N_c = 42.8 - 43.9\,\epsilon$. A "tricritical" line exists which separates a second-order region for low d and large N from a first-order region for large d and small N. Applying $\epsilon = 1$ ($d = 3$), we obtained that $N_c(d = 3) < 0$ for all P. For $P = 2$ we know that this result does not hold and equivalently it does not apply for $P \geq 3$. Loison[38] has done some simulations for $N = 3$ and $N = 4$ with $P = N$ and $P = N - 1$ for the Stiefel $V_{N,P}$ model. He showed that the transition is clearly of first-order using numerical simulations. He generalized this result for all N. We remark that the fully frustrated cubic lattice (generalized Villain lattice) which could have an identical BS also has a first-order transition for Heisenberg spins.[25] In conclusion, the transition in systems with $O(N)/O(N - P)$, the breakdown of symmetry is of first order for $P = N$ and $P = N - 1$ and in particular for the physical case $N = P = 3$, i.e. a non-planar GS for Heisenberg spins.

4.3.3. $Z_2 \otimes SO(N)/SO(N-1)$ *breakdown of symmetry for* $d = 3$

For $N > 2$, this BS corresponds to a collinear GS associated to a BS of the lattice (see Tables 4.2 and 4.3). For example, the stacked Villain lattices with first-neighbor ferromagnetic interaction J_1 and a second-neighbor antiferromagnetic interaction J_2 can have this BS. For $J_2/J_1 < -0.5$, the competition between the two interactions leads to two possible GSs.[22] For $N = 2$ (XY spins), the stacked Villain lattices and the stacked J_1–J_2 model have the same BS as the STA and should display an analogous behavior ("almost second order" for small sizes following by a first-order transition for larger sizes).

We have simulated by MC technique the stacked Villain lattices and the stacked J_1–J_2 lattice which have this kind of BS. We found that the transition seems continuous[39] for small lattices for XY and Heisenberg spins. For XY spins, this is in agreement with the numerical simulation of STA which has an identical BS. We have also studied the Ising-$V_{N,1}$ model, which has the same BS. For $N = 2$ this model is equivalent to the $V_{2,2}$ model, which has a strong first-order behavior as explained previously. For $N = 3$ and $N = 4$, we also found strong first-order transitions. Therefore, it seems that the transition $Z_2 \otimes SO(N)/SO(N-1)$ is of first order for any N even if it could exist as an "almost second-order transition" for small sizes.

4.3.4. $Z_3 \otimes SO(N)/SO(N-1)$ *breakdown of symmetry for* $d = 3$

This BS appears with a collinear GS associated to a three-state Potts symmetry due to the breakdown of lattice symmetries. This corresponds to various lattices where the unit cell is composed of four spins (two parallel spins and two antiparallel spins) like the STA with intermediate NNN interaction ($0.125 < J_2/J_1 < 1$), the fcc, the hcp, the cubic J_1–J_2 model or the pyrochlore (see Tables 4.2 and 4.3).

As seen previously, even the coupling of an Ising symmetry (Z_2) with $SO(N)/SO(N-1)$ can give a first-order transition. Hence, it is expected that the transition with a three-state Potts symmetry (Z_3) also gives a first-order transition, which is even stronger if we consider that the Potts symmetry alone has a first-order transition in $d = 3$ for $q = 3$. Indeed, a strong first-order transition was seen in the fcc,[25,28] hcp,[29] cubic J_1–J_2 model[25,26] pyrochlore[30] and STA with NNN interaction[7,11] for XY and Heisenberg spins.

We have done some simulations for the Potts-$V_{N,1}$ model with the Hamiltonian (4.5) with $q \geq 3$, i.e. the q-state Potts model coupled to an N-component vector. We found a strong first-order transition for any N and q.

In conclusion, the transition for the $Z_q \otimes SO(N)/SO(N-1)$ BS in $d = 3$ is of first order for any $N \geq 2$ and $q \geq 2$.

4.3.5. $Z_q \otimes O(N)/O(N-2)$ *and other breakdown of symmetry in* $d = 3$

This BS appears with a planar GS associated to a three-state Potts symmetry. One example is the STA with NNN interaction $J_2/J_1 > 1$ (see Tables 4.2 and 4.3).

To simulate this BS, we have used a Potts-$V_{N,2}$ model for any q. We found the behavior of a strong first-order transition whatever N and q are. It is not surprising if we consider that the "less frustrated" model Potts-$V_{N,1}$ is already of first order.

We have also simulated the $Z_q \otimes O(N)/O(N-3)$ for $N = 3$, which corresponds to the BS $Z_q \otimes Z_2 \otimes SO(N)$ using a Potts-$V_{N,P}$ model. We found again a strong first-order transition. Similar results are obtained for Ising–Potts-$V_{N,1}$ and Ising–Potts-$V_{N,2}$ models.

To summarize, it seems that the coupling between a Potts or Ising symmetry and a continuous symmetry of vector spins always give rise to a first-order transition.

4.4. Summary

We have studied breakdowns of symmetry of 3D lattices for XY ($N = 2$) and Heisenberg ($N = 3$) spins. The general form of the breakdown of symmetry is $S_{\text{lattice}} \otimes O(N)/O(N-P)$. S_{lattice} is usually a Potts symmetry Z_q (Z_1 corresponds to the identity, Z_2 to an Ising symmetry, Z_3 to a three-state Potts symmetry, etc.). P runs from 1 to 3.

If the frustration is small, the GS can be collinear ($P = 1$) without breaking any symmetry of the lattice $S_{\text{lattice}} = 1$, the BS will be $SO(N)/SO(N-1)$ and the transition will be of second-order like in ferromagnetic systems.

If the frustration is strong enough, we can have a non-collinear GS with the usual planar configuration ($P = 2$) like in the STA or helimagnets and $S_{\text{lattice}} = 1$. Depending on the model and for not "too big" sizes, the transition could appear continuous belonging to a new universality class. For an infinite size, the system will have a first-order transition.

Some more complicated systems could also have a non-planar GS, and for Heisenberg spins, it corresponds to $P = 3$ with usually $S_{\text{lattice}} = 1$. The transition will be of first order with a possible "almost second-order" behavior for small system sizes similar to the $P = 2$ case.

Some even more complicated systems could have a non-collinear GS ($P = 2$ or $P = 3$) with $S_{\text{lattice}} = Z_q$ and $q = 3$. For example, the STA with large NNN interaction has a strong first-order transition.

On the other hand, some frustrated systems have degenerate GS, but by "order-by-disorder", a collinear GS ($P = 1$) is selected. The lattice symmetry could be broken giving an additional BS. $S_{\text{lattice}} = Z_2$ (Ising) for the stacked Villain model, or $S_{\text{lattice}} = Z_3$ in fcc, hcp, STA with intermediate

NNN interaction, and pyrochlore. For the Z_3 symmetry, the transition is strongly of first order even for small sizes. For the Z_2 symmetry, the transition is also of first order for large sizes but looks continuous for small sizes. Therefore, we cannot exclude an "almost second-order" behavior belonging to a new universality class.

In conclusion, frustrated systems have a first-order transition for XY and Heisenberg spins in three dimensions even if for "small" sizes the systems would show a second-order transition. "Small" could mean sizes of thousands of lattice constants and the first-order transition will then not be observable with the actual computer resources.

4.5. $O(N)$ Frustrated Vector Spins in $d = 2$

4.5.1. *Introduction*

The situations in 2D systems are different from those in the three dimensions. The Mermin–Wagner theorem[99] asserts that no magnetization appears at non-zero temperature. However, a transition due to the topological defects[100,101] can appear. The binding–unbinding transition of vortex–antivortex pairs for XY spin systems is a classical example. We note that their exact role for 3D phase transitions is not clear.[102–104] For non-collinear GS, the topological properties of the system differ from those of collinear GS, and new types of transition could appear. In the following, we will briefly review the ferromagnetic case, then the frustrated XY case, applicable to Josephson junction arrays in magnetic fields, and finally the frustrated Heisenberg cases.

4.5.2. *Non-frustrated XY spin systems*

For a non-frustrated XY spin system, the order parameter is $SO(2)$. The topological defects of this group are only the point defect classified by the homotopy group $\Pi_1(SO(2)) = Z$, Z being the topological quantum number of the defect.[100,101] A Kosterlitz–Thouless (KT) transition[105] exists at the critical temperature $T_c \sim 0.9$, driven by the unbounding of vortex–antivortex pairs.

This transition has some special features: for $T < T_c$, the correlation length (ξ) is infinite, while for $T > T_c$, it has an exponential decreasing contrary to the power law behavior in the standard transition.

Several numerical methods exist to calculate the critical temperature and the critical exponents (see Section 4.7).

4.5.3. *Frustrated XY spin systems:* $Z_2 \otimes SO(2)$

A frustrated XY spin system can have many order parameters as previously seen (see Table 4.2). The most studied case is $Z_2 \otimes SO(2)$, which corresponds to a non-collinear GS: triangular lattice,[106–110] Villain lattice or 2D fully frustrated lattice,[4,106,111–118] and related Zig–Zag models.[31,119] In combination to the topological defect point Z, there is a line of Z_2 corresponding to an Ising transition. Therefore, they will have similar properties as the J_1–J_2 model, the $V_{2,2}$ model, or any model where the Z_2 symmetry comes from the lattice. We note that different models can have an identical group and therefore the same topological defects. The following models should have analogous properties: the Villain model on the Villain lattice,[122] The Ising-$V_{2,1} \equiv V_{2,2}$ model,[120,121] the 19-vertex model,[123] the 1D quantum spins,[124] the Coulomb gas representation,[125–128] the $XY - XY$ model,[129,130] or the RSOS model.[131]

This kind of model was widely studied because it is believed to correspond to experimental systems such as Josephson junction arrays of weakly coupled superconducting islands[132,133] or films of Helium ^{3}He.[134–137]

The question is to understand the coupling between the vortex–antivortex and the Ising symmetry. Contrary to the 3D case discussed previously, there are no clear accepted answers to these questions. We will therefore only present the different results in the high-temperature region and in the FSS, giving some clues to understand the physics of these systems.

What are the possible scenarios?

(1) $\boldsymbol{T_c^{\mathrm{KT}} < T_c^{\mathrm{Ising}}}$**:** The KT transition associated to the topological defects appears at a temperature lower than the one associated to the Ising transition even if they can be very close. The two transitions should have standard behaviors ($\nu^{\mathrm{KT}} = 0.5$, $\nu^{\mathrm{Ising}} = 1, \ldots$).

Furthermore, Olsson has proposed that the Ising behavior could be observed only when $L \gg \xi^{\mathrm{KT}}$. Since ξ^{KT} is infinite for $T \leq T_c^{\mathrm{KT}}$ and decreases exponentially for $T > T_c$, it will correspond to a high-temperature region with $T \gg T_c^{\mathrm{KT}}$, and also $T \gg T_c^{\mathrm{Ising}}$ when the two transitions appear in close neighborhood. In particular, the FSS region will have a non-standard Ising behavior for the sizes accessible with actual computer resources. Only for a very large lattice the standard Ising behavior will be visible. We observe the similarity of this hypothesis with the one described previously in three dimensions and we could call the transition in the FSS an "almost new Ising–KT behavior".

(2) $\boldsymbol{T_c^{\mathrm{KT}} \sim T_c^{\mathrm{Ising}}}$: The two transitions appear at the same temperature and display different behaviors compared to the standard one. We remark that for high temperatures $T \gg T_c$, i.e. in the high-temperature region, we cannot assume that the coupling between the Z_2 symmetry and the topological defects would be identical as for $T \sim T_c$. Indeed, with a lot of small Z_2 walls dividing the system, it is not certain that the vortex and the antivortex should play an important role and we could get a standard Ising transition.

(3) $\boldsymbol{T_c^{\mathrm{KT}} > T_c^{\mathrm{Ising}}}$: This hypothesis was advanced by Garel and Doniach in 1980 for the J_1–J_2 model,[138] but numerical MC simulations have shown that it is not the case[22] for this system.

To choose the most probable hypothesis, we will look now at the numerical results:

Some information can be provided by numerical MC simulations in the high-temperature region. By fitting the results, it is possible to get the critical temperature and the critical exponents. Because of the presence of corrections and as a consequence of many free parameters, the extraction of the results is difficult.

For the Ising symmetry, Olsson[122] has found a standard ferromagnetic one with $\nu^{\mathrm{Ising}} \sim 1$ while Jose and Ramirez[114] found $\nu^{\mathrm{Ising}} \sim 0.87$. New simulations should be done to resolve this discrepancy. If $\nu^{\mathrm{Ising}} \neq 1$, it will be a counter-proof to Olsson's claim (first possibility). On the other hand, if $\nu^{\mathrm{Ising}} = 1$, we cannot rule out the second possibility.

The exponent of the KT transition has been calculated by Jose and Ramirez.[114] They found non-standard $\nu^{\mathrm{KT}} \sim 0.3$ to be compared to 0.5 in the ferromagnetic case. If this result is in favor of the second hypothesis, it cannot rule out the first one. Indeed, we can reverse the argument given at the end of the second hypothesis. If at high temperatures the system is composed of many domains separated by walls of Z_2, the correlation between the topological defects could not be the one of the standard one. Therefore, there is no definitive conclusion from the high-temperature simulations.

Now we can look at the results in the FSS.

For the Ising transition, the critical exponents have been precisely determined by Loison and Simon,[22,110] for the J_1–J_2 model using both the FSS and the dynamical properties including the first correction[110] due to the lattice size L. They obtained non-standard critical exponents $\nu^{\mathrm{Ising}} = 0.815(20)$. The "large" error is due to the inclusion of the corrections. See Ref. [110] for a review of all the numerical works. For the triangular lattice, the critical temperature is known with high precision $T_c^{\mathrm{Ising}} = 0.5122(1)$.

The KT transition is more problematic. First, we can use the helicity[105] Υ to find the critical temperature. If we admit the same universal jump at the critical temperature as the ferromagnetic one, we get $T_c^{\mathrm{KT}} = 0.5010(10)$, a temperature much smaller than $T_c^{\mathrm{Ising}} = 0.5122(1)$. But we cannot be sure that $\Upsilon_{\mathrm{jump}}^{\mathrm{frustrated}} = \Upsilon_{\mathrm{jump}}^{\mathrm{ferromagnetic}}$ and this result must be taken with care.

As explained in the previous section, we can use the Binder parameter or the dynamical properties of the model to calculate T_c^{KT} and the critical exponents. Using this method we found very interesting results. First, the Binder parameter shows a power law transition and not an exponential one. In addition, the critical exponents and temperatures calculated by the two methods are in good agreement with $T_c^{\mathrm{KT}} = 0.5102(2)$ just below $T_c^{\mathrm{Ising}} = 0.5122(1)$ for the triangular lattice. The critical exponent $\eta^{\mathrm{KT}} = 0.36(1)$ is very different from the standard one, 0.25. These results are in accord with those on the J_1–J_2 model. What is disturbing is that for the triangular lattice, accepting $T_c^{\mathrm{KT}} = 0.5102(2)$, the helicity jump is smaller than the $\Upsilon_{\mathrm{jump}}^{\mathrm{ferromagnetic}}$, although it is believed that it should be greater.

The situation is puzzling with no definitive conclusions.

4.5.4. *Frustrated XY spin systems: $Z_3 \otimes SO(2)$*

Two-dimensional systems with an order parameter like $Z_3 \otimes SO(2)$ are less numerous than 3D systems. Following Table 4.2, we know at least two systems: the triangular lattice with intermediate NNN interaction and the Potts-$V_{2,1}$ model.

No numerical studies have been done for the former system, but we have studied[140] the latter one. It gives a first-order transition.

Following the reasoning of the previous section, two principal hypotheses appear:

(1) **Olsson hypothesis:** KT transition occurs at a lower temperature following the transition associated to the Potts symmetry Z_3 at a higher temperature. Non-standard critical exponents should appear in the FSS region due to a screening length (ξ^{KT}). The "true" standard second-order behavior will appear only for very large lattices ($L \gg \xi^{\mathrm{KT}}$) or in the high-temperature region.
(2) A new behavior for the KT and Potts transitions is seen at the same critical temperature.

The results are in favor of the second hypothesis. The Potts symmetry has a first-order transition. It means that its correlation length ξ_{Z_3} is finite at the critical temperature and less than ξ_{KT}. There will be no change even if the system size L is much larger and the system will never show a standard three-state Potts second-order transition. Obviously, we cannot directly apply this result to the Ising–KT transition since the two models are not equivalent, although it is nevertheless an argument against the Olsson's hypothesis.

4.5.5. *Frustrated XY spin systems:* $Z_2 \otimes Z_2 \otimes SO(2)$ *and* $Z_3 \otimes Z_2 \otimes SO(2)$

Following Table 4.2, the order parameter $Z_2 \otimes Z_2 \otimes SO(2)$ can appear with a J_1–J_2–J_3 lattice or an Ising-$V_{2,2}$ model. In addition, $Z_3 \otimes Z_2 \otimes SO(2)$ exists in the triangular antiferromagnetic lattice with NNN interaction and in the Potts-$V_{2,2}$ model. No numerical studies have been done on these models.

4.5.6. *Frustrated Heisenberg spin systems:* $SO(3)$

For a non-frustrated Heisenberg spin system, the order parameter will be $SO(3)/SO(2)$ and no topological defects exist in two dimensions, and no phase transition will appear.

For a frustrated system, many order parameters can exist (see Table 4.3), but only the planar GS have received attention. In this case the order parameter is $SO(3)$, and there exist point defects, that is Z_2-vortex–antivortex.[100, 101] These topological defects are different from the Z-vortex present in XY systems. The existence of a critical transition driven by the unbounding of vortex–antivortex was first conjectured by Kawamura and Miyashita.[141]

At low temperatures, the spin waves will dominate the behavior of the system and forbid an infinite correlation length below T_c contrary to XY spins. The behavior should be equivalent to the one present in the 4D ferromagnetic system.[92] This conjecture based on the nonlinear σ model ($d = 2+\epsilon$ expansion) and on symmetry arguments was checked by Southern and Young[93] and Caffarel *et al.*[142]

At higher temperatures, the topological defects will clearly have a role.[142–145]

Caffarel *et al.*[142] have studied numerically two models having identical spin waves but not the same topological defects. One has the topological

Z_2 defects, while the other does not. They showed that the two models are equivalent at low temperatures, but show differences at higher temperatures due to the topological defects.

Southern and Xu,[142] who studied using MC simulations the vorticity associated to the vortex–antivortex, proposed that the vorticity has a jump at the critical temperature similar to that of the KT transition.

Kawamura and Kikuchi[144] using different boundary conditions in MC simulations also observed various phenomena associated to the vorticity which, according to them, are a proof of a phase transition.

Last, Wintel *et al.*[145] have studied theoretically and numerically the region above the suspected critical temperature for the triangular antiferromagnetic lattice ($T_c \sim 0.29$). They claimed that the correlation length and susceptibility must follow a KT law.

There are numerical evidences of the importance of topological defects at finite temperatures and it seems that these systems undergo a kind a topological phase transition. Nevertheless, we have no certainties that the phenomena present at finite sizes would hold for the infinite size.

4.5.7. *Frustrated Heisenberg spin systems:* $Z_2 \otimes SO(3)$, $Z_3 \otimes SO(3), \ldots$

By inspecting Table 4.3, it is not difficult to see that other order parameters exist in frustrated systems. One of the most interesting should be $Z_2 \otimes SO(3)$. It corresponds to a non-planar GS, which could exist in experimental systems. Numerically, the simplest system would be the Stiefel $V_{3,3}$ model or equivalently the Ising-$V_{2,2}$ model. The comparison with the XY case would be very interesting. In particular, the Ising transition could appear near the $SO(3)$ transition (see Section 4.5.6) and the coupling between the Z_2 walls and the vortex could be very instructive. If we follow the second hypothesis of Olsson for the XY case (see Section 4.5.3), the transition should display an "almost second-order transition".

Symmetrically, the coupling of the $SO(3)$ vortex with a Z_3 Potts model, which appear in the Potts-$V_{3,2}$ model, could also disclose interesting properties. Does it show first-order properties as for $SO(2)$ vortex?

Some other breakdowns of symmetry could also appear like $Z_3 \otimes Z_2 \otimes SO(3)$ in the antiferromagnetic triangular lattice with large NNN interaction. Similar questions appear. We note that the $SO(3)/SO(2)$ order parameter does not have a topological transition. The transition of the type $Z_2 \otimes SO(3)/SO(2)$ for the J_1–J_2 model or $Z_3 \otimes SO(3)/SO(2)$ for the triangular lattice with intermediate NNN interaction should have a standard transition (Ising or three-state Potts model).

4.5.8. *Topological defects for $N \geq 4$*

Some other topological defects should exist for an order parameter of the type $SO(N)$ with a GS in $N-1$ dimensions, $Z_2 \otimes SO(N)$ with a GS in N dimensions, and $Z_q \otimes SO(N)$ with a coupling of a GS in $N-1$ dimensions and a lattice symmetry, etc. All questions raised for Heisenberg spins still hold in these cases.

4.6. Conclusions

In this chapter, we have studied the phase transition in frustrated systems between two and four dimensions. We have found various breakdowns of symmetry, contrary to the unique one for ferromagnetic systems.

In three dimensions, the transition is always of first order in the thermodynamic limit. However, for "small" sizes in numerical simulations or for temperatures not "too close" to the transition temperatures in experiments, the system could display an "almost universality class" for an $O(N)/O(N-2)$ breakdown of symmetry. Many compounds studied experimentally are in this class.

The situation is much less clear in two dimensions. Indeed, the topological defects can play a fundamental role and their couplings with a discrete symmetry (Ising or Potts) are unknown. We hope that in the near future the 2D case will be clarified as the 3D one on which our understanding has increased considerably in the last decade.

4.7. Note Added for the Second Edition

Since the first edition of this book in 2005, there has been a considerable progress in the controversial question of the nature of the phase transition in frustrated spin systems.

The most studied system is no doubt the STA with XY spins or Heisenberg spins shown in Fig. 4.1. Chapter 3 by Delamotte *et al.* has given an exhaustive theoretical review on this question and this chapter has provided a review on the principal numerical simulations published till 2003. At the time of the first edition of this book, the non-perturbative RG has already given evidence of a first-order transition in the STA with XY and Heisenberg spins. Numerical simulations also began to find some first-order evidence: Using an improved MC RG scheme to investigate the RG flow of the effective Hamiltonian used in field-theoretical studies for the XY STA, Itakura[8] found that the XY STA exhibits a clear first-order behavior and there are no chiral fixed points of RG flow for $N = 2$.

Since 2003, there have been a number of numerical works that have made further progress toward the final conclusion. In 2004, Peles *et al.*[146] used a continuous model to study the XY STA by MC simulation. They found evidence of a first-order transition. In 2006, Kanki *et al.*,[147] using a microcanonical MC method, found a first-order signature of the XY STA. While these simulations have demonstrated evidence of first-order transition for the XY STA in agreement with the non-perturbative RG analysis, we had to wait for the definite answer in 2008: Using a very high-performance technique for weak first-order transitions, the so-called Wang–Landau flat-histogram method,[148] MC simulations have been recently carried out for the XY STA.[149] A first-order transition in that system has been clearly established, confirming the results of the other authors.

For the Heisenberg case, Itakura[8] found, as in the XY case mentioned above, the absence of chiral fixed points of RG flow. However, he could not find numerical evidence of the first-order transition. He predicted that if the transition is of first order for the Heisenberg spins, it should occur at much larger lattice sizes which he was not able to perform at that time. Finally, using the high performance of the Wang–Landau method, the Heisenberg case was studied with very large sizes[150]: The result shows clearly a first-order transition, putting an end to the 20-year long controversy.

It is interesting to note here that another well-studied frustrated system called Villain's model has also been investigated with the Wang–Landau method: the phase transition was found to be of first order for the Ising,[151] classical XY,[152] and Heisenberg[153] spin models on the Villain's fully frustrated simple cubic lattice shown in Fig. 4.6. The well-known HCP antiferromagnet with Ising and XY spins was also shown to have a first-order transition.[154]

To summarize, it is tempting to conclude that all 3D frustrated spin systems undergo a first-order transition.

Appendix A: Monte Carlo Simulation

The purpose of this section is just to show the fastest algorithm to run MC simulations, the best method to distinguish a first-order transition from a second-order one, and to extract the critical exponents with a reliable estimate of errors.

A.1. *Markov chains and algorithms*

Since we cannot enumerate all the spin configurations, we are forced to use numerical simulations to calculate the physical quantities. The method of choice is the MC method generating a set of phase–space configurations from a Markov chain defined by the transition probability $W[s, s']$ between states $\{s\}$ and $\{s'\}$, where the new configuration depends only on the preceding one. There are many ways to get the transition probability $W[s, s']$.[155] The Metropolis algorithm is the simplest but not very efficient one. A new heat bath algorithm, called Fast Linear Algorithm,[139] for vector spins is more efficient. For example, it is three times faster than the Metropolis algorithm for a 2D triangular antiferromagnetic lattice at the critical temperature. Furthermore, the use of the overrelaxation[155] can reduce strongly the autocorrelation time.[110] The cluster algorithm cannot be used for frustrated spin systems. Indeed, there are two problems. The first is to take into account the competition of different interactions in a plaquette. It means that many spins should be considered in one step when constructing the cluster. Second, even if we were able to construct a cluster, the BS is no more $O(N)/O(N-1)$, but $O(N)/O(N-2)$, for example. Consequently, a symmetry to an $N-1$ plane used in the Wolff's algorithm[156] should be changed.

Thermodynamic averages are taken simply by

$$\bar{A} = (1/N_{\text{MC}}) \sum_{t=1}^{N_{\text{MC}}} A[t], \tag{A.1}$$

where N_{MC} is the number of new configurations generated (MC Steps) and $A[t]$ is the value of the quantity at step t. The great advantage of this method is that the partition function $Z = \sum_{\{s\}} e^{-\beta E[s]}$ need not be calculated.

To estimate the errors of the averaged quantities we have to take into account that each new configuration is correlated with the previous one. We define the autocorrelation time τ by the number of MC steps required to obtain two uncorrelated spin configurations. It is calculated using the autocorrelation function[157] $C(t) = (1/\chi)[\langle A(t)A(0)\rangle - \langle A\rangle^2]$, where time is measured in MC steps (see comments in the appendix of Ref. [22]).

If $N \gg \tau$, then a useful approximation for the error in $\bar{A}$ is given by

$$(\delta\bar{A})^2 = \frac{\chi}{N_{\mathrm{MC}}/(1+2\tau)}, \tag{A.2}$$

$$\chi = \langle A^2\rangle - \langle A\rangle^2. \tag{A.3}$$

This expression is identical to the standard deviation but with an effective number of independent measurements given by

$$\frac{N_{\mathrm{MC}}}{1+2\tau}. \tag{A.4}$$

Problems arise when quantities are a combination of different averages, for example the susceptibility (A.3). We could try to treat $\langle A^2\rangle$ and $\langle A\rangle^2$ as independent quantities and estimate the error by the sum of the errors of the two quantities. But the result will be overestimated due to the correlation between the two elements of the sum. To solve this problem we can use, for example, the Jackknife procedure.[158] An application of this method can be found in the appendix of Ref. [22].

A.2. *The histogram method*

The great advantage of the histogram method in the analysis of MC data is that a run at a single temperature T_1 can be used to extract results for a continuous range of nearby temperatures.[159] In practice, the range of temperatures over which $\langle A\rangle$ may be estimated from a single MC run at T_1 is limited by the range over which reliable statistics can be expected for $H(E)$, the histogram of the energy. A rough guide is $T_a < T < T_b$, where T_a and T_b correspond to the energies $\langle E_a\rangle$ and $\langle E_b\rangle$ at which $H(E) \simeq \frac{1}{2}H_{\mathrm{max}}$. Since $H(E)$ becomes more sharply peaked with system size, the valid temperature range becomes smaller as L increases. Multiple histograms made at a number of nearby temperatures may be combined to increase accuracy.

A.3. *Nature of the transition*

Differentiating a weak first-order from a second-order transition could be difficult. The FSS for a first-order transition has been extensively studied.[160–162] A first-order transition should be identified by the following properties:

(a) The histogram of the energy, $P_T(E)$, has a double peak.
(b) The minimum of the fourth-order energy cumulant W varies as: $W = W^* + b\,L^{-d}$, where W^* is different from 2/3.

(c) The temperatures $T(L)$ at which the specific heat C or the susceptibility χ has a maximum should vary as $T(L) = T_c + a\,L^{-d}$.
(d) The maximum of C and χ are proportional to the volume L^d.

Obviously, a double peak of $P_T(E)$ that becomes more pronounced when the size increases is preferable (way a). If the two peaks are too close, the three other possibilities (b), (c) and (d) can check whether the probability is gaussian (second-order transition) or not (first-order transition) in the limit of infinite size.

A.4. *Second-order phase transition*

For a second-order phase transition, the interesting parameters are the critical temperature and the critical exponents. There are at least three main ways to calculate them:

(a) Consider the high- and low-temperature regions where the correlation length $\xi \ll L$. There, we can fit $\xi = a_\xi(T - T_c)^{-\nu}$ + corrections. The corrections become less important near T_c, but then ξ is very large, and a very big system size is necessary. Furthermore, the autocorrelation time $\tau \sim (T - T_c)^{-z}$ also becomes very large and the error $\delta\xi$ becomes very big even if we can use some tricks to diminish it.[163]
(b) The FSS region is the most powerful method. It is the region with ξ(theoretical) $\gg L$. The best method is to calculate the Binder cumulant[164] $U_L(T) = 1 - \frac{\langle M^4\rangle}{3\langle M^2\rangle^2}$ for different sizes of L.

    (1) We can calculate the critical exponents directly by plotting $\chi L^{-\gamma/\nu}$ as a function of $U_L(T)$ and all the curves must collapse for the correct exponent $-\gamma/\nu = \eta - 2$.[165] Indeed, we can write $\chi L^{\eta-2} = h(U_L)$ with h an unknown function. The other exponents β/ν and ν can be obtained similarly using the magnetization and $V_1 = \frac{\langle ME\rangle}{\langle M\rangle} - \langle E\rangle$. It is the fastest way to get the critical exponents. This method works well even for the Kosterlitz–Thouless transition.[166]
    (2) We can also use $U_L(T)$ to calculate the critical temperature T_c using the crossing of U_L for different sizes. Very good statistics is needed because the evaluation of the first correction is necessary to get a correct value of T_c.
    (3) Having determined T_c, we can calculate the critical exponents using, for example, $\chi \propto L^{+\gamma/\nu}$, etc.

(4) The last mentioned properties are also valid for the maximum (or minimum) of χ and V_1, but we need many simulations at various temperatures to find the maxima because their locations vary as a function of the size L and are different for each quantity.

(c) The dynamical properties[167–171] are not very often used even if they are surely the fastest method available. One has to prepare a state in the GS (i.e. $T = 0$) or randomly (i.e. $T = \infty$) and to observe the dynamical properties of the system at the critical temperature $T = T_c$ for a finite number of MC steps before the equilibrium is reached. For an example, see Ref. [111].

A.5. *Kosterlitz–Thouless (KT) transition*

A KT transition[105] exists for 2D lattices with XY spins. The unbounding of vortex–antivortex pairs appears at the critical temperature, $T_c \sim 0.9$ for a square lattice.

This transition has some special features: for $T < T_c$, the correlation length (ξ) is infinite, while for $T > T_c$ it is exponentially decreasing contrary to the power law behavior in the standard transition.

To find T_c and the critical exponents, we have several ways:

(a) we can fit ξ in the high-temperature region, which is problematic because of the many free parameters and the exponential form.

(b) we can use a method using the behavior of various quantities in the FSS region where $\xi \gg L$.

(1) The first one is to use the universal jump of the helicity Υ at the critical temperature (in the FSS region where $\xi \gg L$), but this method requires the jump of $\Upsilon(L)$ which is known for the non-frustrated case, but not for the frustrated case.

(2) It is therefore interesting to find another method to get the critical exponent, even without calculating T_c. The method (b.1.) introduced previously for a second-order phase transition using the Binder parameter works well for the KT transition.[166] The Binder parameter, contrary to the common belief, crossed around T_c. Plotting $\chi L^{\eta-2}$ as a function of U_L for various sizes, the curves collapse at the correct value of $\eta = 0.25$. This method can be applied whatever the form of ξ is. Therefore, it should be applicable to all phase transitions.

(c) The dynamical behavior of this system can also be used.[170]

For the 2D ferromagnetic XY spin system, all methods mentioned above are in agreement.

Appendix B: Renormalization Group: Landau–Ginzburg Theory, Expansions in Fixed Dimension $d = 3$ and for $d = 4 - \epsilon$ and its Implications for Experiments

In this short section, we would like to give the necessary knowledge to help to understand the concepts of fixed points, flow diagrams, and cross-overs.

First consider a system of Heisenberg spins with a Hamiltonian:

$$H = -J_1 \sum_{(ij)} \mathbf{S}_i . \mathbf{S}_j, \tag{B.1}$$

and the spin is restricted to be of norm $S = 1$. Replacing this constraint by an exponential potential $S = 1 \Leftrightarrow \int_0^\infty e^{u(\mathbf{S}^2-1)^2} \cdot dS$, the Hamiltonian can be written in the form

$$H = K \sum_{\langle ij \rangle} (\mathbf{S}(i) - \mathbf{S}(j))^2 + r \sum_i \mathbf{S}(i)^2 + u_0 \sum_i \left(\mathbf{S}(i)^2\right)^2 . \tag{B.2}$$

Two remarks:

(1) Other terms could be added in the Hamiltonian. These additional terms could be unimportant, which means that they become irrelevant near the fixed point (see later). Or we cannot treat them, which means they are not renormalizable in technical terms. If these neglected terms are important, the method cannot describe the physics of the phase transition. This unfortunate case occurs for frustrated system for low dimensions of space and spins (Chapter 3).

(2) In addition, since the transformation is not exact, the starting value u_0 cannot be known even if it is possible to make a rough guess (Chapter 3).

We have plotted in Fig. B.1 the RG flow for this model.

The system will physically keep its value u_0. But its RG flow will not. Following a series of steps, the value of u will change in approaching the fixed point F. At each step, the correlation length of the system will be divided by a factor $b > 1$. In addition, the flow becomes slower near the fixed point F. Consequently, to reach the neighborhood of F, where the system has a second-order transition, the initial ξ must be large "enough". We know that $\xi(T) \sim (T - T_c)^{-\nu}$ for a second-order phase transition. Hence,

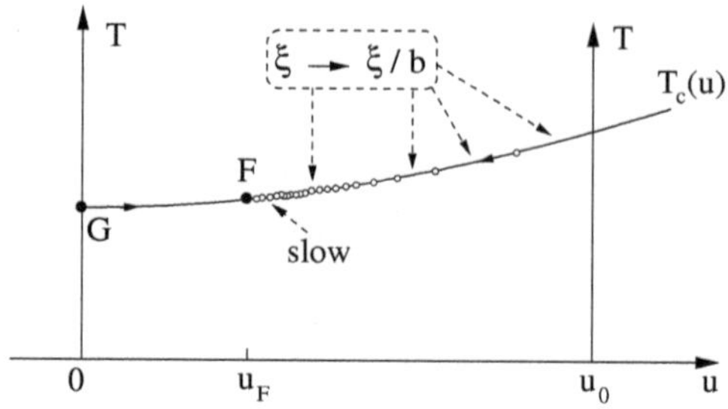

Fig. B.1. RG flow for (B.2). F is the stable fixed point, G is the Gaussian one. Near F the flow is very slow.

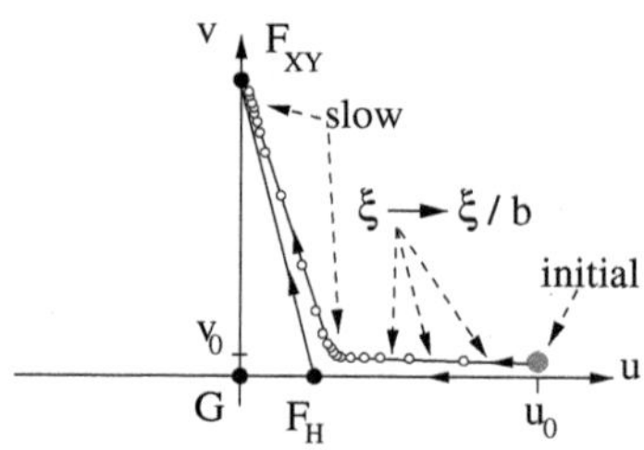

Fig. B.2. Schematic RG flow ($T_c(u,v)$ plane) for (B.2 with $D\sum_i(S_i^z)^2$). F_H is the Heisenberg fixed point, F_{XY} is the XY fixed point, G is the Gaussian one. Near F_H and F_{XY} the flow is very slow.

to obtain numerically or experimentally the value for ξ, which follows a power law (without too many corrections), the temperature must be close "enough" to the critical temperature T_c. Numerically, in the FSS regions where ξ (theoretical is much bigger than the size of the system L, the FSS law (see Section 4.7) will be valid only for L large "enough". We observe that in Fig. 4.9, only the plane $T_c(u,v)$ is plotted.

Consider now a model with an XY anisotropy, for example we add to the Hamiltonian (B.1) a term $D\sum_i(S_i^z)^2$ with $0 < D \ll 1$. We have now two parameters in our Landau–Ginzburg model: u for the length of the spin and v associated to D. The critical plane (u,v) of the flow diagram of this model is plotted in Fig. B.2. From the initial point $(u_0, v_0 \ll 1)$, the flow will go close to the Heisenberg fixed point F_H and then has a cross-over to the F_{XY} fixed point. Near F_H the flow is very slow and needs a lot of steps to escape from the influence of F_H and reaches finally the neighborhood of F_{XY}. Therefore, to observe the "true" XY behavior, the correlation length must be very large. This means that the temperature must be very close to T_c or, if we are in the FSS region, the system size must be very large.

We must know the stability of the fixed point when anisotropies are present. The "quasi"-general rule is that the system goes to the fixed point with the lowest spin symmetry and the biggest space dimensions. For example, consider the quasi 1D system $CsMnBr_3$. This compound will display a 1D behavior before reaching the 3D behavior. Another example is $Cu(HCOO)_2 2CO(ND_2)_2 2D_2O$, which has a small Ising anisotropy. In this case, the system will show a Heisenberg behavior before showing the Ising behavior. In real compounds, small "anisotropies" are almost always present. Therefore, many cross-overs will appear when the temperature approaches the critical temperature. Consequently, interpretations of experiments could be difficult.

To get the critical exponent and the flow diagram, the expansion in fixed dimension d or in $d = 4 - \epsilon$ consists of expanding the exponential $e^{\text{Hamiltonian}(u)}$ around the Gaussian fixed point $u = 0$. Results are in the form of a power series of u. Unfortunately, this series is not convergent, but in the ferromagnetic case it has been shown to be resummable with Pade–Borel techniques. However, there are many ways to resum and a way must be chosen following some criteria.[86] This resummation has been proved to be efficient for ferromagnetic systems; however, it is not certain that it will work for frustrated systems. Indeed, the presence of two vectors ($O(N)/O(N-2)$ BS in this case (compared to the $O(N)/O(N-1)$ BS in the ferromagnetic case) gives rise to a series of power of u and v $(u, v, uv, u^2, \ldots)$ and this double expansion is difficult to resum.

Nomenclature

bct	body-centered-tetragonal,
BS	breakdown of symmetry,
d	dimension,
fcc	face-centered cubic lattice,
FSS	finite size scaling,
GS	ground state,
hcp	hexagonal-close-packed lattice,
KT	Kosterlitz–Thouless,
L	system size,
MC	Monte Carlo,

MCRG	Monte Carlo renormalization group,
N	number of components of the spin **S**,
NN	nearest-neighbors,
NNN	next-nearest-neighbors,
RG	Renormalization Group,
STA	Stacked triangular antiferromagnetic lattices,
STAR	Stacked triangular antiferromagnetic lattices with rigidity,
T	Temperature,
T_c	critical temperature,
$V_{N,P}$	Stiefel model.

Acknowledgments

I am grateful to Prof. K. D. Schotte and Sonoe Sato for their support, discussions, and for critical reading of the manuscript. This work was supported by the DFG SCHO 158/10-1.

References

1. T. Inui, Y. Tanabe, and Y. Onodera, *Group Theory and its Applications in Physics*, Springer Series in Solid–State–Sciences 78 (Springer, Berlin, 1990).
2. S. A. Antonenko and A. I. Sokolov, *Phys. Rev. E* **51**, 1894 (1995).
3. G. Toulouse, *Commun. Phys.* **2**, 115 (1977).
4. J. Villain, *J. Phys. C* **10**, 1717 (1977).
5. H. Kawamura, *J. Phys. Soc. Jpn.* **61**, 1299 (1992); **58**, 584 (1989); **56**, 474 (1987); **55**, 2095 (1986).
6. M. L. Plumer and A. Mailhot, *Phys. Rev. B* **50**, 16113 (1994).
7. E. H. Boubcheur, D. Loison, and H. T. Diep, *Phys. Rev. B* **54**, 4165 (1996).
8. M. Itakura, *J. Phys. Soc. Jpn.* **72**, 74 (2003).
9. M. L. Plumer and A. Mailhot, *J. Phys. Condens. Matter.* **9**, L165 (1997).
10. M. L. Plumer and A. Mailhot, *Phys. Rev. B* **50**, 6854 (1994).
11. D. Loison and H. T. Diep, *Phys. Rev. B* **50**, 16453 (1994).
12. T. Bhattacharya, A. Billoire, R. Lacaze, and Th. Jolicoeur, *J. Phys. I* (*Paris*) **4**, 181 (1994).
13. D. Loison, A. I. Sokolov, B. Delamotte, S. A. Antonenko, K. D. Schotte, and H. T. Diep, *JETP Lett.* **76**, 337 (2000).
14. Th. Jolicoeur, E. Dagotto, E. Gagliano, and S. Bacci, *Phys. Rev. B* **42**, 4800 (1990).
15. C. L. Henley, *J. Appl. Phys.* **61**, 3962 (1987). Also see: M. T. Heinilä and A. S. Oja, *Phys. Rev. B* **48**, 7227 (1993).
16. J. Villain, R. Bidaux, J.-P. Carton, and R. Conte, *J. Phys.* (*Paris*) **41**, 1263 (1980).

17. W. M. Saslow, M. Gabay, and W. M. Zhang, *Phys. Rev. L* **68**, 3627 (1992).
18. H. Kawamura, *Prog. Theo. Phys. Suppl.* **101**, 545 (1990); W.-M. Zhang, W. M. Saslow, and M. Gabay, *Phys. Rev. B* **44**, 5129 (1991); 0204 (1993).
19. H. T. Diep, *Europhys. Lett.* **7**, 725 (1988); *Phys. Rev. B* **39**, 397 (1989).
20. D. Loison, *Physica A* **275**, 207 (1999).
21. R. Quartu and H. T. Diep, *J. Magn. Magn. Mater.* **182**, 38 (1998).
22. D. Loison and P. Simon, *Phys. Rev. B* **61**, 6114 (2000).
23. H. T. Diep, A. Ghazali, and P. Lallemand, *J. Phys. C* **18**, 5881 (1985).
24. P. Lallemand, H. T. Diep, A. Ghazali, and G. Toulouse, *J. Physique-Lettres* **468**, 1087 (1985).
25. J. L. Alonso *et al.*, *Phys. Rev. B* **53**, 2537 (1996).
26. C. Pinettes and H. T. Diep, *J. Appl. Phys.* **83**, 6317 (1998).
27. A. Moreo, E. Dagotto, T. Jolicoeur, and J. Riera, *Phys. Rev. B* **42**, 6283 (1990), J. Ferrer, *Phys. Rev. B* **47**, 8769 (1993).
28. H. T. Diep and H. Kawamura, *Phys. Rev. B* **40**, 7019 (1989).
29. H. T. Diep, *Phys. Rev. B* **45**, 2863 (1992).
30. J. N. Reimers, *Phys. Rev. B* **45**, 7295 (1992); A. Mailhot and M. L. Plumer, *Phys. Rev. B* **48**, 9881 (1993); S. T. Bramwell, M. J. P. Gingras, and J. N. Reimers, *J. Appl. Phys.* **75**, 5523 (1994).
31. E. H. Boubcheur, R. Quartu, H. T. Diep, and O. Nagai, *Phys. Rev. B* **58**, 400 (1998).
32. A. Dobry and H. T. Diep, *Phys. Rev. B* **51**, 6731 (1995).
33. D. Loison and K. D. Schotte, *Eur. Phys. J. B.* **5**, 735 (1998).
34. D. Loison and K. D. Schotte, *Eur. Phys. J. B* **14**, 125 (2000).
35. D. Loison and K. D. Schotte, Submitted to *Eur. Phys. J. B* (unpublished).
36. H. Kunz and G. Zumbach, *J. Phys. A* **26**, 3121 (1993).
37. H. T. Diep and D. Loison, *J. Appl. Phys.* **76**, 6350 (1994).
38. D. Loison, *Eur. Phys. J. B* **15**, 517 (2000).
39. D. Loison, not published.
40. D. Loison, *Phys. Lett. A* **257**, 83 (1999); *Phys. Lett. A* **264**, 208 (1999).
41. R. J. Baxter, *J. Phys. C* **6**, L445 (1973); *Exactly Solved Models in Statistical Mechanics* (Academic Press, London, 1982).
42. P. Peczak and D. P. Landau, *Phys. Rev. B* **39**, 11932 (1989).
43. D. Loison, submitted to *Euro. J. Phys. B.* (unpublished).
44. G. Zumbach, *Nucl. Phys. B* **413**, 771 (1994); *Phys. Lett. A* **190**, 225 (1994); *Phys. Rev. Lett.* **71**, 2421 (1993).
45. A. Z. Patashinskii and V. I. Pokrovskii, The S-matrix method and unitary relations *Fluctuation Theory of Phase Transitions* §VII, vol. 6 (Pergamon press 1979).
46. J. Zinn-Justin, Section 7.4; Real-time quantum field theory and S-matrix, Section 11.8 Dimensional regularization, minimal subtraction: calculation of RG functions *Quantum Field Theory and Critical Phenomena* (Oxford University Press, Oxford, 1996).
47. M. Tissier, D. Mouhanna, and B. Delamotte, *Phys. Rev. B* **61**, 15327 (2000); *ibid* **67**, 134422 (2003).

48. T. E. Mason, Y. S. Yang, M. F. Collins, B. D. Gaulin, K. N. Clausen, and A. Harrison, *J. Magn. Magn. Mater.* **104–107**, 197 (1992).
49. T. E. Mason, M. F. Collins, and B. D. Gaulin, *J. Phys. C* **20**, L945 (1987).
50. Y. Ajiro, T. Nakashima, Y. Unno, H. Kadowaki, M. Mekata, and N. Achiwa, *J. Phys. Soc. Jpn.* **57**, 2648 (1988).
51. B. D. Gaulin, T. E. Mason, M. F. Collins, and J. F. Larese, *Phys. Rev. Lett.* **62**, 1380 (1989).
52. T. E. Mason, B. D. Gaulin, and M. F. Collins, *Phys. Rev. B* **39**, 586 (1989).
53. H. Kadowaki, S. M. Shapiro, T. Inami, and Y. Ajiro, *J. Phys. Soc. Jpn.* **57**, 2640 (1988).
54. J. Wang, D. P. Belanger, and B. D. Gaulin, *Phys. Rev. Lett.* **66**, 3195 (1991).
55. H. Weber, D. Beckmann, J. Wosnitza, H. V. Löhneysen, and D. Visser, *Int. J. Modern Phys. B* **9**, 1387 (1995).
56. T. Goto, T. Inami, and Y. Ajiro, *J. Phys. Soc. Jpn.* **59**, 2328 (1990).
57. R. Deutschmann, H. V. Löhneysen, J. Wosnitza, R. K. Kremer, and D. Visser, *Euro. Phys. Lett.* **17**, 637 (1992).
58. M. F. Collins, O. A. Petrenko, *Can. J. Phys.* **75**, 605 (1997).
59. T. Kato, T. Asano, Y. Ajiro, S. Kawano, T. Ihii, and K. *Iio Physica B* **213–214**, 182 (1995).
60. T. Kato, K. Iio, T. Hoshimo, T. Mitsui, and H. Tanaka, *J. Phys. Soc. Jpn.* **61**, 275 (1992).
61. H. Tanaka, H. Nakamo, and S. Matsumo, *J. Phys. Soc. Jpn.* **63**, 3169 (1994)
62. H. B. Weber, T. Werner, J. Wosnitza, H. V. Löhneysen, and U. Schotte, *Phys. Rev. B* **54**, 15924 (1996)
63. D. A. Tindall, M. O. Steinitz, and M. L. Plumer, *J. Phys. F* **7**, L263 (1977).
64. K. D. Jayasuriya, S. J. Campbell, and A. M. Stewart, *J. Phys. F* **15**, 225 (1985).
65. B. D. Gaulin, M. Hagen, and H. R. Child, *J. Physique Coll.* **49** C8, 327 (1988).
66. T. R. Thurston, G. Helgesen, D. Gibbs, J. P. Hill, B. D. Gaulin, and G. Shirane, *Phys. Rev. Lett.* **70**, 3151 (1993); T. R. Thurston, G. Helgesen, J. P. Hill, D. Gibbs, B. D. Gaulin, and P. J. Simpson, *Phys. Rev. B* **49**, 15730 (1994).
67. P. Du Plessis, A. M. Venter, and G. H. F. *Brits J. Phys.* (*Cond. Mat.*) **7**, 9863 (1995).
68. J. Ecker and G. Shirane, *Solid State Commun.* **19**, 911 (1976).
69. G. Helgesen, J. P. Hill, T. R. Thurston, D. Gibbs, J. Kwo, and M. Hong, *Phys. Rev. B* **50**, 2990 (1994).
70. S. W. Zochowski, D. A. Tindall, M. Kahrizi, J. Genosser, and M. O. Steinitz, *J. Magn. Magn. Mater.* **54–57**, 707 (1986).
71. H. U. Åström and G. Benediktson, *J. Phys. F* **18**, 2113 (1988).
72. F. L. Lederman and M. B. Salomon, *Solid State Commun.* **15**, 1373 (1974).
73. K. D. Jayasuriya, S. J. Campbell, and A. M. Stewart, *Phys. Rev. B* **31**, 6032 (1985).

74. P. Du Plessis, C. F. Van Doorn, and D. C. Van Delden, *J. Magn. Magn. Mater.* **40**, 91 (1983).
75. G. H. F. Brits and P. Du Plessis, *J. Phys. F* **18**, 2659 (1988).
76. E. Loh, C. L. Chien, and J. C. Walker, *Phys. Lett. A* **49**, 357 (1974).
77. K. D. Jayasuriya, A. M. Stewart, S. J. Campbell, and E. S. R. Gopal, *J. Phys. F* **14**, 1725 (1984).
78. O. W. Dietrich and J. Als-Nielsen, *Phys. Rev.* **162**, 315 (1967).
79. C. C. Tang, P. W. Haycock, W. G. Stirling, C. C. Wilson, D. Keen, and D. Fort, *Physica B* **205**, 105 (1995).
80. K. H. Hirota, G. Shirane, P. M. Gehring, and C. F. Majkrzak, *Phys. Rev B* **49**, 11 967 (1994); P. M. Gehring, K. H. Hirota, C. F. Majkrzak, and G. Shirane, *Phys. Rev. L* **71**, 1087 (1993).
81. C. C. Tang, W. G. Stirling, D. L. Jones, A. J. Rolloson, A. H. Thomas, and D. Fort, *J. Magn. Magn. Mater.* **103**, 86 (1992).
82. H. Kadowaki, K. Ubukoshi, K. Hirakawa, J. L. Martinez, and G. Shirane, *J. Phys. Soc. Jpn.* **56**, 4027 (1987).
83. J. Wosnitza, R. Deutschmann, H. V. Löhneysen, and R. K. Kremer, *J. Phys. Condens. Matt.* **6**, 8045 (1994).
84. K. Koyama and M. Matsuura, *J. Phys. Soc. Jpn.* **54**, 4085 (1985).
85. G. C. DeFotis, F. Palacio, and R. L. Carlin, *Physica B* **95**, 380 (1978); G. C. DeFotis and S. A. Pugh, *Phys. Rev. B* **24**, 6497 (1981); G. C. DeFotis and J. R. Laughlin, *J. Magn. Magn. Mater.* **54–57**, 713 (1986).
86. J. C. Le Guillou and J. Zinn-Justin, *J. Phys.* (*Paris*) *Lett.* **46**, L137 (1985).
87. S. A. Antonenko, A. I. Sokolov, and V. B. Varnashev, *Phys. Lett. A* **208**, 161 (1995).
88. B. Kastening, *Phys. Rev. D* **57**, 3567 (1998).
89. S. A. Antonenko and A. I. Sokolov, *Phys. Rev. B* **49**, 15901 (1994).
90. A. Pelisseto, P. Rossi, and E. Vicari, *Phys. Rev. B* **63**, 140414(R) (2001).
91. P. Calabrese, P. Parruccini, and A. I. Sokolov, *Phys. Rev. B* **66**, 180403(R) (2002).
92. P. Azaria, B. Delamotte, F. Delduc, and T. Jolicoeur, *Nucl. Phys. B* **408**, 485 (1993); P. Azaria, B. Delamotte, and T. Jolicoeur, *Phys. Rev. Lett.* **64**, 3175 (1990); *J. App. Phys.* **69**, 6170 (1991).
93. B. W. Southern and A. P. Young, *Phys. Rev. B* **48**, 13170 (1993).
94. D. R. T. Jones, A. Love, and M. A. Moore, *J. Phys. C* **9**, 743 (1976).
95. B. Nienhuis, A. N. Berker, E. K. Riedel, and M. Schick, *Phys. Rev. Lett.* **43**, 737 (1979).
96. M. Nauenberg and B. Nienhuis, *Phys. Rev. Lett.* **33**, 944 (1974).
97. L. Saul, *Phys. Rev. B* **46**, 13847 (1992).
98. H. Kawamura, *J. Phys. Soc. Jpn.* **59**, 2305 (1990).
99. N. D. Mermin and H. Wagner, *Phys. Rev. Lett.* **17**, 1133 (1966).
100. N. D. Mermin, *Rev. Mod. Phys.* **51**, 591 (1979).
101. G. Toulouse and M. Kleman, *J. Physique Lett.* **37**, 149 (1976).

102. G. Kohring, R. E. Shrock, and P. Wills, *Phys. Rev. Lett.* **57**, 1358 (1986).
103. G. A. Williams, *Phys. Rev. Lett.* **59**, 1926 (1987).
104. S. R. Shenoy and B. Chattopadhyay, *Phys. Rev. B* **51**, 9129 (1995); *Phys. Rev. Lett.* **72**, 400 (1994); B. Chattopadhyay, M. C. Mahato, and S. R. Shenoy, *Phys. Rev. B* **47**, 15159 (1993); S. R. Shenoy, *Phys. Rev. B* **42**, 8595 (1990); *Phys. Rev. B* **40**, 5056 (1989).
105. V. L. Berezinskii, *Soviet Phys. JETP* **32**, 493 (1971); J. M. Kosterlitz and D. J. Thouless, *J. Phys. C* **6**, 1181 (1973); J. M. Kosterlitz, *J. Phys. C* **7**, 1046 (1974).
106. J. Lee, J. M. Kosterlitz, and E. Granato, *Phys. Rev. B* **43**, 11531 (1991).
107. S. Miyashita and H. Shiba, *J. Phys. Soc. Jpn.* **53**, 1145 (1984).
108. S. Lee and K. C. Lee, *Phys. Rev. B* **57**, 8472 (1998).
109. L. Caprioti, R. Vaia, A. Cuccoli, and V. Togneti, *Phys. Rev. B* **58**, 273 (1998).
110. D. Loison, cond-mat/0001122, unpublished.
111. S. Teitel and C. Jayaprakash, *Phys. Rev. Lett.* **51**, 1999 (1983); *Phys. Rev. B* **27**, 598 (1983).
112. B. Berge, H. T. Diep, A. Ghazali, and P. Lallemand, *Phys. Rev. B* **34**, 3177 (1986).
113. D. B. Nicolaides, *J. Phys. A* **24**, L231 (1991).
114. J. V. Jose and G. Ramirez-Santiago, *Phys. Rev. Lett.* **77**, 4849 (1996); **68**, 1224 (1992); *Phys. Rev. B* **49**, 9567 (1994).
115. J. M. Thjissen and H. J. F. Knops, *Phys. Rev. B* **42**, 2438 (1990).
116. E. Granato, M. P. Nightingale, *Phys. Rev. B* **48**, 7438 (1993).
117. S. Lee and K. C. Lee, *Phys. Rev. B* **49**, 15184 (1994).
118. H. J. Luo, L. Schülke, and B. Zheng, *Phys. Rev. Lett.* **81**, 180 (1998).
119. M. Benakly and E. Granato, *Phys. Rev. B* **55**, 8361 (1997).
120. M. P. Nightingale, E. Granato, and J. M. Kosterlitz, *Phys. Rev. B* **52**, 7402 (1995).
121. J. Lee, E. Granato, and J. M. Kosterlitz, *Phys. Rev. Lett.* **66**, 2090 (1991); *Phys. Rev. B* **44**, 4819 (1991).
122. P. Olsson, *Phys. Rev. Lett.* **75**, 2758 (1995); **77**, 4850 (1996); *Phys. Rev. B* **55**, 3585 (1997).
123. Y. M. M. Knops, B. Nienhuis, H. J. F. Knops, and J. W. Blöte, *Phys. Rev. B* **50**, 1061 (1994).
124. E. Granato, *Phys. Rev. B* **45**, 2557 (1992).
125. M. Yosefin and E. Domany, *Phys. Rev. B* **32**, 1778 (1985).
126. P. Minnhagen, *Phys. Rev. Lett.* **54**, 2351 (1985); *Phys. Rev. B* **32**, 3088 (1985).
127. G. S. Grest, *Phys. Rev. B* **39**, 9267 (1989).
128. J. R. Lee, *Phys. Rev. B* **49**, 3317 (1994).
129. M. Y. Choi and D. Stoud, *Phys. Rev. B* **32**, 5773 (1985).
130. G. S. Jeon, S. Y. Park, and M. Y. Choi, *Phys. Rev. B* **55**, 14088 (1997).
131. S. Lee, K. C. Lee, and J. M. Kosterlitz, cond-mat/9612242 (unpublished).
132. C. J. Lobb, *Physica B* **152**, 1 (1988).

133. X. S. Ling, H. J. Lezec, J. S. Tsai, J. Fujita, H. Numata, Y. Nakamura, Y. Ochiai, Chao Tang, P. M. Chaikin, and S. Bhattacharya, *Phys. Rev. Lett.* **76**, 2989 (1996).
134. T. C. Halsey, *J. Phys. C* **18**, 2437 (1985).
135. S. E. Korshunov, *J. Stat. Phys.* **43**, 17 (1986).
136. V. Kotsubo, K. D. Hahn, and J. P. Parpia, *Phys. Rev. Lett.* **58**, 804 (1987).
137. J. Xu and C. Crooker, *Phys. Rev. Lett.* **65**, 3005 (1990).
138. T. Garel and S. Doniach, *J. Phys. C* **13**, L887 (1980).
139. The Fast Linear Algorithm developed by D. Loison is three times faster than the Metropolis algorithm at the critical temperature for frustrated two dimensional systems. Submitted to *Euro. J. Phys. B*. Accessible at the home page http://www.physik.fu-berlin.de/~loison/
140. D. Loison, submitted to *Euro. J. Phys. B.* (unpublished).
141. H. Kawamura and S. Miyashita, *J. Phys. Soc. Jpn.* **53**, 1145 (1985).
142. M. Caffarel, P. Azaria, B. Delamotte, and D. Mouhanna, *Phys. Rev. B* **64**, 014412 (2001).
143. B. W. Southern and H. J. Xu, *Phys. Rev. B* **52**, R3836 (1995).
144. H. Kawamura and M. Kikuchi, *Phys. Rev. B* **47**, 1134 (1993).
145. M. Wintel, H. U. Everts, and W. Apel, *Europhys. Lett.* **25**, 711 (1994); *Phys. Rev. B* **52**, 13480 (1995).
146. A. Peles, B. W. Southern, B. Delamotte, D. Mouhanna, and M. Tissier, *Phys. Rev. B* **69**, 220408 (2004).
147. K. Kanki, D. Loison, and K.-D. Schotte, *J. Phys. Soc. Jpn.* **75**, 015001 (2006).
148. F. Wang and D. P. Landau, *Phys. Rev. Lett.* **86**, 2050 (2001); *Phys. Rev. E* **64**, 056101 (2001).
149. V. Thanh Ngo and H. T. Diep, *J. Appl. Phys.* **103**, 07C712 (2008).
150. V. Thanh Ngo and H. T. Diep, *Phys. Rev. E* **78**, 031119 (2008).
151. V. Thanh Ngo, D. Tien Hoang, and H. T. Diep, *J. Phys. Cond. Matt.* **23**, 226002 (2011).
152. V. Thanh Ngo, D. Tien Hoang, and H. T. Diep, *Phys. Rev. E* **82**, 041123 (2010).
153. V. Thanh Ngo, D. Tien Hoang, and H. T. Diep, *Modern Phys. Lett. B* **25**, 929–936 (2011).
154. Danh-Tai Hoang and H. T. Diep, *Phys. Rev. E* **85**, 041107 (2012).
155. M. Creutz, *Phys. Rev. B* **36**, 515 (1987).
156. U. Wolff, *Phys. Rev. Lett.* **62**, 361 (1989); *Nucl. Phys. B* **322**, 759 (1989).
157. H. Müller-Krumbhaar and K. Binder, *J. Stat. Phys.* **8**, 1 (1973).
158. B. Efron, *The Jackknife, The Bootstrap and Other Resampling Plans* (SIAM, Philadelphia, PA, 1982).
159. A. M. Ferrenberg and R. H. Swendsen, *Phys. Rev. Lett.* **61**, (1988) 2635; *ibid* **63**, (1989) 1195; *Comput Phys.* Sept/Oct (1989) 101.
160. V. Privman and M. E. Fisher, *J. Stat. Phys.* **33**, 385 (1983).
161. K. Binder, *Rep. Prog. Phys.* **50**, 783 (1987).

162. A. Billoire, R. Lacaze, and A. Morel, *Nucl. Phys. B* **370**, 773 (1992).
163. G. Prussner, D. Loison, and K. D. Schotte, *Physica A* **299**, 557 (2001).
164. K. Binder, *Z. Phys. B* **43**, 119 (1981).
165. D. Loison, *Physica A* **271**, 157 (1991).
166. D. Loison, *J. Phys. Condens. Matter.* **11**, L401 (1999).
167. H. K. Janssen, B. Schaub, and B. Schmittmann, *Z. Phys. B* **73**, 539 (1989).
168. D. A. Huse, *Phys. Rev. B* **40**, 304 (1989).
169. K. Humayun and A. J. Bray, *J. Phys. A* **24**, 1915 (1991).
170. B. Zhang, *Int. J. Mod. Phys.* **12**, 1419 (1998).
171. H. J. Luo and B. Zheng, *Mod. Phys. Lett. B* **11**, 615 (1997).

CHAPTER 5

TWO-DIMENSIONAL QUANTUM ANTIFERROMAGNETS

Grégoire Misguich*,‡ and Claire Lhuillier†,§

*Institut de Physique Théorique,
CEA, IPhT, F-91191 Gif-sur-Yvette, France
CNRS, URA 2306, F-91191 Gif-sur-Yvette, France
†Laboratoire de Physique Théorique de la matière condensée
Université P. et M. Curie and UMR 7600 of CNRS
Case 121, 4 Place Jussieu, 75252 Paris Cedex, France
‡gregoire.misguich@cea.fr
§claire.lhuillier@upmc.fr

This review presents some theoretical advances in the field of quantum magnetism in 2D systems and, in particular, quantum spin liquids. The first edition published in 2005 has been updated for the present second edition of the book: the section devoted to the kagomé antiferromagnet (Section 5.7) as well as the concluding section (Section 5.8) have been completely rewritten. The other sections (Sections 5.1–5.6) are unchanged from the first edition of the book.

5.1. Introduction

In this review we present some theoretical advances in the field of quantum magnetism in 2D systems. The spin-1/2 nearest-neighbor 2D Heisenberg models on Bravais lattices (square,[1] triangular[2]) are Néel ordered[a] at $T = 0$. Frustration, small coordination number, and competition between

[a]This generic kind of order with a macroscopic sublattice magnetization is called *magnetic* long-range order (LRO), in contrast to other ordered phases where the long-range ordered correlations concern $S = 0$ scalar observables (on dimers, quadrumers, etc.).

interactions can lead to specific quantum phases without magnetic long-range order. This subject has been a highly debated issue in the field of quantum magnetism in the past decade. It was revived by the discovery of high-T_c superconductivity in the doped cuprates and fuelled by numerous experimental studies of 2D antiferromagnetic insulators.[3]

Section 5.2 is devoted to the spin-1/2 Heisenberg model on the square lattice with first- and second-neighbor interactions (J_1–J_2 model). This model is one of the most studied in the field and this section is a short guide to the literature, with a special emphasis on the various methods used for this problem.

Section 5.3 deals with the general properties of valence bond crystals (VBCs) and related states, the simplest phase which is commonly realized in frustrated spin systems without magnetic LRO.

In Section 5.4, we present large-N generalizations of the Heisenberg model. This approach was extensively developed by Read and Sachdev during the early 1990s and has been the first to give an insight into the alternative between VBC and related phases, which have a long-range order in local singlet patterns (whence the name of crystals) and resonating valence bond (RVB) spin liquids (SLs) which are translationally invariant phases where the quantum coherence is a central issue.

Section 5.5 presents some results of quantum dimer models (QDMs). These models are effective approaches to the quantum phases of antiferromagnets which are dominated by short-range valence bonds (or singlets). They have recently received special attention and provide useful insights onto the phenomenology of VBC and RVB SL.

In Section 5.6, we review some results concerning models with multiple-spin exchange (MSE) (also called ring exchange) interactions. These interactions are now recognized to be present in several physical systems and appear to play an important role in the stabilization of RVB liquid ground states (GSs).

The last section is devoted to the Heisenberg model on the kagomé lattice (and related models). Despite currently ongoing important studies on this subject, the understanding of the low-energy physics of the spin-1/2 kagomé antiferromagnet remains a challenging problem and we discuss some of the important results and questions.

We should warn the readers that this review is quite "inhomogeneous" and cannot, of course, replace textbooks.[4–7] While some parts deal with some rather recent works (QDM or MSE for instance), some others are devoted to older results which we think are still of importance for current research (beginning of the section J_1–J_2, large-N). The final part devoted to kagomé reflects our own views and some unpublished material on still unsettled issues. Some parts are intended to be more pedagogical and concrete (QDM and beginning of large-N section) while some others contain more qualitative discussions of the physical issues (end of the section J_1–J_2, VBC, kagomé).

5.2. J_1–J_2 Model on the Square Lattice

We consider the following Heisenberg model on the square lattice:

$$\mathcal{H} = 2J_1 \sum_{\langle ij \rangle} \vec{S}_i \cdot \vec{S}_j + 2J_2 \sum_{\langle\langle ij \rangle\rangle} \vec{S}_i \cdot \vec{S}_j \tag{5.1}$$

where $\langle ij \rangle$ and $\langle\langle ij \rangle\rangle$ denote pairs of nearest and next-nearest-neighbors respectively. Although quite simple in appearance, this spin model realizes several interesting phenomena which are relevant to a large class of 2D frustrated quantum magnets: classical degeneracy, order by disorder, destruction of some long-range order by quantum fluctuations, break down of the spin wave expansion, opening of a spin gap, and (possibly?) spontaneous translation symmetry breaking, etc. For this reason, we start with a general overview of some important results concerning this system. We will focus on the properties related to *frustration*. A review on the non-frustrated model ($J_2 = 0$) can be found in Ref. [1].

5.2.1. *Classical ground state and spin wave analysis*

It is easy to find *some* classical GS of a translation-invariant Heisenberg model on a Bravais lattice because the energy can always be minimized by a planar helix

$$\vec{S}_{\mathbf{r}} = \vec{e}_1 \cos(\mathbf{q} \cdot \mathbf{r}) + \vec{e}_2 \sin(\mathbf{q} \cdot \mathbf{r}) \tag{5.2}$$

provided that the pitch $\mathbf{q}$ minimizes the Fourier transform $J(\mathbf{q})$ of the coupling.[8] In the case of the J_1–J_2 model, one has

$$J(\mathbf{q}) = 2J_1(\cos(q_x) + \cos(q_y)) + 2J_2(\cos(q_x + q_y) + \cos(q_x - q_y)) \tag{5.3}$$

- **$J_2 < 0.5J_1$:** $J(\mathbf{q})$ has a single minimum at (π, π). It corresponds to the "usual" Néel state.

- **$J_2 > 0.5J_1$:** $J(\mathbf{q})$ has two isolated minima at $(0, \pi)$ and $(\pi, 0)$. They correspond to ferromagnetic lines (respectively, columns) arranged in an antiferromagnetic way. These states are sometimes called *collinear* (in *real* space). From these planar helix states, one can build many other GS by rotating globally all the spins of one sublattice with respect to the other. Although this costs no energy for classical spins at zero temperature, it is known (order by disorder, see below) that configurations where both sublattices have their staggered magnetization *collinear in spin space* are selected by thermal or quantum fluctuations.
- **$J_2 = 0.5J_1$:** $J(\mathbf{q})$ has lines of minima around the edges of the Brillouin zone. At this point, the classical GS is highly degenerate: We can write $\mathcal{H} = \mathrm{cst} + J_2 \sum (S_1 + S_2 + S_3 + S_4)^2$ where the sum runs over all square plaquettes and any state where each plaquette has a vanishing total spin minimizes the classical energy.

Even at the lowest order in $1/S$, zero-temperature quantum corrections to the sublattice magnetization (order parameter) diverge around $J_2 = 0.5J_1$.[9] Such large-S approximation usually tends to overestimate the stability of the magnetic phases; therefore, this breakdown around $J_2 \sim 0.5J_1$ is a strong evidence for the existence of quantum disordered phase(s) in this region of parameter space.

5.2.2. *Order by disorder* ($J_2 > J_1/2$)

The concept of "order by disorder" was introduced in 1980 by Villain and co-workers[10] in the study of a frustrated Ising model on the square lattice. In this model, the next-neighbor couplings along the rows are ferromagnetic as well as those on the odd columns (named A in the following). The couplings on the even columns (named B) are antiferromagnetic. At $T = 0$, the GS has no average magnetization and is disordered. This changes when thermal fluctuations are introduced: a B-chain sandwiched between two A-chains with parallel spins has *lower excitations* than a B-chain between two A-chains with anti-parallel spins. This gives a larger Boltzmann weight to ferrimagnetically ordered states. Villain *et al.* have exactly shown that the system is indeed ferrimagnetic at low temperatures. They were also able to show that site dilution (non-magnetic sites) selects the same ordered pattern, whence the name of "order by disorder".

A somewhat less drastic phenomenon has been observed in quantum systems. It is the selection of particular long-range ordered quantum states among a larger family of ordered solutions which are classically degenerate at $T = 0$.[b] Consider a spin system in which the molecular field created by the spins of one sublattice on the other cancels, which is the case when $J_2 > 0.5J_1$. Shender[11] showed that if fluctuations are included, the system will select states in which all spins are collinear to each other. This follows from the fact that (moderate) fluctuations at one site are orthogonal to the mean value of the magnetization at that site and the system can gain some magnetic exchange energy by making such fluctuations coplanar on neighboring sites, that is to make the spins collinear. Such a selection of order by quantum fluctuations (and dilution) was discussed by Henley[12] and also appears quite distinctly in a spin wave expansion.[13]

This selection of the $(\pi, 0)$ or $(0, \pi)$ order spontaneously breaks a four-fold lattice symmetry. An Ising order parameter is thus generated. It takes two values depending on whether the ferromagnetic correlations are locally arranged horizontally or vertically. Chandra and co-workers[14] have studied this mechanism and predicted the existence of a finite-temperature Ising phase transition independent of the subsequent development of a sublattice magnetization. This result has been questioned recently[15] and the transition has not been observed so far in the spin-1/2 model.[15,16] It has however been confirmed by some recent Monte Carlo (MC) simulations of the *classical* Heisenberg model.[17] Very similar phenomena are present in the J_1–J_2 quantum Heisenberg model on the *triangular* lattice.[18–21]

Melzi *et al.*[22,23] have studied a quasi 2D spin-1/2 system which is believed to be a J_1–J_2 square lattice Heisenberg antiferromagnet. They found some evidence (splitting of NMR lines) for a collinear ($(\pi, 0)$ or $(0, \pi)$) magnetic ordering. Several estimates[16,23–25] indeed point to $J_2 > J_1$ in this compound.

5.2.3. *Non-magnetic region* ($J_2 \simeq J_1/2$)

Consider the two classical "Ising states" corresponding to the wave vectors (π, π) and $(\pi, 0)$. They can be taken as (crude) variational states

---

[b]In Villain's model the system is truly disordered at $T = 0$ and an ordered solution is entropically selected at finite temperature. In the quantum J_1–J_2 model discussed above, the classical solutions can adopt various ordered patterns: quantum fluctuations select among these patterns the most ordered one, that is the pattern with the highest symmetry and the smallest degeneracy. The ultimate effect of these quantum fluctuations can be the destruction of the Néel order in favor of a fully quantum GS with $\mathcal{O}(1)$ degeneracy.

for the Hamiltonian equation (5.1). Their energies (per site) are $E_{\pi,\pi} = -J_1 + J_2$ and $E_{\pi,0} = -J_2$. As discussed above, these states cross at $J_2 = \frac{1}{2}J_1$. However, one can also consider any first-neighbor singlet (or valence bond) covering of the lattice as another variational state. In such a completely dimerized state, the expectation value of the energy per site is $E_{\text{dimer}} = -\frac{3}{4}J_1$, which is below the two Ising states around $J_2 \simeq J_1/2$. Of course, this very simple argument does not prove anything since "dressing" these classical states with quantum fluctuations (spin flips in the Néel-like states or valence-bond motions in the dimerized wave functions) will lower the energies of all these trial states and it is absolutely not clear which one may eventually win. Nevertheless, this shows in a simple way why non-magnetic states (i.e. rotationally invariant or spin singlet) such as dimerized states can be a route to minimize the energy in a frustrated magnet.[c]

5.2.3.1. *Series expansions*

High-order series expansions can be a powerful technique to investigate frustrated quantum magnets. The general method to generate zero-temperature perturbation expansions in quantum many-body systems was described by Singh *et al.*[28] and Gelfand *et al.*[29] For instance, one can consider the following anisotropic model:

$$\begin{aligned}\mathcal{H}(\lambda) = 2J_1 \sum_{\langle ij\rangle}&[S_i^z S_j^z + \lambda(S_i^x S_j^x + S_i^y S_j^y)]\\ &+ 2J_2 \sum_{\langle\langle ij\rangle\rangle} [S_i^z S_j^z + \lambda(S_i^x S_j^x + S_i^y S_j^y)],\end{aligned} \tag{5.4}$$

where $\mathcal{H}(\lambda = 0)$ is a classical Ising model the GS of which is known. The series expansion about the Ising limit amounts to compute expectation values in the GS $|\lambda\rangle$ of $\mathcal{H}(\lambda)$ in powers of λ:

$$\frac{\langle\lambda|\hat{O}|\lambda\rangle}{\langle\lambda|\lambda\rangle} = \sum_n a_n \lambda^n \tag{5.5}$$

(energy gaps, dispersion relations, and susceptibilities can also be computed in the same approach). The calculation of a_n requires the enumeration and

[c]Klein[26] introduced a general procedure to generate local and $SU(2)$ symmetric Hamiltonians for which any first-neighbor dimerized state is an exact GS. These Hamiltonians are simply defined as sums of projectors which annihilate all dimer coverings. The Majumdar–Gosh[27] chain is the simplest example of a "Klein model".

evaluation of the *connected clusters* of size $\sim n$, whose number grows exponentially with n. Depending on the quantity $\hat{O}$ and on the model, orders from 7 to 20 can typically be obtained on present computers. The series is then extrapolated to $\lambda = 1$ by standard Padé, Dlog Padé or integrated differential approximations. Such a series expansion about the Ising limit was done by Weihong *et al.*[30] for the first-neighbor square lattice antiferromagnet. Oitmaa and Weihong[31] extended the series to the J_1–J_2 model, where each a_n is now a polynomial in J_1 and J_2. The disappearance of Néel order in the Heisenberg model manifests itself by a vanishing sublattice magnetization as well as some singular behavior of the series for $\lambda_c < 1$. The results indicate the absence of Néel long-range order in the strongly frustrated region $0.4 \leq J_2/J_1 \leq 0.6$. Such an expansion can locate with a reasonable accuracy the phase boundary of the Néel ordered regions but provides no direct information on the nature of the non-magnetic phase.

To study the model around $J_2 \simeq J_1/2$, several other expansions have been carried out. A dimer expansion about an exactly dimerized model was done by Gelfand *et al.*,[32] Gelfand,[33] Singh *et al.*,[34] and Kotov *et al.*[35] In this approach, J_1 and J_2 are set to zero everywhere except on isolated bonds arranged in a columnar way and all the other couplings are treated perturbatively. At zeroth order, the GS is simply a product of singlets. In these calculations, the dimerized phase remains stable in the intermediate region. Singh *et al.*[34] also performed a different kind of zero-temperature series expansion. They started from a model of isolated four-spin plaquettes in order to check a prediction made by Zhitomirsky and Ueda[36] that such plaquettes could spontaneously form around $J_2 \simeq J_1/2$ to produce a state which is invariant under $\pi/4$ lattice rotations. Although the GS energy they obtained is very close to the one obtained from the dimerized limit (within error bars of the extrapolation procedure), they observed an instability in the plaquette scenario (the triplet gap vanishes before reaching the isotropic square lattice model) which suggests that the plaquette order is not the issue (the analysis of the exact numerical spectra for 36 sites confirmed this result[37]).

Sushkov *et al.*[38] (improved numerical results compared to Ref. [39]) computed the susceptibility χ_D associated with the appearance of columnar dimer order in the (π, π) Néel phase by a series expansion about the Ising limit. Such a susceptibility seems to diverge at $J_2/J_1 = g_{c1} \simeq 0.405 \pm 0.04$. On the contrary, the disappearance of the magnetic LRO is observed (through the Néel order parameter or through the anisotropy in spin space of the spin–spin correlations) at $J_2/J_1 = g_{c2} \simeq 0.39 \pm 0.02$.

This point could *a priori* be *different* from g_{c1}. In such a case, the system would first break the $\pi/4$ lattice rotation symmetry at g_{c1}, while magnetic LRO remains (gapless spin waves). Only at $g_{c2} > g_{c1}$ the $SU(2)$ rotation symmetry is restored and the magnetic excitations acquire a gap. From field theoretical arguments based on effective actions valid close to the critical points, Sushkov *et al.*[38] argue that the proximity (or possible equality) of g_{c1} and g_{c2} is a general feature in frustrated magnets which originate from the coupling of triplet and singlet excitations.

Sushkov *et al.*[39] computed the susceptibility χ_P associated to plaquette order by an expansion around the dimerized limit, assuming that the system has columnar dimer LRO. The result shows a divergence of χ_P when $J_2/J_1 \to g_{c3} = 0.5 \pm 0.02$. From these results, Sushkov *et al.* suggested that the translation symmetry along the columns is broken down at g_{c3} (giving rise to an eight-fold degenerate GS in the thermodynamic limit) before the $(\pi, 0) - (0, \pi)$ magnetically ordered phase appears at $g_{c4} \simeq 0.6$. This picture is qualitatively consistent with the spin–spin correlations computed in a 10×10 system with a density matrix renormalization group (DMRG) algorithm.[40]

Due to the relatively short series (typically of order 7) involved and the uncertainties in the extrapolation procedures, such results should be confirmed by other methods, but this succession of quantum phase transitions represents a very interesting scenario. We note that if the model has a fully symmetric liquid GS in some parameter range, it should be difficult to capture from series expansions about limits where some lattice symmetries are explicitly broken.

5.2.3.2. *Exact diagonalizations*

Exact diagonalizations have *a priori* no bias and were used very early in this field.[41–43] Large-size computations and sophisticated finite-size scaling analysis are nevertheless needed to extract significant information. Schulz *et al.*[44] performed extensive exact diagonalizations of the J_1–J_2 model for system sizes up to 36 sites. They analyzed the behavior of several quantities such as structure factors (Néel order parameter), GS energy, spin wave velocities (obtained from the finite-size corrections to the GS energy), spin stiffness, and uniform susceptibility. Their analysis, including quantitative comparisons with nonlinear sigma model predictions,[45] concluded to the absence of Néel long-range order in the strongly frustrated region $0.4 \leq J_2/J_1 \leq 0.6$. There, they show enhanced columnar dimer–dimer correlations as well as chiral ones but the size effects were not clear enough to discriminate between short- or long-range order for these order parameters.

5.2.3.3. *Quantum Monte Carlo*

Quantum Monte Carlo (QMC) methods have been extensively applied to the J_1–J_2 model in the low-frustration regime giving an highly accurate description of the Néel phase (Ref. [46] and references therein). In the non-magnetic and highly frustrated regime, a simple QMC approach is ineffective due to the so-called sign problem. The fixed-node approach is the first answer to this problem: the exact imaginary time propagator $e^{-\tau\mathcal{H}}$ used to filter out the GS from a variational guess $|\psi_g\rangle$ is replaced by an approximate propagator, which has the same nodes as $|\psi_g\rangle$. The quality of the result depends on the quality of the nodal regions of $|\psi_g\rangle$. Various schemes have been used to try to go beyond this limitation: stochastic reconfiguration,[47] eventually associated to a few Lanczos iterations.[37,48] An alternative method has been devised by du Croo de Jongh *et al.*,[40] where the guiding function is replaced by the result of a DMRG calculation.[49,50] Both methods have their own bias. Using the first of the schemes, Capriotti and Sorella[37] concluded that for $J_2/J_1 \sim 0.45$, a Gutzwiller-projected BCS wave function $|p\,BCS\rangle$ was an excellent guiding wave function:

$$|p\,BCS\rangle = \hat{\Pi}|BCS\rangle, \tag{5.6}$$

$$|BCS\rangle = \exp\left(\sum_{i,j} f_{i,j} c^{\dagger}_{i\uparrow} c\dagger_{j\downarrow}\right)|0\rangle, \tag{5.7}$$

where $|0\rangle$ is the fermion vacuum, $c^{\dagger}_{i\uparrow} c\dagger_{j\downarrow}$ creates a valence bond on sites (i, j), and $\hat{\Pi}$ projects out states with double occupancy. The pairing amplitude $f_{i,j}$ (often called gap function Δ_k)[d] is optimized with an MC algorithm in order to minimize the energy. Capriotti and Sorella gave convincing indications that their wave function is quite accurate. The best variational energies are obtained in the frustrated region with a pairing amplitude which mixes $d_{x^2-y^2}$ and d_{xy} symmetries. In particular, it reproduces the correct nodal structure of the GS in the frustrated region at least for moderate system sizes where the variational result can be checked against the exact

[d] After the Gutzwiller projection, Δ_k is no longer the observable gap.[51]

result. This is a subtle and non-trivial information for systems which do not obey the Marshall's sign rule such as this frustrated model. They concluded from these results that the system probably had no long-range order neither in dimer–dimer correlations nor in four-spin plaquette correlations. On the other hand, du Croo de Jongh *et al.* argued in favor of columnar dimerized phase which also breaks the translation symmetry along the columns (plaquette-like correlations similar to those found by series expansions[39]).

The comparison of the results of these different approaches shows that this problem remains a very challenging one. The model in the frustrated regime is probably never very far from a quantum critical point and in these conditions none of the available methods seems to be able to discriminate between a VBC with tiny gaps both in the singlet and triplet sectors, a critical phase with a quasi order in dimers and gapless singlet excitations, and a true SL with gaps in any sector of spin but no long-range order in any observable. As we will explain in the following sections, some other frustrated models are thoroughly embedded in the non-magnetic phases and exhibit quantum phases which are easier to characterize.

5.3. Valence Bond Crystals

5.3.1. *Definitions*

Among the different quantum solutions to overcome frustration, the VBC is the simplest scenario. In this phase, neighboring spins arrange themselves in a regular pattern of singlets: dimers,[e] quadrumers, or $2n$-mers $S = 0$ plaquettes. The stability of this phase comes from the extreme stability of small $S = 0$ clusters (recall that the energy of a singlet of two spins $1/2$ is $-3/4$ to be compared to the energy of two classical (or Ising) spins which is only $-1/4$) and eventually from the fact that frustrated bonds between two different singlets do not contribute to the total energy.

In a VBC phase, there is no $SU(2)$ symmetry breaking, no long-range order in spin–spin correlations, but long-range order in dimer–dimer or larger singlet units. Except at a quantum critical point, all excitations of a VBC are gapped. Depending on the lattice geometry, such a wave function can spontaneously break some lattice symmetry (*spontaneous VBC*) or can remain fully symmetric (*explicit VBC*). In a strict sense, the name VBC

[e]Whence, the name Spin Peierls phase is sometimes given to the VBC phase.

should be reserved for systems with a spontaneous lattice symmetry breaking. However, since these two kinds of systems share many similarities, we will discuss both in this section.

When the Hamiltonian has some inequivalent bonds and an integer spin in the unit cell (even number of spin-1/2 for instance), the system can take full advantage of the strong bonds and minimize the effects of the frustrating ones. In that case, the *explicit* VBC is the "natural" strong coupling solution. One can build a simple Hamiltonian in which the bonds which are not occupied by the singlet objects are turned off. The resulting model is a set of small decoupled clusters (dimers or larger plaquettes) and the GS is a trivial product of singlets. Importantly, this strong coupling limit has the same lattice symmetry as the original one. Going back to the original Hamiltonian, *no quantum phase transition is encountered when going from the trivial singlet product up to a real interacting GS.* Models with an half-odd-integer spin in the unit cell cannot realize a VBC unless they *spontaneously* enlarge their unit cell. In these situations, there is no unique elected position for the $2n$-mers and a symmetry breaking must take place in order to form a VBC. Examples of these two kinds of VBC will be given below.

5.3.2. *One-dimensional and quasi- one-dimensional examples (spin-1/2 systems)*

One of the simplest example of (spontaneous) VBC is observed in the J_1–J_2 model on the chain for $J_2/J_1 > (J_2/J_1)_c \sim 0.24$.[27,52–55] For $J_2/J_1 = 0.5$, the doubly degenerate GS are exact products of dimers[5,27]:

$$|MG_{\pm}\rangle = \prod_{n=1}^{N/2} |(2n, 2n \pm 1)\rangle. \tag{5.8}$$

Here and in the following, we call "dimer" a pair of spins in a singlet state, and note it:

$$|(i,j)\rangle = \frac{1}{\sqrt{2}}[|i,+\rangle|j,-\rangle - |i,-\rangle|j,+\rangle]. \tag{5.9}$$

For all $J_2/J_1 > (J_2/J_1)_c$, the GS are products of dimers, dressed by fluctuations of valence bonds, and dimer long-range order persists in all the range of parameters. This model has gapful excitations which can be described as pairs of scattering spin-1/2 solitons separating the two exact GS[52] (these fractionalized excitations are specific for the 1D chain).

The Heisenberg chain with alternating strong and weak bonds (Spin Peierls instability) indeed has a unique GS where the dimers are mainly located on the strong bonds. In that case, the spin-1/2 excitations are confined by the underlying potential and the true excitations are gapful integer magnons Refs. [56, 57] and references therein). It is an explicit VBC.

A two-leg ladder with AF rung exchange also has a unique VBC GS and gapped magnons as excitations.[58] On the contrary, Nersesyan and Tsvelik[59,60] have proposed an example of frustrated ladder, with a spontaneously dimerized GS and gapful excitations. Excitations of this last model are identified as pairs of singlet and triplet domain walls connecting the two GS, and they form a continuum.

As can be seen from this rapid and non-exhaustive enumeration, VBC GS are relatively frequent in frustrated 1D spin-1/2 models. All these systems are gapful, but the excitations could be of a different nature emerging as modes (associated to integer spin excitations) or as a continuum of pairs of excitations that could be either fractionalized (it is then specific for 1D systems) or not.

5.3.3. *Valence Bond Solids*

The VBS wave function was introduced by Affleck, Kennedy, Lieb, and Tasaki (AKLT).[61,62] It can be constructed whenever the spin S on a site is a multiple of one half the coordination number z: $2S = 0 \bmod z$. Let us consider the simplest case $2S = z$. In that case, the local spin S can be seen as the symmetric combination of $2S$ (fictitious) spin-1/2. Now on each bond of the lattice, one can make a singlet between two fictitious spin-1/2. Such a product of singlets does not belong to the physical Hilbert space of the original spin-S model but to a much larger space. The VBS wave function is defined as the projection of the singlet-product state onto the physical space. This projection amounts to symmetrize (for all lattice sites) the wave function with respect to the fictitious spins to force them into a physical spin-S state. A VBS can be viewed as an *explicit VBC of the fictitious spins*. Simple Hamiltonians with short-range and $SU(2)$-symmetric interactions for which the VBS is an exact GS can be constructed (sum of projectors[61,62]). By construction, the VBS wave function is a spin singlet and *breaks no lattice symmetry*. By extension, we may say that a system is in a VBS *phase* if its GS can be adiabatically transformed into the VBS wave function without crossing a phase transition. Similar to the VBC, models in the VBS phase have a gap to all excitations[f] but their wave functions are slightly more complex and their order parameter is non-local. The order of VBS is associated to long-range singlet–singlet correlations in the *fictitious spins*. Expressing such are observable in terms of the physical spins leads to a non-local quantity called *string order parameter*.[63,64] Contrary to explicit VBC, VBS have fractionalized degrees of freedom at the edges of the system with open boundary conditions. These are simply associated to the unpaired fictitious spins. To our knowledge, these properties have not been explored in quantum 2D systems.

The spin-1 Heisenberg chain is the prototype of VBS in 1D.[g] Such a state has a unique GS, a gap in the excitations, and exponentially decreasing spin–spin and dimer–dimer[65] correlations. See the chapter by P. Lecheminant in this volume for more details about the VBS phase of the spin-1 chain.

A spin-3/2 specific $SU(2)$-invariant model on the honeycomb lattice[61,62] is another example of 2D VBS. The spin-1 Heisenberg model on the kagomé lattice was proposed to realize a VBS-like GS[66] in which singlets form on every hexagon without any spontaneous symmetry breaking (hexagonal singlet solid).[h] A similar approach was carried out for the spin-1 pyrochlore Heisenberg model.[67] In that case a lattice distortion was invoked to lift the degeneracy between the two singlet states of each tetrahedron.

5.3.4. *Two-dimensional examples of VBC*

5.3.4.1. *Without spontaneous lattice symmetry breaking*

Two spin-1/2 experimental examples of 2D (explicit) VBC have recently attracted attention: CaV_4O_9[68–74] and $SrCu_2(BO_3)_2$.[75–84] In CaV_4O_9, the V^{4+} ions are on a 1/5-depleted square lattice, and in $SrCu_2(BO_3)_2$, the exchange couplings between Cn^{2+} ions realize the Shastry–Sutherland

[f] However this may not always be true when the coordination number of the lattice is large.[61] In such cases, the VBS wave function is still a spin singlet but has long-range spin–spin correlations. We do not consider such cases here.

[g] In 1D, some authors call "Haldane systems" all the spin-gapped phases, whatever their true nature: VBC or VBS.

[h] Each kagomé site belongs to two hexagons. Each physical spin-1 can be split into two spin-1/2, each of them being involved in the formation of a singlet on one neighboring hexagon.

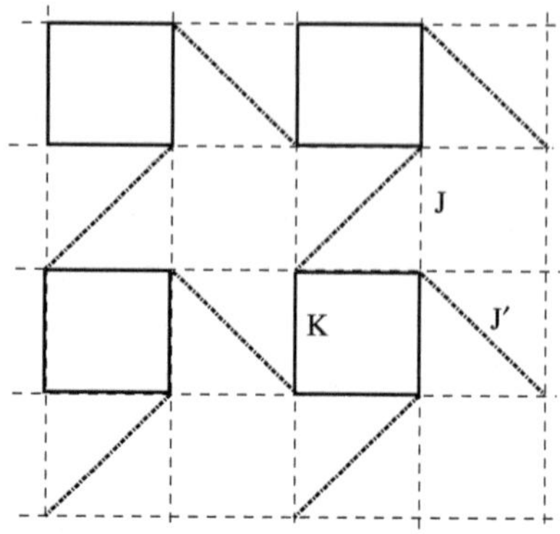

Fig. 5.1. The 1/5-depleted lattice and the Shastry–Sutherland lattice. The strong bonds of the Shastry–Sutherland model are the bonds J' (dotted dashed lines): they can accommodate orthogonal dimers ($J = K$ can be considered as a "perturbation" in the real $SrCu_2(BO_3)_2$). The lattice formed by the strongest exchanges in CaV_4O_9 is obtained with $J = 0$. The phase diagram of this model[82] contains (at least) collinear Néel phases, a dimer, and four-spin plaquette VBC.

model.[85] A lattice embedding the main couplings of these two physical problems is drawn in Fig. 5.1. Interactions are of the Heisenberg type.[i]

VBC are obvious GS in the strong coupling limits of each problem.

The exact GS of the Shastry–Sutherland model is built from singlets on the J' bonds.[77,85] For $J'/J \gtrsim 1.55 \pm 0.05$, the model has a gap in the spectrum of excitations and is in a dimerized VBC phase.[j] For $J'/J \lesssim 1.15 \pm 0.05$, the system is in the (π, π) Néel state of the square lattice (results of zero-temperature series expansion[87]). The possibility of an intermediate phase, possibly with helical short-range correlations, has been actively discussed in the literature.[79,88,89] Miyahara and Ueda have recently written a review of the theory of the orthogonal dimer Heisenberg spin model for $SrCu_2(BO_3)_2$.[84]

The 1/5-depleted Heisenberg square lattice model ($J = 0$) has been studied as a function of the ratio of the two different couplings: bonds within

[i]Small Dzyaloshinsky–Moriya interactions have been identified in $SrCu_2(BO_3)_2$.[86]

[j]Consider a decomposition of the Shastry–Sutherland lattice as edge- and corner-sharing triangles. For $J' = 2J$, the Hamiltonian is accordingly written as a sum of $J(\vec{S}_1+\vec{S}_2+\vec{S}_3)^2$ terms for each triangle (J'-bonds are shared by two triangles) and each such term is minimized by the dimerized state. This shows rigorously that the dimerized state is the GS as soon as $J'/J \gtrsim 2$.

a plaquette (K) and dimer bonds (J') between plaquettes. At isotropic coupling ($J' = K$) collinear long-range Néel order survives the depletion (the order parameter is about 35%[71] of the maximum classical value). A small unbalance in couplings drives the system either in a four-spin plaquette VBC ($K > J'$) or in a dimer VBC ($K < J'$). Both (explicit) VBC phases have a spin gap. A recent generalization of these models by Läuchli *et al.* encompasses both the 1/5-depleted Heisenberg square lattice model and the Shastry–Sutherland model[82] (see Fig. 5.1). Its phase diagram exhibits collinear Néel phases ((π, π) or $(0, \pi)$) separated from the VBC phases by second-order phase transitions. Transition between the two VBC phases which have different symmetries occurs via a first-order phase transition.[k]

Excitations in these models come from the promotion of local singlets to triplet excitations. In 2D, the ordered dimer background provides a confining force for the spin-1/2 excitations. Indeed, separating two unpaired spins (that is two *spinons*) creates a disruption in the ordered pattern all the way from the first to the second. The energy cost is thus proportional to the length of the string of defaults, and both spin-1/2 excitations remain in fact confined. Only integer spin excitations are expected. On the contrary, in these strongly coupled models single–triplet hoppings can be difficult and correlated motions might be important, leading to a large zoology of excited modes (see Ref. [81] and references therein). This potential frustration of the triplet motion favors the appearance of magnetization plateaus in VBC models.[75,90–94] This aspect was briefly discussed in the lecture notes published by the authors.[95]

5.3.4.2. *With spontaneous lattice symmetry breaking*

In the previous models, the (explicit) VBC phases do not break any lattice symmetry. They can be directly related to the geometry and relative strength of the couplings. In more symmetric situations with frustration, spontaneous symmetry-breaking VBC can appear as a way to overcome this frustration by taking full advantage of the quantum fluctuations. This is probably the case in the J_1–J_3 model on the square lattice for intermediate $J_3/J_1 \sim 0.6$,[96] in the J_1–J_2 model on the hexagonal lattice for intermediate $J_2/J_1 \sim 0.4$,[97] and in the Heisenberg model on the checker-board

[k] A recent $Sp(N)$ study of the Shastry–Sutherland model[89] suggests that a spin liquid phase with deconfined spinons might appear in such a model. No evidence of such a phase has emerged from the $SU(2)$ studies so far.

lattice.[98–103] In the two first cases, the GS are dressed columnar VBC of dimers. Translation and C_4 (respectively, C_3 only) symmetries of the lattice are spontaneously broken. The GS is 4 (respectively 3) times degenerate. Spin–spin correlations decrease exponentially with the system size. All excitations are gapped. Contrary to the J_1–J_2 model on the square lattice, exact diagonalizations[96,97,99] give a rather straightforward information on these systems where the correlation lengths are small enough (far enough from the critical points which limit the boundary of the VBC phases).

The spin-1/2 Heisenberg model on the checker board lattice, which can also be seen as a planar lattice of corner-sharing tetrahedrons (see Fig. 5.2), has received atmost importance for different reasons.[98–105] The problem classically has a continuous local degeneracy: the Hamiltonian can be rewritten as the sum of the squares of the total spin of each tetrahedron, and every configuration with a zero spin on each tetrahedron is a GS. Classically, this problem shares this property with the Heisenberg model on the kagomé, pyrochlore,[106,107] garnet,[108] and pyrochlore slab[109] lattices (these lattices made of corner-sharing "simplexes" with two, three or four spins each were dubbed "bisimplex" lattices by Henley[110]).

The quantum spin-1/2 antiferromagnet on the kagomé lattice has been found to be quite specific with a small gap (if any) toward triplet excitations and an anomalous density of gapless low-lying singlets excitations (see Section 5.7).

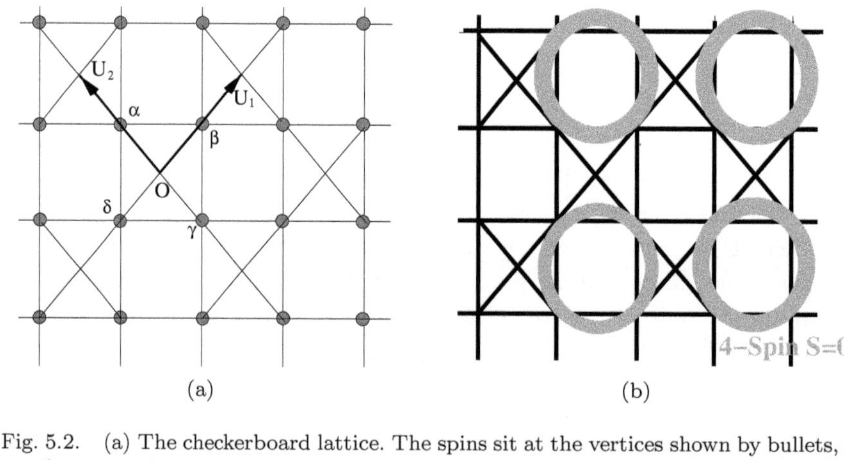

Fig. 5.2. (a) The checkerboard lattice. The spins sit at the vertices shown by bullets, all couplings are identical, $\mathbf{u}_1$, $\mathbf{u}_2$ are the unit vectors of the Bravais lattice. (b) The GS of the Heisenberg model on the checker board lattice are translational symmetry breaking VBC of four-spin plaquettes on the uncrossed squares.

The quantum scenario on the checkerboard lattice is quite different. The GS is a (dressed) product of four-spin $S = 0$ plaquettes on uncrossed squares: this state breaks translational symmetry but not C_4 (the point group being defined at the center of an empty plaquette). It has a two-fold degeneracy in the thermodynamic limit: this is easily seen in the symmetries and finite-size scaling of the low-lying levels of exact spectra.[99] The choice of the 4-spin $S = 0$ states on the uncrossed squares corresponds to the most stable local configuration. The product of $S = 0$ quadrumers is only weakly renormalized in the exact GS.[99]

Excitations of this model have been studied in different approaches, such as exact diagonalizations,[111] series expansions,[101] real space renormalization transformation.[102] All the excitations (singlet and triplet) are gapped. The triplet excitations originate from the triplet excitation of an uncrossed plaquette, they are weakly dispersive with a large gap. The singlet excitations cannot be described so simply: from exact diagonalization data, one can suspect that some of these excitations come from the reorganization of two adjacent triplet on crossed squares. The Contractor Renormalization (CORE) method of Berg *et al.*[102] on the contrary suggests that these excitations are domain walls between the two translated plaquette VBC GSs.

5.3.5. *Methods*

Spin waves[69] and Schwinger bosons[70, 88] are simple approximations to study the phase diagram of a quantum frustrated magnet. But the first method only gives an approximate knowledge on the range of existence of the Néel phases, and it is rather difficult to include the effect of fluctuations beyond the mean-field approximation within the Schwinger–Boson formulation. As discussed in the section devoted to large-N approaches (Section 5.4), spontaneous lattice symmetry breaking are very likely to be due to topological effects (Berry phase of instantons) which cannot be captured by the mean-field state. This probably explains why no spontaneous VBC has ever been found (to our knowledge) in Schwinger–Boson calculations.[1] For unfrustrated problems, as the Heisenberg model on the 1/5-depleted square lattice,[71] QMC is considered as the method which can give benchmarks for other approaches.

---

[1]It seems however that spontaneous VBC naturally arise in large-N approaches with fermionic representation of $SU(N)$ when $1/N$ corrections are considered (see Section 5.5.2).

Although VBC are naturally described with the help of spin-1/2 valence-bonds, the appearance of such states in the low-spin limit can sometimes be anticipated from an appropriate large-S approach. An example is given in the work of Tchernyshyov *et al.*[103] (see also a previous paper by Henley[110]) on the checkerboard Heisenberg model. In this model, when both couplings (square lattice bonds and diagonal bonds) are equal, GS has a continuous local degeneracy. However, leading $1/S$ corrections select collinear states out of this huge manifold.[110] There remains an exponential number of such states and they do not have any magnetic order. However, they exhibit long-range *bond* order and a spontaneous symmetry breaking[103] which is analogous to the one observed numerically in the spin-1/2 case.

For frustrated problems, exact diagonalizations can be useful tools in situations where the system is not too close from a critical point, that is when the correlation length is not too large. Successful applications of exact diagonalizations methods to 2D Heisenberg magnets realizing a VBC include studies of the J_1–J_3 model on the square lattice,[96] J_1–J_2 model on the hexagonal lattice,[97] and Heisenberg model on the checkerboard lattice.[99] In such situations, analysis of the quasi degeneracy of the low-lying levels of the spectra and of their finite-size scaling gives an unbiased and direct information on the symmetry breakings in the thermodynamic limit.[99] However, the boundaries of the phases and the quantum critical points cannot be accurately determined with this method. The series expansions described in the previous section appear to be a powerful approach to determine the phase boundaries.[74, 112] If the spin–spin correlation length is not very short, as is the case in the J_1–J_2 model on the square lattice for $J_2/J_1 \sim 0.5$,[37] it is very difficult to decide the exact diagonalizations between a VBC, a critical phase, or an RVB SL.

Concerning excitations, exact diagonalizations give the gaps in each sector and provide a crude approximation of the dispersion laws of the first excitations. The large-scale nature of the excitations (as for example domain wall excitations) can escape this method. The semi-analytical approaches which can be used for the study of the excitations of the VBC, all use as a basic departure point the excitations of a local cluster of spins conveniently renormalized by perturbation[73, 101] or effective renormalization technique.[102] Contrary to exact diagonalizations, these methods are not limited by finite-size effects, but the results can be biased by the departure point.[101, 102]

5.3.6. *Summary of the properties of VBC phases*

The generic features of VBC (whatever the dimensionality of the lattice) are as follows:

- A spin gap, no $SU(2)$ symmetry-breaking and short-range spin–spin correlations,
- Long-range order in dimer–dimer and/or larger $S = 0$ plaquettes. The coupling of this order to lattice distortions is probable in experimental realizations of spontaneous VBC.
- In *spontaneous* VBC phases, the GS is degenerate. From the theoretical point of view, the discrete symmetry of the order parameter of the VBC which spontaneously breaks a lattice symmetry may give birth to a finite-temperature Ising-like transition.[14] Simultaneity between this transition and a possible structural transition is likely when the couplings of the spins to the lattice degrees of freedom (phonons) are considered.[113]
- VBC have gapped excitations in the $S = 0$ sector as well as in other S sectors. A wide zoology of modes is to be expected as well as continua associated to multi-particle excitations or scattering of domain walls (in the case of a spontaneous symmetry breaking of the GS). In two dimensions, all these excitations have integer spins (the ordered background inducing a confinement of the spin-1/2 excitations)

Frustration on the square lattice or more generally on bipartite lattices is often overcome by VBC phases. The appearance of VBC in triangular geometries is possible in principle but there is up to now no examples of such phases in pure spin-1/2 models (in Section 5.5, examples will be given within the framework of quantum dimer models).

It has been advocated in the large-N approaches (see Section 5.4) that, at least in two dimensions, collinear spin–spin correlations generically lead to VBC or VBS and only non-collinear spin–spin correlations can give birth to RVB SL with unconfined spin-1/2 excitations. The present knowledge of $SU(2)$ phase diagrams supports this prediction. The VBC found so far numerically in $SU(2)$ spin models appears to be in regions of parameter space where the spin–spin correlations are characterized by some short-range collinear order in the large-S limit. The J_1–J_2 model on the honeycomb lattice has a classical incommensurate phase in the regime of high frustration and there are some evidences that in the quantum phase diagram the collinear phase is separated from the columnar VBC phase by an RVB SL.[97] The multiple-spin exchange (MSE) model on the triangular lattice is also believed to be an RVB SL[114] and the corresponding classical GS generically have non-coplanar spin configurations. Capriotti *et al.*[37] argued that the spin-1/2 square lattice J_1–J_2 model could be an RVB SL. If confirmed, this would be the first counter-example to the general rule explained above (the Heisenberg model on the pyrochlore lattice might be an other counter-example[102]).

5.4. Large-N Methods

Introduced by Affleck,[115] Affleck and Marston,[116] Arovas and Auerbach,[117] and Read and Sachdev[118–120] in the context of spin models, large-N approaches are powerful methods to investigate quantum antiferromagnets. When N is taken to infinity, many of these models can be solved by saddle point methods and finite-N corrections can be, at least in principle, explored in a controlled way. A success of these approaches is that they can describe the phenomenology of a large variety of phases encountered in quantum magnets, e.g., Magnetic LRO (possibly with order by disorder selection), as well as phases dominated by short-range valence bonds, e.g., VBC, VBS and RVB liquids. One crucial result (due to Read and Sachdev[118–122]) concerning these three latter phases is that the analysis of finite-N corrections to some large-N formulations ($Sp(N)$ for instance, see below) provides a general criterion to decide which of these three phase should appear in a given model.

This criterion is the following in 2D: if the (large-N equivalent of the) "spin" S at each site matches the lattice coordination number z by $2S = 0 \mod z$, a VBS is to be expected. If it is not the case (as for a spin-1/2 model on the square lattice), one should look at the local spin–spin correlations. If they are reminiscent of a collinear order, a VBC with spontaneous translation symmetry breaking is expected whereas a non-collinear short-range correlation generically gives rise to an RVB phase without any broken symmetry and deconfined spinon excitations. These results are of course based on a large-N generalization of the original spin model, and there is no guaranty at all that these rules should always apply to $SU(2)$ models. To our knowledge, they have however not been manifestly found in error up to now. In the following, we will present some of the important reasoning steps leading to this result.

5.4.1. *Bond variables*

The $SU(2)$ algebra of a spin S at one site can be represented by two species of particles $a^\dagger_\sigma$ (with $\sigma = \uparrow, \downarrow$), provided that the total number of particles on one site is constrained to be $a^\dagger_\uparrow a_\uparrow + a^\dagger_\downarrow a_\downarrow = 2S$. The raising operator S^+ (respectively, S^-) is simply represented by $a^\dagger_\uparrow a_\downarrow$ (respectively, $a^\dagger_\downarrow a_\uparrow$). These particles can be chosen to be fermions (Abrikosov fermions) or bosons (Schwinger bosons). These particles carry a magnetization $\pm\frac{1}{2}$ since $S^z = \frac{1}{2}(a^\dagger_\uparrow a_\uparrow - a^\dagger_\downarrow a_\downarrow)$. For this reason, they are often called *spinons.* The Heisenberg interaction is a quartic interaction for these particles:

$$\vec{S}_i \cdot \vec{S}_j = S^2 - \frac{1}{2} A^\dagger_{ij} A_{ij} \tag{5.10}$$

with the bond operator A_{ij} defined by

$$A_{ij} = a_{j\downarrow} a_{i\uparrow} - a_{j\uparrow} a_{i\downarrow}. \tag{5.11}$$

Acting on the vacuum, $A^\dagger_{ij}$ creates a spin singlet on the bond (ij). Physically, $A^\dagger_{ij} A_{ij}$ measures the number of singlets on that bond and Eq. (5.10) shows that the antiferromagnetic Heisenberg interaction just tries to maximize that number.

The idea of large-N methods is to generalize the $SU(2)$ symmetry of the spin-S algebra to a larger group $SU(N)$ (or $Sp(N)$) by letting the index σ go from 1 to N (or $2N$). The $SU(N)$ (or $Sp(N)$) generalization of the Heisenberg model is solved by a saddle point calculation of the action. The $N = \infty$ limit is very similar to a mean-field decoupling of the four-body interaction of the physical $SU(2)$ model: $A^\dagger_{ij} A_{ij} \simeq A^\dagger_{ij} \langle A_{ij} \rangle + \langle A^\dagger_{ij} \rangle A_{ij} - |\langle A_{ij} \rangle|^2$.

5.4.2. *SU(N)*

The generalization of the Heisenberg model to such a symmetry group depends only on the choice of the irreducible representation of $SU(N)$ according to which the "spin" operators transform (and not on the choice of fermions or bosons to implement the representation). For $SU(2)$, this amounts to specify the magnitude S of the spin. For $SU(N)$, irreducible representations are labeled by Young tableaux. The case of a general rectangular tableau with n_c columns and m rows was discussed by Read and Sachdev[123] and n_c appears to continue to play a role similar to $2S$ in the large-N phase diagrams.[m] In this review, we will focus on a slightly different large-N generalization of the $SU(2)$ model which is able to deal with both frustration and magnetic states.

5.4.3. $Sp(N)$

To perform a large-N extension of *frustrated* Heisenberg models, one has to use either fermions[116] or bosons with an $Sp(N)$ symmetry. The latter seems to produce phase diagrams that closely resemble the $SU(2)$ problems and we will focus on this representation which was introduced by Read and Sachdev.[120] The presentation below is largely inspired from their papers.[118–122]

We now have $2N$ flavors of bosons at each site: $b_{i,\sigma}$ with $\sigma = 1..2N$ and we define an $Sp(N)$-invariant bond operator

$$A_{ij} = \sum_{\sigma,\sigma'=1..2N} \mathcal{J}_{\sigma,\sigma'} b_i^\sigma b_j^{\sigma'}, \tag{5.12}$$

where the $2N \times 2N$ antisymmetric tensor $\mathcal{J}$ is block diagonal

$$\mathcal{J} = \begin{bmatrix} 0 & 1 & & & \\ -1 & 0 & & & \\ & & \ddots & & \\ & & & 0 & 1 \\ & & & -1 & 0 \end{bmatrix}, \tag{5.13}$$

and generalizes the $SU(2)$ antisymmetric tensor ϵ_{ij} ($SU(2)$ is identical to $Sp(1)$). Up to some constant, the $Sp(N)$ Hamiltonian is

$$\mathcal{H} = -\frac{1}{N} \sum_{ij} J_{ij} A_{ij}^\dagger A_{ij} \tag{5.14}$$

---

[m]Taking the limit $N \to \infty$ with m fixed of order 1 and $n_c \sim N$ is most conveniently done with bosons $b_{\alpha p}^\dagger$ where $\alpha = 1 \cdots N$ is the $SU(N)$ index, while $p = 1 \cdots m$ label the different "colors". There are therefore Nm kinds of bosons. On the contrary, it is convenient to use n_c "colors" of fermions (still with an $SU(N)$ index) to deal with the case $n_c \sim \mathcal{O}(1)$ and $m \sim N$. Bosonic representations with $n_c \sim N$ are appropriate to describe magnetically ordered phases[117] but cannot be used for frustrated models (such a representation is not self-conjugate). On the contrary, fermionic representations, such as the $m = N/2$ and $n_c = 1$ one,[116,124] can be used on any lattice but they do not display magnetically ordered phases and tend to favor dimerized states.[116,123,124]

with the constraints

$$\forall i \sum_{\sigma=1}^{2N} b_{i\sigma}^{\dagger} b_{i\sigma} = n_c, \tag{5.15}$$

$n_c = 2S$ in the $SU(2)$ case and $n_c/N = \kappa$ will be kept constant when taking the large-N limit. The partition function can be represented by an imaginary time functional integral

$$Z = \int \mathcal{D}[\lambda_i, b_i^{\sigma}, b_i^{\dagger\sigma}] \exp\left(-\int_0^{\beta} (L_0 + \mathcal{H}) d\tau\right), \tag{5.16}$$

$$L_0 = \sum_{i\sigma} b_i^{\dagger\sigma} (\partial_\tau + i\lambda_i) b_{i\sigma} - iN\kappa \sum_i \lambda_i. \tag{5.17}$$

The $\lambda_i(\tau)$ are Lagrange multipliers that enforce the constraint Eq. (5.15) at every site. Bond degrees of freedom Q_{ij} are introduced in order to decouple the bosons (Hubbard–Stratonovitch). The partition function is now

$$Z = \int \mathcal{D}[Q_{ij}, \bar{Q}_{ij}, \lambda_i, b_i^{\sigma}, b_i^{\dagger\sigma}] \exp\left(-\int_0^{\beta} (L_0 + L_1) d\tau\right) \tag{5.18}$$

with

$$L_1 = \sum_{ij} \left[N \frac{|Q_{ij}|^2}{J_{ij}} - (A_{ij}^{\dagger} Q_{ij} + \text{h.c})\right], \tag{5.19}$$

so that a Gaussian integration of the Q_{ij} gives back Eq. (5.16). The bond variables are $Sp(N)$ invariant, and they can take non-zero expectation values at a mean-field level without breaking the $Sp(N)$ symmetry. As we explain below, they are however not *gauge invariant.*

5.4.3.1. *Gauge invariance*

An important property of Eq. (5.18) is the $U(1)$ gauge invariance associated to the following transformations:

$$b_{i\sigma} \to b_{i\sigma} e^{i\phi_i}, \tag{5.20}$$

$$Q_{ij} \to Q_{ij} e^{i(\phi_i + \phi_j)}, \tag{5.21}$$

$$\lambda_i \to \lambda_i - \partial_\tau \phi_i, \tag{5.22}$$

where $\phi_i(\tau)$ are arbitrary site- and imaginary time-dependent angles. This gauge invariance comes from the conservation of the local boson number and reflects the fact that the magnitude of the spin is constant at each site. If we focus on the phase degrees of freedom of the bond variables, the Eq. (5.18) describes a system of charge-1 bosons coupled to a $U(1)$ lattice gauge theory.[125,126] These gauge degrees of freedom play a crucial role in the analysis of the fluctuations about meanfield solutions.

Effective action for the bond variables — The boson degrees of freedom can be integrated out to give an effective action for the bond variables:

$$Z = \int \mathcal{D}[Q_{ij}, \bar{Q}_{ij}, \lambda_i] \exp\left(-S^{\text{eff}}\right), \tag{5.23}$$

$$S^{\text{eff}} = N \int_0^\beta \left[\sum_{ij} \frac{|Q_{ij}|^2}{J_{ij}} - i\kappa \sum_i \lambda_i \right] - N \text{Tr} \log G, \tag{5.24}$$

where G^{-1} is the quadratic form which couples the bosons in Eq. (5.18) (propagator). It depends on the bond variables and on λ_i. We may write formally $G^{-1} = \partial_\tau - i\lambda - Q$. The term $N\text{Tr} \log G$ is the free energy of the bosons in the presence of the bond fields. By construction, the action S^{eff} is gauge invariant with respect to the transformations of Eqs. (5.20)–(5.22). So far, this is an exact formulation of the original model for arbitrary N.

5.4.3.2. *Mean-field ($N = \infty$ limit)*

Since N factorizes (no flavor index is left in Eq. (5.24), Z is dominated by the saddle point of S^{eff} when N is large. For simple models such as the first-neighbor antiferromagnet on the cubic lattice (any space dimension), the saddle point can be determined analytically. The $N = \infty$ limit is almost equivalent to the Schwinger-boson mean-field theory.[127,128] This can otherwise be done numerically.[n] In this large-N limit, two kinds of mean-field solutions can appear. For large enough κ, the bosons condense at some wave vector, and the spectrum of the mean-field Hamiltonian is gapless. This corresponds physically to the Néel long-range order. On the contrary,

[n] To our knowledge, all the saddle points considered so far are static (expectation values of the Q_{ij} are time independent) and the corresponding $\langle Q_{ij} \rangle$ could all be made real with an appropriate gauge transformation. There is no chiral order and the time-reversal symmetry is unbroken. The (oriented) sum of the complex phases of the bond variables around a plaquette defines a $U(1)$ flux. This flux is related to the solid angle formed by the spins and it vanishes in collinear as well as in coplanar states. In such cases, the phases can be therefore be gauged away and the $\langle Q_{ij} \rangle$ can be made real. For this reason, complex bond variables are usually not observed.[129]

for smaller κ (smaller "spin") the mean field Hamiltonian is gapped and the GS preserves the $Sp(N)$ symmetry. Fluctuations around the saddle point are not expected to change drastically the Néel ordered phases, but they play an important role in the physics of $Sp(N)$ symmetric phases. The following is a brief discussion of the effects of fluctuations in these non-magnetic phases.

5.4.3.3. *Fluctuations about the mean-field solution*

At the mean-field level described above, some Q_{ij} acquire a (static in all known cases) non-zero expectation value: $\langle Q_{ij} \rangle = \bar{Q}_{ij}$. For this reason, such a state spontaneously breaks the local gauge invariance of Eqs. (5.20)–(5.22). However, this does not mean that the gauge degrees are all gapped and do not play any role at a low energy. In fact, as remarked by Read and Sachdev, depending on the *geometry* of the lattice defined by the bonds where $Q_{ij} \neq 0$, some long-wavelength gapless gauge excitations survive and the associated fluctuations must be taken into account. More precisely, the fluctuations of the bond variables about the saddle point are decomposed into an amplitude and a phase

$$Q_{ij} = (\bar{Q}_{ij} + q_{ij}) \exp(i\theta_{ij}) \tag{5.25}$$

and we expand S_{eff} with these new variables. Two cases must then be considered:

(i) The lattice made of the sites connected by non-zero $\bar{Q}_{ij}$ bonds is bipartite. This is automatically the case if the original lattice is defined by bonds where the exchange $J_{ij} \neq 0$ is bipartite. This can also be true if the original lattice is a non-bipartite lattice but some bonds have $\bar{Q}_{ij} = 0$ so that the remaining lattice is bipartite. This is the case, for instance, in the J_1–J_2 model on the square lattice,[120] in some regions of the phase diagram of the Shastry–Sutherland model[89] and in the checkerboard Heisenberg model.[104] Such configurations of the bond variables give *collinear* spin structures: spin–spin correlations can either be long ranged (large κ, Néel phase) or short ranged but in both cases the magnetic structure factor is peaked at a simple wave vector $\mathbf{k}_0$ such that $2\mathbf{k}_0$ is a reciprocal lattice vector ($k_0 = (\pi, 0)$, $k_0 = (0, \pi)$, or $k_0 = (\pi, \pi)$ in square geometries).

(ii) The lattice made of the sites connected by non-zero $\bar{Q}_{ij}$ bonds is not bipartite. This happens in some phases of the J_1–J_2–J_3 model on the square lattice,[120] on the triangular or kagomé lattices,[130] in the Shastry–Sutherland model for some values of the exchange parameters[89] and on an anisotropic triangular lattice.[131] Such mean-field states generically have planar but non-collinear spin–spin correlations.

It is simple to check that case (i) preserves a *global* continuous symmetry while such a symmetry is absent in case (ii). Consider the following *global* gauge transformation in case (i):

$$b_{i\sigma} \to b_{i\sigma} e^{i\phi} \quad b_{j\sigma} \to b_{j\sigma} e^{-i\phi}, \tag{5.26}$$

$$Q_{ii'} \to Q_{ii'} e^{2i\phi} \quad Q_{jj'} \to Q_{jj'} e^{-2i\phi}, \tag{5.27}$$

$$Q_{ij} \to Q_{ij}, \tag{5.28}$$

where i, i' belongs to sublattice A and j, j' belongs to sublattice B. This transformation does not change the mean-field parameters $\bar{Q}_{ij}$. The only fields affected by this transformation are those connecting two sites on the same sublattice.[o] They have a zero expectation value in case (i) (or do not even exist if the physical lattice is itself bipartite). For this reason, it is possible to make low-energy (and long-wavelength) gapless gauge excitations about the saddle point by replacing the global staggered phase shift ϕ of Eq. (5.26) by a slowly varying (staggered) θ_{ij} (Eq. (5.25)). A gradient expansion of the effective action performed at the appropriate points in the Brillouin zone for the phase fluctuations only involves gradients of θ. The corresponding action is that of $U(1)$ lattice gauge theory coupled to charge-1 boson (spinons).

5.4.3.4. *Topological effects — instantons and spontaneous dimerization*

So far, only small fluctuations around the saddle point were considered and the contributions of topologically non-trivial gauge-field configurations were ignored. Consequently, the magnitude of the "spin" $n_c/2$ was a continuous parameter and *the information about the integer or half-integer (for instance) character of S-as disappeared.* From Haldane's work on quantum spin chains and non-linear sigma models,[132] it is well known that Berry

[o]In gauge theory language, the fields $Q_{ii'}$ (respectively $Q_{jj'}$) of Eq. (5.27) transform like a charge-2 scalar for the $U(1)$ gauge field. Instead, from Eq. (5.26), the bosons (spinons) carry a charge 1.

phases in spin systems give rise to topological terms in the low-energy effective action which can play a crucial role depending on the parity of $2S$.

In nonlinear sigma models in 2+1 dimensions, the Berry phase vanishes for configurations which are smooth on the scale of the lattice spacing[133, 134] (unlike the $1+1$-dimensional case). However, "hedgehog" space–time singularities[133] give non-trivial Berry phases. Read and Sachdev found that the closely related instantons of the effective $U(1)$ gauge theory described before[p] also play a crucial role in the physics of the $Sp(N)$ (as well as $SU(N)$) spin models.

The Berry phase associated to such a singular configuration depends on details of the lattice geometry. In the short-range ordered (π, π) phase of the square lattice antiferromagnet, this Berry phase is a multiple of $in_c\pi/2$. Although dealing with a gas of interacting ($1/r$ Coulomb-like potential) instantons is a difficult problem (see Ref. [119] and references therein), we can guess that the physics will depend on $n_c \bmod 4$. A detailed analysis[119] shows that when $n_c \neq 0 \bmod 4$, the instantons condense and *spontaneously break the lattice translation symmetry.* This generates a static electric field for the $U(1)$ gauge field. Since the electric field is coupled to the difference of amplitudes of the bond variables, such a state acquires spatially inhomogeneous expectation values of the bond variables, *it is a VBC and spinons are confined in pairs.* In the J_1–J_2 model, around $J_2/J_1 \simeq 0.5$ the mean-field state is short-range ordered with correlations peaking at (π, π). A columnar dimerized state is predicted by Read and Sachdev from this analysis of the fluctuations, in agreement with a number of numerical works on the $SU(2)$ J_1–J_2 spin-1/2 model.

In a recent work by Harada, Kawashima, and Troyer,[135] the phase diagram of the (first neighbor — unfrustrated) $SU(N)$ antiferromagnet on the square lattice with $n_c = 1$ was found to be in complete agreement with Read and Sachdev's predictions. They showed by QMC simulations that for $N \leq 4$ the system is Néel ordered, whereas it is a columnar VBC for $N > 5$. This provides an additional support to the field theory arguments described above. It also underlines that the mechanism of spontaneous symmetry breaking and formation of a VBC may come from quantum fluctuations only and that frustration is not always required (although it clearly enhances quantum fluctuations).

[p] In the gauge theory associated to the phases of the link variables, an instanton corresponds to a tunneling event during which the total magnetic field piercing the lattice is changed by $\pm 2\pi$.

On the contrary, when $n_c = 0 \bmod 4$, the analysis of Read and Sachdev shows that fluctuations should not bring any broken lattice symmetry. Spinons are also confined and this state closely resembles the valence bond solid (VBS) proposed by Affleck *et al.*[61] as a possible GS when the spin S matches the coordination number z according to $2S = 0 \bmod z$ (see Section 5.3.3).

5.4.3.5. *Deconfined phases*

Now we suppose that, starting from a mean-field solution with collinear correlations (case (i), a parameter of the original spin model is varied so that the mean field solution is changed and some bonds $Q_{ii'}$ (i and i' belong to the same sublattice) acquire a non-zero expectation value (case (ii)). In the framework of square lattice antiferromagnets, a finite third-neighbor coupling (J_3) would be needed.[120] From the point of view of the long-wavelength gauge fluctuations (related to the continuum limit of the phases θ_{ij}) discussed above, the appearance of $\bar{Q}_{ii'} \neq 0$ is equivalent to the condensation of a (Higgs) boson of charge 2. This is a spontaneous break down of the global $U(1)$ staggered symmetry of Eqs. (5.26)–(5.28) down to a $\mathbb{Z}_2$ one since the field Q_{ii} is not invariant under Eq. (5.27) except if $\phi = 0$ or π. Based on the results of Fradkin and Shenker[136] concerning confinement in compact lattice gauge theories coupled to matter, Read and Sachdev argued that this Higgs mechanism suppresses the low-energy gauge fluctuations and liberates the spinons. This confinement transition is described by a $\mathbb{Z}_2$ gauge theory. The suppression of the $U(1)$ gauge fluctuations also forbids the condensation of instantons discussed above and the GS remains unique[q] and *bond variables have uniform expectation values. It is a genuine SL without any broken symmetry and deconfined spinons.*

5.5. Quantum Dimer Models

In Section 5.3, we showed that pairing spins-1/2 into singlets at short distances is a rather natural way to overcome frustration in Heisenberg antiferromagnets. QDM are defined in the Hilbert space of the nearest-neighbor valence bond (or dimer) coverings of the lattice. By construction, these models focus on the dynamics in the singlet space and ignore magnetic (gapped magnons or gapped spinons) excitations. For this reason, they are

[q]Except for a discrete topological degeneracy.

(*a priori*) not appropriate to describe the physics of spin systems where magnetic excitations are gapless.

The Hamiltonian of a QDM usually contains kinetic as well as potential energy terms for these dimers. Such Hamiltonians can often be simpler than their spin parents and are amenable to several analytic treatments because of their close relation to classical dimer problems,[137–139] Ising models, and $\mathbb{Z}_2$ gauge theory.[140–142] These models can offer simple descriptions of VBC[143] as well as RVB liquids.[141,144] It is in particular possible to write down some QDM that have a simple and exact VBC GS with spontaneous broken symmetries (such as Rokhsar and Kivelson's model on the square lattice[143] with attractive potential energy only — in which case the exact GS is very simple). Simple solvable QDM which have a dimer liquid GS can also be constructed.[141]

5.5.1. *Hamiltonian*

The first QDM was introduced by Rokhsar and Kivelson.[143] It is defined by an Hamiltonian acting in the Hilbert space of first-neighbor dimer (valence-bonds) coverings of the square lattice and reads as follows:

$$\mathcal{H} = \sum_{\text{Plaquette}} \left[-J \left(|{\parallel}\rangle \langle{=}| + \text{H.c.} \right) + V \left(|{\parallel}\rangle\langle{\parallel}| + |{=}\rangle\langle{=}| \right) \right]. \tag{5.29}$$

Flipping two parallel dimers around a square plaquette is the simplest dimer move on the square lattice and the J terms precisely represent such dynamics. The V terms are diagonal in the dimer basis and account for an attraction or repulsion between nearest-neighbor dimers. These are the two *most local* terms (respecting all lattice symmetries) which can be considered.[r]

5.5.2. *Relation with spin-1/2 models*

There exist different interesting mappings between frustrated Ising models and QDM.[142,145] Here, however, we focus on the relations between QDM

[r] They were originally derived[143] as the lowest order terms of a formal overlap expansion (see Section 5.5.2) of the Heisenberg. In that calculation $J \sim x^4$ and $V \sim x^8$. Note that a three-dimer kinetic term (extending over two neighboring plaquettes) is generated of order x^6 and is not included in Eq. (5.29).

and $SU(N)$ (or $Sp(N)$) spin models in which dimers are related to singlet valence bonds.

Overlap expansion — A valence bond state (product of two-spin singlets — belongs to the spin-1/2, Hilbert space) can be associated to any dimer covering.[s] Two such valence bond states $|a\rangle$ and $|b\rangle$ are not orthogonal, but, as first discussed by Sutherland,[146] their overlap decays exponentially with the length L of the loops of their transition graphs (defined in Section 5.5.3.1) as $|\langle a|b\rangle| = 2^{1-L/2}$. Rokhsar and Kivelson[143] introduced a formal expansion parameter x and replaced $|\langle a|b\rangle|$ by $2x^L$. Their idea is that although $x = \frac{1}{\sqrt{2}}$ for physical $SU(2)$ spins, the physics of some models may be captured by the first orders of a small x expansion. Truncating this expansion to order x^n gives an effective Hamiltonian which contains *local* terms involving at most n dimers.[t] In this approach, the dimer states of the QDM are in one-to-one correspondence with *orthogonalized* valence bonds in the spin Hilbert space.[u]

Fluctuations about large-N saddle points — From the above argument, it could seem that the connection between spin-1/2 models and QDM relies on a variational approximation: the spin Hilbert space is restricted to the nearest-neighbor valence bond subspace. This connection is in fact probably deeper, as some theories describing fluctuations about some large-N saddle points ($1/N$ corrections) are equivalent to (generalized) QDM. This mapping was discussed by Read and Sachdev[123] for representation with $m = 1$ (number of rows in the Young tableau of the $SU(N)$ representation), $N \to \infty$ and $n_c \sim \mathcal{O}(1)$, it leads to a *generalized* QDM where n_c dimers emanate from each site. A QDM also describes $1/N$ corrections in the case

[s] There is however a sign ambiguity (a valence bond is antisymmetric under the exchange of both spins) that can be fixed by choosing an orientation on every bond.

[t] The *signs* of the non-diagonal (kinetic) terms of the effective QDM obtained by such small x expansion depend on the sign convention which was chosen to map valence-bonds to dimers. An important question is to know whether, at least at the lowest non-trivial order, a sign convention giving the same sign for all kinetic terms exists (as in Eq. (5.29)). This is the case on the square[143] and triangular lattices.[144]

[u] This implicitly assumes that the valence-bond states are linearly independent. This can be demonstrated on the square lattice and appears to be the case on the triangular and kagomé lattices for large-enough sizes. The operator which orthogonalizes the valence-bond basis into the dimer basis is $\Omega^{-1/2}$, where $\Omega_{a,b} = \langle a|b\rangle$ is the overlap matrix.

of the fermionic[v] $SU(N)$ generalization[116,124] of the Heisenberg model

$$\mathcal{H} = \frac{1}{N}\sum_{ij} J_{ij} : B^{\dagger}_{ij} B_{ij} :$$
$$= -\frac{1}{N}\sum_{ij} J_{ij}\ B^{\dagger}_{ij} B_{ij} + \text{cst}, \tag{5.30}$$

where

$$B_{ij} = \sum_{\sigma=1}^{N} c^{\dagger}_{j\sigma} c_{i\sigma}, \tag{5.31}$$

and where the $c_{i\sigma}$ are N flavors of fermions satisfying a constraint similar to Eq. (5.15):

$$\sum_{\sigma=1}^{N} c^{\dagger}_{i\sigma} c_{i\sigma} = N/2. \tag{5.32}$$

Rokhsar showed[124] that in the $N \to \infty$ limit, "dimerized states" (or Peierls states) become exact GS of Eq. (5.30) for a large class of models.[w] Quite naturally, $1/N$ corrections will induce a dynamics into this subspace of dimerized states; it can be described by a QDM (with kinetic energy terms only at this order). At lowest order, on the square lattice, a kinetic term identical to the J term in Eq. (5.29) is generated and naturally favors a columnar or resonating-plaquette crystal (in agreement with a number of works on the spin-1/2 model). The same arguments were discussed for the kagomé lattice.[147] In that case the leading $1/N$ corrections to the fermionic saddle point generate three-dimer moves around hexagons and stabilize a crystal of resonating hexagons.[x] This formalism was also applied to the checkerboard model[98] to conclude to the presence of a VBC phase.

[v] Young tableau with $n_c = 1$ column and $m = N/2$ rows.

[w] Let J_0 be the largest value of the J_{ij}. Each dimerization (no site is left empty) where only bonds where $J_{ij} = J_0$ are occupied is a GS. Here, a "dimer" between to neighbors i and j consists of a $SU(N)$ singlet made with N fermions (one of each flavor) hoping back and forth between i and j. It is constructed from $\prod_{\sigma=1}^{N}(c^{\dagger}_{i\sigma} + c^{\dagger}_{j\sigma})|0\rangle$ by projecting out the components which do not satisfy Eq. (5.32). Notice however that in the $N \to \infty$ limit the relative fluctuations of the total number of fermion on each site are of order $1/\sqrt{N}$ and can be neglected.

[x] To our knowledge however there is no clear evidence of such ordering in the spin-1/2 case.

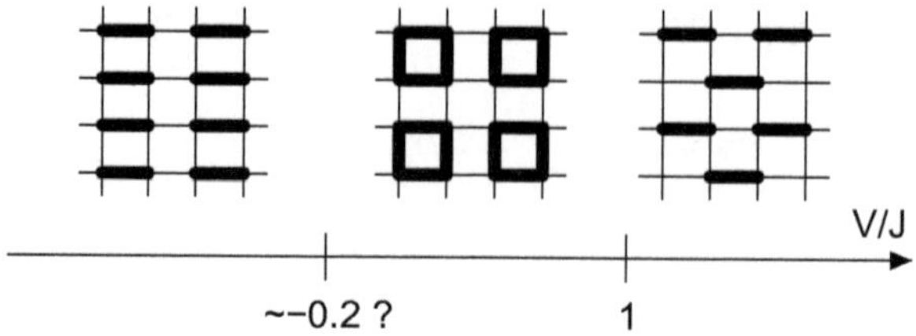

Fig. 5.3. Schematic phase diagram of the square lattice QDM. The possible location of an intermediate plaquette phase is taken from the work of Leung *et al.*[148].

5.5.3. *Square lattice*

The phase diagram of the Rokhsar and Kivelson's square lattice QDM is shown Fig. 5.3. Since a change in the signs of the basis dimer configurations can change J into $-J$ (see Ref. [143]), we will choose $J > 0$ without loss of generality.

5.5.3.1. *Transition graphs and topological sectors*

In order to understand the particularities of QDM on bipartite lattices, it is useful to describe how the set of dimer coverings splits into *topological sectors*. To do so, we first have to define transition graphs: the transition graph of two dimer coverings c and c' is obtained by superimposing c and c' one on top of the other; it defines a set of non-intersecting loops covering the lattice. On each bond where the dimers of c and c' match, a trivial loop of length 2 is obtained. When the lattice is bipartite (two sublattices A and B) these loops can be *oriented* in the following way: Any dimer belonging to c is oriented from A to B and dimers of c' are oriented $B \to A$. The transition graph is then made of loops of the type $A \to B \to A \to B \cdots$. With periodic boundary conditions, two winding numbers[143] $-L/2 \leq \Omega_x, \Omega_y \leq L/2$ are associated to such a transition graph ($L \times L$ sites). Ω_x (respectively, Ω_y) is the net number of topologically non-trivial loops (clockwise minus counterclockwise) encircling the torus in the x (respectively, y) direction.

Dimer coverings can be grouped into different *topological sectors*. By definition, two dimer coverings belong to the same sector if they can be transformed into each other by repeated actions of *local* dimer moves (the transition graph associated to each movement does not wind around the whole system if it has periodic boundary conditions). On the square lattice, two-dimer moves are sufficient to connect any two states in the same sector; that is the Hamiltonian equation (5.29) is ergodic within each topological sector. In a torus geometry, c and c' belong to the same sector if and only if their transition graph has winding numbers $\Omega_x = \Omega_y = 0$. The different topological sectors can be labeled by their winding numbers with respect to some reference columnar configuration. Their number is of order $\mathcal{O}(L^2)$ for a system of linear size L.

5.5.3.2. *Staggered VBC for $V/J > 1$*

When V is sufficiently large, the system tries to minimize the number of parallel dimers. The staggered configuration shown Fig. 5.3 has no such *flippable plaquette.* It is always a zero-energy eigenstate of Eq. (5.29) and becomes a GS for $J \geq V$. It breaks several lattice symmetries (four-fold degenerate) and is a VBC.

The expectation value of the energy per plaquette satisfies $\min(0, V - J) \leq E_0/N_p \leq \max(0, V + J)$. For $V/J > 1$, this gives $0 \leq E_0/N_p$, and any zero-energy state saturates this lower bound and is a therefore a GS. One should however note that it is possible to make zero-energy domain walls in this VBC since the state shown in Fig. 5.4 is also annihilated by the Hamiltonian. No local dimer movement can take place in the staggered VBC (with or without domain walls). Each of these states forms a topological sector with a single configuration which has $|\Omega_x| + |\Omega_y| = L/2$ with respect to a columnar state.

5.5.3.3. *Columnar crystal for $V < 0$*

When parallel dimers attract each other, the system tries to maximize the number of flippable plaquettes. Columnar configurations as shown on the left-hand side of Fig. 5.3 do maximize this number. Such a VBC is exactly realized for $V < 0$ and $J = 0$. Elementary excitations consist of a pair of (say) vertical dimers in a background of vertical columns of horizontal

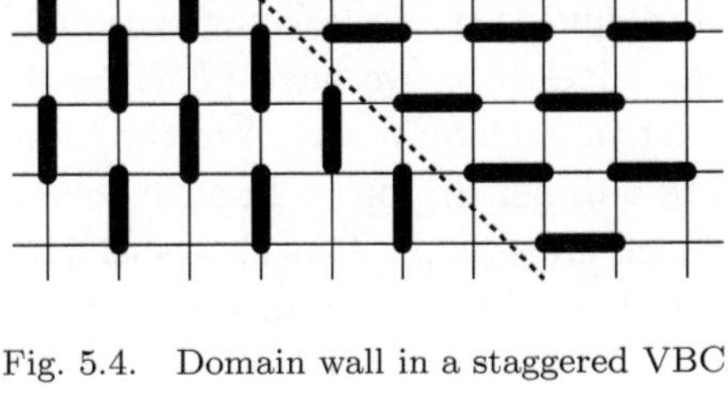

Fig. 5.4. Domain wall in a staggered VBC.

dimers. Such excitations are gapped ($\Delta E = 2|V|$), and this VBC phase will survive to the inclusion of a finite J term. Note that unlike the staggered VBC presented in the previous paragraph, the columnar dimer configuration is not an exact eigenstate when $J \neq 0$. The exact diagonalization data of Leung *et al.*[148] suggest that the columnar phase may disappear around a critical value $V/J \simeq -0.2$.

5.5.3.4. *Plaquette phase*

When the kinetic energy dominates ($|V| \ll J$), the system will try to maximize the number of resonating plaquettes $|\,||\,\rangle + |=\rangle$. This can be achieved through the resonating plaquette crystal shown in Fig. 5.3 and the numerical work (exact diagonalizations up to 8×8 sites) of Leung *et al.*[148] suggests that this phase is realized in an interval $-0.2 \leq V/J \leq 1$. Although this model would not suffer from the fermion sign problem, we are not aware of any QMC simulation on this model.

5.5.3.5. *Rokhsar–Kivelson point*

The point $J = V$ (Rokhsar–Kivelson (RK) point) plays a special role. As remarked by Rokhsar and Kivelson,[143] the equal-amplitude superposition of all dimerizations in a given topological sector is an exact GS. When $J = V$, the Hamiltonian can be written as a sum of projectors

$$\mathcal{H}_{J=V=1} = \sum_p \langle |\Psi_p\rangle\langle\Psi_p|, \tag{5.33}$$

$$|\Psi_p\rangle = |\,||\,\rangle - |=\rangle\,. \tag{5.34}$$

The linear superposition of all dimer coverings belonging to a given sector Ω

$$|0\rangle = \sum_{c\in\Omega} |c\rangle \tag{5.35}$$

is annihilated by Eq. (5.33) and is therefore a GS. The argument is the following. Consider a plaquette p and a configuration $|c\rangle$. If $|c\rangle$ has one or no dimer at all on the edges of p, we have $\langle\Psi_p|c\rangle = 0$. If two dimers are present, then there exists a configuration $|c'\rangle$ in the same sector which only differs from $|c\rangle$ by a two-dimer flip on p. In such a case, the combination $|c\rangle + |c'\rangle$ is again orthogonal to $|\Psi_p\rangle$. This shows that $H|0\rangle = 0$.

When open boundary conditions are considered (this restricts the topological sector to $\Omega_x = \Omega_y = 0$), the RK state is the linear combination of an exponential number of configurations.[y] This is very different from the crystalline states considered so far where some periodic configurations were favored and it closely resembles Anderson's RVB picture. As we shall see, this RK state is not a "true" liquid on the square lattice since dimer–dimer correlations are not of short range but algebraically decay ($\sim 1/r^2$) with distance. The calculation of dimer–dimer correlations in the RK state (Eq. (5.35)) maps onto a *classical* dimer problem solved by Kasteleyn,[137,139] Fisher,[138] and Fisher and Stephenson.[149] From this, Rokhsar and Kivelson[143] constructed gapless excitations (in single-mode approximation) the dispersion relation of which vanishes as $\mathbf{k}^2$ at a small momentum (the origin is taken at (π, π)). Quoting Rokhsar and Kivelson,[143] these excitations (dubbed "resonons") are the *Goldstone mode of the gauge symmetry which allows the phases of the different topological sectors to be varied without changing the energy.* Another mode of gapless excitations (around $(\pi, 0)$ and $(0, \pi)$ in the Brillouin zone), specific to the fact that the GS has critical (algebraic) dimer–dimer correlations, was recently discovered.[150]

The QDM on the square lattice is thus believed to be ordered (VBC) everywhere except at the RK point ($J = V$), where it has quasi long-range (critical) dimer–dimer correlations.

5.5.4. *Hexagonal lattice*

The QDM on the honeycomb lattice was studied by Moessner *et al.*[151] in particular with MC simulations. The phase diagram is very similar to the square lattice-case discussed above. It possesses three crystalline phases (Fig. 5.5) and has algebraically decaying dimer–dimer correlations at the Rokhsar Kivelson point (where the GS in each sector has the same form as Eq. (5.35)). The absence of liquid phase (with exponentially decaying $2n$-mer-$2n$-mer correlations) is believed to prevail in bipartite lattices. This relation between the absence of a deconfined dimer liquid phase[z] and the

---

[y]This is also true for periodic boundary conditions provided the two winding numbers do not scale like the linear size L.

[z]Bipartiteness seems to forbid deconfinement but not short-range dimer–dimer correlations. The 4–8 lattice (squares and octogons) is an example where dimer–dimer correlations are short range. We thank R. Moessner for pointing this to us. On this lattice, the equal-amplitude superposition of all coverings would be similar to an explicit VBC wave function (thus confining). Such situations are only possible when the number of sites is even in the unit cell.

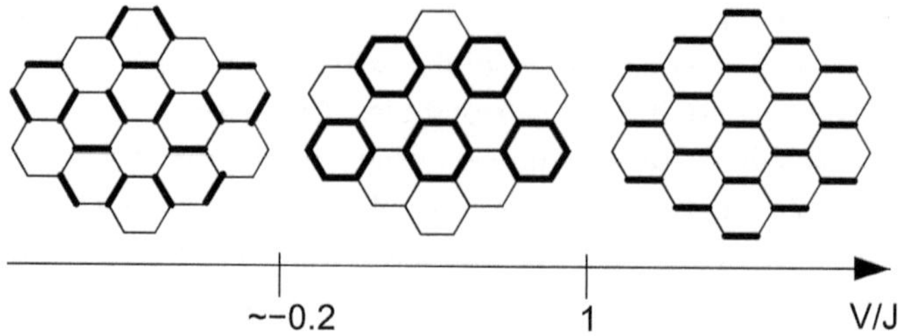

Fig. 5.5. Phase diagram of the hexagonal QDM obtained by Moessner *et al.*[151] Although the VBC shown for $V < 0$ do not have all dimers parallel to the same direction, it is equivalent to the columnar VBC found in the square lattice QDM because it maximizes the number of flippable plaquettes. It also corresponds to the ordering pattern predicted in the large-N theory of Ref. [119]. As for the VBC obtained for $V/J > 1$, it is the hexagonal counterpart of the *staggered* VBC of the square lattice (no flippable plaquette, exact eigenstate, and maximum tilt in a height representation).

bipartite character of the lattice has been discussed by several authors[152] and is related to the possibility of a *height representation*[153] of dimer coverings when the lattice is bipartite.[aa]

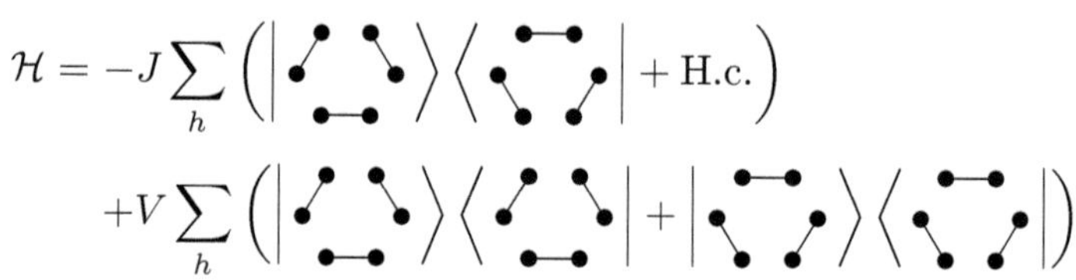

(5.36)

[aa]Consider a bipartite lattice with coordination number z. For each dimer covering, we can associate integers (heights) on the dual lattice by the following rule. Set the height to be zero on a plaquette at the origin. The height is then defined on the whole lattice by turning clockwise (respectively counterclockwise) around sites of the A-sublattice (respectively B-sublattice) and changing the height by $z-1$ when crossing a dimer and by -1 when crossing an empty bond. It is simple to check the difference of heights $\delta h(x) = h_1(x) - h_2(x)$ between two dimerizations is constant inside each loop of their transition graph. Note that the loops of a transition graph can be naturally oriented on a bipartite lattice. Then, $\delta h(x)$ changes by $+z$ (respectively, $-z$) when crossing a clockwise (respectively, counterclockwise) loop of the transition graph. Columnar dimerizations have an averaged height that is flat, and staggered ones have the maximum tilt. The winding numbers (Ω_x, Ω_y) correspond to the average height difference between both sides of the sample. The kinetic energy terms of Eqs. (5.29) and (5.36) change the height of the corresponding plaquette by $\pm z$ and the potential terms ($V > 0$) favor tilted configurations.

Fouet *et al.*[97] studied the spin-1/2 J_1–J_2–J_3 model on the hexagonal lattice by exact diagonalizations and found evidences of a staggered VBC of the type predicted for $V/J > 1$ by Moessner *et al.*[151] in the QDM. Other phases (Néel ordered phase and a possible short-range RVB SL) are also present in the spin-1/2 model.[97]

5.5.5. *Triangular lattice*

The most local dimer Hamiltonian on the triangular lattice contains kinetic and potential two-dimer terms on each rhombus; it was studied by Moessner and Sondhi[144]:

$$\mathcal{H} = -J\sum_r \left(\left|\diagup\diagup\right\rangle \left\langle \substack{\bullet\!-\!\bullet \\ \bullet\!-\!\bullet} \right| + \text{H.c.} \right)$$
$$+V\sum_r \left(\left|\diagup\diagup\right\rangle \left\langle \diagup\diagup \right| + \left| \substack{\bullet\!-\!\bullet \\ \bullet\!-\!\bullet} \right\rangle \left\langle \substack{\bullet\!-\!\bullet \\ \bullet\!-\!\bullet} \right| \right), \quad (5.37)$$

where the sums run over all rhombi r of the lattice (with three possible orientations). This model was shown to possess (at least) three crystalline phases, including a columnar and staggered one (Fig. 5.6) as in the two previous examples. An additional VBC (with resonating diamond plaquettes) with a large unit cell (12 sites) was also predicted around $V = 0$. When $V < 0$ and $J = 0$, the GS is highly degenerate since it is possible, from an ordered columnar configuration, to shift all the dimers along any straight line without changing the number of flippable plaquettes (contrary to the square lattice case). However, an infinitesimal J is expected to lift this degeneracy and to order the GS in a columnar way.

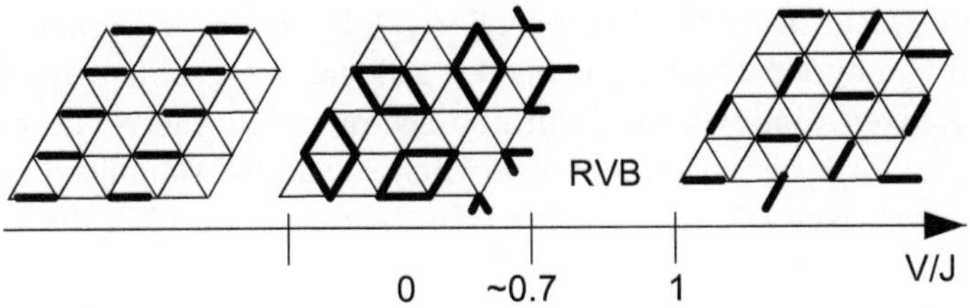

Fig. 5.6. Phase diagram of the triangular lattice QDM obtained by Moessner and Sondhi.[144]

The phenomenology of these ordered phases is that of usual VBC, and we refer to the original paper[144] for details.

5.5.5.1. *RVB liquid at the RK point*

The new physics of this model appears through the existence of a *liquid* phase in the interval $0.7 \lesssim V/J \leq 1$. As for the two previous QDM, the GS are exactly known at the RK point $J = V$. As before, dimer–dimer correlations are obtained exactly at this point by a Pfaffian calculation,[144, 154, 155] but the result shows a *finite* correlation length. From their MC simulations, Moessner *et al.*[144] argued that the spectrum is gapped at the RK point and that this gap persists down to $V/J \simeq 2/3$, that is over a finite range of coupling. This picture is consistent with the exact diagonalizations performed on this model.[156, 157]

5.5.5.2. *Topological order*

When the lattice is not bipartite, the loops of a transition graph can no longer be oriented. The winding numbers Ω_x and Ω_y are now defined as the (positive) number of non-trivial loops around x and y in the transition graph with a reference configuration (say columnar). These two integers are not conserved by local dimer moves. They are however *conserved modulo two*, which leaves only four sectors.[bb] Consequently, *the dimer liquid GS is four-fold degenerate* at the RK point. This degeneracy holds exactly at the RK point even on finite-size samples, but it is expected to hold in the thermodynamic limit in the whole liquid phase ($0.7 \lesssim V/J \leq 1$).

Conventional orders are often associated to a spontaneously broken symmetry and lead to GS degeneracies in the thermodynamic limit. The four-fold degeneracy discussed above is the signature of some kind of order, called topological order.[158] The peculiarity of this order is that it is not associated to any local order parameter: A local observable cannot decide whether a given dimerization is in one sector or another. The existence of topological order is intimately associated to the fractionalized nature of the elementary excitations (see below). In the case of an RVB dimer liquid,

[bb]The two-dimer moves included in Eq. (5.37) are not sufficient to guaranty ergodicity with each of the four sectors. Staggered states (12-fold degenerate — not $\mathcal{O}(L)$ like on the square lattice) have no flippable plaquette but can be connected to other states with four-dimer moves.[144]

these excitations have been known to be *Ising vortices* for a long time[159,160] (dubbed *visons* in the recent literature[161,162]). We will now discuss these excitations in more detail in the framework of a QDM which realizes the same dimer liquid phase but for which not only the GS but also all the eigenstates are known exactly.

5.5.6. *Solvable QDM on the kagomé lattice*

An exactly solvable QDM on the kagomé lattice was introduced by Serban, Pasquier, and one of us.[141] It offers a very simple and explicit realization of the ideas discussed above (visons, topological order, etc.).

5.5.6.1. *Hamiltonian*

The kagomé lattice QDM introduced in Ref. [141] contains only kinetic terms and has no external parameter. The Hamiltonian reads as follows:

$$\mathcal{H} = -\sum_{h} \sigma^x(h), \tag{5.38}$$

where

$$\sigma^x(h) = \sum_{\alpha=1}^{32} |d_\alpha(h)\rangle\langle\bar{d}_\alpha(h)| + \text{H.c.} \tag{5.39}$$

The sum runs over the 32 loops on the lattice which enclose a single hexagon and around which dimers can be moved (see Table 5.1 for the eight inequivalent loops). The shortest loop is the hexagon itself, it involves 3 dimers. 4, 5 and 6-dimer moves are also possible by including 2, 4, and 6 additional triangles (the loop length must be even). The largest loop is the star. For each loop α, we associate the two ways dimers can be placed along that loop: $|d_\alpha(h)\rangle$ and $|\bar{d}_\alpha(h)\rangle$. Note that $\sigma^x(h)$ measures the relative phases of dimer configurations displaying, respectively, the $d_\alpha(h)$ and $\bar{d}_\alpha(h)$ patterns in the wave function.

5.5.6.2. *RK ground state*

As for the QDM discussed previously, the GS of this Hamiltonian is the equal-amplitude superposition of all dimer coverings belonging to a given topological sector (as on the triangular lattice there are four sectors). This

Table 5.1. The eight different classes of loops which can surround a hexagon of the kagomé lattice.

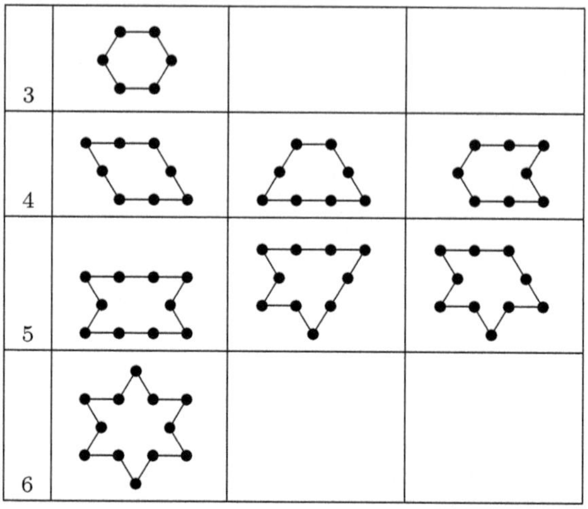

Note: Including all possible symmetries we find 32 possible loops. The first column indicates the number of dimers involved in the coherent motion around the hexagon.

can be readily shown by writing $\mathcal{H}$ as a sum of projectors:

$$\mathcal{H} = -N_h + \sum_h \sum_{\alpha=1}^{32} [|d_\alpha(h)\rangle - |\bar{d}_\alpha(h)\rangle][\langle d_\alpha(h)| - \langle \bar{d}_\alpha(h)|], \tag{5.40}$$

where N_h is the number of hexagons on the lattice. When expanding the products, the diagonal terms give a simple constant since

$$\sum_{\alpha=1}^{32} |d_\alpha\rangle\langle d_\alpha| + |\bar{d}_\alpha\rangle\langle \bar{d}_\alpha| = 1. \tag{5.41}$$

This reflects the fact that, for any dimerization, the dimers on hexagon h match *one and only one* of the 2×32 patterns $\{d_\alpha, \bar{d}_\alpha\}$.

Unlike the square or triangular case, the RK GS $|0\rangle = \sum_{c\in\Omega} |c\rangle$ are not degenerate with some staggered VBC.[cc] This means that the Hamiltonian

[cc]Because resonances loops of length up to 12 are present the dynamics is ergodic in each of the four topological sectors.[141]

of Eq. (5.38) is not at a phase transition to a VBC. As we will explain, it is *inside* a liquid RVB phase.

The RK wave function can be viewed as a dimer condensate. It is similar to the GS of liquid ^{4}He which has the same positive amplitude for any configuration and permuted images.[163] An important difference, however, is that the QDM state is incompressible and cannot sustain acoustic phonons. This can be related to the fact that the $U(1)$ symmetry of the Bose liquid is absent in the QDM on non-bipartite lattices. It is replaced instead by a discrete $\mathbb{Z}_2$ gauge symmetry (see Section 5.5.6.7).

5.5.6.3. *Ising pseudo-spin variables*

The kinetic energy operators σ^x defined in Eq. (5.39) commute with each other. This is obvious when two such operators act on remote hexagons, but it also holds for neighboring ones. This property can easily be demonstrated with the help of the arrow representation of dimer coverings introduced by Zeng and Elser.[164] This mapping of kagomé dimerizations to arrows on the bonds of the honeycomb lattice is illustrated in Fig. 5.7. Each arrow has two possible directions: It points toward the interior of one of the two neighboring triangles. If site i belongs to a dimer (i, j), its arrow must point toward the triangle the site j belongs to. A dimer covering can be constructed from any arrow configuration provided that the number of outgoing arrows is odd (1 or 3) on every triangle.

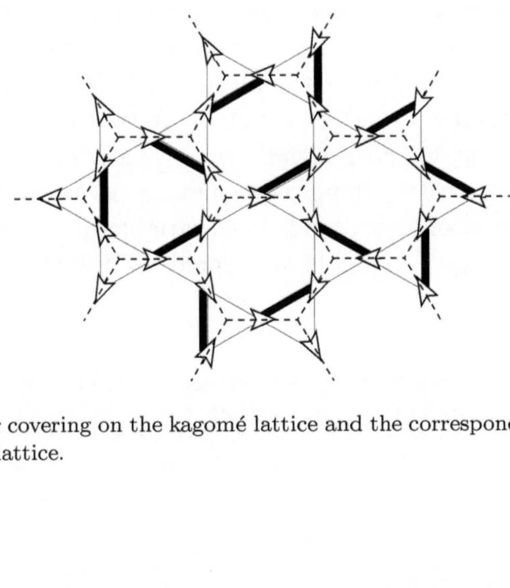

Fig. 5.7. A dimer covering on the kagomé lattice and the corresponding arrows. Dashed lines: honeycomb lattice.

The operators σ^x have a particularly simple meaning in terms of the arrow degrees of freedom: $\sigma^x(h)$ flips the 6 arrows sitting on h.[dd] It is then clear that the σ^x commute and that $\sigma^x(h)^2 = 1$. In fact, these operators can be used as Ising pseudo-spin variables and the Hamiltonian now describes non-interacting pseudo-spins in a uniform magnetic field pointing in the x direction. In the GS, we have $\sigma^x(h) = 1$ on every hexagon.

5.5.6.4. *Dimer–dimer correlations*

The GS is the most possible disordered dimer liquid as the dimer–dimer correlations strictly vanish beyond a few lattice spacings. Such correlations can be computed by the Pfaffian method. On the kagomé lattice, the determinant of the Kasteleyn matrix (which is directly related to the partition function of the classical dimers problem) is exactly constant in Fourier space.[165] Since dimer–dimer correlations are obtained from the Fourier transform of the inverse of this determinant, they turn out to be strictly zero beyond a few lattice spacings (as soon as the two bonds do not touch a common triangle).[141] This result can also be obtained by a simpler argument[141,166] using the σ^x operators. This result is related to the kagomé geometry.[ee] This absence of long-range dimer–dimer correlations demonstrates that the RK state is a dimer liquid and that it breaks no lattice symmetry.

On the triangular lattice, even at high temperature, dimer–dimer correlations decay exponentially with distance but these correlations remain *finite* at any distance. On the square lattice, such correlations are even larger because they decay only as a power law. This means that the infinite hard-core dimer repulsion makes QDM non-trivial even at infinite temperatures; dimers cannot be free when they are fully packed. From this point of view, we see that the kagomé lattice is particular: It is as close as possible to a free dimer gas, except for non-trivial correlations over a few lattice spacings. This is the reason why dimer coverings on the kagomé lattice should be handled with independent pseudo-spin variables and why the RK state on this lattice is the most possible disordered RVB liquid.

[dd]Flipping all the arrows around any closed loop (such as around an hexagon) preserves the local constraint imposed on arrow configurations. Flipping the arrows around a topologically non-trivial loop changes the topological sector.

[ee]The model of Eq. (5.38) can be generalized to any lattice made of corner-sharing triangles.[141]

5.5.6.5. *Visons excitations*

The σ^x operators can be simultaneously diagonalized, but they must satisfy the global constraint $\prod_h \sigma^x(h) = 1$ since this product flips every arrow *twice*. It must therefore leave all dimerizations unchanged. Therefore, the lowest excitations have an energy 4 above the GS, and they are made of a *pair* of hexagons a and b in a $\sigma^x(a) = \sigma^x(b) = -1$ state. a and b are the locations of two Ising vortices (or *visons*[161, 162]). As remarked before, this means that the relative phases of the configurations with $d_\alpha(h)$ and $\bar{d}_\alpha(h)$ patterns have now changed sign. The corresponding wave function is obtained in the following way. Consider a string Ω which goes from a to b (see Fig. 5.8), and let $\Omega(a, b)$ be the operator which measures the parity ± 1 of the number of dimers crossing that string. $\Omega(a, b)$ commutes with all $\sigma^x(h)$, except for the ends of the string: $\sigma^x(a)\Omega(a, b) = -\Omega(a, b)\sigma^x(a)$. A dimer move changes the sign of $\Omega(a, b)$ if and only if the associated loop crosses the string an odd number of times, which can only be done by surrounding one end of the string. This shows that $\Omega(a, b)$ flips the σ^x in a and b.[ff] As the RK GS $|0\rangle$, $\Omega(a, b)|0\rangle$ is a linear combination of all dimer configurations belonging to one sector. However, the amplitudes are now 1 and -1 depending on the number of dimers crossing Ω. This wave function therefore has nodes, and it is at an excited state of energy 4 with two vortices in a and b. It is easy to see that a different choice Ω' for the string connecting a and b gives the same state up to a global sign which depends on the parity of the number of kagomé sites enclosed by $\Omega \cup \Omega'$.

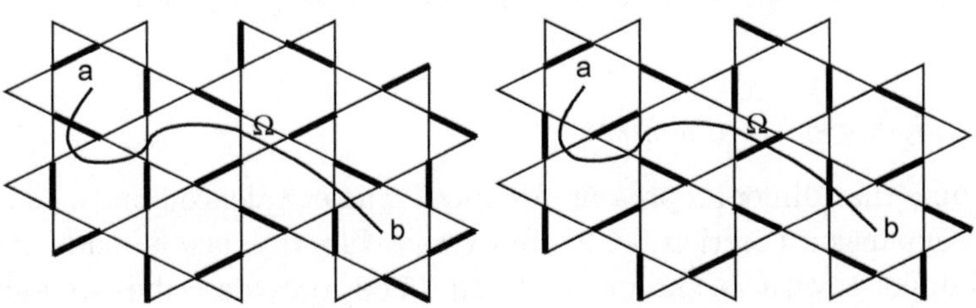

Fig. 5.8. A pair of visons (located in a and b) is created by applying to the RK wave function a factor (-1) for each dimer crossing the string Ω. The dimerization shown on the left appears in the linear superposition of the two-vison state with the sign -1, whereas the one on the right has the sign $+1$.

[ff]Up to a global sign (reference dependent) $\Omega(a, b)$ is equal to $\sigma^z(a)\sigma^z(b)$ where the σ^z operators are those introduced by Zeng and Elser.

These vortex excitations carry a $\mathbb{Z}_2$ charge since attempting to put two vortices on the same hexagon does not change the state. Such excitations are not local in terms of the dimer degrees of freedom. Indeed, determining the sign of a given dimerization in a state with two visons which are far apart requires the knowledge of the dimer locations along the whole string connecting the two vortex cores. In this model, the visons appear to be static and non-interacting. This is a particularity of this solvable model, but the existence of gapped vison excitations is believed to be a robust property of RVB liquids. In more realistic models, the visons will acquire a dynamics and a dispersion relation but will remain gapped.[gg] They will also have some interactions with each other but should remain *deconfined.* This property is particularly clear in the kagomé QDM: Visons are necessarily created by pairs but the energy is independent of their relative distances.

The Ising vortices also offer a simple picture of the topological degeneracy. Consider a GS $|+\rangle$ of the model which lives in the sector where the winding number Ω_y (with respect to some arbitrary but fixed dimerization) is even. Another GS $|-\rangle$ is obtained in the odd-Ω_y sector. Now consider the combination $|0\rangle = |+\rangle + |-\rangle$ and apply the operator $\Omega(0, L_x)$ corresponding to a closed loop surrounding the torus in the x direction. This amounts to creating a pair of nearby visons at the origin, taking one of them around the torus in the x direction and annihilating them. This can also be viewed as the creation of a vison in one hole of the torus (with no energy cost). It is simple to check that $\Omega(0, L_x)|0\rangle = |+\rangle - |-\rangle$ (up to an irrelevant global sign). This provides a simple relation between the vison-pair creation operator and the existence of two topologically distinct GS $|+\rangle + |-\rangle$ and $|+\rangle - |-\rangle$.

5.5.6.6. *Spinons deconfinement*

We assume that dimers represent "dressed" singlet valence bonds, as in the overlap expansion (Section 5.5.2). Since the Hilbert space is made of fully packed dimer coverings, the model of Eq. (5.38) only describes spin-singlet states. However, as any QDM, it can be extended to include static holes or spinons. Configurations with unpaired sites (spinon or holon) are now allowed, but the kinetic terms of the original Hamiltonian through which the loop passes on an empty site give zero value. Consider a system with

[gg]It is possible to add potential energy terms to Eq. (5.38) to drive the system outside of the liquid phase and this transition corresponds to a vison condensation.

two static spinons in x and y. As on the square[143] and triangular lattices,[144] at the RK point, the exact GS $|x, y\rangle$ remains the sum of all dimer coverings and the GS energy is independent of the distance between the two spinons (except at very short distance if they belong to a common hexagon). This is a first indication that RVB *spin* liquid has deconfined spin-1/2 excitations (spinons). In the QDM language, these excitations are simply unpaired sites in a dimer liquid background. Such unpaired sites are necessarily created by pairs, but they can then propagate freely (no attractive potential) when they are sufficiently far apart.

Another calculation allows to test the deconfinement properties of a dimer liquid. We consider the state $|\psi\rangle = \sum_{\mathbf{r}\neq 0} |0, \mathbf{r}\rangle$, where $|0, \mathbf{r}\rangle$ is the (un-normalized) GS with two spinons in 0 and $\mathbf{r}$. The probability to find a spinon in $\mathbf{r}$ in the $|\psi\rangle$ can be obtained by the relatively involved calculation of the monomer correlation[hh] with Pfaffians. One the square lattice, this probability goes to zero as $1/\sqrt{r}$.[149] This shows that the second spinon is (quasi-) confined in the vicinity of the first one on the square lattice because escaping far away represents a large "entropy" cost in the dimer background. On the triangular lattice, it goes exponentially to a constant.[154] This result is a signature of deconfinement. In fact, the same signature can be obtained on the kagomé lattice without any technical calculation since the monomer correlation is exactly 1/4 at any distance.[167]

If unpaired sites are allowed, one can describe spinons or holons. Unfortunately, in the presence of simple kinetic energy terms for these objects, the model can no longer be solved. However, one can consider a static spinon and its interaction with visons: when the spinon is adiabatically taken around a vison, the dimers are shifted along a path encircling the vison. Because the vison wave function is particularly simple in this model, it is easy to check that this multiplies the wave function by a factor -1. This is the signature of a long-range statistical interaction[159, 160] between visons and spinons (or holons). In more realistic models, as long as the visons are gapped excitations, the spinons are expected to be deconfined. On the contrary, if the visons condense their long-range statistical interaction with spinon, then their motion frustrated. This is no longer true if they propagate in *pairs*, in which case they are not sensitive any more to visons (see Ref. [141] for an extension of the present QDM with a vison condensation). This

---

[hh] Ratio of the number of dimer coverings with two holes in 0 and $\mathbf{r}$ to the number of dimer coverings without a hole.

simple physical picture illustrates the relation between vison condensation and spinon confinement.

5.5.6.7. $\mathbb{Z}_2$ *gauge theory*

The forces responsible for confinement are usually associated to gauge fields and their fluctuations. Whereas $U(1)$ compact gauge theories are generically confining in $2+1$ dimensions,[136, 168] $\mathbb{Z}_2$ gauge theories are known to possess deconfined phases.[169] For this reason, some attention has been paid to the connections between $\mathbb{Z}_2$ theories and fractionalized phases in 2D electronic systems.[161]

It is known[140] that QDM can be obtained as special limits of $\mathbb{Z}_2$ gauge theories, the gauge variable being the dimer number on a bond. However, on the kagomé lattice this connection can be made exact and completely explicit since there is a one-to-one correspondence between dimer coverings and physical states (i.e. gauge-invariant) of a $\mathbb{Z}_2$ gauge theory.[141] In this mapping, the gauge fields are Ising variables living on the link of the honeycomb lattice (i.e. kagomé sites) and are constructed from the arrows described previously. As for the constraints of gauge invariance, they correspond to the odd parity of the number of outgoing arrows on every triangle. The σ^x operator used to define a solvable QDM translates into a gauge-invariant plaquette operator for the gauge degrees of freedom (product of the Ising gauge variables around an hexagon). With this mapping, the visons appear to be vortices in the $\mathbb{Z}_2$ gauge field and the solvable model of Eq. (5.38) maps to the deconfined phase of the gauge theory (precisely at infinite temperature).

5.5.7. *A QDM with an extensive ground-state entropy*

So far we have discussed QDM that realize spontaneous VBC, critical states, or RVB liquids. We wish to mention here that these three scenarios may not be the only possible GS for QDM. In particular, a QDM on the kagomé lattice with an extensive GS entropy was recently discussed.[170] This model was introduced from the observation that the dimer kinetic energy terms arising from an overlap expansion (Section 5.5.2) generally have non-trivial *signs* as soon as resonance loops of *different lengths* are considered. It was then realized that such signs (which make the QDM no longer appropriate for QMC simulations) can lead to qualitatively new phases, different from VBC or RVB liquids. The Hamiltonian introduced in Ref. [170] is similar to that of Eqs. (5.38)–(5.39) except that each resonance loop α is now included with a sign$(-1)^{n_\alpha}$, where $n_\alpha = 3, \ldots, 6$ is the number of dimers involved:

$$\mathcal{H} = \sum_h (-1)^{n_\alpha} [|d_\alpha(h)\rangle\langle\bar{d}_\alpha(h)| + |\bar{d}_\alpha(h)\rangle\langle d_\alpha(h)|]. \tag{5.42}$$

These signs are precisely those appearing in the overlap expansion (of the order of one hexagon) of the Heisenberg model on the kagomé lattice. This expansion was carried out by Zeng and Elser[171] in an insightful paper which laid the basis of the analysis of the kagomé antiferromagnet in the first-neighbor valence bond subspace.

Although not exactly solvable, the Hamiltonian of Eq. (5.42) was shown to be a dimer liquid (short-range dimer–dimer correlation) and to have a huge GS degeneracy $\sim 2^{N/6} = 1.122^N$ (N is the number of kagomé sites). In addition, several numerical indications pointed to a critical behavior of this system,[170] with a possible algebraic decay of energy–energy correlations.[ii] It was argued that the effective QDM describing the singlet dynamics of the spin-1/2 Heisenberg antiferromagnet on the kagomé lattice could be *close* (in parameter space) to Eq. (5.42). If correct, this sheds light on the very large density of singlet states observed at low energy in the numerical spectra of that spin model (see Section 5.7).

5.6. Multiple-Spin Exchange Models

5.6.1. *Physical realizations of multiple-spin interactions*

5.6.1.1. *Nuclear magnetism of solid ^{3}He*

Solid ^{3}He was the first magnetic system in which the importance of MSE interactions was recognized.[172–174] Due to the large zero-point motion of the atoms about their mean positions, tunneling events during which 2, 3, or 4 atoms exchange their positions in a cyclic way are frequent. These processes generate an effective interaction between the (nuclear) spins which can be written as follows:

$$\mathcal{H} = \sum_P -J_P(-1)^P(P_{\text{spin}} + P_{\text{spin}}^{-1}), \tag{5.43}$$

where the sum runs over permutations P, $J_P > 0$ is the exchange frequency of the associated tunneling process (in real space), and P_{spin} acts on the

[ii]Note that a 1D analog of Eq. (5.42) can be defined and exactly maps onto the Ising chain in transverse field at its critical point.

Hilbert space of spin-1/2 located on the site of the crystal. The sign $-(-1)^P$ depends of the signature of the permutation P and is a consequence of the Pauli principle. For a *cyclic* permutation involving n spins, this sign is just $(-1)^n$ and is responsible for the ferromagnetic character of processes involving an odd number f spins. For spin-1/2 particles, two- and three-spin exchange terms reduce to the familiar Heisenberg interaction:

$$P_{12} = 2\vec{S}_1 \cdot \vec{S}_2 + \tfrac{1}{2}, \tag{5.44}$$

$$P_{123} + P_{321} = P_{12} + P_{23} + P_{31} - 1, \tag{5.45}$$

but this is no longer true for $n \geq 4$:

$$P_{1234} + P_{4321} = P_{12}P_{34} + P_{14}P_{23} - P_{13}P_{24} + P_{13} + P_{24} - 1 \tag{5.46}$$

which can be expressed (thanks to Eq. (5.44)) as a sum of terms with two and four Pauli matrices.

^{3}He can form solid atomic monolayers with a triangular geometry when adsorbed on a graphite substrate at ultra-low temperatures (milli-Kelvin range). This 2D magnet has been studied for a long time (see Refs. [175,176] and references therein), and the importance of MSE interactions involving up to six atoms has now been recognized.[177,178] The exchange frequencies of the most important processes have been computed by Path Integral Monte Carlo (PIMC)[179–182] (analytic WKB calculations have also been carried out[177,183]) as a function of the density. The proposed MSE Hamiltonian describing the magnetic properties of this 2D quantum crystal reads

$$\mathcal{H} = (J_2 - 2J_3) \sum P_{12} + J_4 \sum (P_{1\ldots4} + \text{H.c.})$$
$$- J_5 \sum (P_{1\ldots5} + \text{H.c}) + J_6 \sum (P_{1\ldots6} + \text{H.c.}), \tag{5.47}$$

where Eq. (5.45) was used to absorb the three-spin terms into an effective first-neighbor Heisenberg exchange $J_2^{\text{eff}} = J_2 - 2J_3$. At high density, the hard-core potential between Helium atoms only leaves three-body exchanges possible ($J_3 \gg J_{n\neq3}$) and Eq. (5.47) reduces to a first-neighbor Heisenberg *ferromagnet*,[184] as observed experimentally for the first time by Franco *et al.*[185] in high-density layers. On the contrary, the second layer solidifies at lower density and higher order exchange terms cannot be ignored.[jj] PIMC simulations[182] and high-temperature fits of the experimental data[178] showed that the relative strength of two- and four-spin terms if roughly $J_2^{\mathrm{eff}}/J_4 \sim -2$ in the low-density second layer solid.

The J_2–J_4 model was studied by exact diagonalizations. It exhibits many distinct phases. There is evidence of a short-range RVB SL phase with no broken symmetry,[114,186,187] and different gapless exotic phases, with $SU(2)$ symmetry breaking but with no on-site magnetization. Exotic order parameters have been evidenced in different parts of the phase diagram, as a chiral current[397] and an octupolar moment.[398]

The most recent ultra-low temperature measurements of specific heat[188] and uniform susceptibility[189] are not incompatible with a gapless phase, but the signature of the octupolar moment remains to be observed.[399]

5.6.1.2. *Wigner crystal*

The Wigner crystal is another fermionic solid with a triangular geometry where MSE interactions can play an important role. At very low density, the Coulomb energy dominates, the crystal is almost classical, and MSE interactions are very small. Exchange frequencies J_P can be computed in this regime by a semi-classical (WKB) approximation[177,190,191] and, as for the high-density solid of ^{3}He, three-body exchanges dominate and give rise to ferromagnetism. However, at higher density and close to melting, PIMC calculations of the exchange frequency[192] showed that the magnetism may be described by an MSE model with parameters (J_2^{eff} and J_4) close to those where the triangular MSE model is expected to be an RVB SL. Unlike the ^{3}He case, the particles (electrons) are charged and an external magnetic field also has an orbital effect; it also introduces complex phases in the exchange energies: $P+P^{-1} \to e^{i\alpha}P+e^{-i\alpha}P^{-1}$ where the angle $\alpha = 2\pi\phi/\phi_0$ is proportional to the magnetic flux ϕ passing through the area enclosed by the exchange trajectory and ϕ_0 is the unit flux quantum. This can give rise to very rich phase diagrams[191,192] where complex MSE terms compete with the Zeeman effect (see Ref. [193] for some early experimental attempts to explore this physics).

[jj] The first layer is then so dense that exchange is strongly suppressed. The first layer can also be replaced by an ^{4}He or HD monolayer.

5.6.1.3. *Cuprates*

The possibility of a significant four-spin exchange around square Cu plaquettes in copper oxide compounds was first suggested by Roger and Delrieu.[194] They interpreted the anomalously large width of Raman scattering spectra as a signature of four-spin exchange in this copper oxide superconductor. The importance of these MSE interactions in CuO_2 planes ($J_4 \sim 0.25J_2$) has then been emphasized by a number of groups and in different materials and by different experimental and theoretical approaches.[195–198] Four-spin plaquette ring exchange also plays a significant role in ladder compounds.[199–202] For instance, exchange parameters with values $J_{\rm rung} = J_{\rm leg} = 110\,{\rm meV}$ and $J_{\rm ring} = 16.5\,{\rm meV}$ were proposed for $La_6Ca_8Cu_{24}O_{41}$ based on the dispersion relation of magnetic excitations.[199,200]

5.6.2. *Two-leg ladders*

Numerous works were devoted to ladder models with four-spin interactions. These include general bi-quadratic interactions as well as models with ring-exchange terms. We will only discuss here the simplest of these MSE models:

$$\mathcal{H} = J\sum_n(\vec{S}_{n,1}\cdot\vec{S}_{n,2} + \vec{S}_{n,1}\cdot\vec{S}_{n+1,1} + \vec{S}_{n,2}\cdot\vec{S}_{n+1,2}) + K\sum_\square (P_{1234} + \text{H.c.}). \tag{5.48}$$

Thanks to several studies,[200,203–205] the phase diagram of this Hamiltonian is now rather well understood and five different phases were identified.

- **Ferromagnetic phase.** The GS is fully polarized. This phase includes the ($J = -1, K = 0$) and the ($J = 0, K = -1$) points.
- **Rung-singlet phase.** This phase includes the GS of the ladder without MSE term ($J = 1, K = 0$). The spectrum is gapped and the GS is unique. A moderate $K/J \gtrsim 0.23 \pm 0.03$ destroys this phase[200,205–207] in favor of the VBC below.
- **Staggered VBC with dimers on the legs.** In one of the two degenerate GS, the dimerized bonds are $(2n,1)-(2n+1,1)$ and $(2n+1,2)-(2n+2,2)$. The VBC disappears for $K/J \gtrsim 0.5$.[205] Such a staggered VBC was first predicted in the framework of a ladder with bi-quadratic interaction by Nersesyan and Tsvelik.[59] Using matrix-product Ansatz, Kolezhuk and Mikeska[60] constructed models which are generalizations of

Eq. (5.48) and which have exact GS with long-range staggered dimer correlations. In this phase, the magnetic excitations are very different from the magnon excitations of the rung-singlet phase above. Here, the excitations do not form well-defined quasi-particles but a continuum made of pairs of domain walls connecting two dimerized GS.[59,60]

- **Scalar chirality phase.** The order parameter is $\langle \vec{S}_{n,1} \cdot (\vec{S}_{n,2} \times \vec{S}_{n+1,2}) \rangle$ and it spontaneously breaks the time-reversal symmetry and translation invariance. The GS is two-fold degenerate up to the next transition at $K/J \simeq 2.8 \pm 0.3$.[205] There exists a duality transformation[204,208] which maps the scalar chirality order parameter onto the dimer order parameter of the VBC above.[kk] Applying such a transformation to the exact VBC GS mentioned above, models with an exactly known GS and scalar chirality LRO can be constructed.[204,208] Although chiral SL have been much discussed in the literature, this is to our knowledge the first realization of such a phase in an SU(2) symmetric spin-1/2 model.
- **Short-range ordered phase with vector-chirality correlations.** The strongest correlations are $\langle (\vec{S}_{n,1} \times \vec{S}_{n,2}) \cdot (\vec{S}_{n',1} \times \vec{S}_{n',2}) \rangle$ but they remain short range. The spectrum is gapped and the GS is unique. This phase includes the pure $K = 1$ model where $J = 0$. This phase is related by the duality transformation discussed above to the rung-singlet phase.[208] This transformation indeed relates the Néel correlations $\langle (\vec{S}_{n,1} - \vec{S}_{n,2}) \cdot (\vec{S}_{n',1} - \vec{S}_{n',2}) \rangle$ (which are the strongest ones in the rung-singlet phase) to the vector-chirality correlations. Close to the ferromagnetic phase ($J < 0$), one observes a crossover to a region where the strongest correlations are ferromagnetic spin–spin correlations along the legs and antiferromagnetic along the rungs.[205]

5.6.3. *MSE model on the square lattice*

The phase diagram of the Hamiltonian equation (5.48) on the *2D square lattice* has been recently studied by Läuchli[209] by exact diagonalizations. Néel, ferromagnetic, columnar VBC, and staggered VBC phases were identified, as in the ladder model above. In addition, a nematic phase characterized by long-range vector chirality correlations (alternating spin currents) was found around the $K = 1$, $J = 0$ point. To our knowledge, this could be the first microscopic realization of a nematic order in a 2D spin-1/2 model.

[kk]The Hamiltonian of Eq. (5.48) is self-dual at $2K = J$.

5.6.4. *RVB phase of the triangular J_2–J_4 MSE*

Because of its relevance to solid ^{3}He films and Wigner crystals, the MSE model on the triangular lattice has been the subject of many studies.[114,186,187,210–213] We will discuss here some properties of the simplest MSE model with up to four-spin cyclic exchange interactions ($J_2 - 2J_3$ and J_4 only in Eq. (5.47)). The classical phase diagram (Fig. 5.9) of this model has been studied by Kubo and collaborators,[210,211] and the quantum one has been roughly scanned in Ref. [187]: We will mainly focus on the short-range RVB spin liquid (see Fig. 5.9), which might be *the first RVB SL encountered in an SU(2)-symmetric spin model.*

5.6.4.1. *Non-planar classical ground states*

It is well-known that a Heisenberg model (with possible second-neighbors, third-neighbors, ... interactions) on a Bravais lattice always admits a planar helical GS at the classical level. This is no longer true when MSE are present, and finding the classical GS for arbitrary J_2 and J_4 is an unsolved problem. A mean-field phase diagram was obtained for the classical model,[210] but very few exact results are known. In the neighborhood of $J_4 = 1$, $J_2 = 0$, the classical GS is known to be a four-sublattice configuration with magnetizations pointing toward the vertices of a regular

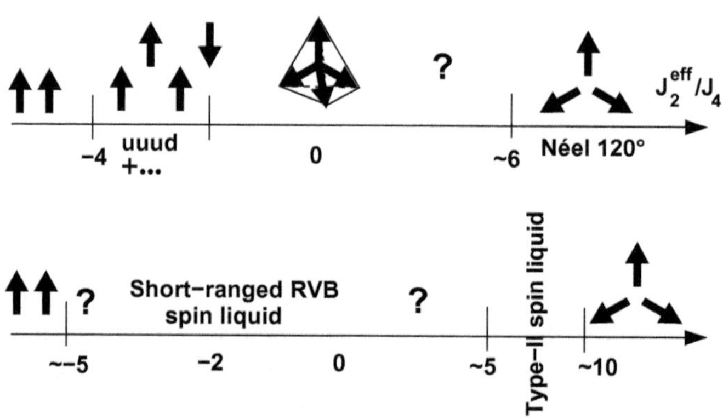

Fig. 5.9. Classical (top) and quantum (bottom) phases of the $J_2 - J_4$ MSE Hamiltonian. The classical model was studied by Kubo and Momoi[210] and is based on a variational approach. The quantum phase diagram is the simplest scenario compatible with the exact diagonalizations data of Refs. [114, 186, 187]. While not completely understood, in the type-II spin liquid region the spectra are characterized by a very large number of singlet excitations below the first triplet state. This is not the case in the RVB phase.

tetrahedron.[210] This is a quite interesting model where the GS spontaneously breaks a discrete Ising symmetry associated to the sign of the triple product $\vec{S}_1 \cdot (\vec{S}_2 \times \vec{S}_3)$ around a triangle. This broken symmetry gives rise to a finite-temperature phase transition which has been observed in MC simulations.[211] This phenomena is similar to the transition predicted in the $(\pi, 0)$ phase of the J_1–J_2 model on the square lattice.[14]

5.6.4.2. *Absence of Néel LRO*

The classical GS at $J_4 = 1, J_2 = 0$ are tetrahedral configurations. Although this phase appears to be stable within the framework of linear spin wave calculations[211] or Schwinger–Boson mean-field theory,[212] exact diagonalizations indicate that the magnetic LRO is washed out by quantum fluctuations.[114, 211] The chiral order predicted to survive at long distances and finite temperatures[211] in the classical system for $J_2 = 0$ is also likely to be washed out by quantum fluctuations.[114]

When $J_2 = 1$, a relatively small amount of $J_4 \sim 0.1$ is sufficient to destroy the three-sublattice Néel LRO realized by the first-neighbor Heisenberg model.[187] The nature of the phase on the other side of this transition is not settled but the finite-size spectra display a large density of singlet excitations at low energy which could be reminiscent of the kagomé situation.[187]

From exact diagonalizations (up to 36 sites), no sign of Néel LRO could be found at $J_4 = 1, J_2 = -2$.[114, 186] In addition, the finite-size analysis showed that the spin–spin correlation length is quite short at $J_2 = -2$ and $J_4 = 1$ and a spin gap of the order of $\Delta \sim 0.8$ exists at this point. Much of the numerical effort to elucidate the nature of the MSE GS was concentrated on this point because it is close to the parameters realized in low-density ^{3}He films (when higher order exchanges are neglected).

5.6.4.3. *Local singlet–singlet correlations — Absence of lattice symmetry breaking*

Having excluded the possibility of a Néel ordered GS at $J_4 = 1, J_2 = -2$ it is natural to look for a possible VBC. Because of the complexity of the MSE Hamiltonian, it is not clear what kind of spatial order should be favored. From the analysis of dimer–dimer correlations (see Fig. 5.10) it appears that parallel valence bonds repel each other at a short distance. This is similar to what is observed in the staggered phase of the J_2–J_4 MSE ladder and square lattice models. For this reason, it appears that a plausible VBC would be the staggered VBC encountered in the triangular

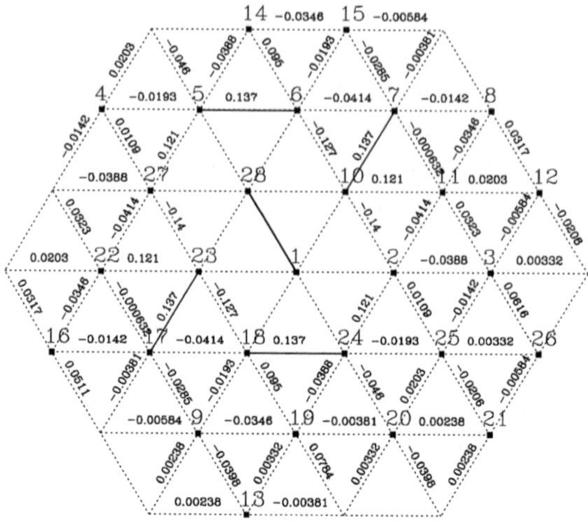

Fig. 5.10. Dimer-dimer correlations in the GS of the J_2–J_4 MSE model on the triangular lattice (28 sites) at $J_2 = -2$, $J_4 = 1$ (from Ref. [114]). Numbers are proportional to $\langle \hat{d}_0 \hat{d}_x \rangle - \langle \hat{d}_0 \rangle \langle \hat{d}_x \rangle$ where the operator $\hat{d}_x$ projects onto the singlet space of the bond x and $\hat{d}_0$ refers to the reference bond $(1, 28)$. These results shows a clear tendency for repulsion between parallel dimers.

QDM for $V > J$ (Section 5.5.5). However, this scenario seems difficult to reconcile with the weakness of dimer–dimer correlations.[114] In addition, the low-energy singlet states and their quantum numbers[214] do not reflect the 12-fold quasi-degeneracy that should be present if the system was to spontaneously break some lattice symmetry according to a staggered VBC pattern. Small systems usually *favor* ordered phases because low-energy and long-wavelength fluctuations that could destabilize an ordered state are reduced compared to larger systems. From the fact that the finite-size spectra do not show the signatures of a staggered VBC symmetry breaking, it is unlikely that the MSE model could develop a VBC of this kind in the thermodynamic limit.

5.6.4.4. *Topological degeneracy and Lieb–Schultz–Mattis theorem*

Because no VBC phase could be identified in the MSE model at $J_4 = 1$, $J_2 = -2$, the numerical data were compared with the predictions of an RVB liquid scenario.

In one dimension, a famous theorem due to Lieb, Schultz, and Mattis[215] (LSM) states that in a 1D spin system with an half-integer spin in the unit cell there is at least on excited state collapsing to the GS in the thermodynamic limit (periodic boundary conditions). There are in fact several arguments suggesting that this theorem might, at least to some extent, also apply to higher dimensions.[157, 216–218, 362] If that is the case, a gapped system with an odd integer spin in the unit cell must have a degenerate GS. The simplest scenario to explain this degeneracy is a translation symmetry breaking. One could think that this would rule out the possibility of any (translation invariant) RVB liquid in such models. This is incorrect because a GS degeneracy can have a topological origin on a system with periodic boundary conditions, as we discussed in the framework of QDM (Section 5.5). Such a phase is characterized by a four-fold topological GS degeneracy when the system is on a torus. That degeneracy allows the system to fulfill LSM's requirement without any spontaneous translation symmetry breaking.[157]

On a finite-size system, the topological degeneracy is only approximate but some constraints exist for the quantum numbers (momentum in particular) of the quasi-degenerate multiplet.[157] A system with periodic boundary conditions with an even number of sites but an *odd number of rows* is expected to have two GS that differ by a momentum π in the direction parallel to the rows, in close analogy to the LSM theorem in dimension one. The numerical spectra of the MSE model exhibit a set of three singlet energy levels collapsing onto the GS when the system size is increased[114] and their quantum number turns out to be consistent with the constraints derived from the general RVB picture.[157]

5.6.4.5. *Deconfined spinons*

The SL phase described above is expected to have deconfined spinons ($S = 1/2$ excitations). These excitations should show up as an incoherent continuum in the spin–spin dynamical structure factor. However, such a feature would probably be rather difficult to observe on small 2D lattices, in particular due to the small number of inequivalent **k**-vectors in the Brillouin zone. On the other contrary, the binding energy of two spinons can be evaluated by comparing the GS energy and the first magnetic excitation energy on even and odd samples. In the case of the MSE model at $J_4 = 1, J_2 = -2$ the results show the existence of a bound state (it is more favorable to put two spinons in the same small sample than in separate ones, which is not surprising) but this *does not mean that the spinons are confined* (contrary to the conclusions of Ref. [114]). Interestingly, this binding energy seems to go to zero for the largest available sizes (Fig. 5.11): This might indicate

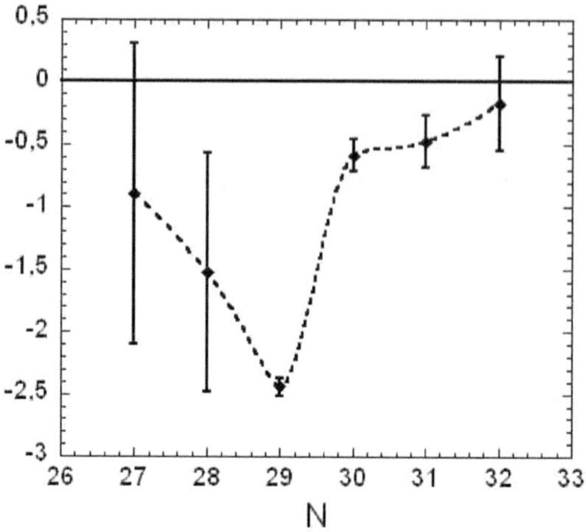

Fig. 5.11. Spinons binding energy as a function of the system size in the MSE model at $J_2 = -2$ and $J_4 = 1$. The vertical bars correspond to the range of values found for different sample shapes.

the absence of attraction between spinons for large enough separation and an asymptotic deconfinement.

It is important to stress here that the RVB SL phase discussed here (and its QDM counterparts of Sections 5.5.5 and 5.5.6) is not the only way to spinon deconfinement in 2D. There is at least another scenario, inherited from one dimension, which is the *sliding Luttinger liquid*. Indeed, the Luttinger liquid behavior and the 1D critical behavior of magnetic chains seem to be robust to small (or moderate) *frustrating transverse couplings between chains*, as observed both in theoretical[219–225] and numerical approaches.[100] This regime between one and two dimensions which may have been observed in Cs_2CuCl_4[226] is the subject of a number of recent studies.[222, 223]

5.6.5. *Other models with MSE interactions*

Multiple-spin interactions are present in a number of models that were found to exhibit fractionalization or an RVB liquid GS. Well-known examples of MSE interactions with an Ising symmetry are $\mathbb{Z}_2$ gauge theories, where the gauge-invariant plaquette term is a product of Pauli matrices $\prod_i \sigma_i^z$. Such theories have a deconfined phase in 2+1 dimension,[169] and their relevance to fractionalized phases of 2D electronic systems has been pointed out by Senthil and Fisher.[161] The connection between $\mathbb{Z}_2$ gauge theories and

QDM was mentioned in Section 5.5.6.7. Some MSE spin models with an Ising symmetry and a fractionalized GS were discussed by Kitaev,[227] Nayak, and Shtengel.[228] In the other limit of a $U(1)$ (or XY) symmetry, several models have been studied. Recent examples are based upon the spin-1/2 four-spin XY ring exchange interaction[229–232]

$$\mathcal{H} = -K \sum_{\langle ijkl \rangle} (S_i^+ S_j^- S_k^+ S_l^- + \text{H.c.}) \tag{5.49}$$

which is the XY analog of the $SU(2)$ MSE interaction $P_{1234} + \text{H.c.}$

5.7. Antiferromagnets on the Kagomé Lattice

The spin-1/2 nearest-neighbor Heisenberg model on the kagomé lattice has attracted much attention in the long quest of spin liquids. The first studies go back to the end of the eighties[171, 244–247] and more than 300 papers have appeared in journals since this period. This model possesses the two ingredients that are considered important to obtain a liquid GS: frustrated geometry (non-bipartiteness) and a low coordination number which enhance the quantum effects. Moreover, the classical model is at a high degeneracy point in parameter space, where several phases meet. The first breakthrough in the study of this tough quantum problem has been done thanks to large-scale exact diagonalizations:[171, 247–249, 251] These studies could not detect any form of LRO, and the model was therefore an interesting candidate for a spin liquid. But the subject remained controversial as the sizes of the clusters in exact diagonalizations were very limited (36 spins during nearly two decades, 48 very, recently,[252] and a large number of competing hypotheses were considered; (Section 5.7.5). A new impetus was recently provided by the discovery of various Cu compounds exhibiting spin liquid behaviors (Section 5.7.6). In the following, we will first describe the properties of simpler "parent" models (classical model Section 5.7.2, Ising model Section 5.7.1, quantum dimer model Section 5.7.3). We will then give the 2012 results of state-of-the-art DMRG numerical approaches (Section 5.7.4) for the pure nearest-neighbor spin-1/2 Heisenberg model, describe various proposals of nearby phases (in parameter space), and provide a brief account of the experimental procedures.

5.7.1. *Ising model*

The classical model remains disordered at all temperatures.[234, 235] The system fails to order even at $T = 0$ and has a large finite entropy per site:

$S_{\rm kag}^{\rm Ising} = 0.502$, more than half the independent spin value, much larger than the triangular lattice value $S_{\rm tri}^{\rm Ising} = 0.323$, and of the order of Pauling approximation for independent triangles $S_{\rm Pauling} = 0.501$.[233] Moessner and Sondhi have studied this Ising model in a transverse magnetic field (the simplest way to include some quantum fluctuations): the model fails to order for any transverse field, at any temperature.[145, 236]

5.7.2. *Classical Heisenberg models on the kagomé lattice*

The nearest-neighbor classical $O(3)$ Heisenberg model on the kagomé lattice also has a huge GS degeneracy. This property holds on different lattices with corner-sharing units such as the checkerboard lattice or the 3D pyrochlore lattice.[106, 107] On all these lattices, the nearest-neighbor Heisenberg Hamiltonian can be written as the sum of the square of the total spin $\vec{S}_\alpha$ of individual units α (a triangle for the kagomé lattice and a tetrahedron in the $2d$ checkerboard or $3d$ pyrochlore cases), which share only one vertex. Classical GS are obtained whenever $\forall \alpha \vec{S}_\alpha = \vec{0}$. This condition fixes the relative positions of the three classical spins of a triangle at 120° from each other in a plane. But it does not fix the relative orientation of the plane of a triad with respect to the planes of the triads on neighboring triangles: The model has a continuous local degeneracy[235, 237] at $T = 0$.[ll]

The classical model has a large density of low-lying excitations at a low temperature.[243] Thermal fluctuations select coplanar configurations because they have the largest phase space for low-energy fluctuations and are therefore entropically favored.[235, 237, 239] The plausibility of long-range order in spin–spin correlations or other observables at a very low temperature has also been discussed.[235, 240, 291, 292] The recent progresses made in numerical simulations point to a selection of the so-called $\sqrt{3} \times \sqrt{3}$ order at ultra-low temperatures.[331, 375]

[ll]Counting the *planar* GS amounts to determine in how many ways one can associate one of the three letters A, B, and C to each site so that each triangle has spins along the three different orientations. This already represents an extensive entropy.[235, 238] In such a planar GS, one may look for a closed loop involving only two spin directions, say $A - B - A - B - \cdots$. By construction, all the spins connected to this loop point in direction C. It is then possible to rotate simultaneously all the spins of the loop by some arbitrary angle about the C axis, without any energy cost. This creates a non-planar GS.

The $T = 0$ instabilities of the classical model to infinitesimal perturbations have been studied very early.[281] Depending on its sign, a second-neighbor Heisenberg coupling J_2 leads to two different coplanar phases with unit cells of 3 or 9 spins. They correspond, respectively, to the sol-called "$q = 0$" and the "$\sqrt{3} \times \sqrt{3}$" magnetic structures in reference to the order wave vector. The Dzyaloshinsky–Moriya interaction (with the DM vector perpendicular to the spin plane) favors the "$q = 0$" phase.[241,242] It has been recently shown that an infinitesimal antiferromagnetic interaction J_3 between third neighbors across hexagons favors a non-planar chiral order with a 12-spin unit cell.[289] Under the action of quantum fluctuations, these classical orders can be destabilized to give birth to distinct classes of spin liquids with different fluxes around close contours.[286,376]

A new algebraic method[290] has been used to enlarge our knowledge of the zero-temperature phase diagram of the classical J_1–J_2–J_3 model, as shown in Fig. 5.12. This method, which is based on symmetry arguments, allows a classification of all the regular classical magnetic orders that can be harbored on a lattice whatever the details of the $SU(2)$ invariant spin Hamiltonian.

5.7.3. *Nearest-neighbor RVB description of the spin-1/2 kagomé antiferromagnet*

Assuming that the GS of the spin-1/2 Heisenberg model has a large Enough spin-gap, an hypothesis supported by early exact diagonalisations,[247] Zeng and Elser[171] proposed a variational description of the GS and low-lying excitations of the kagomé antiferromagnet in the basis of nearest-neighbor valence bonds. They analyzed in this context the dimer dynamics and showed on a $N = 36$ sample that the hexagon VBC — favored by the shortest (three-)dimer moves — melts when introducing higher-order tunneling. Mila and Mambrini[250,253] confirmed that this reduced Hilbert space of next-neighbor valence bonds captures some of the perplexing features of this magnet and specifically the absence of (measurable) gap in the singlet sector of small samples.[249]

Some recent numerical results[254] (in the full spin-1/2 Hilbert space as well as in the RVB subspace) showed that (static) non-magnetic impurities (holes) experience an unexpected *repulsion* in this system and that no significant magnetic moment is created in the vicinity of the impurities. These non-magnetic impurities provide a valuable probe of spinon confinement

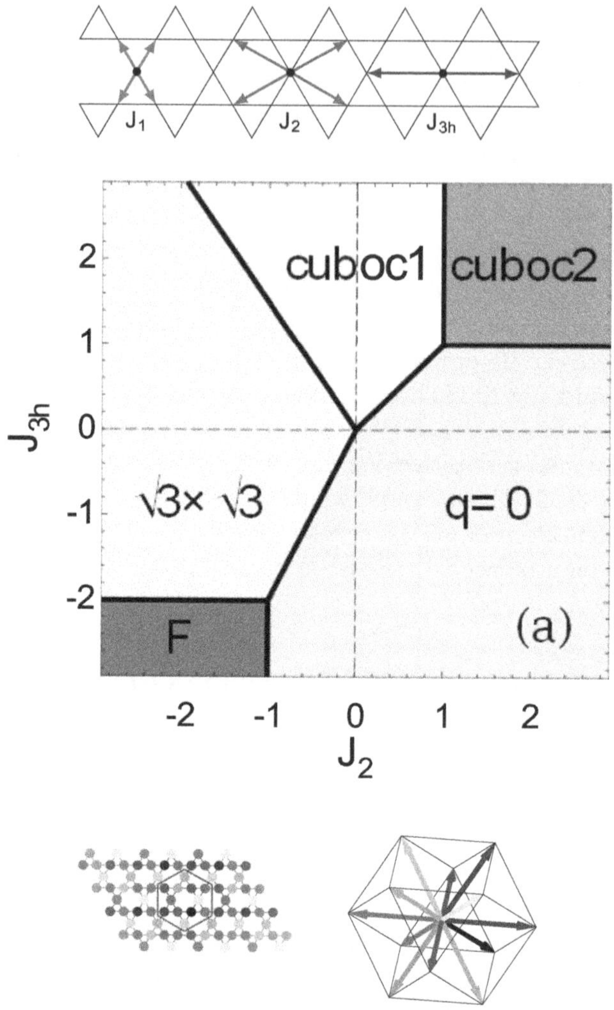

Fig. 5.12. Phase diagram of the model with up to three neighbor interactions at $T = 0$ and $J_1 = 1$. Top: definition of the interactions J_1, J_2, and J_{3h}. Middle: exact classical phase diagram. The point $(J_2, J_{3h}) = (0, 0)$ is a tricritical point. Bottom: description of the *cuboc1* order. Left: on the kagomé lattice each color corresponds to a different magnetic sublattice. The thick line indicates the 12-site unit cell. Right: arrows are the spin orientations with the same color coding as in the left figure. The black lines connecting the end of the vectors form a cuboctahedron. On each triangle, the spins are coplanar at 120 degrees; for opposite sites on each hexagon, spins are anti-parallel. The triple products (determinant) of three spins of the hexagons, either first or second neighbors, are non-zero and measure the chirality of the phase. They change sign in a mirror symmetry or in a spin flip.

in 2D antiferromagnets[271] and the absence of magnetic moment formation suggest's that they are *deconfined.*

5.7.4. *Spin-1/2 Heisenberg model on the kagomé lattice: numerics*

Understanding the nature of the GS of the nearest-neighbor spin-1/2 Heisenberg model on the kagomé lattice is clearly a difficult problem: classical degeneracy, numerous competing phases at the quantum level, and sign problem for QMC methods. The first "unbiased" tool used to approach this problem has been exact diagonalizations.[248, 249, 293, 294, 338] This approach has shown the absence of Néel or VBC LRO,[313] but the sizes up to $N \leq 36$ were insufficient to give a definite answer on the nature of the spin liquid: A careful analysis has even shown that both a critical gapless spin liquid as well as a gapped one remained plausible on the basis of the $N = 36$ spectra.[293] In a numerical *tour de force*, Läuchli *et al.* have very recently obtained a part of the low-lying levels of the exact $N = 48$ spectrum.[252] A refined analysis of these data will probably give some valuable information in a near future.

In the meanwhile, a breakthrough in large-scale DMRG calculations on very large and long cylinders has brought more insight on the GS of this puzzling system, as we will explain below.[295–298]

5.7.4.1. *Ground-state energy per spin*

All these results agree on the extremely low energy per bond of the GS ($\langle 2\vec{S}_i \cdot \vec{S}_j \rangle \simeq -0.438$) $\sim 87\%$ of the energy per bond in an isolated triangle. On this lattice, the energy per bond of the spin-1/2 system is much lower than the classical energy, $\frac{E_{qu.}}{E_{cl.}} \sim 1.74$, a ratio much larger than in any other 2D magnet, that can only be compared to the value obtained for the Bethe chain (1.77). The kagomé lattice is the 2D lattice which offers the largest stabilization due to quantum fluctuations.

5.7.4.2. *Correlations*

Early exact diagonalizations[247] as well as the latest DMRG simulations[298] found spin–spin and dimer–dimer correlations to be short ranged, in agreement with the early series expansion.[256] The scalar chirality correlators $\langle (\vec{S}_1 \cdot (\vec{S}_2 \times \vec{S}_3))(\vec{S}_{1'} \cdot (\vec{S}_{2'} \times \vec{S}_{3'})) \rangle$ between elementary triangles have been shown to be short ranged.[245, 298] Depenbrock *et al.* added that: "Chiral correlators for other loop types and sizes decay even faster" and will come back to this point later when discussing a putative chiral topological spin liquid.[302]

5.7.4.3. *Spin gap*

The spin gap of the symmetric $N = 36$ toroidal cluster is 0.14, but the data on such small systems are difficult to extrapolate safely to the thermodynamic limit. The most recent results of DMRG on large and very long cylinders (with diameter up to ~ 16 lattice spacings) point to a value of the spin gap ~ 0.13.[249,295–298] The 2012 belief is thus that a nearest-neighbor Heisenberg model on the kagomé lattice is indeed a gapped spin liquid.

5.7.4.4. *Singlet gap*

The extension of the Lieb–Shultz–Mattis theorem to two dimensions[217,362] guarantees that the spin-1/2 kagomé antiferromagnet (as any spin model with conserved S^z, short-range interactions, and a half-odd integer spin per unit cell) cannot simultaneously be gapped and have a unique GS on a closed surface of genius larger or equal to one. If we admit that spin excitations are indeed gapped, this implies that the spectrum is either gapless in the singlet sector, or the GS is degenerate in the thermodynamic limit. If the singlet excitations are also gapped, the GS must be degenerate. If this degeneracy is not the consequence of some conventional lattice symmetry breaking (valence bond crystal for instance), it is said to be a "topological degeneracy" (see Sections 5.5.5.2 and 5.5.6.7), and this kind of SL is now called a topological spin liquid.

From the presence of a spin gap and absence of long-range dimer–dimer correlations, a topological liquid is thus presently the most natural scenario for the spin-1/2 Heisenberg problem on the kagomé lattice. But the structure of the singlet sector is not completely unveiled at the moment. In a simple $\mathbb{Z}_2$ topological spin liquid, we expect a four-fold degeneracy on a two-torus and a two-fold degeneracy on a cylinder at the thermodynamic limit. In fact, small systems up to 36 sites exhibit a quasi continuum of $S = 0$ excitations and the absence of a sizeable gap in the singlet sector.[249,293][mm]

[mm]This feature, inconsistent with simple pictures of gapped spin liquids, remained puzzling for a long time and may have recently received an explanation.[302]

There is an interesting evolution of the spectrum between the sizes $N = 36$ and $N = 48$,[252] but a coherent picture for the spectrum of low-lying excitations obtained in ED remains to be constructed.

On the contrary, the first excited singlet state in DMRG is found at a finite energy of about 0.04 or 0.05 above the GS,[296] which seems contradictory with the Lieb–Shultz–Mattis theorem. There might be different explanations to this: either the splitting induced by the finite cylinder circumference is larger than expected or the DMRG method is biased toward stabilizing a linear combination of the (quasi-degenerate) GS with the smallest entanglement entropy.[297]

5.7.4.5. *Entanglement entropy and signature of a $\mathbb{Z}_2$ liquid*

The entanglement entropy (EE) is a powerful tool to detect the presence of "topological order" in states without any conventional local order parameter. In a generic wave function with short-range correlations, the EE S_A of a subsystem A has a leading contribution corresponding to the "area law" $S_A \simeq a l_A^{d-1}$, where l_A is the typical linear size of A and d is the space dimension. This contribution simply originates from the short-range correlations taking place "across" the boundary of the subsystem A. Hamma *et al.*,[364] Kitaev and Preskill,[365] and Levin and Wen[366] discovered that, in the presence of topological order, the EE may also contain a sub-leading contribution of order $\mathcal{O}(l^0)$: $S_A \simeq a l_A^{d-1} - \gamma$. This contribution γ indicates the presence of some non-local form of quantum entanglement. What makes this sub-leading contribution conceptually interesting and practically very useful is the fact that it is "universal": it depends on the topological phase (through the so-called total quantum dimension), but not on the microscopic or short-distance properties of the wave functions. For a $\mathbb{Z}_2$ liquid, the simplest topological phase, we have $\gamma = \ln(2)$.

Extracting γ from a given wave function is a non-trivial problem and the original suggestion[365, 366] was to compute some EE difference between suitably chosen subsystems so that the leading term cancels. This subtraction scheme has been implemented numerically in a few cases.[367–369] It is however easier and more accurate to obtain γ from the EE in a cylinder geometry. There, the translation invariance of the boundary of A allows to extract γ simply from a fit to (at least) two-system size. The cylinder geometry approach to topological EE has been used successfully in several numerical works,[370–373] including some on the kagomé antiferromagnet with first-neighbor coupling only,[298] or with additional next-nearest neighbor J_2.[297] For the first–neighbor model, the DMRG results of Ref. [298] gave γ consistent with ln(2) with a $\simeq$ 10% accuracy. With a further neighbor interaction (a point sitting deeper in a possibly distinct liquid phase[302]), Ref. [297] obtained $\gamma \simeq \ln(2)$ with an accuracy of the order of 1%.

5.7.4.6. *Spin liquids on the kagomé lattice and projective symmetry groups*

Wen was the first to develop a constructive approach of fermionic mean-field theories of spin liquids on the square lattice.[314] The general idea is the following.

In order to be able to describe liquid states with fractionalized excitations (spin-1/2 excitations), it is natural to start with a description in terms of Schwinger bosons or fermions (Section 5.4). Since the microscopic degrees of freedom are instead spin-1 objects, the spinons are inevitably coupled to a gauge field. This gauge field describes the quantum fluctuations of the singlet bonds.

If the effective gauge theory is non-confining, the system has spin-1/2 fractionalized excitations and is a spin liquid. This spin liquid can thus be characterized by the symmetries of the effective theory: space symmetry group (point group and lattice translations), the time reversal symmetry, and the local gauge transformations. Whereas a spin liquid does not break any lattice symmetry, its effective theory in terms of spinons and gauge fields may break the space symmetries insofar as the effect of each space symmetry can be compensated by a gauge transformation, thus achieving an invariant physical spin liquid state. The projective representation of the symmetry group (in short PSG) is defined as the set of all combinations of space and gauge symmetry operations that leave an effective theory (or mean-field Hamiltonian) invariant. The above requirements allow the determination of the distinct mean-field theories describing a fermionic or bosonic spin liquid on a given lattice. This number is finite but may be large (288 fermionic mean-field structures on the square lattice).[314] This construction does not depend on the details of spin Hamiltonian, except for its symmetries. It gives a first general classification of the spin liquids, at least valid in the regime of small gauge fluctuations. It also provides a direct way to determine the low-energy gauge group for a given mean-field state (a particular subset of the PSG, called invariant gauge group

(IGG)).[nn] For a detailed discussion in the context of the kagomé lattice, see Ref. [324].

The PSG approach is a first step toward a classification of spin liquid phases, but many questions remain unanswered.[oo] What is the domain of validity of the PSG classification beyond mean field? How to "measure" the PSG in a spin wave function?[pp] Can two different PSGs correspond to the same phase once fluctuations are included? What is the relation (duality?) between fermionic and bosonic $\mathbb{Z}_2$ liquids?[qq] Is there some spin liquids that cannot be reached by these approaches?

On the kagomé lattice, Wang *et al.* studied the PSG of an SU(2)-invariant Hamiltonian in the Schwinger boson representation and found eight different Schwinger boson mean-field ansatz of $\mathbb{Z}_2$ SLs which preserve all space symmetries (4 of them only have non-zero amplitude on nearest-neighbor bonds).[311] Lu *et al.*[312] have done the same analysis in the Schwinger fermion representation and found 20 different spin liquids (with only five gapped spin liquids with non-zero gauge field on NN bonds).[312] The relation between the 20 $\mathbb{Z}_2$ SLs in Schwinger fermion representation and the 8$\mathbb{Z}_2$ SLs in Schwinger boson representation has not been clarified up to now. The precise gauge structure of the spin liquid exhibited in recent DMRG approaches on the kagomé lattice is not known, although two proposals were made in Refs. [312] and [302].

5.7.5. *Competing phases*

The classical model is known to be at a high degeneracy point in parameter space (see Section 5.7.2) and we can expect that perturbations larger than the spin gap and/or larger than the vison gap will destabilize the $\mathbb{Z}_2$ spin

[nn]Using the Schwinger bosons as a starting point, one generically finds that the low-energy gauge group is $U(1)$ on bipartite lattices and $\mathbb{Z}_2$ on non-bipartite lattices. The former situation leads to confinement (monopole proliferation in the $U(1)$ gauge theory), whereas the situation may or may not — depending if the $\mathbb{Z}_2$ gauge theory is in its confined or deconfined phase — correspond to a liquid with deconfined spinons.

[oo]See the oral communications (by Hermele, Wen, Ran, and others) at the KITP Program: "Frustrated Magnetism and Quantum Spin Liquids: From Theory and Models to Experiments" (August 13–November 9, 2012). Available online at: http://online.kitp.ucsb.edu/online/fragnets12.

[pp]This question is tackled in Ref. [376], where it is proposed to determine a PSG through the measurement of fluxes which are physical observables that can be expressed in terms of spin operators.

[qq]This has been analyzed in detail in one case on the honeycomb lattice.[318]

liquid in favor of different competing phases. It has been shown for instance that a large enough DM coupling (larger than about one tenth the Heisenberg coupling constant) drives the quantum spin liquid toward the $q = 0$ semi-classical Néel long-range order.[317,352] We expect a large enough antiferromagnetic second-neighbor coupling to have the same effect. Other transitions toward the semi-classical $\sqrt{3}\times\sqrt{3}$ and cuboc1 phases upon increasing ferromagnetic second-neighbor or antiferromagnetic third-neighbor interactions are also expected. But the true nature of these transitions and/or the presence of spin liquid phases in between are today unknown. As explained above, the number of possibilities is considerable.

We thus give below a rapid survey of the various states that have been proposed as plausible spin-1/2 GS of the Heisenberg problem, as they could be interesting for nearby phases.

5.7.5.1. *Valence bond crystals*

The "simplest" crystal[264] has a unit cell of 12 spins and is made of resonating *stars* with 6 dimers. The second VBC, with a unit cell of 36 spins, is made of resonating (trimerized) *hexagons*. It was discussed as the most reasonable crystal for the pure J_1 model by Marston and Zeng[147] and Zeng and Elser[171] and rediscussed more recently.[265–267] In both scenarios, it is the energy gained by *local resonances* (involving respectively 6 and 3 valence bonds) which drives the system toward a VBC. From the energy point of view, the star VBC is probably less realistic since it involves a much longer resonance loop. In the pure Heisenberg model, the resonance loop involving 6 valence bonds around a star has a vanishing amplitude at the lowest non-trivial order of the overlap expansion in the RVB subspace, as was shown by Zeng and Elser.[171] In the approximation where only the shortest resonance loops are present, the model was indeed found to be in the *hexagon* VBC phase. A crucial (numerical) result of Zeng and Elser[171] is however that at the pure nearest-neighbor Heisenberg point, this VBC *melts* when higher order resonances loops are included. Extra couplings would be needed to stabilize this VBC phase. For further reading concerning possible VBC instabilities as well as variational approaches (tensor networks, RVB subspace, and projected fermionic wave functions), a see Refs.[303–307]

5.7.5.2. $U(1)$ *Dirac spin liquid*

Ran *et al.*[308] have constructed a variational wave function of a gapless $U(1)$ spin liquid. At the mean-field level, it corresponds to fermionic spinons with a conical dispersion relation at the Fermi level (Dirac fermions). After Gutzwiller projection, this wave function turns out to have a very low energy, even though it has no adjustable parameter. Many correlation functions have an algebraic decay with distance.

The present DMRG results[296,298] strongly indicate a spin gap and therefore dismiss this Dirac $U(1)$ scenario, but the latter may be a good candidate to describe a critical point or even an extended gapless phase nearby. A full study of this interesting phase can be found in Refs.[308–310,315,316,325,326]

From a completely different starting point (easy plane model), Ryu *et al.*[327] have developed a theory of the singlet sector of a gapless algebraic *vortex liquid* theory in a XY model on the kagomé lattice.

5.7.5.3. *Spontaneously breaking the time-reversal symmetry, "chiral" spin liquids*

The simplest $\mathbb{Z}_2$ topological SL does not break any lattice symmetries nor time reversal. A chiral spin liquid does not break $SU(2)$ symmetry nor translations, but it breaks time reversal symmetry and usually some discrete lattice point-group symmetry. The idea that the kagomé lattice can harbor a chiral spin liquid goes back to the end of the eighties.[288,299] This idea has revived recently, Wen *et al.*[300] have studied chiral spin liquids as instability of the Hubbard model and Chua *et al.*[301] have exhibited an exact gapless spin liquid with stable spin Fermi surface in a Kitaev model on the kagomé lattice.

This year, Messio *et al.*[302] have suggested that a chiral $\mathbb{Z}_2$ spin liquid could be stabilized in the pure J_1 Heisenberg model and in the J_1–J_3 model on the kagomé lattice. In the classical phase diagram of the J_1–J_3 model, an infinitesimal J_3, third-neighbor coupling across the hexagons lifts the degeneracy of the pure J_1 model in favor of a 12-sublattice LRO named *cuboc1*, where the order parameter has the symmetry of a cuboctahedron and the neighboring spins are at 120°. This magnetic structure and its time-reversal counterpart (spin inversion) cannot be transformed into each other by a global spin rotation in $SO(3)$. In that sense, it is a *chiral* magnetic order.

Three spins on a nearest-neighbor triangle form a plane, but the spins of a hexagon are not planar: the triple product of second-neighbor spins around an hexagon (scalar chirality) is non-zero ($\langle \vec{S}_i \cdot (\vec{S}_j \times \vec{S}_k) \rangle = 0.0148$). Messio's SL state, the spin liquid descendant of this cuboc1 classical structure, is — at the Schwinger boson mean-field level — slightly more stable than the plain $\mathbb{Z}_2$ topological SL. It does not break any translation symmetry, but, since its ordered parent is chiral, it breaks time reversal (there is a non-trivial flux on the hexagons — i.e. not 0 nor π) and one reflection symmetry. The above-mentioned chirality has been computed in ED in the GS of samples with an odd number of spins ($N = 21$ and 27) and found to be sizeable (0.817 of the classical value).[249] The 48-site spectrum obtained by Laüchli *et al.*[252] may be compatible with the eight-fold quasi-degeneracy expected for such a chiral $\mathbb{Z}_2$ liquid. DMRG or larger size ED calculations of the associated 6-point correlation functions would be very valuable. It has been conjectured that, due to gauge fluctuations (singlet bond fluctuations), the expectation value of the flux along large loops would obey a perimeter law in a chiral spin liquid,[286] but this has not been tested numerically.[rr] Concerning the topological entanglement entropy, the values measured by Jiang *et al.*[297] and Depenbrock *et al.*[298] are compatible with a (non-chiral) RVB $\mathbb{Z}_2$ SL but the values to be expected in the chiral SL scenario remains to be worked out precisely. If a chiral SL phase does exist at $T = 0$, it should undergo a finite-temperature phase transition above which the time-reversal symmetry is restored. The classical problem has been studied, showing a rich interplay of $\mathbb{Z}_2$ vortices and chirality domain walls,[344,363] but the quantum case has not been studied.

5.7.6. *Experiments in compounds with kagomé-like lattices*

The first experiments on compounds with kagomé-like lattices were done on compounds with spin 3/2, 5/2, or larger. We have no place to recall all these results, but it may be remembered that "non-classical" consequences of frustration were very clearly observed in these compounds. $SrCr_{9p}Ga_{12-9p}O_{19}$, with the spin-3/2 Cr ions on a kagomé lattice (or a bilayer of pyrochlore) has been one of the compounds which has been studied through numerous of techniques. The magnetic excitations of this compound as seen by muons spectroscopy were described as itinerant spins 1/2 in a "sea of singlets".[272] The non-linear spin susceptibility of $SrCr_9Ga_{12}0_{19}$ exhibits a very large increase at about 5 K, reminiscent of spin glasses, but neutrons and muons show that a very significant fraction of the spins are not frozen below this

[rr]Such a hypothesis on the correlations could perhaps explain the puzzling structure of the exact spectra of small samples as well as the presence of a non-zero Chern number for spin-1/2 excitations.[249]

temperature and still exhibit very rapid fluctuations.[273] The same phenomena have been observed in two jarosites that are equally good models of kagomé antiferromagnets with half-odd-integer spin per unit cell.[274,275] The low-temperature specific heat of these spin systems is unusually large, with a double–peak structure. In $SrCr_{9p}Ga_{12-9p}O_{19}$, it was shown that the low-temperature peak was insensitive to large magnetic fields and apparently dominated by singlet states.[262] This might have some relationship with the results of numerical calculations performed on the spin-1/2 model.[251]

The theoretical interest in spin-1/2 compounds on the kagomé lattice has lead the experimentalists to a worldwide effort to synthesize new compounds. Volborthite synthesized at the ISSP by Hiroi's group has been the first success. Unhappily, it is not a "perfect" representative of the *spatially isotropic* Hamiltonian and its modelization is still disputed. This compound has attracted a lot of attention. For a partial bibliography, see Refs. [319–323] and references therein.

Herbertsmithite, with the chemical formula $ZnCu_3(OH)_6Cl_2$, a rare mineral identified in 2004 and named after G. F. Herbertsmith (1872–1957) was first synthesized in 2005 in MIT.[345] The spin-1/2 Cu^{2+} ions form 2D kagomé lattices. Inelastic neutron scattering, NMR, and μSR consistently show that, in zero or very low magnetic field, Herbertsmithite remains fluctuating down to 20 mK.[342,346–349,354] This compound is undoubtedly in an SL state but not a perfect realization of the nearest-neighbor Heisenberg model as its gap, if it exists, is smaller than $J/8{,}000$ (where J, the nearest-neighbor coupling constant, is about 170 K). In fact, the lack of inversion center on the magnetic bounds allows for DM interaction of spin orbit origin.[334,335] These couplings have been estimated to be of the order of $0.08J$,[351,355] not far from the theoretically expect ed quantum critical point.[352] The role of impurities in such a material has also been theoretically discussed.[333,337,353] Single crystals of Herbertsmithite are now available and we can expect that future inelastic neutron scattering experiments will uncover new specificities of this spin liquid and will check more deeply the different theoretical scenarios.[308,310,339]

Many spin-1/2 compounds with the kagomé geometry turn out to be ferromagnets.[343,356,357] Domenge *et al.* suggested in 2005 that competing interactions (ferromagnetic between nearest neighbors and antiferromagnetic between second neighbors and farther) could lead classically to a non-magnetic phase with a non-planar 12-sublattice magnetic unit cell (called cuboc2) as well as chiral properties.[358] At $T = 0$, this classical phase breaks time reversal and spin inversion symmetry and a finite temperature is needed to restore these discrete symmetries. Due to the presence of $\mathbb{Z}_2$ vortices, the phase transition is not Ising-like but weakly first order.[344] Messio has recently shown that this phase is the GS of J_1–J_2–J_3 model for a large domain of parameters.[290] The nature of the spin-1/2 phase of this model is for the moment unknown.

In 2008, a new compound called kapellasite, polymorphous to Herbertsmithite, was synthesized in London.[359,360] In this compound, the kagomé planes are well separated from each other by non-magnetic Zn^{2+} and Cl^- ions. The absence of freezing down to very low temperature as well as NMR and inelastic neutron scattering spectra lead experimentalists to the conclusion that it is a spin liquid.[361] *Ab initio* calculations suggested that the J_1–J_2–J_3 parameters describing this compound were such that in the classical limit the GS was a chiral and non-planar magnetic structure with the symmetry of a cuboctahedron (in this *cuboc1* state, the spins are at 120° from each other and define a plane on each small triangle: spins around an hexagon are however not coplanar (chirality)).[289] The inelastic neutron spectra dismissed this hypothesis and suggested that the parent classical phase is instead the *cuboc2* phase (this state also has a 3D order parameter with the symmetry of a cuboctahedron but neighboring spins are at 60 degrees from each other; there is chirality on the small triangles and the spins around distinct hexagons define distinct planes).[290] A high-temperature series analysis of the spin susceptibility and specific heat confirms this hypothesis.[380] The model extracted from these data allows a precise description of the μSR relaxation rate. In spite of a strong disorder, there is no signature of "defects". A theoretical explanation for the continuum of excitations seen in this compound is however still lacking.

Two other compounds have been synthesized recently: vesignieite[379] and haydeeite.[360] As Herbertsmithite and — in spite of its slight distortion — vesignieite seem rather good models of the AF Heisenberg model, they are found to partially order at $T_c \sim J/6$.[377,378] This might be mainly due to the presence of DM interactions relatively larger than those in Herbertsmithite. It has been suggested that these two compounds are respectively below (Herbertsmithite) and above (vesignieite) the quantum critical point separating the spin liquid phase from the $q = 0$ Nel ordered phase.[352] Haydeeite, $\alpha - Cu_3Mg(OD)_6Cl_2$, is isostructural to kapellasite $\alpha - Cu_3Zn(OD)_6Cl_2$, but the relative weight of the first-neighbor ferromagnetic coupling compared to second- and third-neighbor antiferromagnetic coupling is apparently stronger than in kapellasite and the system is a ferromagnet.[380]

5.8. Conclusions

We conclude by summarizing some properties — and related open questions — of the different phases discussed in this review.

The properties of these phases are summarized in Table 5.2. Semi-classical phases with Néel long-range order, magnons as gapless excitations, do exist in spin-1/2 2D systems with moderate frustration: The Heisenberg model on the triangular lattice is the most explicit example, with a sub-lattice magnetization about one half of the classical value.[2] On a large but finite system, the GS manifold is a tower of states including a number of eigenstates which is a power of the number of sites.

An increased frustration, lower coordination number, or smaller spin lead to quantum phases, with a GS of higher symmetry, no long-range order in spin–spin correlations, a spin gap, and the restored $SU(2)$ symmetry. Two main alternatives are then opened: the VBC or VBS phases on the one hand, the RVB SL on the other hand. These states first require the formation of short-range singlets. When a particular local resonance pattern dominates the dynamics of the Hamiltonian, the system will try to maximize the number of occurrences of this pattern. This is usually achieved by a regular arrangement, that is a VBC. When no such pattern dominates, the system may form a translation-invariant RVB SL. In the first case, the GS can be qualitatively described by one ordered configuration of singlets dressed by small fluctuations. In the RVB SL, the amplitudes of the wave function are distributed over an exponentially large number of configurations. These GS lead to very different excitations: $\Delta S = 1$ gapped magnons in the first case (and $\Delta S = 0$ domain-wall excitations), gapped $\Delta S = 0$ visons, and gapped $\Delta S = \frac{1}{2}$ unconfined spinons in the second case.

The bosonic large-N results (Section 5.4) indicate that VBC are expected in situations where the short-range magnetic order is collinear, whereas SL appear when these correlation are non-collinear. Although this is verified in many cases (kagomé in particular), some exceptions seem to exist also. For instance, the recent DMRG simulations on the J_1–J_2 model on the square lattice[382] indicate a $\mathbb{Z}_2$ SL GS in the vicinity of $J_2/J_1 \sim 0.5$.

These states obey the 2D extension of LSM theorem[217,362,388]: if $2S$ (per unit cell) is odd and if excitations are gapped there must be

Table 5.2. Different phases encountered in $SU(2)$-symmetric frustrated models in 2D.

Phase	$2S$/cell	Order	Degeneracy	Broken sym.	Excitations	Thermo.	Examples
Néel AF p-sublattice	Any	Spin–Spin LRO	$\mathcal{O}(N^p)$	$SU(2)$ Translations Point group	Gapless magnons (spin waves)	$C_v \sim T^2$ $\chi \sim$ cst	Spin-1/2 triangular Heisenberg AF
VBC (Section 5.3) (spontaneous)	Odd	Singlet–Singlet LRO colli. spin–spin SRO	>1	Translations Point group	Gapped magnons	C_v and χ activated	Honeycomb J_1–J_2 Checkerboard Square J_1–J_2?
VBC (Section 5.3) (explicit)	Even	(1)	1	None	Gapped magnons	C_v and χ activated	$SrCu_2(BO_3)_2$ CaV_4O_9
VBS (Section 5.3.3)	Even	"String" LRO	1	None	Gapped magnons Edge excitations	C_v and χ activated	AKLT Hamiltonians $S = 1$ kagomé AF ?
RVB SL (Sections 5.5.5–5.6.4)	Odd	Topological non-colli. SRO	4 (torus)	None	Gapped spinons Gapped visons	C_v and χ activated	MSE (Section 5.6) Kagomé AF

Notes: This table summarizes the properties of some important phases encountered in 2D frustrated magnets. S is the value of the spin on each site. "Order" refers to the nature of the long-range correlations (if any). The ground-state degeneracy in the limit of an infinite system (with periodic boundary conditions) is indicated in the fourth column, except for RVB SL, where it is related to the spontaneously broken symmetries mentioned in the next column. Elementary excitations and the low-temperature behavior of the specific heat (C_v) and uniform susceptibility (χ) are given in column six. The last column gives some examples of theoretical or experimental realizations of these phases. The five families of systems presented here of course do not exhaust all the possibilities. Some authors classify all the systems with gapped excitations in a loose category of "quantum disordered systems", alluding to the absence of Néel long-range order. It is a rather unhappy appellation for VBC (which obviously have some order) and in fact for most of the quantum systems with a gap. In classical statistical physics, "disorder" is associated to some extensive entropy, which is not the case in the quantum systems at $T = 0$. (1) The presence of order in an explicit VBC wave function is somewhat a question of definition: while the singlet–singlet correlation $\langle(\vec{S}_1 \cdot \vec{S}_2)(\vec{S}_{1'} \cdot \vec{S}_{2'})\rangle$ should show some long-range modulations when the bond (12) and (1′2′) are far apart, its *connected* counterpart $\langle(\vec{S}_1 \cdot \vec{S}_2)(\vec{S}_{1'} \cdot \vec{S}_{2'})\rangle$-$\langle\vec{S}_1 \cdot \vec{S}_2\rangle\langle\vec{S}_{1'} \cdot \vec{S}_{2'}\rangle$ remains short range.

ground-state degeneracy in the thermodynamic limit (with periodic boundary conditions). However, the origin of the degeneracy differs in the two types of states. In a VBC, the degeneracy is associated with spontaneously broken translation symmetry, whereas in the RVB SL, the degeneracy has a topological origin.[ss] In the VBS (or explicit VBC), the GS is unique but $2S$ is even (in one unit cell). The situation of the hexagonal lattice with respect to this theorem and possible topological degeneracies is interesting: since the lattice has two sites per unit cell, the LSM theorem does not apply. Still, on this lattice, there is no obvious way to construct a gapped and $SU(2)$-symmetric spin-1/2 wave function without any conventional or topological order (see however Ref. [389]). Although not covered in this chapter, we mention that some recent quantum Monte Carlo simulations indicating an insulating SL in the (unfrustrated!) Hubbard model at half-filling on the hexagonal lattice[390] (see also Ref. [396]) have triggered a large number of studies concerning frustrated spin models on this lattice.[391–393, 395]

These paradigms are relatively well understood, at least on the qualitative level. They also appear naturally in the broader context of the classification of Mott insulators.[285] However, several kinds of 2D frustrated magnets do not fall in these simple classes and many open questions remain.

This review was restricted to the $SU(2)$-invariant Hamiltonians. Although the Ising limit has been much studied, the differences between quantum XY and Heisenberg models have received much less attention.

Chiral SL have not been thoroughly discussed in this review. They are characterized by a broken time-reversal symmetry. This possibility has been studied intensely since the 1980s.[36, 286–288] There might be a revival of this line of research in the future.[302, 361]

The issue of quantum phase transitions in a frustrated antiferromagnet is also an active topic that is not presented in this review. Many properties of these critical points are still unknown, not to mention the fascinating problems associated with (quenched) disorder.

Limited by place (and competence), we have not discussed in detail the works done on spatially anisotropic models. This field which is in between 1D (review by Lecheminant in this book) and 2D is extremely flourishing and tackled by bosonization and large-N methods.

[ss]On the kagomé lattice (three spins in the unit cell) a precursor of this topological degeneracy might have been seen in the $N = 48$-sites spectrum.[252]

Ten years ago, we wrote in the first edition: "To conclude we would like to emphasize that new analytical and/or numerical methods are highly desirable to proceed in the analysis of the two emblematic problems by which we have opened and closed this review: the J_1–J_2 model on the square lattice and the spin-1/2 Heisenberg model on the kagomé (and pyrochlore) lattices. In both of these problems a consensus remains to be obtained."

During the last few years, 2D DMRG studies have allowed a breakthrough which has lead to a complete revisiting of these two problems.[295–298, 382] As explained in this rewritten section, the full understanding and consensus is not yet reached but a very important step forward has been achieved.

References

1. E. Manousakis, *Rev. Mod. Phys.* **63**, 1 (1991).
2. B. Bernu, P. Lecheminant, C. Lhuillier, and L. Pierre, *Phys. Rev. B* **50**, 10048 (1994).
3. Proceedings of the *Highly Frustrated Magnetism 2000* conference, published in *J. Can. Phys.* **79**, (2001).
4. E. Fradkin, *Field Theories of Condensed Matter Systems* (Addison-Wesley, 1998).
5. A. Auerbach, *Interacting Electrons and Quantum Magnetism* (Springer-Verlag, Berlin Heidelberg New York, 1994).
6. A. M. Tsvelik, *Quantum Field Theory in Condensed Matter Physics* (Cambridge University Press, 1996).
7. S. Sachdev, *Quantum Phase Transitions* (Cambridge Universtiy Press, New York, 1999).
8. J. M. Luttinger and L. Tisza, *Phys. Rev.* **70**, 954 (1946).
9. P. Chandra and B. Douçot, *Phys. Rev. B* **38**, 9335 (1988).
10. J. Villain, R. Bidaux, J. P. Carton, and R. Conte, *J. Phys. (Paris)* **41**, 1263 (1980).
11. E. Shender, *Sov. Phys. JETP* **56**, 178 (1982).
12. C. L. Henley, *Phys. Rev. Lett.* **62**, 2056 (1989).
13. A. Moreo, E. Dagotto, T. Jolicœur, and J. Riera, *Phys. Rev. B* **42**, 6283 (1990).
14. P. Chandra, P. Coleman, and A. Larkin, *Phys. Rev. Lett.* **64**, 88 (1990).
15. R. R. P. Singh, W. Zheng, J. Oitmaa, O. P. Sushkov, and C. J. Hamer, *Phys. Rev. Lett.* **91**, 017201 (2003).
16. G. Misguich, B. Bernu, and L. Pierre, *Phys. Rev. B* **68**, 113409 (2003).
17. C. Weber, L. Capriotti, G. Misguich, F. Becca, M. Elhajal, and F. Mila, *Phys. Rev. Lett.* **91**, 177202 (2003).
18. T. Jolicœur, E. Dagotto, E. Gagliano, and S. Bacci, *Phys. Rev. B* **42**, 4800 (1990).

19. A. Chubukov and T. Jolicœur, *Phys. Rev. B* **46**, 11137 (1992).
20. S. E. Korshunov, *Phys. Rev. B* **47**, 6165 (1993).
21. P. Lecheminant, B. Bernu, C. Lhuillier, and L. Pierre, *Phys. Rev. B* **52**, 6647 (1995).
22. R. Melzi *et al.*, *Phys. Rev. Lett.* **85**, 1318 (2000).
23. R. Melzi *et al.*, *Phys. Rev. B* **64**, 024409 (2001).
24. H. Rosner *et al.*, *Phys. Rev. Lett.* **88**, 186405 (2002).
25. H. Rosner *et al.*, *Phys. Rev. B* **67**, 014416 (2003).
26. D. J. Klein, *J. Phys. A: Math. Gen.* **15**, 661 (1982).
27. C. K. Majumdar and D. K. Ghosh, *J. Math. Phys.* **10**, 1399 (1969).
28. R. R. P. Singh, M. P. Gelfand, and D. A. Huse, *Phys. Rev. Lett.* **61**, 2484 (1988).
29. M. P. Gelfand, R. R. P. Singh, and D. A. Huse, *J. Stat. Phys.* **59**, 1093 (1990).
30. Z. Weihong, J. Oitmaa, and C. J. Hamer, *Phys. Rev. B* **43**, 8321 (1991).
31. J. Oitmaa and Z. Weihong, *Phys. Rev. B* **54**, 3022 (1996).
32. M. P. Gelfand, R. R. P. Singh, and D. A. Huse, *Phys. Rev. B* **40**, 10801 (1989).
33. M. P. Gelfand, *Phys. Rev. B* **42**, 8206 (1990).
34. R. R. P. Singh, Z. Weihong, C. J. Hamer, and J. Oitmaa, *Phys. Rev. B* **60**, 7278 (1999).
35. V. N. Kotov, J. Oitmaa, O. Sushkov, and Z. Weihong, *Phil. Mag. B* **80**, 1483 (2000).
36. M. E. Zhitomirsky and K. Ueda, *Phys. Rev. B* **54**, 9007 (1996).
37. L. Capriotti, F. Becca, A. Parola, and S. Sorella, *Phys. Rev. Lett.* **87**, 097201 (2001).
38. O. P. Sushkov, J. Oitmaa, and Z. Weihong, *Phys. Rev. B* **66**, 054401 (2002).
39. O. P. Sushkov, J. Oitmaa, and Z. Weihong, *Phys. Rev. B* **63**, 104420 (2001).
40. M. S. L. du Croo de Jongh, J. M. J. Van Leeuwen, and W. Van Saarloos, *Phys. Rev. B* **62**, 14844 (2000).
41. E. Dagotto and A. Moreo, *Phys. Rev. B* **39**, 4744 (1989), *Phys. Rev. Lett.* **63**, 2148 (1989).
42. F. Figueirido *et al.*, *Phys. Rev. B* **41**, 4619 (1990).
43. D. Poilblanc, E. Gagliano, S. Bacci, and E. Dagotto, *Phys. Rev. B* **43**, 10970 (1991).
44. H. J. Schulz and T. A. L. Ziman, *Europhys. Lett.* **18**, 355 (1992); T. Einarsson and H. J. Schulz, *Phys. Rev. B* **51**, 6151 (1995); H. J. Schulz, T. A. L. Ziman, and D. Poilblanc, *J. Physique I* **6**, 675 (1996).
45. S. Chakravarty, B. I. Halperin, and D. R. Nelson, *Phys. Rev. B* **39**, 2344 (1989); H. Neuberger and T. Ziman, *Phys. Rev. B* **39**, 2608 (1989); D. Fisher, *Phys. Rev. B* **39**, 11783 (1989); T. Einarsson and H. Johannesson, *Phys. Rev. B* **43**, 5867 (1991); P. Hasenfratz and F. Niedermayer, Z. *Phys. B. Condens. Matter* **92**, 91 (1993).
46. A. W. Sandvik, *Phys. Rev. B* **56**, 11678 (1997).
47. S. Sorella, *Phys. Rev. Lett.* **80**, 4558 (1998).

48. S. Sorella, *Phys. Rev. B* **64**, 024512 (2001).
49. S. White, *Phys. Rev. Lett.* **69**, 2863 (1992).
50. S. White, *Phys. Rev. B* **48**, 10345 (1993).
51. T. Giamarchi and C. Lhuillier, *Phys. Rev. B* **43**, 12943 (1991).
52. B. Shastry and B. Sutherland, *Phys. Rev. Lett.* **47**, 964 (1981).
53. F. Haldane, *Phys. Rev. B* **25**, 4925 (1982).
54. I. Affleck, J. *Phys. Cond. Matt.* **1**, 3047 (1989).
55. H. Yokoyama and Y. Saiga, J. *Phys. Soc. Jpn.* **66**, 3617 (1997).
56. T. Nakamura and S. Takada, *Phys. Rev. B* **55**, 14413 (1997).
57. D. Augier, E. Sorensen, J. Riera, and D. Poilblanc, *Phys. Rev. B* **60**, 1075 (1999).
58. E. Dagotto and T. M. Rice, *Science* **271**, 618 (1996).
59. A. A. Nersesyan and A. M. Tsvelik, *Phys. Rev. Lett.* **78**, 3939 (1997).
60. A. K. Kolezhuk and H.-J. Mikeska, *Int. J. Mod. Phys. B* **12**, 2325 (1998); *Phys. Rev. Lett.* **80**, 2709 (1998); *Phys. Rev. B* **56**, 11380 (1997).
61. I. Affleck, T. Kennedy, E. Lieb, and H. Tasaki, *Phys. Rev. Lett.* **59**, 799 (1987).
62. I. Affleck, T. Kennedy, E. H. Lieb, and H. Tasaki, *Commun. Math. Phys.* **115**, 477 (1988).
63. M. den Nijs and K. Rommelse, *Phys. Rev. B* **40**, 4709 (1989).
64. T. Kennedy and H. Tasaki, *Phys. Rev. B* **45**, 304 (1992).
65. T. Jolicœur, private communication.
66. K. Hida, *J. Phys. Soc. Jpn.* **69**, 4003 (2000).
67. Y. Yamashita and K. Ueda, *Phys. Rev. Lett.* **85**, 4960 (2000).
68. N. Katoh and M. Imada, *J. Phys. Soc. Jpn.* **63**, 4529 (1994); M. Troyer, H. Tsunetsugu, and D. Wuertz, *Phys. Rev. B* **50**, 13515 (1994); S. Taniguchi *et al.*, *J. Phys. Soc. Jpn.* **64**, 2758 (1995); K. Kodama *et al.*, *J. Phys. Soc. Jpn.* **65**, 1941 (1996); Y. Fukumoto and A. Oguchi, *J. Phys. Soc. Jpn.* **65**, 1440 (1996); K. Kodama *et al.*, *J. Phys. Soc. Jpn.* **66**, 793 (1997); T. Miyasaki and D. Yoshioka, *J. Phys. Soc. Jpn.* **65**, 2370 (1996); T. Ohama, H. Yasuoka, M. Isobe, and Y. Ueda, *J. Phys. Soc. Jpn.* **66**, 23 (1997).
69. K. Ueda, H. Kontani, M. Sigrist, and P. A. Lee, *Phys. Rev. Lett.* **76**, 1932 (1996).
70. M. Albrecht and F. Mila, *Phys. Rev. B* **53**, 2945 (1996).
71. M. Troyer, H. Kontani, and K. Ueda, *Phys. Rev. Lett.* **76**, 3822 (1996).
72. S. Sachdev and N. Read, *Phys. Rev. Lett.* **77**, 4800 (1996).
73. O. A. Starykh *et al.*, *Phys. Rev. Lett.* **77**, 2558 (1996).
74. Z. Weihong *et al.*, *Phys. Rev. B* **55**, 11377 (1997).
75. H. Kageyama *et al.*, *Phys. Rev. Lett.* **82**, 3168 (1999).
76. H. Nojiri *et al.*, *J. Phys. Soc. Jpn.* **68**, 2906 (1999).
77. S. Miyahara and K. Ueda, *Phys. Rev. Lett.* **82**, 3701 (1999).
78. H. Kageyama *et al.*, *Phys. Rev. Lett.* **84**, 5876 (2000).
79. A. Koga and N. Kawakami, *Phys. Rev. Lett.* **84**, 4461 (2000).
80. E. Müller-Hartmann, R. R. P. Singh, C. Knetter, and G. S. Uhrig, *Phys. Rev. Lett.* **84**, 1808 (2000).

81. K. Totsuka, S. Miyahara, and K. Ueda, *Phys. Rev. Lett.* **86**, 520 (2001).
82. A. Läuchli, S. Wessel, and M. Sigrist, *Phys. Rev. B* **66**, 014401 (2002).
83. K. Kodama *et al.*, *Science* **298**, 395 (2002).
84. S. Miyahara and K. Ueda, *J. Phys.: Condens. Matter* **15**, R327–R366 (2003).
85. B. Shastry and B. Sutherland, *Physica B (Amsterdam)* **108**, 1069 (1981).
86. O. Cépas, T. Ziman, *Proceedings of the Conference in Fukuoka*, November (2001). Fukuoka University Press [cond-mat/0207191]; O. Cepas, T. Sakai, and T. Ziman, *Progr. Theor. Phys. Suppl.* **145**, 43 (2002).
87. W. Zheng, J. Oitmaa, and C. J. Hamer *Phys. Rev. B* **65**, 014408 (2002).
88. M. Albrecht and F. Mila, *Europhys. Lett.* **34**, 145 (1996).
89. C. H. Chung, J. B. Marston, and S. Sachdev, *Phys. Rev. B* **64**, 134407 (2001).
90. Y. Fukumoto and A. Oguchi, *J. Phys. Soc. Jpn.* **68**, 3655 (1999).
91. T. Momoi and K. Totsuka, *Phys. Rev. B* **61**, 3231 (2000).
92. T. Momoi and K. Totsuka, *Phys. Rev. B* **62**, 15067 (2000).
93. S. Miyahara and K. Ueda, *Phys. Rev. B* **61**, 3417 (2000).
94. G. Misguich, T. Jolicœur, and S. M. Girvin, *Phys. Rev. Lett.* **87**, 097203 (2001).
95. C. Lhuillier and G. Misguich, in C. Berthier, L. Levy, and G. Martinez (eds.), *High Magnetic Fields* (Springer, Berlin, 2002), pp. 161–190, [cond-mat/0109146].
96. P. W. Leung and N. Lam, *Phys. Rev. B* **53**, 2213 (1996).
97. J.-B. Fouet, P. Sindzingre, and C. Lhuillier, *Eur. Phys. J. B* **20**, 241 (2001).
98. B. Canals, *Phys. Rev. B* **65**, 184408 (2002).
99. J.-B. Fouet, M. Mambrini, P. Sindzingre, and C. Lhuillier, *Phys. Rev. B* **67**, 054411 (2003).
100. P. Sindzingre, J. B. Fouet, and C. Lhuillier, *Phys. Rev. B* **66**, 174424 (2002).
101. W. Brenig and A. Honecker, *Phys. Rev. B* **65**, 140407R (2002).
102. E. Berg, E. Altman, and A. Auerbach, *Phys. Rev. Lett.* **90**, 147204 (2003).
103. O. Tchernyshyov, O. A. Starykh, R. Moessner, and A. G. Abanov, *Phys. Rev. B* **68**, 144422 (2003).
104. R. Moessner, Oleg Tchernyshyov, and S. L. Sondhi, cond-mat/0106286.
105. S. Palmer and J. .T. Chalker, *Phys. Rev. B* **64**, 094412 (2002).
106. R. Moessner and J. T. Chalker, *Phys. Rev. B* **58**, 12049 (1998).
107. R. Moessner and J. T. Chalker, *Phys. Rev. Lett.* **80**, 2929 (1998).
108. O. A. Petrenko and D. McK. Paul, *Phys. Rev. B* **63**, 024409 (2001).
109. H. Kawamura and T. Arimori, *Phys. Rev. Lett.* **88**, 077202 (2002)
110. C. L. Henley, *Can. J. Phys. (Canada)* **79**, 1307 (2001). [cond-mat/0009130]
111. J.-B. Fouet, Ph.D. thesis, Université Cergy Pontoise, 2003.
112. M. P. Gelfand *et al.*, *Phys. Rev. Lett.* **77**, 2794 (1996).
113. F. Becca and F. Mila, *Phys. Rev. Lett.* **89**, 037204 (2002).
114. G. Misguich, C. Lhuillier, B. Bernu and C. Waldtmann, *Phys. Rev. B* **60**, 1064 (1999).
115. I. Affleck, *Phys. Rev. Lett.* **54**, 966 (1985).
116. I. Affleck and J. Marston, *Phys. Rev. B* **37**, 3774 (1988).

117. D. Arovas and A. Auerbach, *Phys. Rev. B* **38**, 316 (1988).
118. N. Read and S. Sachdev, *Phys. Rev. Lett.* **62**, 1694 (1989).
119. N. Read and S. Sachdev, *Phys. Rev. B* **42**, 4568 (1990).
120. N. Read and S. Sachdev, *Phys. Rev. Lett.* **66**, 1773 (1991).
121. S. Sachdev and N. Read, *Int. J. Mod. Phys.* **5**, 219 (1991).
122. S. Sachdev, in Y. Lu, S. Lundqvist, and G. Morand (eds.), *Low Dimensional Quantum Field Theories for Condensed Matter Physicists* (World Scientific, Singapore, 1995), cond-mat/9303014.
123. N. Read and S. Sachdev, *Nucl. Phys. B* **316**, 609 (1989).
124. D. Rokhsar, *Phys. Rev. B* **42**, 2526 (1991).
125. K. G. Wilson, *Phys. Rev. D* **10**, 2445 (1974).
126. R. Balian, J. M. Drouffe, and C. Itzykson, *Phys. Rev. D* **10**, 3376 (1974); *Phys. Rev. D* **11**, 2098 (1975); *Phys. Rev. D* **11**, 2098 (1975).
127. D. P. Arovas and A. Auerbach, *Phys. Rev. B* **38**, 316 (1988); A. Auerbach and D. P. Arovas, *Phys. Rev. Lett.* **61**, 617 (1988); J. E. Hirsch and S. Tang, *Phys. Rev. B* **39**, 2850 (1989).
128. H. A. Ceccatto, C. J. Gazza, and A. E. Trumper, *Phys. Rev. B* **47**, 12329 (1993); A. E. Trumper, L. O. Manuel, C. J. Gazza, and H. A. Ceccatto *Phys. Rev. Lett.* **78**, 2216 (1997).
129. We thank O. Tchernyshyov for pointing us this physical interpretation of the $U(1)$ flux.
130. S. Sachdev, *Phys. Rev. B* **45**, 12377 (1992).
131. C. H. Chung and J. B. Marston and Ross H. McKenzie, J. *Phys.: Condens. Matter* **13**, 5159 (2001).
132. F. D. M. Haldane, *Phys. Lett. A* **93**, 464 (1983); *Phys. Rev. Lett.* **50**, 1153 (1983).
133. F. D. M. Haldane, *Phys. Rev. Lett.* **61**, 1029 (1988).
134. X. G. Wen and A. Zee, *Phys. Rev. Lett.* **61**, 1025 (1988); E. Fradkin and M. Stone, *Phys. Rev. B* **38**, 7215 (1988); T. Dombre and N. Read, *Phys. Rev. B* **38**, 7181 (1988).
135. K. Harada, N. Kawashima, and M. Troyer, *Phys. Rev. Lett.* **90**, 117203 (2003).
136. E. Fradkin and S. H. Shenker, *Phys. Rev. D* **19**, 3682 (1979).
137. P. W. Kasteleyn, *Physica* **27**, 1209 (1961).
138. M. E. Fisher, *Phys. Rev.* **124**, 1664 (1961).
139. P. W. Kasteleyn, *J. Math. Phys.* **4**, 287 (1963).
140. R. Moessner, S. L. Sondhi, and E. Fradkin, *Phys. Rev. B* **65**, 024504 (2002).
141. G. Misguich, D. Serban, and V. Pasquier, *Phys. Rev. Lett.* **89**, 137202 (2002).
142. R. Moessner and S. L. Sondhi, *Phys. Rev. B* **68**, 054405 (2003).
143. D. S. Rokhsar and S. A. Kivelson, *Phys. Rev. Lett.* **61**, 2376 (1988).
144. R. Moessner and S. L. Sondhi, *Phys. Rev. Lett.* **86**, 1881 (2001).
145. R. Moessner, S. L. Sondhi, and P. Chandra, *Phys. Rev. Lett.* **84**, 4457 (2000).
146. B. Sutherland, *Phys. Rev. B* **37**, 3786 (1988).
147. J. B. Marston and C. Zeng, *J. Appl. Phys.* **69**, 5962 (1991).
148. P. W. Leung, K. C. Chiu and K. J. Runge, *Phys. Rev. B* **54**, 12938 (1996).

149. M. E. Fisher and J. Stephenson *Phys. Rev.* **132**, 1411 (1963).
150. R. Moessner and S. L. Sondhi, *Phys. Rev. B* **68**, 184512 (2003); See Appendix B concerning the square-lattice QDM.
151. R. Moessner, S. L. Sondhi, and P. Chandra, *Phys. Rev. B* **64**, 144416 (2001).
152. D. A. Huse, W. Krauth, R. Moessner, and S. L. Sondhi, *Phys. Rev. Lett.* 91, 167004 (2003) and references therein.
153. L. S. Levitov, *Phys. Rev. Lett.* **64**, 92 (1990); C. L. Henley, *J. Stat. Phys.* **89**, 483 (1997).
154. P. Fendley, R. Moessner and S. L. Sondhi, *Phys. Rev. B* **66**, 214513 (2002).
155. A. Ioselevich, D. A. Ivanov, and M. V. Feigelman, *Phys. Rev. B* **66**, 174405 (2002).
156. L. B. Ioffe, M. V. Feigel'man, A. Ioselevich, D. Ivanov, M. Troyer, and G. Blatter, *Nature* **415**, 503 (2002).
157. G. Misguich, C. Lhuillier, M. Mambrini, P. Sindzingre, *Eur. Phys. J. B* **26**, 167 (2002).
158. X. G. Wen, *Phys. Rev. B* **44**, 2664 (1991).
159. S. Kivelson, *Phys. Rev. B* **39**, 259 (1989).
160. N. Read and B. Chakraborty, *Phys. Rev. B* **40**, 7133 (1989).
161. T. Senthil and M. P. A. Fisher, *Phys. Rev. B* **62**, 7850 (2000).
162. T. Senthil and M. P. A. Fisher, *Phys. Rev. Lett.* **86**, 292 (2001); *Phys. Rev. B* **63**, 134521 (2001).
163. R. P. Feynman, *Phys. Rev.* **90**, 1116 (1952); *Phys. Rev.* **91**, 1291 (1953).
164. V. Elser and C. Zeng, *Phys. Rev. B* **48**, 13647 (1993).
165. A. J. Phares and F. J. Wunderlich, *Nuovo Cimento B* **101**, 653 (1988).
166. See Section V.E.6 of Ref. [170]
167. This follows from the independence of the arrow variables, see Section V.B of Ref. [170]
168. A. M. Polyakov, *Gauge Fields and Strings* (Harwood Academic, New York, 1987).
169. J. B. Kogut, *Rev. Mod. Phys.* **51**, 659 (1979).
170. G. Misguich, D. Serban, and V. Pasquier, *Phys. Rev. B* **67**, 214413 (2003).
171. C. Zeng and V. Elser, *Phys. Rev. B* **51**, 8318 (1995).
172. D. J. Thouless, *Proc. Phys. Soc. London* **86**, 893 (1965).
173. M. Roger, J. H. Hetherington, and J. M. Delrieu, *Rev. Mod. Phys.* **55**, 1 (1983).
174. M. C. Cross and D. S. Fisher, *Rev. Mod. Phys.* **57**, 881 (1985).
175. D. S. Greywall and P. A. Busch, *Phys. Rev. Lett.* **62**, 1868 (1989); D. S. Greywall, *Phys. Rev. B* **41**, 1842 (1990).
176. H. Godfrin and R. E. Rapp, *Adv. Phys.* **44**, 113 (1995).
177. M. Roger, *Phys. Rev. B* **30**, 6432 (1984).
178. M. Roger, C. Bäuerle, Yu. M. Bunkov, A.-S. Chen, and H. Godfrin, *Phys. Rev. Lett.* **80**, 1308 (1998).
179. D. M. Ceperley and G. Jacucci, *Phys. Rev. Lett.* **58**, 1648 (1987).
180. B. Bernu, D. Ceperley, and C. Lhuillier, *J. Low Temp. Phys.* **89**, 589 (1992).
181. D. M. Ceperley, *Rev. Mod. Phys.* **67**, 279 (1995).

182. B. Bernu and D. Ceperley, in M. P. Nightingale and C. J. Umrigar (eds.), *Quantum Monte Carlo Methods in Physics and Chemistry* (Kluwer, Dordrecht, The Netherlands, 1999).
183. H. Ashizawa and D. S. Hirashima, *Phys. Rev. B* **62**, 9413 (2000).
184. J. M. Delrieu, M. Roger, and J. H. Hetherington, *J. Low Temp. Phys.* **40**, 71 (1980).
185. H. Franco, R. E. Rapp, and H. Godfrin, *Phys. Rev. Lett.* **57**, 1161 (1986).
186. G. Misguich, B. Bernu, C. Lhuillier, and C. Waldtmann, *Phys. Rev. Lett.* **81**, 1098 (1998).
187. W. LiMing, G. Misguich, P. Sindzingre, and C. Lhuillier, *Phys. Rev. B* **62**, 6372 (2000).
188. K. Ishida, M. Morishita, K. Yawata, and H. Fukuyama, *Phys. Rev. Lett.* **79**, 3451 (1997).
189. E. Collin *et al.*, *Phys. Rev. Lett.* **86**, 2447 (2001).
190. M. Katano and D. S. Hirashima, *Phys. Rev. B* **62**, 2573 (2000).
191. D. S. Hirashima and K. Kubo, *Phys. Rev. B* **63**, 125340 (2001).
192. B. Bernu, L. Candido, and D. M. Ceperley *Phys. Rev. Lett.* **86**, 870 (2001).
193. T. Okamoto and S. Kawaji, *Phys. Rev. B* **57**, 9097 (1998).
194. M. Roger and J. M. Delrieu, *Phys. Rev. B* **39**, 2299 (1989).
195. S. Sugai *et al.*, *Phys. Rev. B* **42**, 1045 (1990).
196. R. Coldea *et al.*, *Phys. Rev. Lett.* **86**, 5377 (2001).
197. E. Müller-Hartmann and A. Reischl, *Eur. Phys. J. B* **28**, 173 (2002).
198. A. A. Katanin, and A. P. Kampf, *Phys. Rev. B* **67**, 100404R (2003).
199. M. Matsuda *et al.*, *Phys. Rev. B* **62**, 8903 (2000).
200. S. Brehmer *et al.*, *Phys. Rev. B* **60**, 329 (1999).
201. K. P. Schmidt, C. Knetter, and G. S. Uhrig, *Europhys. Lett.* **56**, 877 (2001).
202. A. Gößling *et al.*, *Phys. Rev. B* **67**, 052403 (2003).
203. M. Müller, T. Vekua, and H.-J. Mikeska, *Phys. Rev. B* **66**, 134423 (2002).
204. T. Hikihara, T. Momoi, and X. Hu, *Phys. Rev. Lett.* **90**, 087204 (2003).
205. A. Läuchli, G. Schmid, and M. Troyer, *Phys. Rev. B* **67**, 100409R (2003).
206. Y. Honda and T. Horiguchi, cond-mat/0106426.
207. K. Hijii and K. Nomura, *Phys. Rev. B* **65**, 104413 (2002).
208. T. Momoi, T. Hikihara, M. Nakamura, and X. Hu, *Phys. Rev. B* **67**, 174410 (2003).
209. A. Läuchli, talk given at the *Highly Frustrated Magnetism 2003* Conference, Grenoble, France, August 2003.
210. K. Kubo and T. Momoi, Z. *Phys. B* **103**, 485 (1997).
211. T. Momoi, K. Kubo, and K. Niki, *Phys. Rev. Lett.* **79**, 2081 (1997).
212. G. Misguich, B. Bernu, and C. Lhuillier, *J. Low Temp. Phys.* **110**, 327 (1998).
213. K. Kubo, H. Sakamoto, T. Momoi, and K. Niki, *J. Low Temp. Phys.* **111**, 583 (1998).
214. A 16- and a 32-site triangular lattices which do not frustrate the staggered VBC were investigated. In both cases some of the irreducible representations

of the space group required to break the appropriate lattice symmetries are very high in the spectrum. C. Lhuillier and G. Misguich (unpublished).
215. E. H. Lieb, T. D. Schultz, and D. C. Mattis., *Ann. Phys. (N.Y)* **16**, 407 (1961).
216. I. Affleck and E. Lieb, *Lett. Math. Phys.* **12**, 57 (1986).
217. M. Oshikawa, *Phys. Rev. Lett.* **84**, 1535 (2000).
218. M. Oshikawa, *Phys. Rev. Lett.* **90**, 236401 (2003).
219. A. A. Nersesyan, A. O. Gogolin, and F. H. L. Essler, *Phys. Rev. Lett.* **81**, 910 (1998).
220. P. Azaria *et al.*, *Phys. Rev. Lett.* **81**, 1694 (1998).
221. V. J. Emery, E. Fradkin, S. A. Kivelson, and T. C. Lubensky, *Phys. Rev. Lett.* **85**, 2160 (2000).
222. M. Bocquet, F. Essler, A. M. Tsvelik, and A. O. Gogolin, *Phys. Rev. B* **64**,094425 (2001); M. Bocquet, *Phys. Rev. B* **65**, 1884415 (2001).
223. A. Vishwanath and D. Carpentier, *Phys. Rev. Lett.* **86**, 676 (2001).
224. S. Sachdev and K. Park, *Ann. Phys. (N.Y.)* **58**, 298 (2002).
225. O. A. Starykh, R. R. P. Singh, and G. C. Levine, *Phys. Rev. Lett.* **88**, 167203 (2002).
226. R. Coldea, D. A. Tennant, A. M. Tsvelik, and Z. Tylczynski, *Phys. Rev. Lett.* **86**, 1335 (2001).
227. A. Kitaev, *Ann. Phys.* **303**, 2 (2003). [quant-ph/9707021]
228. C. Nayak and K. Shtengel, *Phys. Rev. B* **64**, 064422 (2001).
229. L. Balents, M. P. A. Fisher, and S. M. Girvin, *Phys. Rev. B* **65**, 224412 (2002).
230. A. Paramekanti, L. Balents, and M. P. A. Fisher, *Phys. Rev. B* **66**, 054526 (2002).
231. A. W. Sandvik, S. Daul, R. R. P. Singh, and D. J. Scalapino, *Phys. Rev. Lett.* **89**, 247201 (2002).
232. T. Senthil and O. Motrunich, *Phys. Rev. B* **66**, 205104 (2002).
233. L. Pauling, *The Nature of the Chemical Bond* (Cornell University Press, Ithaca, 1938).
234. K. Kano and S. Naya, *Prog. in Theor. Phys.* **10**, 158 (1953).
235. D. Huse and A. Rutenberg, *Phys. Rev. B* **45**, 7536 (1992).
236. R. Moessner and S. L. Sondhi, *Phys. Rev. B* **63**, 224401 (2001).
237. J. Chalker, P. C. W. Holdsworth, and E. F. Shender, *Phys. Rev. Lett.* **68**, 855 (1992).
238. R. J. Baxter, J. Math. *Phys.* **11**, 784 (1970).
239. I. Richtey, P. Chandra, and P. Coleman, *Phys. Rev. B* **47**, 15342 (1993).
240. J. Reimers and A. Berlinsky, *Phys. Rev. B* **48**, 9539 (1993).
241. M. Elhajal, Ph.D. thesis, Université Joseph Fourier, Grenoble, France, 2002.
242. M. Elhajal, B. Canals, and C. Lacroix, *Phys. Rev. B* **66**, 014422 (2002).
243. A. Keren, *Phys. Rev. Lett.* **72**, 3254 (1994).
244. V. Elser, *Phys. Rev. Lett.* **62**, 2405 (1989).
245. J. Chalker and J. Eastmond, *Phys. Rev. B* **46**, 14201 (1992).

246. S. Sachdev, *Phys. Rev. B* **45**, 12377 (1992).
247. P. Leung and V. Elser, *Phys. Rev. B* **47**, 5459 (1993).
248. P. Lecheminant *et al.*, *Phys. Rev. B* **56**, 2521 (1997).
249. C. Waldtmann *et al.*, Eur. *Phys. J. B* **2**, 501 (1998).
250. F. Mila, *Phys. Rev. Lett.* **81**, 2356 (1998).
251. P. Sindzingre *et al.*, *Phys. Rev. Lett.* **84**, 2953 (2000).
252. A. M. Läuchli, oral communication at the KITP Program: "Frustrated Magnetism and Quantum Spin Liquids: From Theory and Models to Experiments" (August 13 – November 9, 2012). Available online at: http://online.kitp.ucsb.edu/online/fragnets12/laeuchli. Unpublished.
253. M. Mambrini and F. Mila, *Eur. Phys. J. B* **17**, 651 (2001).
254. S. Dommange, M. Mambrini, B. Normand, and F. Mila, *Phys. Rev. B* **68**, 224416 (2003).
255. C. Zeng and V. Elser, *Phys. Rev. B* **42**, 8436 (1990).
256. R. Singh and D. Huse, *Phys. Rev. Lett.* **68**, 1766 (1992).
257. K. Hida, *J. Phys. Soc. Jpn.* **70**, 3673 (2001).
258. D. C. Cabra, M. D. Grynberg, P. C. W. Holdsworth, and P. Pujol, *Phys. Rev. B* **65**, 094418 (2002).
259. N. Eltsner and A. P. Young, *Phys. Rev. B* **50**, 6871 (1994).
260. T. Nakamura and S. Miyashita, *Phys. Rev. B* **52**, 9174 (1995).
261. P. Tomczak and J. Richter, *Phys. Rev. B* **54**, 9004 (1996).
262. A. P. Ramirez, B. Hessen, and M. Winkelmann, *Phys. Rev. Lett.* **84**, 2957 (2000).
263. P. Mendels *et al.*, *Phys. Rev. Lett.* **85**, 3496 (2000).
264. A. V. Syromyatnikov and S. V. Maleyev, *Phys. Rev. B* **66**, 132408 (2002).
265. P. Nikolic and T. Senthil, *Phys. Rev. B* **68**, 214415 (2003).
266. R. R. P. Singh and D. A. Huse, *Phys. Rev. B* **76**, 180407(R) (2007).
267. R. R. P. Singh and D. A. Huse, *Phys. Rev. B* **77**, 144415 (2008).
268. C. Mondelli *et al.*, *Physica B* **284**, 1371 (2000).
269. A. Georges, R. Siddhartan, and S. Florens, *Phys. Rev. Lett.* **87**, 277203 (2001).
270. C. Lhuillier and P. Sindzingre, in Y. Ajiro and J. P. Boucher, (eds.), *Quantum Properties of Low Dimensional Antiferromagnets* (Kyushu University Press, Fukuoka, Japan, 2002), p. 111.
271. S. Sachdev and M. Vojta, in *Proceedings of the XIII International Congress on Mathematical Physics, July 2000, London*, A. Fokas, A. Grigoryan, T. Kibble, and B. Zegarlinski (eds.) (International Press, Boston, 2001), [cond-mat/0009202].
272. Y. Uemura *et al.*, *Phys. Rev. Lett.* **73**, 3306 (1994).
273. S.-H. Lee *et al.*, *Europhys. Lett.* **35**, 127 (1996).
274. A. Keren *et al.*, *Phys. Rev. B* **53**, 6451 (1996).
275. A. S. Wills *et al.*, *Europhys. Lett.* **42**, 325 (1998).
276. K. Awaga *et al.*, *Phys. Rev. B* **49**, 3975 (1994).
277. N. Wada *et al.*, *J. Phys. Soc. Jpn.* **66**, 961 (1997).

278. I. Watanabe *et al.*, *Phys. Rev. B* **58**, 2438 (1998).
279. B. Canals and C. Lacroix, *Phys. Rev. Lett.* **80**, 2933 (1998).
280. A. B. Harris, A. J. Berlinsky, and C. Bruder, *J. Appl. Phys.* **69**, 5200 (1991).
281. A. B. Harris, C. Kallin, and A. J. Berlinsky, *Phys. Rev. B* **45**, 2899 (1992).
282. H. Tsunetsugu, *J. Phys. Soc. Jpn.* **70**, 640 (2001).
283. H. Tsunetsugu, *Phys. Rev. B* **65**, 024415 (2002).
284. J. V. Delft and C. L. Henley, *Phys. Rev. Lett.* **69**, 3236 (1992); *Phys. Rev. B* **48**, 965 (1993).
285. S. Sachdev, *Ann. Phys.* **303**, 226 (2003) [cond-mat/0211027] and *Rev. Mod. Phys.* **75**, 913 (2003).
286. X. Wen, F. Wilczek, and A. Zee, *Phys. Rev. B* **39**, 11413 (1989).
287. V. Kalmeyer and R. Laughlin, *Phys. Rev. Lett.* **59**, 2095 (1987); *Phys. Rev. B* **39**, 11879 (1989).
288. K. Yang, L. Warman, and S. M. Girvin, *Phys. Rev. Lett.* **70**, 2641 (1993).
289. O. Janson, J. Richter, and H. Rosner, *Phys. Rev. Lett.* **101**, 106403 (2008).
290. L. Messio, C. Lhuillier, and G. Misguich, *Phys. Rev. B* **83**, 184401 (2011).
291. M. E. Zhitomirsky, *Phys. Rev. B* **78**, 094423 (2008).
292. C. L. Henley, *Phys. Rev. B* **80**, 180401 (2009).
293. P. Sindzingre and C. Lhuillier, *EPL* **88** 27009 (2009).
294. A. M. Läuchli, J. Sudan, and E. S. Sørensen, *Phys. Rev. B* **83**, 212401 (2011).
295. H. C. Jiang, Z. Y. Weng, and D. N. Sheng, *Phys. Rev. Lett.* **101**, 117203 (2008).
296. S. Yan, D. A. Huse, and S. R. White, *Sci. Magaz.* **332**, 1173 (2011).
297. H.-C. Jiang, Z. Wang, and L. Balents, *Nat. Phys.* **8**, 902 (2012).
298. S. Depenbrock, I. P. McCulloch, and U. Schollwoeck, *Phys. Rev. Lett.* 109, 067201 (2012).
299. V. Kalmeyer and R. B. Laughlin, *Phys. Rev. B* **39** 11879 (1989).
300. J. Wen, A. Rüegg, C.-C. Wang, and G. A Fiete, *Phys. Rev. B* **82**, 075125, (2010).
301. V. Chua, Victor H. Yao, and G.A. Fiete, *Phys. Rev. B* **83**, 180412 (2011).
302. L. Messio, B. Bernu, and C. Lhuillier, *Phys. Rev. Lett.* **108**, 207204 (2012).
303. D. Poilblanc, M. Mambrini, and D. Schwandt, *Phys. Rev. B* **81**, 180402 (2010).
304. G. Evenbly and G. Vidal, *Phys. Rev. Lett.* **104**, 187203 (2010).
305. D. Schwandt, M. Mambrini, and D. Poilblanc, *Phys. Rev. B* **81**, 214413 (2010).
306. Y. Iqbal, F. Becca, and D. Poilblanc, *Phys. Rev. B* **83**, 100404 (2011).
307. D. Poilblanc and G. Misguich, *Phys. Rev. B* **84**, 214401 (2011).
308. Y. Ran, M. Hermele, P. A.Lee, and X.-G. Wen, *Phys. Rev. Lett.* **98**, 117205 (2007).
309. M. Hermele, T. Senthil and M. P. A. Fisher, *Phys. Rev. B* **72**, 104404 (2005).
310. M. Hermele, Y. Ran, P. A. Lee and X.-G. Wen, *Phys. Rev. B* **77**, 224413 (2008).
311. F. Wang and A. Vishwanath, *Phys. Rev. B* **74**, 174423 (2006).

312. Y.-M. Lu, Y. Ran, and P. A. Lee, *Phys. Rev. B* **83**, 224413 (2011).
313. G. Misguich and P. Sindzingre, *J. Phys. Cond. Matt.* **19**, 145202 (2007) and Errata: condmat/0607764v3.
314. X.-G. Wen, *Phys. Rev. B* **65**, 165113 (2002).
315. W.-H. Ko, Z.-X. Liu, T-K. Ng, and P. A. Lee, *Phys. Rev. B* **81**, 024414 (2010);
316. W.-H. Ko and P. A. Lee, *Phys. Rev. B* **84**, 125102 (2011).
317. L. Messio, O. Cépas, and C. Lhuillier, *Phys. Rev. B* **81**, 064428, (2010).
318. Y.-M. Lu and Y. Ran, *Phys. Rev. B* **84**, 024420 (2011).
319. Z. Hiroi, M. Hanawa, N. Kobayashi, M. Nohara, H. Takagi, Y. Kato, and M. Takigawa, *J. Phys. Soc. Japan,* **70**, 3377 (2001).
320. Y. Okamoto, M. Tokunaga, H. Yoshida, A. Matsuo, K. Kindo, and Z. Hiroi, *Phys. Rev. B* **83**, 180407 (2011).
321. G. J. Nilsen, F. C. Coomer, M. A. de Vries, J. R. Stewart, P. P. Deen, A. Harrison, and H. M. Rønnow, *Phys. Rev. B* **84**, 172401, (2011).
322. T. Yavors'kii, W. Apel, and H.-U. Everts, *Phys. Rev. B* **76**, 064430 (2007).
323. Fa Wang, A. Vishwanath, and Y. B. Kim, *Phys. Rev. B* **76**, 094421 (2007).
324. Y. Huh, M. Punk, and S. Sachdev, *Phys. Rev. B* **84**, 094419 (2011).
325. T. Tay and O.-I. Motrunich, *Phys. Rev. B* **84**, 020404 (2011).
326. Y. Iqbal, F. Becca, and D. Poilblanc, *Phys. Rev. B* **84**, 020407 (2011).
327. S. Ryu, O. I. Motrunich, J. Alicea, and M. P. A. Fisher, *Phys. Rev. B* **75**, 184406 (2007).
328. T. H. Han, J. S. Helton, S. Chu, A. Prodi, D. K. Singh, C. Mazzoli, P. Müller, D. G. Nocera and Y. S. Lee, *Phys. Rev. B* **83**, 100402 (2011).
329. J. S. Helton, K. Matan, M. P. Shores, E. A. Nytko, B. M. Bartlett, Y. Qiu, D. G. Nocera, and Y. S. Lee, *Phys. Rev. Lett.* **104**, 147201 (2010).
330. O. Götze, D. J. J. Farnell, R. F. Bishop, P. H. Li, and J. Richter, *Phys. Rev. B* **84**, 224428 (2011).
331. O. Cépas and A. Ralko, *Phys. Rev. B* **84**, 020413 (2011).
332. G. Misguich and B. Bernu, *Phys. Rev. B* **71**, 014417 (2005).
333. G. Misguich and P. Sindzingre, *Eur. Phys. J. B* **59**, (2007).
334. M. Rigol and R. R. P. Singh, *Phys. Rev. Lett.* **98**, 207204 (2007).
335. M. Rigol and R. R. P. Singh, *Phys. Rev. B* **76**, 184403 (2007).
336. R. Chitra and M. J. Rozenberg, *Phys. Rev.B* **77**, 052407 (2008).
337. I. Rousochatzakis, S. Manmana, A. M. Läuchli, B. Normand, and F. Mila, *Phys. Rev. B* **79** 214415 (2009).
338. A. Laeuchli and C. Lhuillier, arXiv:09011065v1.
339. Z. Hao and O. Tchernyshyov, *Phys. Rev. B* **81**, 214445 (2010).
340. F. Bert, S. Nakamae, F. Ladieu, D. L'Hôte, P. Bonville, F. Duc, J.-C. Trombe, and P. Mendels, *Phys. Rev. B* **76**, 132411 (2007).
341. E. Kermarrec, P. Mendels, F. Bert, R. H. Colman, A. S. Wills, P. Strobel, P. Bonville, A. Hillier, and A. Amato, *Phys. Rev. B* **84**, 100401 (2011).
342. M. Jeong, F. Bert, P. Mendels, F. Duc, J. C. Trombe, M. A. de Vries, and A. Harrison, *Phys. Rev. Lett.* **107**, 237201 (2011).

343. Y. Narumi, K. Katsumata, Z. Honda, J.-C. Domenge, P. Sindzingre, C. Lhuillier, Y. Shimaoka, T. C. Kobayashi, and K. Kindo, *Europhys. Lett.* **65** 705 (2004).
344. J.-C. Domenge, C. Lhuillier, L. Messio, L. Pierre, and P. Viot, *Phys. Rev. B* **77**, 172413 (2008).
345. M. P. Shores, E. A. Nytko, B. M. Bartlett, and D. G. Nocera, *J. Am. Chem. Soc.* **127**, 13462 (2005).
346. P. Mendels, F. Bert, M. A. de Vries, A. Olariu, A. Harrison, F. Duc, J. C. Trombe, J. S. Lord, A. Amato, and C. Baines, *Phys. Rev. Lett.* **98**, 077204 (2007).
347. J. S. Helton, K. Matan, M. P. Shores, E. A. Nytko, B. M. Bartlett, Y. Yoshida, Y. Takano, A. Suslov, Y. Qiu, J.-H. Chung, D. G. Nocera, and Y. S. Lee, *Phys. Rev. Lett.* **98** 107204 (2007).
348. A. Olariu, P. Mendels, F. Bert, F. Duc,J. C. Trombe, M. A. de Vries, and A. Harrison, *Phys. Rev. Lett.* **100** 087202 (2008).
349. T. Imai, E. A. Nytko, B. M. Bartlett, M. P. Shores, and D. G. Nocera, *Phys. Rev. Lett.* **100** 077203 (2008).
350. M. Jeong, F. Bert, P. Mendels, F. Duc, J. C. Trombe, M. A. de Vries and A. Harrison, *Phys. Rev. Lett.* **107**, 237201 (2011).
351. A. Zorko, S. Nellutla, J. van Tol, L. C. Brunel, F. Bert, F. Duc, J. C. Trombe, M. A. de Vries, A. Harrison and P. Mendels, *Phys. Rev. Lett.* **101**, 026405 (2008).
352. O. Cépas, C. M. Fong, P. W. Leung, and C. Lhuillier, *Phys. Rev. B.* **78**, 140405 (2008).
353. M. A. de Vries, D. Wulferding, P. Lemmens, J. S. Lord, A. Harrison, P. Bonville, F. Bert, and P. Mendels, *Phys. Rev. B.* **85**, 014422 (2012).
354. M. A. de Vries, J. R. Stewart, P. P. Deen, J. O. Piatek, G. J. Nilsen, H. M. Rønnow, A. Harrison, *Phys. Rev. Lett.* **103**, 237201 (2009).
355. S. El Shawish, S. O. Cépas, and S. Miyashita, *Phys. Rev. B.* **81**, 224421 (2010).
356. H. Kageyama, T. Nakajima, M. Ichihara, F. Sakai, and Y. Ueda, in Y. Ajiro and J. P. Boucher (eds.), *Spin Frustration in Two-Dimensional Compounds* (Kyushu University Press, Fukuoka, Japan, 2002), p. 135.
357. Y. Karaki, Y. M. Kou, A. Yamaguchi, M. Kubota, H. Ishimoto, Z. Honda, and K. Yamada, *J. Low Temp Phys.* **158**, 653 (2010).
358. J.-C. Domenge, P. Sindzingre, C. Lhuillier, and L. Pierre, *Phys. Rev. B* **72**, 024433 (2005).
359. R. H. Colman, C. Ritter, and A. S. Wills, *Chem. Mater.* **20**, 6897 (2008).
360. R. H. Colman, A. Sinclair, and A. S.Wills, *Chem. Mater.* **22**, 5774 (2010).
361. B. Fak, E. Kermarrec, L. Messio, B. Bernu, B. C. Lhuillier, F. Bert, P. Mendels, B. Koteswararao, F. Bouquet, J. Ollivier, A. D. Hillier, A. Amato, R. H. Colman, and A. S. Wills, *Phys. Rev. Lett.* **109**, 037208 (2012).
362. M. B. Hastings, *Phys. Rev. B* **69**, 104431 (2004).

363. L. Messio, J.-C. Domenge, C. Lhuillier, L. Pierre, P. Viot, and G. Misguich, *Phys. Rev. B* **78**, 054435 (2008).
364. A. Hamma, R. Ionicioiu, and P. Zanardi, *Phys. Lett. A* **337**, 22 (2005).
365. A. Kitaev and J. Preskill, *Phys. Rev. Lett.* **96**, 110404 (2006).
366. M. Levin and X.-G. Wen, *Phys. Rev. Lett.* **96**, 110405 (2006).
367. S. Furukawa and G. Misguich, *Phys. Rev. B* **75**, 214407 (2007).
368. S. V. Isakov, M. B. Hastings, and R. G. Melko, *Nat. Phys.* **7**, 772 (2011).
369. Yi Zhang, T. Grover, and A. Vishwanath *Phys. Rev. B* **84**, 075128 (2011).
370. A. M. Läuchli, E. J. Bergholtz, and M. Haque, *New J. Phys.* **12**, 075004 (2010).
371. J.-M. Stéphan, G. Misguich, and V. Pasquier, *J. Stat. Mech.* P02003 (2012).
372. H. C. Jiang, H. Yao, and L. Balents, *Phys. Rev. B* **86**, 024424 (2012).
373. Yi Zhang, T. Grover, A. Turner, M. Oshikawa, and A. Vishwanath, *Phys. Rev. B* **85**, 235151 (2012).
374. O. Cépas and A. Ralko, *Phys. Rev. B* **84**, 020413 (2011).
375. G.-W. Chern and R. Moessner, arXiv:1207.4752.
376. L. Messio, C. Lhuillier, and G. Misguich, in preparation.
377. R. H. Colman, F. Bert, D. Boldrin, A. D. Hillier, P. Manuel, P. Mendels, and A. S. Wills, *Phys. Rev. B* **83**, 180416 (2011).
378. J. A. Quilliam, F. Bert, R. H. Colman, D. Boldrin, A. S. Wills, and P. Mendels, *Phys. Rev. B* **84**, 180401 (2011).
379. Y. Okamoto, H. Yoshida, and Z. Hiroi, *J. Phys. Soc. Jpn.* 78, 033701 (2009).
380. B. Bernu, E. Kermarrec, L. Messio, C. Lhuillier, F. Bert, and P. Mendels, arXiv:1210.2549.
381. R. Coldea, D. A. Tennant, A. M. Tsvelik, and Z. Tylczynski, *Phys. Rev. Lett.* **86**, 1335 (2001).
382. Hong-Chen Jiang, Hong Yao, and Leon Balents, *Phys. Rev. B* **86**, 024424 (2012).
383. F. D. M. Haldane, *Phys. Rev. Lett.* **61**, 2015 (1988).
384. C. L. Kane, and E. J. Mele, *Phys. Rev. Lett.* **95**, 146802 (2005).
385. Y. Hatsugai, *Phys. Rev. Lett.* **71**, 3697 (1993).
386. X.-L. Qi, Y.-S. Wu, and S.-C. Zhang, *Phys. Rev. B* **74**, 085308 (2006).
387. M. Udagawa and Y. Motome, *Phys. Rev. Lett.* **104**, 106409 (2010).
388. B. Nachtergaele and R. Sims, *Commun. Math. Phys.* **276**, 437 (2007).
389. I. Kimchi, S. A. Parameswaran, A. M. Turner, and A. Vishwanath, arXiv:1207.0498.
390. Z. Y. Meng, T. C. Lang, S. Wessel, F. Assaad, and A. Muramatsu, *Nature* **464**, 847 (2010).
391. F. Wang, *Phys. Rev. B* **82**, 024419 (2010).
392. A. F. Albuquerque, D. Schwandt, B. Hetényi, S. Capponi, M. Mambrini, and A. M. Läuchli, *Phys. Rev. B* **84** 024406 (2011).
393. D. C. Cabra, C. A. Lamas, and H. D. Rosales, *Phys. Rev. B* **83** 094506 (2011).
394. B. K. Clark, D. A. Abanin, and S. L. Sondhi, *Phys. Rev. Lett.* **107**, 087204 (2011).

395. H-Y. Yang, A. F. Albuquerque, S. Capponi, A. M. Läuchli, and K. P. Schmidt, *New J. Phys.* **14**, 115027 (2012).
396. S. Sorella, Y. Otsuka, and S. Yunoki, arXiv:1207.1783.
397. A. Läuchli, J. C. Domenge, C. Lhuillier, P. Sindzingre, and M. Troyer, *Phys. Rev. Lett.* **95**, 137206 (2005).
398. T. Momoi, P. Sindzingre, and N. Shannon, *Phys. Rev. Lett.* **97**, 257204 (2006).
399. H. Fukuyama, oral communication at the KITP Conference: "Exotic Phases of Frustrated Magnets" (October 8–12, 2012). Available online at: http://online.kitp.ucsb.edu/online/fragnets-c12/fukuyama/. Unpublished.

CHAPTER 6

ONE-DIMENSIONAL QUANTUM SPIN LIQUIDS

P. Lecheminant

Laboratoire de Physique Théorique et Modélisation, CNRS UMR 8089, Université de Cergy-Pontoise,5 mail Gay-Lussac, Neuville sur Oise, 95301 Cergy-Pontoise Cedex, France

This chapter is intended as a brief overview of some of the quantum spin liquid phases with unbroken $SU(2)$ spin symmetry available in one dimension. The main characteristics of these phases are discussed by means of the bosonization approach. A special emphasis is laid on the interplay between frustration and quantum fluctuations in one dimension.

6.1. Introduction

A central issue in the study of strongly correlated systems is the classification of all possible Mott insulating phases at zero temperature. The general strategy for describing the possible phases associated with the spin degrees of freedom (the so-called quantum magnetism) of Mott insulators consists of analyzing localized spin models such as the antiferromagnetic (AF) Heisenberg model:

$$\mathcal{H} = J \sum_{\langle i,j \rangle} \vec{S}_i \cdot \vec{S}_j + \cdots , \tag{6.1}$$

where $\vec{S}_i$ are quantum spin-S operators on sites i of a lattice, and the sum $\langle i,j \rangle$ is over the nearest-neighbor sites with an AF exchange ($J > 0$); the ellipsis represents additional terms like second-neighbor competing exchange interaction or ring exchange that can be eventually added to the Heisenberg Hamiltonian (6.1). Generically, the ground state (GS) of the model (6.1) without these additional terms displays long-range Néel ordering ($\langle \vec{S}_i \rangle \neq \vec{0}$) which spontaneously breaks the $SU(2)$ symmetry of the lattice Hamiltonian (6.1). The low-energy excitations are gapless spin waves (i.e. magnons) as expected when a continuous symmetry is spontaneously broken.

A central focus of quantum magnetism over the years, starting from the proposal made by Anderson,[1] has been the search for a spin liquid behavior i.e. a phase with no magnetic long-range Néel order. Spin liquid phases are expected to be stabilized in low dimensions or in the presence of frustration, i.e. in situations where quantum fluctuations can strongly suppress magnetism. It gives rise to rich physics with exotic low-energy excitations which require in most cases the use of non-perturbative techniques to fully determine their properties. In the 2D case, different spin liquid phases have been found and they display bond ordering or topological ordering (for a review see for instance Refs. [2, 3] and references therein). An interesting attempt to develop a systematic classification of 2D spin liquids has also recently been explored by Wen.[4]

In one dimension, the quantum analog of the Mermin–Wagner[5] theorem shows that quantum fluctuations always disorder the GS in systems with a continuous symmetry. The generic situation in one dimension is thus to have a spin liquid phase. The 1D case is also extremely favorable since several non-perturbative techniques are available to fully characterize the physical properties of different spin liquids. These powerful techniques include integrability, conformal field theory (CFT),[6,7] the bosonization approach,[8–10] and numerical calculations such as the density matrix renormalization group (DMRG) approach.[11] The determination of the properties of spin liquids is not a purely academic problem since many physical realizations of spin chains have been synthesized over the years. The reader may consult for instance the recent reviews[12,13] on experiments in quasi-1D spin systems.

The main questions in the study of 1D spin liquids are in order: What are the different spin liquid phases available at zero-temperature? What are the different physical properties of these spin liquids? What is the nature of the quantum phase transition[14] between two different spin liquids? Last but not least, the problem of dimensional cross-over, i.e. the fate of the spin liquid as the 2D case is reached, is also a central topic. In particular, is it possible to stabilize a spin liquid phase in $d = 2$ starting from the $d = 1$ case?

In this work, we shall mainly be concerned with the first two questions for spin liquids with unbroken $SU(2)$ spin symmetry. Using the bosonization approach, we shall discuss some of the physical properties of spin liquids that have been found in spin chains and spin ladders with or without frustration. An important question related to this problem is how can one classify all these 1D spin liquid phases at $T = 0$? A first distinction concerns the existence or absence of a spectral gap in the model. In a gapless system (critical spin liquid), one has quasi-long-range Néel order. The leading asymptotics of spin–spin correlation functions display a power law behavior with an exponent characterized by a well-defined underlying CFT.[15] In contrast, in a spin liquid phase with a spectral gap, spin–spin correlation functions have an exponential decay with the distance due to the existence of a finite correlation length ξ: $\langle \vec{S}_i \cdot \vec{S}_j \rangle \sim \exp(-|i-j|/\xi)$. At this point, it is worth stressing that two gapful spin liquid phases sharing the same thermodynamic properties (typically a thermal activation law at low T) may display very different behaviors in other physical quantities of interest such as, for instance, dynamical or optical properties. It is thus necessary to make a scrutiny analysis of the properties of the phase before elaborating a complete classification. Let us first consider the GS of a gapful spin liquid phase. A (discrete) symmetry might be spontaneously broken in the GS resulting on a GS degeneracy. The spin liquid phase may display also an hidden topological order,[16] which manifests itself in a GS degeneracy that depends on the nature of the boundary conditions (BC) used, i.e. periodic or open BC. The resulting chain-end (edge) excitations in the open BC case can be observed in the NMR profile of a spin chain compound doped with non-magnetic impurities like Zn or Mg. Another important distinction between the two spin liquids relies on the quantum number carried by elementary excitations. The fractionalized or integer nature of this quantum number has crucial consequences in the dynamical structure factor of the system which directly probes the elementary nature of the spin-flip. A sharp spectral peak, characteristic of a well-defined $S = 1$ mode, or a broad (incoherent background) feature in the dynamical structure factor can be seen in inelastic neutron scattering experiments. Finally, an interesting question is the existence or absence of bound states in the energy spectrum below the two-particle continuum. Spin liquid phases may then be distinguished at this level, which is also an important fact from the experimental point of view since observation of magnetic singlet bound states can be realized through light scattering experiments.[13] This classification is certainly far from being complete, but it enables us to distinguish between several spin liquid phases and investigate the nature of the quantum phase transition between them.

This chapter is organized as follows: We present, in Section 6.2, the different spin liquid phases that occur in unfrustrated spin chains and spin ladders. The main effects of frustration in 1D spin liquids are described in Section 6.3. In particular, we shall observe that frustration plays its trick by allowing deconfined spinons (carrying fractional $S = 1/2$ quantum number) as elementary excitations and it provides a non-trivial source of incommensurability. Finally, our concluding remarks are given in Section 6.4.

6.2. Unfrustrated Spin Chains

The paradigmatic model to investigate the properties of 1D quantum magnets in the absence of frustration is the AF Heisenberg spin chain given by the Hamiltonian

$$\mathcal{H} = J \sum_i \vec{S}_i \cdot \vec{S}_{i+1}, \tag{6.2}$$

where $\vec{S}_i$ is a spin-S operator at the ith site of the chain and the exchange interaction is AF: $J > 0$. In the classical limit, the model displays Néel long-range order: $\vec{S}_i \sim S(-1)^i \vec{n}$, $\vec{n}$ being an unit vector with an arbitrary fixed orientation in spin space. This solution breaks the $SU(2)$ symmetry of the model (6.2) down to $U(1)$. Corresponding to this symmetry breaking scheme, there are two Goldstone modes that propagate with the same velocity (Lorentz invariance in the low-energy limit). In the language of CFT,[7] it corresponds to a field theory with central charge $c = 2$ ($c = 1$ being the central charge of a free massless boson field). Physically, these Goldstone modes are nothing but a doublet of gapless spin wave modes associated with slow modulations in the orientation of the vector $\vec{n}$, which represents the order parameter of the Néel magnetic structure. In the quantum case, this Néel solution in one dimension is destabilized by strong quantum fluctuations no matter how large S is: a spin liquid phase is formed.

6.2.1. *Spin-1/2 Heisenberg chain*

In the ultra quantum limit, i.e. $S = 1/2$, the nature of the spin liquid phase can be fully determined since the model (6.2) is exactly solvable for

$S = 1/2$ by means of the Bethe ansätze approach. In particular, starting with Bethe's seminal work,[17] a host of exact results have been obtained over the years for GS properties,[18] magnetic susceptibility,[19] thermodynamics,[20] excitation spectrum,[21,22] and correlation functions.[23]

The model displays quantum criticality properties which belong to the Wess–Zumino–Novikov–Witten (WZNW) $su(2)_1$ universality class.[15] The low-energy limit of the spin-1/2 AF Heisenberg chain (6.2) is described by the $su(2)_1$ WZNW CFT with central charge $c = 1$ perturbed by marginally irrelevant current–current interaction.[15,24] The resulting Hamiltonian density reads as follows:

$$\mathcal{H}_{\text{eff}} = \frac{2\pi v}{3}\left(\vec{J}_L^2 + \vec{J}_R^2\right) + \lambda \vec{J}_R \cdot \vec{J}_L, \tag{6.3}$$

where $\vec{J}_R$ and $\vec{J}_L$ are, respectively, the right and left $su(2)_1$ currents that generate the $su(2)_1$ WZNW CFT;[25] these currents satisfy the $su(2)_1$ Kac–Moody commutation relations:

$$\left[J^a_{R,L}(x), J^b_{R,L}(y)\right] = \mp \frac{i\delta^{ab}}{4\pi}\delta'(x-y) + i\epsilon^{abc} J^c_{R,L}(x)\,\delta(x-y). \tag{6.4}$$

In Eq. (6.3), v is the spin velocity and $\lambda < 0$ so that the last contribution is a marginally irrelevant term that renormalizes to zero in the far-infrared (IR) limit. This perturbation accounts for logarithmic corrections in the spin–spin correlation,[26] which is exactly known[27] in the long-distance limit:

$$\langle \vec{S}_0 \cdot \vec{S}_r \rangle \simeq_{r \gg 1} \frac{(-1)^r}{(2\pi)^{3/2}} \frac{(\ln r)^{1/2}}{r}. \tag{6.5}$$

The most striking feature of this spin liquid phase stems from the nature of its elementary excitations. They have been elucidated by Faddeev and Takhtajan[22] within the Bethe anšatze approach and consist of fractional $S = 1/2$ massless excitations called spinons. The lowest excitations are fourfold degenerate and correspond to a triplet ($S = 1$) and a singlet ($S = 0$). The resulting energy spectrum is a continuum in (k, ω) space between a lower boundary (the des Cloizeaux–Pearson dispersion relation[21]) $\omega_{\text{dcp}} = \pi J|\sin k|/2$ $(-\pi < k \leq \pi)$ and an upper boundary $\omega_u = \pi J|\sin(k/2)|$ (see Fig. 6.1). The central point of the analysis is that this continuum can be interpreted as being made up of two spin-1/2

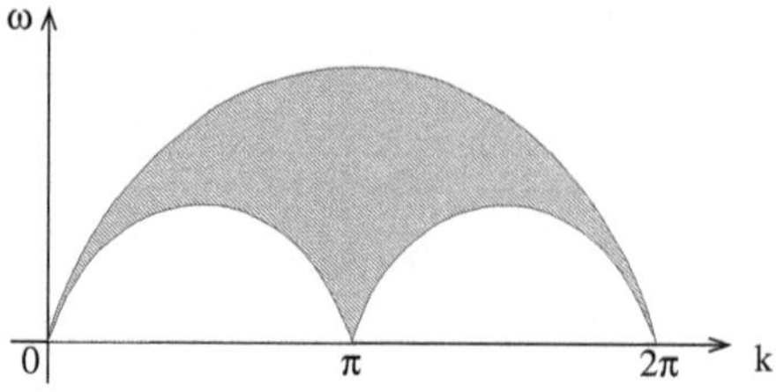

Fig. 6.1. Two-spinon continuum of the spin-1/2 AF Heisenberg chain.

excitations (spinons) with the dispersion

$$\omega_{\text{spinon}} = \frac{\pi J}{2} \sin k, \tag{6.6}$$

with $0 < k < \pi$. A spinon has thus a wave vector restricted to only half of the Brillouin zone. A triplet (or singlet) excitation with momentum k is then described by two spinons with momenta k_1 and k_2 $(0 < k_1 \leq k_2 < \pi)$, such that: $\omega(k) = \omega_{\text{spinon}}(k_1) + \omega_{\text{spinon}}(k_2)$ with $k = k_1 + k_2$ (respectively, $k = k_1 + k_2 - 2\pi$) if $0 < k \leq \pi$ (respectively, $-\pi < k < 0$). It corresponds to a two-parameter continuum $\omega(k) = \pi J \sin(k/2) \cos(k/2 - q_1)$ with $0 < q_1 < k/2$ for $0 < k < \pi$ and $\pi + k < q_1 < \pi + k/2$ for $-\pi < k < 0$ which identifies with the continuum of the spin-1/2 AF Heisenberg chain.

A magnon excitation carrying $S = 1$ quantum number is thus no longer an elementary excitation in this model but is fractionalized into two spinons. One important experimental signature of these spinons is the absence of any sharp peak in the dynamical susceptibility structure factor since a single spin-flip generates a triplet excitation which is not elementary here but made up of two spinons. The spectral density is then a convolution of the spectral densities of the individual spinons. These spinons have fascinating properties, and in particular, they obey semion statistics intermediate between bosons and fermions[28] as can be seen by the study of the elementary excitations of the Haldane–Shastry model[29] which belongs to the same universality class as the spin-1/2 AF Heisenberg chain. A heuristic way to describe the spinons is to consider an anisotropic XXZ version of the Heisenberg model (6.2)

$$\mathcal{H}_{XXZ} = J_z \sum_i S_i^z S_{i+1}^z + \frac{J_\perp}{2} \sum_i \left(S_i^+ S_{i+1}^- + S_{i+1}^+ S_i^- \right). \tag{6.7}$$

This model is still integrable and the GS is two-fold degenerate for $J_z > J_\perp$ as the result of the spontaneous breaking of a discrete Z_2 symmetry. In this regime, a spectral gap is opened and the elementary excitations are massive spinons. A simple way to understand these properties is to consider the Ising limit when $J_z \gg J_\perp$ where the GS reduces to the two Néel ordered states: $\uparrow\downarrow\uparrow\downarrow\uparrow\downarrow$.. and $\downarrow\uparrow\downarrow\uparrow\downarrow\uparrow$... The elementary excitations are massive kinks that carry $\Delta S^z = \pm 1/2$ quantum number and interpolate between these two GSs: $\uparrow\downarrow\uparrow\downarrow\uparrow \times \uparrow\downarrow\uparrow$.. The composite nature of a magnon excitation can then be readily seen by starting from a ground state and applying a single spin-flip on it to obtain the configuration: $\uparrow\downarrow\uparrow\downarrow\downarrow\downarrow\uparrow\downarrow\uparrow\downarrow$. Applying the exchange $J_\perp$ interaction of the model (6.7) on this state, this magnon excitation with $\Delta S^z = -1$ is in fact made up of two kinks of the Néel order, i.e. two spinons: $\uparrow\downarrow \times \downarrow\uparrow\downarrow\uparrow\downarrow \times \downarrow\uparrow\downarrow$. At the transition $J_z = J_\perp$ (isotropic limit), the ground state is disordered by "condensation" of these kinks and the low-lying excitations carrying $\Delta S^z = \pm 1/2$ are transmuted from gapful kinks to gapless spinons. Magnons are still built as pairs of these spinons and have a gapless energy spectrum.

This remarkable production of two particles from a single spin-flip has been explored in inelastic neutron scattering measurements on quasi-1D compounds that are good experimental realizations of the model (6.2). The first clear evidence of spinons came from neutron experiments[30] on $KCuF_3$ that found a continuum of magnetic excitations consistent with that expected from unbound spinons pair. The spinon continuum has also been observed in several other realizations of the spin-1/2 AF Heisenberg chain as the copper benzoate $Cu(C_6D_5COO)_2.3D_2O$,[31] $BaCu_2Si_2O_7$,[32] and $Cu(C_4H_4N_2)(NO_3)_2$.[33]

6.2.2. *Haldane's conjecture*

After the elucidation of the elementary excitations of the spin-1/2 AF Heisenberg chain by Faddeev and Takhtajan[22] in the early eighties, a central question of 1D quantum magnetism concerns the stability of the spinons. What is the fate of these exotic excitations upon switching on small deviations from the model (6.2)? Do these additional terms lead to a confinement of the spinons or do they still remain deconfined?

A simple modification of the spin-1/2 AF Heisenberg chain (6.2) is not to add perturbations but to change the Hilbert space, namely to consider the general spin-S case. The resulting model is no longer integrable, and one has to resort to approximate methods to describe its physical properties.

This question has lead to one of the most profound result in the field of 1D quantum magnetism: the difference between integer and half-integer spins AF Heisenberg chains predicted by Haldane.[34] Integer spin chains are incompressible spin liquids with a finite gap (the so-called Haldane gap) in the energy spectrum. In that case, the spinons are confined and the elementary excitations of the model are massive (optical) $S = 1$ magnons. In contrast, for half-integer spins, the model displays quantum criticality with similar universal properties as for the $S = 1/2$ case that are described by the $su(2)_1$ WZNW universality class. The elementary excitations are still, in this case, gapless spinons.

This remarkable distinction between integer and half-integer spins, historically called Haldane's conjecture, has been obtained by means of a semi-classical analysis of the model with a special emphasis on the topological nature of the order parameter fluctuations. For a review of this approach, the reader may consult the reviews[24,35] and the books.[14,36,37] In the large S limit, the Euclidean action that describes the low-energy properties of the spin-S AF Heisenberg chain reads as follows:

$$\mathcal{S}_{\text{eff}} = \frac{v}{2g} \int d\tau dx \left[\frac{1}{v^2} \left(\partial_\tau \vec{n}\right)^2 + \left(\partial_x \vec{n}\right)^2 \right] + i\, 2\pi S \mathcal{Q}\left(\vec{n}\right), \tag{6.8}$$

where $v = 2JS$, $g = 2/S$, and $\vec{n}$ is the order parameter of the Néel collinear state. The second contribution in Eq. (6.8) is a topological term since $\mathcal{Q}\left(\vec{n}\right) = \int d\tau dx\ \vec{n} \cdot (\partial_\tau \vec{n} \wedge \partial_x \vec{n})/4\pi$ is an integer, called the Pontryagin index, which measures the number of times the spin configuration $\vec{n}(x,\tau)$ covers the surface of the unit sphere S^2. In more mathematical terms, the configurations $\vec{n}(x,\tau)$, with fixed boundary conditions at infinity, are mappings of the sphere S^2 onto S^2 with homotopy classes classified by an integer $\Pi_2(S^2) = Z$, which is nothing but the Pontryagin index Q. For integer spin chains, the term $2\pi S \mathcal{Q}\left(\vec{n}\right)$ in Eq. (6.8) has no effect on the path integral of the model and can be discarded. The effective action (6.8) then reduces to the one of the 2D O(3) nonlinear sigma model which is a massive integrable field theory.[38,39] The exact low-energy spectrum of this field theory consists of a massive bosonic triplet with mass $m \sim e^{-2\pi/g}$ and it exhibits no bound states. The AF Heisenberg chain with integer spin is thus expected to have low-lying triplet excitations with a gap that scales as $\Delta \sim v e^{-\pi S}$ when S is large.

In the case of half-integer spins, the topological term in Eq. (6.8) manifests itself in the path integral through a phase factor $(-1)^{\mathcal{Q}}$ which gives rise to quantum interference between topologically distinct paths in space–time of the order parameter field $\vec{n}(x,\tau)$. It turns out that this process protects the model from a dynamically generated mass gap by quantum fluctuations and, in contrast, it leads to a non-perturbative massless flow toward an IR conformally invariant fixed point[40–42] that belongs to the $su(2)_1$ WZNW universality class. Since this fixed point also describes the universal properties of the spin-1/2 AF Heisenberg chain, all half-integer spin chains should exhibit the same type of emerging quantum criticality with for instance spin–spin correlations that decay according to the power law behavior (6.5).

6.2.3. *Haldane spin liquid: Spin-1 Heisenberg chain*

The simplest incompressible 1D spin liquid phase corresponds to the spin-1 AF Heisenberg chain. The existence of a Haldane gap in this model has generated an intense activity over the years after its prediction. It has been confirmed numerically from exact diagonalizations on finite samples,[43] quantum Monte Carlo (MC) methods,[44] transfer matrix computations,[45] and finally from DMRG calculations.[11,46] In particular, this latter technique predicts a gap $\Delta = 0.41050(2)J$ and a correlation length $\xi \simeq 6.03(1)$ lattice spacings. From the experimental point of view, several Haldane compounds have been synthesized over the years (see for instance Ref. [12] for a recent review). The two most studied compounds are $CsNiCl_3$ (see Ref. [2]) [47] and $Ni(C_2H_8N_2)_2NO_2(ClO_4)$ (NENP),[48] where spin-1 local moments are provided by the Ni^{2+} ions. In particular, inelastic neutron scattering experiments on NENP[48,49] confirm the existence of the Haldane gap. This energy gap has also been observed in several other quasi-1D spin-1 materials as $(CH_3)_4NNi(NO_2)_3$,[50] $AgVP_2S_6$,[51] and Y_2BaNiO_5,[52,53] which is probably the best realization of the spin-1 AF Heisenberg chain. Finally, it is worth noting that the recently identified compounds $Ni(C_5D_{14}N_2)_2N_3(PF_6)$(NDMAP)[54] and $Ni(C_5H_{14}N_2)_2N_3(ClO_4)$ (NDMAZ)[55] enable to investigate experimentally the high-magnetic field properties of the spin-1 Heisenberg chain.[56] From the theoretical point of view, several methods have been introduced to shed light on the nature of the mechanism of the Haldane-gap phenomena and to determine the main characteristics of this incompressible spin liquid phase. These approaches are the nonlinear sigma model field theory obtained in the large spin limit,[34,35] the valence-bond state (VBS) description,[57] the Majorana fermions method,[58] and the restricted Hilbert space approach.[59,60]

Probably, the simplest and most appealing approach to study the physical properties of the spin-1 AF Heisenberg chain consists of adding a biquadratic interaction to the model (6.2)

$$\mathcal{H}_\beta = J \sum_i [\vec{S}_i \cdot \vec{S}_{i+1} + \beta(\vec{S}_i \cdot \vec{S}_{i+1})^2]. \tag{6.9}$$

One of the main interest of this extended Heisenberg model is that for $\beta = 1/3$, the so-called AKLT point,[57] the model has an exactly solvable ground state which captures the main characteristics of the Haldane spin liquid phase. At this special point, the Hamiltonian (6.9) is equivalent to the sum of projection operators $P_2(i, i+1)$ that project onto the spin-2 contribution for every pair of nearest-neighbor (NN) spins:

$$\mathcal{H}_{AKLT} = \mathcal{H}_{\beta=1/3} = 2J \sum_i \left[P_2\,(i, i+1) - \frac{1}{3} \right]. \tag{6.10}$$

From this structure, the GS of this model can be constructed exactly using NN valence bonds. Following Affleck and coworkers pioneer work,[57] each original $S = 1$ spin is written as two spin-1/2 variables in a triplet state. The GS is then obtained by coupling into a singlet state all NN spin-1/2, thus forming a crystalline pattern of valence bonds (see Fig. 6.2). This state is called the valence-bond-solid (VBS) state. For periodic BC and an even number of sites, this VBS state is the unique GS and does not break the translation symmetry of the Hamiltonian (6.10). It can also be shown rigorously[57] that the AKLT model has a spectral gap whose value has been computed numerically: $\Delta \simeq 0.7J$.[61] In addition, the authors of Ref. [57] have also been able to show that the spin–spin correlation function decays exponentially with the distance with a finite correlation length equal to $\xi = 1/\ln 3$ lattice spacings. The VBS state is therefore the simplest incompressible spin liquid phase with a unique GS and a gap to all excitations.

Despite the exponential decay of correlations in this spin liquid phase, there is a subtle form of hidden AF ordering. This hidden order was first

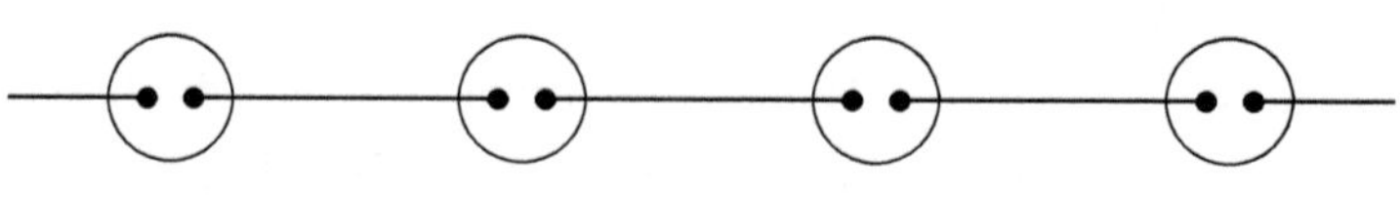

Fig. 6.2. VBS state in a chain with periodic boundary conditions.

discovered by den Nijs and Rommelse[62] using an equivalence of the spin-1 chain to a 2D restricted solid-on-solid model and also later by Tasaki[63] by means of a geometrical approach. A simple way to exhibit this hidden AF ordering is to look at the VBS state written in the conventional S^z representation. A typical configuration has the following structure: ..0 ↑ 0..0 ↓↑↓↑ 0 ↓ 0..0 ↑ 0.., i.e. each ↑ ($S_n^z = +1$ state) is followed by ↓ ($S_n^z = -1$ state) with an arbitrary number of 0 states ($S_n^z = 0$ state) between and vice versa. If we ignore the spins with $S_n^z = 0$ state, then the remaining spins display a perfect long-range spin-1/2 Néel ordering: ↑↓↑↓↑↓ ... The VBS state thus has a perfect dilute AF Néel order. Because of the arbitrary number of $S_n^z = 0$ sites inserted between $S_p^z = \pm 1$ sites, the long-range AF order is invisible in the spin–spin correlation function which has an exponential decay. However, this hidden order becomes manifest in the non-local string order parameter introduced by den Nijs and Rommelse[62] and defined by

$$\mathcal{O}^{\alpha}_{\text{string}} = - \lim_{|i-j| \to +\infty} \left\langle S_i^{\alpha} \exp \left(i\pi \sum_{k=j+1}^{i-1} S_k^{\alpha} \right) S_j^{\alpha} \right\rangle , \tag{6.11}$$

with $\alpha = x, y, z$. In the VBS GS, this order parameter can be computed exactly[62]: $\mathcal{O}^{\alpha}_{\text{string}} = 4/9$, i.e. this object exhibits long-range order in the VBS phase. The existence of such a hidden order is essential to the basic mechanism of the Haldane gap: breaking this topological order to create an excitation costs a finite energy gap. The properties of this dilute AF order was further elucidated by Kennedy and Tasaki.[64] They were able to show, using a non-local unitary transformation, that the long-range order $\mathcal{O}^{\alpha}_{\text{string}} \neq 0$ and the Haldane gap are related to a spontaneous breaking of a hidden $Z_2 \times Z_2$ symmetry. In particular, the string order parameter becomes the usual local ferromagnetic order parameter under this non-local unitary transformation. Another important consequence of this broken $Z_2 \times Z_2$ symmetry is the existence of a quasi-degeneracy of the four lowest energy levels for a finite-size chain. This can also be understood within the VBS state since for open BC the AKLT model (6.10) has exactly four GSs with the existence of two spin-1/2 degrees of freedom that are unpaired at each end of the chain (see Fig. 6.3). Three of these states constitute a spin triplet, whereas the last one is a spin singlet.

Beside the fact that only the ground state of the Hamiltonian (6.10) is known exactly, the AKLT approach is also very useful to describe low-lying

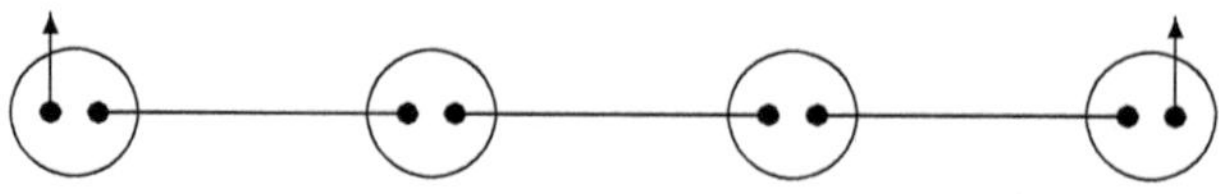

Fig. 6.3. VBS state with open boundary conditions; unpaired bonds are left at the boundaries resulting into two free spin-1/2 objects at the edges.

excitations of the model through a variational approach starting from the VBS state. A first approach[65] consists of creating a "magnon" excitation by applying a spin-flip on the VBS GS ($|\mathrm{VBS}\rangle$):

$$|k\rangle = \frac{1}{\sqrt{N}} \sum_{j=1}^{N} \exp\left(ikj\right) S_j^+ |\mathrm{VBS}\rangle, \tag{6.12}$$

N being the system size. It is very tempting to interpret this excitation as a standard $S = 1$ magnon, i.e. an excitation which modifies only locally the AF Néel order. In fact, the trial wave function (6.12) does not describe a magnon excitation of the perfect hidden Néel order of the VBS state, but it has instead a solitonic nature.[61] Indeed, in the S^z representation, a typical configuration that appears in $S_j^+|\mathrm{VBS}\rangle$ has the following structure: ..0 $\uparrow$ 0..0 $\downarrow\uparrow$ 0 $\uparrow$ 0 $\downarrow$ 0..0 $\uparrow$ 0.. Removing the 0 states, this configuration becomes $\uparrow\downarrow\uparrow\uparrow\downarrow\uparrow \ldots$, i.e. a soliton or domain wall which non-locally destroys the AF Néel long-range order. An alternative approach to describe the elementary excitations of the AKLT (6.10) model is to start from the VBS state and to promote the spins of the link $(j, j+1)$ into a triplet state $|\Phi_j^a\rangle$.[61,66,67] This low-lying excitation carrying momenta k describes a moving triplet bond, usually called crackion, and it is defined by

$$|\Phi^a\left(k\right)\rangle = \frac{1}{\sqrt{N}} \sum_{j=1}^{N} \exp\left(ikj\right) |\Phi_j^a\rangle. \tag{6.13}$$

In fact, the two approaches (6.12) and (6.13) are related and describe the same dispersion relation for the triplet excitation.[61] The solitonic nature of the excitation (6.13) becomes manifest by considering open boundary conditions and the non-local unitary transformation introduced by Kennedy and Tasaki.[64] In that case, the trial excited wave function corresponds to a domain wall that interpolates between two of the four possible GSs.[61]

One of the main success of this VBS approach stems from the fact that the properties of the VBS state are not special to the AKLT point (6.10) but have a larger extent. This model turns out to be smoothly connected to the spin-1 Heisenberg chain in the sense that they share the same physical properties. In this respect, numerical investigations[68–74] of the phase diagram of the bilinear–biquadratic spin-1 Hamiltonian (6.9) have revealed that this model with $|\beta| < 1$ belongs to the same phase, the Haldane phase, whose main characteristics are well described by the VBS state. In particular, the origin of the Haldane gap in the spin-1 Heisenberg chain is a consequence of a hidden $Z_2 \times Z_2$ broken symmetry as for the AKLT model (6.10). At the Heisenberg point $\beta = 0$, the hidden AF Néel ordering is no longer perfect as at the AKLT point. It is only locally destroyed so that there is still long-range order at $\beta = 0$ with a non-zero string order parameter (6.11) but smaller than at the AKLT point: $\mathcal{O}^{\alpha}_{\text{string}} \simeq 0.374$.[46,70,71] A second consequence of the spontaneous breaking of the $Z_2 \times Z_2$ symmetry is the two almost free $S = 1/2$ degrees of freedom at the ends of a finite chain with open BC. At the Heisenberg point, exact diagonalizations[68] of finite open samples with an even number of sites have shown that the ground state is a singlet and the existence of an exponentially low-lying triplet state (the so-called Kennedy triplet) in the Haldane gap. This leads to a four-fold GS degeneracy in the thermodynamic limit which can be simply understood by means of the VBS description of Fig. 6.3 as described above. The existence of these two $S = 1/2$ edge states is a robust and striking property of the Haldane spin liquid phase, and they are present in the whole region $|\beta| < 1$ of the bilinear–biquadratic spin-1 model.[73,74] The physical properties of these $S = 1/2$ chain-boundary excitations in the open spin-1 chain have been investigated in detail numerically[46,73–80] and also analytically.[81–83] Remarkably enough, this liberation of fractional spin-1/2 degrees of freedom in a spin-1 Heisenberg chain has been observed experimentally in the spin-1 compound NENP cut by non-magnetic impurities (Zn^{2+} or Mg^{2+})[84,85] and also doped with magnetic ions (Cu^{2+}).[86] In particular, these spin-1/2 chain-boundary excitations reveal themselves as satellite peaks in the NMR profile of the Mg-doped Y_2BaNiO_5 (see Ref. [87]). A very recent inelastic neutron scattering experiments[88] further probed the microscopic structure of these edge states through the wave vector dependence of the Zeeman resonance.

6.2.4. *General spin-S case*

For higher spin-S value ($S > 1$), the difference between integer-spin and half-integer AF Heisenberg chains has been explained within the semiclassical analysis pioneered by Haldane[34,35] and also by means of the Abelian[89] and non-Abelian[40,90] bosonization approaches. The Haldane's conjecture for $S > 1$ has also been investigated numerically and experimentally over the years.

The numerical studies[91–94] of the $S = 3/2$ AF Heisenberg chain report that the model displays a critical behavior that belongs to the same universality class as the $S = 1/2$ Heisenberg chain. Several quasi-1D AF compounds with spin-3/2 have been found like $CsVBr_3$, $CsVCl_3$,[95,96] and $AgCrP_2S_6$.[97] In particular, the uniform susceptibility of the spin-3/2 $AgCrP_2S_6$ material measured experimentally[97] is consistent with the result for spin-1/2, i.e. confirming the universal behavior of half-integer spins. In the $S = 2$ case, the existence of a Haldane gap has been established in various quantum MC approaches[98] and in DMRG calculations.[99–101] In the thermodynamic limit, a finite gap of $\Delta = 0.085(5)J$ and a finite correlation length $\xi = 49(1)$ lattice spacings have been found in DMRG.[100] Several Haldane spin-2 compounds have been discovered over the years (see Ref. [12] for a review) like $CsCrCl_3$ and (2,2′-bipyridine)trichloromanganase(III) ($MnCl_3$(bipy)).[102] Experiments[102] on this last compound report the existence of a gap $\Delta/J = 0.07 \pm 0.02$, which is in good agreement with the numerical results.

Integer $S > 1$ spin chains share similar properties as the $S = 1$ case. In particular, there is still a hidden AF order in the integer $S > 1$ case. A non-local string order parameter that measures this topological order has been proposed by Oshikawa[103]:

$$\mathcal{O}^{\alpha}_{\text{string}}\left(\frac{\pi}{S}\right) = -\lim_{|i-j|\to+\infty} \left\langle S^{\alpha}_i \exp\left(i\pi/S \sum_{k=j+1}^{i-1} S^{\alpha}_k\right) S^{\alpha}_j \right\rangle, \tag{6.14}$$

where $\alpha = x, y, z$. This order parameter can be exactly computed in spin-S generalization of the VBS state[103,104] and numerically for the Heisenberg chain.[99,100,105] For instance, in the spin-2 case, DMRG calculations[100] found $\mathcal{O}^{\alpha}_{\text{string}}(\pi/2) = 0.726(2)$. However, the full characterization of the hidden AF order for integer spin $S > 1$ is less clear than that for $S = 1$ where, as seen above, it is related to a broken $Z_2 \times Z_2$ symmetry. In fact, the nature of the hidden symmetry breaking is still an open problem for $S > 1$. Insight on this issue might be gained by studying the structure of edge states of open integer spin-S chains. Higher-S VBS states with open BC have $(S+1)^2$-fold degenerate GS with the presence of two free $S/2$ degrees of freedom at each end of the chain in close parallel to the $S = 1$

case (see Fig. 6.3). A nonlinear sigma model approach[106] of the open spin-S Heisenberg chain reveals the existence of spin-$S/2$ edge states in the integer spin case. These results lead us to expect that $Z_{2S} \times Z_{2S}$ might be a good candidate to describe the hidden symmetry breaking scheme in the integer spin-S case. In the $S = 2$ case, the approximate nine-fold degeneracy of the GS of finite open chains have been observed numerically[77,99,100] by DMRG. The presence of these spin-1 chain-end excitations has also been confirmed experimentally in the ESR study of the spin-2 Haldane compound $CsCrCl_3$ doped with non-magnetic Mg^{2+} ions.[107]

The liberation of these fractional spin-$S/2$ degrees of freedom when non-magnetic impurities are introduced in integer spin-S AF Heisenberg chain is a remarkable fact. It stems from the presence of a non-trivial topological ordering in the GS of the spin-S chain. In this respect, the existence of these edge states opens the possibility to make a distinction between spin-S Haldane liquid phases. These massive spin liquid phases share the same thermodynamics properties but have, for instance, different chain-end ESR response due to the special nature of their edge states. In fact, this topological distinction between gapped spin liquid phases can also be applied to the massless case. All half-integer AF Heisenberg spin chains have the same bulk universal properties described by an IR fixed point which belongs to the $su(2)_1$ WZNW universality class. However, the GS of half-integer spin chains still sustains a topological ordering. Indeed, by means of a nonlinear sigma model approach, Ng (see Ref. 106) has given evidence that edge states with fractionalized quantum number $(S - 1/2)/2$ exist in open half-integer spin-S AF Heisenberg chain. This interesting result has been confirmed numerically by examining open spin chains with spins up to $S = 9/2$ by DMRG.[77,108] Recently, a numerical investigation[109] of the non-local string order parameter (6.14) directly reveals the presence of a topological order in half-integer spin-S AF Heisenberg chains. The existence of this hidden topological order (or the related chain-end excitations) leads to a topological distinction between gapless spin liquid phases with the same emerging quantum criticality.

6.2.5. *Two-leg spin ladder*

A further impetus for the study of low-dimensional spin systems was given by the discovery of spin ladder materials one decade ago.[110] These systems consist of a finite number (n_{leg}) of spin-1/2 AF Heisenberg chains coupled by a transverse exchange interaction $J_\perp$. The Hamiltonian of this model reads as follows:

$$\mathcal{H}_{n_{\text{leg}}} = J_{\parallel} \sum_i \sum_{a=1}^{n_{\text{leg}}} \vec{S}_{a,i} \cdot \vec{S}_{a,i+1} + J_{\perp} \sum_i \sum_{a=1}^{n_{\text{leg}}-1} \vec{S}_{a,i} \cdot \vec{S}_{a+1,i}, \tag{6.15}$$

where $\vec{S}_{a,i}$ is a spin-1/2 operator at site i on the ath chain and the in-chain exchange $J_{\parallel}$ is AF ($J_{\parallel} > 0$). One of the main interest of these spin ladders stems from the fact that they are intermediate objects between 1D and 2D systems. Moreover, these quasi-1D systems share many properties with the cuprates and, because of their relative simplicity, may provide some insights on mechanisms behind high-T_c superconductivity.

Despite this original motivation to study spin ladders, these systems display striking properties which make them interesting by themselves in the field of 1D quantum magnetism. In particular, in close parallel to the qualitative difference between integer and half-integer AF Heisenberg spin chains, the universal properties of spin ladders (6.15) with open boundary conditions in the transverse direction strongly depend on the parity of the number of legs n_{leg}.[111–115] Ladders with even number of legs are spin liquids with a finite gap in the excitation spectrum and exponentially decaying spin–spin correlations. The original massless spinons of the spin-1/2 chains get confined by the interchain coupling to form coherent optical spin-1 magnon excitations. Upon doping, these ladders display quasi-long-range superconducting pairing correlations with an approximate d-wave symmetry.[110] In contrast, odd-legged ladders have a gapless spectrum which is characterized by a central charge $c = 1$ corresponding to an effective $S = 1/2$ AF Heisenberg spin chain. Upon doping, odd-legged ladders exhibit a metallic behavior typical of a Luttinger liquid.[116] This difference on the energy spectrum of the spin ladders depending on the parity of n_{leg} is strongly reminiscent of the Haldane conjecture for AF Heisenberg spin chains with $S = n_{\text{leg}}/2$. It can be simply understood from the fact that in the strong ferromagnetic rung coupling limit, the spin ladder (6.15) is indeed equivalent to a single $S = n_{\text{leg}}/2$ AF Heisenberg chain.

Several spin ladder materials have been discovered over the years. The family of compounds $Sr_{n-1}Cu_{n+1}O_{2n}$ approximately realize ladders with $n_{\text{leg}} = (n+1)/2, n = 3, 5, 7, \ldots$ legs.[111] The even–odd scenario in ladders was confirmed experimentally from susceptibility and muons resonance measurements on this compound with $n = 3$ and $n = 5$.[117,118] The existence of a spin gap in the two-leg case ($n_{\text{leg}} = 2$) was found in experiments on the two-leg compounds $A_{14}Cu_{24}O_{41}$ ($A_{14} \equiv La_6Ca_8$, $(Sr,La,Ca)_{14}$),[119]

$Cu_2(C_5H_{12}N_2)_2Cl_4$,[120, 121] and $(C_5H_{12}N)_2CuBr_4$ (see Ref. [122]) and the organic compound BIP-BNO.[123] However, it is worth noting that additional interactions are required to fully describe these materials such as, for instance, a non-negligible ring exchange for the compound $A_{14}Cu_{24}O_{41}$ (see Refs. [124] and [125]) or a small diagonal interchain exchange interaction for $Cu_2(C_5H_{12}N_2)_2Cl_4$.[120, 121]

The existence of a spin gap in the two-leg case can be anticipated by considering a strong-coupling analysis of the model (6.15) with $n_{\rm leg} = 2$. In the limit of strong AF interchain coupling $J_\perp \gg J_\parallel$, the ground state consists of singlet bonds formed across the rungs of the ladder, with triplet excitations separated by a large energy gap of order $J_\perp$. Perturbations in small $J_\parallel$ will cause these singlet bonds to resonate and the triplet excitations form a band with bandwidth $\sim J_\parallel$ but the spectral gap survives.[126] In the strong ferromagnetic interchain $-J_\perp \gg J_\parallel$ coupling case, local spins $S = 1$ associated with each rung of the ladder are formed leading thus to a spin-1 AF Heisenberg chain with a non-zero Haldane gap in the energy spectrum. The cross-over between strong and weak coupling limits has been carefully analyzed in numerical calculations.[127–130] It turns out that the GS in the two strong-coupling limits $|J_\perp|/J_\parallel \gg 1$ evolves adiabatically with increasing $J_\parallel$. The spin gap survives for arbitrarily large $J_\parallel/|J_\perp|$ and finally vanishes for $J_\perp = 0$, i.e. when the two spin-1/2 AF Heisenberg chains are decoupled. The critical point $J_\perp = 0$ separates thus two strong-coupling massive phases: a rung-singlet phase for $J_\perp > 0$ and a phase with $J_\perp < 0$ which is smoothly connected to the Haldane phase of the spin-1 chain.

The opening of the spin gap upon switching on the interchain coupling can be investigated by means of the bosonization approach.[89, 131, 132] In particular, the field theory that accounts for the massless spinon–optical magnon transmutation, when $J_\perp$ is small, corresponds to an SO(3) $\times$ Z_2 symmetric model of four massive Majorana (real) fermions,[132, 133] i.e. four off-critical 2D Ising models (for a review, see for instance, Ref. [10]). The resulting low-energy field theory is described by the following Hamiltonian density[132, 133]:

$$\begin{aligned}\mathcal{H}_{n_{\rm leg}=2} &= -\frac{iv_t}{2}\left(\vec{\xi}_R \cdot \partial_x \vec{\xi}_R - \vec{\xi}_L \cdot \partial_x \vec{\xi}_L\right) - im_t\, \vec{\xi}_R \cdot \vec{\xi}_L \\ &\quad - \frac{iv_s}{2}\left(\xi_R^0 \partial_x \xi_R^0 - \xi_L^0 \partial_x \xi_L^0\right) - im_s \xi_R^0 \xi_L^0 + \mathcal{H}_{\rm marg},\end{aligned} \tag{6.16}$$

where $\vec{\xi}_{R,L}$ are a triplet of right- and left-moving Majorana fermions that describe the $S = 1$ low-lying excitations of the two-leg spin ladder and the

Majorana fermion $\xi^0_{R,L}$ accounts for the singlet excitation. The masses of these real fermions are given by the weak coupling limit: $m_t = J_\perp \lambda^2/2\pi$ and $m_s = -3J_\perp\lambda^2/2\pi$, λ being a non-universal constant. These triplet and singlet massive Majorana fermions are weakly coupled by a marginal perturbation associated to the last term of Eq. (6.16):

$$\mathcal{H}_{\text{marg}} = \frac{g_1}{2}(\vec{\xi}_R \cdot \vec{\xi}_L)^2 + g_2\, \vec{\xi}_R \cdot \vec{\xi}_L\, \xi^0_R \xi^0_L, \tag{6.17}$$

with $g_1 = -g_2 = \pi a_0 J_\perp/2$, a_0 being the lattice spacing. The model (6.16), derived in the weak coupling limit $|J_\perp| \ll J_\parallel$, is expected to have a larger extent and to capture in fact the low-energy properties of the two-leg spin ladder for arbitrary $J_\perp$ with a suitable redefinition of the masses $m_{s,t}$, velocities $v_{s,t}$, and coupling constants $g_{1,2}$. This stems from the continuity between the weak- and strong-coupling limits in the two-leg spin ladder observed numerically.[127–130] In addition, in the strong ferromagnetic rung limit $-J_\perp \gg J_\parallel$, the singlet excitations described by the Majorana fermion $\xi^0_{R,L}$ are frozen ($|m_s| \to \infty$) so that the low-energy properties of the model (6.16) are governed by the triplet magnetic excitations corresponding to the fermions $\vec{\xi}_{R,L}$. The resulting model coincides with the low-energy field theory of the spin-1 AF Heisenberg chain obtained by Tsvelik[58] by perturbing around the Babujian–Takhtajan[134] integrable point of the bilinear–biquadratic spin-1 chain (6.9) with $\beta = -1$. The Hamiltonian (6.16) correctly thus captures the low-energy properties of the two-leg spin ladder in the limit $-J_\perp \gg J_\parallel$ where it reduces to the spin-1 AF Heisenberg chain.

Despite its apparent simplicity, the model (6.16) is not an integrable field theory and takes the form of a massive SO(3) $\times$ Z_2 Gross–Neveu model.[135] However, the leading behavior of the physical quantities of the two-leg spin ladder can be determined by treating the marginal contribution (6.17) perturbatively. As shown in Ref. [132], this term leads to a renormalization of the masses and velocities so that the low-energy description (6.16) simply reduces to four independent massive Majorana fermions, i.e. four decoupled off-critical 2D Ising models. This mapping onto off-critical 2D Ising models can then be exploited to derive the low-energy properties of the two-leg spin ladder such as, for instance, the spectrum of elementary excitations, low-T thermodynamics and the leading asymptotics of spin–spin correlation functions. To this end, one needs to express the spin operators $\vec{S}_{a,i}$ in terms of the Ising fields. In the continuum limit, the spin densities separate into the smooth and staggered parts[40]: $\vec{S}_a(x) = \vec{J}_{aL}(x) + \vec{J}_{aR}(x) + (-1)^{x/a_0}\vec{n}_a(x)$. The chiral $su(2)_1$ currents, uniform parts of the spin densities, can be written locally in terms of the Majorana fermions[132, 136]:

$$\vec{J}_{aR,L} = -\frac{i}{4}\,\vec{\xi}_{R,L} \wedge \vec{\xi}_{R,L} + i\frac{\tau_a}{2}\,\vec{\xi}_{R,L}\xi^0_{R,L}, \tag{6.18}$$

with $\tau_1 = +1$ and $\tau_2 = -1$. In contrast, the staggered magnetizations $\vec{n}_a$ are non-local in terms of the underlying fermions and expressed in terms of the order (σ_a) and disorder (μ_a) of the Ising models[58, 132, 136]:

$$\begin{aligned} \vec{n}_+ &\sim (\mu_0\mu_1\sigma_2\sigma_3, \mu_0\sigma_1\mu_2\sigma_3, \mu_0\sigma_1\sigma_2\mu_3), \\ \vec{n}_- &\sim (\sigma_0\sigma_1\mu_2\mu_3, \sigma_0\mu_1\sigma_2\mu_3, \sigma_0\mu_1\mu_2\sigma_3), \end{aligned} \tag{6.19}$$

where $\vec{n}_\pm = \vec{n}_1 \pm \vec{n}_2$. Due to the existence of non-zero masses $m_{s,t}$ for the fermions in Eq. (6.16), all related Ising models are non-critical. A spectral gap is thus present in the two-leg spin ladder for all signs of the interchain interaction $J_\perp$ and it opens linearly with $J_\perp$ in the weak coupling limit. In addition, the Ising description (6.18) and (6.19) of the spin densities allows the calculation of the leading asymptotics of spin–spin correlations using exact results of the two-point function of a non-critical Ising model. Since the signs of the triplet and singlet masses are always opposite in Eq. (6.16), it can be shown[132] that the dynamical spin susceptibility displays a sharp single-magnon peak near $q = \pi/a_0$ and $\omega = |m_t|$, which reflects the elementary nature of the triplet excitations for all signs of $J_\perp$.

In this respect, the spin liquid phase of the two-leg spin ladder is very similar to the Haldane phase of the spin-1 chain. However, it has been stressed recently that the phase with $J_\perp < 0$, smoothly connected to the Haldane phase of the spin-1 chain, and the rung-singlet phase for $J_\perp > 0$ are, in fact, topologically distinct gapped phases.[137] The distinction is intimately related to the short-range valence bond structure of their ground states. In close parallel to the classification of short-range valence bond configurations on a 2D square lattice,[138] two different topological classes can be defined in the 1D case by counting the number Q_y of valence bonds crossing an arbitrary vertical line.[137] In the case of the rung-singlet phase, Q_y is always even while it is odd for the Haldane phase with $J_\perp < 0$. Two different non-local string order parameters can then be defined in connection to this topological distinction[129, 130, 137]:

$$\begin{aligned}\mathcal{O}^{\alpha}_{\text{even}} &= -\lim_{|i-j|\to+\infty}\langle (S^{\alpha}_{1,i+1}+S^{\alpha}_{2,i})e^{i\pi\sum_{k=j+1}^{i-1}(S^{\alpha}_{1,k+1}+S^{\alpha}_{2,k})} \\ &\quad\times (S^{\alpha}_{1,j+1}+S^{\alpha}_{2,j})\rangle, \\ \mathcal{O}^{\alpha}_{\text{odd}} &= -\lim_{|i-j|\to+\infty}\langle (S^{\alpha}_{1,i}+S^{\alpha}_{2,i})e^{i\pi\sum_{k=j+1}^{i-1}(S^{\alpha}_{1,k}+S^{\alpha}_{2,k})} \\ &\quad\times (S^{\alpha}_{1,j}+S^{\alpha}_{2,j})\rangle,\end{aligned} \tag{6.20}$$

with $\alpha = x, y, z$. The Haldane and rung-singlet phases are then characterized by $\mathcal{O}^{\alpha}_{\text{odd}} \neq 0$, $\mathcal{O}^{\alpha}_{\text{even}} = 0$, and by $\mathcal{O}^{\alpha}_{\text{even}} \neq 0$, $\mathcal{O}^{\alpha}_{\text{odd}} = 0$, respectively. The order parameters (6.20) also reveal the difference nature of the hidden AF Néel order in the Haldane phase (non-zero triplet states along the rung) and in the rung-singlet phase (non-zero triplet states along the diagonal).[129, 130, 137] This topological difference can be discussed in light of the Ising description (6.19) of the two-leg spin ladder. The string order parameters (6.20) can be expressed in terms of the order and disorder Ising operators:[139] $\mathcal{O}^{z}_{\text{odd}} \sim \langle\sigma_1\rangle^2\langle\sigma_2\rangle^2$ and $\mathcal{O}^{z}_{\text{even}} \sim \langle\mu_1\rangle^2\langle\mu_2\rangle^2$. For a ferromagnetic (respectively, AF) interchain coupling, the Ising models in the triplet sector are in their ordered (respectively, disordered) phases so that $\mathcal{O}^{z}_{\text{odd}} \neq 0$ and $\mathcal{O}^{z}_{\text{even}} = 0$ (respectively, $\mathcal{O}^{z}_{\text{odd}} = 0$ and $\mathcal{O}^{z}_{\text{even}} \neq 0$). In this Ising description, the phases with $J_\perp > 0$ and $J_\perp < 0$ are thus simply related by a Kramers–Wannier duality transformation on the underlying Ising models.

The topological distinction between these two phases becomes manifest when analyzing the ground-state degeneracy depending on the nature of BC used. In the open BC case, as noted by the authors of Ref. [137], GSs of gapped spin liquid states characterized by an odd value of Q_y have spin-1/2 edge states, while these end states disappear when Q_y is even. The existence of these $S = 1/2$ chain-end degrees of freedom leads to a GS degeneracy in a two-leg spin ladder with open BC. In the Haldane phase of the ladder with $J_\perp < 0$, finite open chains with an even number of sites have a singlet GS with an exponentially low-lying triplet in the spin gap resulting on a fourfold GS degeneracy in the thermodynamic limit. In contrast, the GS in the rung-singlet phase is always unique irrespective of whether open or periodic BC are used: no low-lying triplet states are found inside the spin gap in open ladder with $J_\perp > 0$ and an even number of sites. In this respect, the rung-singlet phase with $J_\perp > 0$, in contrast to the $J_\perp < 0$ case, is not equivalent to the Haldane phase characterized by

$S = 1/2$ chain-end excitations even though they share similar properties such as the presence of a spin gap and a non-zero string order parameter. The existence or absence of spin-1/2 edge states in the open two-leg spin ladder can also be discussed within the Ising model description (6.16).[83] In particular, the four-fold GS degeneracy of the open two-leg ladder with a ferromagnetic interchain coupling can be obtained within this approach. Indeed, for $J_\perp < 0$, the three Ising models for the triplet sector are all in their ordered phases while the Ising model for the singlet degrees of freedom belongs to its disorder phase so that $\langle \sigma_i \rangle \neq 0$ $(i = 1, 2, 3)$ and $\langle \sigma_0 \rangle = 0$. In that case, each Ising model in the triplet sector has a doubly degenerate GS which gives thus an eight-fold degeneracy. However, there is a redundancy in the Ising description since the triplet Hamiltonian in Eq. (6.16) and the total uniform and staggered magnetizations in Eqs. (6.18) and (6.19) are all invariant under the transformation: $\xi^i_{R,L} \rightarrow -\xi^i_{R,L}, \mu_i \rightarrow \mu_i$, and $\sigma_i \rightarrow -\sigma_i$, $i = 1, 2, 3$. This leads to a physical four-fold GS degeneracy as it should be in the Haldane phase. Another advantage of this Ising model description is to make explicit the spontaneous breaking of a hidden $Z_2 \times Z_2$ symmetry associated to the GS degeneracy for $J_\perp < 0$. The existence of this hidden symmetry can also be revealed with help of a non-local unitary transformation on the lattice spins of the two-leg ladder[140] as in the spin-1 AF Heisenberg chain.[64]

Finally, the Haldane and rung-singlet phases of the two-leg spin ladder can also be distinguished by their optical properties. In the rung-singlet phase, the existence of singlet or triplet two-magnon bound states below the two-magnon continuum has been predicted theoretically by a number of groups.[141–144] Recently, the singlet two-magnon bound state has been observed in the optical conductivity spectrum of the compound $(La,Ca)_{14}Cu_{24}O_{41}$,[145] which contains layers with Cu_2O_3 two-leg spin ladder with $J_\perp > 0$. On the contrary, it is expected that for a ferromagnetic interchain coupling no bound states are present below the two-particle continuum as in the spin-1 AF Heisenberg chain.[46] The existence of bound states in the two-leg spin ladder can be discussed in the context of the field theory approach (6.16). Indeed, the marginal term (6.17) plays its trick by giving rise to an effective interaction between the magnon excitations. In the AF interchain case, the contribution (6.17) is in fact marginally relevant and the interaction between magnons are attractive. The presence of bound states in this case can be argued qualitatively as a result of this attractive interaction. The Hamiltonian (6.17) takes the form of a SO(4) Gross–Neveu model up to irrelevant contributions and a duality transformation $\xi_R^0 \to -\xi_R^0$ in the singlet sector. The latter model is integrable[38] and, for $J_\perp > 0$, the low-lying excitations are massive kinks and anti-kinks interpolating between the two degenerate GS resulting from the spontaneous symmetry breaking of the Z_2 symmetry: $\xi_R^a \xi_L^a \to -\xi_R^a \xi_L^a$, $a = 0, \ldots, 3$. The mass terms in the full Hamiltonian (6.16) break explicitly this Z_2 symmetry and the GS degeneracy is lifted so that kink configurations are no longer asymptotic states of the field theory. The situation is in close parallel to the 2D Ising model in its low-temperature phase upon switching on a magnetic field. The mass terms are then expected to induce a linear confining potential between the kinks giving rise to a sequence of bound states. In contrast, when $J_\perp < 0$, the perturbation (6.17) is a marginally irrelevant contribution and the spectrum of the model (6.16) consists of massive fermions and their multiparticle excitations.

6.2.6. *Non-Haldane spin liquid*

The Haldane and rung-singlet phases of the two-leg spin ladder are not the only possible gapped spin-liquid states available in 1D unfrustrated quantum magnets. In this respect, Nersesyan and Tsvelik[133] have discussed the example of a gapped spin liquid phase without any coherent magnon excitations. The spectral function of this state displays a broad feature rather than a single sharp magnon peak. Such a spin liquid can be stabilized by the introduction of a four-spin interchain interaction that couples two spin-1/2 AF Heisenberg chains:

$$\mathcal{H}_{\text{so}} = J \sum_i \sum_{a=1}^{2} \vec{S}_{a,i} \cdot \vec{S}_{a,i+1} + K \sum_i (\vec{S}_{1,i} \cdot \vec{S}_{1,i+1})(\vec{S}_{2,i} \cdot \vec{S}_{2,i+1}). \quad (6.21)$$

The biquadratic interaction represents an interchain coupling for the spin-dimerization operators ($\epsilon_{a,i} \sim (-1)^i \vec{S}_{a,i+1} \cdot \vec{S}_{a,i}$) of each chain which can be effectively generated by spin–phonon interaction. A second motivation to investigate the effect of this four-spin interaction stems from orbital degeneracy. In most of transition metal compounds, in addition to the usual spin degeneracy, the low-lying electron states are also characterized by orbital degeneracy.[146] A starting point to study magnetic properties of magnetic insulators with Jahn–Teller ions is the two-band Hubbard-like models. At quarter-filling (one electron per atom), this system is a Mott insulator in the limit of strong Coulomb interaction and the state of each ion can be characterized by a spin degrees of freedom $\vec{S}_{1,i}$ and an orbital state described by a pseudo-spin-1/2 $\vec{S}_{2,i}$. In the large Coulomb repulsion limit, the simplest Hamiltonian that describes the competition between spin and orbital degrees of freedom in one dimension reduces to the spin–orbital model (6.21). In particular, this model has been introduced by Pati *et al.*[147] to explain the unusual magnetic properties of the quasi-1D spin gapped material $Na_2Ti_2Sb_2O$.[148]

The spin–orbital model (6.21) is unfrustrated for $K < 0$, whereas the frustration manifests itself in the AF case $K > 0$ only in the intermediate regime $K \simeq J$ as we shall see later. The Hamiltonian (6.21) is invariant under independent SU(2) rotations in the spin ($\vec{S}_1$) and orbital ($\vec{S}_2$) spaces. For generic couplings, the model (6.21) is thus SU(2) × SU(2) × Z_2 symmetric, the additional Ising symmetry being the exchange between the spins $\vec{S}_1$ and $\vec{S}_2$. In the weak coupling limit $|K| \ll J$, this underlying SO(4) symmetry of the spin–orbital model (6.21) is reflected in the form of its low-energy Hamiltonian density which is described in terms of four massive Majorana fermions[133, 149]:

$$\mathcal{H}_{\text{so}} \simeq -\frac{iv}{2}\sum_{a=0}^{3}\left(\xi_R^a \partial_x \xi_R^a - \xi_L^a \partial_x \xi_L^a\right) - im\sum_{a=0}^{3}\xi_R^a \xi_L^a, \tag{6.22}$$

with $m = \alpha K$, α being a positive non-universal constant. In Eq. (6.22), a marginal contribution, similar to Eq. (6.17), has been neglected. In contrast to the low-energy description (6.16) of the standard two-leg spin ladder, the triplet and singlet Majorana modes become equally important here as a consequence of the SO(4) symmetry of the spin-orbital model (6.21). The resulting low-energy properties of the model can then be determined by this mapping onto four off-critical 2D Ising models. As the staggered magnetizations (6.19), the spin dimerization fields $\epsilon_{1,2}$ can be expressed in terms of the order and disorder operators of the underlying Ising models:

$$\epsilon_a \sim \mu_1\mu_2\mu_3\mu_4 \pm \tau_a \sigma_1\sigma_2\sigma_3\sigma_4. \tag{6.23}$$

For an AF biquadratic coupling ($K > 0$), the four Ising models belong to their ordered phases so that the model (6.22) enters a spontaneously dimerized phase with a finite gap and $\langle\epsilon_1\rangle = -\langle\epsilon_2\rangle = \pm|\epsilon_0|$. The GS is thus two-fold degenerate and dimerizes with an alternating pattern as shown in Fig. 6.4(a). In contrast, in the ferromagnetic $K < 0$ case, the dimerization is now in-phase between the two chains: $\langle\epsilon_1\rangle = \langle\epsilon_2\rangle = \pm|\epsilon_0|$ as depicted by Fig. 6.4(b). A first distinction between this gapped spin liquid state and the Haldane and rung-singlet phases of the two-leg spin ladder stems from the

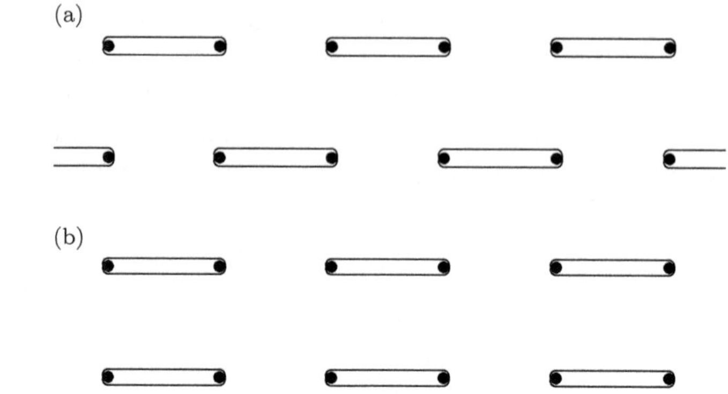

Fig. 6.4. GS of the spin–orbital model in the weak coupling limit $|K| \ll J$. The bonds indicate a singlet pairing between the spins. (a) Alternating dimerization for $K > 0$; (b) in-phase dimerization for $K < 0$. In each case, a second GS is obtained by applying the one-step translation symmetry on each chain to the states depicted here.

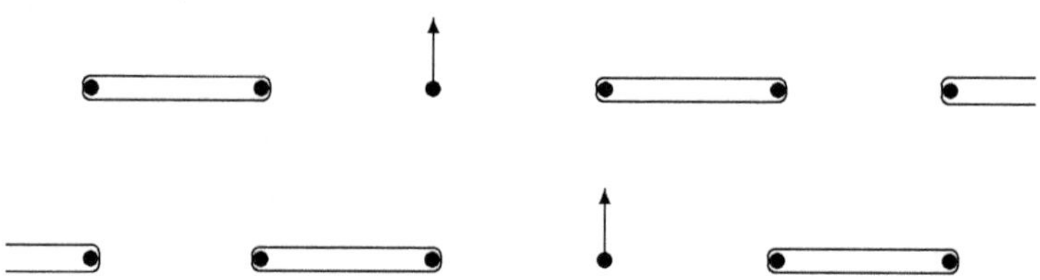

Fig. 6.5. Triplet excitation of the alternating dimerization described in terms of a pair of dimerization kinks of each chain.

GS degeneracy and the fact that the lattice translation symmetry is spontaneously broken in the GSs of Fig. 6.4. A more drastic difference appears at the level of the low-lying excitations. Indeed, the elementary excitations of the spin–orbital model are neither optical magnons nor massive spinons but a pair of propagating massive triplet or singlet kinks connecting two spontaneously dimerized GSs.[133] An example of such a triplet excitation is described in Fig. 6.5 for the staggered dimerization phase with $K > 0$. The composite nature of the $S = 1$ excitation reveals itself in the dynamical structure factor of the model, which can be determined using the Ising description (6.19) of the total and relative staggered magnetizations.[133] The dynamical magnetic susceptibility displays a two-particle threshold instead of a sharp magnon peak near $q = \pi/a_0$ as in the two-leg spin ladder or the spin-1 chain. This incoherent background in the dynamical structure factor leads to a new gapped spin liquid phase in unfrustrated quantum magnetism without any coherent magnon excitations. In this respect, this state has been called non-Haldane spin liquid by Nersesyan and Tsvelik.[133] The existence of this spin liquid phase has been confirmed non-perturbatively at two special points in the phase diagram of the spin–orbital model (6.21). At $J = 3K/4$, the GS is exactly known[150] and is a product of checkerboard-ordered spin and orbital singlets as in Fig. 6.4(a) obtained in the weak coupling limit. At $K = -4J$, the model is exactly solvable and the GS energy and triplet energy gap have been determined exactly.[151] The system has two spontaneously dimerized GSs and belongs to the class of non-Haldane spin liquid.

Finally, it is worth noting that the alternating and in-phase dimerization phases can be distinguished in close parallel to the topological distinction between the two gapped phases of the two-leg spin ladder. Using the topological criterion of Kim *et al.*,[137] one observes from Fig. 6.4 that the number Q_y of valence bonds crossing an vertical line is odd (respectively, even) for the staggered (respectively, in-phase) dimerization. The two non-Haldane spin liquid phases with $K > 0$ and $K < 0$ belong thus to two different topological classes. This difference becomes manifest when investigating the structure of the edge states of the two phases with open BC. The physical properties of these boundary excitations of the semi-infinite spin–orbital model (6.21) can be determined by means of the Ising mapping (6.22) of the weak coupling limit similar to the cut two-leg ladder.[83] For $K < 0$, such edge states are absent, whereas in the $K > 0$ case, two spin-1/2 chain-end excitations are expected.[152]

6.3. Frustration Effects

In this section, we shall review some of the main aspects of the interplay between frustration and quantum fluctuations in AF spin chains and spin ladders. The natural question is whether frustration can stabilize new types of spin liquid phases with exotic spin excitations not encountered in Section 6.2. The paradigmatic model to analyze the effect of frustration in spin chains is the J_1–J_2 spin-1/2 Heisenberg chain with Hamiltonian:

$$\mathcal{H} = J_1 \sum_i \vec{S}_i \cdot \vec{S}_{i+1} + J_2 \sum_i \vec{S}_i \cdot \vec{S}_{i+2}, \tag{6.24}$$

where the next-nearest-neighbor (NNN) coupling $J_2 > 0$ is a competing AF interaction which introduces frustration. This model can also be viewed as a

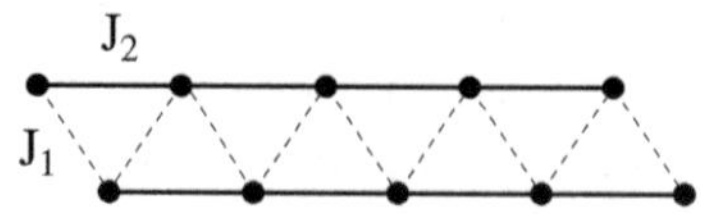

Fig. 6.6. Two-leg zigzag ladder.

frustrated two-leg spin ladder where the spin chains are coupled in a zigzag way as shown in Fig. 6.6. Let us first discuss some of the main characteristics of frustration in 1D spin systems by means of a semi-classical approach.

6.3.1. *Semi-classical analysis*

Classically, one of the main effect of frustration is to favor non-collinear magnetic ordering where the spins lie in a plane rather than along a single direction as in unfrustrated spin systems:

$$\vec{S}_i = S\cos(\vec{Q}\cdot\vec{r}_i)\vec{n}_1 + S\sin(\vec{Q}\cdot\vec{r}_i)\vec{n}_2, \tag{6.25}$$

where $\vec{r}_i$ is the spatial location of the site i and $\vec{n}_{1,2}$ are two mutually orthogonal fixed vectors in spin space with unit length. The magnetic structure corresponding to the ordering (6.25) is a circular spiral with a pitch angle related to the wave vector $\vec{Q}$ which is, in general, incommensurate. In the case of the J_1–J_2 chain (6.24), the spiral GS is stabilized when $J_2/J_1 > 1/4$. The spins are arranged in a canted configuration in which each spin makes an angle α with its predecessor such that $\cos\alpha = -J_1/4J_2$. The classical GS of the model is doubly degenerate since the spin configurations (6.25) can turn clockwise and counterclockwise with the same energy along the spiral axis $\vec{n}_3 = \vec{n}_1 \wedge \vec{n}_2$. A Z_2 discrete symmetry characterized by this right- and left-handed chirality is thus spontaneously broken in this helical structure. A corresponding chiral ordering[153] can be defined and detected by the chiral order parameter: $\langle(\vec{S}_i \wedge \vec{S}_{i+1})\cdot\vec{n}_3\rangle \neq 0$. In contrast to the Néel state of unfrustrated magnets, the SU(2) spin rotation of the Heisenberg Hamiltonian is completely broken in the non-collinear state (6.25). Three Goldstone or spin wave modes are thus expected here from this spontaneous symmetry breaking scheme. It leads to a CFT with central charge $c = 3$ with extended criticality in comparison to the unfrustrated case ($c = 2$, see Section 6.2). A second distinction stems from the spin wave calculation[154] for non-collinear ordered states which reports the presence of two different spin wave velocities. Therefore, the low-energy field theory describing canted magnets is expected to be non-Lorentz invariant. To identify this effective field theory, one needs to define a suitable order parameter for canted magnets. The long-wavelength fluctuations of the spin configurations (6.25) can be captured by introducing a two-component complex field $z_a, a = 1, 2$ of unit modulus ($|z_1|^2 + |z_2|^2 = 1$) such that[2]:

$$\vec{n}_1 + i\vec{n}_2 = \epsilon_{ac} z_c \vec{\sigma}_{ab} z_b, \tag{6.26}$$

where ϵ_{ab} is the standard antisymmetric tensor and $\vec{\sigma}$ is a vector formed by the Pauli matrices. It is indeed straightforward to see that this identification (6.26) correctly reproduces the constraints: $\vec{n}_1^2 = \vec{n}_2^2 = 1$ and $\vec{n}_1 \cdot \vec{n}_2 = 0$ if the two-component complex field z_a belongs to the 3D surface of the unit sphere in four dimensions denoted by S^3. Moreover, it is worth noting that the field $z = (z_1, z_2)$ transforms like a $S = 1/2$ spinor under spin rotations. The representation (6.26) is double-valued since z and $-z$ describe the same non-collinearly ordered state. The order parameter space corresponding to canted magnets (6.25) is thus S^3/Z_2.

A crucial consequence of this identification is that the order parameter allows topologically non-trivial vortices having a 2π circulation since $\pi_1(S^3/Z_2) = Z_2$.[155] Upon encircling such a Z_2 vortex, called now a vison,[156] by a closed loop, the value of the spinor complex field $z = (z_1, z_2)$ changes smoothly from z to $-z$, i.e. two points diametrically opposed on the surface of a sphere S^3. The existence of these stable vison defects reveals itself in the semiclassical description of frustrated spin chains like the $J_1\!-\!J_2$ Heisenberg chain (6.24). In particular, in sharp contrast to unfrustrated spin chains, no Pontryagin topological term has been found in the effective action that describes the low-energy properties of the $J_1\!-\!J_2$ Heisenberg chain[157,158] and 2D frustrated Heisenberg AF[159] in the large spin limit. This stems from the very special topological nature of the order parameter for canted magnets which has a trivial second homotopy group: $\pi_2(S^3/Z_2) = 0$.[155] However, there is still a subtle distinction between half-integer and integer spins in the $J_1\!-\!J_2$ Heisenberg chain[157,160] related to the existence of stable visons. A Berry phase calculation[157,160–162] shows that tunneling between sectors with different Z_2 topological number is possible for integer spins but not for half-integer spins due to cancellation between pairs of paths. The GS of the $J_1\!-\!J_2$ model is thus non-degenerate for integer spins and two-fold degenerate (two different sectors non-coupled by tunneling effect) for half-integer spins. The quantum number carried by elementary excitations also depends strongly on the nature of the spins.[160] In the half-integer case, there is an energy gap toward the creation of Z_2 visons so that these topological defects are strongly suppressed. The low-energy excitations are then described by the spinor field $z = (z_1, z_2)$ of Eq. (6.26) which is free to propagate in absence of visons and carries a $S = 1/2$ quantum number under spin rotations. Neglecting anisotropy, the Euclidean effective action that governs the leading low-energy properties of the J_1–J_2 chain for half-integer spins is the $SU(2) \times SU(2)$ principal chiral model defined by

$$\mathcal{S}_G = \frac{v}{2g} \int dx d\tau \; \mathrm{Tr}\left(\partial_\mu G^\dagger \partial_\mu G\right), \tag{6.27}$$

where G is a $SU(2)$ matrix formed by the two complex fields $z_{1,2}$. This field theory[163, 164] is integrable and its low-energy spectrum consists of massive excitation with $S = 1/2$ quantum number, i.e. massive deconfined spinons. In contrast, in the integer spin case, the visons now proliferate in the GS so that the spinor field $z = (z_1, z_2)$ can no longer be defined as single-valued configurations. The suitable fields that describe the low-energy excitations of the $J_1 - J_2$ chain with integer spins are bilinear with respect to z and correspond to the $\vec{n}_{1,2}$ fields of Eq. (6.26). In that case, neglecting anisotropy, the effective field theory that captures the leading low-energy properties of the frustrated spin chain with integer spins is the nonlinear sigma model SO(3) $\times$ SO(3) with action

$$\mathcal{S}_R = \frac{v}{2g} \int dx d\tau \; \mathrm{Tr}\left(\partial_\mu R^{-1} \partial_\mu R\right), \tag{6.28}$$

where R is a rotation matrix made of the triplet vectors $\vec{n}_{1,2}$ and $\vec{n}_3 = \vec{n}_1 \wedge \vec{n}_2$. The low-lying spectrum of this field theory can be determined by a large N approach[157, 158] or by means of a strong-coupling analysis[160]: it consists of a triplet of $S = 1$ massive magnons. In the integer spin case, the spinons are thus expected to be confined to optical magnons.

An important prediction, obtained in this semi-classical description, is thus the existence of a gapped spin liquid phase for half-integer spins with a two-fold degenerate ground state and massive deconfined spinons. Such a state is stabilized by frustration and represents a spin liquid phase not encountered in Section 6.2. We shall now consider the ultra-quantum case with $S = 1/2$, where a combination of numerical and field theoretical techniques can be used to fully determine the main characteristics of this spin liquid phase.

6.3.2. *Spin liquid phase with massive deconfined spinons*

The phase diagram of the spin-1/2 J_1-J_2 Heisenberg chain (6.24) has been studied extensively over the years after the bosonization analysis of Haldane.[165] This problem is not a purely academic question since inorganic compounds such as $CuGeO_3$ (see Refs. [166] and [167]) or LiV_2O_5 (see Ref. [168]) can be considered as prototypes of the spin-1/2 J_1-J_2 chain (6.24). In particular, values such as $J_1 \simeq 160$ K and $J_2/J_1 \simeq 0.36$ have been proposed for $CuGeO_3$.[169] In addition, the quasi-1D compound $SrCuO_2$ contains a collection of spin-1/2 Heisenberg chains assembled pairwise in an array of weakly coupled zigzag ladders.[170,171] These zigzag ladders are built from corner-sharing Cu–O chains with an exchange J_2 staked pairwise in edge-sharing geometry. The frustrating interaction J_1 between the chains (see Fig. 6.6) stems from the nearly 90° Cu–O–Cu bonds and is expected to be weak.[170] Finally, two other possible realizations of the spin-1/2 J_1-J_2 chain (6.24) have been recently proposed: the compound $(N_2H_5)CuCl_3$ which can be described as a two-leg zigzag spin ladder with $J_1/J_2 \simeq 0.25$[172] and Cu[2-(2-aminomethyl)pyridine]Br_2 with a ratio $J_2/J_1 \simeq 0.2$.[173]

A starting point for investigating the phase diagram of the spin-1/2 J_1-J_2 chain (6.24) is to consider the weak coupling limit when $J_2 \ll J_1$. This enables us to study the stability of the massless spinons of the spin-1/2 AF Heisenberg chain upon switching on a small NNN frustrating interaction J_2. In this weak-coupling regime, the low-energy effective Hamiltonian density of the model reads as follows[165]:

$$\mathcal{H}_{\text{eff}} = \frac{2\pi v}{3}(\vec{J}_L^2 + \vec{J}_R^2) + \gamma \vec{J}_L \cdot \vec{J}_R, \tag{6.29}$$

where $\vec{J}_{L,R}$ are the left and right su$(2)_1$ currents that generate the su$(2)_1$ quantum criticality of the spin-1/2 AF Heisenberg chain. The low-energy physics of the model are thus mainly determined by the marginal current–current interaction of Eq. (6.29) with coupling constant $\gamma \simeq J_2 - J_{2c}$, J_{2c} being a non-universal positive constant. For a small value of J_2, one has $\gamma < 0$ so that the interaction in Eq. (6.29) is a marginally irrelevant contribution. The low-energy physics of the J_1-J_2 chain is thus identical to that of the spin-1/2 AF Heisenberg chain which is governed by the $su(2)_1$ fixed point with additional logarithmic corrections introduced by the current–current interaction. In this respect, frustration plays no important role in the regime $J_2 < J_{2c}$ and the spin-1/2 Heisenberg phase with massless spinons is stable upon switching on a small value of J_2. However, for $J_2 > J_{2c}$ ($\gamma > 0$), the current–current interaction becomes marginally relevant and a strong coupling regime develops with a dynamically generated spectral gap. A phase transition of Berezinskii–Kosterlitz–Thouless (BKT) type[174] occurs at $J_2 = J_{2c}$ which separates the gapless spin-1/2 Heisenberg phase from a fully massive region. The actual value of the transition has been determined numerically by the different groups to be $J_{2c} \simeq 0.2411 J_1$.[175–178] At this point, it is worth noting that the quasi-1D material Cu[2-(2-aminomethyl)pyridine]Br_2 can be described by a spin-1/2 J_1-J_2 Heisenberg chain with $J_2/J_1 \simeq 0.2$ (see Ref. [173]) and should thus belong to the critical Heisenberg phase with $J_2 < J_{2c}$. The absence of a spin gap for this compound has been reported experimentally.[173]

The main characteristics of the strong coupling massive phase with $J_2 > J_{2c}$ can be determined from Eq. (6.29). The field theory (6.29) is indeed integrable and corresponds to a chiral Gross–Neveu model or a non-Abelian version of the Thirring model with a $SU(2)$ symmetry. The low-energy excitations of the model (6.29) are a massive doublet (massive spinons) with mass $m \sim a_0^{-1} \exp(-2\pi/\gamma)$.[179, 180] A striking effect of frustration is thus the formation of a mass for the spinons of the spin-1/2 AF Heisenberg chain without confining them into $S = 1$ excitations as in the two-leg spin ladder. A second consequence of frustration is the presence of spontaneous dimerization in the model. A simple way to exhibit this dimerization is to use the bosonization approach to express the Hamiltonian (6.29) in terms of a $\beta^2 = 8\pi$ sine-Gordon model[165]:

$$\mathcal{H}_{\text{eff}} = \frac{v}{2}[(\partial_x \Phi)^2 + (\partial_x \Theta)^2] + \frac{\gamma}{2\pi} \partial_x \Phi_R \partial_x \Phi_L - \frac{\gamma a_0^2}{4\pi^2} \cos \sqrt{8\pi} \Phi, \qquad (6.30)$$

where $\Phi_{R,L}$ are the chiral components of the bosonic field Φ ($\Phi = \Phi_R + \Phi_L$) and Θ is its dual field ($\Theta = \Phi_L - \Phi_R$). The $SU(2)$ invariance of the model (6.30) is hidden in the structure of the interaction with a single coupling constant and the fact that the bosonic field Φ is compactified on a circle with a special radius $R = 1/\sqrt{2\pi}$ consistent with the $SU(2)$ symmetry. This compactification leads to the following identification:

$$\Phi \sim \Phi + 2\pi R n = \Phi + n\sqrt{2\pi}, \qquad (6.31)$$

n being integer. In the phase with $J_2 > J_{2c}$, one has $\gamma > 0$ so that the bosonic field is pinned at one of its minima: $\langle \Phi \rangle = p\sqrt{\pi/2}$, p being-integer. However, by taking into account the identification (6.31), the $\beta^2 = 8\pi$

sine-Gordon model (6.30) has only two inequivalent GSs with $\langle\Phi\rangle = 0$ and $\langle\Phi\rangle = \sqrt{\pi/2}$. This two-fold degeneracy can be interpreted as resulting from the spontaneous breaking of a discrete Z_2 symmetry. This symmetry identifies with the one-step translation symmetry ($\mathcal{T}_{a_0}$) which is described by the following shift on the bosonic field[24]: $\Phi \to \Phi + \sqrt{\pi/2}$. This symmetry is spontaneously broken in the phase $J_2 > J_{2c}$ and the two GS field configurations are connected by this translation symmetry. An order parameter designed to characterize this phase is the spin dimerization operator $\epsilon_n = (-1)^n \vec{S}_n \cdot \vec{S}_{n+1}$ which admits the following bosonic representation in the continuum limit $\epsilon \sim \cos\sqrt{2\pi}\Phi$.[10,24] This operator changes sign under the lattice translation symmetry and has a non-zero expectation value $\langle\epsilon\rangle \neq 0$ in the two GSs of the $\beta^2 = 8\pi$ sine-Gordon model (6.30).

For $J_2 > J_{2c}$, frustration thus stabilizes a gapful spontaneously dimerized phase which is characterized by a two-fold degenerate GS and a spontaneous breaking of the lattice translation symmetry. The elementary excitations of this phase are massive spinons that carry $S = 1/2$ quantum number and identify with the kinks of the underlying dimerization. A simple way to understand the emergence of these deconfined spinons is to consider the Majumdar–Ghosh (MG)[181] point at $J_2 = J_1/2$, where the GS of the lattice model (6.24) is exactly known. This MG point for the spontaneously dimerized phase plays a role similar to that of the AKLT point (6.10) for describing the main properties of the Haldane phase in the phase diagram of the spin-1 bilinear–biquadratic chain (6.9). For $J_2 = J_1/2$, the Hamiltonian (6.24) takes the following form up to a constant:

$$\mathcal{H}_{\mathrm{MG}} = \frac{3J_1}{4} \sum_i P_{3/2}\left(i-1, i, i+1\right), \tag{6.32}$$

where $P_{3/2}\left(i-1, i, i+1\right)$ projects the total spin of the three spins located at sites $i-1$, i, and $i+1$ onto the $S = 3/2$ subspace. For an even number of sites N and periodic BC, the GS of the MG model (6.32) is two-fold degenerate and corresponds to the two singlet states:

$$\begin{aligned} |\Phi_1\rangle &= [1,2]\,[3,4]\cdots[N-1,N]\,, \\ |\Phi_2\rangle &= [2,3]\,[4,5]\cdots[N,1]\,, \end{aligned} \tag{6.33}$$

where $[i, i+1]$ denotes a singlet bond between the spins at the i and $i+1$ sites. The two GSs (6.33) are represented in Fig. 6.7. The lattice translation symmetry $\mathcal{T}_{a_0}$ is broken in the two GS (6.33) and it exchanges them: $\mathcal{T}_{a_0}|\Phi_{1,2}\rangle = |\Phi_{2,1}\rangle$. The existence of a gap in the MG model (6.32) has been

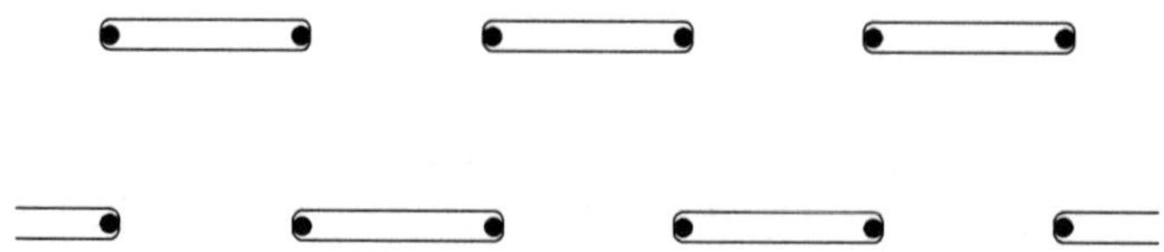

Fig. 6.7. The two dimerized GS at the MG point; (a) $|\Phi_1\rangle$, (b) $|\Phi_2\rangle$.

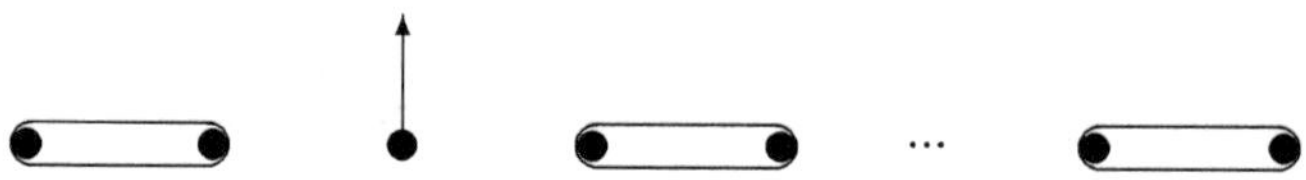

Fig. 6.8. Dimerization kink (spinon) with $S = 1/2$ quantum number.

shown rigorously by Affleck *et al.*[57] and the spin–spin correlation function can be determined exactly at the MG point[182]:

$$\langle \vec{S}_i \cdot \vec{S}_j \rangle = \frac{3}{4}\delta_{i,j} - \frac{3}{8}\delta_{|i-j|,1}. \tag{6.34}$$

This correlation function is thus zero for distance larger than one lattice spacing. The GS degeneracy and the spontaneous breaking of a discrete Z_2 symmetry suggest the existence of topological excitations which interpolate between the two GSs (6.33). These kink excitations can be viewed as the insertion of a spin-1/2 in a sea of singlet valence bond states as shown in Fig. 6.8. This spin-1/2 excitation (spinon) is nothing but a domain wall between the two states (6.33). For periodic BC, these kink states always appear in pairs (see Fig. 6.9) and the low-lying excitations of the MG model can be built starting from the state[182]:

$$\begin{aligned} |p, m\rangle = [1,2]..[2p-3, 2p-2]\alpha_{2p-1}[2p, 2p+1]\cdots \\ [2m-2, 2m-1]\alpha_{2m}[2m+1, 2m+2]\cdots[N-1, N], \end{aligned} \tag{6.35}$$

where α_{2p-1} and α_{2m} denote spin-1/2 states located at the sites $2p-1$ and $2m$, respectively. A variational approach of the low-lying excitations[182, 183] can then be done by considering a linear combination of states (6.35) and it leads to a triplet and singlet continuum spin excitations. In particular, the dispersion relation of a massive spinon, obtained within this approach,

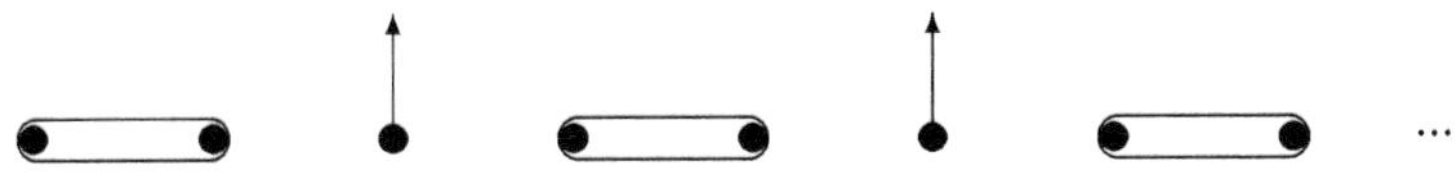

Fig. 6.9. Spin-1 excitation built from two spinons.

takes the form[182]:

$$\epsilon(k) = \frac{5J_1}{8} + \frac{J_1}{2}\cos(2ka_0). \tag{6.36}$$

This dispersion relation has been verified numerically by exact diagonalizations.[184] Finally, as shown by Caspers and Magnus,[185] there are additional exact singlet and triplet bound-states at momentum $k = \pi/2a_0$, which are degenerate with energy $E = J_1$. Exact diagonalization and DMRG calculations[184] have confirmed the existence of these bound-states for a small range of momenta close to $k = \pi/2a_0$ in the region $J_2 \geq J_1/2$.

In summary, the spontaneously; dimerized phase for $J_2 > J_{2c}$ represents a distinct spin liquid phase stabilized by frustration. The main distinction, from the spin liquid phases of the two-leg spin ladder and spin–orbital model, originates in the fractionalized nature of the quantum number carried by the elementary excitations. In particular, instead of a sharp $S = 1$ magnon peak as in the two-leg spin ladder, the dynamical structure factor of the model in this dimerized phase displays an incoherent background with additional bound-states features. The main difference between the dimerized phase of Fig. 6.7 and the staggered dimerization phase of the spin–orbital model (6.21) stems from the nature of the low-lying excitations. In the latter phase, the excitations are a pair of propagating massive triplet kinks, as described above, whereas here the elementary excitations are massive spinons.

This dimerized phase, with deconfined massive spinon excitations, extends in the entire region with $J_2 > J_{2c}$ as it has been shown numerically. Figure 6.10 represents the evolution of the spin gap Δ, computed by DMRG,[178] as a function of the NNN interaction J_2. As depicted by Fig. 6.10, the spin gap is maximum for $J_2/J_1 \simeq 0.6$ and decreases to zero in the large J_2 limit. The absence of a spin gap in this regime can be easily understood since for $J_2 \gg J_1$, the $J_1 - J_2$ spin chain can be better viewed as a two-leg zigzag ladder (see Fig. 6.6). In the limit $J_2 \to +\infty$, the two spin-1/2 Heisenberg chains are decoupled so that the model is

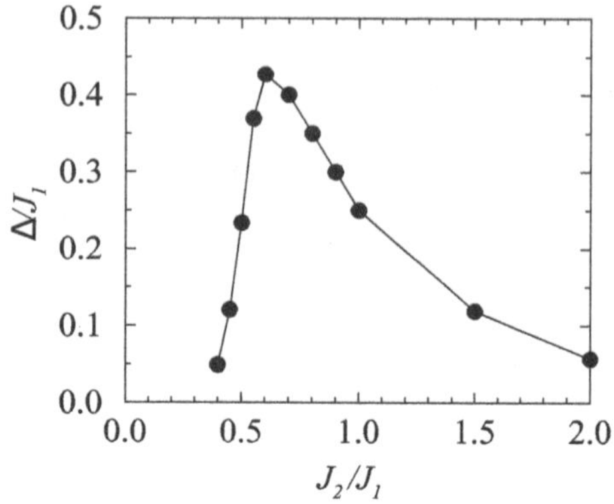

Fig. 6.10. Evolution of the spin gap, computed by DMRG, as a function of J_2 for $J_2 > J_{2c}$ (taken from Ref. [178]).

critical: $\Delta = 0$. From the results of Fig. 6.6, it is very tempting to conclude that the whole region with $J_2 > J_{2c}$ describes a single phase. However, a subtle qualitative change occurs in the model at the level of spin correlation functions after the MG point in close parallel to the Haldane phase of the spin-1 bilinear–biquadratic chain (6.9) just after the AKLT point (6.10).[72, 186, 187] Indeed, an incommensurate behavior develops in the real-space spin–spin correlation function for $J_2 > J_1/2$.[176, 178, 188–190] This incommensurability can be interpreted as a quantum signature of the spiral structure of the classical ground state of the $J_1 - J_2$ spin chain for $J_2 > J_1/4$. After the MG point, the leading asymptotic of the spin–spin correlation function behaves as

$$\langle \vec{S}_i \cdot \vec{S}_{i+r} \rangle \sim \frac{1}{r^{1/2}} \cos\left(q\, r a_0\right) \exp\left(-r a_0/\xi\right), \tag{6.37}$$

where the oscillation factor depends on J_2/J_1. In parallel to the classification of commensurate–incommensurate transitions in Ising spin systems,[191] the MG point, as the AKLT point (6.10), is a disorder point of the first kind.[72] The momentum $q_{\max}$ which maximizes the static structure factor (Fourier transform of the spin–spin correlation function) becomes incommensurate not at the MG point but slightly after $J_{2L} \simeq 0.5206 J_1$ (see Ref. [188]) at a Lifshitz point.[72] For $J_2 > J_{2L}$, the static structure factor displays a double-peak structure rather than a single peak. The low-lying excitations are then characterized by an incommensurate momenta $q_{\max} \neq \pi/a_0$. In contrast to the classical case, it is worth noting that the

BKT phase transition between the critical and dimerized phases and the onset of incommensurability at the Lifshitz point occur at different points of the phase diagram, $J_{2c} \simeq 0.2411 J_1$ and $J_{2L} \simeq 0.5206 J_1$, respectively. Though no real phase transition occurs for $J_2 > J_{2c}$ in the model, the dimerized phase in the region $J_2 > J_{2L}$ can be distinguished from that at $J_{2c} < J_2 < J_{2L}$ due to the existence of this incommensurability. In this respect, the spontaneously dimerized phase for $J_2 > J_{2L}$ with deconfined massive spinons and incommensurate correlations represents a remarkable 1D spin liquid phase stabilized by frustration.

This incommensurability induced by frustration is, in fact, quite general in 1D and not restricted to the spin-1/2 case. In particular, this phenomenon also appears in the spin-1 case. The zero-temperature phase diagram of the spin-1 J_1–J_2 Heisenberg chain (6.24) has been investigated numerically[158, 192–196] and a spectral gap exists for all finite values of J_2. A first-order transition occurs at $(J_2/J_1)_T \simeq 0.744$ (see Ref. [194]) separating a Haldane phase from a double-Haldane phase.[194–196] The quantum number carried by elementary excitations is $S = 1$ in full agreement with the semi-classical approach[160] described above. Frustration plays its trick here by giving rise to a similar onset of incommensurability as in the spin-1/2 case. A disorder point of the first kind occurs at $(J_2/J_1)_D \simeq 0.284$ together with a Lifshitz point $(J_2/J_1)_L \simeq 0.3725$[194] after which the static structure factor develops a two-peak structure.

6.3.3. *Field theory of spin liquid with incommensurate correlations*

Frustration thus represents a novel mechanism in one dimension for generating incommensurability as external magnetic fields or Dzyaloshinskii–Moriya interaction.[197] The zigzag ladder representation (Fig. 6.6) of the J_1–J_2 model enables one to investigate, by a weak coupling approach $J_1 \ll J_2$, the main characteristics of frustration and to shed light on the mechanism that gives rise to incommensurate correlations in the large J_2 regime. In the spin-1/2 case, the onset of incommensurability and the presence of deconfined massive spinons might be understood starting from the limit where the two spin-1/2 AF Heisenberg chains are decoupled.

The continuum limit of the $S = 1/2$ two-leg zigzag ladder has been analyzed by several groups.[136, 178, 198–202] The interacting part of the Hamiltonian density of the low-energy field theory reads as follows:

$$\mathcal{H}_{\text{int}} \simeq g_1(\vec{J}_{1L} \cdot \vec{J}_{2R} + \vec{J}_{2L} \cdot \vec{J}_{1R}) + g_2\, \vec{n}_1 \cdot \partial_x \vec{n}_2, \tag{6.38}$$

where $\vec{J}_{a,L,R}$ $(a = 1, 2)$ are the left–right $su(2)_1$ currents corresponding to the continuum description of the ath spin-1/2 AF Heisenberg chain. In Eq. (6.38), $\vec{n}_a$ denotes the staggered magnetization of the spin density of the chain with index $a = 1, 2$. In the continuum limit, this $\vec{n}_a$ field identifies with the vector part of the primary field, with scaling dimension $\Delta_n = 1/2$, of the $SU(2)_1$ WZNW CFT transforming according to the fundamental representation of SU(2).[10, 15, 24, 40] In this continuum description, frustration suppresses geometrically the standard backscattering contribution $\vec{n}_1 \cdot \vec{n}_2$ which governs the low-energy physics of the two-leg spin ladder. This term represents a strongly relevant perturbation of scaling dimension 1. It corresponds to the energy operators of the underlying Ising models of the low-energy description (6.16) of the two-leg spin ladder.[89, 132] As seen in Section 6.2, this backscattering contribution leads to the formation of a spin gap in this model as well as the confinement of the original massless spinons into optical magnons. In the presence of frustration, the interacting part (6.38) is now only marginal relevant and consists of two terms of different natures. The first one, with coupling constant g_1, is a current–current interaction similar to that which appears in the effective field theory (6.29) of the spin-1/2 $J_1 - J_2$ Heisenberg chain in the weak coupling limit $J_2 \ll J_1$. The second marginal contribution in Eq. (6.38), called twist term,[199] $\mathcal{O}_{\text{twist}} = \vec{n}_1 \cdot \partial_x \vec{n}_2$ is a novel parity-breaking perturbation which contains the staggered magnetizations of each chain and also a spatial derivative. In contrast to the current–current perturbation, this twist term is not a scalar under the Lorentz transformation in $1 + 1$ dimensions but behaves as a vector. In the CFT jargon, this kind of perturbation is characterized by a non-zero conformal spin $S = \pm 1$ together with its scaling dimension $\Delta = 2$. In fact, this twist perturbation has been forgotten in the first bosonization analysis of the two-leg zigzag spin ladder.[136, 178] It has been discovered by Nersesyan *et al.*[199] and independently by Allen.[198] It is interesting to observe that a similar operator appears in the effective field theory approach to the spin-1 two-leg zigzag ladder in the large J_2 limit.[203] This twist perturbation is thus the hallmark of frustration in the low-energy description of frustrated spin ladders with zigzag interchain interaction.

The effect of such a non-zero conformal spin perturbation is non-trivial since the usual irrelevant versus relevant criterion of perturbative field theory does not hold for such a non-scalar contribution (see, for instance, the discussion in Ref. [10]). In the case of a $S = \pm 1$ perturbation, it is expected that the generic effect of this contribution is the onset of incommensurability. In this respect, a simple example of incommensurability arising from such a perturbation is the spin-1/2 XXZ Heisenberg chain in a magnetic field along the z-axis. The Hamiltonian density of this model, obtained within the bosonization approach, reads as follows [204]:

$$\mathcal{H} = \frac{v}{2}\left[(\partial_x\Phi)^2 + (\partial_x\Theta)^2\right] - g\cos\beta\Phi - h\,\partial_x\Phi. \tag{6.39}$$

Here, the $S = \pm 1$ perturbation is described by the uniform part of the spin density $\partial_x\Phi$. It is well known that in this model, when the magnetic field h is increased, a commensurate–incommensurate phase transition takes place with the appearance of an incommensurate phase with critical correlation functions.[205–208] This transition occurs for a finite or vanishing magnetic field depending on the relevance or irrelevance of the $\cos\beta\Phi$ term. A similar incommensurability generated by a $S = \pm 1$ conformal spin perturbation occurs in the spin-1/2 XXZ Heisenberg chain in a transverse magnetic field,[209–211] in a model of two spinless Luttinger chains weakly coupled by a single-particle interchain hopping,[209, 212, 213] and in the quantum axial next-to-nearest neighbor Ising (ANNNI) chain in a transverse magnetic field.[214] Finally, it is important to note that there exist some exactly solvable CFT models perturbed by $S = \pm 1$ conformal spin term for which the presence of incommensurability can be shown non-perturbatively.[215, 216]

All these results suggest that the twist term $\mathcal{O}_{\text{twist}} = \vec{n}_1 \cdot \partial_x \vec{n}_2$, as proposed by Nersesyan *et al.*,[199] should be at the origin of the incommensurability found numerically in the spin-1/2 $J_1 - J_2$ Heisenberg chain (6.24) when $J_2 > 0.52 J_1$. To this end, this twist term and the current–current interaction of Eq. (6.38) can be expressed in terms of the four Majorana fermions $\xi^a_{R,L}, a = 0,1,2,3$ of the continuum limit (6.16) of the standard two-leg spin ladder.[198, 199] In particular, the current–current interaction is built from of the following tensor: $\mathcal{O}^{ab}_{\text{cc}} = \xi^a_R \xi^a_L \xi^b_R \xi^b_L$ with $a, b = 0,1,2,3$ and $a \neq b$. The twist perturbation is also local in terms of these Majorana fermions but with a different structure. A typical term that enters its expression is $\mathcal{O}^{abcd}_{\text{twist}} = \xi^a_R \xi^b_L \xi^c_L \xi^d_L$ with $a,b,c,d = 0,1,2,3$ and $a \neq b \neq c \neq d$. The non-zero conformal spin of this twist term is reflected here by the different number of right and left fermions in this expression. The field theory (6.38), expressed in terms of these four Majorana fermions, turns out not to be integrable. The renormalization group (RG) flow analysis reveals that the current–current interaction and the twist term are equally important in the IR limit. They reach a strong coupling regime simultaneously with a fixed ratio.[198,201,202] The nature of this strong coupling regime is still an open problem since the emerging IR field theory is not integrable and one cannot disentangle the effect of the two interactions of Eq. (6.38). However, it is very tempting to explain the main characteristics of the spontaneously dimerized phase with incommensurate correlation of the spin-1/2 J_1–J_2 Heisenberg chain in the large J_2 limit from these two contributions. On the one hand, the current–current contribution of Eq. (6.38), equivalent to two decoupled $SU(2)$ Thirring model, has similar properties as the field theory (6.29) obtained in the weak coupling regime $J_2 \ll J_1$. The spontaneous dimerization and the existence of massive deconfined spinons should result from this interaction. In fact, this current–current interaction appears alone, without a twist term, in the continuum limit of a frustrated two-leg spin ladder with crossings along a special line of the couplings.[201] It has been shown that this model, in the weak coupling limit, is characterized by a weak spontaneous dimerization with massive spinons as elementary excitations.[201] Finally, as already stressed, the twist perturbation of Eq. (6.38) should be at the origin of the incommensurability in the large J_2 limit.[199] In this respect, the effect of the twist term of the two-leg zigzag ladder has been analyzed within an RPA approach, and it leads indeed to some incommensurability behavior in the spin–spin correlation function.[217]

A simple way to disentangle the effects of the current–current interaction and the twist perturbation is to introduce an exchange anisotropy which makes the twist contribution more relevant in the RG sense.[199] The extreme case is the XY version of the spin-1/2 J_1–J_2 Heisenberg chain with Hamiltonian:

$$\mathcal{H}_{XY} = J_1 \sum_n (S_n^x S_{n+1}^x + S_n^y S_{n+1}^y) + J_2 \sum_n (S_n^x S_{n+2}^x + S_n^y S_{n+2}^y). \quad (6.40)$$

In the ladder limit when $J_2 \gg J_1$, the low-energy physics of this model can be analyzed by means of the bosonization approach. Introducing two bosonic fields $\Phi_\pm$, the resulting bosonic Hamiltonian density of the model reads as follows[199]:

$$\mathcal{H}_{XY} \simeq \frac{v}{2} \sum_{a=\pm} \left((\partial_x \Phi_a)^2 + (\partial_x \Theta_a)^2 \right) + g\, \partial_x \Theta_+ \sin(\sqrt{2\pi}\Theta_-), \quad (6.41)$$

where $\Theta_\pm$ are the dual fields associated to the bosonic fields $\Phi_\pm$. The Hamiltonian (6.41) describes a non-trivial field theory with a relevant $S = \pm 1$ conformal spin twist perturbation with scaling dimension $\Delta_g = 3/2$. In this strong anisotropic XY case, the current–current perturbation is less dominant and can be safely neglected to derive the nature of the phase of the model when $J_2 \gg J_1$. The presence of incommensurability in the system can then be found using a mean-field analysis of the model by decoupling the two pieces of the twist term of Eq. (6.41). In particular, the leading asymptotic behavior of the transverse spin–spin correlation functions of the model obtained by this mean-field approach is given by[199]:

$$\langle S_1^\dagger (x) \, S_a^- (0) \rangle \sim \frac{\exp{(iqx)}}{|x|^{1/4}}, \quad a = 1, 2, \tag{6.42}$$

which displays an incommensurate critical behavior with an oscillating factor: $q - \pi/a_0 \sim (J_1/J_2)^2$. The physical picture that emerges from this mean-field analysis is the existence of a critical spin nematic phase that preserves the $U(1)$ and time-reversal symmetries and displays long-range chiral ordering in its GS: $\langle (\vec{S}_{n+1} \wedge \vec{S}_n)_z \rangle \neq 0$. It is important to note that this spin nematic phase does not break the time-reversal symmetry but spontaneously breaks a Z_2 symmetry of the model which is a tensor product of site-parity and link-parity symmetries on the two chains: $P_L^{(1)} \times P_S^{(2)}$. The z-component of the spin current J_{as}^z associated to the ath spin-1/2 XY chain ($a = 1, 2$) takes a non-zero expectation value in the ground state of the mean field Hamiltonian[199]:

$$\langle J_{1s}^z \rangle = \langle J_{2s}^z \rangle = -\frac{v}{\sqrt{2\pi}} \langle \partial_x \Theta_+ \rangle \neq 0. \tag{6.43}$$

As a result, this produces a picture of local non-zero spin currents polarized along the z-anistropy axis circulating around the triangular plaquettes of the two-leg zigzag spin ladder. This phase has been first found numerically by Nishiyama,[218] who has investigated the existence of chiral order of the Josephson ladder with half a flux quantum per plaquette. It has also been obtained by Hikihara *et al.*[219] by means of the DMRG approach and by Sacramento and Vieira[220] using exact diagonalization calculations. This critical incommensurate phase is the quantum analogue of the classical spiral phase with chiral ordering.[153] It is thus natural to expect that this phase is not restricted to the spin-1/2 case but should exist in the range of parameters of the model (6.40) in the general spin case. Indeed, the existence of this critical spin nematic phase in the spin-S case has been shown numerically for $S = 1, 3/2, 2$,[219, 221] by means of a semiclassical method,[222] and finally by the Abelian bosonization approach[223] of a general spin S introduced by Schulz.[89]

6.3.4. *Extended criticality stabilized by frustration*

In addition to incommensurability, a second characteristic of classical canted magnets is the presence of three gapless spin wave modes instead of two as in the Néel collinear state. An interesting question is whether frustration can lead to a new type of emerging quantum criticality not encountered in unfrustrated spin chains or ladders. In this last case, the critical behavior of half-integer AF Heisenberg spin chains or odd-legged spin ladders is characterized by one gapless bosonic mode, i.e. by a CFT with central charge $c = 1$. As it has already been pointed out in the previous section, a striking effect of frustration in the continuum limit stems from the fact that the low-energy effective field theory is mainly governed by marginal interactions. New type of IR critical behaviors may result from the delicate balance between these marginal perturbations. In the following, we shall give some examples of critical phases and quantum critical points with extended criticality stabilized by frustration in 1D spin systems.

6.3.4.1. *Critical phases with SU(N) quantum criticality*

A first example of a system with extended criticality is the 1D spin–orbital model (6.21) in a regime where frustration reveals itself $K \sim J$. At the special point $J = K/4$, the Hamiltonian (6.21) can be expressed in terms of a product of two-body permutation operator in S_1 and S_2 subspaces

$$\begin{aligned}\mathcal{H}_{\mathrm{so}} &= J \sum_i \left(2\vec{S}_{1,i} \cdot \vec{S}_{1,i+1} + \frac{1}{2}\right)\left(2\vec{S}_{2,i} \cdot \vec{S}_{2,i+1} + \frac{1}{2}\right) \\ &= J \sum_i P_{i,i+1}^{(S_1=1/2)} \, P_{i,i+1}^{(S_2=1/2)}. \end{aligned} \tag{6.44}$$

Since this Hamiltonian exchanges both $\vec{S}_{1,i}$ and $\vec{S}_{2,i}$ spins at the same time, the spin–orbital model at $J = K/4$ is not only $SU(2) \times SU(2)$ symmetric but also has an enlarged $SU(4)$ symmetry which unifies the spin and orbital degrees of freedom.[224–226] More precisely, the Hamiltonian (6.44) can be recasted as an AF Heisenberg spin chain with $SU(4)$ spins up to a constant

$$\mathcal{H}_{\mathrm{SU}(4)} = J \sum_i \sum_{A=1}^{15} T_i^A T_{i+1}^A, \tag{6.45}$$

where T^A are the 15 generators belonging to the fundamental representation of $SU(4)$. This model is exactly solvable by means of the Bethe anšatze approach[227] and its low-energy spectrum consists of three gapless spinons with wave vectors $\pm\pi/2a_0$ and π/a_0.[227,228] As shown by Affleck,[229] the critical theory corresponds to the $su(4)_1$ WZNW model with central charge $c = 3$ (three massless bosonic modes). The existence of this quantum critical point with an $SU(4)$ symmetry allows us to study the spin–orbital model (6.21) by a continuum description in an intermediate coupling regime $J \sim K$ where frustration shows off. In this respect, it is important to notice that, according to the Zamolodchikov's c theorem,[230] the $SU(4)$ critical point with central charge $c = 3$ cannot be reached by an RG trajectory starting from the decoupling limit ($K = 0$) of two spin-1/2 AF Heisenberg chains with total central charge $c = 2$. Stated differently, the physics in the neighborhood of the $SU(4)$ point cannot be understood in terms of weakly coupled $S = 1/2$ Heisenberg chains. The strategy to tackle this intermediate coupling regime is to start from the $SU(4)$ Hubbard chain at quater-filling and apply the bosonization approach to obtain the continuum description of the spin densities $\vec{S}_{1,2}(x)$ at the $SU(4)$ point in the large Coulomb repulsion limit. The low-energy field theory which describes small deviations from the $SU(4)$ symmetric point can then be derived.[231,232] The resulting effective Hamiltonian density associated to the symmetry breaking scheme $SU(4) \to SU(2) \times SU(2)$ of the spin–orbital model (6.21) reads as follows[231]:

$$\begin{aligned}\mathcal{H}_{\text{eff}} &= -\frac{iv_s}{2}(\vec{\xi}_{sR}\cdot\partial_x\vec{\xi}_{sR} - \vec{\xi}_{sL}\cdot\partial_x\vec{\xi}_{sL}) - \frac{iv_o}{2}(\vec{\xi}_{oR}\cdot\partial_x\vec{\xi}_{oR} - \vec{\xi}_{oL}\cdot\partial_x\vec{\xi}_{oL})\\ &\quad + g_1(\vec{\xi}_{sR}\cdot\vec{\xi}_{sL})^2 + g_2(\vec{\xi}_{oR}\cdot\vec{\xi}_{oL})^2 + g_3(\vec{\xi}_{sR}\cdot\vec{\xi}_{sL})(\vec{\xi}_{oR}\cdot\vec{\xi}_{oL}), \quad (6.46)\end{aligned}$$

where $\vec{\xi}_{sR,L}$ and $\vec{\xi}_{oR,L}$ are two triplets of Majorana fermions which act, respectively, in the spin and orbital sectors. In Eq. (6.46), we have considered a more general situation than the model (6.21) by allowing the exchange in spin (J_1) and orbital (J_2) channels to be different: the three coupling constants g_i are independent. The interaction of the low-energy field theory (6.46) is marginal and describes two $SO(3)$ Gross–Neveu[135] models marginally coupled. This field theory is not integrable so that one has to recourse to perturbation theory to elucidate the phase diagram of the spin–orbital model in the vicinity of the $SU(4)$ symmetric point. In fact, the all-order beta functions of the field theory (6.46) can be determined using the approach of Gerganov *et al.*[233] These authors have computed the all-order beta functions of a general model with anisotropic current–current interactions in a special minimal scheme prescription. The model (6.46) can be viewed as a current–current interaction corresponding to the symmetry breaking scheme $SO(6) \to SO(3) \times SO(3)$. The application of the general formula given in Ref. [233] confirms the conclusions derived from the one-loop calculation[231,232] which reveals the existence of two phases with different remarkable properties.

A first phase, for $J_1 \simeq J_2 > K/4$, has a spectral gap and the GS has a similar staggered dimerization as the weak-coupling phase of Fig. 6.4. Provided the anisotropy is not too large $J_1 \neq J_2$ in the vicinity of the $SU(4)$ symmetric point, the massive phase displays an approximate $SO(6) \sim SU(4)$ enlarged symmetry. Indeed, by neglecting the velocity anisotropy in Eq. (6.46), the RG equations reveal a flow to strong coupling with an attraction along a special direction with enlarged symmetry described by the following Hamiltonian density[231]:

$$\mathcal{H}_{\text{IR}} \simeq -\frac{iv_*}{2} \sum_{a=1}^{6} (\xi_R^a \partial_x \xi_R^a - \xi_L^a \partial_x \xi_L^a) + g_* \left(\sum_{a=1}^{6} \xi_R^a \xi_L^a \right)^2 , \tag{6.47}$$

with $g_* > 0$. This Hamiltonian is identified as the $SO(6)$ Gross–Neveu model, which is a massive integrable field theory. Its spectrum is known and consists of the fundamental fermion, with mass M, together with a kink and anti-kink with mass $m_{\text{kink}} = M/\sqrt{2}$.[38] The initial model (6.46), $SU(2) \times SU(2)$ symmetric, acquires thus in the IR limit an enlarged $SO(6)$ symmetry. In more physical terms, the spin and orbital degrees of freedom are unified and described by the same multiplet with six components. A similar example of symmetry resaturation by interactions is the emergence of a SO(8) symmetry in a weakly coupled two-leg Hubbard ladder at half-filling[234] and in the $SU(4)$ Hubbard chain at half-filling.[235]

The second phase with $J_1 \simeq J_2 < K/4$ has striking properties. All couplings in Eq. (6.46) flow to zero in the IR limit and the interaction is marginally irrelevant. The six Majorana fermions are thus massless and the phase displays extended quantum criticality characterized by a central charge $c = 3$. Spin–spin correlation functions decay algebraically with exponent $3/2$ and exhibit a four-site periodicity ($2k_F = \pi/2a_0$). The critical behavior at the $SU(4)$ point (6.45) extends to a finite region of the phase diagram of the spin–orbital model. In this respect, this remarkable gapless phase with extended quantum criticality $c = 3$ represents a new universality class, stabilized by frustration, in spin chains and spin ladders. It is worth

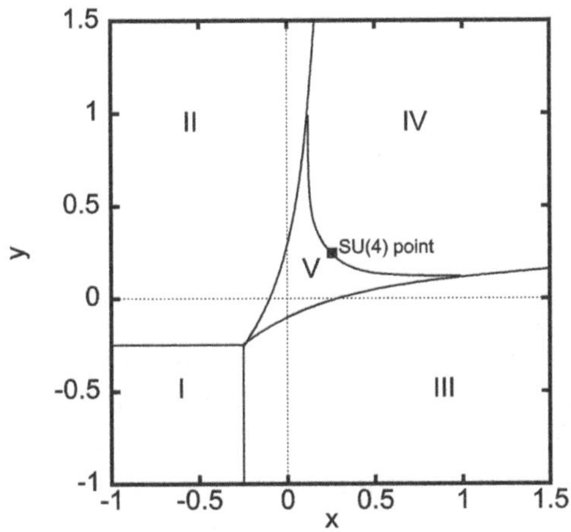

Fig. 6.11. Phase diagram of the spin–orbital model obtained by DMRG; $x = J_2/K$ and $y = J_1/K$; taken from Ref. [232].

noting that this phase has a maximum of gapless modes allowed by the classical structure of a spiral (three Goldstone modes).

The phase diagram of the anisotropic spin–orbital model (6.21) with $J_1 \neq J_2$ has been investigated numerically by means of the DMRG technique.[147, 232, 236] Figure 6.11 represents the resulting $T = 0$ phase diagram obtained in Refs. [232] and [236] In phase I, both spin and orbital degrees of freedom are in fully polarized ferromagnetic states. In phase II, the orbital degrees of freedom are still in a ferromagnetic state, whereas the spin degrees of freedom are now critical (criticality of a spin-1/2 AF Heisenberg chain) and vice versa in phase III. Phase IV corresponds to the staggered dimerized phase of Fig. 6.4 while phase V is the $c = 3$ gapless phase with an extended quantum criticality. In fact, this critical phase has been first discovered only along a special line $-K/4 < J_1 = J_2 < K/4$ in the DMRG calculation of Pati *et al.*[147] Additional DMRG works[232, 236] have shown the existence of this gapless phase in an extended region (phase V) of the phase diagram (Fig. 6.11).

A second model which displays extended quantum criticality induced by frustration is the spin-1 bilinear–biquadratic chain (6.9) with a biquadratic coupling constant $\beta > 1$. In this regime, frustration manifests itself and the classical GS is an incommensurate spiral. The Haldane phase of the spin-1 bilinear–biquadratic chains (6.9) with $|\beta| < 1$ ends at the integrable point with $\beta = 1$, the so-called Uimin–Lai–Sutherland,[227, 237] point. This model can be expressed in terms of $SU(3)$ spins and takes the form of a $SU(3)$ AF Heisenberg spin chain:

$$\mathcal{H}_{SU(3)} = J \sum_i \sum_{A=1}^{8} T_i^A T_{i+1}^A, \tag{6.48}$$

where T^A are the eight generators belonging to the fundamental representation of $SU(3)$. This model is exactly solvable by means of the Bethe ansätze approach[227, 237] and its low-energy spectrum consists of two gapless spinons with wave vectors $\pm 2\pi/3a_0$.[227] As shown by Affleck,[229] the critical theory corresponds to the $su(3)_1$ WZNW model with central charge $c = 2$ (two massless bosonic modes). The existence of this quantum critical point with an $SU(3)$ symmetry allows to study the spin-1 bilinear–biquadratic chain (6.9) in the vicinity of $\beta \simeq 1$ by means of a field theory approach.[238] The low-energy effective field theory is described by marginal current–current interactions associated to the symmetry breaking scheme $SU(3) \to SU(2)$.[238] The one-loop RG flow near $\beta \simeq 1$ reveals the existence of two distinct phases: on the one hand, a phase with a dynamical mass generation when $\beta < 1$ which signals the onset of the Haldane phase, and on the other hand, a massless phase for $\beta > 1$ where interactions are marginally irrelevant. The $SU(3)$ quantum criticality of the Uimin–Lai–Sutherland point, with two gapless bosonic modes, extends to a finite region of the phase diagram of the generalized spin-1 Heisenberg chain (6.9). This massless phase has been first predicted in a numerical investigation of this model by Fáth and Sólyom.[239] This massless phase with approximate $SU(3)$ symmetry, stabilized by frustration, is similar in spirit to the previous $SU(4)$ massless phase of the spin–orbital model (6.21). Finally, it is worth noting that this $c = 2$ massless phase also appears in the phase diagram of the so-called spin tube model, a three-leg spin ladder with frustrated periodic interchain interaction, in a magnetic field.[240]

6.3.4.2. *Chirally stabilized critical spin liquid*

Frustration can also induce an exotic quantum critical point in frustrated spin ladders. A possible lattice realization of this phenomenon is a model (see Fig. 6.12) of three $S = 1/2$ AF Heisenberg spin chains weakly coupled by on-rung $J_\perp$ and plaquette-diagonal $J_\perp$ interchain interactions.[241] The

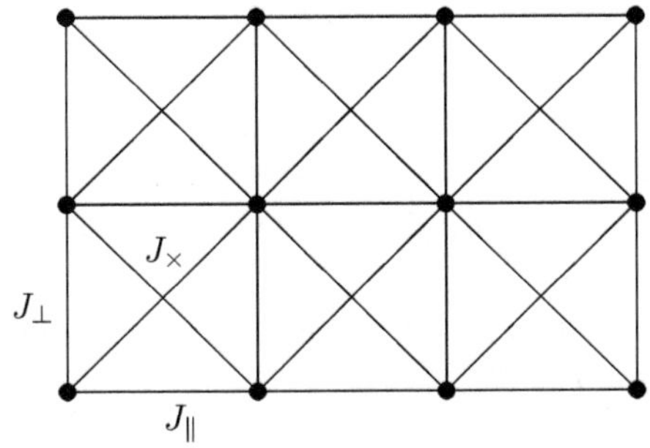

Fig. 6.12. Three-leg spin ladder with crossings.

Hamiltonian of this model reads as follows:

$$\mathcal{H}_\times = J_\parallel \sum_i \sum_{a=1}^{3} \vec{S}_{a,i} \cdot \vec{S}_{a,i+1} + J_\perp \sum_i \vec{S}_{2,i} \cdot (\vec{S}_{1,i} + \vec{S}_{3,i})$$
$$+ J_\times \sum_i [(\vec{S}_{1,i} + \vec{S}_{3,i}) \cdot \vec{S}_{2,i+1} + (\vec{S}_{1,i+1} + \vec{S}_{3,i+1}) \cdot \vec{S}_{2,i}]. \quad (6.49)$$

In the continuum limit with $J_\perp, J_\times \ll J_\parallel$, the Hamiltonian of the lattice model (6.49) takes the form:

$$\mathcal{H}_\times = \frac{2\pi v}{3} \sum_{a=1}^{3} (\vec{J}_{aR}^2 + \vec{J}_{aL}^2) + \tilde{g}\, \vec{n}_2 \cdot (\vec{n}_1 + \vec{n}_3)$$
$$+ g\, [\vec{J}_{2R} \cdot (\vec{J}_{1L} + \vec{J}_{3L}) + \vec{J}_{2L} \cdot (\vec{J}_{1R} + \vec{J}_{3R})], \quad (6.50)$$

with $g = a_0(J_\perp + 2J_\times)$ and $\tilde{g} = a_0\,(J_\perp - 2J_\times)$. In Eq. (6.50), $\vec{J}_{aR,L}$ ($a = 1, 2, 3$) are the right and left chiral $su(2)_1$ currents that account for the low-energy description of the uniform part of the spin density of the ath spin-1/2 chain. It is interesting to note that in Eq. (6.50), there is no marginally relevant twist perturbation $\vec{n}_1 \partial_x \vec{n}_2$ which appears in the continuum description of the two-leg spin ladder with a small zigzag interchain coupling (6.38). The two interaction terms in Eq. (6.50) are of different nature. On the one hand, the first contribution, with coupling constant $\tilde{g}$, is a strongly relevant perturbation with scaling dimension $\Delta_{\tilde{g}} = 1$. On the other hand, the second term is a current–current interaction term which is only marginal and, as long as $\tilde{g}$ is not too small, can be discarded. As a result, for generic values of g and $\tilde{g}$, the low-energy physics of the model (6.49) will be essentially that of the standard three-leg ladder, and frustration will play no role (except for renormalization of mass gaps and velocities). The important point here is that in contrast to non-frustrated ladders, the two coupling constants $g, \tilde{g}$ can vary independently, and there exists a vicinity of the line $J_\perp = 2J_\times$ ($\tilde{g} = 0$) where the low-energy properties of the model are mainly determined by current–current interchain interaction.

Along the special line $J_\perp = 2J_\times$, the low-energy effective Hamiltonian (6.50) simplifies as follows[241]:

$$\mathcal{H}_\times = -\frac{iv}{2}(\xi_R^0 \partial_x \xi_R^0 - \xi_L^0 \partial_x \xi_L^0) + \frac{2\pi v}{3}(\vec{J}_{2R}^2 + \vec{J}_{2L}^2) + \frac{\pi v}{2}(\vec{I}_R^2 + \vec{I}_L^2) + g\,[\vec{J}_{2R} \cdot \vec{I}_L + \vec{J}_{2L} \cdot \vec{I}_R], \tag{6.51}$$

where $\xi_{R,L}^0$ are Majorana fermions which are associated to the Z_2 ($1 \rightarrow 3$) discrete interchange symmetry between the surface chains labeled $a = 1, 3$ in the lattice model (6.49); the total chiral current of the surface chains is noted $\vec{I}_{R,L} = \vec{J}_{1R,L} + \vec{J}_{3R,L}$ and corresponds to a $su(2)_2$ WZNW current. The Hamiltonian (6.51) has an interesting structure. First, the Majorana fermions $\xi_{R,L}^0$ do not participate in the interaction and remain thus critical in the IR limit. These massless degrees of freedom can be interpreted as an effective 2D Ising model at $T = T_c$ which accounts for singlet excitations between the surface chains. All non-trivial physics of the model is incorporated in the current–current dependent part of the Hamiltonian (6.51), denoted by $\mathcal{H}_{cc}$, which describes marginally coupled $su(2)_1$ and $su(2)_2$ WZNW models. Moreover, this Hamiltonian separates into two commuting and *chirally asymmetric* parts: $\mathcal{H}_{cc} = \mathcal{H}_1 + \mathcal{H}_2$, ($[\mathcal{H}_1, \mathcal{H}_2] = 0$), where

$$\mathcal{H}_1 = \frac{\pi v}{2}\,\vec{I}_R^2 + \frac{2\pi v}{3}\,\vec{J}_{2L}^2 + g\,\vec{I}_R \cdot \vec{J}_{2L}, \tag{6.52}$$

and $\mathcal{H}_2$ is obtained from $\mathcal{H}_1$ by inverting chiralities of all the currents. The model (6.52) is integrable by means of the Bethe ansätze approach.[163, 242] A simple RG analysis shows that, at $g > 0$, the interaction is marginally relevant. Usually, the development of a strong coupling regime is accompanied by a dynamical mass generation and the loss of conformal invariance in the strong coupling limit. However, here due to the chiral asymmetry of $\mathcal{H}_1$, it turns out that the effective interaction flows towards an intermediate fixed point where conformal invariance is recovered with a smaller central charge. This critical behavior has been identified as the universality class of chirally stabilized fluids, introduced by Andrei *et al.*[242] In the case of the model (6.52), the symmetry of the IR fixed point, obtained from the Bethe ansätze analysis,[242] turns out to be $su(2)_1|_R \times Z_2|_L$. This result can also be derived using a Toulouse point approach.[241] As a whole, taking into account the contribution of the Majorana fermions $\xi^0_{R,L}$, the model (6.51) displays critical properties characterized by a fixed point with an $su(2)_1 \times Z_2 \times Z_2$ symmetry and a $c = 2$ central charge. This emerging quantum criticality can be interpreted as the criticality resulting from an effective $S = 1/2$ AF Heisenberg chain and two decoupled critical Ising models. Frustration, introduced by the diagonal exchange interaction $J_\times$, shows up in a non-trivial way by stabilizing a quantum critical point with central charge $c = 2$. This IR behavior differs from the standard $c = 1$ quantum criticality in unfrustrated three-leg spin ladders by the two gapless non-magnetic, singlet, degrees of freedom described by the Ising models. The physical properties of the model at this $c = 2$ critical point can be determined by a Toulouse point approach of the model (6.51).[241] The slowest spin–spin correlation functions of the model at the IR critical point correspond to the staggered correlations between the spins of the surface chains which decay with a universal exponent 3/2. As a consequence, the low-temperature dependence of the NMR relaxation rate scales as $1/T_1 \sim \sqrt{T}$ in contrast to $1/T_1 \sim$ const for the spin-1/2 AF Heisenberg chain.[10] The $c = 2$ quantum critical point, induced by frustration, describes thus a new universality class in spin ladders. In addition, the Toulouse point analysis enables us to investigate the effect of the neglected backscattering term $\vec{n}_2 \cdot (\vec{n}_1 + \vec{n}_3)$ of the model (6.50) in the vicinity of the line $\tilde{g} = 0$. The main effect of this operator is to open a spectral gap in the Ising degrees of freedom of the $c = 2$ fixed point but has no dramatic effect on the effective $S = 1/2$ AF Heisenberg chain. The $c = 2$ IR fixed point is thus unstable with respect to the interchain backscattering term and the model (6.50) will display IR critical properties governed by the $c = 1$ fixed point of the three-leg spin ladder or the $S = 1/2$ AF Heisenberg spin chain.

Finally, it is worth noting that the zero-temperature phase diagram of the three-leg spin ladder with crossings (6.49) has been recently investigated by means of the DMRG approach.[243] Two critical phases have been found corresponding to the $S = 3/2$ AF Heisenberg chain and the three-leg spin ladder. Frustration, introduced through the diagonal interchain interaction $J_\times$, induces a quantum phase transition between these two critical phases belonging to the standard $c = 1$ universality class. At large $J_\times$, the transition has been found to be of first order. In the weak coupling regime, the transition occurs at $J_\perp = 2J_\times$ when $J_\times < 0.8J_\parallel$ is in full agreement with the field theoretical description (6.50) of the model. Extended DMRG calculations are required to fully characterize the nature of the phase boundary at $J_\perp = 2J_\times$ in the weak coupling regime in particular to verify the chiral spin liquid behavior with $c = 2$ criticality proposed in Ref. [241].

6.4. Concluding Remarks

The suppression of magnetism by quantum fluctuations in one dimension gives rise to a large variety of zero-temperature spin liquid phases with strikingly different physical properties. The canonical examples are the emerging quantum criticality of the spin-1/2 AF Heisenberg chain, controlled by gapless elementary excitations with fractional quantum numbers (spinons), and the formation of an incompressible Haldane spin liquid phase in the spin-1 case with optical $S = 1$ magnon excitations. Several other types of incompressible spin liquid phases can be stabilized with different dynamical or optical properties. A distinct spin liquid behavior may be revealed in the form of the dynamical structure factor with the presence or absence of a sharp spectral peak which directly probes the elementary nature of the triplet excitation of the phase. In this respect, the staggered dimerized phase of the two-leg spin ladder with a biquadratic interchain exchange represents a non-Haldane spin liquid phase even though a spectral gap is formed by quantum fluctuations. The main difference between this phase and the Haldane phase of the spin-1 Heisenberg chain stems from the composite nature of the spin-flip $S = 1$ excitation due to the presence of a two-particle threshold in the dynamical structure factor instead of a sharp magnon peak. Two-gapped spin liquid phases can also be distinguished at the level of the topology of the short-range valence bond description of their GSs. A spin liquid phase may exhibit a GS degeneracy depending on the nature of the boundary conditions used. In particular, for open boundary conditions, chain-end spin excitations can result from this GS degeneracy leading to well-defined satellite peaks in the NMR profile of the system doped with non-magnetic impurities. An example of this topological distinction between two gapped phases is provided by the two-leg spin ladder. In the Haldane phase of the open two-leg ladder with a ferromagnetic rung interchain $J_\perp < 0$, $S = 1/2$ edge states are formed, whereas these chain-end excitations disappear in the rung singlet phase of the ladder with $J_\perp > 0$.

Frustration, i.e. the impossibility to satisfy simultaneously every pairwise interaction, can induce additional types of spin liquid phases with exotic properties. In the spin-1/2 case, one of the most striking effect of frustration is the stabilization of a spontaneously dimerized phase with massive deconfined spinon excitations. Frustration thus represents a direct route to fractionalization. The existence of these deconfined spinon excitations has important consequences, as for instance, in the study of doping effects of such a spin liquid phase with non-magnetic impurities. On general grounds, it is expected that no free spin degrees of freedom are generated around non-magnetic impurities due to the presence of deconfined spinons.[244] In the case of the spontaneously dimerized phase of the spin-1/2 J_1–J_2 Heisenberg chain, the absence of induced free spin degrees of freedom or edge states has been shown recently.[245] A second remarkable effect of frustration is the onset of incommensurability in spin chains whose classical ground state has a spiral structure. In this respect, frustration is a novel mechanism in one dimension for generating incommensurate behavior as external magnetic fields or Dzyaloshinskii–Moriya interaction. An example of such a spin liquid phase with incommensurate correlation is the quantum $J_1 - J_2$ Heisenberg chain in the large NNA limit. When the $SU(2)$ symmetry of this model is explicitely broken, an incommensurate phase is produced by frustration with local non-zero spin currents, polarized along the anisotropic axis, circulating around triangular plaquettes. This phase is the quantum signature of the classical spiral phase induced by frustration. Finally, a last effect of frustration, discussed in this review, is the possible realization of a new type of emerging quantum criticality in spin chains and ladders. In the continuum description, the low-energy effective field theory of a frustrated spin chain is mainly governed by marginal interactions. An extended IR critical behavior characterized by a CFT with central charge $c > 1$ may result from the delicate balance between these marginal contributions.

Regarding perspectives, a natural question to raise is the existence of a novel 1D spin liquid phase with unbroken $SU(2)$ spin symmetry which displays physical properties not described in this review. In particular, an interesting possibility is the realization of a 1D version of the chiral spin liquid phase[246] which breaks spontaneously the time-reversal symmetry. In fact, such a phase has been identified recently in the phase diagram of the two-leg spin ladder with a four-spin cyclic exchange.[247, 248] In the large ring-exchange limit, a spin liquid phase characterized by a non-zero scalar-chirality operator $\langle \vec{S}_{1,i} \cdot (\vec{S}_{2,i} \wedge \vec{S}_{1,i+1}) \rangle$, breaking both parity and time-reversal symmetries, has been found in DMRG calculations.[247,248] Using duality arguments, the authors of Ref. [248] have also exhibited several spin ladder models with exact ground states which display spontaneous breaking of the time-reversal symmetry. Finally, a central issue is the possibility to stabilize a 2D $SU(2)$-invariant spin liquid phase with exotic properties starting from the 1D limit. Recently, a model of 2D frustrated spin system with $SU(2)$ spin symmetry and strong spatially anisotropy has been introduced by Nersesyan and Tsvelik.[249] This model can also be interpreted as a collection of weakly coupled spin chains. It has been shown that, in the limit of infinite number of chains, this system displays a spin liquid phase with massive deconfined fractional spin-1/2 excitations (spinons).[249] The existence of spinons in 2D spin systems has also been found in a crossed-chain model[250] which is a 2D anisotropic version of the pyrochlore lattice. Remarkably enough, for a range of coupling constants,[250,251] this system is a non-dimerized spin liquid with deconfined gapless spinons as elementary excitations. This spin liquid phase, stabilized by frustration, is an example of a $SU(2)$ version of a sliding Luttinger liquid phase.[252] We hope that other 2D spin liquid phases with exotic properties will be reported in the near future.

6.5. Note Added for the Second Edition

Since the writing of this review in 2004, there have been many important developments in 1D spin chain problems both theoretically and experimentally. It will be too long to review these recent progresses here. Let us, however, mention briefly two advances that have attracted much interest in recent years.

On the theory side, one outstanding breakthrough is the recent complete classification of all 1D gapful phases using some mathematical tools in algebraic topology. This progress is related to the understanding of topological phases which do not break any symmetry and cannot be characterized by local order parameters as it has been discussed in this review. Though all 1D gapful phases have short-range entanglement, a topological phase can still be defined in 1D by the presence of a symmetry which protects the phase.[253] A topological phase with such symmetry is called a symmetry-protected topological (SPT) phase.[253,254] Otherwise, it can be smoothly connected to a trivial gapful phase without any phase transition.[255] The Haldane phase of the spin-1 Heisenberg chain displays striking properties which makes it a paradigmatic example of an SPT phase in 1D. As reviewed in Section 2.3, there is a finite gap above a unique ground state with periodic boundary conditions while spin-1/2 edge states are liberated in open chains. This phase exhibits non-local string long-range ordering as well. These properties have been shown to be protected by the presence of three global symmetries: the dihedral group of π rotations along the x, y, z axes, time-reversal and inversion symmetries.[256] In addition, one important consequence of the latter work is the prediction that the Haldane phases with even-integer spins turn out not to be topological phases in stark contrast to the odd-integer spins case.[256] The complete classification of SPT of 1D spin systems has been obtained recently.[257] It relies on the determination of the projective representations of the underlying symmetry group which can be obtained using group cohomology.[257] Remarkably, together with the CFT approach,[258] which governs the 1D gapless Lorentz invariant phases, the cohomology approach leads to the full classification of 1D quantum phases.

On the experimental side, one important direction of research is to find new experimental realizations of the exotic 1D quantum phases using artificial solids. These phases are usually extremely hard to be investigated with the standard compounds of the solid-state physics. However, these phases might be implemented and studied experimentally using the precision of quantum optics and other technologies. In this respect, ultracold atoms loading into optical lattices is a very important field which emerged in 2002.[259] For instance, Heisenberg spin chain problems have been implemented theoretically using cold atoms, trapped ions, or polar molecules.[260] Other discussed possibilities involve coupled cavities, Josephson junctions, and cavity QED experiments.[261] It is likely that, in the future, these experiments will be performed and some phases discussed in this review will be observed.

Acknowledgments

We would like to thank D. Allen, P. Azaria, E. Boulat, B. Douçot, F. H. L. Essler, Th. Giamarchi, Th. Jolicoeur, A. A. Nersesyan, E. Orignac, T. T. Truong, and A. M. Tsvelik for helpful discussions and their interest in this work.

References

1. P. W. Anderson, *Mater. Res. Bull.* **8**, 153 (1973); *Science* **235**, 1196 (1987).
2. S. Sachdev and K. Park, *Ann. Phys.* **298**, 58 (2002); S. Sachdev, *ibid.* **303**, 226 (2003).

3. C. Lhuillier and G. Misguich, Frustrated quantum magnets, in C. Berthier, L. P. Lévy, and G. Martinez (eds.), *High Magnetic Fields*, Lecture Notes in Physics Vol. 595, (Springer, 2002), p.161; arXiv: cond-mat/0109146; See also, G. Misguich and C. Lhuillier, Frustration in two-dimensional quantum antiferromagnets, in this volume, arXiv: cond-mat/0310405.
4. X. G. Wen, *Phys. Rev. B* **65**, 165113 (2002).
5. N. D. Mermin and H. Wagner, *Phys. Rev. Lett.* **17**, 1307 (1968).
6. A. A. Belavin, A. M. Polyakov, and A. B. Zamolodchikov, *Nucl. Phys. B* **241**, 333 (1984).
7. P. Di Francesco, P. Mathieu, and D. Sénéchal, *Conformal Field Theory* (Springer-Verlag, New York, 1997).
8. A. Luther and I. Peschel, *Phys. Rev. B* **9**, 2911 (1974).
9. S. Coleman, *Phys. Rev. B* **11**, 2088 (1975).
10. A. O. Gogolin, A. A. Nersesyan, and A. M. Tsvelik, *Bosonization and Strongly Correlated Systems* (Cambridge University Press, Cambridge, 1998).
11. S. R. White, *Phys. Rev. Lett.* **69**, 2863 (1992); *Phys. Rev. B* **48**, 10345 (1993).
12. M. Yamashita, T. Ishii, and H. Matsuzaka, *Coord. Chem. Rev.* **198**, 347 (2000).
13. P. Lemmens, G. Güntherodt, and C. Gros, *Phys. Rep.* **375**, 1 (2003).
14. S. Sachdev, *Quantum Phase Transitions* (Cambridge University Press, Cambridge, 1999).
15. I. Affleck, *Phys. Rev. Lett.* **55**, 1355 (1985); *Nucl. Phys. B* **265**, 409 (1986).
16. X. G. Wen, *Adv. Phys.* **44**, 405 (1995).
17. H. Bethe, *Z. Phys.* **71**, 205 (1931).
18. L. Hulthén, *Ark. Mat., Astron. Fys.* **26A**, 1 (1938).
19. R. B. Griffiths, *Phys. Rev.* **133**, A768 (1964); C. N. Yang and C. P. Yang, *Phys. Rev.* **150**, 327 (1966); S. Eggert, I. Affleck, and M. Takahashi, *Phys. Rev. Lett.* **73**, 332 (1994); S. Lukyanov, *Nucl. Phys. B* **522**, 533 (1998); A. Klümper and D. C. Johnston, *Phys. Rev. Lett.* **84**, 4701 (2000).
20. M. Takahashi, *Thermodynamics of One-Dimensional Solvable Models* (Cambridge University Press, Cambridge, 1999).
21. J. des Cloizeaux and J. J. Pearson, *Phys. Rev.* **128**, 2131 (1962).
22. L. D. Faddeev and L. A. Takhtajan, *Phys. Lett. A* **85**, 375 (1981).
23. A. Luther and I. Peschel, *Phys. Rev. B* **9**, 2911 (1974); A. H. Bougourzi, M. Couture, and M. Kacir, *ibid.* **54**, 12669 (1996); A. Abada, A. H. Bougourzi, and B. Si-Lakhal, *Nucl. Phys. B* **497**, 733 (1997); N. Kitanine, J. M. Maillet, and V. Terras, *ibid.* **554**, 647 (1999).
24. I. Affleck, *Field Theory Methods and Quantum Critical Phenomena*, in Les Houches, session XLIX, *Champs, Cordes et Phénomènes critiques* (Elsevier, New-York, 1989).
25. V. G. Knizhnik and A. B. Zamolodchikov, *Nucl. Phys. B* **247**, 83 (1984).

26. I. Affleck, D. Gepner, H. J. Schulz, and T. Ziman, *J. Phys. A* **22**, 511 (1989); R. R. P. Singh, M. E. Fisher, and R. Shankar, *Phys. Rev. B* **39**, 2562 (1989); T. Giamarchi and H. J. Schulz, *ibid.* 4620 (1989).
27. I. Affleck, *J. Phys. A* **31**, 4573 (1998); T. Hikihara and A. Furusaki, *Phys. Rev. B* **58**, R583 (1998); S. Lukyanov and V. Terras, *Nucl. Phys. B* **654**, 323 (2003).
28. F. D. M. Haldane, *Phys. Rev. Lett.* **66**, 1529 (1991).
29. F. D. M. Haldane, *Phys. Rev. Lett.* **60**, 635 (1988); B. S. Shastry, *ibid.* 639 (1988).
30. S. E. Nagler, D. A. Tennant, R. A. Cowley, T. G. Perring, and S. K. Satija, *Phys. Rev. B* **44**, 12361 (1991); D. A. Tennant, T. G. Perring, R. A. Cowley, and S. E. Nagler, *Phys. Rev. Lett.* **70**, 4003 (1993); D. A. Tennant, R. A. Cowley, S. E. Nagler, and A. M. Tsvelik, *Phys. Rev. B* **52**, 13368 (1995).
31. D. C. Dender, D. Davidović, D. H. Reich, C. Broholm, K. Lefmann, and G. Aeppli, *Phys. Rev. B* **53**, 2583 (1996).
32. I. Tsukada, Y. Sasago, K. Uchinokura, A. Zheludev, S. Maslov, G. Shirane, K. Kakurai, and E. Ressouche, *Phys. Rev. B* **60**, 6601 (1999).
33. P. R. Hammar, M. B. Stone, D. H. Reich, C. Broholm, P. J. Gibson, M. M. Turnbull, C. P. Landee, and M. Oshikawa, *Phys. Rev. B* **59**, 1008 (1999).
34. F. D. M. Haldane, *Phys. Rev. Lett.* **50**, 1153 (1983); *Phys. Lett. A* **93**, 464 (1983).
35. I. Affleck, *J. Phys. Condens. Matter* **1**, 3047 (1989).
36. E. Fradkin, *Field Theories of Condensed Matter Systems* (Addison-Wesley, 1991).
37. A. Auerbach, *Interacting Electrons and Quantum Magnetism* (Springer-Verlag, New York, 1994).
38. A. B. Zamolodchikov and Al. B. Zamolodchikov, *Ann. Phys.* **120**, 253 (1979).
39. P. B. Wiegmann, *Phys. Lett. B* **152**, 209 (1985).
40. I. Affleck and F. D. M. Haldane, *Phys. Rev. B* **36**, 5291 (1987).
41. R. Shankar and N. Read, *Nucl. Phys. B* **336**, 457 (1990).
42. A. B. Zamolodchikov and Al. B. Zamolodchikov, *Nucl. Phys. B* **379**, 602 (1992).
43. R. Botet and R. Jullien, *Phys. Rev. B* **27**, 613 (1983); R. Botet, R. Jullien, and M. Kolb, *Phys. Rev. B* **28**, 3914 (1983); M. Kolb, R. Botet, and R. Jullien, *J. Phys. A* **16**, L673 (1983); U. Glaus and T. Schneider, *Phys. Rev. B* **30**, 215 (1984); J. B. Parkinson and J. C. Bonner, *ibid* **32**, 4703 (1985); H. J. Schulz and T. A. L. Ziman, *ibid.* **33**, 6545 (1986); A. Moreo, *ibid.* **35**, 8562 (1987); T. Sakai and M. Takahashi, *ibid.* **42**, 1090, 4537 (1990); O. Golinelli, T. Jolicoeur, and R. Lacaze, *ibid.* **46**, 10854 (1992); S. Haas, J. Riera, and E. Dagotto, *Phys. Rev. B* **48**, 3281 (1993); M. Kaburagi, I. Harada, and T. Tonegawa, *J. Phys. Soc. Jpn.* **62**, 1848 (1993); T. Sakai and M. Takahashi, *J. Phys. Soc. Jpn.* **63**, 755 (1994).

44. M. P. Nightingale and H. W. Blöte, *Phys. Rev. B* **33**, 659 (1986); M. Marcu and J. Müller, *Phys. Lett. A* **119**, 469 (1987); M. Takahashi, *Phys. Rev. B* **38**, 5188 (1988); *Phys. Rev. Lett.* **62**, 2313 (1989); K. Nomura, *Phys. Rev. B* **40**, 2421 (1989); S. Liang, *Phys. Rev. Lett.* **64**, 1597 (1990); J. Deisz, M. Jarrell, and D. L. Cox, *Phys. Rev. B* **42**, 4869 (1990); S. V. Meshkov, *ibid.* **48**, 6167 (1993); S. Miyashita and S. Yamamoto, *ibid.* 913 (1993); J. Deisz, M. Jarrell, and D. L. Cox, *ibid.* 10227 (1993).
45. K. Kubo and S. Takada, *J. Phys. Soc. Jpn.* **55**, 438 (1986); K. Betsuyaku, *Phys. Rev. B* **34**, 8125 (1986); K. Kubo, *ibid.* **46**, 866 (1992).
46. S. R. White and D. A. Huse, *Phys. Rev. B* **48**, 3844 (1993); E. S. Sorensen and I. Affleck, *Phys. Rev. Lett.* **71**, 1633 (1993); *Phys. Rev. B* **49**, 15771 (1994).
47. W. J. L. Buyers, R. M. Morra, R. L. Armstrong, M. J. Hogan, P. Gerlach, and K. Hirakawa, *Phys. Rev. Lett.* **56**, 371 (1986); M. Steiner, K. Kakurai, J. K. Kjems, D. Petitgrand, and R. Pynn, *J. Appl. Phys.* **61**, 3953 (1987); R. M. Morra, W. J. L. Buyers, R. L. Armstrong, and K. Hirakawa, *Phys. Rev. B* **38**, 543 (1988); Z. Tun, W. J. L. Buyers, R. L. Armstrong, K. Hirakawa, and B. Briat, *ibid.* **42**, 4677 (1990); I. A. Zaliznyak, L. P. Regnault, and D. Petitgrand, *ibid.* **50**, 15824 (1994).
48. J. P. Renard, M. Verdaguer, L. P. Regnault, W. A. C. Erkelens, J. Rossat-Mignod, and W. G. Stirling, *Europhys. Lett.* **3**, 949 (1987).
49. J. P. Renard, M. M. Verdaguer, L. P. Regnault, W. A. C. Erkelens, J. Rossat-Mignod, J. Ribas, W. G. Stirling, and C. Vettier, *J. Appl. Phys.* **63**, 3538 (1988); L. P. Regnault, C. Vettier, J. Rossat-Mignod, and J. P. Renard, *Physica B* **156–157**, 247 (1989); *ibid.* **180–181**, 188 (1992); S. Ma, C. Broholm, D. H. Reich, B. J. Sternlieb, and R. W. Erwin, *Phys. Rev. Lett.* **69**, 3571 (1992); L. P. Regnault, I. Zaliznyak, J. P. Renard, and C. Vettier, *Phys. Rev. B* **50**, 9174 (1994); S. Ma, D. H. Reich, C. Broholm, B. J. Sternlieb, and R. W. Erwin, *ibid.* **51**, 3289 (1995).
50. V. Gadet, M. Verdaguer, V. Briois, A. Gleizes, J. P. Renard, P. Beauvillain, C. Chappert, T. Goto, K. Le Dang, and P. Veillet, *Phys. Rev. B* **44**, 705 (1991).
51. H. Mutka, C. Payen, P. Molinié, J. L. Soubeyroux, P. Colombet, and A. D. Taylor, *Phys. Rev. Lett.* **67**, 497 (1991).
52. J. Darriet and J. P. Renard, *Solid State Commun.* **86**, 409 (1993); T. Sakaguchi, K. Kakurai, T. Yokoo, and J. Akimitsu, *J. Phys. Soc. Jpn.* **65**, 3025 (1996).
53. G. Xu, J. F. Ditusa, T. Ito, K. Oka, H. Takagi, C. Broholm, and G. Aeppli, *Phys. Rev. B* **54**, R6827 (1996).
54. Z. Honda, H. Askawa, and K. Katsumata, *Phys. Rev. Lett.* **81**, 2566 (1998); Y. Chen, Z. Honda, A. Zheludev, C. Broholm, K. Katsumata, and S. M. Shapiro, *ibid.* **86**, 1618 (2001).
55. Z. Honda, K. Katsumata, H. Aruga Katori, K. Yamada, T. Ohishi, T. Manabe, and M. Yamashita, *J. Phys. Condens. Matter* **9**, L83 (1997).

56. A. Zheludev, Z. Honda, K. Katsumata, R. Feyerherm, and K. Prokes, *Europhys. Lett.* **55**, 868 (2001); A. Zheludev, Z. Honda, Y. Chen, C. L. Broholm, K. Katsumata, and S. M. Shapiro, *Phys. Rev. Lett.* **88**, 077206 (2002).
57. I. Affleck, T. Kennedy, E. H. Lieb, and H. Tasaki, *Phys. Rev. Lett.* **59**, 799 (1987); *Commun. Math. Phys.* **115**, 477 (1988).
58. A. M. Tsvelik, *Phys. Rev. B* **42**, 10499 (1990).
59. G. Gómez-Santos, *Phys. Rev. Lett.* **63**, 790 (1989).
60. H.-J. Mikeska, *Europhys. Lett.* **19**, 39 (1992).
61. G. Fáth and J. Sólyom, *J. Phys. Condens. Matter* **5**, 8983 (1993).
62. M. den Nijs and K. Rommelse, *Phys. Rev. B* **40**, 4709 (1989).
63. H. Tasaki, *Phys. Rev. Lett.* **66**, 798 (1991).
64. T. Kennedy and H. Tasaki, *Phys. Rev. B* **45**, 304 (1992); *Commun. Math. Phys.* **147**, 431 (1992).
65. D. P. Arovas, A. Auerbach, and F. D. M. Haldane, *Phys. Rev. Lett.* **60**, 531 (1988).
66. S. Knabe, *J. Stat. Phys.* **52**, 627 (1988).
67. R. Scharf and H.-J. Mikeska, *J. Phys. Condens. Matter* **7**, 5083 (1995).
68. T. Kennedy, *J. Phys. Condens. Matter* **2**, 5737 (1990).
69. K. Chang, I. Affleck, G. W. Hayden, and Z. G. Soos, *J. Phys. Condens. Matter* **1**, 153 (1989); E. S. Sorensen and A. P. Young, *Phys. Rev. B* **42**, 754 (1990); J. Deisz, *ibid.* **46**, 2885 (1992).
70. S. M. Girvin and D. P. Arovas, *Phys. Scr. T* **27**, 156 (1989); D. P. Arovas and S. M. Girvin, *Recent Progress in Many-Body Theories*, Vol 3 (Plenum Press, New York, 1992), p. 315.
71. Y. Hatsugai and M. Kohmoto, *Phys. Rev. B* **44**, 11789 (1991); K. Kubo, *ibid.* **46**, 866 (1992); F. Alcaraz and Y. Hatsugai, *ibid.* 13914 (1992);
72. U. Schollwöck, T. Jolicoeur, and T. Garel, *Phys. Rev. B* **53**, 3304 (1996); O. Golinelli, T. Jolicoeur, and E. S. Sorensen, *Eur. Phys. J B* **11**, 199 (1999).
73. E. Polizzi, F. Mila, and E. S. Sorensen, *Phys. Rev. B* **58**, 2407 (1998).
74. S.-W. Tsai and J. B. Marston, *Phys. Rev. B* **62**, 5546 (2000).
75. S. Miyashita and S. Yamamoto, *Phys. Rev. B* **48**, 913, 9528 (1993); *ibid.* **51**, 3649 (1995).
76. M. Kaburagi, I. Harada, and T. Tonegawa, *J. Phys. Soc. Jpn.* **62**, 1848 (1993).
77. S. Qin, T. K. Ng, and Z. B. Su, *Phys. Rev. B* **52**, 12844 (1995).
78. Ö. Legeza, G. Fáth, and J. Sólyom, *Phys. Rev. B* **55**, 291 (1997); Ö. Legeza and J. Sólyom, *ibid.* **59**, 3606 (1999).
79. C. D. Batista, K. Hallberg, and A. A. Aligia, *Phys. Rev. B* **58**, 9248 (1998); *ibid.* **60**, R12553 (1999).
80. F. Alet and E. S. Sorensen, *Phys. Rev. B* **62**, 14116 (2000).
81. P. Mitra, B. Halperin, and I. Affleck, *Phys. Rev. B* **45**, 5299 (1992).
82. E. S. Sorensen and I. Affleck, *Phys. Rev. B* **49**, 15771 (1994).
83. P. Lecheminant and E. Orignac, *Phys. Rev. B* **65**, 174406 (2002).

84. S. H. Glarum, S. Geschwind, K. M. Lee, M. L. Kaplan, and J. Michel, *Phys. Rev. Lett.* **67**, 1614 (1991).
85. T. Goto, S. Satoh, Y. Matsumata, and M. Hagiwara, *Phys. Rev. B* **55**, 2709 (1997).
86. M. Hagiwara, K. Katsumata, I. Affleck, B. I. Halperin, and J. P. Renard, *Phys. Rev. Lett.* **65**, 3181 (1990).
87. F. Tedoldi, R. Santachiara, and M. Horvatić, *Phys. Rev. Lett.* **83**, 412 (1999).
88. M. Kenzelmann, G. Xu, I. A. Zaliznyak, C. Broholm, J. F. DiTusa, G. Aeppli, T. Ito, K. Oka, and H. Takagi, *Phys. Rev. Lett.* **90**, 087202 (2003).
89. H. J. Schulz, *Phys. Rev. B* **34**, 6372 (1986).
90. D. C. Cabra, P. Pujol, and C. von Reichenbach, *Phys. Rev. B* **58**, 65 (1998).
91. A. Moreo, *Phys. Rev. B* **35**, 8562 (1987); C. C. Alcaraz and A. Moreo, *ibid.* **46**, 2896 (1992).
92. T. Ziman and H. J. Schulz, *Phys. Rev. Lett.* **59**, 140 (1987).
93. K. Hallberg, X. Q. G. Wang, P. Horsch, and A. Moreo, *Phys. Rev. Lett.* **76**, 4955 (1996).
94. J. Lou, J. Dai, S. Qin, Z. Su, and L. Yu, *Phys. Rev. B* **62**, 8600 (2000).
95. M. Niel, C. Cros, G. Le Flem, M. Pouchard, and P. Hagenmuller, *Physica B* **86–88**, 702 (1977).
96. S. Itoh, K. Kakurai, Y. Endoh, and H. Tanaka, *Physica B* **213–214**, 161 (1995).
97. H. Mutka, C. Payen, and P. Molinié, *Europhys. Lett.* **21**, 623 (1993).
98. N. Hatano and M. Suzuki, *J. Phys. Soc. Jpn.* **62**, 1346 (1993); S. V. Meshkov, *Phys. Rev. B* **48**, 6167 (1993); J. Deisz, M. Jarell, and D. L. Cox, *ibid.* **48**, 10227 (1993); G. Sun, *ibid.* **51**, 8370 (1995); S. Yamamoto, *Phys. Rev. Lett.* **75**, 3348 (1995).
99. Y. Nishiyama, K. Totsuka, N. Hatano, and M. Suzuki, *J. Phys. Soc. Jpn.* **64**, 414 (1995).
100. U. Schollwöck and T. Jolicoeur, *Europhys. Lett.* **30**, 493 (1995); U. Schollwöck, O. Golinelli, and T. Jolicoeur, *Phys. Rev. B* **54**, 4038 (1996).
101. S. Qin, Y.-L. Liu, and L. Yu, *Phys. Rev. B* **55**, 2721 (1997); S. Qin, X. Wang, and L. Yu, *ibid.* **56**, R14251 (1997).
102. G. E. Granroth, M. W. Meisel, M. Chaparala, T. Jolicoeur, B. H. Ward, and D. R. Talham, *Phys. Rev. Lett.* **77**, 1616 (1996).
103. M. Oshikawa, *J. Phys. Condens. Matter* **4**, 7469 (1992).
104. K. Totsuka and M. Suzuki, *J. Phys. Condens. Matter* **7**, 1639 (1995).
105. Y. Hatsugai, *J. Phys. Soc. Jpn.* **61**, 3856 (1992).
106. T.-K. Ng, *Phys. Rev. B* **50**, 555 (1994).
107. H. Yamazaki and K. Katsumata, *Phys. Rev. B* **54**, R6831 (1996).
108. J. Lou, S. Qin, T.-K. Ng, and Z. Su, *Phys. Rev. B* **65**, 104401 (2002).
109. S. Qin, J. Lou, L. Sun, and C. Chen, *Phys. Rev. Lett.* **90**, 067202 (2003); J. Lou, S. Qin, and C. Chen, *Phys. Rev. Lett.* **91**, 087204 (2003); C. Chen, *Phys. Lett. A* **317**, 135 (2003).

110. For reviews, see E. Dagotto and T. M. Rice, *Science* **271**, 618 (1996); T. M. Rice, *Z. Phys. B* **103**, 165 (1997); E. Dagotto, *Rep. Prog. Phys.* **62**, 1525 (1999).
111. S. Gopolan, T. M. Rice, and M. Sigrist, *Phys. Rev. B* **49**, 8901 (1994).
112. S. R. White, R. M. Noack, and D. J. Scalapino, *Phys. Rev. Lett.* **73**, 886 (1994).
113. B. Frischmuth, B. Ammon, and M. Troyer, *Phys. Rev. B* **54**, R3714 (1996); B. Frischmuth, S. Haas, G. Sierra, and T. M. Rice, *ibid.* **55**, R3340 (1997).
114. A. G. Rojo, *Phys. Rev. B* **53**, 9172 (1996).
115. G. Sierra, *J. Phys. A* **29**, 3299 (1996).
116. For a review, see for instance: J. Voit, *Rep. Prog. Phys.* **58**, 977 (1995); H. J. Schulz, G. Cuniberti, and P. Pieri, *Lecture notes of the Chia Laguna summer school*, (Springer Verlag, 1997); arXiv: cond-mat/9807366.
117. M. Azuma, Z. Hiroi, M. Takano, K. Ishida, and Y. Kitaoka, *Phys. Rev. Lett.* **73**, 3463 (1994).
118. K. Kojima, A. Keren, G. M. Luke, B. Nachumi, W. D. Wu, Y. J. Uemura, M. Azuma, and M. Takano, *Phys. Rev. Lett.* **74**, 2812 (1995).
119. S. A. Carter, B. Batlogg, R. J. Carla, J. J. Krajewski, W. F. Peck Jr., and T. M. Rice, *Phys. Rev. Lett.* **77**, 1378 (1996); R. S. Eccleston, M. Uehara, J. Akimitsu, H. Eisaki, N. Motoyama, and S.-I. Uchida, *ibid.* **81**, 1702 (1998); T. Imai, K. R. Thurber, K. M. Shen, A. W. Hunt, and F. C. Chou, *ibid.* 220 (1998); M. Takigawa, N. Motoyama, H. Eisaki, and S. Uchida, *Phys. Rev. B* **57**, 1124 (1998); L. P. Regnault, A. H. Moudden, J. P. Boucher, E. Lorenzo, A. Hiess, A. Vietkin, and A. Revcolevschi, *Physica B* **259–261**, 1038 (1999); A. Glozar, G. Blumberg, B. S. Dennis, B. S. Shastry, N. Motoyama, H. Eisaki, and S. Uchida, *Phys. Rev. Lett.* **87**, 197202 (2001).
120. C. A. Hayward, D. Poilblanc, and L. P. Lévy, *Phys. Rev. B* **54**, R12649 (1996); G. Chaboussant, P. A. Crowell, L. P. Lévy, O. Piovesana, A. Madouri, and D. Mailly, *ibid.* **55**, 3046 (1997); G. Chaboussant, M.-H. Julien, Y. Fagot-Revurat, L. P. Lévy, C. Berthier, M. Horvatić, and O. Piovesana, *Phys. Rev. Lett.* **79**, 925 (1997); G. Chaboussant, Y. Fagot-Revurat, M.-H. Julien, M. E. Hanson, C. Berthier, M. Horvatić, L. P. Lévy, and O. Piovesana, *ibid.* **80**, 2713 (1998).
121. P. R. Hammar, D. H. Reich, C. Broholm, and F. Trouw, *Phys. Rev. B* **57**, 7846 (1998).
122. B. C. Watson *et al.*, *Phys. Rev. Lett.* **86**, 5168 (2001).
123. K. Katoh, Y. Hosokoshi, K. Inoue, and T. Goto, *J. Phys. Soc. Jpn.* **69**, 1008 (2000); K. Katoh, Y. Hosokoshi, K. Inoue, M. I. Bartashevich, H. Nakano, and T. Goto, *J. Phys. Chem. Sol.* **63**, 1277 (2002).
124. S. Brehmer, H.-J. Mikeska, M. Müller, N. Nagaosa, and S. Uchida, *Phys. Rev. B* **60**, 329 (1999); M. Matsuda, K. Katsumata, R. S. Eccleston, S. Brehmer, and H.-J. Mikeska, *ibid.* **62**, 8903 (2000).
125. T. S. Nunner, P. Brune, T. Kopp, M. Windt, and M. Grüninger, *Phys. Rev. B* **66**, 180404 (2002).

126. T. Barnes, E. Dagotto, J. Riera, and E. S. Swanson, *Phys. Rev. B* **47**, 3196 (1993).
127. K. Hida, *J. Phys. Soc. Jpn.* **60**, 1347 (1991); *ibid.* **64**, 4896 (1995).
128. H. Watanabe, *Phys. Rev. B* **50**, 13442 (1994); *ibid.* **52**, 12508 (1995).
129. Y. Nishiyama, N. Hatano, and M. Suzuki, *J. Phys. Soc. Jpn.* **64**, 1967 (1995).
130. S. R. White, *Phys. Rev. B* **53**, 52 (1996).
131. S. P. Strong and A. J. Millis, *Phys. Rev. Lett.* **69**, 2419 (1992); *Phys. Rev. B* **50**, 9911 (1994).
132. D. G. Shelton, A. A. Nersesyan, and A. M. Tsvelik, *Phys. Rev. B* **53**, 8521 (1996).
133. A. A. Nersesyan and A. M. Tsvelik, *Phys. Rev. Lett.* **78**, 3969 (1997); *Errata* **79**, 1171 (1997).
134. L. Takhtajan, *Phys. Lett. A* **87**, 479 (1982); J. Babujian, *ibid.* **90**, 479 (1982).
135. D. J. Gross and A. Neveu, *Phys. Rev. D* **10**, 3235 (1974).
136. D. Allen and D. Sénéchal, *Phys. Rev. B* **55**, 299 (1997).
137. E. H. Kim, G. Fáth, J. Sólyom, and D. J. Scalapino, *Phys. Rev. B* **62**, 14965 (2000); G. Fáth, Ö. Legeza, and J. Sólyom, *ibid.* **63**, 134403 (2001).
138. N. Read and B. Chakraborty, *Phys. Rev. B* **40**, 7133 (1989).
139. M. Nakamura, *Physica B* **329–333**, 1000 (2003).
140. S. Takada and H. Watanabe, *J. Phys. Soc. Jpn.* **61**, 39 (1992).
141. K. Damle and S. Sachdev, *Phys. Rev. B* **57**, 8307 (1998).
142. O. P. Sushkov and V. N. Kotov, *Phys. Rev. Lett.* **81**, 1941 (1998); V. N. Kotov, O. P. Sushkov and R. Eder, *Phys. Rev. B* **59**, 6266 (1999).
143. C. Jurecka and W. Brenig, *Phys. Rev. B* **61**, 14307 (2000).
144. S. Trebst, H. Monien, C. J. Hamer, Z. Weihong, and R. R. P. Singh, *Phys. Rev. Lett.* **85**, 4373 (2000); W. Zheng, C. J. Hamer, R. R. P. Singh, S. Trebst, and H. Monien, *Phys. Rev. B* **63**, 144410 (2001).
145. M. Windt, M. Grüninger, T. Nunner, C. Knetter, K. P. Schmidt, G. S. Uhrig, T. Kopp, A. Freimuth, U. Ammerahl, B. Büchner, and A. Revcolevschi, *Phys. Rev. Lett.* **87**, 127002 (2001).
146. K. I. Kugel and D. I. Khomskii, *Sov. Phys. Usp.* **25**, 231 (1982).
147. S. K. Pati, R. R. P. Singh, and D. I. Khomskii, *Phys. Rev. Lett.* **81**, 5406 (1998).
148. E. Axtell, T. Ozawa, S. Kauzlarich, and R. R. P. Singh, *J. Solid State Chem.* **134**, 423 (1997).
149. E. Orignac, R. Citro, and N. Andrei, *Phys. Rev. B* **61**, 11533 (2000).
150. A. K. Kolezhuk and H.-J. Mikeska, *Phys. Rev. Lett.* **80**, 2709 (1998); A. K. Kolezhuk, H.-J. Mikeska, and U. Schollwöck, *Phys. Rev. B* **63**, 064418 (2000).
151. M. J. Martins and B. Nienhuis, *Phys. Rev. Lett.* **85**, 4956 (2000).
152. P. Lecheminant and E. Orignac, unpublished.
153. J. Villain, *J. Phys. Fr* **38**, 385 (1977); Chiral order in Helimagnets, proceedings of the 13th IUPAP Conference on Statistical Physics, edited by C. Weil, D. Cabid, C. G. Kuper, and I. Riess (1978).
154. T. Jolicoeur and J. C. Le Guillou, *Phys. Rev. B* **40**, 2727 (1989).

155. See for instance, N. D. Mermin, *Rev. Mod. Phys.* **51**, 591 (1979).
156. T. Senthil and M. P. A. Fisher, *Phys. Rev. B* **62**, 7850 (2000).
157. S. Rao and D. Sen, *Nucl. Phys. B* **424**, 547 (1994); *J. Phys. Condens. Matter* **9**, 1831 (1997).
158. D. Allen and D. Sénéchal, *Phys. Rev. B* **51**, 6394 (1995).
159. T. Dombre and N. Read, *Phys. Rev. B* **39**, 6797 (1989).
160. F. D. M. Haldane, talk at the workshop New Theoretical Approaches to Strongly Correlated Systems, Cambridge (2000); F. D. M. Haldane, *Phys. Rev. Lett.* **61**, 1029 (1988).
161. J. von Delft and C. L. Henley, *Phys. Rev. Lett.* **69**, 3236 (1992); *Phys. Rev. B* **48**, 965 (1992).
162. D. Loss, D. P. DiVincenzo, and G. Grinstein, *Phys. Rev. Lett.* **69**, 3232 (1992).
163. A. M. Polyakov and P. B. Wiegmann, *Phys. Lett. B* **131**, 121 (1983); *ibid.* **141**, 223 (1984).
164. E. Ogievetsky, N. Reshetikhin, and P. B. Wiegmann, *Nucl. Phys. B* **280**, 45 (1987).
165. F. D. M. Haldane, *Phys. Rev. B* **25**, 4925 (1982); Errata *ibib.* **26**, 5257 (1982).
166. M. Hase, I. Terasaki, and K. Uchinokura, *Phys. Rev. Lett.* **70**, 3651 (1993).
167. For a review, see for instance: J. P. Boucher and L. P. Regnault, *J. Phys. I* **6**, 1939 (1996).
168. M. Isobe and Y. Ueda, *J. Phys. Soc. Jpn.* **65**, 3142 (1996); N. Fujiwara, H. Yasuoka, M. Isobe, Y. Ueda, and S. Maegawa, *Phys. Rev. B* **55**, R11945 (1997).
169. G. Castilla, S. Chakravarty, and V. J. Emery, *Phys. Rev. Lett.* **75**, 1823 (1995); J. Riera and A. Dobry, *Phys. Rev. B* **51**, 16098 (1995).
170. M. Matsuda and K. Katsumata, *J. Mag. Mag. Mat.* **140–144**, 1671 (1995); M. Matsuda, K. Katsumata, K. M. Kojima, M. Larkin, G. M. Luke, J. Merrin, B. Nachumi, Y. J. Uemura, H. Eisaki, N. Motoyama, S. Uchida, and G. Shirane, *Phys. Rev. B* **55**, R11953 (1997).
171. N. Motoyama, H. Eisaki, and S. Uchida, *Phys. Rev. Lett.* **76**, 3212 (1996).
172. N. Maeshima, M. Hagiwara, Y. Narumi, K. Kindo, T. C. Kobayashi, and K. Okunishi, *J. Phys. Condens. Matter* **15**, 3607 (2003).
173. H. Kikuchi, H. Nagasawa, Y. Ajiro, T. Asano, and T. Goto, *Physica B* **284–288**, 1631 (2000).
174. V. L. Berezinskii, *Sov. Phys. JETP* **34**, 610 (1972); J. M. Kosterlitz and D. J. Thouless, *J. Phys. C* **6**, 1181 (1973); J. M. Kosterlitz, *ibid.* **7**, 1046 (1974).
175. K. Okamoto and K. Nomura, *Phys. Lett. A* **169**, 433 (1992); K. Nomura and K. Okamoto, *J. Phys. A* **27**, 5773 (1994).
176. R. Chitra, S. Pati, H. R. Krishnamurthy, D. Sen, and S. Ramasesha, *Phys. Rev. B* **52**, 6581 (1995).
177. S. Eggert, *Phys. Rev. B* **54**, 9612 (1996).
178. S. White and I. Affleck, *Phys. Rev. B* **54**, 9862 (1996).

179. N. Andrei and J. H. Lowenstein, *Phys. Rev. Lett.* **43**, 1698 (1979); *Phys. Lett. B* **90**, 106 (1980).
180. A. A. Belavin, *Phys. Lett. B* **87**, 117 (1979).
181. C. K. Majumdar and D. K. Ghosh, *J. Math. Phys.* **10**, 1388, 1399 (1969).
182. B. S. Shastry and B. Sutherland, *Phys. Rev. Lett.* **47**, 964 (1981).
183. W. J. Caspers, K. M. Emmett, and W. Magnus, *J. Phys. A* **17**, 2687 (1984).
184. E. Sorensen, I. Affleck, D. Augier, and D. Poilblanc, *Phys. Rev. B* **58**, R14701 (1998).
185. W. J. Caspers and W. Magnus, *Phys. Lett. A* **88**, 103 (1982).
186. G. Fáth and A. Süto, *Phys. Rev. B* **62**, 3778 (2000).
187. K. Nomura, *J. Phys. Soc. Jpn.* **72**, 476 (2003).
188. R. Bursill, G. A. Gehring, D. J. J. Farnell, J. B. Parkinson, T. Xiang, and C. Zeng, *J. Phys. Condens. Matter* **7**, 8605 (1995).
189. S. Watanabe and H. Yokoyama, *J. Phys. Soc. Jpn.* **68**, 2073 (1999).
190. A. A. Aligia, C. D. Batista, and F. H. L. Essler, *Phys. Rev. B* **62**, 3259 (2000).
191. J. Stephenson, *Can. J. Phys.* **47**, 2621 (1969); *ibid.* **48**, 1724, 2118 (1970); *J. Math. Phys.* **12**, 420 (1970).
192. T. Tonegawa, M. Kaburagi, N. Ichikawa, and I. Harada, *J. Phys. Soc. Jpn.* **61**, 2890 (1992).
193. S. Pati, R. Chitra, D. Sen, H. R. Krishnamurthy, and S. Ramasesha, *Europhys. Lett.* **33**, 707 (1996).
194. A. Kolezhuk, R. Roth, and U. Schollwöck, *Phys. Rev. Lett.* **77**, 5142 (1996); *Phys. Rev. B* **55**, 8928 (1997).
195. R. Roth and U. Schollwöck, *Phys. Rev. B* **58**, 9264 (1998).
196. A. Kolezhuk and U. Schollwöck, *Phys. Rev. B* **65**, 100401 (2002).
197. I. Dzyaloshinskii, *J. Phys. Chem. Solids* **4**, 241 (1958); T. Moriya, *Phys. Rev. Lett.* **4**, 228 (1960); *Phys. Rev.* **120**, 91 (1960).
198. D. Allen, PhD thesis, Sherbrooke university, 1998.
199. A. A. Nersesyan, A. O. Gogolin, and F. H. L. Essler, *Phys. Rev. Lett.* **81**, 910 (1998).
200. D. C. Cabra, A. Honecker, and P. Pujol, *Eur. Phys. J. B* **13**, 55 (2000).
201. D. Allen, F. H. L. Essler, and A. A. Nersesyan, *Phys. Rev. B* **61**, 8871 (2000).
202. C. Itoi and S. Qin, *Phys. Rev. B* **63**, 224423 (2001).
203. D. Allen and D. Sénéchal, *Phys. Rev. B* **61**, 12134 (2000).
204. R. Chitra and T. Giamarchi, *Phys. Rev. B* **55**, 5816 (1997).
205. G. I. Dzhaparidze and A. A. Nersesyan, *JETP Lett.* **27**, 334 (1978); *J. Low. Temp. Phys.* **37**, 95 (1979).
206. V. L. Pokrovsky and A. L. Talapov, *Phys. Rev. Lett.* **42**, 65 (1979); *Sov. Phys. JETP* **51**, 134 (1980).
207. H. J. Schulz, *Phys. Rev. B* **22**, 5274 (1980).
208. Y. Okwamoto, *J. Phys. Soc. Jpn.* **49**, 8 (1980).
209. A. A. Nersesyan, A. Luther, and F. V. Kusmartsev, *Phys. Lett. A* **176**, 363 (1993).

210. D. V. Dmitriev, V. Ya. Krivnov, and A. A. Ovchinnikov, *Phys. Rev. B* **65**, 172409 (2002); D. V. Dmitriev, V. Ya. Krivnov, A. A. Ovchinnikov, and A. Langari, *JETP* **95**, 538 (2002).
211. A. Dutta and D. Sen, *Phys. Rev. B* **67**, 094435 (2003).
212. A. A. Nersesyan, *Phys. Lett. A* **153**, 49 (1991); A. A. Nersesyan, A. Luther, and F. V. Kusmartsev, *JETP Lett.* **55**, 692 (1992).
213. P. Fendley and C. Nayak, *Phys. Rev. B* **63**, 115102 (2001).
214. D. Allen, P. Azaria, and P. Lecheminant, *J. Phys. A* **34**, L305 (2001).
215. J. L. Cardy, *Nucl. Phys. B* **389**, 577 (1993).
216. A. M. Tsvelik, *Nucl. Phys. B* **612**, 479 (2001).
217. D. Allen, private communication and unpublished.
218. Y. Nishiyama, *Eur. Phys. J. B* **17**, 295 (2000).
219. T. Hikihara, M. Kaburagi, and H. Kawamura, *Phys. Rev. B* **63**, 174430 (2001); *Prog. Theor. Phys. Suppl.* **145**, 58 (2002).
220. P. D. Sacramento and V. R. Vieira, *J. Phys. Condens. Matter* **14**, 591 (2002); V. R. Vieira, N. Guihery, J. P. Rodriguez, and P. D. Sacramento, *Phys. Rev. B* **63**, 224417 (2001).
221. M. Kaburagi, H. Kawamura, and T. Hikihara, *J. Phys. Soc. Jpn.* **68**, 3185 (1999); T. Hikihara, M. Kaburagi, H. Kawamura, and T. Tonegawa, *ibid.* **69**, 259 (2000).
222. A. K. Kolezhuk, *Phys. Rev. B* **62**, R6057 (2000); *Prog. Theor. Phys. Suppl.* **145**, 29 (2002).
223. P. Lecheminant, T. Jolicoeur, and P. Azaria, *Phys. Rev. B* **63**, 174426 (2001); T. Jolicoeur and P. Lecheminant, *Prog. Theor. Phys. Suppl.* **145**, 23 (2002).
224. Y. Q. Li, M. Ma, D. N. Shi, and F. C. Zhang, *Phys. Rev. Lett.* **81**, 3527 (1998).
225. Y. Yamashita, N. Shibata, and K. Ueda, *Phys. Rev. B* **58**, 9114 (1998).
226. B. Frischmuth, F. Mila, and M. Troyer, *Phys. Rev. Lett.* **82**, 835 (1999).
227. B. Sutherland, *Phys. Rev. B* **12**, 3795 (1975).
228. Y.-Q. Li, M. Ma, D.-N. Shi, and F.-C. Zhang, *Phys. Rev. B* **60**, 12781 (1999).
229. I. Affleck, *Nucl. Phys. B* **305**, 582 (1988).
230. A. B. Zamolodchikov, *JETP Lett.* **43**, 730 (1986).
231. P. Azaria, A. O. Gogolin, P. Lecheminant, and A. A. Nersesyan, *Phys. Rev. Lett.* **83**, 624 (1999); P. Azaria, E. Boulat, and P. Lecheminant, *Phys. Rev. B* **61**, 12112 (2000).
232. C. Itoi, S. Qin, and I. Affleck, *Phys. Rev. B* **61**, 6747 (2000).
233. B. Gerganov, A. Leclair, and M. Moriconi, *Phys. Rev. Lett.* **86**, 4753 (2001); A. Leclair, *Phys. Rev. B* **64**, 045329 (2001).
234. H.-H. Lin, L. Balents, and M. P. A. Fisher, *Phys. Rev. B* **58**, 1794 (1998).
235. R. Assaraf, P. Azaria, E. Boulat, M. Caffarel, and P. Lecheminant, arXiv: cond-mat/0310090.
236. Y. Yamashita, N. Shibata, and K. Ueda, *J. Phys. Soc. Jpn.* **69**, 242 (2000).
237. G. V. Uimin, *JETP Lett.* **12**, 225 (1970); C. K. Lai, *J. Math. Phys.* **15**, 1675 (1974).

238. C. Itoi and M.-H. Kato, *Phys. Rev. B* **55**, 8295 (1997).
239. G. Fáth and J. Sólyom, *Phys. Rev. B* **44**, 11836 (1991); *ibid.* **47**, 872 (1993).
240. R. Citro, E. Orignac, N. Andrei, C. Itoi, and S. Qin, *J. Phys. Condens. Matter* **12**, 3041 (2000).
241. P. Azaria, P. Lecheminant, and A. A. Nersesyan, *Phys. Rev. B* **58**, R8881 (1998); P. Azaria and P. Lecheminant, *Nucl. Phys. B* **575**, 439 (2000).
242. N. Andrei, M. Douglas, and A. Jerez, *Phys. Rev. B* **58**, 7619 (1998).
243. X. Wang, N. Zhu, and C. Chen, *Phys. Rev. B* **66**, 172405 (2002).
244. S. Sachdev and M. Vojta, in *Proceedings of the XIII International Congress on Mathematical Physics*, edited by A. Fokas, A. Grigoryan, T. Kibble, and B. Zegarlinski, (International Press, Boston, 2001); arXiv: cond-mat/0009202.
245. B. Normand and F. Mila, *Phys. Rev. B* **65**, 104411 (2002).
246. X. G. Wen, F. Wilczek, and A. Zee, *Phys. Rev. B* **39**, 11413 (1989).
247. A. Läuchli, G. Schmid, and M. Troyer, *Phys. Rev. B* **67**, 100409(R) (2003).
248. T. Hikihara, T. Momoi, and X. Hu, *Phys. Rev. Lett.* **90**, 087204 (2003); T. Momoi, T. Hikihara, M. Nakamura, and X. Hu, *Phys. Rev. B* **67**, 174410 (2003).
249. A. A. Nersesyan and A. M. Tsvelik, *Phys. Rev. B* **67**, 024422 (2003); F. A. Smirnov and A. M. Tsvelik, *ibid.* **68**, 144412 (2003); M. J. Bhaseen and A. M. Tsvelik, *ibid.* **68**, 094405 (2003).
250. O. A. Starykh, R. R. P. Singh, and G. C. Levine, *Phys. Rev. Lett.* **88**, 167203 (2002); R. R. P. Singh, O. A. Starykh, and P. J. Freitas, *J. Appl. Phys.* **83**, 7387 (1998).
251. P. Sindzingre, J.-B. Fouet, and C. Lhuillier, *Phys. Rev. B* **66**, 174424 (2002).
252. V. J. Emery, E. Fradkin, S. A. Kivelson, and T. C. Lubensky, *Phys. Rev. Lett.* **85**, 2160 (2000); A. Vishwanath and D. Carpentier, *Phys. Rev. Lett.* **86**, 676 (2001); R. Mukhopadhyay, C. L. Kane, and T. C. Lubensky, *Phys. Rev. B* **63**, 081103 (2001); *ibid.* **64**, 045120 (2001); S. L. Sondhi and K. Yang, *ibid.* **63**, 054430 (2001).
253. Z.-C. Gu and X.-G. Wen, *Phys. Rev. B* **80**, 155131 (2009).
254. X. Chen, Z.-C. Gu, and X.-G. Wen, *Phys. Rev. B* **82**, 155138 (2010).
255. F. Pollmann, A. M. Turner, E. Berg, and M. Oshikawa, *Phys. Rev. B* **81**, 064439 (2010).
256. F. Pollmann, E. Berg, A. M. Turner, and M. Oshikawa, *Phys. Rev. B* **85**, 075125 (2012).
257. X. Chen, Z.-C. Gu, and X. G. Wen, *Phys. Rev. B* **83**, 035107 (2011); *Phys. Rev. B* **84**, 235128 (2011); N. Schuch, D. Pérez-García, and I. Cirac, *Phys. Rev. B* **84**, 165139 (2011); L. Fidkowski and A. Kitaev, *Phys. Rev. B* **83**, 075103 (2011).
258. A. A. Belavin, A. M. Polyakov, and A. B. Zamolodchikov, *Nucl. Phys. B* **241**, 333 (1984).
259. M. Lewenstein, A. Sanpera, and V. Ahufinger, *Ultracold Atoms in Optical Lattices: Simulating Quantum Many-Body Systems* (Oxford University Press, 2012).

260. J. J. Garcia-Ripoll, M. A. Delgado, and J. I. Cirac, *Phys. Rev. Lett.* **93**, 250405 (2004); D. Porras and J. I. Cirac, *Phys. Rev. Lett.* **92**, 207901 (2004); A. V. Gorshkov *et al.*, *Phys. Rev. A* **84**, 033619 (2011).
261. Z.-X. Chen *et al.*, *Phys. Rev. A* **81**, 022303 (2010); L.-H. Du *et al.*, *Phys. Rev. A* **86**, 032302 (2012); Z.-X. Liu, Z.-B. Yang, Y.-J. Han, W. Yi, and X.-G. Wen, *Phys. Rev. B* **86**, 195122 (2012).

CHAPTER 7

SPIN ICE

Steven T. Bramwell*, Michel J. P. Gingras†,‡
and Peter C. W. Holdsworth§

* *University College London, Department of Chemistry, 20 Gordon Street, London WC1H 0AJ, UK*
† *Department of Physics, University of Waterloo, Waterloo, Ontario, N2L 3G1, Canada*
‡ *Canadian Institute for Advanced Research, 180 Dundas Street West, Toronto, Ontario, M5G 1Z8, Canada*
§ *Laboratoire de Physique, Ecole Normale Supérieure, 46 Avenue d'Italie, F-69634 Lyon Cedex 07, France*

Pauling's model of hydrogen disorder in water ice represents the prototype of a frustrated system. Over the years, it has spawned several analogous models, including Anderson's model antiferromagnet and the statistical "vertex" models. Spin ice is a 16-vertex model of "ferromagnetic frustration" that is approximated by real materials, most notably the rare earth pyrochlores $Ho_2Ti_2O_7$, $Dy_2Ti_2O_7$, and $Ho_2Sn_2O_7$. These "spin ice materials" have the Pauling zero-point entropy and in all respects represent almost ideal realizations of Pauling's model. They provide experimentalists with unprecedented access to a wide variety of novel magnetic states and phase transitions that are located in different regions of the field–temperature phase diagram. They afford theoreticians the opportunity to explore many new features of the magnetic interactions and statistical mechanics of frustrated systems. This chapter is a comprehensive review of the physics — both experimental and theoretical — of spin ice. It starts with a discussion of the historic problem of water ice and its relation to spin ice and other frustrated magnets. The properties of spin ice are then discussed in three sections that deal with the zero-field spin ice state, the numerous field-induced states (including the recently identified "kagomé ice"), and the magnetic dynamics. Some materials related to spin ice are briefly described, and the chapter is concluded with a short summary of spin ice physics.

7.1. Introduction

Pauling's model of the hydrogen structure of water ice, with its characteristic property of zero-point entropy, stands as a canonical model of disorder in condensed matter.[1] Magnetic models with similar properties were developed half a century ago by Wannier[2] and Anderson,[3] who demonstrated how zero-point entropy could arise in Ising model antiferromagnets with triangular[2] or tetrahedral geometry.[3] These antiferromagnetic systems are now described as "geometrically frustrated" as their lattice geometry indicates that they are unable to satisfy the local minimum energy conditions (although the concept of frustration first arose from spin glass research where it is usually associated with random chemical disorder[6]). However, despite the historic role of Pauling's model in magnetism research, it is only relatively recently that its magnetic equivalent has been realized in the laboratory. "Spin ice"[4,5] — a microscopic mapping of Pauling's model onto a *ferromagnetic*, rather than an antiferromagnetic system — is approximated almost ideally by the so-called spin ice materials. These systems finally provide concrete experimental examples of the basic models of magnetic frustration envisaged by Wannier and Anderson and of the more generalized ice-type models that have been studied theoretically for many years.[7,8]

The prototypical spin ice materials $Ho_2Ti_2O_7$ and $Dy_2Ti_2O_7$ are oxides with the pyrochlore structure in which the Ti^{4+} ions are non-magnetic and the trivalent rare earth ions occupy a face-centered cubic lattice of corner-linked tetrahedra, as shown in Fig. 7.1. They are highly crystalline electrical insulators with band gaps of the order of optical frequencies. The crystal field on the rare earth ions is much stronger than the (effectively ferromagnetic) dipole interaction, which in turn largely dominates the superexchange (thought to be weakly antiferromagnetic). The single-ion ground state (GS) is a doublet, composed almost entirely of free ion sub-levels with the maximum $M_J = \pm J$,[a] with the first group of excited crystal field states lying

[a]The total angular momentum of the free rare earth ion is characterized by two quantum numbers J and M_J, which takes the values $J, J-1 \cdots -J$. The observable magnetic moment is $\mu_z = -g_J \mu_B M_J$, where g_J is the Landé g-factor and z denotes the quantization axis. $M_J = \pm J$ therefore corresponds to a moment $\mp g_J \mu_B J$.

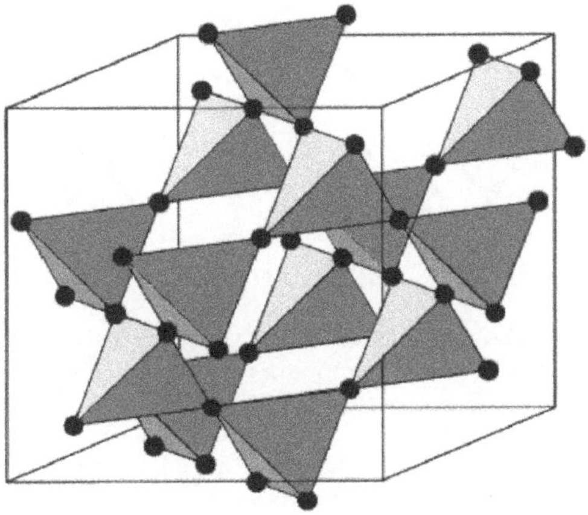

Fig. 7.1. The pyrochlore lattice occupied by the rare earth ions in the spin ice materials.

some 300 K above the GS. This gives rise to a strong single-ion anisotropy that at low temperature constrains the atomic magnetic moment (henceforth "spin") parallel to its local quantization axis, a member of the $\langle 111 \rangle$ set.[b] Consequently, in terms of its static properties, each moment behaves like a classical Ising spin lying parallel to the axis that connects the center of the elementary tetrahedron of the pyrochlore lattice to its vertex.

When one adds ferromagnetic spin–spin coupling to this model of $\langle 111 \rangle$ Ising spins, the GS of a single tetrahedron is determined by the condition that two spins must point "in" and two spins must point "out" of each tetrahedron. This "two in, two out" rule minimizes only four of the six pairwise interactions, which is the best that can be achieved, given the tetrahedral geometry. Hence, spin ice is frustrated. The GS rules for spin ice are equivalent to the rules that control local hydrogen arrangements in Pauling's model of water ice. In fact, the mapping of the $\langle 111 \rangle$ Ising ferromagnet onto Pauling's model is precise if one considers cubic ice, the spins playing the role of proton displacement vectors located at the midpoints of the oxide–oxide lines of contact (see Fig. 7.2). The remarkable fact is that this scenario actually occurs in the spin ice materials, to an excellent approximation. Similar to water ice, they have zero-point entropy, and there are also analogies in the dynamics. However, unlike water ice, the spin ice

[b]We adopt the standard crystallographic convention that $[\cdots]$ refers to the indices of a single lattice direction and $\langle \cdots \rangle$ refers to the set of all equivalent lattice directions.

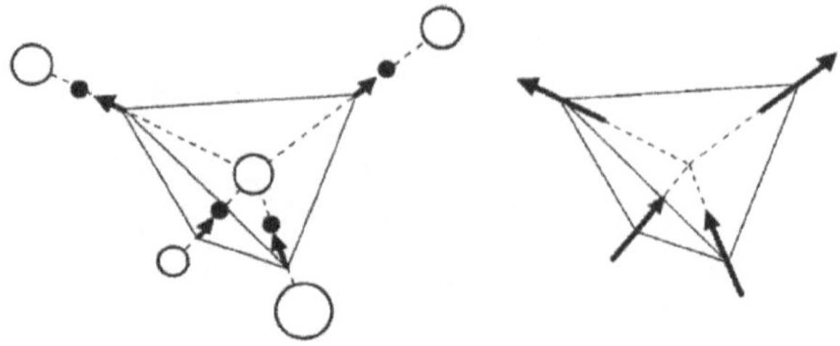

Fig. 7.2. Illustration of the equivalence of the water ice rule "two protons near, two protons far" and the spin ice rule "two spins in, two spins out". The diagram (left) illustrates a water molecule in the tetrahedral coordination of the ice structure (oxides are denoted by open circles, protons by filled circles). The proton positions are located by displacement vectors that occupy a lattice of linked tetrahedra. In spin ice (right), the displacement vectors are replaced by rare earth moments ("spins") occupying the pyrochlore lattice (Fig. 7.1), which is the dual lattice (i.e. the lattice formed by the mid-points of the bonds) of the oxide lattice in cubic ice.

materials are ideal systems for both experimental and theoretical studies: samples are easy to handle; the Hamiltonian is simple, well-established, and tractable; magnetic fields can be applied that far exceed in relative magnitude any electric fields that can be applied to water ice; spin dynamics are fast compared to proton dynamics, and so forth. In fact, studies of spin ice materials have started to reveal much new physics that might in principle be observed in ice itself, were it more amenable to experiment. It could thus be argued that spin ice has already replaced water ice as the canonical realization of an "ice-type" model.

Whatever its ultimate significance, the spin ice phenomenon provides an ideal system for the combined experimental and theoretical approach to the statistical mechanics of frustrated systems, as well as an interesting new type of magnetic material. The purpose of this review is to describe the current state of experimental and theoretical work on spin ice, to identify the main trends for future research, and to assess the significance of the new physics revealed. It is not primarily a comparison of spin ice and water ice, although we do describe the historic problem of water ice in Section 7.2.1. Nor is it an attempt to put spin ice in the general context of frustrated magnetism, although we briefly allude to this in Section 7.2.5. It is hoped that this review will be of interest to those who know nothing of spin ice as well as workers in the field who are seeking an overview. We note two previous reviews of spin ice[9,10] and that spin ice has been discussed in the generally broader context of frustrated materials and related phenomena.[11,12]

7.2. From Water Ice to Spin Ice

7.2.1. *Pauling's model*

Ice is the paradigm for disordered systems, not least because of the experimental evidence it shows for the apparent violation of the third law of thermodynamics. A series of quite remarkable experiments terminated, in 1936, with Giauque and Stout's paper *The Entropy of Water and the Third Law of Thermodynamics. Heat Capacity of Ice from 15 to 273 K.*[13] In their paper, Giauque and Stout reported calorimetric measurements on water through the freezing and vaporization transitions up to the high-temperature gas phase. The entropy was calculated by integrating the heat capacity and adding the latent heat contributions; the low-temperature measurements being extrapolated to zero, using Debye's then recently derived formula.[14] The resulting entropy, $S_1 = 44.28\,\text{Cal/deg}\cdot\text{mol}$, was compared with a second estimate, $S_2 = 45.10\,\text{Cal/deg}\cdot\text{mol}$, calculated using the statistical mechanics of an ideal gas, with internal energy levels taken from spectroscopic measurements. The two values were found to disagree by $S_0 = 0.82 \pm 0.05\,\text{Cal/deg}\cdot\text{mol}$. The difference was famously accounted for, to an excellent approximation, by Pauling,[1] who showed that ice possesses a finite entropy (estimated as $0.81\,\text{Cal/deg}\cdot\text{mol}$) at zero temperature due to the configurational disorder intrinsic to the proton system.

The ice problem is a classic example of how the separation of energy scales can leave certain classes of internal degrees of freedom as irrelevant and frustrated bystanders in the energy minimization process. Here, the chemical binding energy of the water molecule is so strong (221 kCal/mol) that it is left essentially unchanged in the solid condensed phase. As a result, the GS configuration does not minimize the electrostatic energy of a neutral ensemble of O^{2-} and H^+ ions. Rather, in the hexagonal ("wurtzite") and cubic ("sphalerite") ice phases,[15] the oxygen ions form an open tetrahedral structure, whose 109° angles conform closely to that encompassing the H–O–H bonds of an isolated water molecule. In the wurtzite phase, the O–O bond length is 2.76 Å, while the O–H bond in the water molecule measures only 0.96 Å. As the internal H_2O molecular structure is maintained, the minimum energy position is not half way between the two oxide ions. Instead, there are two equivalent positions for a proton lying on the O–O line of contact.[1] With the four-fold oxygen coordination, there is one proton per O–O "bond". The constraint of H_2O molecular integrity therefore leads to the "ice rules" for a GS configuration: for each oxygen, two protons must be in the near position and two in the far position. From an electrostatic point of view, the protons would like to be as far apart as possible. The ice rules therefore frustrate the effective proton–proton interaction.

The ice rules were enunciated by Bernal and Fowler in 1933.[16] In that period, X-ray diffraction could only determine the oxygen structure, but Bernal and Fowler argued that the hydrogen ions must lie along the oxygen–oxygen lines of contact. They proposed an ordered proton structure, undoubtedly presuming that this would be the case. However, around that time, Giauque and co-workers first established the probable existence of a zero-point entropy.[17] This led Pauling to his proposal (published in 1935) that the open tetrahedral structure of ice affords many equivalent ways of satisfying the ice rules. Pauling went on to approximately calculate the associated configurational entropy in the following way: consider one mole of ice, containing a number N of O^{2-} ions. Each O–O bond can be taken as having two proton positions, which gives 2^{2N} possible configurations. Out of the 16 possible configurations associated with each oxygen, 10 are energetically unfavorable: the OH_4^{2+} configuration, the four OH_3^+ configurations, the 4 OH^- configurations, and the O^{2-} configuration. This leaves the six configurations that satisfy the Bernal–Fowler rules as the allowed local proton configurations around each oxygen ion. An upper bound on the number of GS configurations is therefore estimated as $\Omega \leq 2^{2N}(6/16)^N = (3/2)^N$. The corresponding configurational entropy, $S_0 = k_B \ln(\Omega) = N k_B \ln(3/2) = 0.81\,\text{Cal/deg}\cdot\text{mol}$ is in remarkable agreement with the missing entropy in Giauque and Stout's experiment, $0.82 \pm 0.05\,\text{Cal/deg}\cdot\text{mol}$. Although Pauling's calculation neglects both the global constraint on the number of protons and the local constraints coming from closed loops on the Wurtzite lattice, it has been subsequently shown to be accurate to 1–2%.[18]

7.2.2. *Why is the zero-point entropy not zero?*

Pauling's model assumes then that the protons remain in disordered configurations consistent with the Bernal–Fowler rules. The proton disorder has been confirmed by neutron scattering on deuterated or heavy ice.[19] Remarkably, it seems that the calorimetry results do not require an ergodic exploration of the GS configurations. Even in Giauque and Stout's original paper, it is remarked that the response time of the sample goes through a maximum at around 100 K, corresponding to a freezing of the proton degrees of freedom. It seems then that the system settles essentially into a single GS and does not evolve in time over the manifold of possible states. However, as there are no long-range correlations, it is self-averaging. That is, no information is transmitted over macroscopic distances and the sample behaves itself as a Gibbs ensemble of equivalent states. This interpretation is in strong contrast with the case of spin glasses, where despite high (quasi-)GS degeneracy, each state requires collective organization.

This might seem paradoxical: if ice (or spin ice) freezes into a particular ice-rules state, this seems just as definite a molecular (spin) arrangement as an ordered state. So why is the GS entropy not zero in both cases? The question can be answered at several levels: if the sample were cut into many sub-samples, subsequent measurements on one of these would give no information about the microscopic state of the other sub-samples. The opposite is true of a system that has collectively broken symmetry to find a long-range ordered state. The excitation spectrum above the GSs, constituting the specific heat, must reflect this difference in information: for ice, this is confirmed by Giauque and Stout's experiment. Another very simple example is an Ising paramagnetic system in zero magnetic field. All 2^N configurations have zero energy. At all temperatures, C_H would be zero and all entropy measurements would give $\Delta S = 0$. The GS entropy is then very trivially $S_0 = Nk_{\rm B}\ln(2)$. The simplest models of ice and spin ice are somewhere between this trivial example and an ordered system: they allow zero energy excitations, not of single particles, but of a small number of particles. What about the lack of dynamics? Given the self-averaging, dynamics are not needed: an ergodic system and a Gibbs ensemble give identical results. Discussions of the definition of entropy in this context can be found in Rushbrooke[20] and Mayer and Mayer.[21]

Following Giauque and Stout's experiment and Pauling's explanation, one might still ask the question, can the third law really be violated in practice? A highly degenerate GS structure would be expected to be extremely susceptible to perturbations. Hence, there should always be some interaction — further neighbor exchange, electric or magnetic dipole, external fields, quantum fluctuations (i.e. proton or spin tunneling), etc. — that lifts the degeneracy, giving finally a single preferred state and no anomaly. This is certainly true and we will see an explicit example of it later, in the case of the spin ice model, when dipolar interactions are considered. However, experimentally this is not what is observed either in water ice or in the magnetic analogues discussed in this review. The evidence is that the third law does not apply to these systems on the time scale of any conceivable experiment. Rushbrooke has therefore described the observed zero-point entropy as a "practical entropy", to logically distinguish it from the entropy of the expected (but unobserved) equilibrium state.[20]

We can now recognize such degeneracy as a standard consequence of having frustrated degrees of freedom. Pauling's model of two proton positions per bond is Ising-like and the frustration can be described by a nearest-neighbor exchange term. Many such model systems have been studied, as described in Section 7.2.3 and elsewhere in this book. The recently discovered spin ice materials[4] provide an important experimental connection between these models and water ice. This is well illustrated by the calorimetry experiment by Ramirez *et al.* on the spin ice material $Dy_2Ti_2O_7$,[22] in which Pauling's GS entropy was measured using a procedure analogous to that used by Giauque and Stout.

7.2.3. *Generalizations of Pauling's model*

7.2.3.1. *Wannier's model*

In an elegant 1950 paper, Wannier[2] investigated the statistical mechanics of antiferromagnetically coupled Ising spins on a triangular lattice, using the then recently developed methods of Kaufmann and Onsager. His main conclusion, that *The system is disordered at all temperatures and possesses no Curie point.* probably represents the first published recognition of frustrated magnetism. The GS of the model is macroscopically degenerate: it has two spins up and one down (or vice versa) per triangle and has zero-point entropy 0.323 Nk_B, where N is the number of spins.

7.2.3.2. *Anderson's model*

Anderson[3] considered the problem of cation ordering in inverse spinels and magnetic ordering in normal spinels. To a first approximation, both these problems map onto the antiferromagnetic Ising model on the B-site spinel lattice, which is identical to the pyrochlore lattice, Fig. 7.1. The task is to place + and − signs on this lattice in the ratio 1:1, such that the number of +/− pairs is maximized. This condition is fulfilled with two + and two − signs per tetrahedron. The tetrahedra of the spinel lattice are arranged on the same lattice as the oxide ions in cubic ice. The spinel lattice consists of alternate "up" and "down" tetrahedra, so the Pauling ice rules are recovered if a + sign represents "proton in" on an up tetrahedron and "proton out" on a down tetrahedron, and vice versa for a "minus" sign. The model therefore has the Pauling zero-point entropy and represents one of the first examples of what is now called a geometrically frustrated antiferromagnet, predated only by Wannier's model (see above). As regards to the spinel systems, Anderson calculated the Coulomb (Madelung) energies of ordered phases and concluded that the energy advantage ΔU of adopting one of these was only a few percent of the near-neighbor energy, implying a transition at a temperature $T_c \propto \Delta U/\Delta S$ (where $\Delta S \approx S_0$, the zero-point entropy). Because T_c is a low temperature, the experimental dynamics might be too slow to equilibrate the system, leading to observable ionic disorder. This is born out experimentally both in the spinels and the pyrochlore fluorides of general formula $AXYF_6$ (e.g. $CsMnFeF_6$), which have a fully disordered pyrochlore lattice of X and Y ions even in highly annealed samples. It is interesting to note that Anderson also considered different ratios of X and Y ions, including the ratio 3:1, a situation that maps onto the "kagomé ice" model discussed in Section 7.4.5 as well as onto Wannier's model.[2]

In the context of frustration, Anderson's model is important as a paradigm for geometrical frustration on lattices built on corner-sharing simplex units (triangles or tetrahedra), typified by the pyrochlore and kagomé lattices. The idea was revived 20 years later by Villain,[23] who discussed the properties of the pyrochlore Heisenberg antiferromagnet, heralding many years of research into frustrated antiferromagnetic materials. As remarked by Anderson himself, this model shares with the Ising antiferromagnet the property of a disordered GS: in this case, a spin liquid or cooperative paramagnet.[23,24] The Heisenberg antiferromagnet model is realistic, being approximated by several real magnetic materials, for example $CsNiCrF_6$ and $CsMnFeF_6$.[25,26] In contrast, there are no known realizations of Anderson's Ising antiferromagnet. The reasons for this are discussed in Section 7.2.3.4.

7.2.3.3. *Vertex models*

In the 1960s and 1970s, models of ferroelectrics derived from the Pauling ice model became of great interest, as, in two dimensions at least, their properties could often be calculated exactly.[7,8] These became known as "vertex models". A vertex is a local configuration of displacement vectors that has a definite energy: for example "four protons in" would constitute a single vertex. The phase diagram of the models could be mapped out in terms of the weighting of these vertices, which for real systems are determined by their Gibbs–Boltzmann probabilities (which can be controlled to some extent by the application of an external field). A landmark calculation was Baxter's exact solution of the eight-vertex model in two dimensions, which is equivalent to a Pauling model that includes the OH_4^{2+} and O^{2-} configurations along with the six ice-rule states, to make eight in total.

Some mappings of vertex models onto Ising models have previously been proposed. For example, the 2D eight-vertex model can be mapped alternatively onto Ising models with two and four spin interactions or Ising models with further neighbor interactions: see Ref. [7] for details.

7.2.3.4. *Possibility of realizing magnetic vertex models*

From the above discussion, it should be clear that it is usually possible to map an ice-type or vertex model onto a magnetic system. The problem is that the magnetic models so derived are often rather unrealistic. This is exemplified by the absence of any reported examples of Anderson's antiferromagnet in nearly 50 years since its publication.[3] This fact may be understood by the following argument. The occurrence of Ising-like variables in real magnetic materials is always associated with magnetocrystalline anisotropy arising from the combined effect of the local crystal field and spin–orbit coupling. The local magnetic moments therefore always have a symmetry compatible with the local point symmetry of the crystal structure, which means that practical Ising magnets require a unique crystallographic axis. The simplest model for Ising-like spins in the cubic pyrochlore symmetry is with *local* easy axes defined by the four local axes of point symmetry, each a different member of the $\langle 111 \rangle$ set. In this case, the crystal fields themselves frustrate the exchange interaction as they inhibit parallel or antiparallel nearest-neighbor spin configurations. Within this restricted phase space, the physics resulting from ferromagnetic and antiferromagnetic couplings are reversed, compared to simple (uniaxial) Ising models.[4,5,27,28] Counterintuitively the antiferromagnet behaves as an Ising ferromagnet and does *not* have a macroscopically degenerate GS.[5] It is rather *ferromagnetically* coupled spin ice that maps onto Pauling's model[1] and hence onto Anderson's model.[3] In this sense, spin ice *is* the experimental realization of Anderson's antiferromagnet. However, it should be noted that ferromagnetic spin ice has a much richer experimental phenomenology than the antiferromagnetic Ising model: This is because a laboratory field can couple to many order parameters in the ferromagnetic system, which is not possible in its antiferromagnetic counterpart. Thus, the interesting physics of spin ice in an applied field (described in Section 7.4) is not an immediately obvious property of the Ising antiferromagnet. Hence, we have an example where two equivalent statistical mechanical model systems (spin ice and Anderson's antiferromagnet) are very different from the experimental point of view.

The symmetry argument used above does not apply to Wannier's antiferromagnet (the Ising model on the triangular lattice)[2] because the normal to the triangular net defines a unique axis. Although there are examples of Ising-like triangular antiferromagnets, such as $CsCoCl_3$,[29] it appears that no realizations of the Wannier zero-point entropy have yet been found. However, we describe later in Section 7.4.5.1, how spin ice in a magnetic field applied on [111] maps on to the GS of Wannier's model antiferromagnet[2] and may thus be considered its experimental realization.

It is also evident that spin ice provides the first truly ideal realization of a vertex model in magnetism: the previous mappings of vertex models onto magnetic Ising models are either not very realistic or not very well represented experimentally. In fact, spin ice is a 16-vertex model with the 16 vertices corresponding to the 2^4 spin configurations of a single tetrahedron. The beauty of spin ice is that the experimentalist has significant power to selectively control the statistical weights of the vertices, via the application of an external field along several of the principal symmetry axes, for example [100], [110], [111], and [211]. This remarkable feature of spin ice, which is explored further in Section 7.4, means that one finally has a system that connects comprehensively to the theory of vertex models. It is hoped that this property might re-invigorate interest in exact analysis of such systems (see for example, Ref. [30]).

7.2.4. *Spin ice*

As often happens, the appropriate labeling of a new scientific phenomenon can contribute to the process of unravelling its origin and understanding it. This appears to have been the case for "spin ice". The name "spin ice" and the concept of frustrated ferromagnetism was initially proposed in Ref. [4]. Specifically, Ref. [4] introduced the "spin ice model", that is the $\langle 111 \rangle$ Ising pyrochlore ferromagnet (now known as the near-neighbor spin ice model) in order to explain diverse experiments on holmium titanate, $Ho_2Ti_2O_7$. It is relevant in the context of this review to first discuss why the experimental results demanded the invention of the spin ice model.

7.2.4.1. *Definition of the spin ice model and its application to* $Ho_2Ti_2O_7$

The first evidence of magnetic frustration in $Ho_2Ti_2O_7$ came from a muon spin relaxation (μSR) experiment performed at the ISIS facility in 1996[31] (and reported briefly as part of Ref. [32]). The initial hypothesis being tested was that $Ho_2Ti_2O_7$ was a geometrically frustrated antiferromagnet, which might not magnetically order down to the lowest measurable temperatures. μSR is a sensitive measure of magnetic order as an ordered spin structure will often produce a well-defined and narrow distribution of dipolar fields at the muon stopping site, giving rise to characteristic precessional oscillations in the relaxation signal. In zero applied magnetic field, $Ho_2Ti_2O_7$ showed no such oscillations; but after the application and removal of a field, oscillations were found to be "frozen" into the signal. This effect (see Section 7.3.1.2) was reminiscent of spin glass behavior and consistent with the hypothesis of frustrated antiferromagnetism. However, careful magnetization measurements on single-crystal samples performed in order to estimate the Curie–Weiss temperature, θ_{CW}, found that θ_{CW} depended strongly on sample shape, indicative of a large demagnetization factor. Estimation of the latter, combined with measurements on quasi-ellipsoidal samples of different aspect ratio,[33] fixed the Curie–Weiss temperature accurately at $\theta_{CW} = 1.9 \pm 0.1\,\text{K}$.[4,32] The surprise was that this value is *ferromagnetic*: the Curie–Weiss temperature is a robust measure of the sum of pairwise coupling energies experienced by a spin, so $Ho_2Ti_2O_7$ could be thought of as having predominantly ferromagnetic spin–spin interactions. At first sight, this was inconsistent with the μSR result, so in order to get a more detailed microscopic picture of $Ho_2Ti_2O_7$, several neutron scattering experiments were performed on a large single crystal. In zero applied field, these measurements confirmed the μSR result of no magnetic order down to 50 mK, but the build up of diffuse scattering near the Brillouin zone centers in reciprocal space was again more characteristic of ferromagnetic than antiferromagnetic correlations. In an applied field, the dramatic growth of Bragg peaks indicated the restoration of magnetic order. The relaxation of these peaks confirmed the history dependence already noted in the μSR experiment, while their positions at the $Q = 0$ and $Q = X$ points of the Brillouin zone were again curiously indicative of both ferromagnetic and antiferromagnetic ordering patterns in a field.

In the mean time, an analysis of powder magnetization measurements had confirmed, as found by previous authors (see Section 7.2.4.3), that Ho^{3+}

could be represented to be a good approximation by an Ising GS doublet with large moment. By analogy with $LiHoF_4$ (see Ref. [34]) and the single-molecule magnets[35] (which have large Ising-like spins), this suggested a picture of Ising spins fixed along the local quantization axis: in this case, the trigonal $\langle 111 \rangle$ axes that connect the vertices of each tetrahedron to its center.

Thus, at the most basic level of approximation, the experiments demanded a model with $\langle 111 \rangle$ spins and near-neighbor ferromagnetic coupling.[4] Of course, this is a phenomenological approach that does not inquire as to the nature of the coupling: subsequent refinements have taken a more microscopic starting point that invokes realistic dipolar and exchange interactions[36,37] (see Section 7.3.2). For the $\langle 111 \rangle$ Ising ferromagnet, a simple calculation showed that the GS of a single tetrahedron of spins is the configuration with two spins pointing in and two pointing out. As Anderson had shown the pyrochlore lattice to be the dual lattice (i.e. the lattice formed by the mid points of the bonds) of cubic ice, "two spins in, two spins out" was therefore found to correspond to the ice rules, and the nearest-neighbor $\langle 111 \rangle$ Ising ferromagnet was mapped directly onto the Pauling ice model. Hence, it was christened the "spin ice" model.[4] This model not only provided a natural explanation of the unexpected absence of magnetic order in ferromagnetic $Ho_2Ti_2O_7$ but also gave a good description of its detailed magnetic properties. In particular, it was demonstrated in Ref. [4] that numerical simulations of the model were consistent with both the bulk magnetization measurements and the zero-field diffuse neutron scattering arising from the disordered spin structure. Furthermore, the field-induced ordering patterns matched well with the predictions of the model. At $T \sim 0.3\,\mathrm{K}$, the zero-field magnetic state of $Ho_2Ti_2O_7$ was thus established to approximate an ice-rules state, and the implication was that it should have the Pauling zero-point entropy.[4]

7.2.4.2. *Identification of spin ice materials*

Following the initial proposal of spin ice in $Ho_2Ti_2O_7$,[4] it was suggested,[38] on the basis of similar magnetization properties, that $Dy_2Ti_2O_7$ and $Yb_2Ti_2O_7$ were also spin ices. This prediction was later born out for the Dy material[22] but not for the Yb material.[40,41] An early and important experimental result was the direct measurement of the predicted zero-point entropy in $Dy_2Ti_2O_7$ by Ramirez and co-workers,[22] discussed further in Section 7.3.1.1. Some confusion subsequently arose from the suggestion[36,39] that $Ho_2Ti_2O_7$ may have a partially ordered GS and may not be spin ice. This idea arose as a theoretical interpretation of a Schottky-like increase of the heat capacity below 0.6 K. It implied that the neutron scattering experiments of Ref. [4], that had ruled out such a state, were in error. However, later neutron measurements[42,43] confirmed the correctness of Ref. [4] and again ruled out the ordered spin structure proposed in Ref. [36] below 0.6 K. In fact the low-temperature specific heat rise had previously been observed in isostructural Ho_2GaSbO_7 and attributed to the freezing out of the nuclear spins.[44] Analysis of the specific heat with subtraction of the nuclear contribution[42,45,46] finally established the Pauling entropy for $Ho_2Ti_2O_7$ (see Section 7.3.1.1). It is now clear from the combination of bulk and microscopic measurements discussed in this review, that both $Ho_2Ti_2O_7$ and $Dy_2Ti_2O_7$ as well as $Ho_2Sn_2O_7$ (see Ref. [47]) are spin ice materials with the Pauling zero-point entropy and that their similarities are much greater than their differences. The principal difference in fact relates to the hyperfine effect, which is much stronger in the Ho materials than in the Dy materials (see next section). Other possible realizations of spin ice are mentioned in Section 7.6.

7.2.4.3. *Basic properties of the spin ice materials*

The spin ice titanates or stannates can be prepared as high-quality powder samples by ceramic techniques.[48,49] The titanates, but not the stannates, can also be prepared as large single crystals either by flux methods[50] or by the floating zone technique.[51,52] The flux-grown crystals have an octahedral habit, typically slightly flattened along [111], while the higher quality crystals are of a translucent brown to amber color.

The spin ice materials adopt the cubic pyrochlore structure, space group $Fd\bar{3}m$. This structure has the rare earth ion on the 16d site, Fig. 7.1, Ti or Sn on 16c (forming another pyrochlore lattice), and an oxygen on 48f and another on 8b. The cubic lattice parameter is about 10.1 Å for $Ho_2Ti_2O_7$ and $Dy_2Ti_2O_7$ and about 10.4 Å for $Ho_2Sn_2O_7$. For the latter compound, the x parameter for the 48f site is 0.3368(5).[47] The point symmetry at the rare earth site is defined by the local trigonal axis, a member of the $\langle 111 \rangle$ cubic direction set, that connects the center of the elementary tetrahedron with its vertex. An 8b oxygen lies at the center of each tetrahedron. The rare earth is coordinated by two such oxygen ions that lie along the trigonal axis. The coordination shell is then completed by six more oxygen ions at a slightly greater distance from the central ion that form a puckered hexagon with its rotation axis along the local trigonal axis.

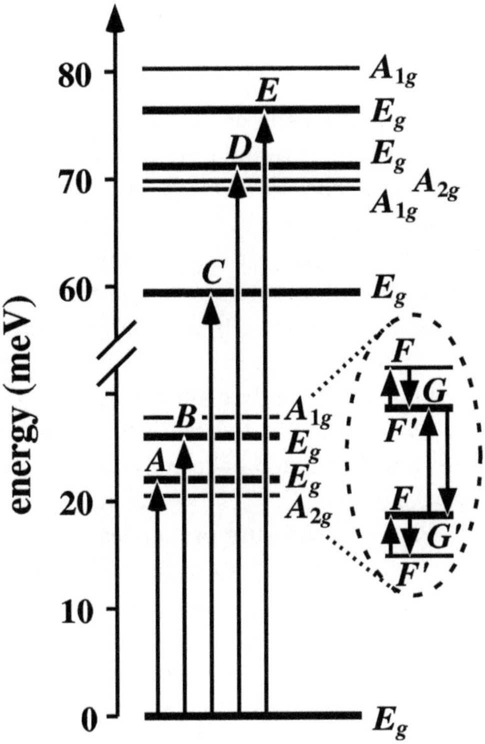

Fig. 7.3. $Ho_2Ti_2O_7$: Single-ion energy-level scheme (from Ref. [39]). The arrows indicate transitions observed by neutron scattering.

The crystal fields of the spin ice materials have been considered in various works.[4,39,44,53–57] A detailed neutron scattering investigation by Rosenkranz *et al.* established the energy-level scheme for Ho^{3+} in $Ho_2Ti_2O_7$ (see Fig. 7.3, Ref. [39]). Most authors agree that the GS of Ho^{3+} in $Ho_2Ti_2O_7$ is an almost pure $|\pm 8\rangle$ doublet and that of Dy^{3+} in $Dy_2Ti_2O_7$ is an almost pure $|\pm 15/2\rangle$ Kramers doublet. However, the crystal field symmetry allows admixture of some other $|J, M_J\rangle$ states (with $M_J \neq J$) into the GS doublet. Estimates of the degree of admixture differ. Reference [55] reports, on the basis of fitting low-field susceptibility curves between 77 K and 300 K, that the GS doublet of Ho^{3+} in $Ho_2Ti_2O_7$ includes 12% of other states. However, the crystal field parameters derived in Ref. [55] deviate substantially from those found by Rosenkranz *et al.* using inelastic neutron scattering techniques,[39] which suggest a smaller admixture and should probably be considered to be the more accurate results. There is no available neutron scattering study of $Dy_2Ti_2O_7$, but magnetization studies[54] find evidence of a slightly lower magnetic moment than expected for pure $|\pm 15/2\rangle$. Rosenkranz *et al.*[39] and Qiu[57] have estimated the crystal field scheme for $Dy_2Ti_2O_7$ using the crystal field parameters of $Ho_2Ti_2O_7$ with a correction for the different radial integrals of Dy^{3+}.[57] This approach finds that the GS doublet is 99% $|\pm 15/2\rangle$. In all the spin ice materials, the first excited crystal field levels are of the order of 300 K above the GS. At elevated temperatures, $Ho_2Ti_2O_7$ and $Dy_2Ti_2O_7$ show significant magnetostriction[53] and quite large dielectric constants. However, there is no evidence of anomalous effects at the low temperatures where spin ice behavior occurs. The dielectric properties have been studied in detail by Mamsurova *et al.*,[58] who conclude that anomalies observed at ~40 K and ~80 K are associated with defects: it seems likely that samples prepared at high temperature are slightly oxygen deficient and contain a small amount (~0.25%) of magnetic Ti^{3+} or Ti^{2+}. Once again, these do not seem to significantly affect the low-temperature properties.

Natural Dy has a mixture of isotopes while Ho is a single isotope ^{165}Ho. Nuclear–electronic hyperfine coupling is particularly strong in the Ho materials (~0.35) K,[42,44,56] but very small (<50 mK) in $Dy_2Ti_2O_7$, as evidenced by the absence of a contribution to the heat capacity.[22] The specific heat behavior of the spin ice materials is discussed further in Section 7.3.1.1. That of $Dy_2Ti_2O_7$ was previously considered in Refs. [44] and [59]; it is interesting to note that Blöte *et al.*[44] noted a shortfall in entropy, now understood to result from the macroscopic degeneracy of the spin ice state.

The basic magnetic characterization of powder samples of the spin ice materials is described in Refs. [44], [54] and [60–63]. The magnetic susceptibility in general shows a very small Curie–Weiss temperature, which must be carefully corrected for the strong demagnetizing factor: for a spherical or powder sample, this correction amounts to adding 1.4 K to the value estimated from the raw data. Note that some experimental reports do not make this important correction. Accurate estimates on single-crystal samples of the corrected values are $\theta_{CW} = 1.9$ K for $Ho_2Ti_2O_7$[4] and 1.2 K for $Dy_2Ti_2O_7$.[54,64] The value 1.8 K for $Ho_2Sn_2O_7$ (see Ref. [65]) has been estimated from measurements on a powder sample. The dipolar interaction alone, assuming a full classical moment, would yield ~3 K, so the shortfall has been ascribed to antiferromagnetic superexchange coupling.[37] Assuming that this hypothesis is correct (there may be other possible reasons for the shortfall including lower than expected magnetic moments), one can obtain a self-consistent description of the mean magnetic coupling energy in the spin ice materials. For example, the antiferromagnetic superexchange can alternatively be estimated by specific heat methods, as described in Section 7.3.1.1.

7.2.5. *Spin ice as a frustrated magnet*

7.2.5.1. *Frustration and underconstraining*

The macroscopically degenerate GSs that are characteristic of frustrated systems can be understood in terms of underconstraining arguments.[66] That is, in terms of the difference between the number of constraints necessary to determine a GS and the number of degrees of freedom that the system possesses. In an unfrustrated system, the two differ by a microscopic number at most. For example, in a classical ferromagnet, fixing the two independent directions of the first spin, S_1^x, S_1^y, the GS configuration is satisfied only by satisfying the $2(N-1)$ microscopic constraints $S_i^x = S_1^x$, $S_i^y = S_1^y$. Once satisfied, the GS is uniquely defined. For discrete systems, enumeration of the degeneracy can be cast in a similar language. For example, for Anderson's Ising pyrochlore antiferromagnet, the GS condition is underconstrained, demanding only that the total magnetization of the four Ising spins on each tetrahedron be zero. Six of the $2^4 = 16$ possible spin configurations satisfy this condition. Counting 2^4 configurations for each tetrahedron gives, for a system of N spins and $N/2$ tetrahedra, a total number of microstates, $\tilde{\Omega}_T = 2^{2N}$, which greatly overestimates the exact total, $\Omega_T = 2^N$. This is because the spins are shared between two tetrahedra, so that the configurations on each tetrahedron are not independent. Following a Pauling-like argument; allocating $2^2 = 4$ states per tetrahedron and assuming that 6/16 of them satisfy the constraint, leads to a GS degeneracy $\Omega_0 = \{2^2(6/16)\}^{N/2} = (3/2)^{N/2}$. The corresponding entropy $S_0 = k_{\rm B}\ln(\Omega_0) = (Nk_{\rm B}/2)\ln(3/2)$ is of course just Pauling's original result. In the mapping from ice to Anderson's antiferromagnet, or alternatively to spin ice, N oxygen ions lead to $2N$ spins and we recover the Pauling entropy.

The macroscopic degeneracy can therefore be expressed in terms of the number of unconstrained degrees of freedom, or in the case of discrete symmetry, in terms of a local, unconstrained phase space. These Pauling-type arguments that neglect collective constraints coming, for example, from closed loops in the lattice, give an upper bound to the GS entropy. For corner sharing, or tetrahedral systems, as has been shown both experimentally and numerically, this upper bound is an excellent approximation to the true GS entropy. A more detailed discussion, applied to continuous spin systems can be found in Ref. [66].

7.2.5.2. ⟨111⟩ *Pyrochlore models*

As discussed above, the pyrochlore structure is remarkable in that its "natural" magnetic Ising variables (i.e. those consistent with point symmetry) are fully frustrated only when the the near-neighbor interactions are *ferromagnetic.* With antiferromagnetic coupling, these variables remain unfrustrated by the interaction and the system orders magnetically. This situation is the opposite to the case of the Heisenberg model on the pyrochlore lattice, where the antiferromagnet is highly frustrated and the ferromagnet is not.[5,27,66]

The contrasting behavior of the Ising and Heisenberg models makes it interesting to inquire as to the behavior of the "intermediate" case of continuous spins, with local easy-axis (Ising-like) anisotropy. The case of ferromagnetic coupling has been discussed in Ref. [67]. The macroscopic spin ice degeneracy is found to be lifted in favor of a particular state with $q = 0$ long-range order, where q is the reciprocal lattice vector of the underlying Bravais lattice. Relaxing the spins continuously away from the ⟨111⟩ crystal field directions, one gains exchange energy at the cost of crystal field energy. As the gain goes linearly with the relaxation, while the loss goes quadratically, the modified minimum energy states are distorted away from the $q = 0$ spin ice state. This maximizes the energy gain, and hence, the distorted state becomes the unique GS. This model is discussed further in Section 7.4.1.1.

In the case of easy axis crystal fields and antiferromagnetic coupling, the system has the ordered GS of the ⟨111⟩ Ising model, the so-called FeF_3 ("all in–all out") structure, observed experimentally in this material.[68,69] This state has global two-fold symmetry, and for strong crystal fields, there is a second order Ising-like phase transition.[27] For weaker fields there is a first-order phase transition. This is separated from the second-order transition by a tri-critical point with critical exponents that differ from those of the three dimensional Ising model.[70]

One can also consider the case of easy-plane or XY spins, i.e. spins that lie in the planes perpendicular to the local ⟨111⟩ trigonal axes.[71–73] This case is closer to the Heisenberg model in that the antiferromagnet has a highly degenerate GS. However, the GS entropy is sub-extensive and there is a fluctuation-driven, or "order by disorder"[23] transition to a long-range ordered state, observed in the material $Er_2Ti_2O_7$.[71,72] With ferromagnetic coupling, one finds a transition to an ordered ferromagnetic state with canted spins.[73]

7.3. Properties of the Zero-Field Spin Ice State

7.3.1. *Experimental properties*

7.3.1.1. *Heat capacity: Zero-point entropy*

The key thermodynamic characteristic of the spin ice state is its non-zero Pauling entropy, S_0, with $S_0 \approx (Nk_B/2)\ln(\frac{3}{2})$.[1] Ramirez and co-workers[22] directly investigated this property via experimental measurements of the heat capacity of $Dy_2Ti_2O_7$.

The experimental procedure used in Ref. [22] closely followed the one used by Giauque in 1935 in his determination of the residual entropy of water ice.[13] At temperatures below 10 K, the specific heat contribution from phonons in $Dy_2Ti_2O_7$ is sufficiently small to allow one to measure the magnetic contribution that develops as the paramagnetic regime $T < \theta_{CW}$ is entered upon cooling. Figure 7.4(a) shows the magnetic specific heat of $Dy_2Ti_2O_7$ from a temperature $T = 7\,\text{K}$ down to 300 mK. One observes a rise of the specific heat upon cooling with a broad maximum around $T_{peak} \sim 1.2\,\text{K}$. This growth of the specific heat indicates the build-up of magnetic correlations and the removal of magnetic entropy as the system is cooled. The broad maximum and the absence of a sharp feature in $C_H(T)$ is a strong indication that the system does not develop long-range magnetic order via a thermodynamic phase transition. The magnetic specific heat drops rapidly below T_{peak}, essentially vanishing below $T_{min} \sim 400\,\text{mK}$. This indicates that the spins freeze as the temperature is lowered below T_{peak} and that all thermal spin fluctuations have ceased below T_{min}. The entropy removed from the system upon cooling can be determined from the heat capacity measurements using the thermodynamic relationship

$$\delta S \equiv S(T_2) - S(T_1) = \int_{T_1}^{T_2} \frac{C_H(T)}{T} dT, \tag{7.1}$$

where δS is the entropy difference between temperatures T_2 and T_1. Since there are two possible spin orientations per degree of freedom, the molar spin entropy in the high-temperature paramagnetic regime $(T \gg \theta_{CW})$

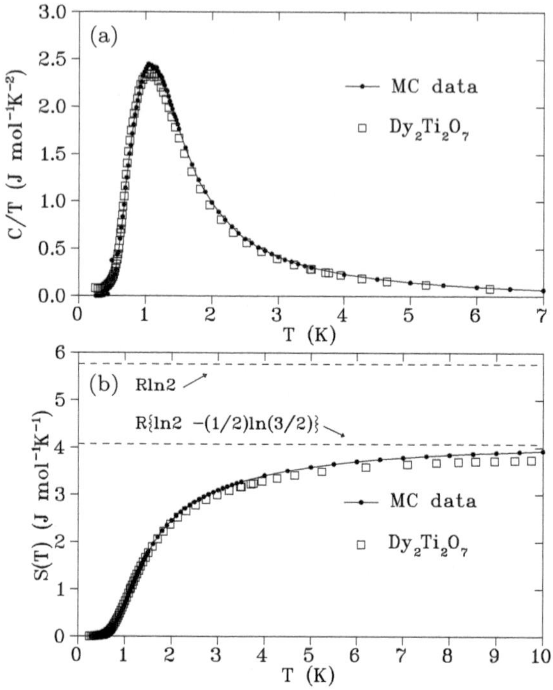

Fig. 7.4. $Dy_2Ti_2O_7$: (a) Specific heat and (b) entropy versus temperature, measured by Ramirez *et al.* (Ref. [22]). This quantity corresponds to δS of Eq. (7.1) and shows evidence of the Pauling zero-point entropy. The data are compared with those calculated by Monte Carlo simulations of the dipolar spin ice model (Ref. [37]) with exchange $J_{\rm NN} = -1.24\,\rm K$ and dipolar coupling $D_{\rm NN} = 2.35\,\rm K$ (see Section 7.3.2).

is $R\ln(2)$ (R is the molar gas constant, $R = 8.3145\,\rm J\,mol^{-1}\,K^{-1}$). Figure 7.4(b), taken from Ref. [22], shows the temperature dependence of δS as T is increased from 400 mK to 10 K. Figure 7.4(b) shows that the entropy recovered upon warming the system is about $3.9\,\rm J\,mol^{-1}\,K^{-1}$, a value that is considerably less than the value $R\ln(2) \approx 5.76\,\rm J\,mol^{-1}\,K^{-1}$. The difference, $1.86\,\rm J\,mol^{-1}\,K^{-1}$, is close to the Pauling estimate for the GS entropy associated with the extensive GS degeneracy of spin ice: $S_0 = (R/2)\ln(3/2) = 1.68\,\rm mol^{-1}\,K^{-1}$, which includes a factor of 2 to account for the mapping of 1 mole of ice onto 2 moles of spin ice (see Section 7.2.5.1). This result is consistent with the interpretation that the state reached upon cooling from the paramagnetic phase down to $T \sim 400\,\rm mK$ in $Dy_2Ti_2O_7$ carries extensive entropy close to S_0 and obeys the ice rules.

The experiment of Ramirez *et al.*, apart from the cryogenics involved, is thus somewhat simpler than the original experiment on ice: given the separation of energy scales between the effective coupling, $\sim O(1)$ K, and the crystal field strength, $\sim O(100)$ K, one can assume, to a good approximation, that at $\sim O(10)$ K, there are uncorrelated Ising degrees of freedom, with molar entropy $S = R\ln(2)$. Unlike water ice, the spin ice state is reached from the high-temperature uncorrelated phase without passing through gas–liquid or liquid–solid transitions. Furthermore, spin ice offers a second test of the zero-point entropy, unavailable in water ice: the application of a magnetic field to restore long-range order. For example, a field applied along the [100] direction lifts the degeneracy in favor of a non-collinear ferromagnetic state.[38] Repeating the experiment in a field, one would therefore expect to recuperate the missing entropy. Despite performing the experiment on a powder sample of $Dy_2Ti_2O_7$ and of the loss of ergodicity at low temperature (which one might now expect to hinder the process of the selection of a unique state), this change in behavior was qualitatively confirmed.[22]

A demonstration, via heat capacity measurements, that the low-temperature disordered state attained by $Ho_2Ti_2O_7$ (see Ref. [4]) indeed has zero-point entropy, proved to be less straightforward than for $Dy_2Ti_2O_7$,[45] causing some confusion as to the true nature of the low-temperature state attained.[36,74] The temperature dependence of $C_H(T)$ for $Ho_2Ti_2O_7$ (Fig. 7.5) is significantly different from that of $Dy_2Ti_2O_7$ (Fig. 7.4). First, there is no broad maximum, and at a temperature close to $T \sim 0.6$ K, the specific heat rises, followed by a sudden freezing-out and loss of thermal equilibrium.[36] The behavior of $Ho_2Ti_2O_7$ was interpreted by some authors as an indication for some kind of partial ordering transition, rather than spin ice behavior.[36,74] However, the holmium ion is somewhat atypical in that it possesses a particularly large hyperfine coupling between the nuclear spins and electronic spins. The importance of this term had previously been commented on by Blöte *et al.* in earlier heat capacity measurements on isostructural Ho_2GaSbO_7.[44] In fact, Ho^{3+} materials commonly have large nuclear–electronic coupling[75,76]; for example, this has proved to be of importance in the quantitative calculation of the transverse field versus temperature phase diagram in the dipolar Ising ferromagnet $LiHoF_4$.[77] The hyperfine interaction leads to a Shottky nuclear specific heat anomaly at $T_{\rm nuc}$. To isolate the electronic component from the low-temperature $C_H(T)$ in $Ho_2Ti_2O_7$, one therefore has to first remove the nuclear contribution[45]

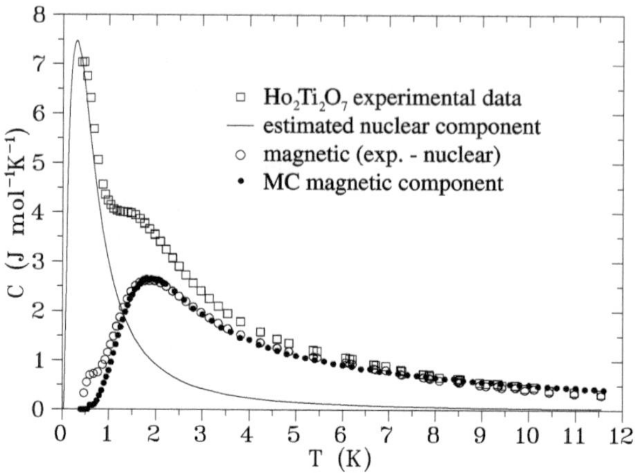

Fig. 7.5. $Ho_2Ti_2O_7$: The total specific heat is shown by the empty squares and the expected nuclear contribution by the solid line. The electronic contribution has been estimated by subtracting these two curves (open circles). Near to 0.7 K, this subtraction is prone to a large error (see text). Dipolar spin ice simulation results are indicated by the filled circles (from Ref. [42]).

(see also Ref. [55]). In a different study,[42] the specific heat of $Ho_2Ti_2O_7$ was measured on a single crystal rather than a powder sample, making thermal equilibration easier at low temperatures.[36] The low-temperature magnetic specific heat (with phonon contribution subtracted) of a $Ho_2Ti_2O_7$ single crystal is shown by the open squares in Fig. 7.5. One can discern a shoulder in the data around a temperature of $T_{\rm cr} \approx 1.5\,$K. This shoulder is indicative of a change of regime in the nature of the thermal excitations from $T > T_{\rm cr}$ to $T < T_{\rm cr}$. The nuclear hyperfine contribution, which can be estimated from measurements on Ho_2GaSbO_7,[44] is shown by the solid line. Subtracting this from the above data gives an estimate for the electronic contribution. This is shown by the open circles in Fig. 7.5, which now bears a strong resemblance to the $C_H(T)$ data above for $Dy_2Ti_2O_7$ in Fig. 7.4(a). One can see a kink in the subtracted data around a temperature of 600 mK. The kink has been argued to be an artifact of the data manipulation (i.e. subtraction of estimated hyperfine contribution to the specific heat) and not an indication of any real rapid change in the electronic spin configuration. This conclusion is supported by neutron scattering results at 50 mK, which show no sign of the spin–spin correlations building up above those found in MC simulations.[42] The temperature integration of the electronic contribution of $C_H(T)$ for $Ho_2Ti_2O_7$ in Fig. 7.6 shows convincingly a residual

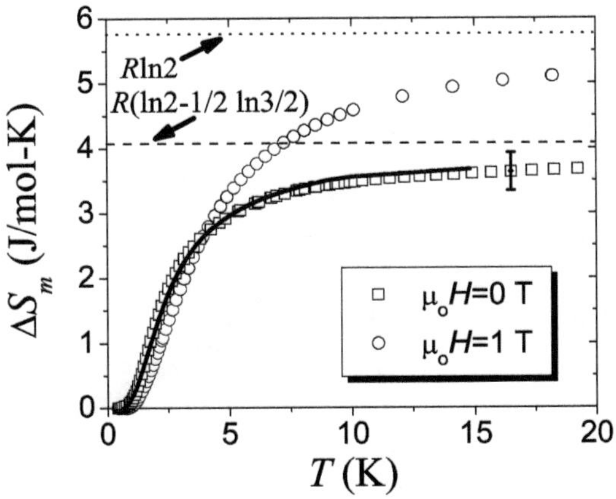

Fig. 7.6. $Ho_2Ti_2O_7$: magnetic entropy versus temperature (from Ref. [46]), showing the existence of a Pauling zero-point entropy and its removal by an applied field. The solid line is the prediction of the near-neighbor spin ice model of Ref. [5].

entropy close to the Pauling value (S_0).[46] Similarly, the zero-point entropy is released by the application of a magnetic field of 1 T. These results confirm that the low-temperature disordered magnetic state of $Ho_2Ti_2O_7$ is indeed a spin ice state, just as in the case of $Dy_2Ti_2O_7$.

7.3.1.2. *Low field magnetic susceptibility: Spin freezing*

The first magnetization study of the spin ice regime was performed by Matsuhira *et al.* on polycrystalline samples of $Ho_2Sn_2O_7$ and $Ho_2Ti_2O_7$ at temperatures down to T = 100 mK.[65] Below the spin ice freezing temperature of $Ho_2Sn_2O_7$, there is a very pronounced divergence between the field-cooled (FC) and zero-field-cooled (ZFC) magnetization (see Fig. 7.7). In the history-dependent regime, there is strong magnetic hysteresis (Fig. 7.7) and a remanent moment can be frozen into the sample. This remanence had previously been observed on single crystalline $Ho_2Ti_2O_7$ using μSR[32] and neutron scattering.[4] It is clear from these studies that the spin ice state is frozen, at least as far as major magnetic reorientations are concerned (however, see Section 7.5). Very recently, low-temperature hysteresis curves have been reported for powder samples of $Dy_2Ti_2O_7$, again confirming the frozen nature of the spin ice state.[78] It should be noted, as discussed in Section 7.5, that the static frozen nature of spin ice is firmly established only as regards major reorientations of the magnetization: there is evidence that some local spin dynamics persists to the lowest measured temperatures.[79,80]

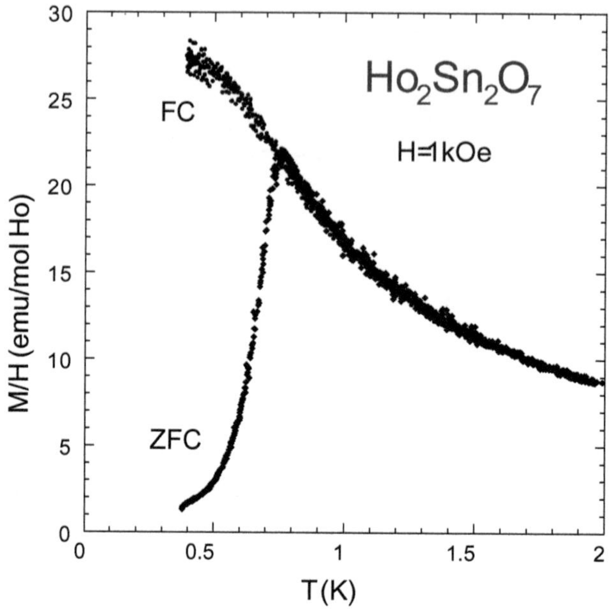

Fig. 7.7. $Ho_2Sn_2O_7$: magnetization versus temperature for a polycrystalline sample, showing a pronounced splitting between field-cooled (FC) and zero-field-cooled (ZFC) data (Ref. [65]).

7.3.1.3. *Spin arrangement observed by neutron scattering*

The neutron scattering from all the spin ice materials is, at low temperatures, within the energy resolution of most conventional neutron spectrometers (but see Section 7.5.3.2). The measured scattering patterns therefore measure the static spin correlations. Zero-field neutron scattering[4, 42, 43] on $Ho_2Ti_2O_7$ in the millikelvin range reveals broad diffuse features. In the *hhl* plane of reciprocal space, these are centered around (001), (003) (110), and $(3/2, 3/2, 3/2)$ (see Fig. 7.8). These features are predicted qualitatively by the near-neighbor spin ice model,[4] but there are clear differences of detail, particularly as regards the shape of the features. A very recently reported study of $Dy_2Ti_2O_7$ (see Ref. [81]) has similarly confirmed the qualitative relevance of the near-neighbor model, but again differences in detail. For a description of these differences, one must turn to the dipolar spin ice model discussed in Section 7.3.2. $Ho_2Sn_2O_7$ has only been studied as a powder, but again, the necessity of considering the dipolar interaction is evident.[47]

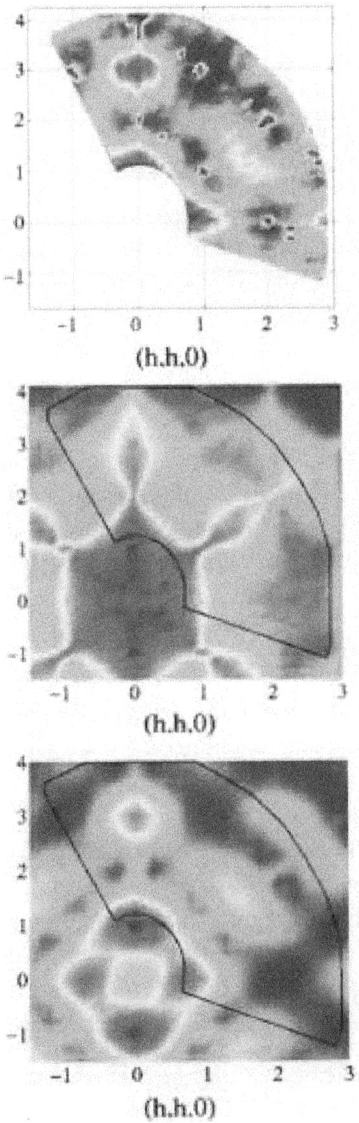

Fig. 7.8. $Ho_2Ti_2O_7$: Neutron scattering in the *hhl* plane showing experimental data (upper panel; the sharp spots are nuclear Bragg scattering with no magnetic component), compared with Monte Carlo simulations of the near-neighbor spin ice model (middle panel) and dipolar model (lower panel) (Ref. [42]). Blue indicates the weakest and red-brown the strongest intensity.

7.3.2. *Microscopic theories and experimental tests*

7.3.2.1. *Near-neighbor spin ice model: Successes and failures*

From the above discussion, we have seen that the near-neighbor model,[4,5] when applied to $Ho_2Ti_2O_7$, $Dy_2Ti_2O_7$, and $Ho_2Sn_2O_7$, correctly predicts the existence of the spin ice state, its frozen nature, and the existence and magnitude of its zero-point entropy. Considering the low-field experiments, bulk measurements do not demand a higher level theory. However, the need arises when one considers the neutron scattering experiments of Fig. 7.8, for example. Similarly, we shall see later that the near-neighbor model qualitatively captures the behavior of spin ice in an applied field remarkably well, but for a detailed quantitative understanding, a more accurate magnetic Hamiltonian is required. The main improvement that has been studied to date is the inclusion of the full dipolar interaction between two idealized magnetic moments, with the addition of a phenomenological near-neighbor exchange term. This model was first suggested by Siddharthan *et al.*[36] and has become known as the "dipolar spin ice model". It is to this model that we now turn.

7.3.2.2. *The problem of treating the dipolar interaction*

Consider the simple theoretical model of $\langle 111 \rangle$ Ising spins interacting via nearest-neighbor ferromagnetic exchange. We emphasized above how the simplicity of this model gives rise to Bernal–Fowler ice rules for the Ising spins. This leads to underconstrained conditions for the minimum energy states, which consequently results in a macroscopic GS degeneracy. As discussed above, the experimental evidence for spin ice behavior has so far been established in rare earth compounds which possess ideal Ising-like spins with large magnetic moments of approximately $\mu \approx 10\mu_B$. These interact via long-range dipole–dipole interactions. The strength of the dipole–dipole energy, D, for nearest-neighbor spins can be estimated using the formula

$$D = \frac{\mu_0}{4\pi} \frac{\mu^2}{r_{NN}^3}, \tag{7.2}$$

where r_{NN} is the nearest-neighbor distance. Taking $r_{NN} = (a/4)\sqrt{2}$, where $a \sim 10\,\text{Å}$ is the size of the conventional cubic unit cell, we get $D \sim 1.4\,\text{K}$. The Curie–Weiss temperature, θ_{CW}, as determined via magnetic susceptibility measurements, is a measure of the overall magnetic energy scale in a system. In view of the experimental fact that θ_{CW} in the spin ice materials is or the order of $\sim 10^0$ K, the dipolar interactions are therefore a sizeable contribution to the overall energy scale. The Hamiltonian for magnetic dipole–dipole interactions, $H_{\rm dd}$ is

$$H_{\rm dd} = Dr_{\rm NN}^3 \sum_{i>j} \frac{\boldsymbol{S}_i^{\hat{z}_i} \cdot \boldsymbol{S}_j^{\hat{z}_j}}{|\boldsymbol{r}_{ij}|^3} - \frac{3(\boldsymbol{S}_i^{\hat{z}_i} \cdot \boldsymbol{r}_{ij})(\boldsymbol{S}_j^{\hat{z}_j} \cdot \boldsymbol{r}_{ij})}{|\boldsymbol{r}_{ij}|^5}, \tag{7.3}$$

where $\boldsymbol{S}_i = S_i \hat{z}_i$ are three-component classical spin vectors of unit length $|\boldsymbol{S}_i| = 1$ pointing along the local trigonal $\langle 111 \rangle$ axes with $\boldsymbol{\mu}_i = \mu \boldsymbol{S}_i$. The size of the magnetic moment, μ, is absorbed in the definition of D which sets the energy scale for the dipolar interactions. In general, one expects that in magnetic rare earth insulators, there exist exchange interactions as well as magnetic dipolar interactions. A suitable starting point is therefore the Hamiltonian, H, where both long-range dipolar interactions and nearest-neighbor exchange with exchange constant, J, are considered

$$H = -J \sum_{\langle ij \rangle} \boldsymbol{S}_i^{\hat{z}_i} \cdot \boldsymbol{S}_j^{\hat{z}_j} + Dr_{\rm NN}^3 \sum_{i>j} \frac{\boldsymbol{S}_i^{\hat{z}_i} \cdot \boldsymbol{S}_j^{\hat{z}_j}}{|\boldsymbol{r}_{ij}|^3} - \frac{3(\boldsymbol{S}_i^{\hat{z}_i} \cdot \boldsymbol{r}_{ij})(\boldsymbol{S}_j^{\hat{z}_j} \cdot \boldsymbol{r}_{ij})}{|\boldsymbol{r}_{ij}|^5}. \tag{7.4}$$

The Hamiltonian H is presumably more realistic and relevant to real rare earth materials than the nearest-neighbor spin Hamiltonian defined in Ref. [4], so the first question to understand is under what circumstances, if any, it allows for spin ice behavior, as signaled by a zero-point entropy.

First consider the contribution to H coming from first nearest-neighbor spins. For nearest-neighbor $\langle 111 \rangle$ Ising spins $\boldsymbol{S}_i$ on the pyrochlore lattice, $\boldsymbol{S}_i \cdot \boldsymbol{S}_j = (-1/3)\sigma_i\sigma_j$ and $(\boldsymbol{S}_i \cdot \boldsymbol{r}_{ij})(\boldsymbol{S}_j \cdot \boldsymbol{r}_{ij}) = -(2/3)\sigma_i\sigma_j$ for $\boldsymbol{r}_{ij} = \boldsymbol{r}_{\rm NN}$, where we have now introduced Ising variables $\sigma_i = \pm 1$ to represent the spin $\boldsymbol{S}_i$ points along or antiparallel to the member of the $\langle 111 \rangle$ set that defines the local point symmetry (see also Ref. [27]). Rewriting H in terms of σ_i and σ_j for the combined nearest-neighbor exchange and dipoles at nearest neighbor, we have

$$H_{\rm eff} = \sum_{\langle ij \rangle} J_{\rm eff} \sigma_i \sigma_j, \tag{7.5}$$

where $J_{\rm eff} \equiv J_{\rm NN} + D_{\rm NN}$. The effective Ising nearest-neighbor exchange is $J_{\rm NN} \equiv J/3$ and the effective Ising nearest-neighbor dipole coupling is $D_{\rm NN} \equiv 5D/3$, with $D > 0$.

To avoid confusion, we briefly comment on the sign convention in Eq. (7.5), as well as the definitions of $J_{\rm NN}$, $D_{\rm NN}$, and $J_{\rm eff}$. Eq. (7.5) is a mapping of the ferromagnetic spin ice Hamiltonian of Refs. [4] and [5] on to an antiferromagnetic Ising model (equivalent to Anderson's[3]). A mapping of this sort was discussed by Moessner.[27] In the case of Eq. (7.5), the absence of the usual minus sign in front of Hamiltonian is intentional: it allows one to preserve the positive (ferromagnetic) sign of the interaction $J_{\rm eff}$. Furthermore, elimination of the vector spins of the spin ice model and their replacement with global scalar Ising variables, σ_i, absorbs the full magnitude of the interaction energy into $J_{\rm eff}$, allowing a transparent discussion of its magnitude (however, it should be noted that Eq. (7.5) is of no utility when discussing field-induced states, where it is essential to preserve the vector nature of the spins).

From Eq. (7.5), one can construct a very simple nearest-neighbor criterion for spin ice behavior. For $J_{\rm eff} > 0$, one has an effective antiferromagnetic nearest-neighbor model for the σ_i variables with extensive degeneracy. For $J_{\rm eff} < 0$, i.e. for J antiferromagnetic and $J < -5D$, one has two unique GSs related by global symmetry that corresponds to all spins pointing in or out of a tetrahedral unit.[5,27,72] This corresponds to the magnetic structure of FeF_3 alluded to above, and which orders at finite temperature.[37] This simple argument therefore predicts that materials with antiferromagnetic exchange J ($J < 0$) should display spin ice behavior provided $J > -5D$. For example, as we will discuss in more detail below, $Dy_2Ti_2O_7$ has $J \approx -3.72\,{\rm K}$, $D \approx 1.41\,{\rm K}$ ($J_{\rm eff} \approx 1.1\,{\rm K}$) while $Ho_2Ti_2O_7$ has $J \approx -1.65\,{\rm K}$, $D \approx 1.41$ ($J_{\rm eff} \approx 1.8\,{\rm K}$), hence satisfying the nearest-neighbor spin ice criterion, $J_{\rm eff} > 0$. As discussed in the next section, the criterion $J_{\rm eff} = J/3 + 5D/3 > 0$ for spin ice behavior is not exact when considering long-range dipole–dipole interactions. In that case, the condition for spin ice behavior is $J_{\rm NN}/D_{\rm NN} > -0.905$, giving $J_{\rm eff}/D_{\rm NN} > 0.095$.[37]

Let us now return to the dipolar contribution ($H_{\rm dd}$ in Eq. (7.3)) and comment on two important properties of the magnetic dipolar interaction. First, it is long ranged and the truncation at nearest neighbor would seem to be a gross approximation. Indeed, the magnitude of the dipolar coupling decays very slowly as $1/r_{ij}^3$, i.e. as the third power of the distance $r_{\rm ij} \equiv |\boldsymbol{r}_{\rm ij}|$ separating two magnetic moments $\boldsymbol{\mu}_i$ and $\boldsymbol{\mu}_j$. Second, dipolar interactions are strongly anisotropic: they depend on the relative orientation of the interacting moments with respect to the separation vector $\boldsymbol{r}_{ij}$ via the $\boldsymbol{\mu}_i \cdot \boldsymbol{r}_{ij}$ terms in Eq. (7.3). Dipolar couplings are complicated interactions because of the combination of these two features. In contrast, spin ice behavior and extensive degeneracy are intimately tied to the existence of Bernal–Fowler ice rules, which themselves arise from the overwhelming simplicity of a nearest-neighbor model.

These considerations lead to a first and very interesting question regarding the origin of spin ice behavior in dipole-coupled rare earth materials: how is it that the seemingly "complicated" dipolar interactions allow for the emergence of ice rules and consequent extensive degeneracy? From a more technical perspective, one does generally expect that long-range interactions beyond nearest neighbor will combine with spin anisotropy to select a unique ordered GS. For example, in the context of highly frustrated antiferromagnets, it is known that exchange interactions beyond nearest neighbor select a unique GS in the classical Heisenberg antiferromagnet on the 2D kagomé lattice.[82] Similarly, they typically select a unique ordering wave vector in classical Heisenberg pyrochlore antiferromagnets.[83] Of course, in all such cases, there exist combinations of the distance, r, dependent exchange, $J(r)$, where the classical GS selection is not complete.[82,83] However, this situation requires a fine-tuning of the $J(r)$ that rarely occurs in real materials. It has been found that weak dipolar interactions in the pyrochlore lattice with antiferromagnetically coupled Heisenberg spins lead to a unique ordered GS (apart from a finite number of discrete symmetry-related states).[84,85]

Having mentioned a number of examples where interactions beyond nearest neighbor can give rise to ordered GS selection, we return to the question of the distance dependence of the dipolar contribution to Hamiltonian equation (7.4). For example, the second nearest-neighbor distance is $\sqrt{3}$ times larger than the nearest-neighbor distance, and one then has a second nearest-neighbor energy scale, $D_{\mathrm{NNN}} \sim 0.2D_{\mathrm{NN}}$. This implies an important perturbation compared to $J_{\mathrm{eff}} = J_{\mathrm{NN}} + D_{\mathrm{NN}}$, especially in the context of antiferromagnetic (negative) J_{NN}, where $J_{\mathrm{eff}} < D_{\mathrm{NN}}$. Specifically, for $Dy_2Ti_2O_7$, the second nearest-neighbor energy scale is a large proportion of the total energy. One would therefore naturally expect dipolar interactions beyond the nearest neighbor to cause the different ice-rule states to have different energies. This would break the degeneracy of the spin ice manifold and cause long-range order below a critical temperature $T_N \sim O(D_{\mathrm{NN}})$. Thus, there arises the original puzzle posed by the spin ice materials that can be summarized by two questions:

- Is the experimental observation of spin ice behavior consistent with dominant long-range dipolar interactions?

- If so, then why is spin ice behavior maintained in the presence of these interactions when one would expect them to lead to long-range order below a temperature T_c of order $D_{\rm NN}$?

This issue was first investigated by Siddharthan and collaborators who used classical MC simulations to investigate the thermodynamic properties of Hamiltonian H, Eq. (7.4), with the dipolar interaction truncated at the fifth[36] and 12th[74] nearest neighbor. We recall that at the nearest-neighbor level, a positive $J_{\rm eff}$ gives rise to a two-in/two-out GS on a single tetrahedron, which results in a net magnetic moment for an isolated tetrahedron. It is therefore not obvious *a priori* that the long range part of the dipolar interactions in Hamiltonian equation (7.4) should not stabilize a ferromagnetic GS. In contrast, it is known from experiments that the spin ice state does not carry a net (bulk) magnetic moment at a sufficiently large length scale. If the GS of H is not ferromagnetic, the precise cancellation of the moment at large enough length scale is a delicate matter that arises from the large distance behavior of the dipolar interactions. Hence, precise handling of the long-distance dependence is crucial, and in such a situation, the dipolar interactions clearly need to be summed up to infinite distance. It is well known that dipole lattice summations are difficult to handle and are conditionally convergent. A standard technique to handle slowly convergent $1/r^3$ dipolar interactions is the Ewald summation method.[86–89] This method has been succesfully used to investigate the properties of dipolar fluids[90,91] as well as to understand the nature of critical phenomena in ferromagnets subject to long-range magnetic dipole–dipole interactions.[92,93] Very roughly speaking, the Ewald summation method implements a mathematical construct that is akin to the Madelung construction used to calculate the electrostatic binding energy of ionic crystals. In the next section, we describe the results obtained by MC simulations of the dipolar spin ice Hamiltonian H (Eq. (7.4)) that uses the Ewald method.

7.3.2.3. *The Ewald Monte Carlo*

The reader is referred to Refs. [86–89] for a discussion of the Ewald method as implemented in a conventional MC simulation. In essence, for a finite-size cubic system of linear size L that contains L^3 conventional pyrochlore cubic unit cells, hence $16L^3$ spins, the Ewald summation method recasts the original direct real space dipole–dipole interaction in Eq. (7.4) by an effective interaction where the image dipoles of those contained in the cell of size

L are resummed to infinity. In this sense, the Ewald method enhances the tendency toward periodic long-range order for "small" L. Once the Ewald interactions between Ising spins σ_i and σ_j that belong to the simulation cell have been calculated (and stored in a data structure), the MC simulation proceeds using the conventional Metropolis algorithm.

To make quantitative contact with experimental results, one needs to know the numerical values of the parameters entering the Hamiltonian equation (7.4). Since the lattice structure is known from X-ray or neutron diffraction, the size a of the conventional cubic unit cell is known, and one knows the distance r_{ij} between spins. The only unknowns in Eq. (7.4) are therefore the value of the nearest-neighbor exchange J and the magnitude of the magnetic moment μ. For the rare earth spin ice materials, the value of μ is close to its maximum possible value given the Landé g-factor g_J and the total angular momentum quantum number J of the ion.[39] For example, for Dy^{3+} $g = 4/3$ and $J = 15/2$ while for Ho^{3+} $g = 5/4$ and $J = 8$, giving a value $\mu \approx 10\mu_B$ in both cases. This gives an energy scale D close to 1.41 K for both $Ho_2Ti_2O_7$ and $Dy_2Ti_2O_7$ using a nearest-neighbor distance $a\sqrt{2}/4$ with $a \approx 10.1$ Å. This value of D (and $D_{NN} = 5D/3 = 2.35\,\text{K}$) is the value used in previous Monte Carlo simulations.[37,42]

Simulations on the dipolar spin ice model are carried out using the standard single-spin-flip Metropolis algorithm. To mimic the experimental conditions pertinent to real materials, the simulation sample is cooled slowly. At each temperature step, the system is equilibrated carefully, with various thermodynamic quantities of interest calculated. The single-spin-flip Metropolis algorithm is able to map out three different regions of the T vs J_{NN}/D_{NN} phase diagram shown in Fig. 7.9. Thermodynamic data indicate that when the nearest-neighbor exchange is antiferromagnetic and sufficiently large compared to the dipolar interactions ($J_{NN} < 0$ and $|J_{NN}| \gg D_{NN}$), the system undergoes a second-order phase transition to an all-in–all-out GS characterized by wave vector $\boldsymbol{q} = 0$. In the spin ice regime, $J_{\text{eff}}/D_{NN} = 1 + J_{NN}/D_{NN} > 0.095$, each specific heat data set for different J_{NN} shows qualitatively the same broad peak as observed in the nearest-neighbor model[38] (see Fig. 7.10). The height, C_{peak}, and peak temperature position, T_{peak}, show very little dependence on system size, for simulation cells of $L = 2$–6. However, both C_{peak} and T_{peak} are found to depend strongly on the ratio of J_{NN}/D_{NN}, as illustrated in Fig. 7.11.

By re-scaling the temperature scale for the specific heat corresponding to a number of different interaction parameters J_{NN} and D_{NN}, one can

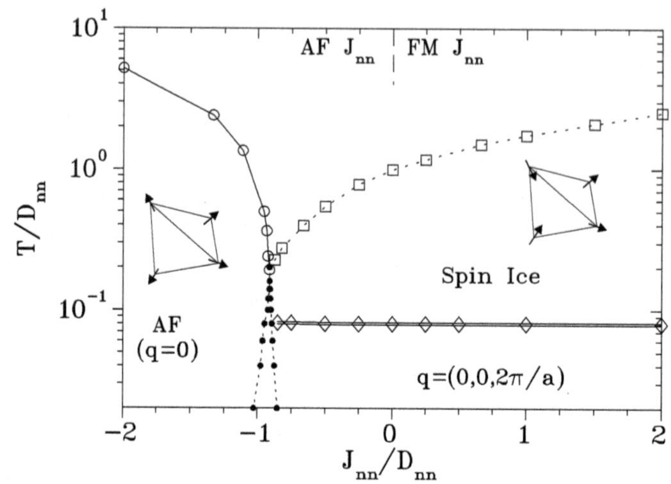

Fig. 7.9. Dipolar spin ice model: the phase diagram. The antiferromagnetic (AF) GS is an all-spins-in or all-spins-out configuration for each tetrahedron for $J_{\mathrm{NN}}/D_{\mathrm{NN}} < -0.905$. For $J_{\mathrm{NN}}/D_{\mathrm{NN}} > -0.905$, the system exhibits spin ice phenomenology as revealed by a broad specific heat peak at a temperature T_{peak} shown by the dashed lines (open squares). A loop MC algorithm, which allows for better thermal equilibration than conventional Metropolis, reveals a transition to a long-range ordered spin ice state with ordering wave vector $\boldsymbol{q} = (0, 0, 2\pi/a)$ at a temperature $T_c \approx 0.07D_{\mathrm{NN}}$. The region encompassed between the almost vertical dotted lines displays hysteresis in the long-range ordered state selected ($\boldsymbol{q} = 0$ vs. $\boldsymbol{q} = (0, 0, 2\pi/a)$) as $J_{\mathrm{NN}}/D_{\mathrm{NN}}$ is varied at fixed temperature T (from Refs. [94] and [95]).

expose more clearly the dependence of the specific heat on the competition between the nearest-neighbor exchange J_{NN} and the dipole–dipole interactions. In Fig. 7.10, this dependence is illustrated in the regime $J_{\mathrm{NN}}/D_{\mathrm{NN}} > 0$. The data show that in terms of an effective energy scale, $J_{\mathrm{eff}} \equiv J_{\mathrm{NN}} + D_{\mathrm{NN}}$, the medium- to long-range effects of the dipolar interactions are in some sense "screened", and one recovers qualitatively the short-range physics of the nearest-neighbor spin ice model. As J_{NN} becomes antiferromagnetic (see C_{peak} vs $J_{\mathrm{NN}}/D_{\mathrm{NN}}$ for $J_{\mathrm{NN}}/D_{\mathrm{NN}} < 0$ in Fig. 7.11), the approximate collapse onto a single-energy scale becomes less accurate, with the specific heat becoming dependent on $J_{\mathrm{NN}}/D_{\mathrm{NN}}$. It is within this regime that both $Ho_2Ti_2O_7$ and $Dy_2Ti_2O_7$ are expected to be realized, as we now discuss.

Since D_{NN} is calculated from Eq. (7.2), J_{NN} must be determined from experimental data. By fitting either the height, C_{peak}, or the peak temperature, T_{peak}, of the specific heat curves of the MC simulation to the

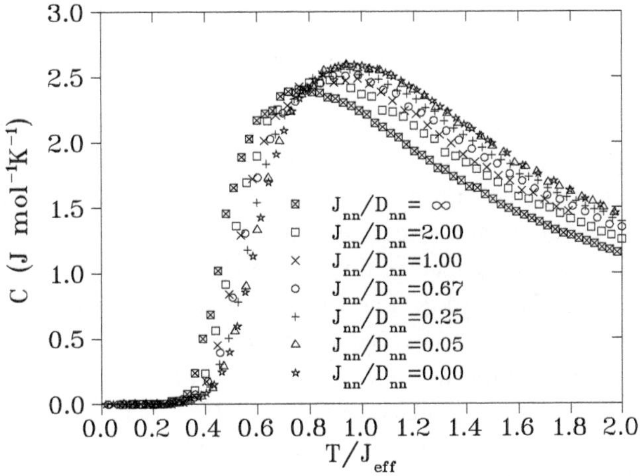

Fig. 7.10. Dipolar spin ice model: specific heat for system size $L = 2$, with temperature, T, re-scaled into units of the effective nearest-neighbor interaction $J_{\rm eff} \equiv J_{\rm NN} + D_{\rm NN}$. $J_{\rm NN}/D_{\rm NN} = 0$ corresponds to purely dipolar interactions, while $J_{\rm NN}/D_{\rm NN} = \infty$ corresponds to nearest-neighbor ferromagnetic exchange only. Simulation runs for $L = 4$ were also performed, but revealed no important finite-size effects (from Ref. [95]).

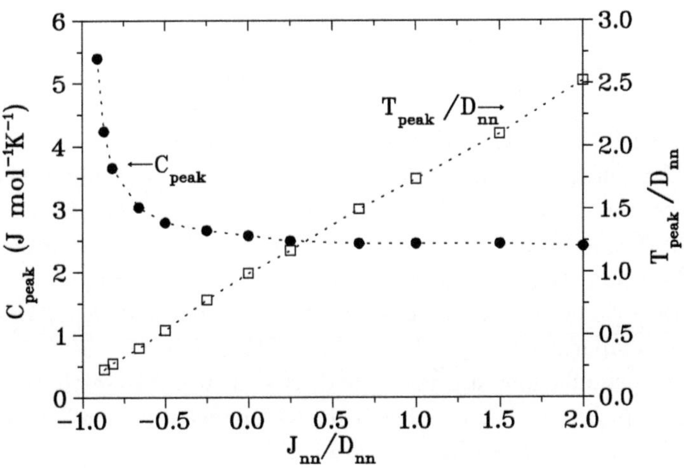

Fig. 7.11. Dipolar spin ice model: dependence of the simulated specific heat peak height $C_{\rm peak}$ and temperature $T_{\rm peak}$ on exchange and dipole–dipole interaction parameters. In this figure, $D_{\rm NN}$ is set to 2.35 K (from Refs. [37] and [95]).

experimental results,[22] one finds $J_{\rm NN} = -1.24\,\rm K$ for $Dy_2Ti_2O_7$. The results of this fitting are illustrated in the top panel of Fig. 7.4(a).[37] A fitting of the height or peak temperature of the experimental magnetic specific heat for $Ho_2Ti_2O_7$ gives $J_{\rm NN} = -0.52\,\rm K$ for this material.[42] As discussed above, a good fit of the electronic specific heat for $Ho_2Ti_2O_7$ can also be achieved once the hyperfine contribution has been removed (see Fig. 7.5).[42,45,46]

As discussed above, numerical integration of the specific heat divided by temperature can be performed to determine the entropy of both $Ho_2Ti_2O_7$ and $Dy_2Ti_2O_7$ (see Eq. (7.1)). The results for $Dy_2Ti_2O_7$ are illustrated in Fig. 7.4(b). The entropy recovered between $T = 0.4\,\rm K$, where $C_H(T)$ is very small, up to a temperature $T = 10\,\rm K$, is $S(T = 10\,\rm K) - S(T \approx 0) \approx 3.930\,\rm J\,mol^{-1}\,K^{-1}$. As we can see in Fig. 7.4(b), the Monte Carlo data for $S(T)$ at $T = 10\,\rm K$ is slightly below the Pauling's value $R\{\ln(2) - (1/2)\ln(3/2)\}$. To perform the calculation of the recovered entropy between $T = 10\,\rm K$ up to $T = \infty$, one can extrapolate the temperature dependence of the specific heat $C_H(T)$ for $T > 10\,\rm K$ by matching the Monte Carlo value of $C_H(T)$ at $T = 10\,\rm K$ with the $1/T^2$ high-temperature paramagnetic temperature regime, $C_H(T) = C_\infty/T^2$. This gives a value $C_\infty = 29.015\,\rm J\,mol^{-1}\,K$, and an extra entropy of $S(T = \infty) - S(T = 10) = 0.145\,\rm J\,mol^{-1}\,K^{-1}$, hence a value $S(T = \infty) - S(T \approx 0) = 4.075\,\rm J\,mol^{-1}\,K^{-1}$ is in exceedingly close agreement with Pauling's value, $4.077\,\rm J\,mol^{-1}\,K^{-1}$. Hence, the simulation with the appropriate experimental parameters retains Pauling's entropy $S_0 = R/2\ln(3/2)$, as is found experimentally for $Dy_2Ti_2O_7$ and $Ho_2Ti_2O_7$.

While the above conventional MC simulations of the model Hamiltonian for the spin ice compounds, Eq. (7.4), yields a reasonably successful quantitative theory of bulk spin ice behavior, there still remains the second question raised earlier in Section 7.3.2.2 as to why dipolar interactions, despite their anisotropic and long-range nature, do not appear to lift the macroscopic degeneracy associated with the ice rules and select an ordered state at a temperature $T \sim D_{\rm NN}$. As explained in the Section 7.3.2.4, as far as the simulations are concerned, the answer lies in the use of the single-spin-flip dynamics MC algorithm which becomes inefficient at low temperatures deep in the spin ice state.

7.3.2.4. *Mean-field theory*

The key question left unanswered by the above MC simulations is in some sense the simplest one of all: are there preferred spin–spin correlations in dipolar spin ice within the ice rule obeying low-temperature state below T_{peak} that lead, in principle, to a long-range ordered state?

The simplest approach to address this question is to calculate the wave vector ($\boldsymbol{q}$) dependence of the static susceptibility, $\chi(\boldsymbol{q})$, in the paramagnetic phase.[83] There are several ways to do this. The simplest way is to perform a high-temperature series expansion of $\chi(\boldsymbol{q})$.[82] A more expeditive approach is to use mean-field theory (or the random phase approximation). In this context, mean-field theory can be interpreted as a partial sum of an infinite number of diagrams in the high-temperature series expansion.[82] In the limit of very high temperature, the leading $1/T$ correction of $\chi(\boldsymbol{q})$ beyond the Curie behavior is the same in both high-temperature series expansion and in mean-field theory. The aim of such a calculation is to identify the periodicity, $\boldsymbol{q}_o$, of the spatial modulation of the magnetization density that first becomes unstable as the temperature is decreased. Except for sufficiently simple spin Hamiltonians, such a calculation will not give information about the GS. A well-known example is that of the ANNNI (anisotropic next-nearest-neighbor Ising) model where there exist regions in exchange coupling parameter space where incommensurate soft modes first appear upon cooling at T_c, but which immediately give way to a cascade of phase transitions to distinct states until one ultimately reaches a "simple" GS.[96] If from the mean-field calculation, one can identify an ordering wave vector $\boldsymbol{q}_o$ (and eigenvectors of the interacting susceptibility) for which the corresponding magnitude of the spins at each lattice site $\boldsymbol{r}$, $\boldsymbol{S}(\boldsymbol{r}) = 1$, (the so-called permanent moment structure), then one has identified not only the critical mode at T_c but also the GS (unless the transition is pre-empted by a first-order transition at a different $\boldsymbol{q}$ than $\boldsymbol{q}_o$ and a temperature $T > T_c(\boldsymbol{q}_o)$).

The key step in the mean-field calculation is the determination of the Fourier transform of the spin–spin interactions $\mathcal{J}^{ab}_{\alpha\beta}(i,j)$. Indices i and j label the sites $\boldsymbol{r}_i$ and $\boldsymbol{r}_j$ occupied by spins $\boldsymbol{S}(\boldsymbol{r}_i)$, α and β label the spin components ($\alpha = x, y, z$), and $\mathcal{J}^{ab}_{\alpha\beta}(i,j)$ incorporates both exchange and dipolar interactions. The dependence on the spin components is due to anisotropy, as arises from the dipolar interactions. Finally, one introduces sub-lattice indices a and b that label the sublattices of the tetrahedral basis. Here, there are four Bravais sublattices, chosen to have their origins on either an "up" or "down" tetrahedron (see Fig. 7.1). As a result of this, there are four branches for each $\boldsymbol{q}$ value. Details of the calculations are presented in Refs. [97] and [98]. The mean-field critical temperature,

$T_c^{\rm mf}$, is given by the largest eigenvalue, $\lambda^r(\boldsymbol{q})$, of $\mathcal{J}_{\alpha\beta}^{ab}(\boldsymbol{q})$ over the whole Brillouin zone (r labels the four eigenvalues/modes at each $\boldsymbol{q}$). More compactly, $T_c = \max_{\boldsymbol{q}}\{\lambda^{\max}(\boldsymbol{q})\}$, where $\{\lambda^{\max}(\boldsymbol{q})\}$ is the largest of the four eigenvalues ($r = 1, 2, 3, 4$) at wave vector $\boldsymbol{q}$, and $\max_{\boldsymbol{q}}$ indicates the global maximum of the spectrum of $\{\lambda^{\max}\}$ over all $\boldsymbol{q}$. The value of $\boldsymbol{q}$ for which $\lambda^r(\boldsymbol{q})$ is maximum is the ordering wave vector $\boldsymbol{q}_o$.

It was shown in Ref. [97] that truncating the dipolar lattice sums in Eq. (7.4) at ever-increasing cut-off distance leads to the development of a very flat quasi-dispersionless spectrum for $\lambda^{\max}(\boldsymbol{q})$ as the dipolar cut-off goes to infinity. Treating dipolar interactions via the Ewald method, where the long-range dipolar interactions are not truncated, one finds a completely smooth and quasi-degenerate soft mode spectrum with a global maximum (critical mode) at $\{001\}^{\rm c}$: see Fig. 7.12.[98] Inspection of the corresponding eigenvectors of $\mathcal{J}_{\alpha\beta}^{ab}(\boldsymbol{q})$ associated with the doubly degenerate critical mode at $\{001\}$ indicates a two-in–two-out spin ice structure, where the spins on sub-lattices $a = 1, 3$ point opposite to those on sub-lattices $a = 2, 4$ in a tetrahedral unit. Reference [98] discusses how these soft modes can be used to reconstruct the long-range ordered (permanent moment) structure, where the thermal local magnetization is the same on all sites, and which corresponds to the long-range ordered spin ice state that develops at low temperature if ergodicity can be maintained in MC simulations (see Section 7.3.2.5).

From mean-field theory, one can also determine the value of $J_{\rm NN}/D_{\rm NN}$ at which the ordering changes from an "all-in–all-out" $\boldsymbol{q} = 0$ state (i.e. from large negative antiferromagnetic $J_{\rm NN}$) to the (001) long-range ordered spin ice state. It is found that the transition between the two states occurs at $J_{\rm NN}/D_{\rm NN} = -0.905$. This agrees with the value found in MC simulations results for the transition between all-in–all-out $\boldsymbol{q} = 0$ ordering and spin ice regime[37] and agrees as well with zero-temperature energy calculations of the all-in–all-out GS vs. the (001) state mentioned above.

Having obtained strong evidence from mean-field theory that there exists a well-defined, unique ordering wave vector in the long-range dipolar spin ice model at the Gaussian mean-field level, one can proceed with the search for a transition to long-range order in the model that is dynamically inhibited when using conventional single-spin-flip MC simulations.

---

[c] $\{\cdots\}$ represents the set of symmetry-related reciprocal lattice points.

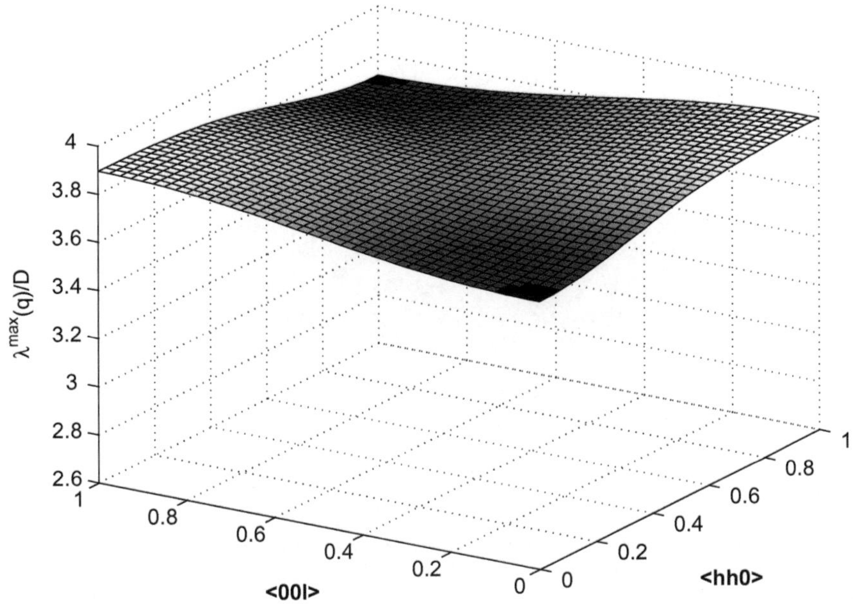

Fig. 7.12. Dipolar spin ice model: the scaled maximum eigenvalues, $\lambda^{\max}(\boldsymbol{q})/D$, in the (hhl) plane. The dipole–dipole interactions are treated with the Ewald approach. Here, the exchange coupling J was set to zero (from Ref. [98]).

7.3.2.5. *The loop Monte Carlo*

In order to investigate the low-temperature ordering properties of dipolar spin ice, one needs a MC algorithm with non-local updates that effectively bypass the energy barriers that separate nearly degenerate states and allow the simulation to explore the restricted ice-rule phase space that prevents ordering in the model.[94, 95, 99] Such an algorithm can then be used to explore and characterize the long-range ordered state that arises out of ice-rule manifold and which is slightly energetically favored by the long-range dipolar interactions over an arbitrary ice rule obeying state.

One first identifies the true zero-energy modes that can take the near-neighbor spin ice model from one spin ice state to another exactly degenerate ice state. An example of these zero modes, or loops, is shown in Fig. 7.13. An initial working hypothesis is that in the dipolar spin ice model, with interactions beyond nearest neighbor, the system freezes into a state that obeys the ice rules. This is indeed what is found in simulations: systems simulated using conventional single spin-flips always freeze out in an ice-rules

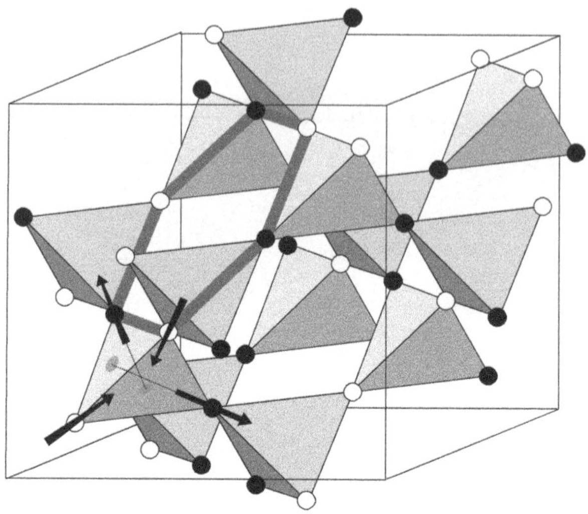

Fig. 7.13. The pyrochlore lattice, illustrating a spin ice state and a possible "loop move". The lower left "downward" tetrahedron shows Ising spins as arrows. Each spin axis is along the local ⟨111⟩ quantization axis, which goes from one site to the center of the opposing triangular face (as shown by the disks) and meets with the three other ⟨111⟩ axes in the middle of the tetrahedron. For clarity, black and white circles on the lattice points denote other spins. White represents a spin pointing into a downward tetrahedron while black is a spin pointing outward. The entire lattice is shown in an ice-rules state (two black and two white sites for every tetrahedron). The hexagon (thick gray line) shows a minimal size loop move, which corresponds to reversing all colors (spins) on the loop to produce a new ice-rules state (from Ref. [99]).

state with no "defects" that violate the ice rules. With interactions beyond nearest neighbor, these loop moves become low-energy excitations (quasi-zero modes) that can take the dipolar spin ice model from one ice-rules state to another without introducing spin defects. This allows all of the quasi-degenerate spin ice states to be sampled ergodically and facilitates the development of a long-range ordered state at low temperatures.

Barkema and Newman[100, 101] proposed a loop algorithm that was originally designed for 2D square ice models within an MC simulation. This scheme can be readily adapted to the three dimensional pyrochlore lattice. In the context of square ice, two types of loop algorithm, the so-called long- and short-loop algorithm, were investigated.[94] In the square ice model, each vertex on a square lattice has four spins associated with it. The vertices are analogous to tetrahedron centers. The ice rules correspond to "two spins pointing in, two spins pointing out" at each vertex. In the Barkema and

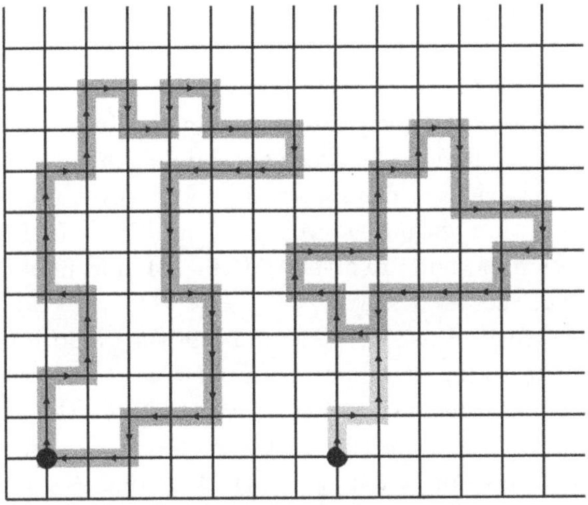

Fig. 7.14. Long and short loops formed by the Newman and Barkema algorithm (Refs. [100] and [101]) on a square ice lattice. Vertices are represented by points where lattice lines cross. Each vertex has two spins pointing in and two spins pointing out; however, for clarity, only spins which are included in the loops are shown. Starting vertices are indicated by large black dots. On the left is an example of a long loop, which is completed when it encounters its own starting vertex. On the right is a short loop, which is complete when it crosses itself at any point. Dark gray lines outline completed loops. The excluded tail of the short loop is shown in light gray (from Refs. [94] and [95]).

Newman algorithm, a loop is formed by tracing a path through ice-rules vertices, alternating between spins pointing into and spins pointing out of the vertices. A "long loop" is completed when the path closes upon the same spin from which it started. A "short loop" is formed whenever the path traced by the loop encounters any other vertex (tetrahedron) already included in the loop — excluding the dangling tail of spins (Fig. 7.14).

The implementation of the Barkema and Newman loop algorithm for the three dimensional pyrochlore lattice spin ice problem proceeds as follows. In this system, the smallest complete loop that is a zero mode on the pyrochlore lattice consists of six spins (see Fig. 7.13). Such a loop in spin ice was previously identified in Ref. [5] and also by Pauling[1] in the context of water ice and by Anderson[3] in the context of the spinel lattice. However, using the above loop algorithm, much larger loops are possible. When used with the pyrochlore lattice (Fig. 7.13), such a loop must pass through two spins on each tetrahedron. A loop always "enters" a tetrahedron through an inward-pointing spin and "leaves" a tetrahedron through an outward-pointing spin. The periodic boundary conditions of the lattice create no difficulties. If one forms a closed loop in this manner, each spin on the loop is reversed, and the whole system stays in an ice-rules state. However, small increases or decreases in the dipole–dipole energy may arise due to small energy differences between the old and the new ice-rules state. These small energy changes caused by the loop moves are evaluated via a Metropolis algorithm within the MC program.[100,101] Full details of the loop MC simulation method are given in Refs. [94] and [95]. Here, we only give a summary of the key results.

It was discussed in Refs. [94], [95] and [99] that loop MC simulations are successful in restoring ergodicity in simulations of dipolar spin ice. A main consequence of this restoration of ergodicity is the observation of a low-temperature phase transition in the model. For example, as a first indication, a MC loop simulation of $Dy_2Ti_2O_7$ (using $J_{NN} = -1.24\,K$ and $D_{NN} = 2.35\,K$) now shows a sharp peak in the specific heat at a temperature significantly below the spin ice peak (see Fig. 7.15). The feature in the

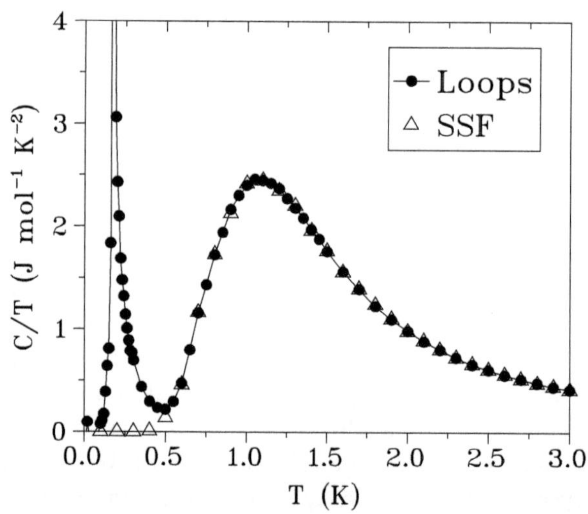

Fig. 7.15. Dipolar spin ice model: the low-temperature magnetic specific heat calculated by numerical simulation for a system size $L = 4$ and with simulation parameters set for $Dy_2Ti_2O_7$. Closed circles are simulation data run with the short-loop algorithm (Loops), open triangles are data obtained using the single-spin-flip (SSF) Metropolis algorithm. The broad feature in the specific heat at $T \approx 1\,K$ indicates the rapid development of the spin ice rule obeying states. The sharp feature at T_c is the appearance of a phase transition to a GS being made dynamically accessible via the non-local loop dynamics (from Refs. [94] and [95]).

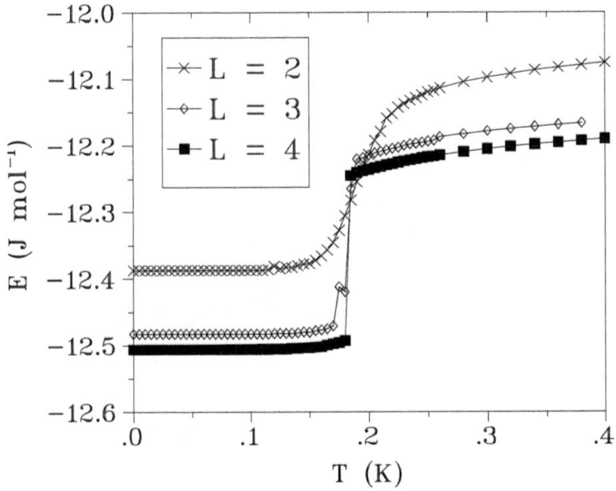

Fig. 7.16. Dipolar spin ice model: details of the loop Monte Carlo simulation energy near the transition for different system sizes (from Refs. [94] and [95]).

specific heat and the abrupt drop in energy (see Fig. 7.16) at the same temperature give compelling preliminary evidence that the loop algorithm is successful in allowing a phase transition to occur at a temperature of $T_c \approx 0.18\,\mathrm{K}$. Also, the energy curve shows a discontinuous drop at T_c (i.e. latent heat) for large lattice sizes, suggesting a first-order phase transition (Fig. 7.16). In the remainder of this section, we discuss the nature of the low-temperature ordered state of dipolar spin ice and the phase transition that leads to it.

Direct inspection of the spin directions at $T < 0.18\,\mathrm{K}$ reveals that the long-range ordered state obeys the ice rules, has zero magnetic moment per unit cell and is commensurate with the cubic pyrochlore unit cell (see Fig. 7.17). This state corresponds to the critical mode found in the mean-field calculation discussed above.[97,98] There are 12 symmetrically equivalent spin configurations for the GS as explained below, two for each cubic axis ($\langle 100 \rangle$) direction and their spin-reversed states. The ordering wave vector $\boldsymbol{q}_o$ lies parallel to one of the cubic axis directions, specifically $\boldsymbol{q}_o = (0, 0, 2\pi/a)$ or one of its symmetrically equivalent (star) directions. To construct the ordered state, first consider a starting tetrahedron with its six possible ice-rules states. For a given ordering wave vector $\boldsymbol{q}_o$, this tetrahedron selects one of the four possible spin configurations (two independent configurations and their spin-reversals, $\boldsymbol{S}_i^a \to -\boldsymbol{S}_i^a$) with a total magnetic

(a)

(b)

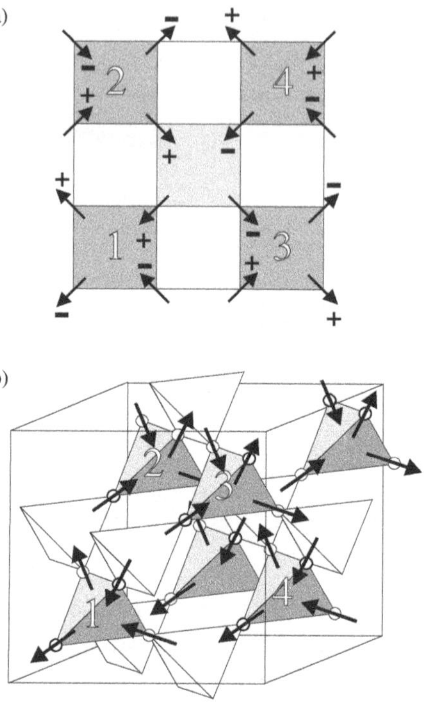

Fig. 7.17. Dipolar spin ice model: the long-range ordered $\boldsymbol{q} = (0, 0, 2\pi/a)$ GS projected down the z axis (a); the four tetrahedra making up the cubic unit cell appear as dark gray squares. The light gray square in the middle does not represent a tetrahedron, but its diagonally opposing spins are in the same lattice plane. The component of each spin parallel to the z axis is indicated by a + and − sign. In perspective (b), the four tetrahedra of the unit cell are numbered to enable comparison with (a) (from Ref. [99]).

moment for the tetrahedron perpendicular to $\boldsymbol{q}_o$. The entire ordered state may be described by planes (perpendicular to $\boldsymbol{q}_o$) of such tetrahedra. The wavelength defined by this $\boldsymbol{q}_o$ physically corresponds to antiferromagnetically stacked planes of tetrahedra, which means that a given plane has tetrahedra of reversed-spin configuration with respect to the planes above and below it (i.e. those displaced by half of cubic unit cell constant). To describe this state, we define the multi-component order parameter

$$\boldsymbol{\Psi}_{\alpha}^{m} = \frac{1}{N} \left| \sum_{j=1}^{N/4} \sum_{a=1}^{4} \sigma_j^a e^{(i\phi_a^m + i\boldsymbol{q}_\alpha \cdot \boldsymbol{R}_j)} \right|. \tag{7.6}$$

This type of labeling is natural given that the pyrochlore lattice can be viewed as a face-centered cubic (FCC) lattice with a "downward" tetrahedral basis (Fig. 7.1). Thus, j labels the FCC lattice points of the pyrochlore lattice, and the index a sums over the four spins that make the basis connected to each j. The index α labels the three possible symmetry-related $\boldsymbol{q}_o$ ordering wave vectors. For a given $\boldsymbol{q}_\alpha$, as described above, there are two ice-rules configurations and their reversals which can each form a GS. Thus, $m = 1, 2$ labels these possibilities with the phase factors $\{\phi_a^m\}$, describing the given configurations m. Each Ising variable σ_j^a has a value $+1$ or -1 when a spin points into or out of its downward tetrahedron j, respectively.

$\boldsymbol{\Psi}_\alpha^m$ has six degenerate components, each of which can take on a value between 0 and 1. Upon cooling through the transition, the system selects a unique ordered configuration, causing the corresponding component of $\boldsymbol{\Psi}_\alpha^m$ to rise to unity and the other five to fall to zero (provided the finite-size system is simulated below the critical temperature over a time scale less than the ergodic time scale over which full spin symmetry would be restored). The component which rises to unity is equally likely to be any one of the six that are randomly selected by spontaneous symmetry breaking.

The average magnitude of the multicomponent order is defined as

$$\langle \Psi \rangle = \sqrt{\sum_{m=1}^{2} \sum_{\alpha=1}^{3} (\boldsymbol{\Psi}_\alpha^m)^2}. \tag{7.7}$$

Figure 7.18 is a plot of this quantity versus temperature for three different system sizes. For $T < T_c$, the different lattice sizes produce identical order parameters. By contrast, $\langle \Psi \rangle$ for the smaller lattice size displays pronounced rounding near T_c and an increased residual value for large T. The larger lattice size produces an order parameter with a clear discontinuity at T_c. This discontinuity in the order parameter combined with the discontinuity of the total energy in Fig. 7.16 can be viewed as strong evidence for a first-order transition.

The loop MC simulation allows an estimate of the latent heat at the transition. Referring to Fig. 7.16, the energy discontinuity, ΔE, as read off this graph for $L = 4$, is

$$\Delta E \approx 0.248\,\mathrm{J\,mol^{-1}}. \tag{7.8}$$

This behavior is also consistent with the transition being of first order. Below, we compare this ΔE to the latent heat associated with the entropy

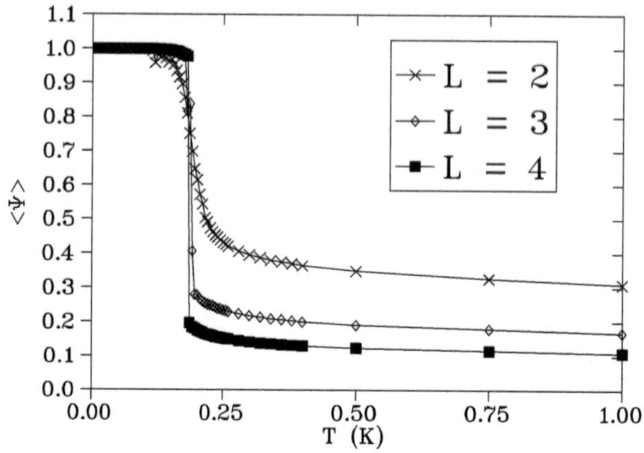

Fig. 7.18. Dipolar spin ice model: the $\boldsymbol{q} = (0, 0, 2\pi/a)$ order parameter. Curves are shown for system size $L = 2$, $L = 3$, and $L = 4$ (from Ref. [99]).

jump at the transition and calculated through the finite-size scaling of the height of the specific heat[102]:

$$C_{\text{peak}} \approx \frac{(\Delta E)^2}{4k_{\text{B}}T_c^2}L^d + a, \tag{7.9}$$

where L^d is the system volume as before and a is the y-intercept of the finite-size scaling graph of $C_{\text{peak}}(L)$ *vs* L^3. To achieve good statistics of the finite-size scaling behavior of the transition to long-range order, one has to employ histogram methods and reweighting techniques discussed in Refs. [103–105] and employed in the present context in Refs. [94], [95] and [99].

The L^3 dependence of $C_{\text{peak}}(L)$ in Fig. 7.19 shows that the finite-size scaling is consistent with that expected for a first-order transition. Also, as a second estimator of the latent heat, the slope of the L^3 fit allows to determine ΔE using Eq. (7.9). One deduces a latent heat of

$$\Delta E = 0.245\,\text{J}\,\text{mol}^{-1}, \tag{7.10}$$

consistent with the value obtained in Fig. 7.16 (for the $L = 4$ system) from reading directly off of the energy graph (see Eq. (7.8)). From the estimate of the latent heat, one finds that the entropy jump at the transition at $T_c \sim 0.18\,\text{K}$ is approximately $\Delta E/T_c \approx 1.37\,\text{J/mol}\,\text{K}$. Adding to this

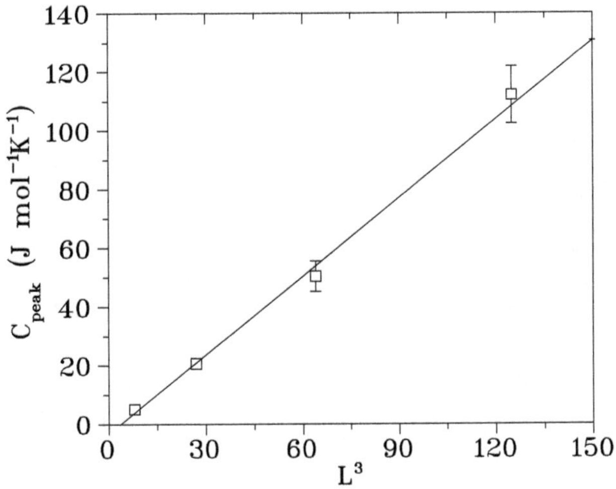

Fig. 7.19. Dipolar spin ice model: finite-size scaling fit for the specific peak heights of the ordering transition observed in the loop Monte Carlo simulations. Data points represent the mean C_{peak} value for a given L averaged over 10 simulations for each L. Error bars show one standard deviation (from Ref. [99]).

entropy jump the (small) amount of entropy removed between $0 < T < T_c$, and between $T_c < T = \infty$, one finds within 2% the full entropy expected for such an Ising system, that is $R\ln(2) = 5.765\,\text{Jmol}^{-1}\,\text{K}^{-1}$ (see Fig. 7.20).

Taking all indicators together, one finds that that the transition to long-range order at 180 mK recovers all residual Pauling entropy of the dipolar spin ice model. Thus, one can assert that the degeneracy associated with the spin ice model, and the corresponding value of zero-point entropy, is "naturally" lifted due to the internal perturbations beyond nearest neighbor within the long-range dipole–dipole interactions, if equilibrium can be maintained at sufficiently low temperatures.

To summarize the results presented in this section, we refer the reader to the dipolar spin ice MC phase diagram, Fig. 7.9. As illustrated in the figure, the transition between the spin ice phase (which retains Pauling's entropy) and the $\boldsymbol{q}_o = (0, 0, 2\pi/a)$ ordered phase is independent of the strength of J_{NN}. One finds that this first-order line also partially follows the boundary between the antiferromagnetic ordered phase and the higher temperature paramagnetic phase and that a tricritical point separates these two regions of the line, which occurs near the value $J_{\text{NN}}/D_{\text{NN}} \sim -1.1$.

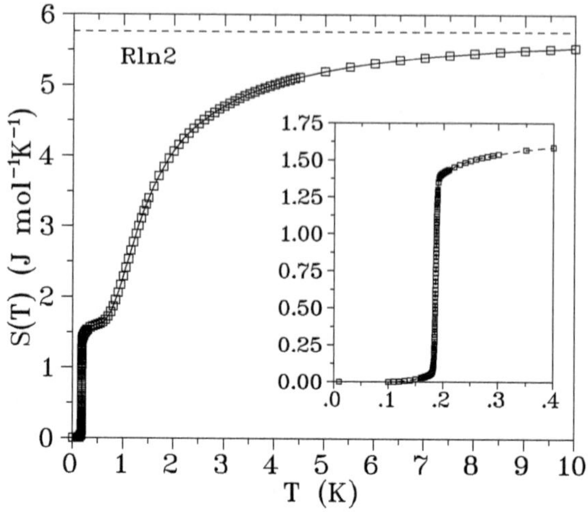

Fig. 7.20. Dipolar spin ice model: the entropy calculated by integrating the specific heat simulated by the loop Monte Carlo method. The entire value of entropy for the system ($R\ln 2$) is recovered in the high-temperature limit. The inset shows the details of the entropy recovered by the transition to long-range order (from Refs. [94] and [95]).

7.3.2.6. *Application of the dipolar model to neutron scattering results*

As mentioned in Section 7.3.1.3, elastic neutron scattering is a sufficiently detailed microscopic probe to identify corrections to the near-neighbor spin ice model that are not required to describe the bulk properties. Referring to Fig. 7.8, of the several characteristic "blobs" of scattering observed in $Ho_2Ti_2O_7$, those centered at $(0,0,3)$ are rounder in reality than predicted by the near-neighbor model[106] and those centered at $(3/2,3/2,3/2)$ are more elongated along $[\bar{1}\bar{1}1]$. As shown in the figure, such differences are well accounted for by the dipolar model.[42] The description is quantitative,[42] as has been further confirmed by the neutron measurements of Kanada *et al.*,[43] who fitted the experimental scattering along $hh0$ and $00l$ at temperatures between 0.4 K and 50 K to the predictions of the dipolar model. Likewise, Kadowaki *et al.* fitted the experimental powder diffraction from $Ho_2Sn_2O_7$ with the prediction of the dipolar model between $T = 0.4$ K and 50 K.[47]

$Dy_2Ti_2O_7$ does not lend itself well to neutron studies as several of the natural Dy isotopes are strongly neutron absorbing. This problem has been solved by Fennell *et al.*[81,107] who have prepared a single crystal of $^{162}Dy_2Ti_2O_7$, with greatly reduced absorption. A very recently reported comparison of experiment and theory has shown that the spin ice state at 0.3 K is reasonably well accounted for by the dipolar model, but that the differences between experiment an theory are much greater than those for the Ho materials. In particular, there is enhanced magnetic scattering around the Brillouin zone boundaries that is not well accounted for by the dipolar model.[81]

7.3.2.7. *How realistic is the dipolar model?*

We have shown above how the dipolar model with single-spin-flip dynamics provides a good description of the elastic neutron scattering and bulk properties of the spin ice materials. It is also evident — especially from the latest neutron scattering results of Fennell *et al.*[81] — that the model is not perfect. As improvements to the model one could consider, for example, further neighbor exchange or a more accurate quantum mechanical description of the admixing of excited M_J levels into the GS doublet.[77]

For example, it could be argued that it is better to consider, rather than Ising spins, Heisenberg spins with a strong uniaxial anisotropy D to force them along $\langle 111 \rangle$.[98] This model with ferromagnetic near-neighbor exchange coupling was considered by Champion *et al.*,[67] who found that it has an ordering transition to a ferromagnetic state with the spins canted off the local trigonal axis toward ferromagnetic alignment (the canting decreasing to zero as $D \to 0$). The transition is first order for $D/J > 5$ and second order for $D/J < 5$. Another related model, which combines dipolar interactions with g-factor anisotropy (but no exchange) has recently been considered by Chamberlain *et al.*,[108] who numerically calculated the $T = 0$ GSs as a function of anisotropy. For the Ising limit, the ordered state identified by Melko *et al.*[99] is recovered, while for the Heisenberg limit, a GS similar to Champion's canted ferromagnetic state is found. It thus seems likely that the combination of dipolar interactions and non-Ising magnetic moments might result in a competition between these two GSs. However, it should be emphasized that to date, only the dipolar spin ice model has received significant experimental support.

It is then relevant to ask, is the dipolar model sufficiently accurate to have identified a "true" GS in the spin ice materials? Thus, as discussed in detail above, Melko *et al.*[99] showed that taking into account the correct long-range dipolar interaction and implementing a special dynamics, the degeneracy is lifted, the system remains ergodic and the missing entropy is liberated through a low-temperature phase transition. However, although thermodynamics as we know it is clearly saved in this way, experimentally there is no evidence of such a transition in the spin ice materials. Loss of ergodicity means that the spin ice materials behave, in this regard, much more like the idealized near-neighbor spin ice model[4] than this essentially exact numerical calculation. It is of course possible that minor terms in the spin Hamiltonian are not yet correctly identified,[81] causing this disagreement between experiment and theory.

7.4. Field-Induced Phases

In Section 7.1, we remarked how the (near-neighbor) spin ice model is a 16-vertex model in which the statistical weights of the vertices can be controlled by the application of a magnetic field along different symmetry directions. In experimental terms, this gives the possibility of a remarkable degree of control over the physics of spin ice materials, and in-field studies indeed have revealed many interesting new phenomena. In this section, we review both the theoretical and experimental aspects of spin ice behavior in an applied field.

7.4.1. *Theory*

7.4.1.1. *Near-neighbor model*

The application of a magnetic field to the near-neighbor spin ice model shows several interesting features. MC calculations of magnetization curves for the three symmetry axes, [100] [110], and [111], are shown in Fig. 7.21.[38] The magnetization per spin saturates at slightly different values depending on the field direction: $1/\sqrt(3)$, $1/\sqrt(6)$, and $1/2$ of the spin length for [100], [110], and [111], respectively.

With the field applied along the four-fold axis [100], all spins couple to the field and are saturated in a canted ferromagnetic state at high fields. Barely visible on the simulated magnetization field plot is a small jump in the magnetization near to saturation. This is in fact a first-order phase transition, which is more clearly visible as a spike in the specific heat. The relevant data, and the corresponding magnetic phase diagram are shown in Figs. 7.22 and 7.23. There are two lines of first-order phase transitions terminating in a critical end point (note that this might still be first order[109]). The phase transition involves a jump in magnetization between two phases of the same symmetry. In this case, it is analogous to the liquid–gas transition, which involves a jump in density. Such symmetry-sustaining phase

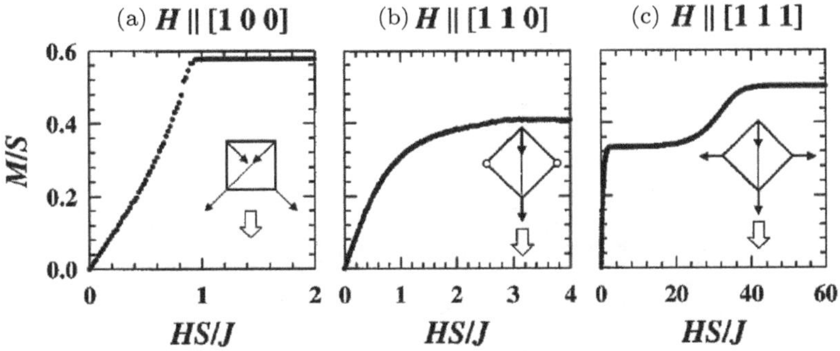

Fig. 7.21. Near-neighbor spin ice model: MC simulation results for the magnetization in an applied field for the three symmetry axes [100] [110], and [111] (Ref. [38]).

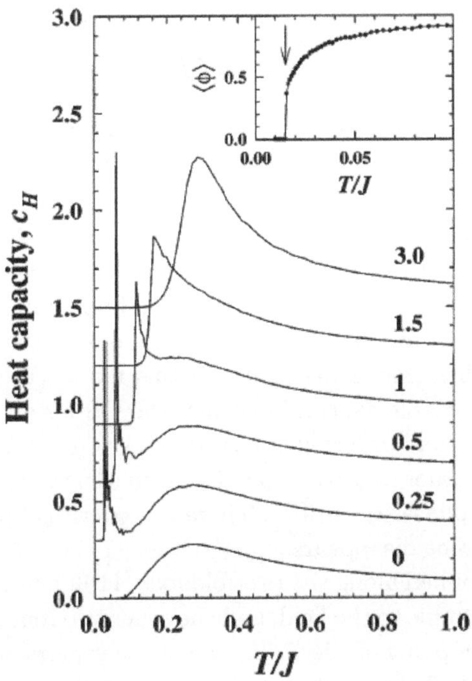

Fig. 7.22. Near-neighbor spin ice model: MC simulation results in a fields applied on [100], revealing a specific heat spike that indicates a line of first-order phase transitions (see Fig. 7.23). The inset shows the "liquid–gas order parameter" $\langle\Phi\rangle = m_l - m_g$, where m_l and m_g are the magnetizations above and below the critical line, respectively (Ref. [38]).

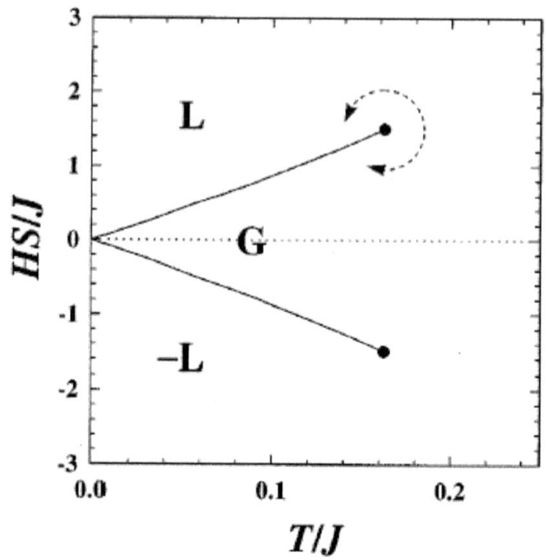

Fig. 7.23. Near-neighbor spin ice model: magnetic phase diagram for a field applied along [100], emphasizing the analogy with a liquid–gas system (Ref. [38]).

transitions are known in ferroelectrics,[110] and the general shape can be justified by the Landau theory with a negative fourth-order term in the free energy expansion. An extension of the near-neighbor spin ice model studied by Champion *et al.*[67] involves the use of continuous spins with strong $\langle 111 \rangle$ anisotropy D, rather than Ising spins. In this model, the [100] ferromagnetic state (with variable canting) becomes the GS for all finite D and the corresponding zero-field ordering transition crosses over from second order at small D to first order at a triple point that marks the confluence of three lines of first-order phase transitions. There is a tricritical point at $D/J \approx 5$ where the triple point disappears.

With a field applied along the two-fold axis [110], now only two spins on each tetrahedron couple to the field, and the other two remain perpendicular (see inset of middle panel of Fig. 7.21). The former pairs are aligned by the field into collinear "α" chains of canted spins, such that one spin points into and one out of each tetrahedron. Application of the ice rule means that the perpendicular spin pair must also be "one in, one out" and hence align in collinear so-called β chains of canted spins oriented perpendicular to the field. However, at the level of the near-neighbor model, these perpendicular chains are not energetically coupled together.

With the field along the three-fold axis [111], there is an initial and rapid saturation of the magnetization to a "plateau" of 1/3 of the spin length, which is then followed by a second rise to full saturation (Fig. 7.21(c)). In this orientation, one spin is parallel to the field while the other three can all couple to it. Hence, the lowest energy state has three spins in and one out (or vice versa, depending on whether it is an "up" or "down" tetrahedron). Hence, the second rise in magnetization involves breaking the spin ice rules. As pointed out in Ref. [111], the plateau state has one spin aligned and the other three fixed by the ice rules. This leads to the very interesting property of retaining an extensive zero-point entropy, as discussed further in Section 7.4.5.

7.4.1.2. *Dipolar model*

The simple nearest-neighbor model allows us to understand the constraints introduced by the applied magnetic field at the single-tetrahedron level. To discuss quantitatively the development of spin–spin correlations in an applied field, it is necessary to discuss the long-range dipolar interactions.

Recent numerical work on the dipolar model in a field applied along [110] has found a transition to long-range order at low temperatures for exchange and dipole–dipole coupling parameters appropriate for $Dy_2Ti_2O_7$.[112,113] Yoshida *et al.* have described the corresponding phase diagram.[106] Below 0.15 T, the GS is the theoretical zero-field ordered state identified by Melko *et al.*[99] in which the α and β chains (defined above) are three-dimensionally ordered in an antiferromagnetic way. However, above μ_0H = 0.15 T, the GS is the $Q = X$ state discussed by Harris *et al.*,[4] which has ferromagnetic α chains. In the dipolar model, these are found to persist even above the temperature (~1 K) where the β chains disorder for $\mu_0 H \to \infty$. Thus, the principal difference between the near-neighbor and dipolar models relates to the ordering of the β chains in the latter but not in the former.

7.4.2. *Magnetization measurements above $T = 1$ K*

It was first shown by Cornelius and Gardner[46] that the magnetization curves of $Ho_2Ti_2O_7$ near 2 K bear a striking qualitative resemblance to the predictions of the near-neighbor model,[38] although the saturation magnetizations measured in Ref. [46] were not in good agreement with the theory. This has recently been corrected by Petrenko *et al.*,[114] who re-measured the magnetization curves along the major symmetry directions [100], [110], and

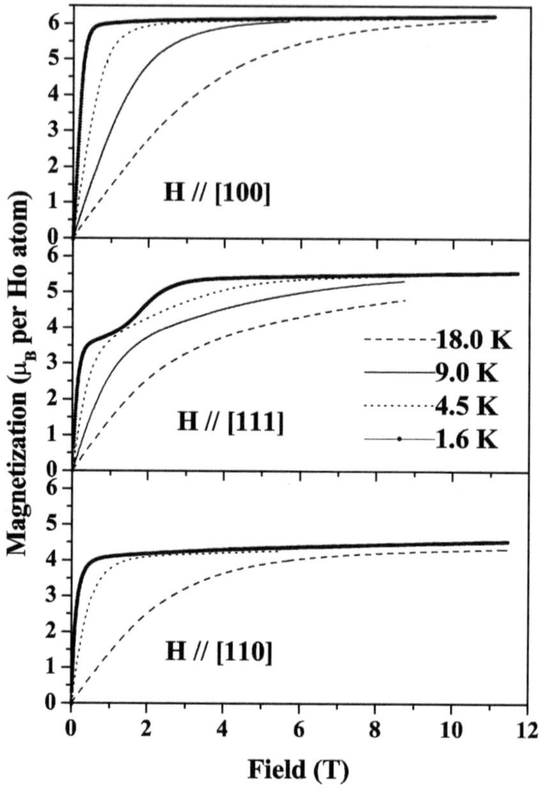

Fig. 7.24. $Ho_2Ti_2O_7$: Magnetization vs. field curves (Ref. [114]), showing qualitative agreement with the theoretical predictions of Fig. 7.21.

[001]. They found very good agreement with both the qualitative shape and saturation magnetization values of the near-neighbor spin ice model (see Fig. 7.24). In the mean time, Fukuzawa *et al.*[64] measured magnetization on single crystals of $Dy_2Ti_2O_7$ (Fig. 7.25) and found excellent agreement with the predictions of the dipolar model which, at the temperatures considered, are qualitatively similar to those of the near-neighbor model. The development of the [111] plateau state is a particularly noticeable feature of these curves. The plateau for $Dy_2Ti_2O_7$ in Fig. 7.25 at 1.8 K is less visible than that for $Ho_2Ti_2O_7$ in Fig. 7.24 at 1.6 K since, as discussed above, $Ho_2Ti_2O_7$ enters the spin ice regime at $T_{\rm peak} \sim 1.9$ K while $Dy_2Ti_2O_7$ is not yet really in the spin ice state at 1.8 K since it enters the spin ice regime below $\sim T_{\rm peak} \sim 1.2$ K.

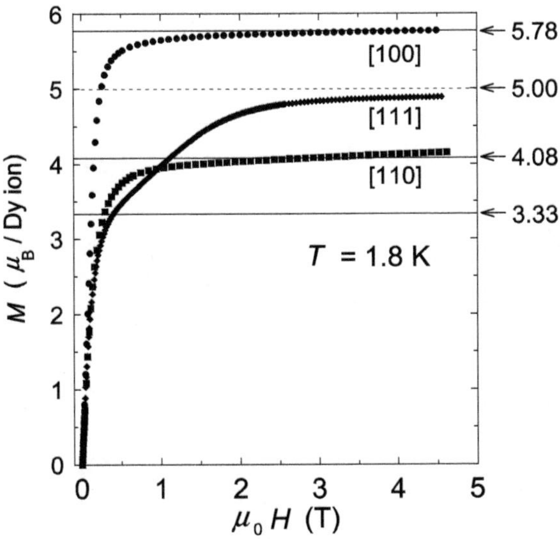

Fig. 7.25. $Dy_2Ti_2O_7$: Magnetization versus field curves (Ref. [64]), showing qualitative agreement with the theoretical predictions of Fig. 7.21. The curves have been justified in detail by a comparison with the dipolar spin ice model.

7.4.3. *Bulk measurements at low temperature*

7.4.3.1. [111] *Direction*

The most comprehensive bulk measurements in this regime are on the specific heat.[111, 115, 116] The application of a magnetic field generally results in a shift of the Schottky-like peak in the heat capacity to higher temperature as the energy cost of a spin-flip is increased by the Zeeman splitting. The magnetic entropy can be determined by integrating the specific heat into the paramagnetic regime, as in Ref. [22]. This is particularly interesting when a field is applied along [111] such that the sample is in the plateau state (in reference to Fig.7.21(c)). Here one observes the entropy of "kagomé ice", discussed in Section 7.4.5.1. At high magnetic field (~1 T), it has been confirmed that all the entropy is eliminated, regardless of the direction of application of the field.[116] However, some anomalous effects have been noticed in the specific heat below 300 mK, which require further investigation.[117, 118]

A significant paper by Sakakibara *et al.*[119] has examined the evolution of the bulk magnetization with the field applied along [111] at millikelvin temperatures. Remarkably, the increase in magnetization out of the

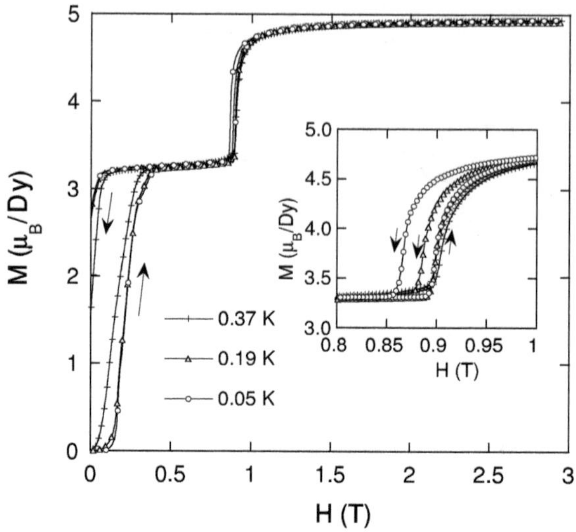

Fig. 7.26. $Dy_2Ti_2O_7$: Field applied along [111]. Magnetization vs. field isotherms in the millikelvin temperature range showing the evolution of a first-order phase transition between the magnetization plateau and saturation (Ref. [119]). Inset shows the field hysteresis around 0.9 T for the three temperatures that correspond to the main panel.

plateau state to full saturation evolves into a first-order transition below $T_c = 0.36\,\mathrm{K}$, as shown in Fig. 7.26. As in the case of the predicted transition with field along [100],[38] this transition with field along [111] is symmetry sustaining and can be considered analogous to the liquid–gas transition. There is a line of first-order transitions on the magnetization field phase diagram with a weak hysteresis. This line disappears at a critical point with divergent susceptibility at $T_c = 0.36(2)\,\mathrm{K}$, $H_c = 0.93$ T. The first-order line can be extrapolated to zero temperature at a finite field (0.905 T) (see Fig. 7.27). The authors found that the phase transition lies outside the spin ice regime of very strong history dependence. It is this feature that makes the transition easily observable: although Harris *et al.*[38] predicted a liquid–gas type transition for the field applied on [100] rather than [111], it seems that any transition in the [100] direction would be affected by strong history dependence and thus be hard to establish with certainty (see Section 7.4.3.3). Further experimental results regarding this phase transition are discussed in Section 7.4.5.3.

The temperature and field dependence of the [111] phase transition may be calculated by a simple thermodynamic argument.[115] The molar entropy (per Dy) of the plateau state is $S_k = 0.65\ \mathrm{J\,K^{-1}\,mol^{-1}}$ (see below)

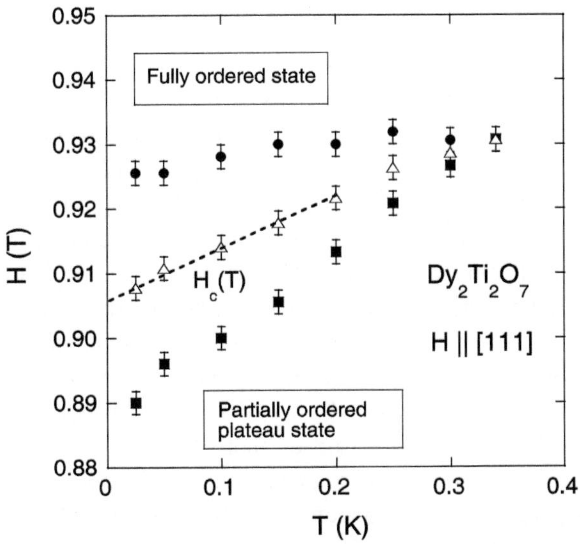

Fig. 7.27. $Dy_2Ti_2O_7$: Field applied along [111]. Magnetic phase diagram showing a line of first-order phase transitions terminating in a critical end point (Ref. [119]).

while its molar field energy is $U_1 = -(10/3)\mu_B H N_A \, \text{J mol}^{-1}$ (here N_A is Avogadro's number). For the saturated state, the molar entropy is zero and $U_s = -(10/2)\mu_B H N_A \, \text{J mol}^{-1}$, respectively. There is also an energy change in passing through the transition, arising from the combined exchange and dipolar couplings: $J_{\text{eff}} = J_{\text{NN}} + D_{\text{NN}} = 2.35\,\text{K} - 1.24\,\text{K} = 1.1\,\text{K}$ equivalent to $U_0 = 1.1\,\text{K} k_B N_A$ (J mol^{-1}). The molar Gibbs functions of the kagomé (k) and saturated (s) states are

$$G_k = U_0 + U_1 - TS_k, \tag{7.11}$$

$$G_s = U_s. \tag{7.12}$$

Solving these equations gives a zero-temperature transition at $\mu_0 H =$ 0.98 T and a slope of the first-order line of $\mu_0 dH_c/dT = 0.070\ \text{TK}^{-1}$, in good agreement with the experimental values 0.905 and 0.08, respectively. This argument suggests that the phase transition might be contained in the near-neighbor spin ice model even though it was not observed in Ref. [38]. This point is discussed further in Section 7.4.5.3.

We finally mention a very recent report that has found another susceptibility peak at slightly higher fields than the first-order line in $Dy_2Ti_2O_7$; currently, the origin of this peak is not clear.[117,118]

7.4.3.2. [110] *Direction*

A specific heat study of $Dy_2Ti_2O_7$ in a magnetic field applied along [110] has been reported.[120] In this direction, as explained above, the system forms two sets of canted ferromagnetic chains: one in which the chains run parallel to the field and one in which the chains run perpendicular to it. These chains are conveniently labeled α and β, respectively.[120] The spins in the α chains have a large component on the applied field and are pinned by it for sufficiently strong fields and at a sufficiently low temperature. The spins in the β chains are strictly perpendicular to the applied field. Hiroi *et al.*[120] argue that the β chains behave to a first approximation as an assembly of independent Ising ferromagnetic chains. In an applied field of $\mu_0 H = B_0 = 1.5\,T$, they observe two peaks: one at ~7 K, which is ascribed to freezing of the spins on the α chains, and a second at just above 1 K, which is argued to arise from the development of correlations among the spins on the β chains. However, this peak is much sharper than the Schottky-like peak expected for a 1D Ising model, even with the inclusion of further neighbor interactions. An experimentally deduced phase diagram[120] is shown in Fig. 7.28. Note that none of the phases indicated on the experimental phase diagram are fully ordered phases. This contrasts with the theoretical phase diagram

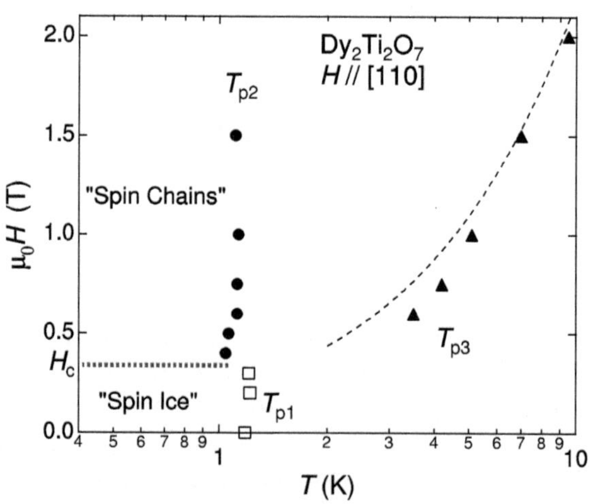

Fig. 7.28. $Dy_2Ti_2O_7$: phase diagram for a field applied along [110] showing three lines of specific heat maxima: T_{p1}, T_{p2}, and T_{p3} arising from spin ice freezing and formation of β and α chains, respectively (Ref. [120]). Note that none of the phases indicated are fully ordered phases.

for the dipolar spin ice model recently determined by Yoshida *et al.*[112] and discussed in Section 7.4.1.2.

7.4.3.3. [100] *Direction*

Higashinaka *et al.* have reported a study of the specific heat of $Dy_2Ti_2O_7$ in a magnetic field applied on [100].[121] Unlike the numerical results of the near-neighbor model (Fig. 7.22), the experimental data revealed only one peak at all values of the applied field (0–5 T) in the temperature range $T > 0.35\,\mathrm{K}$. A single Schottky-like peak was observed to shift to higher temperatures with increasing field, corresponding to a shift of the dominant energy scale from nearest-neighbor ferromagnetic interactions to the Zeeman energy of a single spin. Despite the absence of any observed phase transition, it is perhaps premature to completely rule out the possibility of a "liquid–gas" phase transition of the type predicted for the near-neighbor model (and actually observed with the field along [111] — see above) until these data have been extended to a lower temperature.

7.4.3.4. [211] *Direction*

The [211] direction bisects a triangular face of the elementary tetrahedron of the pyrochlore lattice. Therefore, for strong fields applied along [211], for each tetrahedron, three spins are frozen by the field and one spin is perpendicular to the field. One might again expect in this case a low-temperature transition since 1/4 spins are decoupled from the external applied field. Such a field configuration was studied in Ref. [121]. No sign of a transition above 0.3 K was observed: only a broad specific heat anomaly around 1.7 K. Very recent theoretical work on the dipolar model by Ruff *et al.*[113] confirms that while a [211] field is perpendicular to one sublattice, there is an internal exchange plus dipolar field that enforces the ice rules and which results in a Schottky anomaly at 1.7 K, as observed experimentally.[121] The same work[113] suggests that a detuning of the large field from perfect perpendicular alignment causes cancellation of this zero-temperature internal exchange plus dipolar field, giving rise to a sharp crossover to a dipolar ferromagnetic GS at a temperature of 0.6 K in the limit of large field.[113] This feature is a crossover and not a true phase transition since the component of the detuned field away from perfect alignment overcompensates the exchange plus dipolar field at all non-zero temperatures. Consequently, the global symmetry is broken at all temperatures with the magnetization along the small detuned field component. The addition of a small third nearest-neighbor antiferromagnetic exchange coupling, $J_3 = -0.02\,\mathrm{K}$, brings the transition down to 0.35 K, the temperature at which a feature is observed in powder-specific heat experiments.[22] More experimental single-crystal work is needed to demonstrate the validity of this detuned field idea and also its relevance to the powder-specific heat experiment[22] discussed in the next section.

7.4.3.5. *Powder measurements*

In a sufficiently homogenous powder, an applied field couples to crystallites that collectively represent all crystallographic directions. Therefore, powder measurements give some qualitative guide to the overall behavior. Three notable features were found in powder-specific heat measurements on $Dy_2Ti_2O_7$ in a field.[22] First, a sharp peak at 0.35 K that develops below 0.75 T remains at the same temperature and to a large extent is unaffected in its shape up to fields of 7 T. Second, another feature at a temperature of 0.5 K first develops for a field of 1 T and has essentially disappeared above 3 T. Finally, a third feature is observed at 1.2 K that first develops at 0.5 T and remains noticeable up to 6 T, although it is never as sharp as the 0.35 K sharp peak.

It was proposed in Ref. [22] that some of these three specific heat features might originate from microcrystals of the powder sample that are oriented along, for example, the [110] and [211] directions, such that a finite proportion of all the spins are perpendicular to the magnetic field. It currently appears that the 1.2 K feature in powder-specific heat experiments may indeed be related to fields along $\langle 110 \rangle$ directions,[112,113,120] while the origin of the 0.35 K feature is still in need of investigation.[113] It also seems that that the 0.5 K feature is related to the kagomé ice physics that occur for a field aligned along a $\langle 111 \rangle$ direction (see Section 7.4.5, dedicated to kagomé ice).[117] In particular, a sharp specific heat peak at ~0.5 K is observed for a field of $B_c \sim 1$ T. This field corresponds to the critical field B_c for the zero-temperature transition from the plateau state to the aligned state, which breaks the ice rules (see Section 7.4.3.1).

7.4.4. *Neutron scattering results*

7.4.4.1. [110] *Direction*

As described above, the expectation from the near-neighbor spin ice model is for parallel or α chains to be "pinned" by the field, while uncorrelated perpendicular or β chains develop as a result of the ice rules. For both $Ho_2Ti_2O_7$ and $Dy_2Ti_2O_7$, neutron scattering[4, 57, 107, 122] reveals the growth of Bragg peaks at the Brillouin zone centers ($Q = 0$) arising from the α chains formed when the field is applied. The β chains separately give rise to broader peaks at the $Q = X$ positions. The scattering from a crystal of $^{162}Dy_2Ti_2O_7$ is illustrated in Fig. 7.29. The $Q = X$ peaks from the β chains are narrow (though not quite resolution limited) in the direction parallel to the β chain direction and much broader in the perpendicular direction. This indicates very long range correlations within the chains and much shorter range correlations between chains. The respective correlation lengths can be estimated by fitting the peak shape. They extend for 1200 Å and 16 Å at $T = 0.05$ K, $B_0 = 1.5$ T.[122]

Based on recent numerical work,[112, 113] one would conclude that the correlations between β chains are induced by the long-range dipolar interaction. It is found experimentally that the degree of order can be increased by annealing at elevated temperature ($\sim$1 K) or repetitive field-cycling at 50 mK.[122] However, true long-range order is never established. It has been pointed out[120, 122] that the β chains themselves occupy an almost undistorted triangular lattice, and hence, the absence of a true long-range order might reflect the geometric frustration inherent to this system. At low temperatures ($\sim$50 mK), for both $Ho_2Ti_2O_7$ and $Dy_2Ti_2O_7$, the growth of Bragg peaks in an applied field is discontinuous, with at least two, and possibly several, abrupt jumps in the Bragg intensity as a function of field.[107, 122] These may be indicative of first-order phase transitions. It is also interesting to note that the $Q = 0$ Bragg intensity shows hysteresis: the diffracted intensity is related to the square of the magnetization $\vec{M}$, so one observes something similar to $\vec{M}^2$ vs. B_0. The pristine[d] "magnetization" curve unusually lies outside the hysteresis loop measured after field cycling. Hence, the zero-field-cooled spin ice state is much harder to magnetize than the field-cycled state.

For temperatures above 0.4 K, the neutron scattering of Refs. [57], [107] and [122] and the bulk measurements of Ref. [120] are in reasonable agreement regarding the general features of the phase diagram shown in Fig. 7.28. That is, there are four disordered regimes, as shown in the figure. At higher temperatures ($T > 4$ K), the integrated magnetic intensity of the (220)

[d]That is, the initial magnetization curve in the first quadrant of the hysteresis loop, with field increasing.

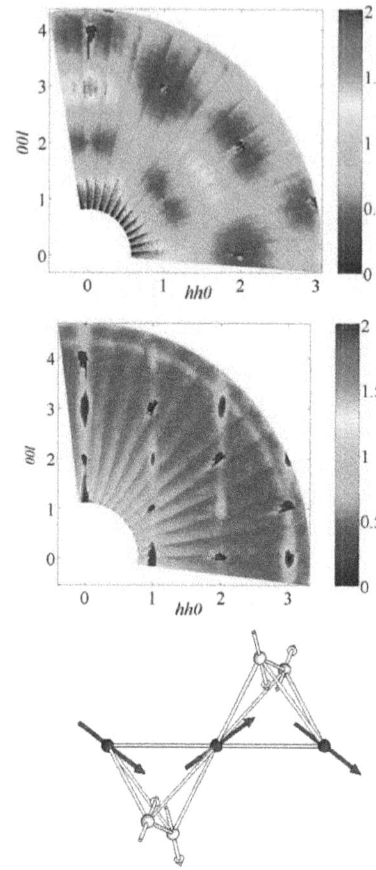

Fig. 7.29. $Dy_2Ti_2O_7$: Neutron scattering in the *hhl* plane in zero field and a field applied parallel to $[1\bar{1}0]$ (i.e. perpendicular to the plane of this picture) (Ref. [122]). Blue is the lowest intensity and brown the highest. In zero field, the Bragg peaks observed have no magnetic contribution. The figure illustrates the broad features of the zero-field spin ice state (upper panel) replaced in an applied field of 1.5 T (middle panel) by strong magnetic Bragg scattering at the $Q = 0$ positions (*hkl* all odd or all even) and broader scattering (extended along $00l$) at the $Q = X$ positions (the remaining positions with integer *hkl*). These correspond to the α and β chains, respectively, as illustrated in the lower panel which is drawn in the same orientation (slightly offset for clarity) as the scattering map. Dark spins represent β chains, and light spins represent α chains.

reflection of $Dy_2Ti_2O_7$ collapses onto a single function of $\mu_0 H/T$ showing that, in this temperature range, the spin–spin coupling is unimportant and the induced moment represents the paramagnetic response of single ions.[57]

7.4.4.2. [100], [111], *and* [211] *directions*

Fennell[122] has recently studied fields applied along the [100] direction, in $Dy_2Ti_2O_7$ and [100], [111], and [211] directions in $Ho_2Ti_2O_7$.

On the application of a magnetic field to $Ho_2Ti_2O_7$ and $Dy_2Ti_2O_7$ along the easy [100] direction, the spins are ordered in a canted ferromagnetic state giving rise to Bragg peaks at the Brillouin zone centers ($Q = 0$). Their intensity saturates above $B_0 = 0.5\,T$. Below $T = 0.3\,K$, hysteresis is observed, and below 1.5 K, one or possibly two distinct jumps in the Bragg intensity develop as a function of field. It is possible that these are related to the predicted first-order liquid–gas type transition,[38] but this is hard to demonstrate convincingly because of the very strong history dependence. With the field along [111], $Ho_2Ti_2O_7$ shows the expected two-step saturation growth of $Q = 0$ Bragg peaks below ∼2 K. With the field applied along [211], one spin is perpendicular to the field and the other three are "pinned" by it into a two-in, one-out configuration. The neutron Bragg intensities are consistent with processes that involve the ordering of the perpendicular spin according to the ice rules as a function of field or temperature.

7.4.5. *Kagomé ice*

The existence and characterization of kagomé ice is one of the most remarkable recent developments in the physics of spin ice. Wills *et al.*[123] were the first to conceive of a two dimensional spin ice model (see below) while Matsuhira *et al.*[111] later formulated a similar model and applied it to the [111] plateau state of the spin ice materials; one should also mention the contribution of Higashinaka *et al.*[116] In the context of this review, it therefore seems appropriate to start with the real pyrochlore materials with field along [111] and show how these relate to the idealized model of Wills *et al.*

7.4.5.1. *Basic kagomé ice model and mappings*

Matsuhira *et al.*[111] pointed out that the [111] plateau state of spin ice, should, according to the ice rules, retain a zero-point entropy. To see this, note that when projected along [111], the pyrochlore lattice can be considered as an assembly of kagomé lattice planes (i.e. planes of corner-linked triangles) decorated by spins alternately above and below the plane to form "up", and "down" tetrahedra (see Fig. 7.30). The plateau state corresponds to the alignment of the apical spin in each tetrahedron, so if the field is "up",

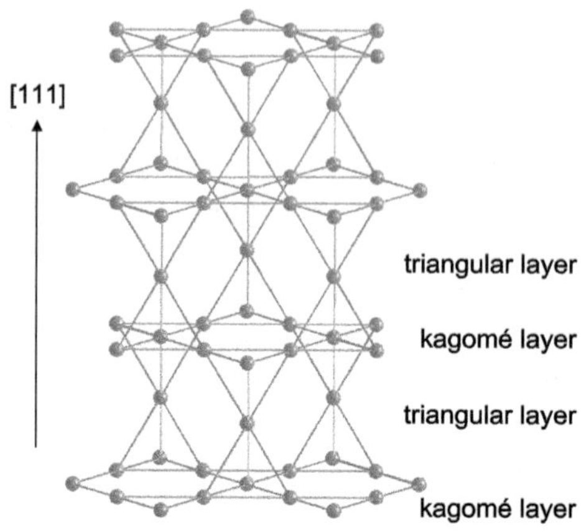

Fig. 7.30. Illustration of how the pyrochlore lattice can be regarded as an assembly of kagomé planes decorated by interstitial spins, when viewed along [111] (Ref. [111]).

then this spin points "out" on an up tetrahedron and "in" on a down tetrahedron. To obey the ice rules, the remaining spins (three per tetrahedron) in the kagomé plane must adopt the configuration in–in–out or in–out–out on up and down tetrahedra, respectively. This leaves an intra-tetrahedron degree of freedom analogous to that in Pauling's model and leads to a zero-point entropy. The "in-plane" spins do not lie precisely in the kagomé plane, leading to a slight energetic preference in the applied field for (say) in–in–out over out–in–in (because, depending on whether one is discussing an "up" or "down" tetrahedron, one of these will have a larger component on the field). However, it is assumed this effect is dominated by the ice rules. A kagomé ice state projected onto the [111] plane is illustrated in Fig. 7.31.

It is clear, therefore, that this situation closely resembles the idealized 2D model defined by Wills *et al.*[123] and termed "kagomé spin ice". In this model, one has spins on a kagomé lattice directed along the bisectors of the equilateral triangles and ferromagnetically coupled, as in spin ice itself. Wills *et al.* also considered the effect of magnetic fields and further neighbor interactions on this model and argued that the model might be relevant to vanadium jarosite materials.

The kagomé spin ice model of Wills *et al.* can be mapped on to the model of two A ions and one B ion occupying the kagomé lattice such that the number of AB interactions is maximized. As mentioned in

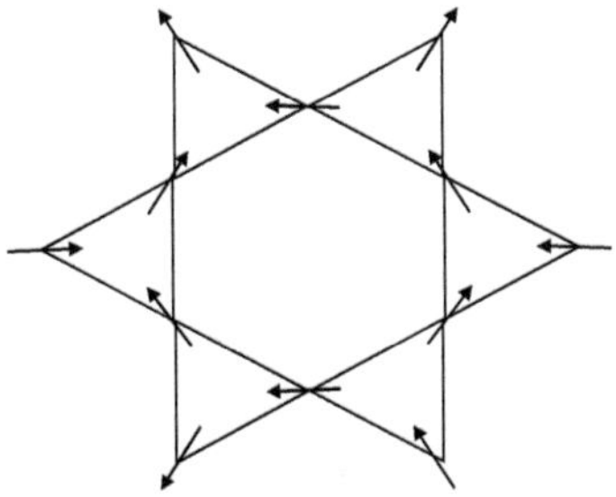

Fig. 7.31. A state of the kagomé spin ice model.

Section 7.2.3.2, this model was considered by Anderson who mapped it onto Wannier's solution of the triangular Ising antiferromagnet. To understand this mapping, place a lattice point at the center of every hexagon in the kagomé lattice and position +/− variables such that each triangle has, at random, either + + − or − + +. The new lattice is a triangular lattice with the +/− variables forming one of the many degenerate GSs of Wannier's antiferromagnet. The kagomé lattice points lie on the mid-points of the bonds of this triangular lattice, so if an +− bond is denoted A and a −− or ++ bond is denoted B, then the kagomé lattice is populated by A/B variables such that each triangle has AAB, as required. The merit of mapping on to Wannier's model is that the latter is exactly solved with intensive molar zero-point entropy 0.32306 R,[2] which translates as 0.1077 R in the model by Wills *et al.* or 0.08077 R for kagomé ice on the pyrochlore lattice. This result can alternatively be obtained by mapping onto a model of dimers that populate the bonds of the honeycomb lattice occupied by the centers of the triangles.[124]

7.4.5.2. *Experimental results: Specific heat*

Matsuhira *et al.*[111] measured the zero-point entropy of $Dy_2Ti_2O_7$ as a function of field applied on [111] by integrating the specific heat. They found that the entropy falls from approximately Pauling's value in zero field to the value 0.8 ± 0.1 JK^{-1} per mole of Dy in the plateau state for applied fields between 0.25 and 0.75 T (see Fig. 7.32). This value was later refined to 0.65 ± 0.1 by more detailed subtraction of a background from the specific heat.[115] This estimate corresponds to an entropy 0.078 R per mole of Dy, which is very close to the theoretical value 0.081 R (see previous section). These results constitute excellent evidence for the existence of the kagomé

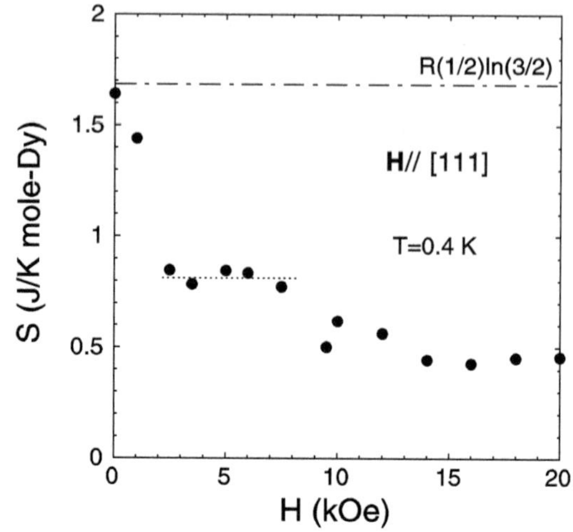

Fig. 7.32. $Dy_2Ti_2O_7$: Specific heat versus applied field along [111] (from Ref. [111]) showing a value in the "plateau state" (dotted line) that is consistent with kagomé ice behavior.

ice state in this material. The transition from the kagomé ice state with increasing magnetic field was discussed earlier in Section 7.4.3.1 and in the next section.

7.4.5.3. *Theory of the kagomé ice state: Kastelyn transition*

It is now well understood that the application of a magnetic field along the [111] direction in the spin ice compounds leads to two magnetization plateaux at low temperatures ($T \ll J_{\text{eff}}$) and as a function of applied magnetic field. The plateau at strong field corresponds to the maximally polarized three-in, one-out state (in reference to an "up" tetrahedron). The entropy of the low-field plateau (the kagomé ice state) is reduced compared to the zero-field spin ice state, but is still extensive. Recent theoretical work by Moessner and Sondhi[125] and Isakov *et al.*[126] have revealed some remarkable features of spin ice in this regime. In Ref. [125], the available degrees of freedom, which in the low-field plateau live on decoupled kagomé planes, are mapped to hard-core dimers on a honeycomb lattice. The resulting 2D kagomé ice state is found to be critical. Small tilts of the applied field at low temperature away from perfect [111] alignment lead to a vanishing of the entropy and the termination of the critical phase by a "Kasteleyn" transition characterized by highly anisotropic scaling. Reference [126] has considered the crossovers between the different plateau regimes from a point of view of (entropically) interacting defects. It was found in an analytical theory for the nearest-neighbor spin ice model, which covers most of the magnetization curve, that the entropy is non-monotonic as a function of applied field and exhibits a giant spike at the low-field plateau to high-field plateau transition. In other words, the entropy becomes much larger than its kagomé ice value (even exceeding the Pauling value) before finally falling to zero.

In very recent experimental work, the existence of an entropy spike has been confirmed. Aoki *et al.*[127] have used magnetocaloric effect measurements to make a very accurate determination of the entropy of $Dy_2Ti_2O_7$ as a function of field as one crosses the critical line in Fig. 7.27. The experimental data are illustrated in Fig. 7.33. Although the observed spike is large (reaching 1 J/K · mol Dy at T = 500 mK), it is smaller than that predicted theoretically (2.4 J/K · mol Dy) for the near-neighbor model and falls with decreasing temperature, also in contrast to the theory. As described in Section 7.4.3.1, the experimental transition is first order, while in the near-neighbor model, it appears to be a crossover. Thus, one could speculate

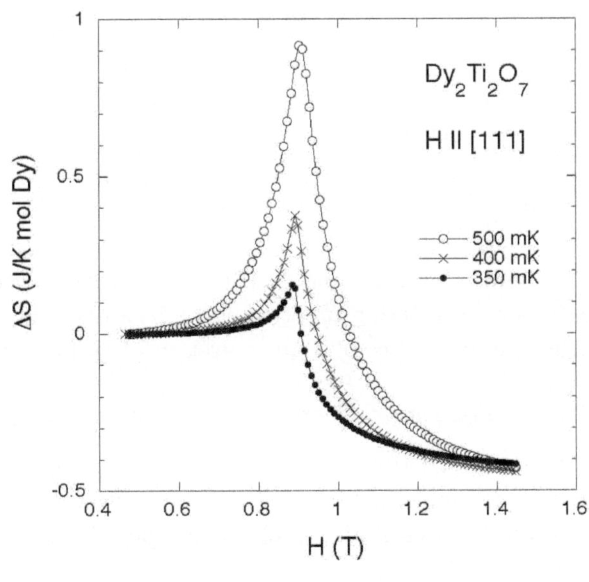

Fig. 7.33. $Dy_2Ti_2O_7$: change in entropy with a field along ⟨111⟩ illustrating the giant entropy spike discussed in the text (Ref. [127]).

that here too, the differences between experiment and theory may be a consequence of long-range dipolar couplings.

7.5. Spin Dynamics of the Spin Ice Materials

From an experimental perspective, the spin ice materials display extremely interesting dynamical properties, which are the subject of this section. We begin with some background to the experimental quantities of interest, before proceeding to describe the main experimental results.

7.5.1. *Experimental quantities of interest*

The purpose of this section is to briefly review, for the benefit of the non-expert reader, the quantities that are typically measured when probing the spin dynamics.

7.5.1.1. *Correlation functions and neutron scattering*

The thermal fluctuations of an equilibrium system may be specified by time- and distance-dependent correlation functions or equivalently by Fourier-transformed correlation functions, depending on wave vector $\mathbf{Q}$ and angular frequency ω. The canonical experimental probe of a spin system is therefore neutron scattering, which in its conventional form can directly measure the spin correlation function $S^{\alpha\beta}(\mathbf{Q},\omega)$, known in scattering theory as the dynamic scattering function (here $\alpha\beta = x, y, z$ specify the spin components). This function is defined as

$$S^{\alpha\beta}(\mathbf{Q},\omega) = \frac{1}{2\pi N}\int \langle S^{\alpha}\left(-\mathbf{Q},0\right) S^{\beta}\left(\mathbf{Q},t\right)\rangle \exp\left(i\omega t\right)dt, \qquad (7.13)$$

where $\langle\cdots\rangle$ is a thermal average and $S^{\alpha}(\mathbf{Q},t)$ a Fourier spin component at time t. The dynamic scattering function contains a detailed microscopic description of the spin dynamics of a given system at equilibrium.

7.5.1.2. *Fluctuation–dissipation theorem and AC susceptibility*

The diffuse part of $S(\mathbf{Q},\omega)$ (for compactness we have dropped the labels α,β) can be related by the fluctuation–dissipation theorem to the imaginary part of the generalized susceptibility $\chi(\mathbf{Q},\omega)$, by the relation

$$S_{\text{diffuse}}(\mathbf{Q},\omega) = \frac{\hbar}{\pi g^2\mu_{\text{B}}^2}\frac{1}{1-\exp(-\hbar\omega/k_{\text{B}}T)}\chi''(\mathbf{Q},\omega). \qquad (7.14)$$

This relationship expresses the very important fact that the linear response of a system magnetization $M(\mathbf{Q}, \omega)$ to weak applied fields $h(\mathbf{Q}, \omega)$ is completely determined by the equilibrium correlation functions. Physically, this arises because, to first order in perturbation theory, the system does not distinguish between a perturbation to the spin structure caused by a thermal fluctuation and one caused by an external field. This is useful as it means that the equilibrium fluctuations can also be probed by the direct application of a laboratory field $h(0, \omega)$. Typically, one performs frequency-dependent (AC) susceptibility measurements, which directly yield both $\chi'(0, \omega)$ and $\chi''(0, \omega)$. These two functions are inter-related by the Kramers–Kronig relations and so contains a complete description of the response of the Fourier component $M(0, \omega)$. The advantage of AC susceptibility is that it can probe a frequency range inaccessible to neutron scattering ($<10^6$ Hz) while its limitation is that it is a bulk measurement: information on the spatial correlations are averaged out of the problem.

7.5.1.3. *Spectral shape function*

In practical terms, it is convenient to separate statics and dynamics and consider a wave vector-dependent susceptibility $\chi(\mathbf{Q})$ and a normalized spectral shape function $F(\mathbf{Q}, \omega)$. The quantity $\chi(\mathbf{Q})$ may be measured by collecting scattered neutrons of all energies, which under suitable conditions is equivalent to integrating $S(\mathbf{Q}, \omega)$ over ω, while $F(\mathbf{Q}, \omega)$ can be measured by examining the spectrum of scattered neutrons at fixed scattering vector $\mathbf{Q}$. The diffuse scattering function is written

$$S_{\text{diffuse}}(\mathbf{Q}, \omega) = \frac{kT}{g^2 \mu_B^2} \chi(\mathbf{Q}) \frac{\hbar\omega}{1 - \exp(-\hbar\omega / k_B T)} F(\mathbf{Q}, \omega). \tag{7.15}$$

The Fourier transform of $F(\mathbf{Q}, \omega)$ is the quantity $F(\mathbf{Q}, t)$, a measure of how a spin component $S(\mathbf{Q}, t)$ relaxes after a perturbation. It is equal to the "intermediate scattering function", $S(\mathbf{Q}, t)$ divided by its value at $t = 0$, and can be determined directly by the technique of neutron spin echo (NSE).

7.5.1.4. *Exponential relaxation*

The simplest type of relaxation occurs when a field is turned off at time $t = 0$ and $S(\mathbf{Q}, t)$ relaxes exponentially. With single-ion dynamics, there is no wave vector dependence, and the following relations apply:

$$F(\mathbf{Q},t) = \exp(-\Gamma t), \tag{7.16}$$

$$S(Q,\omega) \approx k_{\mathrm{B}}T\chi(\mathbf{Q})\frac{\Gamma}{\omega^2+\Gamma^2} \quad (\text{assuming } k_{\mathrm{B}}T \gg h\omega), \tag{7.17}$$

$$\chi'(0,\omega) = \chi(0)\frac{\Gamma^2}{\omega^2+\Gamma^2}, \tag{7.18}$$

$$\chi''(0,\omega) = \chi(0)\frac{\omega\Gamma}{\omega^2+\Gamma^2}. \tag{7.19}$$

Exponential relaxation of this sort is often assumed in phenomenological approaches and arises typically in diffusion problems; for example, the case of Arrhenius relaxation across a barrier with attempt frequency $\Gamma = \Gamma_0 \exp(-\Delta/k_{\mathrm{B}}T)$. It also arises in the analogous case of the electrical field response of non-interacting dipolar liquids (Debye relaxation). Departure from this form may be revealed by a Cole–Cole plot of χ' vs. χ'' which should be semicircular for pure exponential relaxation. A phenomenological extension of this concept is the Davidson–Cole formalism,[128, 129] which assumes $\chi = \chi(0)/(1+i\omega t_c)^\beta$. The case $\beta = 1$ corresponds to a single exponential relaxation, while $0 < \beta < 1$ corresponds to a distribution of exponential relaxation times with t_c being an upper cut off.

7.5.2. *Differences between* $Ho_2Ti_2O_7$ *and* $Dy_2Ti_2O_7$

Early approaches to the spin dynamics of the spin ice materials were complicated by an apparent difference between $Ho_2Ti_2O_7$ and $Dy_2Ti_2O_7$. It is now clear that the behavior of the two materials is qualitatively the same, except perhaps in the very low temperature range where hyperfine coupling becomes important (see Section 7.5.5.2). Hence, in the following, we organize the discussion in a logical sense as a function of temperature range, rather than discussing the various spin ice materials individually.

7.5.3. *Relaxation at high temperature, $T \sim 15$ K and above*

7.5.3.1. *AC susceptibility (AC-χ)*

It came as a surprise in 2001 when two papers — by Matsuhira *et al.*[130] and Snyder *et al.*[131] — reported a peak in AC susceptibility of polycrystalline $Dy_2Ti_2O_7$ near 15 K, suggestive of spin freezing, but at a temperature scale well above that expected for spin ice. This feature is henceforth referred to as "the 15 K peak", although it should be noted that its peak temperature

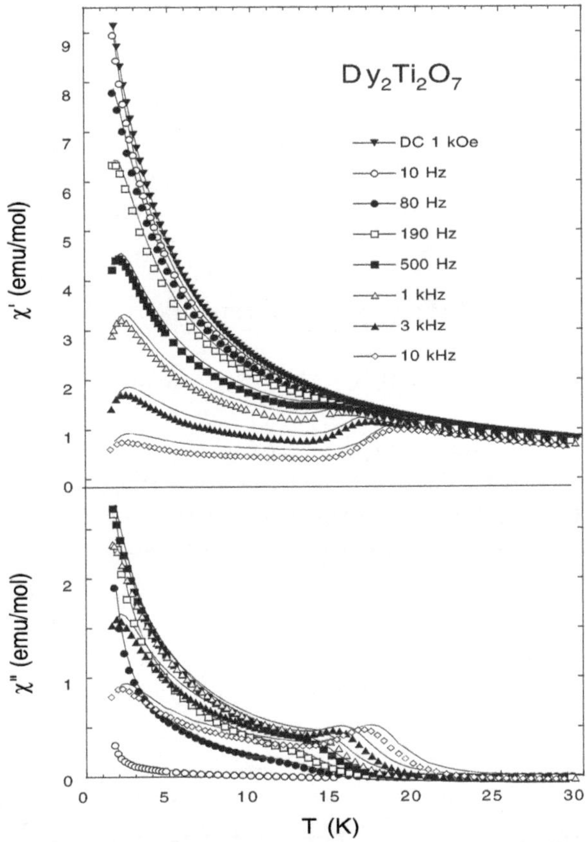

Fig. 7.34. $Dy_2Ti_2O_7$: AC susceptibility (upper panel: χ', lower panel χ'') as a function of temperature at several frequencies, illustrating the "15 K peak" (Ref. [130]).

shifts with frequency as illustrated in Fig. 7.34. The frequency shift of the 15 K peak can be described by an Arrhenius law, with activation energy $E_a = 220\,\text{K}$.[130, 131] It is accompanied by a semicircular Cole–Cole plot corresponding to a single exponential relaxation. An initial interpretation[131] was that the cooperative spin ice freezing was already occurring at this high temperature, but this interpretation was later shown to be inconsistent with direct microscopic measurement of the spin correlations[79] (see below). The 15 K peak is not clearly visible in the case of $Ho_2Ti_2O_7$, but it can be revealed by the application of a magnetic field to single-crystal samples,[79] giving an activation barrier of $\sim 250\,\text{K}$. The reason for the difference between $Ho_2Ti_2O_7$ and $Dy_2Ti_2O_7$ is discussed further in Section 7.34.

7.5.3.2. *Neutron spin echo (NSE)*

Ehlers *et al.* used NSE to determine the microscopic origin of the high-temperature spin dynamics of a polycrystalline sample of $Ho_2Ti_2O_7$, measured over a temperature range 0.3 K–200 K.[79] Further experimental work extended this range to 800 K.[132] As expected from,[130, 131] these experiments revealed a single exponential decay with approximate Arrhenius activation[79]:

$$F(\mathbf{Q},t) = (1-A)\exp(-\Gamma t), \tag{7.20}$$

with $A = 0.09(1)$, $\Gamma = 21.1(2) \times 10^{11} \exp(-\Delta/k_{\rm B}T)$ Hz, and $\Delta/k_{\rm B} = 293(12)$ K.

The Arrhenius behavior of the correlation function $F(Q,t)$ at $t = 1$ ns is illustrated in Fig. 7.35. The neutron scattering was found to be Q-independent ($Q = |\mathbf{Q}|$), which establishes that the single exponential decay is a single-ion phenomenon. In view of its energy scale Δ matching that of the first set of excited crystal field levels (see Fig.7.3), it is almost certainly a spin reversal that involves promotion of a spin to these states followed by de-excitation. It would seem a reasonable assumption, as suggested in Ref. [79], that an analogous process occurs in $Dy_2Ti_2O_7$.

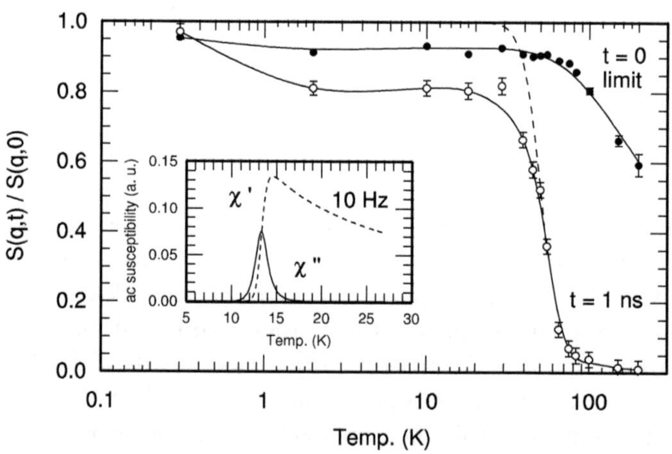

Fig. 7.35. $Ho_2Ti_2O_7$: Spin relaxation measured by neutron spin echo (NSE) (Ref. [132]) at $t = 0$ and $t = 1$ ns. Note the approximately athermal regime in the $t = 1$ ns data below T = 15 K. The dashed line represents Arrhenius kinetics and the inset the AC susceptibility that corresponds to this approximation.

It is tempting to regard this process as activation over a quasi-classical barrier in a manner analogous to the large-spin, single-molecule magnet "Mn_{12}".[35] In this system, the crystal field gives rise to a Stark-like[e] splitting of the $|J, M_J\rangle$ levels giving a ladder of $|\pm M_J\rangle$ states characterized by $M_J = J, J-1, \ldots, 0$. These states form what may be regarded as a quasi-classical barrier in that a classical spin reorientation involves surmounting the ladder of states in a sequential fashion. The spin ice materials share the property of an almost pure doublet GS: $|\pm 8\rangle$ and $|\pm 15/2\rangle$ for Ho and Dy, respectively. However, the excited states are generally complicated admixtures of other M_J levels, so at this point the analogy breaks down. Nevertheless, it should be realistic to regard the spin ice relaxation as a promotion to an excited level, a quantum spin rotation within that level and then de-excitation to give the net result of a spin-flip. In this regard, it would be useful to use the crystal field scheme of Rosenkranz *et al.*[39] to calculate the relaxation of the single-ion system. It should also be noted that the physical origin of the spin dynamics (i.e. the transition matrix element in the Fermi Golden rule) has not yet been established, although magnetoelastic interactions would appear to be a likely cause.

7.5.3.3. *Origin of the 15 K AC susceptibility peak*

As described above, the 15 K peak is observed in the low field AC susceptibility of $Dy_2Ti_2O_7$ but not clearly in that of $Ho_2Ti_2O_7$. This alternate appearance and non-appearance of the 15 K peak in the spin ice materials contains a clue to its origin. The relaxation function of $Ho_2Ti_2O_7$ determined by NSE in the frequency range 10^9–10^{12} Hz can be transformed to an AC susceptibility (see Section 7.5.1.4) and extrapolated to $Q = 0$ and into the AC susceptibility frequency window ($\leq 10^5$ Hz). In that case, it does predict a peak at 15 K and a rapid drop to zero below that temperature (see inset, Fig. 7.35). Therefore, both $Ho_2Ti_2O_7$ and $Dy_2Ti_2O_7$ exhibit a freezing of this "high-temperature" relaxation process near to 15 K.

Why then does this freezing process give rise to a susceptibility peak in $Dy_2Ti_2O_7$ but not in $Ho_2Ti_2O_7$? We first observe that, mathematically, it is perfectly possible to formulate two relaxation functions of the same functional form but with slightly different parameters, one of which does give rise to a susceptibility peak and the other of which does not. In the

[e] The Stark effect of atomic spectroscopy involves the splitting of free-ion terms into $|\pm M_J\rangle$ levels under the influence of a uniform electric field.

present case, one should imagine the exponential decay function of the "high-temperature" process "cut off' rather abruptly above some characteristic time t_c, which is approximately temperature independent. As the temperature is lowered, the exponential decay rate of the high-temperature process decreases. If $1/t_c$ lies to the low-frequency side of the experimental (AC-χ) frequency window, there is a strong susceptibility peak because all the observed response is due to the high-temperature process. However, if $1/t_c$ lies to the high-frequency side, there is no peak as all the response is due to the low-temperature process, characterized by t_c. Thus, to get a "no-peak scenario", the form $F(Q,t) = \exp(-E_a/k_{\mathrm{B}}T)$ must be cut off in the range of frequencies that fall between the neutron and AC-χ experimental windows, and one can conclude that there is most likely an approximately temperature-independent relaxation process active in both $Ho_2Ti_2O_7$ and $Dy_2Ti_2O_7$ but with slightly different characteristic times. In agreement with this scenario, it has been found that the application of a magnetic field to $Ho_2Ti_2O_7$ results in the clear appearance of the 15 K peak in this material.[79] The implication is that the applied field changes the rate of the low-temperature process to bring t_c into the experimental window.

To summarize the argument, there are two distinct relaxation processes in the spin ice materials. These can be thought of as a high-temperature thermally activated process and a lower temperature approximately athermal process. The slowing down of the high-temperature process is observable as a "15 K peak" if the low-temperature process is sufficiently slow to allow it to be visible. The rate of the low-temperature process is characterized by a "cut off" in the spin relaxation function. The cut-off time scale t_c can be adjusted experimentally either by the application of a field, which drives it to lower frequency, revealing the 15 K peak, or possibly by doping with impurities (see below) which drives it to higher frequency, obscuring the 15 K peak. The simplest model consistent with these facts is a dynamical process that is approximately temperature independent near 15 K, which is suppressed by an applied field.[79] At 15 K, spin ice is strictly paramagnetic, with weak spin ice correlations starting to develop, but all spin configurations are still highly populated. It has been suggested[79] that many spin configurations put an instantaneous strong transverse dipolar field on a given spin, giving rise to quantum mechanical tunneling dynamics, which take over from the Arrhenius dynamics below 15 K and are finally extinguished near 1 K, where the appropriate configurations are frozen out.

The NSE furnishes direct evidence for a quantum tunneling regime in that $S(q, t = 1\text{ns})$ (Fig. 7.35) becomes temperature independent below $T = 15\,\text{K}$ before rising again (though not quite to unity — see below) as the build up of spin ice correlations occurs near 2 K. Thus, it is established that the spin relaxation at 1 ns is approximately athermal over a broad temperature range, which is evidence for a tunneling process. If this interpretation stands the test of time, it represents a remarkably high temperature for such tunneling dynamics to be observed and also a remarkable case of tunneling induced by the dipolar interaction itself. The temperature scale is far in excess of that of Mn_{12} (see Ref. [35]) or such systems as $LiHoF_4$.[34] The reason, of course, is the enormous crystal field splitting in the spin ice materials. However, if spin tunneling is involved, it is not yet clear as to whether it can be thought of in terms of direct resonant transitions between essentially pure $|\pm J\rangle$ states or in terms of quantum rotations mediated by admixtures of other M_J states even in the absence of external or dipolar fields. Further evidence for tunneling is discussed in Section 7.5.4.2.

7.5.4. *Relaxation in the range* $1\,K \leq T \leq 15\,K$

In many ways, this is the most interesting temperature range for the spin dynamics of the spin ice materials. As the temperature is lowered, single-ion dynamics should give way to collective dynamics as the system organizes itself into the spin ice state. In this temperature range, the spin-flip dynamics are mainly out of the dynamical range of neutron scattering techniques and one must rely on AC susceptibility. However, it is clear from neutron scattering that the static spin ice correlations develop strongly below about 1 K. These are characterized in the dynamics by a second susceptibility peak in this temperature range for $Ho_2Ti_2O_7$, $Ho_2Sn_2O_7$, and $Dy_2Ti_2O_7$, as illustrated in Fig. 7.36.

7.5.4.1. *AC susceptibility: Phenomenological model*

Assuming phenomenological Arrhenius dynamics, the frequency shift of the temperature of the "1 K peak" corresponds to an activation energy of the order of 20 K in all three materials with an attempt frequency of the order 10^{10} Hz. However, it is clear that below 15 K, there are multiple relaxation times in the samples, with strong deviations from a semicircular Cole–Cole plot. For $Dy_2Ti_2O_7$, Matsuhira *et al.*[130] used the Davidson–Cole formalism to reconstruct a distribution of relaxation times. These distributions

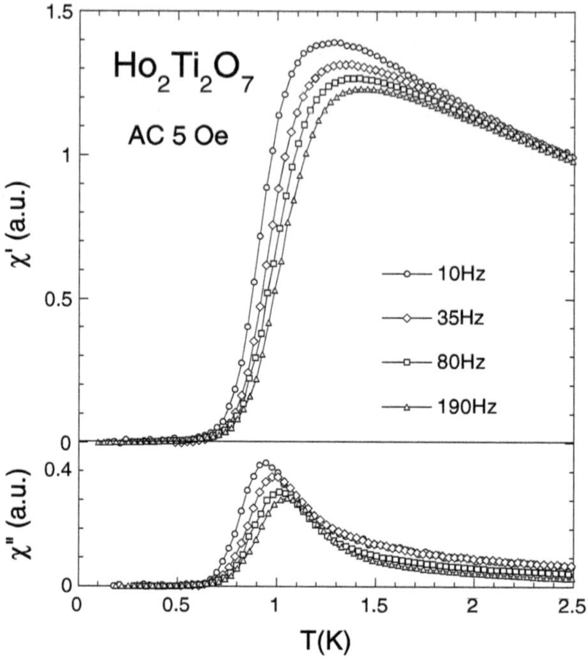

Fig. 7.36. $Ho_2Ti_2O_7$: AC susceptibility measured on a powder sample showing the spin ice dynamical freezing (Ref. [65]).

broaden below $T = 15\,\mathrm{K}$, with a range of frequencies below a rather sharp peak. This might indicate the relaxation of clusters, rather than single spins. However, it should be noted that while this phenomenological approach establishes that $F(0,t)$ can be represented as a sum of exponentials, it does not necessarily mean that there really is a set of non-interacting Arrhenius-type processes in the sample. It has been pointed out by Cornelius that the crossover between a single time scale Arrhenius decay and multiple relaxation times is analogous to a transition from "strong" to "fragile" behavior in glass-forming liquids.[133]

7.5.4.2. *AC susceptibility: Toward a microscopic model*

Important evidence for the existence of the temperature regime of quantum tunneling suggested by Ehlers *et al.*[79] has recently been provided by the discovery of a remarkable feature in the AC susceptibility of polycrystalline $Dy_2Ti_2O_7$.[134] Initially, Snyder *et al.*[134] estimated the average relaxation time τ in polycrystalline $Dy_2Ti_2O_7$ from the maximum in χ'' and confirmed

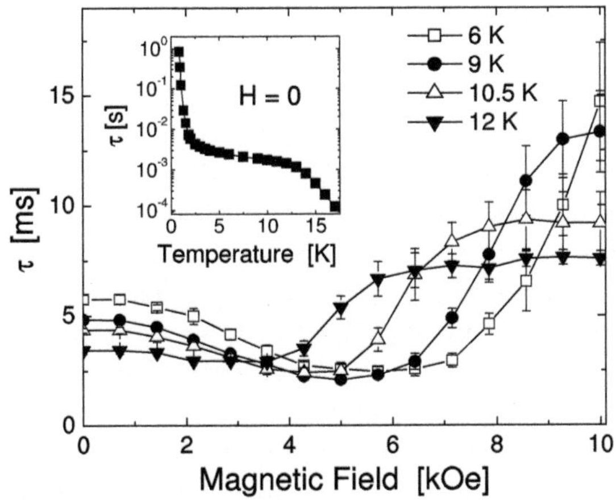

Fig. 7.37. $Dy_2Ti_2O_7$: Field dependence of the characteristic relaxation time measured by AC susceptibility, illustrating a minimum at finite applied field. The inset shows the estimated relaxation time versus temperature, illustrating a weakly temperature-dependent regime ascribed to quantum tunneling (Ref. [134]).

that this is approximately temperature independent below $T = 15\,\mathrm{K}$ (see inset, Fig. 7.37). As expected from Ref. [79], τ increases (i.e. the tunneling slows) in strong applied fields due to Zeeman splitting. However, at weak fields, τ was found to *decrease*, reaching a temperature-dependent minimum at applied fields of the order of a few kOe (see Fig. 7.37). This initial increase in tunneling rate has been interpreted as due to the presence of transverse field components as well as to the partial cancellation of the longitudinal field components. The latter effect moves toward restoring the degeneracy to the $|\pm 15/2\rangle$ states.

It is clear that there is great scope for more experimental work on the spin dynamics in this regime, particularly by examining the effect of field direction on single-crystal samples. There is also a need for a microscopic theory, particularly one that can justify the effective activation energy scale of the order of 20 K, which although clearly connected with the dipolar forces, is not yet justified in detail.

7.5.5. *Spin dynamics in the spin ice regime below* 1 *K*

7.5.5.1. *Slow relaxation*

It is now virtually certain that electron spin-flip dynamics are exponentially frozen out below about 1 K in all the spin ice materials. This is dramatically confirmed by the DC susceptibility measurement of Matsuhira *et al.* on polycrystalline $Ho_2Sn_2O_7$ shown in Figure 7.7. Here, one observes a massive splitting between field-cooled (FC) and zero-field-cooled (ZFC) susceptibility and the corresponding magnetic hysteresis loops. Magnetic hysteresis is also observed in $Ho_2Ti_2O_7$ and $Dy_2Ti_2O_7$.[78] As mentioned above, after the application of a magnetic field, remanent moments can be frozen into the spin ice samples. Snyder *et al.*[78] have reported a comparison of the thermoremanent and isothermal-remnant magnetization of $Dy_2Ti_2O_7$.[78] Thermoremanent magnetization was prepared by cooling in field H and slowly reducing the field to zero. The isothermal remanent magnetization was prepared by cooling in a field, applying a field H, waiting to obtain almost complete relaxation, and then removing the field. In both cases, the remanent magnetization was then measured as a function of time to determine its equilibrium value. The resulting differences (Fig. 7.38) show that the frozen spin ice state possesses the property of memory of its preparation in a manner similar to that displayed by spin glasses. Indeed, the data of Fig. 7.38 look strikingly similar to those for the spin glass system $AuFe_{0.05}$ (see Ref. [135]) (even though the spin ice freezing is certainly distinct from that of a spin glass as emphasized in Ref. [78]). Future studies of the slow relaxation or aging properties of the spin ice systems promise to be very interesting.

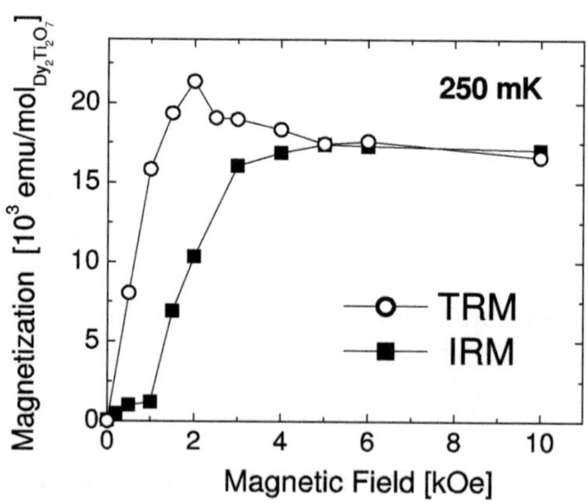

Fig. 7.38. $Dy_2Ti_2O_7$: Thermoremanent magnetization (TRM) and isothermal remanent magnetization (IRM) as a function of applied field at 250 mK (Ref. [78]).

7.5.5.2. *Evidence for residual dynamics in the frozen state*

There is some evidence of residual dynamics in the spin ice materials existing down to the lowest measured temperatures. The NSE results mentioned above suggest a small component of very fast relaxation that is manifest in a finite constant A in Eq. (7.20). μSR experiments also suggest spin dynamics in both $Ho_2Ti_2O_7$ and $Dy_2Ti_2O_7$.[136] Since there are no spin-flip dynamics at these temperatures, it seems most likely that all residual dynamics can be attributed to hyperfine coupling to the nuclear spins which can induce a fast "wobble" or incoherent oscillation of the combined electronic–nuclear spin around its local quantization axis.[79,136]

Bertin *et al.*[80] have suggested the remarkable possibility that the specific heat behavior of $Dy_2Ti_2O_7$ is indicative of electron spin dynamics persisting to the lowest temperatures in this material. Basically, the anomalous absence of a nuclear specific heat in $Dy_2Ti_2O_7$ is attributed to electronic spin dynamics. However, this argument would imply that $Ho_2Ti_2O_7$ (which has a full nuclear contribution[42]) is purely static, in apparent contravention of the NSE and μSR results. It is clear that more work is needed to fully understand these low-temperature dynamical effects.

7.5.6. *Doped spin ice*

The removal of protons in water ice by doping with KOH leads to a long-range ordered state.[137] One interpretation of this is that proton dynamics are strongly restored in the doped samples allowing the system to locate its "true" GS. Of course, this idea cannot be proved or disproved as KOH-doped ice is not ice itself. It does, however, make it particularly interesting to examine the spin dynamics of spin ice materials in which spins have been removed by doping with non-magnetic impurities.

Snyder *et al.*[131,138] have reported AC susceptibility measurements on polycrystalline $Dy_{2-x}Y_xTi_2O_7$ with x up to 0.4. The doping of spin ice has, broadly speaking, the opposite effect of applying a field leading to the "disappearance" of the 15 K peak, which was originally interpreted as a cooperative process. However, as shown in Ref. [79], the increase in susceptibility below 15 K on doping can only be interpreted as a decrease in $F(Q,t)$ or a speeding up of dynamics at a given temperature. The fact that there is no change in the susceptibility at temperatures above the 15 K peak shows that the process freezing out at that temperature is itself not affected by doping: it is a single-ion Arrhenius process, as confirmed by the fact that the parameters describing this peak do not change with x. As argued in Section 5.2.3, the "disappearance" of the peak is due to the speeding up of a lower temperature relaxation process, which is now thought to involve quantum mechanical spin tunneling.[79,134] Very recently, Snyder *et al.*[139] have reported an extension of their study to include both Y and Lu dopants and values of x up to $x = 1.98$ (i.e a magnetically very dilute sample). The notable conclusion of this work is that the 15 K peak, suppressed for $x < 0.4$, actually re-emerges for higher doping levels, whereupon it shifts up to ~22 K ($x = 1.98$). This suggests that the lower temperature quantum relaxation process is absent in very dilute samples and that the crystal field changes with doping. It is a remarkable fact that the sample with $x = 1.98$ (Y) appears practically static to AC susceptibility below $T \approx 20$ K. The latest results are argued to confirm the single-ion nature of the higher temperature relaxation process.

The speeding up of the low-temperature dynamics on doping has also been observed on doped polycrystalline $Ho_{2-x}La_xTi_2O_7$ with $x = 0.05$.[79] Here, the lower temperature susceptibility peak shifts in a manner consistent with a 100-fold increase in relaxation rate. However, such a dramatic speeding up is not observed in Yttrium-doped $Ho_2Ti_2O_7$,[140] which suggests that lattice distortions (or perhaps defects) play a role: La^{3+} is a much larger ion than Ho^{3+}, whereas Y^{3+} is about the same size. In this system, weak doping does not stabilize long-range order at millikelvin temperatures nor does it significantly affect the entropy per spin.[79,140]

The possibility of a dilution-induced transition to long-range order was investigated by Melko[94] using conventional spin-flip MC simulations. However, this preliminary result did not reveal any sign of a dilution-induced transition to long-range order that might have been made accessible to single-spin-flip dynamics via the reduction of energy barriers. This result is consistent with the experimental observations described above. For sufficiently large x, one expects the degeneracy to be lifted by the random dilution[23] and the GS of the system to become that of a dipolar Ising spin glass, akin to that of $LiHo_{1-x}Y_xF_4$.[141]

7.5.7. *Spin ice under pressure*

Water ice under under applied pressure has fascinating physics, yielding many phases.[142] This makes the application of pressure to spin ice particularly interesting. Mirebeau and Goncharenko[143] have reported an elastic neutron scattering investigation of polycrystalline $Ho_2Ti_2O_7$, in applied pressures of up to 6 GPa at temperatures down to 1.4 K. At the lowest temperatures, the dipolar spin ice correlations are clearly visible in the powder scattering pattern. However, there is no discernible change at high pressure: the conclusion is that dipolar spin ice is stable under applied pressure. This contrasts with the behavior of the spin liquid material $Tb_2Ti_2O_7$, which orders magnetically under the application of pressure.[144]

7.6. Spin Ice-Related Materials

The discussion in this chapter has been highly focused on the currently best characterized spin ice materials, $Ho_2Ti_2O_7$, $Dy_2Ti_2O_7$, and $Ho_2Sn_2O_7$, for which strong evidence of spin ice phenomenology has been established based on bulk measurements and neutron scattering experiments. In addition, there are two other obvious candidates for spin ice behavior: $Dy_2Sn_2O_7$,[62] which shows very similar AC susceptibility behavior to $Dy_2Ti_2O_7$, and Ho_2GaSbO_7, which (as already mentioned) has very similar specific heat properties as $Ho_2Ti_2O_7$.[44] We also note that Matsuhira *et al.*[62] have suggested that $Pr_2Sn_2O_7$ might be a spin ice as judged by its bulk properties. If proved to be correct, this is particularly interesting as the dipole interaction is weak in this material: hence, it might be an ideal realization of near neighbor, rather than dipolar spin ice.

Throughout this article, we have used the term "spin ice material" to refer to materials that closely approximate the spin ice model as originally defined in Ref. [4]. But we have also explored the relationship between this model and the Heisenberg model antiferromagnet (see Section 7.2), which in one sense can be seen as an extension, by Anderson[3] and Villain,[23] of the concept originally introduced by Pauling,[1] Similarly, holmium and dysprosium titanates are themselves part of a larger series of rare earth titanates that have been extensively studied in recent years as model highly frustrated antiferromagnets (the stannates have been less studied). Hence, in this section, we briefly summarize the properties of these and other materials that are closely related to the spin ice family.

7.6.1. *Rare earth titanates*

Among the titanates, there is a gamut of materials that exhibit interesting magnetic behavior that depend on the choice of the rare earth trivalent ion, as different rare earth ions possess different types of anisotropy. $Gd_2Ti_2O_7$ has $L = 0$ and $S = 7/2$ and will therefore have only very weak single-ion anisotropy. This system should be well described to a good approximation by a Heisenberg antiferromagnet, where the crystal field is a weaker perturbation to the free ion state than are the long-range dipole–dipole interactions.[145] $Gd_2Ti_2O_7$ exhibits two transitions in zero applied field, at 1.1 K and 0.7 K, which have been seen in neutron scattering[146] and bulk measurements,[145,147] as well as a number of phase transitions in applied magnetic field.[147,148] Considering only antiferromagnetic exchange and dipolar interactions, the GS of this idealized model is an ordered state with propagation vector $k = 0$ (see Refs. [84] and [85]) that is robust to quantum spin fluctuations.[85] However, this is not the ordered state observed in $Gd_2Ti_2O_7$.[146] Recent neutron scattering experiments[149] have firmly established that the zero-field state of $Gd_2Ti_2O_7$, described by the propagation vector $\boldsymbol{k} = 1/2, 1/2, 1/2$, is partially ordered between 0.7 K and 1.0 K, and fully ordered below 0.7 K in a four-wave vector structure. Recent theoretical calculations[150,151] find consistency with the structure between 0.7 K and 1.0 K, but the 4-k structure below 0.7 K has not yet been justified theoretically. There are open questions as to the dynamics of the Gd moments in $Gd_2Ti_2O_7$.[80]

Except for $Gd_2Ti_2O_7$, the other heavy rare earth $R_2Ti_2O_7$ materials have strong single-ion anisotropy. $Tb_2Ti_2O_7$ is an antiferromagnetic material with a measured Curie–Weiss temperature θ_{CW} of approximately -20 K. If treated as a $\langle 111 \rangle$ Ising antiferromagnet (see Section 7.3.2.2) it has $J \approx -2.64$ K and $D \approx 0.48$ K, giving $J_{\text{eff}} \approx -0.08$ K and should in principle order at approximately 1.2 K[37]: but in fact, it does not display long-range order down to a temperature of 60 mK[152] and has been advanced as one of the paradigm materials displaying spin liquid (or collective paramagnetic) behavior in three dimensions. There is, on the other hand, a report of glassy behavior setting in below 70 mK.[153] Such glassy behavior has also been reported in the related $Tb_2Nb_2O_7$ compound below 0.8 K.[154] At the present time, the explanation as to why $Tb_2Ti_2O_7$ fails to order down to 60–70 mK thus remains a puzzle.[155,156] This puzzle is made even more intriguing when it is noted that $Tb_2Ti_2O_7$ orders at approximately 2 K under applied pressure in excess of 2 GPa.[144]

$Er_2Ti_2O_7$ and $Yb_2Ti_2O_7$ have XY spins, which approximately lie perpendicular to the local trigonal axis. $Er_2Ti_2O_7$ is an antiferromagnet, while $Yb_2Ti_2O_7$ has a small positive ferromagnetic θ_{CW}. $Er_2Ti_2O_7$ has a transition to long-range order at 1.17 K, probably stabilized by an "order-by-disorder" mechanism.[71] $Yb_2Ti_2O_7$ displays a sharp first-order transition in the specific heat at approximately 0.24 K where static order was not initially observed.[40] However, this conclusion has recently been questioned, as neutron diffraction evidence for a transition to long-range ferromagnetic order has been found.[41]

7.6.2. *Other pyrochlores related to spin ice*

$Ho_2Ru_2O_7$ has been advanced as another candidate spin ice material, but one in which the transition metal Ru^{4+} is magnetic, unlike Ti^{4+} in $Ho_2Ti_2O_7$.[157] However, recent neutron studies[158] have revealed that $Ho_2Ru_2O_7$ has two magnetic transitions at $T \sim 95$ K and $T \sim 1.4$ K to long-range ordered states involving the Ru and Ho sublattices, respectively. Between these transitions, the Ho^{3+} moments form short-ranged ordered spin clusters. The internal field provided by the ordered S=1 Ru^{4+} moments disrupts the fragile spin ice state of the Ho^{3+} moments and drives them to develop long-range order below 1.4 K.[158]

Departing from insulating materials and considering the manifestation of spin ice phenomenology within metallic pyrochlores, two materials are worth noting. In $Nd_2Mo_2O_7$, the spin ice-like correlations on the Nd^{3+} sublattice perturb the metallic behavior on the Mo sublattice to give a gigantic anomalous Hall effect.[159] We also note a very recent theoretical study of a "double-exchange spin ice model" that may be relevant to this system.[160] Another example is $Dy_2Ir_2O_7$, which is also metallic, but where the Dy spins appear to be antiferromagnetically coupled, leading to a transition to long-range order.[161]

There are many other chemical modifications that can be made on the rare earth titanate structure, by substituting two ions, with charges 3+ and 5+, on the Ti site. However, few of these have yet been tested for spin ice behavior. One exception is $Ho_2(HoSb)O_7$,[162] which shows complex non-collinear order below $T_N \approx 2$ K.

7.7. Conclusions

We have tried to present a relatively comprehensive review of spin ice research which is, we hope, up to date at the time of writing; we apologize to any authors whose work we have inadvertently missed or misrepresented. The field is developing quickly, particularly in the area of spin dynamics and field-dependent properties, and it is anticipated that some parts of this review will quickly become out of date! However, we hope it will be a useful account of the first 7 years of spin ice research. We have mainly restricted our attention to the best characterized spin ice materials $Ho_2Ti_2O_7$, $Dy_2Ti_2O_7$, and $Ho_2Sn_2O_7$. We have not elaborated on many possible avenues for future research, which might include, for example, studies of epitaxial films: we note that these have already been prepared with other pyrochlores.[163]

Let us conclude by summarizing some of the main features of spin ice physics. The near-neighbor spin ice model is a 16-vertex model of frustrated ferromagnetism that is realized in several real materials. The model — and the materials — have zero-point entropy in the GS and the experimentalist has remarkable control of the statistical weightings of the "vertices" by applying a field to single crystals. This leads to many interesting field-induced states including kagomé ice, with zero-point entropy and the "spin chain" state, which in simple terms can be thought of as a 1D Ising ferromagnet. Several interesting phase transitions are also observed, including "liquid–gas" type transitions and possibly the Kastelyn transition. Through various mappings, the spin ice materials can be considered ideal experimental realizations of some basic models of frustration: Anderson's Ising antiferromagnet, Wannier's triangular antiferromagnet, and, of course, Pauling's ice model. The spin ice state represents a particular kind of frozen magnetic state with correlations of limited range. The high-temperature dynamics include a crossover from thermally activated to quantum tunneling dynamics at a particularly high temperature (15 K), with the spin ice freezing process near 1 K apparently distinct from that of a conventional spin glass. The search for a detailed microscopic understanding of the spin ice materials has uncovered several new features of the dipolar interaction. For all these reasons, spin ice has proved to be a fruitful playground in which to explore, develop, and challenge our overall understanding of frustrated spin systems.

7.8. Note Added for the Second Edition

There have been significant developments in spin ice physics since the publication of our review in 2005; more than can be successfully reviewed in an update covering a few pages. Around the time of publication, interest was turning to the inherent correlations in the ensemble of low-temperature states satisfying the ice rules. The ice rules constitute a local topological constraint such that the magnetic degrees of freedom map onto an emergent lattice gauge field satisfying everywhere a discrete divergence-free condition.

The resulting, so-called Coulomb phase has dipolar correlations and characteristic diffuse neutron scattering patterns with narrow "pinch points" at the Brillouin zone centers. Breaking the ice rules means breaking the constraints, producing topological excitations which, within the approximation that the ice rules states form a true flat band, are freely moving, deconfined objects. Remarkably, it turns out that, thanks to the underlying dipolar interaction, the topological objects also carry effective magnetic charge, so that the excitations in spin ice appear as mobile "magnetic monopoles". This picture of Coulomb phase and monopoles has been confirmed in a number of different experimental set ups. Monopole dynamics and the possibility of magnetic current are intriguing subjects that have received considerable attention and generated invigorating discussion in the literature. Here, the problematic centers on the fact that "vacuum" from which the monopoles are excited is unusual, as it conserves large internal magnetic fields. A spin-off bonanza is followed in the field of artificial spin ice, where arrays of 2D nanomagnets have also been shown to generate and deconfine effective magnetic charge with analogous dynamics. Most recently, attention has begun turning away from classical and artificial spin ice systems to materials and models where quantum fluctuations play a key role. Here, the emergence goes beyond effective electrostatics or magnetostatics, to a full analogy with quantum electrodynamics with conjugate magnetic and pseudo-electric charge and photon-like excitations travelling at an effective "speed of light" fixed by the magnetic interactions.

For recent short reviews on classical and quantum spin ice in the above context, see:

- C. Castelnovo, R. Moessner, and S. Sondhi, *Ann. Rev. Cond. Matt. Phys.* **3**, 35 (2012).
- L. Savary and L. Balents, *Phys. Rev. Lett.* **108**, 037202 (2012).
- N. R. Hayre, K. A. Ross, R. Applegate, T. Lin, R. R. P. Singh B. D. Gaulin, and M. J. P. Gingras, *Cond. Matt.* (2012), arXiv:1211.5934.

Acknowledgments

S.T.B. acknowledges funding from the EPSRC and thanks the E.N.S. Lyon for a Loadenbach Visiting Professorship (2003). M. J. P. G. acknowledges the following sources of funding: NSERC of Canada, Canada Research Chair Program, Research Corporation, Canada Foundation for Innovation,

Canadian Institute for Advanced Research, Ontario Innovation Trust, Ontario Challenge Fund, and Materials and Manufacturing Ontario.

We thank Andrew Cornelius, Georg Ehlers, Jason Gardner, Kazuyuki Matsuhira, Oleg Petrenko, Stephan Rosenkranz, Toshiro Sakakibara, and Peter Schiffer for supplying figures and for permission to reproduce them. We thank Colin Broholm and Yiming Qiu for supplying a copy of his Ph.D. thesis (Ref. [57]).

It is a pleasure to thank our recent collaborators in the area of spin ice physics: Simon Banks, Daniel Cabra, Benjamin Canals, Dickon Champion, Andrew Cornelius, Adrian del Maestro, Byron den Hertog, Maxime Dion, Sarah Dunsiger, Georg Ehlers, Matt Enjalran, Bjorn Fåk, Tom Fennell, Mark Field, Hideto Fukazawa, Jason Gardner, Bruce Gaulin, John Greedan, Mark Green, Mark Harris, Ryuji Higashinaka, Ying-Jer Kao, Rob Kiefl, Philip King, Jorge Lago, Yoshi Maeno, Des McMorrow, Roger Melko, Isabelle Mirebeau, Roderich Moessner, Hamid Molavian, Martin Orendáč, Bashir Ouladdiaf, Ivan Parkin, Oleg Petrenko, Pierre Pujol, Jacob Ruff, Zaher Salman, Fahad Shahabazi, Masae Shirai, Jeff Sonier, Ross Stewart, Andrea Taroni, Andrew Wills, and Taras Yavorsk'ii. S. T. B. would particularly like to thank Mark Harris for an initial collaboration in this field. We have also benefited from collaborations and discussions with many other people, too numerous to mention here; we thank all of them.

References

1. L. Pauling, *J. Am. Chem. Soc.* **57**, 2680 (1935).
2. G. H. Wannier, *Phys. Rev.* **79**, 357 (1950); erratum *Phys. Rev. B* **7**, 5017 (1973).
3. P. W. Anderson, *Phys. Rev.* **102**, 1008 (1956).
4. M. J. Harris, S. T. Bramwell, D. F. McMorrow, T. Zeiske, and K. W. Godfrey, *Phys. Rev. Lett.* **79**, 2554 (1997).
5. S. T. Bramwell and M. J. Harris, *J. Phys. Condens. Matt.* **10**, L215 (1998).
6. G. Toulouse, *Commun. Phys.* **2**, 115 (1977).
7. R. J. Baxter, *Exactly Solved Models in Statistical Mechanics* (Academic Press, 1982).
8. E. H. Lieb and F. Y. Wu in C. Domb and M. S. Green (eds.), *Phase Transitions and Critical Phenomena* (Academc Press, New York, 1972).
9. S. T. Bramwell and M. J. P. Gingras, *Science* **294**, 1495 (2001).
10. B. S. Shastry, *Physica B* **329**, 1024 (2003).
11. A. P. Ramirez, C. L. Broholm, R. J. Cava, and G. R. Kowach, *Physica B* **280**, 290 (2000).
12. A. P. Ramirez, *Nature* **421**, 483 (2003).
13. W. F. Giauque and J. W. Stout, *J. Am. Chem. Soc.* **58**, 1144 (1936).

14. P. Debye, *Ann. der Physik* **39**, 789 (1912).
15. J. M. Ziman, *Models of Disorder: The Theoretical Physics of Homogeneously Disordered Systems* (Cambridge University Press, 1976).
16. J. D. Bernal and R. H. Fowler, *J. Chem. Phys.* **1**, 515 (1933).
17. W. F. Giauque and M. F. Ashley, *Phys. Rev.* **43**, 81 (1933).
18. J. F. Nagle, *J. Math. Phys.* **7**, 1484 (1966).
19. E. O. Wolllan, W. L. Davidson, and C. G. Schull, *Phys Rev.* **75**, 1348 (1949).
20. G. S. Rushbrooke, *Introduction to Statistical Mechanics* (Oxford University Press, 1949).
21. J. E. Mayer and M. G. Mayer, *Statistical Mechanics*, 2nd edn. (J. Wiley, 1977).
22. A. P. Ramirez, A. Hayashi, R. J. Cava, R. Siddharthan, and B. S. Shastry, *Nature* **399**, 333 (1999).
23. J. Villain, *Z. Phys. B* **33**, 31 (1979).
24. R. Moessner and J. T. Chalker, *Phys. Rev. Lett.* **80**, 2929 (1998).
25. M. J. Harris, M. P. Zinkin, Z. Tun, B. M. Wanklyn, and I. P. Swainson, *Phys. Rev. Lett.* **73**, 189 (1994).
26. M. J. Harris, M. P. Zinkin, and T. Zeiske, *Phys. Rev. B* **52**, R707 (1995).
27. R. Moessner, *Phys. Rev. B* **57**, R5587 (1998).
28. A. J. Garcia-Adeva and D. L. Huber, *Phys. Rev. B* **64**, 172403 (2001).
29. N. Mekata, *J. Phys. Soc. Japan* **42**, 76 (1977).
30. G. I. Watson, *J. Stat. Phys.* **94**, 1045 (1999).
31. M. J. Harris, S. T. Bramwell, and P. J. C. King, unpublished.
32. M. J. Harris, S. T. Bramwell, T. Zeiske, D. F. McMorrow, and P. J. C. King, *J. Magn. Magn. Mater.* **177**, 757 (1998).
33. S. T. Bramwell, unpublished.
34. D. Bitko, T. F. Rosenbaum, and G. Aeppli, *Phys. Rev. Lett.* **77**, 940 (1996).
35. J. R. Friedman, M. P. Sarachik, J. Tejada, and R. Ziolo, *Phys. Rev. Lett.* **76**, 3830 (1996).
36. R. Siddharthan, B. S. Shastry, A. P. Ramirez, A. Hayashi, R. J. Cava, and S. Rosenkranz, *Phys. Rev. Lett.* **83**, 1854 (1999).
37. B. C. den Hertog and M. J. P. Gingras, *Phys. Rev. Lett.* **84**, 3430 (2000).
38. M. J. Harris, S. T. Bramwell, P. C. W. Holdsworth, and J. D. M. Champion, *Phys. Rev. Lett.* **81**, 4496 (1998).
39. S. Rosenkranz, A. P. Ramirez, A. Hayashi, R. J. Cava, R. Siddharthan, and B. S. Shastry, *J. Appl. Phys.* **84**, 5914 (2000).
40. J. A.Hodges, P. Bonville, A. Forget, A. Yaouanc, P. Dalmas de Reotier, G. André, M. Rams, K. Krolas, C. Ritter, P. C. M. Gubbens, C. T. Kaiser, P. J. C. King, and C. Baines, *Phys. Rev. Lett.* **88**, 077204 (2002).
41. Y. Yasui, M. Soda, S. Iikubo, M. Ito, M. Sato, N. Hamaguchi, T. Matsushita, N. Wada, T. Takeuchi, N. Aso, and K. Kakurai, *J. Phys. Soc. Japan* **72**, 3014 (2003).
42. S. T. Bramwell, M. J. Harris, B. C. den Hertog, M. J. P. Gingras, J. S. Gardner, D. F. McMorrow, A. R. Wildes, A. L. Cornelius, J. D. M. Champion, R. G. Melko, and T. Fennell, *Phys. Rev. Lett.* **87**, 047205 (2001).

43. M. Kanada, Y. Yasui, Y. Kondo, S. Iikubo, M. Ito, M. Harashina, M. Sato, H. Okumura, K. Kakurai, and H. Kadowaki, *J. Phys. Soc. Japan* **71**, 313 (2002).
44. H. W. J. Blöte, R. F. Wielinga, and W. J. Huiskamp, *Physica* **43**, 549 (1969).
45. B. C. den Hertog, M. J. P. Gingras, S. T. Bramwell, and M. J. Harris, cond-mat/9912220 (1999).
46. A. L. Cornelius and J. S. Gardner, *Phys. Rev. B* **64**, 060406 (2001).
47. H. Kadowaki, Y. Ishii, K. Matsuhira, and Y. Hinatsu, *Phys. Rev. B* **65**, 144421 (2002).
48. R. S. Roth, *J. Res. Natl. Bur. Stand.* **56**, 17 (1956).
49. O. Knop, F. Brisse, L. Castelliz, and Sutarno, *Can J. Chem.* **43**, 2812 (1965).
50. B. M. Wanklyn *et al.*, *J. Mater. Sci.* **3**, 395 (1968).
51. G. Balakrishnan, O. A. Petrenko, M. R. Lees, and D. McK. Paul, *J. Phys. Condens. Matter* **10**, L723 (1998).
52. J. S. Gardner, B. D. Gaulin, and D. McK. Paul, *J. Cryst. Growth* **191**, 740 (1998).
53. L. G. Mamsurova, K. K. Pukhov, N. G. Trusevich, and L. G. Shcherbakova, *Sov. Phys. Solid State* **27**, 1214 (1985).
54. S. T. Bramwell, M. N. Field, M. J. Harris, and I. P. Parkin, *J. Phys. Condens. Matter* **12**, 483 (2000).
55. Y. M. Jana and D. Ghosh, *Phys. Rev. B* **61**, 9657 (2000).
56. Y. M. Jana, A. Sengupta, and D. Ghosh, *J. Magn, Magn, Mater.* **248**, 7 (2002).
57. Y. Qiu, Ph.D. Thesis, John Hopkins University, 2002.
58. L. G. Mamsurova, K. S. Pigal'skii, N. G. Trusevich, and L. G. Shcherbakova, *Sov. Phys. Solid State* **27**, 978 (1985).
59. D. J. Flood, *J. Appl. Phys.* **45**, 4041 (1974).
60. J. D. Cashion, A. H. Cooke, M. J. M. Leask, T. L. Thorpe, and M. R. Wells, *J. Mater. Sci.* **3**, 402 (1968).
61. V. Bondah-Jagalu and S. T. Bramwell, *Can J. Phys.* **79**, 1381 (2001).
62. K. Matsuhira, Y. Hinatsu, K. Tenya, H. Amitsuka, and T. Sakakibara, *J. Phys. Soc. Japan* **71**, 1576 (2002).
63. T. Katsufuji, *Phys. Rev. B* **69**, 064422 (2004).
64. H. Fukazawa, R. G. Melko, R. Higashinaka, Y. Maeno, and M. J. P. Gingras, *Phys. Rev. B* **65**, 054410 (2002).
65. K. Matsuhira, Y. Hinatsu, K. Tenya, and T. Sakakibara, *J. Phys. Condens. Matter* **12**, L649 (2000).
66. R. Moessner and J. T. Chalker, *Phys. Rev. B* **58**, 12049 (1998).
67. J. D. M. Champion, S. T. Bramwell, P. C. W. Holdsworth, and M. J. Harris, *Europhys. Lett.* **57**, 93 (2002).
68. G. Ferey, R. de Pape, M. Leblanc, and J. Pannetier, *Rev. Chim. Minerale* **23**, 474 (1986).
69. J. N. Reimers, J. E. Greedan, and M. Bjorgvinsson, *Phys. Rev. B* **45**, 7295 (1992).
70. P. C. W. Holdsworth, unpublished.

71. J. D. M. Champion, M. J. Harris, P. C. W. Holdsworth, A. S. Wills, G. Balakrishnan, S. T. Bramwell, E. Čižmár, T. Fennell, J. S. Gardner, J. Lago, D. F. McMorrow, M. Orendáč, A. Orendáčová, D. M[c]K. Paul, R. I. Smith, M. T. F. Telling, and A. Wildes, *Phys. Rev. B* **68**, 020401 (2003).
72. S. T. Bramwell, M. J. P. Gingras, and J. N. Reimers, *J. Appl. Phys* **75**, 5523 (1994).
73. J. D. M. Champion, Ph.D. Thesis, London University, 2001.
74. R. Siddharthan, B. S. Shastry, and A. P. Ramirez, *Phys. Rev. B* **63**, 184412 (2001).
75. B. Bleaney, F. N. H. Robinson, and M. R. Wells, *Proc. R. Soc. Lond. A* **362**, 179 (1978).
76. K. Ram and K. K. Sharma, *J. Phys. C.: Solid State Phys.* **18**, 619 (1985).
77. P. B. Chakraborty, P. Henelius, H. Kjonsberg, A. W. Sandvik, and S. M. Girvin, cond-mat/0402051 (2004).
78. J. Snyder, B. G. Ueland, J. S. Slusky, H. Karunadasa, R. J. Cava, and P. Schiffer, *Phys. Rev. B* **69**, 064414 (2004).
79. G. Ehlers, A. L. Cornelius, M. Orendáč, M. Kajnakova, T. Fennell, S. T. Bramwell, and J. S. Gardner, *J. Phys. Condens. Matter* **15**, L9 (2003).
80. E. Bertin, J. A. Hodges, J.-P. Bouchaud, J. P. Sanchez, P. Bonville, and P. Vuillet, *Eur. Phys. J.* B **27**, 347 (2002).
81. T. Fennell, O. A. Petrenko, B. Fåk, S. T. Bramwell, M. Enjalran, T. Yavors'kii, M. J. P. Gingras, R. G. Melko, and G. Balakrishnan, cond-mat/0404594 (2004).
82. A. B. Harris, C. Kallin, and A.J. Berlinsky, *Phys. Rev. B* **45**, 2899 (1992).
83. J. N. Reimers, A. J. Berlisnky, and A. C. Shi, *Phys. Rev. B* **43**, 865 (1991).
84. S. E. Palmer and J. T. Chalker, *Phys. Rev. B* **62**, 4888 (2000).
85. A. G. Del Maestro and M. J. P. Gingras, *J. Phys. Condens. Matter* **16**, 3339 (2004).
86. P. P. Ewald, *Ann. Physik* **64**, 253 (1921).
87. S. W. de Leeuw, J. W. Perram, and E. R. Smith, *Ann. Rev. Phys. Chem.* **37**, 245 (1986).
88. M. Born and Huang, *Dynamical Theory of Crystal Lattices* (Oxford University Press, 1968).
89. J. M. Ziman, *Principles of the Theory of Solids*, 2nd edn. (Cambridge University Press, Cambridge, 1972).
90. P. J. Camp, J. C. Shelley, and G. N. Patey, *Phys. Rev. Lett.* **84**, 115 (2000).
91. G. Ayton, M. J. P. Gingras, and G. N. Patey, *Phys. Rev. Lett.* **75**, 2360 (1995).
92. A. Aharony and M. E. Fisher, *Phys. Rev. B* **8**, 3323 (1973).
93. Domb and Green (eds.), *Phase Transitions and Critical Phenomena*, Vol. 3 (Academic Press, London, 1974).
94. R. G. Melko, M. Sc. Thesis, University of Waterloo (2001).
95. R. G. Melko, M. Enjalran, B. C. den Hertog, and M J. P. Gingras, cond-mat/0308282 (2003).
96. W. Selke, *Phys. Rep.* **170**, 213 (1988).

97. M. J. P. Gingras and B. C. den Hertog, *Can. J. Phys.* **79**, 1339 (2001).
98. M. Enjalran and M. J. P. Gingras. cond-mat/0307151 (2003).
99. R. G. Melko, B. C. den Hertog, and M. J. P. Gingras, *Phys. Rev. Lett.* **87**, 067203 (2001).
100. G. T. Barkema and M. E. J. Newman, *Phys. Rev. E* **57**, 1155 (1998).
101. M. E. J. Newman and G. T. Barkema, *Monte Carlo Methods in Statistical Physics* (Clarendon Press, Oxford, 1999).
102. K. Binder and D. W. Heermann, *Monte Carlo Simulation in Statistical Physics* (Springer, Germany, 1997).
103. B. A. Berg and T. Neuhaus, *Phys. Rev. Lett.* **68**, 9 (1992).
104. U. H. E. Hansmann and Y. Okamoto, *Physica A* **212**, 415 (1994).
105. A. M. Ferrenberg and R. H. Swendsen, *Phys. Rev. Lett.* **61**, 2635 (1998).
106. S. Yoshida, K. Nemoto, and K. Wada, *J. Phys. Soc. Japan* **71**, 948 (2002).
107. T. Fennell, O. A. Petrenko, G. Balakrishnan, S. T. Bramwell, J. D. M. Champion, B. Fåk, M. J. Harris, and D. McK. Paul, *Appl. Phys. A: Mat. Sci. Proc.* **74**, S889 (2002).
108. S. L. Chamberlain, S. T. Hess, and L. R. Corruccini, *Phys. Lett. A* **323**, 310 (2004).
109. M. E. Fisher and A. N. Berker, *Phys. Rev. B* **26**, 2507 (1982).
110. T. Mitsui, I. Tatsuzaki, and E. Nakamura, *An Introduction to the Physics of Ferroelectrics and Related Phenomena* (Gordon and Breach Science Publishers, 1976).
111. K. Matsuhira, Z. Hiroi, T. Tayama, S. Takagi, and T. Sakakibara, *J. Phys. Condens. Matter* **14**, L559 (2002).
112. S. Yoshida, K. Nemoto, and K. Wada, cond-mat/0403762 (2004).
113. J. P. C. Ruff, R. G. Melko, and M. J. P. Gingras, unpublished.
114. O. Petrenko, G. R. Lees, and G. Balakrishnan, *Phys. Rev. B* **68**, 012406 (2003).
115. Z. Hiroi, K. Matsuhira, S. Takagi, T. Tayama, and T. Sakakibara, *J. Phys. Soc. Japan* **72**, 411 (2003).
116. H. Fukazawa, R. Higashinaka, and Y. Maeno, *Phys. Rev. B* **68**, 014415 (2003).
117. R. Higashinaka, H. Fukazawa, K. Deguchi, and Y. Maeno, *J. Phys. Condens. Matter* **14**, S679 (2004).
118. R. Higashinaka, H. Fukuzawa, K. Deguchi, and Y. Maeno, cond-mat/0406307 (2004).
119. T. Sakakibara, T. Tayama, Z. Hiroi, K. Matsuhira, and S. Takagi, *Phys. Rev. Lett.* **90**, 207205 (2003).
120. Z. Hiroi, K. Matsuhira, and M. Ogata, *J. Phys. Soc. Japan* **72**, 3045 (2003).
121. R. Higashinaka, H. Fukazawa, and Y. Maeno, *Physica B* **329**, 1040 (2003).
122. T. Fennell, Ph.D. Thesis, University of London (2003).
123. A. S. Wills, R. Ballou, and C. Lacroix, *Phys. Rev. B* **66**, 144407 (2002).
124. M. Udagawa, M. Ogata, and Z. Hiroi, *J. Phys. Soc. Japan* **71**, 2365 (2002).
125. R. Moessner and S. L. Sondhi, *Phys. Rev. B* **68**, 064411 (2003).

126. S. V. Isakov, K. S. Raman, R. Moessner, and S. L. Sondhi, cond-mat/0404417 (2004).
127. H. Aoki, T. Sakakibara, K. Matsuhira, and Z. Hiroi, *J.Phys. Soc. Japan* (to be published, 2004).
128. D. W. Davidson and R. H. Cole, *J. Chem. Phys.* **18**, 1417 (1950).
129. D. W. Davidson and R. H. Cole, *J. Chem. Phys.* **19**, 1484 (1951).
130. K. Matsuhira, Y. Hinatsu, and T. Sakakibara, *J. Phys. Condens. Matter* **13**, L737, 2001.
131. J. Snyder, J. S. Slusky, R. J. Cava, and P. Schiffer, *Nature* **413**, 48 (2001).
132. G. Ehlers, A. L. Cornelius, T. Fennell, M. Koza, S. T. Bramwell, and J. S. Gardner, *J. Phys. Condens. Matter* **16**, S635 (2004).
133. A. L. Cornelius, private communication (2004).
134. J. Snyder, B. G. Ueland, J. S. Slusky, H. Karunadasa, R. J. Cava, A. Mizel, and P. Schiffer, *Phys. Rev. Lett.* **91**, 107201 (2004).
135. J. L. Tholence and R. Tournier, *J. Phys. (Paris)* **35**, C4 (1974).
136. J. Lago and S. R. Dunsiger, unpublished.
137. Y. Tajima, T. Matsuo, and H. Sugo, *Nature* **299**, 1810 (1982).
138. J. Snyder, J. S. Slusky, R. J. Cava, and P. Schiffer, *Phys. Rev. B* **66**, 064432 (2002).
139. J. Snyder, B. G. Ueland, A. Mizel, J. S. Slusky, H. Karunadasa, R. J. Cava, and P. Schiffer, cond-mat/0405233 (2004).
140. M. Orendáč *et al.*, to be published (2004).
141. D. H. Reich, B. Ellman, J. Yang, T. F. Rosenbaum, G. Aeppli, and D. P. Belanger, *Phys. Rev. B* **42**, 4631 (1990).
142. R. F. Whitworth and V. F. Petrenko (eds.), *Physics of Ice* (Oxford University Press, 2002).
143. I. Mirebeau and I. Goncharenko, *J. Phys. Condens. Matter* **16**, S653 (2004).
144. I. Mirebeau, I. N. Goncharenko, P. Cadavez-Pares, S. T. Bramwell, M. J. P. Gingras, and J. S. Gardner, *Nature* **420**, 54 (2002).
145. N. P. Raju, M. Dion, M. J. P. Gingras, T. E. Mason, and J. E. Greedan, *Phys. Rev. B* **59**, 14489 (1999).
146. J. D. M. Champion, A. S. Wills, T. Fennell, S. T. Bramwell, J. S. Gardner, and M. A. Green, *Phys. Rev. B* **64**, 140407(R) (2001).
147. A. P. Ramirez, B. S. Shastry, A. Hayashi, D. A.Huse J. J. Krajewsk, and R. J. Cava, *Phys. Rev. Lett.* **89**, 067202 (1999).
148. O. Petrenko, G. R. Lees, G. Balakrishnan, and D. McK. Paul, cond-mat/0309405 (2004).
149. J. R. Stewart, G. Ehlers, A. S. Wills, S. T. Bramwell, and J. S. Gardner, *J. Phys. Condens. Matter* **16**, L321 (2004).
150. M. Enjalran and M. J. P. Gingras, cond-mat/0307152 (2003).
151. O. Cépas and B. Sriram Shastry, *Phys. Rev. B* **69**, 184402 (2004).
152. J. S. Gardner, A. Keren, G. Ehlers, C. Stock, E. Segal, J. M. Roper, B. Fåk, M. B. Stone, P. R. Hammar, D. H. Reich, and B. D. Gaulin, *Phys. Rev. B* **68**, 180401 (2003).

153. G. Luo, S. T. Hess, and L. R. Corruccini, *Phys. Lett. A* **291**, 306 (2001).
154. Y. M. Jana, Y. M. O. Sakai, R. Higashinaka, H. Fukazawa, Y. Maeno, P. Dasgupta, and D. Ghosh, *Phys. Rev. B* **68**, 174413 (2003).
155. M. Enjalran, M. J. P. Gingras, Y.-J. Kao, A. Del Maestro, and H. R. Molavian, *J. Phys. Condens. Matter* **16**, S673 (2004).
156. Y.-J. Kao, M. Enjalran, A. del Maestro, H. L. Malavian, and M. J. P. Gingras, *Phys.Rev. B* **68**, 172407 (2003).
157. C. Bansal, H. Kawanaka, H. Bando, and Y. Nishihara, *Phys. Rev. B* **420**, 052406 (2002).
158. C. R. Wiebe, J. S. Gardner, S.-J. Kim, G. M. Luke, A. S. Wills, B. D. Gaulin, J. E. Greedan, I. Swainson, Y. Qiu, and C. Jones, cond-mat/0404235 (2004).
159. Y. Taguchi, Y. Oohara, H. Yoshizawa, H. Yoshizawa, N. Nagaosa, and Y. Tokura, *Science* **291**, 2573 (2001).
160. Y. Shimomura, S. Miyahara, and N. Fukuzawa, cond-mat/0406366 (2004).
161. D. Yanagishima and Y. Maeno, *J. Phys. Soc. Japan* **70**, 2880 (2001).
162. T. Fennell, S. T. Bramwell, and M. A. Green, *Can J. Phys.* **79**, 1415 (2001).
163. J. Nishimura, T. Fukumura, M. Ohtani, Y. Taguchi, M. Kawasaki, I. Ohkubo, H. Koinuma, H. Ohguchi, K. Ono, M. Oshima, and Y. Tokura, *Appl. Phys. Lett.* **82**, 1571 (1998).

CHAPTER 8

EXPERIMENTAL STUDIES OF FRUSTRATED PYROCHLORE ANTIFERROMAGNETS

Bruce D. Gaulin*,‡ and Jason S. Gardner†

**Department of Physics and Astronomy,*
Canadian Institute for Advanced Research, McMaster University
Hamilton, ON, L8S 4M1, Canada
†Physics Department, Brookhaven National Laboratory,
Upton, New York 11973-5000, USA &
NIST Center for Neutron Research,
National Institute of Standards and Technology,
Gaithersburg, Maryland 20899-8562, USA
‡gaulin@mcmaster.ca

A lot of recent experimental progress has been made in the study of magnetic materials made up of antiferromagnetically coupled magnetic moments residing on networks of corner-sharing tetrahedra. Such networks of magnetic moments display phenomena broadly known as geometric frustration. They are found in nature in a variety of cubic pyrochlore, spinel- and Laves-phase materials, with the magnetic moments arising from either rare earth or transition metal electrons. Typically, the rare earth moments are relatively large and are only weakly coupled by exchange, such that the competition between antiferromagnetic exchange and dipolar interactions is important. Antiferromagnetic exchange interactions are stronger in transition metal-based materials, and the manifestation of geometrical frustration is different. This chapter focuses on experimental progress in this area from the last 10 years primarily due to neutron scattering studies.

8.1. Introduction

It is now well appreciated that the combination of antiferromagnetic interactions and certain lattice symmetries leads to phenomena broadly known as geometrical frustration.[1] The system under question cannot simultaneously satisfy all of its near-neighbor pairwise exchange interactions and the resulting ground state can have a large degeneracy. This situation is illustrated schematically in Figs. 8.1(a) and 8.1(b). Figure 8.1(a) shows two antiferromagnetically coupled spins on a triangle and the frustration that arises if their ground-state configuration requires nearest neighbour spins to be antiparallel. The tetrahedron, shown in Fig. 8.1(b) is to three dimensions what the triangle is to two dimensions, and a related degeneracy is relevant to finding a ground-state arrangement for four near-neighbor spins at the vertices of a tetrahedron.

Such assemblies of interacting spins with triangular and tetrahedral coordination are in fact common in nature. In two dimensions, one can form a network of edge-sharing triangles, giving rise to the familiar triangular lattice, which forms the basal plane of all hexagonal crystals. One can also form a network of corner-sharing triangles resulting in the *kagomé* lattice, which while relatively rare, is still manifest in a variety of quasi-2D crystal structures. One can also imagine more complicated assemblies of triangular plaquettes, such as the "Maple Leaf" lattice,[2] wherein every spin has five-fold coordination in two dimensions; however, to date such structures have not been exhibited in real materials.

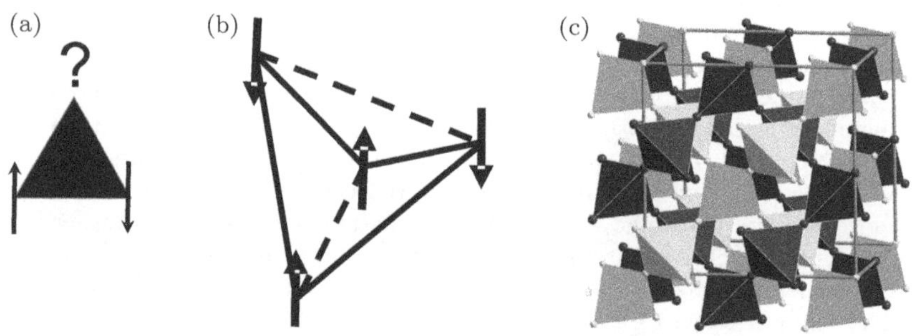

Fig. 8.1. (a) Two of three Ising spins located on a triangular unit can form an up-down pair, while the spin at the apex is frustrated. (b) Ising spins on a tetrahedron showing four satisfied (antiferromagnetic) and two frustrated (ferromagnetic) bonds. (c) The oxide pyrochlore unit cell showing the A (light) and B (dark) metal sublattices, each of which independently forms a network of corner-sharing tetrahedra.

In three dimensions, a network of edge-sharing tetrahedra results in the familiar face-centered cubic lattice, while networks of corner-sharing tetrahedra constitute sublattices of the cubic pyrochlore, spinel- and Laves-phase materials, all of which are relatively common.

In the context of magnetism in solids, this type of problem was considered theoretically first by Wannier[3] who showed that the thermodynamics of the classical Ising antiferromagnet on a two-dimensional triangular lattice was qualitatively different from that of the corresponding ferromagnet. Theoretical treatment of the problem of classical spins with antiferromagnetic interactions on a lattice of corner-sharing tetrahedra was considered by both Anderson[4] and Villain,[5] the latter of whom coined the term "cooperative paramagnet" to describe the resulting low-temperature state.

This type of phenomena, however, is not restricted to antiferromagnetism or even to magnets, but arises much more generally in interacting cooperative systems. It arises, for example both in hydrogen ordering in solid H_2O[6] and in the corresponding problem in magnetism, spin ice,[7] which arises in the presence of *ferromagnetism* and tetrahedrally coordinated spins (phenomena considered separately in this volume by Bramwell, Gingras and Holdsworth).

8.2. Pyrochlore Lattices

The cubic oxide pyrochlores crystallize into the space group $Fd\bar{3}m$[8] and have the chemical composition $A_2B_2O_7$, where the A sublattice is often occupied by a trivalent rare earth ion and the B sublattice by a tetravalent transition metal ion. Both the A and B sublattices, independently, form a network of corner-sharing tetrahedra, as shown in Fig. 8.1(c). The A and B networks are shifted relative to each other along the cubic [111] direction.

As either the rare earth or the transition metal site may possess a magnetic moment, it is possible for pyrochlores to exist in which either the A sublattice is magnetic, as occurs in $Tb_2Ti_2O_7$[9] for example, or the B sublattice is, as occurs in $Y_2Mo_2O_7$.[10] It is also possible for both sublattices to be simultaneously magnetic, as occurs in $Tb_2Mo_2O_7$,[11] or for neither to be magnetic, as occurs in $Y_2Ti_2O_7$[12] for example.

The local chemical environment around the A and B sites is not the same, with the A site being eight-fold coordinated by a distorted cube of O^{2-} ions, while the B site is coordinated within BO_6 octahedra.[8] This local structure can profoundly influence the magnetism in these compounds, particularly for the rare earth moment at the A site. The nature of the distortion of the oxygen environment determines the crystalline electric field splittings relevant to the rare earth ion, which in turn determine the nature of the rare earth magnetic moment and its anisotropies. The transition metal moment tends to be spin only, due to quenching of its orbital magnetic moment.

The rare earth magnetic moments at the A site of the pyrochlores interact with each other via rather weak exchange in insulating pyrochlores. This is due to the fact that the 4f electrons are relatively localized and shielded from the neighboring electronic environment by their 5s and 5p electrons. The electronic overlaps which drive superexchange interactions are therefore weak and near-neighbour exchange is typically ∼10 K or less in such rare earth magnets. The orbitally quenched moments in transition metal-based pyrochlores interact much more strongly as the d-electrons are more extended and the overlap and resulting exchange can be an order of magnitude or more stronger than in the insulating rare earth-based pyrochlores.

Metallic pyrochlores also exist, and some of these relax the condition that the A site be occupied by a rare-earth atom. For example, much recent attention has focused on $Cd_2Re_2O_7$[13] and $Cd_2Os_2O_7$,[14] as $Cd_2Re_2O_7$ has been found to be superconducting at low temperatures (∼1–2 K) and $Cd_2Os_2O_7$ undergoes an intriguing metal–insulator transition at relatively high temperatures. There is also very interesting metal-insulating behavior observed across the $RE_2Mo_2O_7$ series, where RE is a rare earth ion.[15] In this family of materials, the conductivity changes from metallic to insulating as the RE atom moves from the light rare earths to the heavy rare earths. Very recently, superconductivity has also been discovered in the so-called β-pyrochlores KOs_2O_6[16] and $RbOs_2O_6$,[17] at ∼9.6 and 6.3 K, respectively. These metals also crystallize in the cubic space group $Fd\bar{3}m$ with Os occupying sites on a network of corner-sharing tetrahedra.

The network of corner-sharing tetrahedra also forms within the cubic spinels[18] which also crystalize in the space group $Fd\bar{3}m$. These materials often have the chemical composition AB_2O_4, and the B site sublattice also takes on the pyrochlore lattice. In insulating spinels where the A site is divalent, such as Zn^{2+} or Mg^{2+}, the B site can be a trivalent magnetic transition metal ion, such as Cr^{3+} in $ZnCr_2O_4$. These materials are Mott insulators and, as in the case of the transition metal ions on the pyrochlore lattice, the antiferromagnetic interactions between the B site ions are relatively strong.

In the case where the A site is monovalent, such as Li^{1+} in LiV_2O_4,[19] a narrow band metal results within non-integral valence on the transition metal at the B site, in this example the V site. Although the straightforward conditions for geometrical frustration due to near-neighbor antiferromagnetic interactions are not obviously relevant to the metallic magnet, interesting heavy fermion behavior is observed in this material which has been proposed as a consequence of the frustrated magnetic interactions.[20]

At low temperatures, the insulating cubic spinel magnets often distort to tetragonal symmetry,[21] and thus, the "full" frustration of the pyrochlore lattice can be partially relieved. Recent neutron scattering work has proposed that this distortion is in fact driven by the geometric frustration in a manner akin to the spin-Peierls transitions[22] in quasi-1D $s = 1/2$ antiferromagnetic materials.

The coupling between lattice and magnetic degrees of freedom in pyrochlore antiferromagnets has been examined theoretically by Tchernyshyov *et al.*[23] Their work shows that a Jahn–Teller-like distortion of the tetrahedra can lower the overall energy of the coupled lattice and magnetism and result in appropriately dimerized phases similar to experimental observations.[22]

The same pyrochlore sublattice also appears in cubic Laves-phase materials such as YMn_2,[24] where the Mn atoms reside on it in its high-temperature, paramagnetic phase. This metal undergoes a strongly first-order magnetic phase transition to a long-period modulated magnetic structure with an accompanying lattice distortion. The phase transitions can be suppressed with minor alloying, of Sc at the Y site[25] for example, or with applied pressure.[26]

In what follows, we will review progress that has been made primarily through neutron scattering studies on some of the simplest such systems described by antiferromagnetically coupled spins on the pyrochlore lattice. These will include the insulating pyrochlores $Tb_2Ti_2O_7$ and $Y_2Mo_2O_7$, as well as the insulating spinel $ZnCr_2O_4$. This survey is clearly not comprehensive, but rather selective on systems for which our experimental understanding is relatively mature, and representative of the richness in low-temperature ground states that are displayed under the umbrella of geometrical frustration.

8.3. Neutron Scattering Techniques

The complete experimental characterization of magnetic materials requires that an array of advanced experimental techniques be applied to the new materials. Beyond basic structural characterization (typically X-ray diffraction), susceptibility, magnetization and heat capacity measurements are usually required to establish the fundamental magnetic properties of the materials. Metallic materials and materials where itinerant charge degrees of freedom are of interest also require an understanding of transport and Fermi surface properties. To make further progress, microscopic probes of magnetism must be employed, including NMR, μSR and neutron scattering.

Of these experimental probes, neutron scattering is the only one capable of providing information on the static and dynamic behavior of the magnetic system as a function of wave vector spanning the entire Brillouin zone, and as a function of frequency covering spin dynamics on time scales from 10^{-9} to 10^{-13}. For this reason, neutron scattering is often regarded as an essential probe in the understanding of magnetic systems.

The neutron is, of course, electrically neutral, but carries a spin-1/2 magnetic moment. As such, neutron scattering is a weak probe of matter, in the sense that neutrons penetrate deeply into most materials and large volumes of material are often needed to carry out neutron experiments, especially inelastic neutron scattering experiments. It is also a weak probe of matter in the sense that the scattering cross section can be described within the Born Approximation and Fermi's Golden Rule. This is a significant advantage as it means that theoretical calculations can be compared directly to the relevant neutron scattering measurements.

The double, differential scattering cross section describing the relative magnetic neutron scattering into an element of solid angle dΩ and with the scattered neutron having a final energy between E_f and $E_f + dE_f$ can be written as

$$\frac{d^2\sigma}{d\Omega_f dE_f} = \frac{N}{\hbar}\frac{k_f}{k_i}p^2 e^{-2W}\Sigma_{\alpha\beta}(\delta_{\alpha,\beta} - \hat{Q}_\alpha\hat{Q}_\beta)S^{\alpha\beta}(\mathbf{Q},\omega) \tag{8.1}$$

Here, k_i and k_f are the initial and scattered neutron wave vectors; $p = \frac{\gamma r_0}{2}gf(\mathbf{Q})$, where γ is the gyromagnetic ratio of the neutron, r_0 is the classical electron radius, g is the Lande g factor, and $f(\mathbf{Q})$ is the magnetic form factor appropriate to the magnetic moment density of the atom; and e^{-2W} is the Debye–Waller factor.

The physics of the spins, and more generally the magnetic moments, are contained in Eq. (8.2) as follows:

$$S^{\alpha\beta}(\mathbf{Q},\omega) = \frac{1}{2\pi}\int_{-\infty}^{\infty} dte^{-i\omega t}\Sigma_l e^{(i\mathbf{Q}\cdot\mathbf{r}_l)}\langle S_0^\alpha S_l^\beta(t)\rangle \tag{8.2}$$

The $\Sigma_{\alpha\beta}(\delta_{\alpha,\beta} - \hat{Q}_\alpha \hat{Q}_\beta) S^{\alpha\beta}(\mathbf{Q}, \omega)$ term in Eq. 8.1 ensures that the neutron scattering cross section is sensitive to only those components of moment which lie in a plane perpendicular to $\mathbf{Q} = \mathbf{k}_i - \mathbf{k}_f$; therefore, the neutron scattering experiment directly measures the Fourier transform in space and time of this spin-pair correlation function.

Integrating over all energy levels and over a complete Brillouin zone gives rise to the sum rule as follows:

$$\int_{-\infty}^{\infty} d\omega \int_{\mathrm{BZ}} d\mathbf{Q} S^{\alpha\beta}(\mathbf{Q}, \omega) \sim S(S+1)\delta_{\alpha\beta} \tag{8.3}$$

This relation implies that the total magnetic scattering scales as the size of the moment with $S(S+1)$ and that systems with suppressed magnetic ordering and therefore suppressed ordered magnetic moments necessarily exhibit enhanced inelastic scattering so as to satisfy the above sum rule. Excellent comprehensive reviews of neutron scattering exist and are referenced for the interested reader.[27]

8.4. Cooperative Paramagnetism in $Tb_2Ti_2O_7$

The insulating rare earth pyrochlore $Tb_2Ti_2O_7$ is a well-studied system where the A-site Tb^{3+} ions are magnetic due to the partially filled 4f shell, while the B-site Ti^{4+} ion is non-magnetic. Like most of the $A_2B_2O_7$ pyrochlores, the conventional cubic unit cell, shown in Fig. 8.1(c), has cube edge dimensions of $a \sim 10.1$ Å. The crystallographic structure of this material has been very carefully studied[28] and it is characterized as an (almost) disorder-free cubic pyrochlore.

It is often the case that new materials are first studied in polycrystalline form, as that is the form in which they are first prepared. If the material is of sufficient interest, an effort will generally be made to produce high-quality single crystals, although this may ultimately prove to be difficult or impossible. The experimental study of the cubic pyrochlore $Tb_2Ti_2O_7$ follows this path, with initial measurements performed on polycrystalline materials and later more sophisticated measurements made on single crystals.

Powder neutron diffraction measurements on $Tb_2Ti_2O_7$[9] taken at 2.5 K are shown Fig. 8.2(a) and these results clearly feature two distributions of scattering. One of these consists of sharp nuclear Bragg peaks, narrow in Q, due to the crystalline order in the material. But a second diffuse, liquid-like component is also clearly visible and is attributed to magnetic short-range order.

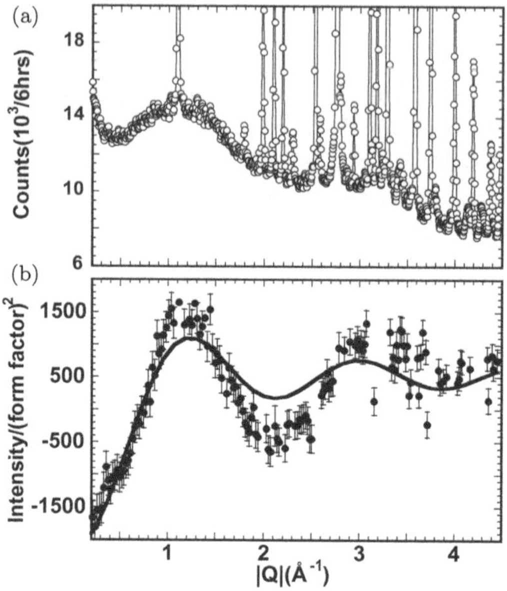

Fig. 8.2. Neutron powder diffraction data taken on $Tb_2Ti_2O_7$ at 2.5 K are shown.[9] Panel (a) shows the raw data while the panel (b) shows the same data after a data set at $T = 100$ K (representing the nuclear scattering background) has been subtracted from it (see text).

Although the general trend of the diffuse scattering is clear in the raw data, the isolation of this magnetic scattering is not completely straightforward. In principle, the best method for accomplishing this is to use neutron polarization analysis.[29] However, this method involves large trade-offs in intensity and is not routinely employed for that reason. A substitute method is to subtract a high-temperature data set from a low-temperature data set. Assuming that the magnetic scattering is Q-independent at high temperatures and that the nuclear scattering shows little temperature dependence, the resulting net scattered intensity can be attributed to magnetism.

Figure 8.2(b) shows just such a subtraction with a T = 100 K data set subtracted from the $T = 2.5$ K data set that is shown in Fig. 8.2(a). These data have been corrected for the magnetic form factor appropriate to Tb^{3+} and fit to a theoretical model for short-range *isotropic* spin correlations over a few near, neighbor distances. The theoretical form[30] used in the fit to the data in Fig. 8.2(b) is appropriate to spin correlations existing over near-neighbor distances only. It is clear from Fig. 8.2(b), that this simple form for the scattering places the peaks and valleys in $S(Q)$ at the correct positions in Q, but that the quantitative fit is imperfect. Nonetheless, as the width of the scattering is also roughly accounted for by this model with only near, neighbor spin correlations, we can conclude that the spin correlations at $T = 2.5$ K in this material are of very short range.

These magnetic $S(Q)$ results obtained here on $Tb_2Ti_2O_7$ are similar to the results of the magnetic $S(Q)$ derived from measurements on powder $Tb_2Mo_2O_7$ samples,[31] wherein both the Tb^{3+} and Mo^{4+} sites are magnetic. In the $Tb_2Mo_2O_7$ case, a short-range order peak is present near 2 Å^{-1}, in addition to those seen in Fig. 8.2 near 1 and 3 Å^{-1}. This indicates an interaction between the two A and B frustrated pyrochlore sublattices. Single, crystal measurements of the magnetic $S(\mathbf{Q})$ along high-symmetry directions in $CsNiCrF_6$[32] also possess very similar characteristics to those seen in Fig. 8.2, although in this $CsNiCrF_6$ case, two magnetic species, Ni^{2+} and Cr^{3+}, are randomly distributed over a single pyrochlore sublattice.

Inelastic scattering experiments were also carried out on polycrystalline $Tb_2Ti_2O_7$[33] and a typical scan at $Q = 2$ Å^{-1} is shown in Fig. 8.3. Three bands of excitations are clearly observed near 0.37, 2.53, and 3.50 THz. As the excitations in this powder sample are relatively sharp in energy, they must originate from relatively dispersionless excitations. Their dispersion as a function of modulus of Q was explicitly mapped[33] out and is shown in Fig. 8.4. Indeed, at temperatures above 30 K, there is very little dispersion

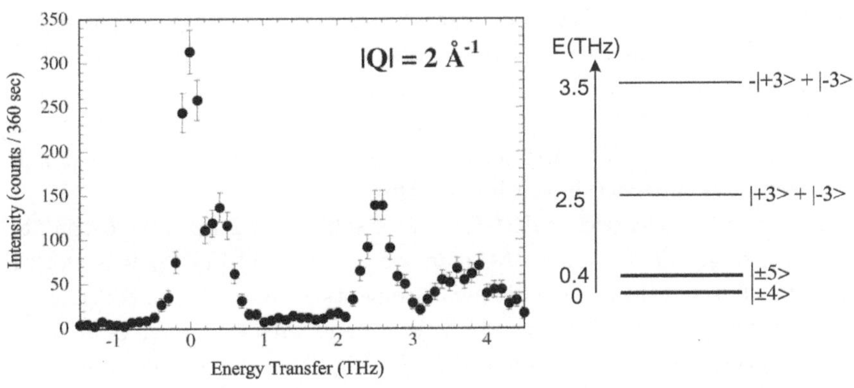

Fig. 8.3. A constant $|\mathbf{Q}|$ scan (2 Å^{-1}) at 11 K from polycrystalline $Tb_2Ti_2O_7$ is shown. At ~3.8 THz, the magnetic scattering is contaminated by an optical phonon branch. The proposed crystalline electric field level scheme and the major components of the eigenstates derived from these measurements are shown on the right.[33]

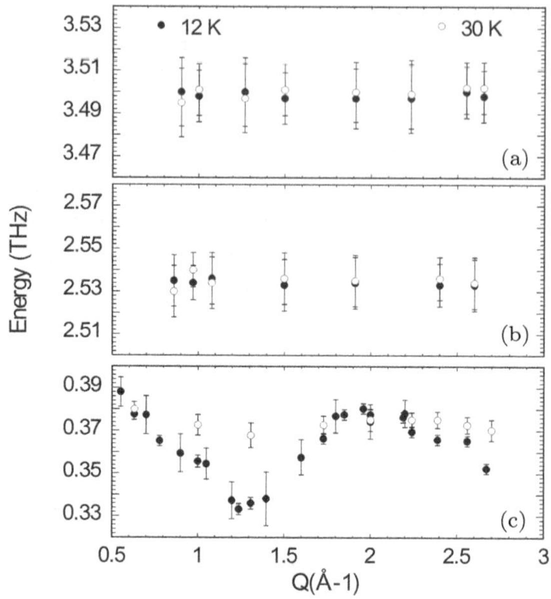

Fig. 8.4. The dispersion relation for the three lowest lying excitations in $Tb_2Ti_2O_7$. In panel (c), an unusual partial softening of the lowest energy excitation at $|\mathbf{Q}| \sim 1.2$ Å^{-1} is seen below 20 K.[33]

in any of the three branches of magnetic excitations, a characteristic of crystalline electric field levels associated with the rare earth site. A weak, but very interesting dispersion develops in the lowest energy band near 0.37 THz ~20 K, at temperatures below ~20 K, as can be seen in Fig. 8.4(c). This dispersion is such that the energy of the excitation is lowest at those wave vectors for which the magnetic structure factor, $S(Q)$, is maximum. This behaviour is very reminiscent of the roton minimum in the single-particle excitation spectrum of liquid ^{4}He.[34]

A crystalline electric field (CEF) level scheme appropriate to the $^7F_6$ configuration of Tb^{3+} in the A site environment of $Tb_2Ti_2O_7$ was worked out on the basis of these inelastic measurements.[35] With $J = 6$, the ground-state and first excited states are both doublets, with two singlets at much higher energies, as shown schematically in Fig. 8.3. The ground-state doublet is made up primarily of J^z eigenstates $|\pm 4\rangle$ and $|\pm 1\rangle$ and the first excited doublet of J^z eigenstates $|\pm 5\rangle$ and $|\pm 2\rangle$. One of the other two singlets corresponds exclusively to $|\pm 6\rangle$ and $|\pm 3\rangle$, while the remaining singlet corresponds to $|\pm 6\rangle$, $|\pm 3\rangle$, and $|0\rangle$. Note that within this CEF scheme the ground state is connected to all these low-lying excited states by $\langle A|S^{+}|0\rangle$ and $\langle A|S^{-}|0\rangle$ matrix elements, and thus, all these transitions are expected to be visible with neutron spectroscopy, which obeys such a dipole selection rule.

Such a CEF scheme implies that the *ground-state moment* (as opposed to the full paramagnetic moment) of the Tb^{3+} is $\sim$5 μ_B. Also, were the excited-state doublet at Δ $\sim$20 K, separated from the ground state by a significantly greater energy, then the moment would be considered to be an extreme Ising moment, i.e. capable of pointing only along local $\langle 111\rangle$ directions; that is either into or out of the tetrahedra. We will return to this issue later.

The CEF excitations provide information as to the size and anisotropy of the local Tb^{3+} moments. However, the nature of the magnetic interactions between the moments must be derived from magnetic susceptibility measurements, as shown in Fig. 8.5[35] for both $Tb_2Ti_2O_7$ and $(Tb_{0.02}Y_{0.98})_2Ti_2O_7$. As can be seen, Curie–Weiss fits to the high-temperature susceptibility given a paramagnetic moment of 9.6 μ_B/Tb^{3+}, very close to the 9.72 value appropriate to the $^7F_6$ free ion value. The Curie–Weiss temperature which parametrizes the average magnetic interactions is obtained from the T-intercept of the Curie–Weiss fit at high temperatures: this is clearly negative for both $Tb_2Ti_2O_7$ and $(Tb_{0.02}Y_{0.98})_2Ti_2O_7$ indicating antiferromagnetic exchange interactions. Magnetically dilute

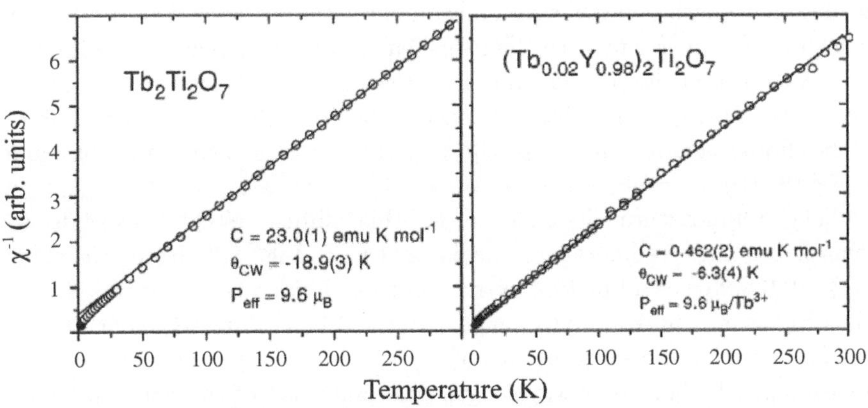

Fig. 8.5. The temperature dependence of the inverse susceptibility of the $Tb_2Ti_2O_7$ and a magnetically diluted sample, $(Tb_{0.02}Y_{0.98})_2Ti_2O_7$, intended to isolate Tb^{3+} magnetic moments.[35]

$(Tb_{0.02}Y_{0.98})_2Ti_2O_7$ would be expected to be representative of a non-interacting system, and consequently, the Curie–Weiss temperature derived from its analysis is attributed to crystal field and non-exchange contributions to Θ_{CW}. For that reason, the Θ_{CW}, which we attribute to magnetic interactions in $Tb_2Ti_2O_7$, is the difference between the Θ_{CW} for each of $Tb_2Ti_2O_7$ and $(Tb_{0.02}Y_{0.98})_2Ti_2O_7$; that is

$$\Theta_{CW}(Tb_2Ti_2O_7) - \Theta_{CW}(Tb_{0.02}Y_{0.98})_2Ti_2O_7 \sim -13\,\mathrm{K} \qquad (8.4)$$

Together with an estimate for the contribution of the magnetic dipole interactions to Θ_{CW}, an estimate for the exchange contribution to Θ_{CW} is in the range from −10.6 to −14.2 K and it is antiferromagnetic.

Relatively large single crystals of $Tb_2Ti_2O_7$ could be grown using floating-zone image furnace techniques,[36] and the resulting single crystals enabled a series of experiments which could probe the 4D dynamic structure factor $S(\mathbf{Q}, \hbar\omega)$. Figure 8.6(a) shows the difference between $S(\mathbf{Q}) = \int S(\mathbf{Q}, \hbar\omega) d\omega$ measured at $T = 9\,\mathrm{K}$ and that measured at $T = 100$ K,[33] within the (h,h,l) plane in reciprocal space. Once again, a difference between a low- and high-temperature data set is employed to isolate the magnetic structure factor. For cubic crystals, this (h,h,l) plane in reciprocal space is often employed in neutron scattering studies as it contains the (0,0,l), (h,h,0), and (h,h,h) high-symmetry directions.

One clearly observes a "checkerboard" pattern to the scattering, with diffuse scattering covering entire Brillouin zones. This indicates spin correlations on a length scale that is small compared to the dimensions of the unit cell ∼10.2 Å, consistent with the near-neighbor spin correlations estimated from the data on the powder samples. However, the single-crystal data are clearly *not* isotropic in reciprocal space, explaining the quantitative failure of the simple isotropic model for the structure factor shown as the solid line in Fig. 8.2(b).

The temperature dependence to this diffuse $S(\mathbf{Q})$[33] extends to remarkably high temperatures as is shown in Fig. 8.7. Here, we show $S(\mathbf{Q}=(0,0,2.25))$ as a function of temperature. This wave vector was chosen for study as it is quite close to the peak in the diffuse scattering near (0,0,2), but far enough removed to be free of any spurious scattering effects due to, for example, higher order wavelength contamination of the neutron beam. One sees a dramatic upturn to the temperature dependence below ∼25 K, but the diffuse scattering continues to evolve with temperature until at least 100 K. It was for this reason that the $T = 100$ K data set was used in the

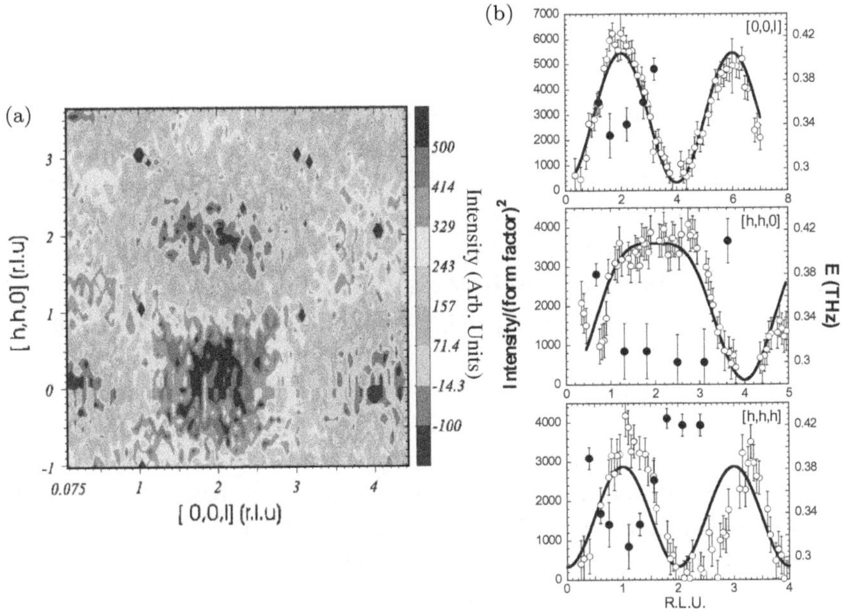

Fig. 8.6. (a) The diffuse magnetic scattering from a single crystal of $Tb_2Ti_2O_7$ in the (h,h,l) plane is shown. (b) Cuts through the diffuse scattering (open circles) along the high-symmetry directions (0,0,l), (h,h,0) and (h,h,h) are shown along with the dispersion of the lowest lying magnetic excitation (closed circles) mode at 4 K. The diffuse scattering intensity scale is on the panel (a), while the energy scale of the lowest magnetic excitation is on the panel (b).[33]

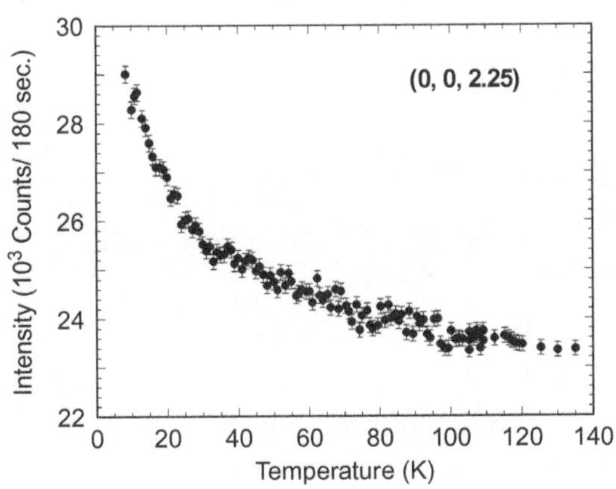

Fig. 8.7. The temperature dependence of the diffuse scattering as measured at (0,0,2.25), close to the peak in the magnetic diffuse scattering at $\mathbf{Q} = (0,0,2)$.[33]

subtraction of the data sets shown in Fig. 8.6. The temperature scale of ~25 K corresponds to the energy scale on which the excited-state doublet is populated and develops the dispersion seen in Fig. 8.4(c).

With a high-quality single crystal, the dispersion of the low-lying $\hbar\omega$ ~0.33 THz magnetic excitation can be studied as a function of **Q**,[33] as opposed to modulus of Q derived from powder samples. One sees the development of dispersion for temperatures below ~25 K, as was observed in powders.[9] This low-temperature dispersion is plotted along the three high-symmetry directions within the (h,h,l) plane in Fig. 8.6, on which are overlaid cuts of $S(\mathbf{Q})$ along (0,0,l) (top), (h,h,0) (middle) and (h,h,h) (bottom). We see that the minima in the dispersion of this low-lying magnetic mode correspond exactly with peaks in $S(\mathbf{Q})$, even though $S(\mathbf{Q})$ is anisotropic in **Q**; that is, even though it displays the "checkerboard" pattern within the (h,h,l) plane of reciprocal space.

Related $S(\mathbf{Q})$ and $S(\mathbf{Q}, \hbar\omega)$ measurements have been carried out to much lower temperatures, as low as 0.07 K.[9,37] Remarkably, the magnetic $S(\mathbf{Q})$ remains largely the same as the checkerboard pattern observed at 9 K and shown in Fig. 8.6. As we will discuss subsequently, the absence of a phase transition down to these very low temperatures is surprising, and ultimately enigmatic.

Neutron spin echo[37,39] and μSR[9,40] measurements have also been carried out to similarly low temperatures to probe the fluctuating nature of the magnetic moments. The normalized intermediate scattering function $S(Q,t)/S(Q,0)$ measured in neutron spin echo from powder samples is shown in Fig. 8.8.[37] Above ~0.3 K, $S(Q,t)/S(Q,0)$ decays with time on

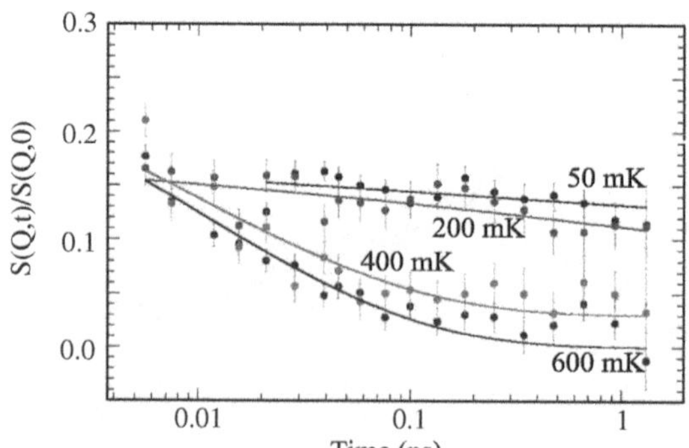

Fig. 8.8. The normalized, intermediate scattering function measured with neutron spin echo techniques at several temperatures from polycrystalline $Tb_2Ti_2O_7$.[37]

the time scale of 0.01–1 ns, consistent with $\frac{1}{T_1}$ μSR measurements which show fluctuating moments down to at least 0.05 K.[9] Below T = 0.3 K, the time dependence of $S(Q,t)/S(Q,0)$ flattens out, indicating a population of frozen magnetic moments. However, a large majority of the magnetic moments remain fluctuating below 0.3 K as the normalized $S(Q,t)/S(Q,0)$ is not observed any higher than ~0.2 for any times within the neutron spin echo time window. Susceptibility measurements also indicate freezing below ~0.3 K.[37,41] We attribute this partial spin freezing to a small subset of magnetic moments distributed near a small number of defects.

Evidence for spin freezing was also found independently in neutron scattering experiments performed on other single-crystal samples.[42] In this case, hysteresis is observed in the scattering near (0,0,2) at temperatures less than ~1.7 K. However, as in the samples described above, most of the scattering remains diffuse in nature even below 1.7 K.

From an experimental point of view, $Tb_2Ti_2O_7$ is well characterized, at least in zero applied magnetic field. It is appropriate to then ask how consistent is the information which has been derived from the different measurements. That is, how well do we understand the low-temperature properties of this frustrated magnetic material?

The most striking experimental feature of $Tb_2Ti_2O_7$ is that most of the Tb^{3+} moments remain dynamic to temperatures as low as 0.05 K. As discussed at length by Kao *et al.*,[38] this is not what one would conclude for antiferromagnetically coupled spins with strong [111] anisotropy. In this case, the ground state is in fact not degenerate, and a non-collinear Néel state is expected with all spins pointing into a tetrahedron, and all spins pointing out of the neighboring tetrahedra. Interestingly, this is in contrast to the "spin ice" ferromagnet with strong local [111] anisotropy, which does display a macroscopic degeneracy at low temperatures. Given estimates for Θ_{CW} and contributions from dipolar interactions in $Tb_2Ti_2O_7$, an ordering temperature of $T_N \sim 1$ K is expected into a non-collinear Néel state.

However, the measured magnetic structure factor $S(\mathbf{Q})$ shown in Fig. 8.6 *cannot* be accounted for by antiferromagnetically coupled spins in the presence of a strong local $\langle 111 \rangle$ anisotropy. This is explicitly shown in Fig. 8.9(a) wherein the calculation of $S(\mathbf{Q})$ including appropriate antiferromagnetic exchange, long-range dipolar interactions and strong local $\langle 111 \rangle$ anisotropy is presented. Clearly, this calculation clearly does not describe the experimental measurement on $Tb_2Ti_2O_7$ shown in Fig. 8.6.

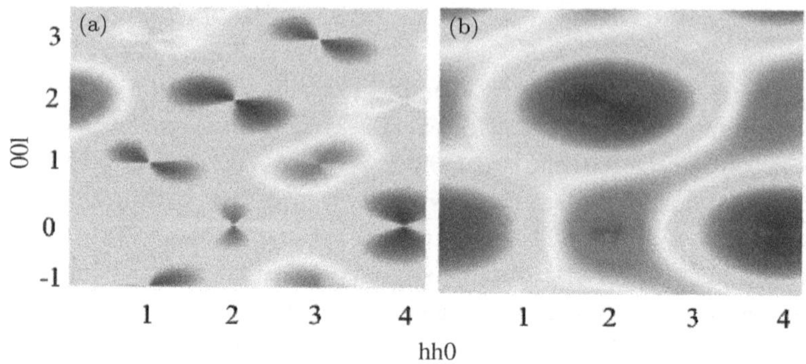

Fig. 8.9. Two theoretical models[38] of the diffuse magnetic scattering in the (h,h,l) plane of $Tb_2Ti_2O_7$. Panel (a) shows a calculation assuming extreme Ising moments, while the panel (b) incorporates the finite gap to the lowest lying magnetic excitations which tends to restore spin isotropy to the system. The calculation in panel (b) is a good descriptor of the measured $S(\mathbf{Q})$ shown in Fig. 8.6.

In fact, Gardner *et al.*[33] have shown that a remarkably simple calculation on the basis of extreme near-neighbor correlations only can reproduce most of the features of the experimentally determined $S(\mathbf{Q})$. As this calculation assumes no spin direction, it implies an isotropy to the local spin structure at odds with the local $\langle 111 \rangle$ anisotropy assumed to be relevant on the basis of the CEF measurements shown in Fig. 8.6. Yasui *et al.*[42] also attempted to model their measured $S(\mathbf{Q})$ on the basis of proposed spin correlations over two interconnected tetrahedra connected by a single, central spin (that is, over a seven-spin cluster). These results were also at odds with a picture in which the spins are restricted to points along local $\langle 111 \rangle$ directions.

However, Kao *et al.*,[38] went much further, and incorporated the finite and relatively small energy gap ($\sim$20 K) separating the ground-state doublet from the excited-state doublet in $Tb_2Ti_2O_7$ using a random phase approximation. The inclusion of the effect of this low-lying excited state restores the spin isotropy to the problem and gives rise to a calculated $S(\mathbf{Q})$, which now includes the effects of the antiferromagnetic exchange, the long-range dipolar interactions as well as the details of the CEF scheme to produce the calculated magnetic $S(\mathbf{Q})$ shown in Fig. 8.9(b). This new calculation is clearly an excellent descriptor of the measured $S(\mathbf{Q})$ and represents a significant step forward in our detailed understanding of this material. However, it still leaves the problem as to why $Tb_2Ti_2O_7$ fails to achieve long-range order near 1 K?

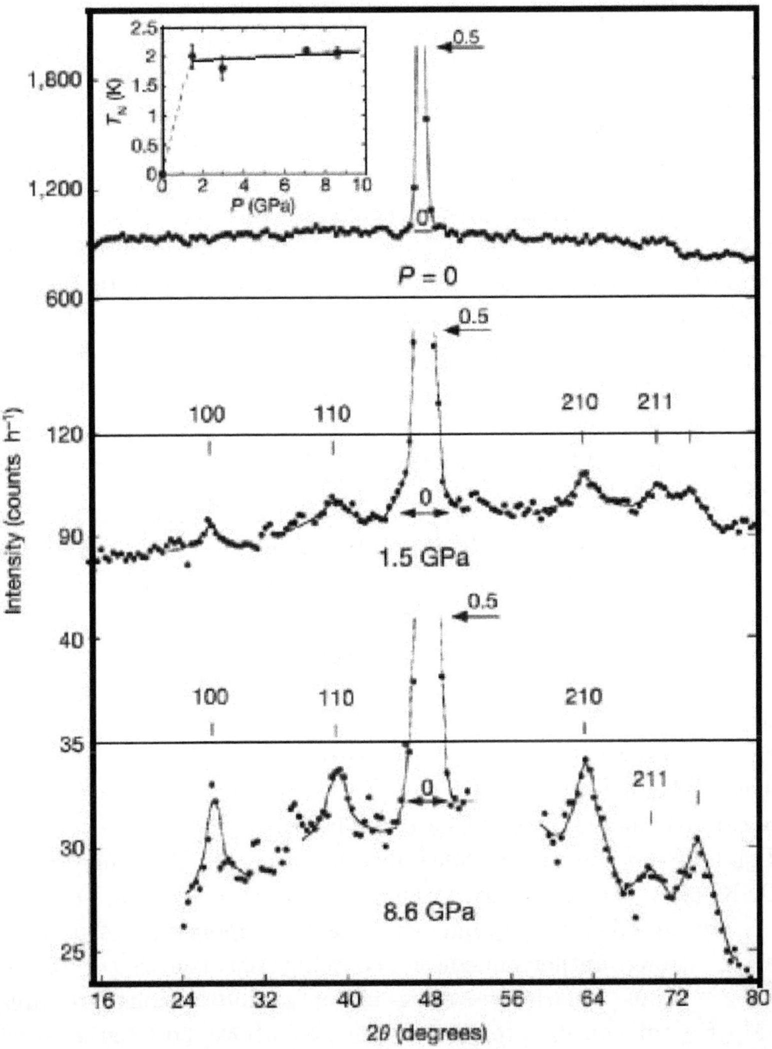

Fig. 8.10. Pressure-induced magnetic Bragg scattering from polycrystalline $Tb_2Ti_2O_7$ at <2 K. The data show the pressure evolution of static correlations in the cooperative paramagnet. The inset depicts the pressure dependence of the transition temperature.[43]

Indeed, recent measurements by Mirebeau *et al.*[43] have shown that polycrystalline $Tb_2Ti_2O_7$ does indeed achieve long-range magnetic order near 2 K, but only under hydrostatic pressure exceeding ~1 GPa. Powder neutron diffraction data as a function of pressure are shown in Fig. 8.10.

Weak magnetic Bragg peaks can be seen to arise out of a flat background at applied pressures of 1.5 GPa and above. This remarkable experiment also shows that even under the conditions of applied hydrostatic pressure, the antiferromagnetic order coexists with spin liquid-like fluctuations; that is, considerable diffuse scattering remains even in the presence of the magnetic Bragg peaks. This study indicates that $Tb_2Ti_2O_7$ is close to an instability to Néel order, as theoretical analysis suggests it should be. However, in the absence of an external perturbation it remains in an enigmatic spin liquid to very low temperatures. Clearly, future work looking at the stability of the spin liquid state to other external perturbations such as unidirectional stress and magnetic fields will be important for further progress in understanding this remarkable material.

8.5. The Spin Glass Ground State in $Y_2Mo_2O_7$

The pyrochlore $Y_2Mo_2O_7$ is a well-studied system[10, 44–51], wherein magnetic Mo^{4+} moments reside at the B-site in the pyrochlore structure.[8] As was the case with $Tb_2Ti_2O_7$, the conventional unit cell shown in Fig. 8.1(c) has dimensions such that a cube edge is ~10.2 Å. The structure of $Y_2Mo_2O_7$ has complexities, which we will discuss subsequently. However, these are relatively subtle, and our starting point in the description of $Y_2Mo_2O_7$ is of antiferromagnetically coupled Mo^{4+} moments which interact on a network of corner-sharing tetrahedra. To date, $Y_2Mo_2O_7$ has been produced only as polycrystalline material, and thus, a more limited set of measurements can be carried out than was the case in $Tb_2Ti_2O_7$.

High-temperature susceptibility measurements, in the range from 300 to 1000 K, on $Y_2Mo_2O_7$ yield a large Curie–Weiss temperature of $\Theta_{CW} \sim -200$ K and an effective paramagnetic magnetic moment of 2.55 μ_B/Mo indicating strong antiferromagnetic coupling between $S = 1$ Mo^{4+} moments.[44] Thus, relatively strong exchange interactions are present in $Y_2Mo_2O_7$, in contrast to those discussed above between rare earth Tb^{3+} moments in $Tb_2Ti_2O_7$.[9] The paramagnetic moment in $Y_2Mo_2O_7$ is close to that expected for orbitally quenched Mo^{4+} magnetic moments: $g\sqrt{S(S+1)} = 2.8\mu_B$.

As shown in of Fig.8.11(a), the DC magnetic susceptibility shows a clear break between the field-cooled (FC) and zero-field-cooled (ZFC) susceptibilities below $T_g \sim 22$ K. This difference between the magnetic susceptibilities as measured under different protocols is typical of chemically disordered spin glasses, such as CuMn, EuSrS, and CdMnTe.

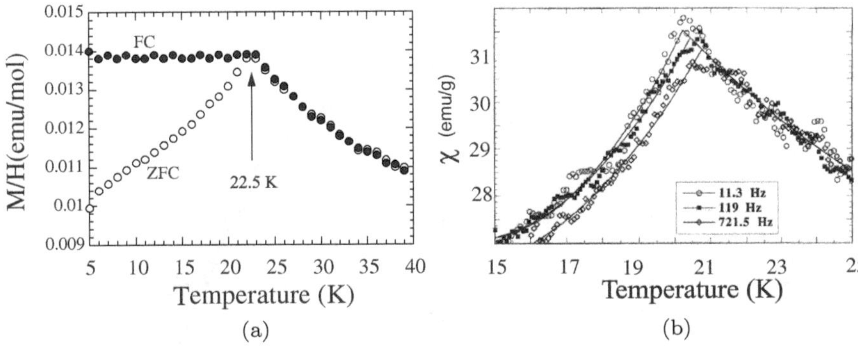

Fig. 8.11. (a) Static magnetic susceptibility of $Y_2Mo_2O_7$ revealing the history dependent signature of a spin-glass-like transition.[45] (b) The frequency dependence of the AC susceptibility is shown.[46]

Further study of the nonlinear DC-susceptibility[47] as well as AC-susceptibility measurements[46] confirm the spin glass nature of the low-temperature ground state in $Y_2Mo_2O_7$. AC-susceptibility data are shown in Fig. 8.11(b) at frequencies ranging from 11.3 to 721 Hz. These data show a clear peak whose position moves to lower temperatures with decreasing frequency, again typical of chemically disordered spin glasses. (These data indicate a slightly lower T_g, likely due to the absolute calibration of the thermometry.)

It is well appreciated that chemically disordered spin glasses display critical divergences associated with their *nonlinear* susceptibility,[52] as opposed to the divergences which characterize the susceptibility or staggered susceptibility in conventional ferromagnets and antiferromagnets undergoing critical phase transitions. Nonlinear susceptibility measurements on $Y_2Mo_2O_7$[47] shows such static critical behavior, parameterized in terms of critical exponents $\gamma \sim 2.8$, $\beta \sim 0.8$, and $\delta \sim 4.7$ and consistent with an underlying thermodynamic phase transition to the spin glass state near 22 K. These critical exponents are typical of 3D spin glasses. The scaling analysis of the nonlinear DC-susceptibility data is shown in Fig. 8.12.

Neutron scattering measurements[45] were performed on the SPINS triple-axis spectrometer located on a cold neutron guide at the NIST Center for Neutron Research. The $|Q|$ dependence of the elastic neutron scattering from the powder sample is shown in Fig. 8.13(a). These measurements employed neutrons with constant scattered energy of 5.1 meV. Figure 8.13(a) shows the difference between low-temperature elastic

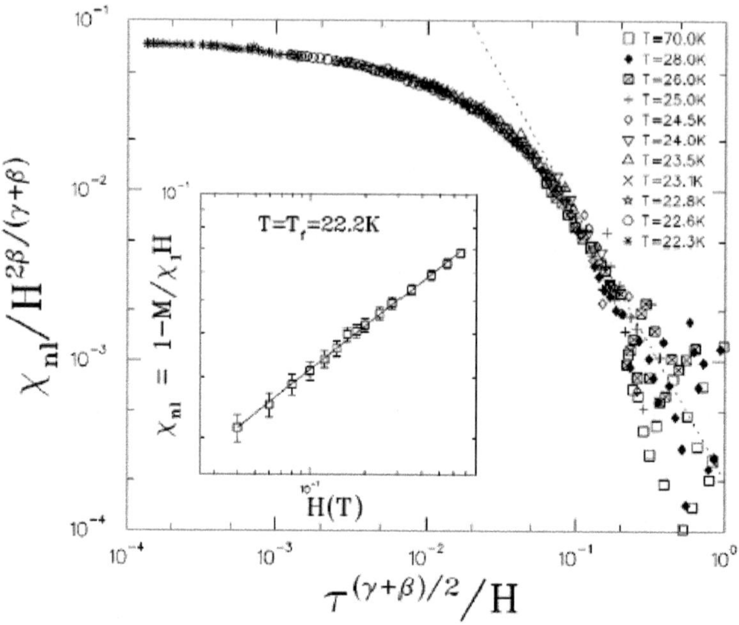

Fig. 8.12. The scaling behavior of the nonlinear DC-susceptibility of $Y_2Mo_2O_7$ is shown as a function of both magnetic field strength and temperature. The inset shows the field dependence of the nonlinear DC-susceptibility at $T_f = T_g = 22.2K$.[47]

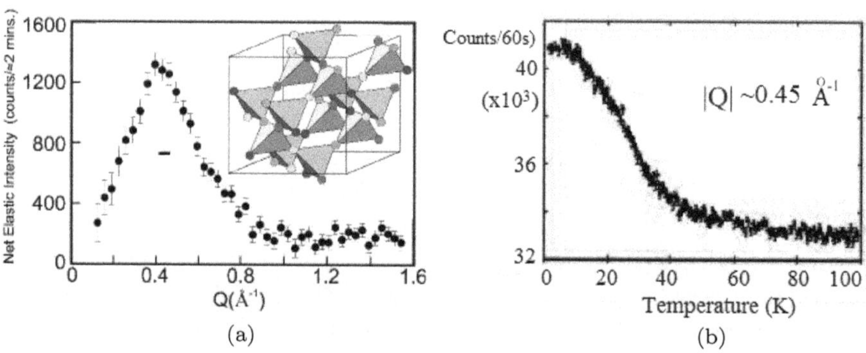

Fig. 8.13. (a) The $|Q|$ dependence of the elastic magnetic scattering from $Y_2Mo_2O_7$ at 1.5 K, obtained by subtracting a 50 K data set from a 1.5 K data set. The proposed four-sublattice spin arrangement for short-range correlations in $Y_2Mo_2O_7$, consistent with the data, is shown in the inset. (b) The temperature dependence of the peak of the elastic scattering at $|Q|$=0.45 Å^{-1} is shown.[45]

neutron scattering taken at T=1.4 K and that taken at a relatively high temperature, 50 K. One clearly observes a diffuse peak in the elastic scattering centered at $|Q| \sim 0.44$ Å^{-1}. For reference, a horizontal bar on the same plot indicates the resolution width of Bragg peaks measured using the SPINS spectrometer in this configuration. The breadth of the diffuse scattering clearly exceeds the resolution width, indicative of a finite and relatively short correlation length. The energy resolution of the nominally elastic scattering experiment sets the time scale on which the measured spin correlations are static. For the present set of measurements, this is $\tau > \frac{\hbar}{\Delta E} = 3ps$.

This diffuse scattering is quite different from that observed in $Tb_2Ti_2O_7$, where the first peak in the diffuse magnetic scattering appears at $|Q| \sim 1.2$ Å^{-1} (see Fig. 8.2). In the case of $Y_2Mo_2O_7$, the peak in the diffuse scattering appears at $Q \sim 0.44$ Å^{-2}. We understand this position of the peak in the diffuse scattering as being given by $\sim 2\pi/d(110)$, where $d(110) \sim \sqrt{2} \times 10.2$ Å is the face diagonal of a conventional cubic unit cell. Such a local periodicity permits a four-sublattice local structure along [110], which is shown schematically in the inset to Fig. 8.13(a).

The moment direction associated with each of the four sublattices cannot be known given the only information available is that associated with the peak of the diffuse magnetic scattering. However, the constraint $\sum_i \mathbf{S}_i=0$ is implied by the vanishing of the magnetic scattering as Q goes to zero. We therefore represent each of the four sublattices by a different color and enforce the $\sum_i \mathbf{S}_i=0$ constraint by requiring that each tetrahedron is color neutral.

The half width at half maximum of this diffuse magnetic scattering gives rise to a correlation length which we set as $\xi \sim 1/\text{HWHM} \sim 5$ Å. This is consistent with a "cluster glass" spin configuration below $T_g \sim 22$ K, in which locally ordered conventional unit cells of the type shown in the inset Fig. 8.13(a) are incoherently correlated with each other.

The fraction of the total magnetic moment which participates in this spin glass state can be estimated from this neutron scattering experiment. The magnitude of the frozen, staggered magnetization is derived by integrating the elastic magnetic scattering data and comparing it to nuclear scattering whose intensity can be calculated. This analysis produces the estimate that 2/3 of the Mo^{4+} moment is frozen at low temperatures. This is an unusually small fraction of the total spin population participating in the ordered state of a 3D magnet in the limit $T/\Theta_{\text{CW}} \to 0$.

The temperature dependence of the elastic scattering at $|Q| \sim 0.44$ Å^{-1} is shown in Fig. 8.13(b). Interestingly, the temperature dependence falls off as expected as the temperature is raised; however, T_g is given roughly by the inflection point on this curve, not by where the intensity becomes temperature independent. This approximately temperature-independent regime only occurs for temperatures greater than ~ 50 K, more than twice T_g.

Inelastic neutron scattering measurements were also performed at the maximum in the diffuse scattering, $|Q| \sim 0.44$ Å^{-1}. These data are shown in both a color contour map in Fig. 8.14(a), where the intensity scale is linear, and in Fig. 8.14(b), where a subset of the same data is plotted on a semilog

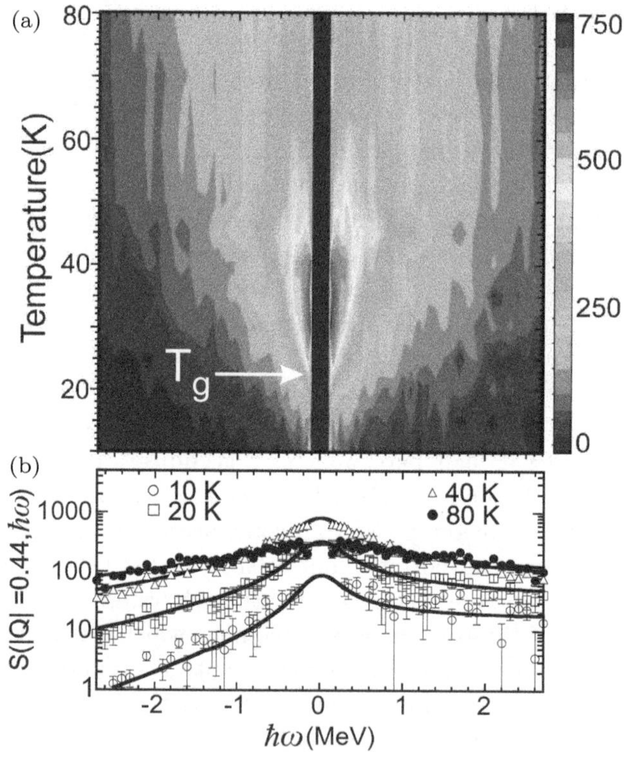

Fig. 8.14. The normalized, background-subtracted inelastic scattering data for $S(Q,\omega)$ from $Y_2Mo_2O_7$. The contour plot is produced from 12 data sets between 10 and 80 K similar to those shown in panel (b). Panel (b) shows the selected data sets from those in panel (a), but on a semilog scale and plotted with fits to the data which parameterize the inelastic scattering and which are described in detail in the text.[45]

scale. The color contour map shows clearly that the inelastic scattering collapses into the elastic position at $T_g \sim 22$ K, indicating that the spins have frozen on the time scale of the neutron measurements, even though the spatial correlations correspond to a correlation length of only ~5 Å. What is also striking about the color contour map in Fig. 8.14(a), is that strong temporal correlations begin as high as 50 K $\sim 2 \times T_g$, which would represent a very wide "critical" regime for conventional critical phenomena near continuous phase transitions. This is similar to what was concluded on the basis of the elastic magnetic scattering alone (Fig.8.13(b)).

It is straightforward to integrate the inelastic scattering, thus measuring $\int_{-3\text{meV}}^{3\text{meV}} S(Q = 0.44\ \text{Å}, \hbar\omega)\, d\omega$ in $Y_2Mo_2O_7$, and this is shown in Fig. 8.15(a). This integral excludes the elastic scattering, and it is pleasing to see that its temperature dependence assumes the complement of the nominally elastic magnetic scattering in $Y_2Mo_2O_7$ shown in Fig. 8.13. Again in this case, T_g is given by the inflection point on the temperature dependence of this inelastic scattering.

This energy dependence of the inelastic scattering could be fit to a simple form of $S(Q = 0.44\ \text{Å}^{-1}, \hbar\omega)$ and consequently a simple form of the

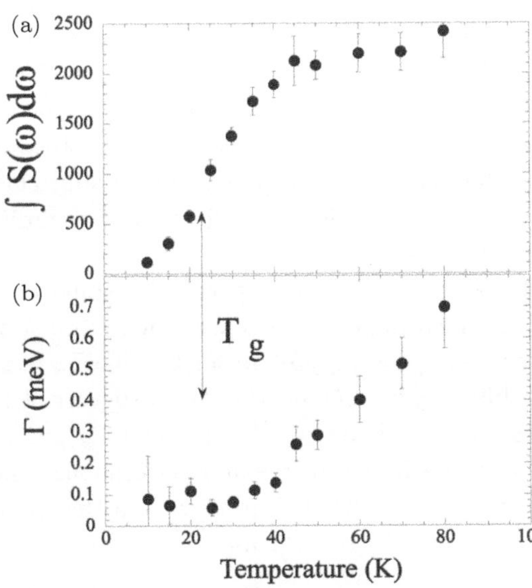

Fig. 8.15. Results from the fits shown in Fig. 8.14(b) to the inelastic data as described in the text. Panel (a) shows the integrated inelastic scattering as a function of temperature. Panel (b) shows the temperature dependence of the relaxation rate.[45]

imaginary part of the dynamic susceptibility, $\chi''(\hbar\omega)$,

$$S(Q = 0.44\mathring{A}, \hbar\omega) = \frac{1}{\pi}\chi''(\hbar\omega)[1 + n(\hbar\omega)] \tag{8.5}$$

with

$$\chi''(\hbar\omega) = \chi_o \arctan\left(\frac{\hbar\omega}{\Gamma}\right). \tag{8.6}$$

The fits shown in Fig. 8.14(b) incorporate an appropriate instrumental resolution-convolution of this theoretical form for $S(Q{=}0.44\ \text{Å}^{-1}, \hbar\omega)$ to the inelastic scattering data. The description is clearly very good, and this allows us to extract the measured relaxation rate, Γ, from the data. This is shown in Fig. 8.15(b) as a function of temperature. The spin freezing transition at $T_g \sim$ 22 K corresponds to the point at which the relaxation rate goes to zero.

The spin glass ground state in $Y_2Mo_2O_7$ is certainly well characterized but the question as to what drives the formation of such a state remains not fully addressed. In fact, many geometrically frustrated antiferromagnets possess low-temperature states which are spin glass-like, or which possess substantial glassy characteristics. This is true, for example, in the case of the well-studied *kagomé*-like (or "pyrochlore-slab") material $SrCr_{9p}Ga_{12-9p}O_{19}$,[53] as well as for the site-random pyrochlore $CsNiCrF_6$,[54] where Ni and Cr ions randomly populate a single pyrochlore sublattice. However, in these latter two examples, there is rather obvious chemical disorder present, and thus one is not surprised that a spin glass state occurs at low temperatures, as frustration and disorder are traditional building blocks for a spin glass state.

Of course, all real materials are imperfect at some level. In the case of $Y_2Mo_2O_7$, the chemical imperfections are much more subtle than those discussed earlier for $SrCr_{9p}Ga_{12-9p}O_{19}$ or for $CsNiCrF_6$. Standard neutron and X-ray powder diffraction refinements of the structure of $Y_2Mo_2O_7$ show a well-ordered crystallographic lattice,[48] with the only potential indication of inhomogeneity residing in a root mean square displacement parameter for the Mo^{4+} site, which is about 40% greater than that describing atomic displacements in other pyrochlores which are believed to be chemically well ordered.

While it is not easy to improve the crystalline perfection in $Y_2Mo_2O_7$, one can systematically study material containing more disorder. Interestingly, the intentional introduction of disorder to $Y_2Mo_2O_7$ in the form of

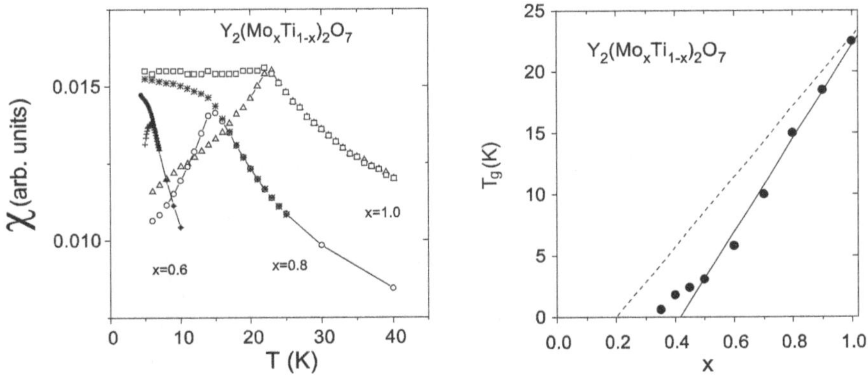

Fig. 8.16. Static susceptibility of $Y_2(Mo,Ti)_2O_7$. It is clear that when the magnetic sublattice is diluted, the freezing temperature, T_g, is reduced.[49]

substituting diamagnetic Ti^{4+} for magnetic Mo^{4+} does not enhance the spin glass state. Rather, it causes the spin glass state to behave as if it were any other ordered magnetic state.[49] As can be seen in Fig. 8.16, T_g drops to zero in an almost linear fashion, near 60% substitution. This percolation threshold for the spin glass state is very close to that expected for site dilution on the 3D pyrochlore lattice, and thus, somewhat counterintuitively, the spin glass state behaves as any other ordered state would behave on dilution.

More sophisticated structural analysis does report evidence for chemical inhomogeneity in $Y_2Mo_2O_7$. This evidence is in the form of both XAFS[50] analysis of Mo^{4+}–Mo^{4+} pair distances, as well as the ^{89}Y NMR[51] spectrum which provides evidence for discrete lattice distortions. The ^{89}Y NMR spectrum from $Y_2Mo_2O_7$ is shown at two temperatures, both well above T_g, in Fig. 8.17. Multiple peaks as a function of frequency are evident, and this work speculates that these are due to subtle lattice distortions which may relieve the magnetic frustration, thereby lowering the overall free energy in $Y_2Mo_2O_7$.

As was previously mentioned, recent theoretical analysis[23] has also investigated the role of magnetoelastic coupling in generating "order by disorder" in pyrochlore antiferromagnets. While not necessarily relevant to the analysis of $Y_2Mo_2O_7$, structural phase transitions are known to occur within the cubic pyrochlores which lower the translation symmetry of the lattice. One such well-studied example is the superconducting pyrochlore $Cd_2Re_2O_7$[55] which transforms to a tetragonal subgroup of

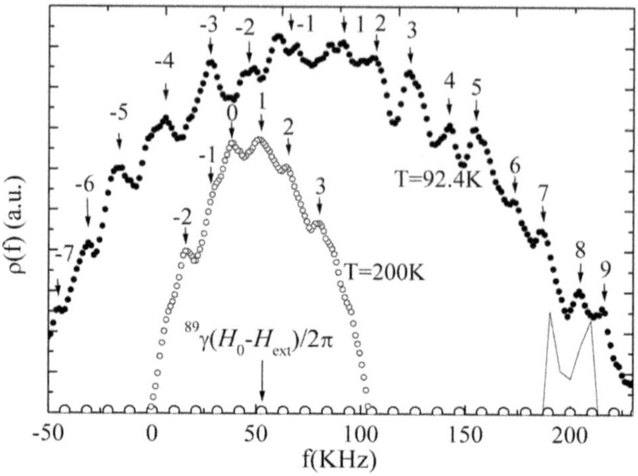

Fig. 8.17. ^{89}Y-NMR data from $Y_2Mo_2O_7$ at 200 and 92 K are shown. The very detailed structure as a function of frequency on top of the usual Gaussian peak is interpreted in terms of local distortions of the Mo^{4+} sublattice.[51]

the $Fd\bar{3}m$ pyrochlore lattice. This material is apparently non-magnetic, but the point remains that lattice and magnetic degrees of freedom are coupled.

8.6. Composite Spin Degrees of Freedom and Spin-Peierls-Like Ground State in the Frustrated Spinel $ZnCr_2O_4$

As mentioned in Section 8.2, the pyrochlore lattice of corner-sharing tetrahedra is also found on the B-sublattice of the cubic spinels with the chemical composition AB_2O_4. In insulators of this form, it is common for the B-site to be occupied by a trivalent transition metal ion, which opens up the possibility of "another flavor" of antiferromagnetism on the pyrochlore lattice. Perhaps the best investigated example of such a compound is $ZnCr_2O_4$.[18] Lee *et al.*[56] have extensively studied this material using neutron scattering techniques, again using the SPINS cold triple-axis spectrometer at NIST. Their results reveal yet another ground state, distinct from either the cooperative paramagnet which $Tb_2Ti_2O_7$ displays or the spin glass ground state which $Y_2Mo_2O_7$ displays as a consequence of geometrical frustration. In the case of the Cr^{3+} magnetic moments in $ZnCr_2O_4$, the ground state is based on nanoscale, hexagonal rings of correlated spins. Lee and coworkers showed that this frustrated spinel undergoes a strongly first-order phase transition at 12.5 K from a cubic spin liquid with a gapless spin excitation spectrum, to a tetragonal Néel state with a local spin resonance at finite energy: 4.5 meV. Lee *et al.* make the reasonable argument that such a phase transition, with coupled spin and lattice degrees of freedom, possesses striking similarities to the spin-Peierls phase transition known to be relevant to quasi-1D, quantum antiferromagnets.

As was the case in the study of $Tb_2Ti_2O_7$, initial measurements on $ZnCr_2O_4$ were performed on polycrystalline materials. Later, when single-crystal samples were available, a more sophisticated analysis was allowed. The initial measurements on polycrystalline materials are surveyed in Fig. 8.18. Panel (b) shows the temperature dependence of the magnetic Bragg scattering on entering the Néel state in $ZnCr_2O_4$ as well as the accompanying lattice strain as the crystal simultaneously distorts from its cubic high-temperature structure to the tetragonal low-temperature structure. The abrupt temperature dependence associated with both the magnetic and

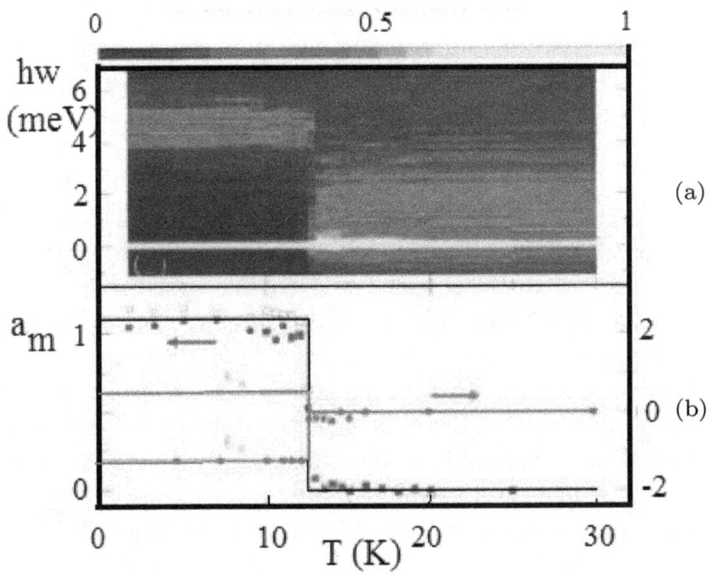

Fig. 8.18. (a) A colour image of constant-Q scans measured on a polycrystalline sample of $ZnCr_2O_4$ using a triple-axis spectrometer as a function of temperature.[56] (b) The temperature dependence of the magnetic Bragg scattering (squares) and the lattice strain along a and c (circles) show abrupt temperature dependencies characteristic of a strongly discontinuous phase transition.

structural phase transitions are as expected for a strongly discontinuous or first-order phase transition.

Figure 8.18(a) shows the inelastic neutron scattering from polycrystalline $ZnCr_2O_4$ at a modulus of wave vector, Q = 1.5 Å^{-1}. Strong quasielastic magnetic neutron scattering is observed near this wave vector above T_N. On lowering the temperature below ~12.5 K, this magnetic scattering becomes localized into an inelastic resonance. Figure 8.18(a) clearly shows that the inelastic magnetic resonance itself develops abruptly, at 4.5 meV upon cooling through the first-order phase transition.

This powder-averaged magnetic inelastic scattering can be quantitatively analyzed and Lee *et al.* showed that the spectral weight of the resonance accounts for roughly 22% of the total fluctuating moment in Cr. Very interestingly, they also showed that the inelastic scattering which was centered at Q =1.5 Å^{-1} was broad in wave vector space, characterized by a half width at half maximum of $\kappa = 0.48$ Å^{-1}. This indicates the presence of a resonating spin cluster, rather than individual spins or single tetrahedra, as the origin for the inelastic magnetic resonance.

This study also pointed out very interesting analogies between the phase transition observed in $ZnCr_2O_4$ and that characterizing the spin-Peierls transition in quasi-1D quantum antiferromagnets such as $CuGeO_3$.[58] There are differences between the phenomena in $ZnCr_2O_4$ and that of spin-Peierls systems; notably the order of the transition and the nature of the low-temperature ground state. However, the similarities related to the coupling between lattice and spin degrees of freedom such that a finite energy magnetic resonance appears below the phase transition and with it a concomitant lowering of the magnetic energy of the system. The analogous phenomena in a spin-Peierls system occurs through a lattice dimerization and the development of a magnetic singlet–triplet energy gap.

A subsequent neutron scattering study on single–crystal $ZnCr_2O_4$ revealed substantially more detail as to the nature of the correlated spin clusters. Lee *et al.*[57] were able to determine the magnetic dynamic form factor within two planes of reciprocal space, work which required single crystals, and which is shown in Figs. 8.19 (a) and (b). These data show the **Q** dependence of the quasielastic scattering at $\hbar\omega = 1$ meV and $T = 15$ K, above T_N, in the (h,k,0) and (h,k,k) planes of reciprocal space. One sees broad maxima at the Brillouin zone boundaries, signaling the emergence of confined nanoscale spin clusters. Unlike the scattering due to single correlated tetrahedra observed in the pyrochlore $Tb_2Ti_2O_7$,[33] the scattering

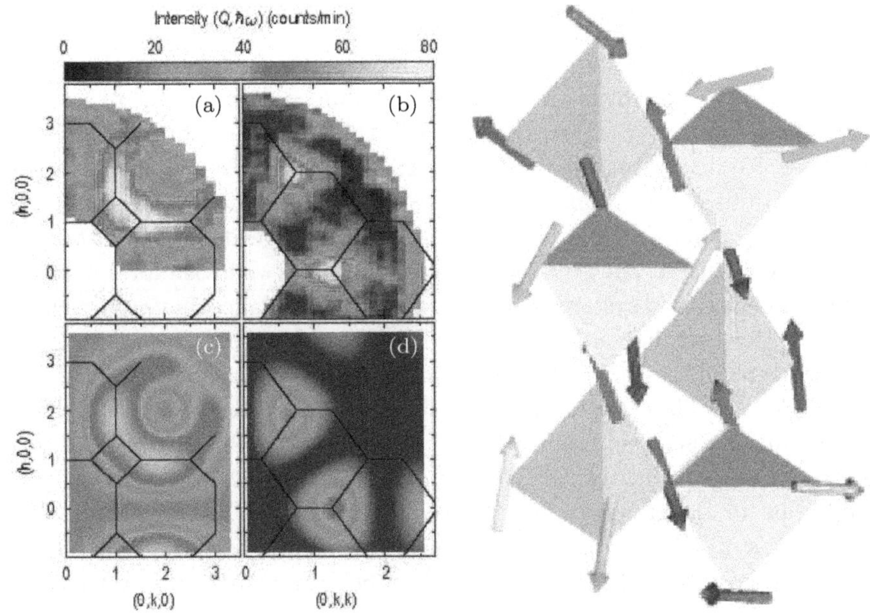

(A) (B)

Fig. 8.19. (A): (a) and (b) Color maps of inelastic scattering at $\hbar\omega = 1$ meV for two high-symmetry planes in a single crystal of $ZnCr_2O_4$ at 15 K. (c) and (d) The calculated form factor for correlated antiferromagnetic hexagons similar to those shown schematically in (B). (B): One hexagon and the associated six tetrahedra are used as the basis for the diffuse scattering calculations on which the form factors are shown on the left-hand side (c) and (d).[57]

shown in Fig. 8.19 requires larger correlated units to account for the details of the **Q**-dependence of this inelastic scattering.

It was found that the scattering was well described by antiferromagnetic hexagonal spin-loops, made up of six tetrahedra where two spins from each tetrahedron occupy the vertices of the loop. The remaining two spins of a given tetrahedron belong to a different hexagon, thereby allowing the assignment of all the spins on the magnetic sublattice to hexagons simultaneously and producing $N/6$ weakly interacting degrees of freedom, from the N original spins. The calculated dynamic form factor for such a correlated nanoscale spin structure is shown in Figures 8.19 (c) and (d) for each of the two planes studied. In this manner, $ZnCr_2O_4$ is convincingly shown at low temperatures to be composed of weakly interacting hexagonal spin-loops.

8.7. Conclusions and Outlook

We have reviewed experimental progress in understanding three of the many magnetic materials which are known to combine antiferromagnetism and tetrahedral local geometries. Even within this subset of three frustrated materials, we find quite different exotic ground states: the enigmatic cooperative paramagnetic ground state of $Tb_2Ti_2O_7$; the spin glass ground state of $Y_2Mo_2O_7$; and the composite spin degrees of freedom and distorted pyrochlore lattice in $ZnCr_2O_4$. Esthetically, we may well wish for a form of universal behavior in systems which, at least at first glance, have so much in common. However, it is also very rewarding to appreciate the richness of the complex ground states which geometrical frustration has enabled in these real materials. There seems little doubt that this richness will continue as a theme in the elucidation of the physical properties of geometrically frustrated magnets in the near future. There is also little doubt that the production of single-crystal materials in pristine form, as well as access to the most advanced instrumentation for neutron scattering and other advanced characterization will continue to be extremely important in our ability to study and ultimately understand such novel materials.

Acknowledgments

This work was supported at McMaster University by NSERC of Canada and by the Canadian Institute for Advanced Research. It was supported at Brookhaven National Laboratory by the U.S. Department of Energy under contract DE-AC02-98CH10886. It is a pleasure to thank our many colleagues and collaborators who have contributed to our involvement in this work. These include D.P. Belanger, A.J. Berlinsky, C.L. Broholm, W. Buyers, M.F. Collins, G. Ehlers, M.J.P. Gingras, I.N. Goncharenko, J.E. Greedan, A. Harrison, C. Kallin, H. Kawamura, A. Keren, S.-H. Lee, G.M. Luke, T.E. Mason, M. Mao, M. Mekata, I. Mirebeau, R. Moessner, N. Raju, A.P. Ramirez, J.N. Reimers, P. Schiffer, C.V. Stager, J.L. Thollence and Y.J. Uemura. Finally, we are indebted to A.A.M. Gaulin and K.M. Hughes for a critical reading of the manuscript.

References

1. H. T. Diep (ed.), *Magnetic Systems with Competing Interactions* (World Scientific, Singapore, 1994); P. Shiffer and A. P. Ramirez, *Commun. Cond. Mat. Phys.* **18**, 21 (1996); M. J. P. Gingras (eds.), *Can. J. Phys.* **79**, (2001).
2. Z. F. Wang and B. W. Southern, *Phys. Rev. B* **68**, 094419 (2003).

3. G. H. Wannier, *Phys. Rev.* **79**, 357 (1950).
4. P. W. Anderson, *Phys. Rev.* **102**, 1008 (1956).
5. J. Villain, *Z. Phys. B* **33**, 31 (1979).
6. L. Pauling, *J. Am. Chem. Soc.* **57**, 2680 (1935).
7. S. T. Bramwell and M. J. P. Gingras, *Science* **294**, 1495 (2001).
8. M. A. Subramanian, G. Aravamudan, and G. V. Subba Rao, *Prog. Solid State Chem.* **15**, 55 (1983).
9. J. S. Gardner, S. R. Dunsiger, B. D. Gaulin, M. J. P. Gingras, J. E. Greedan, R. F. Kiefl, M. D. Lumsden, W. A. MacFarlane, N. P. Raju, J. E. Sonier, I. Swainson, and Z. Tun, *Phys. Rev. Lett.* **82**, 1012 (1999).
10. J. E. Greedan, M. Sato, Y. Xu, and F. S. Razavi, *Solid State Comm.* **59**, 895 (1986).
11. J. E. Greedan, J. N. Reimers, S. L. Penny, and C. V. Stager, *J. App. Phys.* **67**, 5967 (1990).
12. D. Goldschmidt and H. L. Tuller, *Phys. Rev. B* **34**, 5558 (1986).
13. H. Sakai, K. Yoshimura, H. Ohno, H. Kato, S. Kambe, R. E. Walstedt, T. D. Matsuda, and Y. Haga, *J. Phys.: Cond. Matt.* **13**, L785 (2001).
14. D. Mandrus, J. R. Thompson, R. Gaal, L. Forro, J. C. Bryan, B. C. Chakoumakos, L. M. Woods, B. C. Sales, R. S. Fishman, and V. Keppens, *Phys. Rev. B* **63**, 195104 (2001).
15. T. Katsufuji, H. Y. Hwang, and S. W. Cheong, *Phys. Rev. Lett.* **84**, 1998 (2000).
16. S. Yonezawa, Y. Muraoka, Y. Matsushita, and Z. Hiroi, *J. Phys.: Cond. Matt.* **16**, L9 (2004).
17. S. Yonezawa, Y. Muraoka, Y. Matsushita, and Z. Hiroi, *J. Phys. Soc. Jpn.* **73**, 819 (2004).
18. F. Leccabue, C. Pelosi, E. Agostinelli, V. Fares, D. Fiorani, and E. Paparazzo, *J. Crystal Growth* **79**, 410 (1986).
19. S. Kondo, D. C. Johnston, C. A. Swenson, F. Borsa, A. V. Mahajan, L. L. Miller, T. Gu, A. I. Goldman, M. B. Maple, D. A. Gajewski, E. J. Freeman, N. R. Dilley, R. P. Dickey, J. Merrin, K. Kojima, G. M. Luke, Y. J. Uemura, O. Chmaissem, and J. D. Jorgensen, *Phys. Rev. Lett.* **78**, 3729 (1997).
20. S. H. Lee, Y. Qiu, C. Broholm, Y. Ueda, and J. J. Rush, *Phys. Rev. Lett.* **86**, 5554 (2001).
21. R. L. Millard, R. C. Peterson, and B. K. Hunter, *Am. Mineral* **80**, 885 (1995).
22. S. H. Lee, C. Broholm, T. H. Kim, W. Ratcliff, and S. W. Cheong, *Phys. Rev. Lett.* **84**, 3718 (2000).
23. O. Tchernyshyov, R. Moessner, and S. L. Sondhi, *Phys. Rev. B.* **66**, 064403 (2002).
24. R. Cywinski, S. H. Kilcoyne, and C.A. Scott, *J. Phys: Cond. Matt.* **3**, 6473 (1991).
25. R. Ballou, E. LelievreBerna, and B. Fåk, *Phys. Rev. Lett.* **76**, 2125 (1996).
26. G. M. Kalvius, E. Schreier, M. Ekstrom, O. Hartmann, S. Henneberger, A. Kratzer, R. Wappling, E. Martin, F.J. Burghart, R. Ballou, J. Deportes, and C. Niedermayer, *Hyp. Int.* **128**, 275 (2000).

27. L. Price and K. Skold (eds.), *Methods of Experimental Physics, Neutron Scattering*, Vol. 23, Parts A,B,C,D (Academic Press, New York, 1987) and G. Shirane, S. M. Shapiro, and J. M. Tranquada *Neutron Scattering with a Triple Axis Spectrometer* (Cambridge University Press, London, 2002).
28. S. W. Han, J. S. Gardner, and C. H. Booth, *Phys. Rev. B* **69**, 024416 (2004).
29. R. M. Moon, T. Riste, and W. C. Koehler, *Phys. Rev.* **181**, 920 (1969).
30. E. F. Bertaut and P. Burlet, *Solid State Commun.* **5**, 279 (1967).
31. B. D. Gaulin, J. N. Reimers, T. E. Mason, J. E. Greedan, and Z. Tun, *Phys. Rev. Lett.* **69**, 3244 (1992).
32. M. J. Harris, M. P. Zinkin, Z. Tun, B. M. Wanklyn, and I. P. Swainson, *Phys. Rev. Lett.* **73**, 189 (1994).
33. J. S. Gardner, B. D. Gaulin, A. J. Berlinsky, P. Waldron, S. R. Dunsiger, N. P. Raju, and J. E. Greedan, *Phys. Rev. B* **64**, 224416 (2001).
34. H. R. Glyde and E. C. Svensson, *Methods of Experimental Physics*, Vol. 23 (Academic Press, San Diego, 1987), p. 303.
35. M. J. P. Gingras, B. C. den Hertog, M. Faucher, J. S. Gardner, S. R. Dunsiger, L. J. Chang, B. D. Gaulin, N. P. Raju, and J. E. Greedan, *Phys. Rev. B* **62**, 6496 (2000).
36. J. S. Gardner, B. D. Gaulin, and D. M. K. Paul, *J. Crystal Growth* **191**, 740 (1998).
37. J. S. Gardner, A. Keren, G. Ehlers, C. Stock, E. Segal, J. M. Roper, B. Fåk, P. R. Hammar, M. B. Stone, D. H. Reich, and B. D. Gaulin, *Phys. Rev. B* **68**, 134410 (2003).
38. Y. J. Kao, M. Enjalran, A. Del Maestro, H. R. Molavian, and M. J. P. Gingras, *Phys. Rev. B* **68**, 172407 (2003).
39. J. S. Gardner, G. Ehlers, S. T. Bramwell, and B. D. Gaulin, *J. Phys: Condens. Matter* **16**, S643 (2004).
40. A. Keren, J. S. Gardner, G. Ehlers, A. Fukaya, E. Segal, and Y. J. Uemura, *Phys. Rev. Lett.* **92**, 107204 (2004).
41. G. Luo, S. T. Hess, and L. R. Corruccini, *Phys. Lett. A* **291**, 306 (2001).
42. Y. Yasui, M. Kanada, M. Ito, H. Harashina, M. Sato, H. Okumura, K. Kakurai, and H. J. Kadowaki, *Phys. Soc. Jpn.* **71**, 599 (2002).
43. I. Mirebeau, I. N. Goncharenko, P. Cadavez-Peres, S. T. Bramwell, M. J. P. Gingras, and J. S. Gardner, *Nature* **420**, 54 (2002).
44. N. P. Raju, E. Gmelin, and R. K. Kremer, *Phys. Rev. B.* **46**, 5405 (1992) and D. Walton, A. McCleary, C. V. Stager, and N. P. Raju, *Phys. Rev. B* **59**, 135 (1999).
45. J. S. Gardner, B. D. Gaulin, S.-H. Lee, C. Broholm, N. P. Raju, and J. E. Greedan, *Phys. Rev. Lett.* **83**, 211 (1999).
46. J. L. Thollence and B.D. Gaulin, unpublished.
47. M. J. P. Gingras, C. V. Stager, N. P. Raju, B. D. Gaulin, and J. E. Greedan, *Phys. Rev. Lett.* **78**, 947 (1997).
48. J. N. Reimers, J. E. Greedan, and M. Sato, *J. Solid State Chem.* **72**, 390 (1988).
49. N. P. Raju *et al.*, to be published.

50. C. H. Booth, J. S. Gardner, G. H. Kwei, R. H. Heffner, F. Bridges, and M. A. Subramanian, *Phys. Rev. B* **62**, R755 (2000).
51. A. Keren and J. S. Gardner, *Phys. Rev. Lett.* **87**, 177201 (2001).
52. K. Binder and A. P. Young, *Rev. Miod. Phys.* **58**, 801 (1986); J. A. Mydosh, *Spin Glasses: An Experimental Introduction* (Taylor and Francis, Oxford, 1993).
53. X. Obradors, A. Labarta, A. Isalgué, J. Tejada, J. Rodriguez, and M. Pernet, *Solid State Commun.* **65**, 189 (1988).
54. M. J. Harris and M. P. Zinkin, *Mod. Phys. Lett. B* **10**, 417 (1996).
55. J. P. Castellan, B. D. Gaulin, J. van Duijn, M. J. Lewis, M. D. Lumsden, R. Jin, J. He, S. E. Nagler, and D. Mandrus, *Phys. Rev. B* **66**, 134528 (2002).
56. S.-H. Lee, C. Broholm, T. H. Kim, W. Radcliff II, and S.-W. Cheong, *Phys. Rev. Lett.* **84**, 3718 (2000).
57. S.-H. Lee, C. Broholm, W. Radcliff II, G. Gasparovic, Q. Huang, T. H. Kim, and S.-W. Cheong, *Nature* **418**, 856 (2002).
58. K. Uchinokura, *J. Phys.: Condens. Matter* **14**, R195 (2002).

CHAPTER 9

RECENT PROGRESS IN SPIN GLASSES

N. Kawashima[*,‡] and H. Rieger[†,§]

[*] *Department of Physics, Tokyo Metropolitan University, Hachioji, Tokyo 192-0397, Japan*
[†] *Theoretische Physik, Universität des Saarlandes, 66041 Saarbrücken, Germany*
[‡] *nao@phys.metro-u.ac.jp*
[§] *H.Rieger@mx.uni-saarland.de*

We review recent findings on spin glass models. Both the equilibrium properties and the dynamic properties are covered. We focus on progress in theoretical, in particular numerical, studies, while its relationship to real magnetic materials is also mentioned.

The motivation that pulls researchers toward spin glasses is possibly not the potential future use of spin glass materials in "practical" applications. It is rather the expectation that there must be something very fundamental in systems with randomness and frustration. It seems that this expectation has been acquiring a firmer ground as the difficulty of the problem is appreciated more clearly. For example, a close relationship has been established between spin glass problems and a class of optimization problems known to be NP-hard. It is now widely accepted that a good (i.e. polynomial) computational algorithm for solving any one of NP-hard problems most probably does not exist. Therefore, it is quite natural to expect that the Ising spin glass problem, as one of the problem in the class, would be very hard to solve, which indeed turns out to be the case in a vast number of numerical studies. Nonetheless, it is often the case that only numerical studies can decide whether a certain hypothetical picture applies to a given instance. Consequently, the major part of what we present in this chapter is inevitably a description of the current status of spin glass studies, what is being said and done on the issue, rather than an account of established facts. In what follows, we put an emphasis on the theoretical and the numerical progress achieved in the field in the last decade. We mention older theoretical and experimental works only when it is necessary for readers to be able to follow the current topics without reading through many other articles. Many of the important older works and experimental results, therefore, have been left out. For a review on these, the readers are referred to Refs. [1–5].

This chapter is organized as follows. In Section 9.1, we first give a brief overview of two well-known paradigms on spin glasses and summarize the predictions derived from them. Then, in the subsequent Sections 9.2, 9.3, and 9.4, we discuss equilibrium properties of Ising spin glass models. In particular, in Section 9.3, a recent active debate, as to which paradigm is appropriate for realistic short-range spin glass models in three dimensions, is presented. Then, we proceed with the dynamical properties in Section 9.5 to examine these paradigms on a different ground. In this section, an emphasis is put on an aging phenomena while some other non-equilibrium properties are also discussed. Models with continuous degrees of freedom as well as the Potts spin glass models are mentioned in Section 9.6. The effects of a weak disorder discussed in Section 9.7 provide us with a set of interesting issues quite different from those arising from a strong disorder dealt with in the preceding sections. Several exact relations can be found and play an important role in shedding light on the issue under debate. Finally, we see in Section 9.8 what results from the interplay between quantum fluctuations and randomness in spin glasses.

9.1. Two Pictures

The equilibrium properties of Ising spin glass models in finite dimensions have been investigated mainly through developments of working hypotheses and numerical techniques. In principle, we can judge which working hypothesis is correct by numerical computations. The working hypotheses are often called "pictures". Among various pictures, unarguably the most frequently mentioned are the replica-symmetry-breaking (RSB) picture and the droplet picture. In fact, the dichotomy between these two has been the central issue in the spin glass research for more than a decade. However, it is rather difficult to perform a conclusive numerical test in the 3D case. One thing that renders the test so difficult is that the lower critical dimension appears to be close to three. Another factor is, as mentioned in the introduction, that there is no polynomial algorithm for obtaining the ground state of a given instance in three (and higher) dimension. Therefore, the best we can do at present is to summarize on-going arguments, discriminating what has been established from what has not, and presenting the most important numerical evidences. We try to do these in what follows.

It should also be pointed out that most of the "pictures" are asymptotic theories that are supposed to be correct only for large systems. In contrast to most regular systems, the utility of an asymptotic theory for glassy systems cannot be taken for granted. It was pointed out[6] that in most real spin glass materials, the typical length scale of an equilibrated domain within the experimental time scale may not exceed a small number of lattice spacings. If this is the case, considering the numerical evidence of the existence of large corrections to scaling, it can happen that the asymptotically correct theory fails to explain experimental results. The fact that many experiments on spin glasses, those at temperatures below the transition in particular, are not performed in equilibrium implies that an equilibrium theory may not be appropriate. Some attempts to construct a theory for a realistic time and length scales are reviewed in Section 9.5.

9.1.1. *Mean-field picture*

We start with a brief description of the two pictures mentioned above in this section and in Section 9.1.2. We do not, however, intend to repeat the whole history of the development of each picture or discuss every technical detail. The reader may find a complete description of these pictures in a number of review articles.[1,2,7–10]

The RSB picture or the mean-field picture is based on Parisi's solution[11–14] of the Sherrington–Kirkpatric (SK) model[15] and its interpretation in terms of a multitude of thermodynamic states.[16,17] The SK model of N Ising spins, S_i, is defined by the Hamiltonian

$$H = -\sum_{(i,j)} J_{ij} S_i S_j,$$

where the summation is taken over all $N(N-1)/2$ pairs of spins (i, j) with $i \neq j$. The coupling constants J_{ij} are quenched Gaussian random variables with zero mean and variance J/N, where J is a positive constant of $\mathcal{O}(1)$

that sets the energy scale. In Ref. [15], the replica method was used in which n identical copies (replicas) of the system are introduced to perform the disorder average of the free energy: $-\beta F = [\ln Z]_{\rm av} = \lim_{n\to 0}([Z^n]_{\rm av} - 1)/n$. They found a solution of the mean-field equations using a replica symmetric ansätze (see below). This solution has a phase transition and the transition point is located at the temperature $T_{\rm c} = J$. Below this temperature, the order parameter

$$q_{\rm EA} = q^{\alpha\beta} \equiv \frac{1}{N} \sum_i S_i^\alpha S_i^\beta, \tag{9.1}$$

has a non-zero value, where indices α and β specify replicas. The replica symmetric Ansatz means that only solutions to the mean-field equations with $q^{\alpha\beta}$ being independent of α and β (before taking the limit $n \to 0$) are allowed. However, this solution did not describe the low-temperature phase correctly since the entropy predicted by this solution became negative at low temperatures.

In Ref. [18], a different approach was taken for the SK model. They constructed a set of equations (called the TAP equations) that express the equation of state for each bond realization of the SK model in terms of the local magnetization, $m_i = \langle S_i \rangle$. They studied the eigen-modes of the stability matrix of stable solutions of the TAP equations and found that the spectrum of the eigen-values extends down to zero, suggesting that the solutions are only marginally stable in the thermodynamic limit. Furthermore, the stability matrix in the replica space, i.e. the second derivatives of the free energy with respect to the replica order parameters, $q^{\alpha\beta}$, was examined.[19] It turned out that the replica symmetric solution is stable only in the paramagnetic phase and becomes unstable below the phase boundary called the AT line in the T–H plane. Therefore, the SK solution is correct above the AT line, whereas it is not correct below the AT line.

To find the correct solution of the SK model below the AT line, the replica symmetry had to be broken. The way the symmetry is to be broken is, however, highly non-trivial. A solution, called the Parisi solution, was proposed[11–14] based on a novel metric structure of the replica-index space. It has now been proved with acceptable mathematical rigor that the solution is exact.[20] In the Parisi solution, a metric structure is introduced in the replica-index space, and it is assumed that the overlap q between two replicas depends only on the distance between the two in this index space. As a result, the order parameter is actually not a single number, in contrast to the EA order parameter q_{EA}, but a function $q(x)$ $(0 \leq x \leq 1)$, where x is a suitable parameterization of the distance between two replica indices. A remarkable feature of Parisi's solution below the AT line was that $q(x)$ varies continuously as a function of x.

The order-parameter function $q(x)$ first appeared to be a mathematical artifact, but its physical meaning was later clarified[14] by relating $q(x)$ to the probability distribution $P(q)$ of the overlap q between *pure states*, rather than the replicas. A pure state is defined as an extremal equilibrium distribution in the thermodynamic limit that cannot be expressed as a linear combination of any other distributions. Denoting the inverse function of $q(x)$ as $x(q)$, it was found that

$$x(q) = \int_{-\infty}^{q} dq' P(q'). \tag{9.2}$$

Here, the overlap distribution $P(q)$ is formally (and somewhat symbolically) defined as

$$P(q) = \sum_{\alpha,\beta} P_\alpha P_\beta \delta(q - q^{\alpha\beta}). \tag{9.3}$$

Here, the indices α and β specify pure states, not replicas, $q_{\alpha\beta}$ is the overlap between the two pure states α and β, and P_α is the probability for the system to be in the pure state α.

Equation (9.2) may provide a way to investigate the structure of the space of the equilibrium distributions with the replica method. We can conclude, for instance, that more than one pure state exists, and the overlap between two pure states varies depending on the particular pair of states if $q(x)$ is not constant. If $q(x)$ changes continuously in a finite range of x, as in Parisi's solution for the SK model, there must even be an infinite number of pure states. Simply stated, the RSB picture for systems in finite dimensions with short-range interactions consists of the hypothesis that there are infinitely many pure states with varying overlaps and various predictions derived from the hypothesis.

The definition (9.3) of $P(q)$, however, must be taken with caution since the existence of pure states as a well-defined thermodynamic limit is not clear for disordered systems. In fact, it was shown[16,17] that $P_J(q)$ for a bond realization J depends on J even in the thermodynamic limit. As will be discussed later in the chapter, this means[21] that a unique thermodynamic limit of $P_J(q)$ with fixed bond realization J does not exist. Therefore, the order parameter, $q(x)$, is not self-averaging for the SK model, in striking contrast to homogeneous systems.

Because of this subtle nature of the order parameter, the exact implication of the RSB picture for more realistic spin glass models in finite dimensions is somewhat ambiguous and has been a subject of debates.[9,10] One focus is the interpretation of the numerical result for finite systems. The definition of q and $P(q)$ often used in numerical simulations is based on the overlap between two spin configurations, S and S', independently chosen from the equilibrium distribution of the same bond realization. Namely,

$$\begin{aligned} q(S,S') &\equiv \frac{1}{N}\sum_{i=1}^{N} S_i S'_i, \quad P(q;L) \equiv [P_J(q;L)], \\ P_J(q;L) &\equiv \sum_S \sum_{S'} P_J(S;L) P_J(S';L)\delta(q - q(S,S')), \end{aligned} \tag{9.4}$$

where $P_J(S;L)$ is the normalized Boltzmann weight of the spin configuration S in a system of the linear size L with the bond configuration J. Throughout this article, unless stated otherwise, the square bracket $[\cdots]$ denotes the bond-configuration average with the distribution $P(J) = \prod_{(ij)} P_1(J_{ij})$, where the single-bond distribution P_1 is independent of (ij). The following link-overlap and its distribution are also often discussed recently:

$$\begin{aligned} q_l(S,S') &\equiv \frac{1}{N_l}\sum_{(ij)} S_i S_j S'_i S'_j, \quad P(q_l;L) \equiv [P_J(q_l;L)], \\ P_J(q_l;L) &\equiv \sum_S \sum_{S'} P_J(S;L) P_J(S';L)\delta(q_l - q_l(S,S')), \end{aligned} \tag{9.5}$$

where the summation concerning (ij) is over all the pairs of nearest-neighbor spins and N_l is the total number of the pairs in the system.

The following is the list of features that follow from the RSB picture.

(1) The overlap distribution, $P(q;L)$, converges to a function $P(q)$ that has a continuous part in the limit $L \to \infty$. In particular, $P(0;L)$ converges to a finite value $P(0) > 0$.
(2) Some quantities, such as $P_J(q;L)$, are non-self-averaging, i.e. a unique thermodynamic limit does not exist for a fixed bond realization.
(3) The bond-averaged link-overlap distribution $P(q_l;L)$ is not $\delta(q_l)$ in the thermodynamic limit, i.e. its width converges to a finite value.

(4) Global excitations with the energy cost of $\mathcal{O}(1)$ exist.
(5) A change in the boundary conditions generally affects spins located far away from the boundary.

9.1.2. *Droplet picture*

Another well-known picture[22–26] is based on a scaling hypothesis on local excitations and produces a markedly different conclusion for various quantities, in particular, $P(q)$. The basic assumption in this picture is that even below the transition temperature in zero magnetic field, there are only two pure states that are mapped to each other by the total inversion of spins as in the homogeneous ferromagnets. Then, several scaling properties are assumed for compact excitations with varying sizes, which are called *droplets*. To be specific, a droplet of the scale l has a typical excitation free energy $\epsilon_l \sim \Upsilon l^\theta$, where θ is called the droplet exponent. Since this exponent relates the energy scale to the length, similar to the stiffness exponent θ_S that relates the domain wall excitation energy to the system size, the simplest assumption is to identify θ with the stiffness exponent θ_S.

Among various results derived from this picture, of particular importance are the scaling forms of the correlation functions.[23,24] For example, it is predicted for models with a continuous bond distribution and with no magnetic field that the following asymptotic form should apply

$$[\langle S_i S_j\rangle^2] - [\langle S_i\rangle^2][\langle S_j\rangle^2] \sim T/\Upsilon R^\theta, \tag{9.6}$$

at near-zero temperatures, where R is the distance between site i and site j and the thermal average $\langle\cdots\rangle$ is taken in a single pure state (if there are two). It follows that the variance of the distribution function $P(q;L)$ defined in (9.4) (more strictly, the variance of $P(|q|;L)$) decreases as the system size increases:

$$(\Delta q)^2 \equiv [\langle q^2\rangle] - [\langle |q|\rangle]^2 \propto L^{-\theta} \to 0.$$

Therefore, when $P(q)$ is defined as the thermodynamic limit of $P(q;L)$, it consists of a pair of delta peaks at $q = \pm q_\mathrm{EA}$.

For the models with discrete energy levels, such as the $\pm J$ model, the scaling form of the correlation function near $T = 0$ must be modified. However, it is widely believed that the discretized nature of the energy level is not relevant at finite temperatures and the behavior is qualitatively the same as the one for the models with continuous energy levels.

Below, we list the defining properties of the droplet picture together with some predictions derived from it.

(1) The distribution of the excitation free energy of droplets of the scale l, $P(\epsilon_l)$, is continuous down to zero energy.
(2) $P(\epsilon_l)$ has the typical energy scale Υl^θ, with Υ being an $\mathcal{O}(1)$ constant. Specifically, $P(\epsilon_l)$ has the scaling form

$$P(\epsilon_l) \sim \frac{1}{\Upsilon l^\theta} \tilde{P}\left(\frac{\epsilon_l}{\Upsilon l^\theta}\right).$$

(3) $P(q)$ is self-averaging and it consists of a pair of delta peaks below the transition temperature.
(4) Since $P(0; L)$ is proportional to the excitation probability of a droplet that contains approximately one half of all the spins, for a system with a continuous bond distribution, $P(0; L)$ has the following temperature and size dependence near $T = 0$:

$$P(0; L) \sim [\text{Prob } (\epsilon_L < T)] \sim \frac{T}{\overline{\epsilon_L}} \propto TL^{-\theta}. \tag{9.7}$$

(5) For systems with a continuous bond distribution, the bond-averaged link-overlap distribution $P(q_l; L)$ consists of a single delta function in the thermodynamic limit. At low temperatures, its width depends on the size and the temperature as[27]

$$(\Delta q_l)^2 \propto TL^{-\mu} \quad (\mu_l \equiv 2d_S - 2d - \theta), \tag{9.8}$$

where d_S is the fractal dimension of the surface of droplets.
(6) For systems with finite-temperature spin glass transition, θ is positive. Therefore, a global excitation costs an infinite energy.
(7) A change in the boundary conditions does not affect the spins located far away from the boundary, except for a potential simultaneous inversion of all spins.

9.2. Equilibrium Properties of 2D Ising Spin Glasses

Two-dimensional spin glass models are easier to study than 3D models from a technical point of view. However, there are still some open questions. Among others, the possibility of the existence of a phase transition at non-zero temperatures in the discrete distribution models is important.

9.2.1. *Zero-temperature transition?*

One of the arguments that supports the zero-temperature phase transition scenario is based on the numerical estimates of the stiffness exponent that characterizes the size dependence of the domain wall free-energy. The stiffness is simply the difference in the free energy caused by a particular change in the boundary conditions. The most frequent choice in numerical computations is the application of periodic vs. anti-periodic boundary condition. The stiffness exponent θ_S is then determined by the asymptotic size dependence of the stiffness, ΔF:

$$\Delta F \propto L^{\theta_S}.$$

According to Ref. [23], when the distribution of coupling constants is continuous and has a non-zero weight at $J = 0$, whether a phase transition takes place at a finite temperature or not is determined by the sign of the stiffness exponent. If it is positive, the zero-temperature phase is strongly ordered, leading to a finite-temperature phase transition, whereas otherwise, the order is fragile and infinitesimal thermal fluctuations would destroy it, leading to criticality at zero temperature. Numerical computations of the ground states of finite systems in two dimensions were carried out[23, 28, 29] for estimating the stiffness exponent at $T = 0$. For the Gaussian distribution of the bonds, it turned out that $\theta_S \sim -0.3$. At present, one of the most accurate and reliable estimates of the stiffness exponent is given in Ref. [30] as $\theta_{\rm S} = -0.282(2)$ for the Gaussian bond distribution with periodic/anti-periodic boundary conditions. Due to the negative value of the stiffness exponent, it is widely believed that there is no finite-temperature transition in this model and that the system is critical at zero temperature.

The stiffness exponent of $\pm J$ model, on the other hand, is much closer to the marginal value, 0. It was estimated,[31] with the assumption of the power-low dependence of the stiffness on the system size, as

$$\theta_S = -0.060(4),$$

while the possibility of $\theta_S = 0$ was not ruled out. The possibility of the stiffness exponent being zero was strongly suggested by another GS computation.[30]

Recently, it was pointed out that the stiffness exponent has to be interpreted differently for $\pm J$ models. Amoruso *et al.*[32] carried out a renormalization calculation using the Migdal–Kadanoff method. They found that the stiffness exponent, calculated within this approximation, is zero in any dimension lower than the lower critical dimension. This finding suggests that the stiffness exponent being zero does not necessarily imply that the system is exactly at the lower critical dimension. This type of dimensional dependence of the stiffness exponent was observed only for the class of the bond distribution for which the discretized nature of the coupling constant is not smeared out by the renormalization. Since the $\pm J$ bond distribution falls into this class, the above-mentioned numerical estimates, $\theta_S \sim 0$, may only mean that the lower critical dimension is two *or above*. For the $\pm J$ model, however, even the possibility of a finite-temperature phase transition was suggested,[33] as we discuss in greater detail in the subsequent sections.

9.2.2. *Droplet argument for Gaussian-coupling models*

For two dimensional systems with a symmetric ($P(J) = P(-J)$) and continuous bond distribution, it was argued[23] that there is only one independent critical exponent. In other words, all critical indices are related to the stiffness exponent via scaling laws. We consider the following finite-size scaling form of the singular part of the free energy F_s,

$$\Phi(T, H, L) \equiv \beta F_s = -[\log Z] \sim \phi(TL^y, HL^{y_h}).$$

The condition that the total magnetization $[M] \equiv \lim_{H\to 0} \partial F_s/\partial H$ is proportional to $L^{d/2}$ at $T = 0$ relates y to y_h via

$$y_h = y + \frac{d}{2}.$$

Then, by differentiating Φ with respect to T and H, we can express any critical index by y. For example, the magnetization per spin depends on the magnetic field as $m = L^{-d}M \sim= L^{-\frac{d}{2}} \sim H^{\frac{d}{2y_h}}$, which means that the exponent δ appearing in the scaling of the magnetization with the magnetic field is $\delta = 1 - 2y/d$. The nonlinear susceptibility $\chi_2 \equiv L^{-d}\partial^3 M/\partial H^3$ at $H = 0$ depends on the temperature as $\chi_2 \sim T^{-\gamma_2}$ with $\gamma_2 = 3 + \frac{d}{y}$. Since the spin glass susceptibility χ_{SG} at $H = 0$ is related to χ_2 as $\chi_{SG} \equiv L^{-d}\sum_{i,j}[\langle S_i S_j\rangle^2] = T^3\chi_2$, the corresponding exponent is $\gamma_{SG} = d/y$. Similarly, the specific heat exponent can be expressed as $\alpha = -d/y$. As for the exponent η_{SG}, which characterizes the asymptotic form of the two-point spin glass correlation function $[\langle S_i S_j\rangle^2] \sim R_{ij}^{d-2+\eta_{SG}}$, we have $2 - \eta_{SG} = y\gamma_{SG}$, which yields $\eta_{SG} = 2 - d$. This is consistent with the fact that the correlation function does not decay at zero temperature.

9.2.3. *Droplets in Gaussian-coupling models: Numerics*

As mentioned above, one naturally expects[24] that the exponent $-y$ is identical to the stiffness exponent θ_S since both the exponents relate the energy scale to the length scale. However, the estimates of various critical exponents[34–36,38,39] based on the system size up to $L \sim 50$ seem to satisfy the scaling relation with $y \sim 0.5$ rather than $y = -\theta_S \sim 0.3$.

In order to check if y coincides with $-\theta$, a direct numerical estimate of the droplet excitation exponent θ was carried out using a heuristic optimization procedure[40,41] applied to the EA model with a Gaussian bond distribution in two dimensions. For each realization of the model, the GS spin configuration was computed with free boundary conditions. Then, the spins on the boundary were fixed as they were in the original GS, whereas the spin at the center was fixed in the opposite direction. These constraints lead to a new ground state that is identical to the old one on the boundary but differs from it in the vicinity of the central spin by a droplet of flipped spins around the center. The droplet is typically as large as the system itself. The system-size dependence of the droplet volume V and the droplet excitation energy E could be described well by the scaling laws

$$V(L) \propto L^{d_D} \quad \text{and} \quad E(L) \propto (V(L))^{\theta/d_D}$$

with

$$d_D = 1.80(2) \quad \text{and} \quad -\theta = 0.47(5)$$

for the range of the system size $5 \leq L \leq 49$. In particular, the value of $-\theta$ agreed with most of the previous estimates of y. This can also be understood within the droplet theory because all scaling forms derived by droplet arguments are identical to what one can get via the ordinary finite-size scaling by identifying θ and $-y$.

However, a recent computation[42] demonstrates the presence of a cross-over in the droplet excitation energy. They performed essentially the same calculations as the one described above for larger systems (up to $L = 160$) and argued that there may be a correction term due to the self-interaction of the surface of the droplets and that the droplet excitation energy has the form

$$\Delta E(L) \sim AL^{\theta} + BL^{-\omega}$$

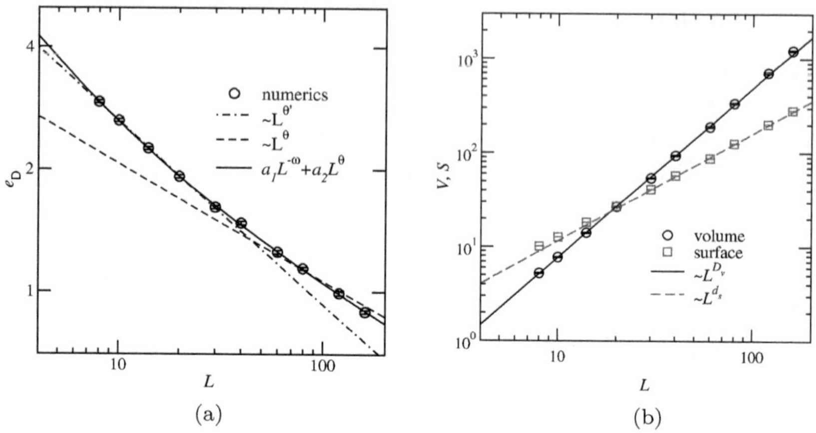

Fig. 9.1. (a) The droplet energy as a function of the system size. The two dashed straight lines represent the algebraic dependences $L^{-0.47}$ for smaller systems and $L^{-0.29}$ for larger ones, respectively. The solid curve is a fitting function (see text). (b) The volume and surface of the droplets (from Ref. [42]).

with $\omega > |\theta_D|$. The numerical data could be well fitted by choosing

$$-\theta_D = 0.29 \quad \text{and} \quad \omega = 0.97(5)$$

as shown in Fig. 9.1(a). The fact that the estimate of the droplet exponent now becomes close to the stiffness exponent indicates the validity of the simplest scenario. However, the result on the fractal dimension of droplets seems to suggest the contrary; their result appears to confirm the previous result $d_D \sim 1.8$ (Fig. 9.1(b)). While this may be another transient behavior, the problem of compactness of the droplets remains open in the 2D system.

The same model was studied in Ref. [43] and then in Ref. [44] at zero temperature using a different approach. For each realization of the model, the ground state for periodic boundary condition was compared with the ground state for another set of boundary conditions, such as anti-periodic and random boundary conditions. (In the latter case, half of the bonds across the boundary are chosen at random and inverted while the other half are kept unchanged.) The authors of Ref. [44] argued that in the RSB picture the change in the boundary condition would induce an excitation whose surface-to-volume ratio should not diminish as the system size increases, whereas in the droplet picture, it should decrease down to zero. Therefore, we can differentiate between the two pictures by applying a boundary condition that induces a domain wall in the system and focusing on a small box located at the center of the system. If the probability of having the domain wall crossing this part is finite even in the thermodynamic limit, it would indicate the validity of the RSB picture. On the other hand, it should decrease to zero if the droplet picture is valid.

They performed a calculation of exact GSs of many instances of the EA model in two dimensions with the Gaussian bond distribution. The system size explored was up to $L = 30$. They considered a box of the linear size of $L_{\text{box}} = 2$ located at the center of the system and measured the distribution of the overlap q_{box} on this box:

$$q_{\text{box}}^{\alpha\beta} \equiv \frac{1}{L_{\text{box}}^d} \sum_{i \in \text{"box"}} S_i^\alpha S_i^\beta,$$

where α and β are the indices specifying real replicas. If the boundary passes the box with a finite probability, the distribution of $q_{\text{box}}^{\alpha\beta}$ should have a finite weight around $q_{\text{box}}^{\alpha\beta} = 0$. This weight turned out to be decreasing with the power of -0.7 as a function of the system size. Based on this result, they concluded that the zero-temperature structure of the 2D EA model is trivial, i.e. only two pure states exist (see Fig. 9.4).

9.2.4. *Finite-temperature transition?*

There are some numerical results[33] of Monte Carlo (MC) simulation on the $\pm J$ model that suggest the existence of a phase transition at a finite temperature. Specifically, the overlap distribution $P(q; L)$ and the binder parameter g were computed. The estimates of the Binder parameter

$$g_q \equiv \frac{1}{2}\left(3 - \frac{[\langle q^4 \rangle]}{[\langle q^2 \rangle]^2}\right) \tag{9.9}$$

as a function of the temperature for various system sizes appeared to have a common crossing point. The finite-size scaling plot seems better when a finite transition temperature $T_c \sim 0.24J$ was assumed instead of zero transition temperature. Their estimates for the critical indices of this finite-temperature transition are $\nu \sim 1.8$ and $\eta \sim 0.2$. The overlap distribution $P(q; L)$ could be nicely scaled with the same η but with a slightly higher temperature $T_c \approx 0.29J$. Similar results were obtained[45] for an asymmetric bond distribution.

Numerical evidences suggest that the stiffness exponent of the two-dimensional $\pm J$ model is non-positive, as discussed in Section 9.2.1, which indicates that the lower critical dimension of the $\pm J$ model is equal to or greater than two. Now, if a spin glass transition takes place at a finite temperature as suggested, it means that the lower critical dimension is not greater than two. Therefore, the only scenario consistent with all the available numerical and analytical results is that the lower critical dimension is exactly two and that the stiffness exponent is exactly zero in two dimensions. More evidences, however, appear to be necessary to settle this issue beyond reasonable doubts.

9.3. Equilibrium Properties of Three-Dimensional Models

The problem of spin glasses in three dimensions is the central topic of the field. Whereas the RSB nature of the low-temperature phase in four dimensions is much less controversial, the nature of the low-temperature phase in three dimensions still remains the subject of an active debate.

In what follows, we present a number of numerical results. Because of the severe technical limitations for three dimensional cases, all the results obtained are for small systems, typically up to the linear size of about 10 lattice spacings or slightly more. Therefore, many important issues are left open, in particular the question as to what picture yields the correct description of the low-temperature phase of the EA model, which is the main subject of Sections 9.3.3–9.3.8.

9.3.1. *Finite-temperature transition?*

Even the very existence of a phase transition at finite temperature was not easy to establish. An evidence of the existence was obtained through the calculation of the domain wall energy at zero temperature,[29] in which a positive estimate for the stiffness exponent was obtained. However, the system size was rather limited ($L \leq 8$) and the estimated value of the stiffness exponent was small ($\theta \sim 0.2$). Therefore, this finding about the stiffness exponent alone was not sufficient to establish the existence of a finite-temperature phase transition. Finite-temperature approaches could not settle the issue, either. For example, while MC simulations[46,47] strongly suggested the existence of a transition at a finite temperature, they could not rule out the possibility of zero-temperature singularity with an exponentially diverging correlation length; it was suggested[48] that all existing data at that time (namely, 1994) were consistent with both hypotheses: $T_c > 0$ and $T_c = 0$. In particular, the simulation results of Ref. [48] for the spin-glass susceptibility could be fitted by a functional form consistent with a

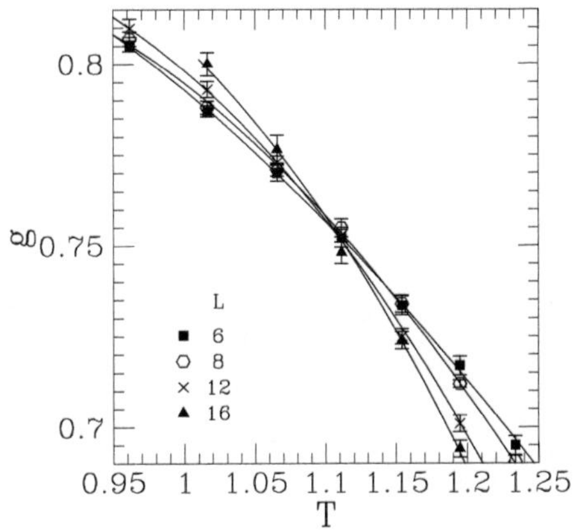

Fig. 9.2. The Binder parameter vs. temperature for the $\pm J$ model in three dimensions (from Ref. [49]).

zero-temperature singularity, $\chi_{SG} = 1 + A/(T - T_c)^\gamma$, just as well as the one consistent with a finite-temperature transition $\chi_{SG} = A(e^{(B/T)^p} - 1) + C$.

However, another set of MC results[49] showed beyond reasonable doubts the existence of a finite-temperature transition in the $\pm J$ model in three dimensions. Specifically, it demonstrated that the Binder parameter curves for different system sizes cross at the same point (Fig. 9.2) near

$$T \sim 1.1 \quad \text{and} \quad g_{\text{sg}} \sim 0.75.$$

With the assumption of the algebraic singularity at the critical point, the critical temperature and the critical indices were estimated as

$$T_c = 1.11(4), \quad \nu = 1.7(3), \quad \text{and} \quad \eta = -0.35(5).$$

These estimates were consistent with previous ones[46] and confirmed by other simulations as shown in Table 9.1.

A MC simulation for larger system sizes[50] clarified the issue further. The results of the correlation length and the spin glass susceptibility were extrapolated to the infinite system-size limit via finite-size scaling and were well fitted by a curve representing an algebraic divergence at a finite temperature. While the existence of a finite-temperature phase transition could be concluded from this result, the nature of the singularities at the critical point could not unambiguously be settled. (The singularity could be an essential rather than an algebraic singularity.) In Section 9.3.2, we present other numerical results concerning the nature of the critical point.

Table 9.1. The estimates of the critical temperature and the exponents for the 3D spin glass models.

Authors	Material/Model	T_c/J	ν	η	β	γ	z
Gunnarsson *et al.*[56]	FeMnTiO	—	1.7	−0.35	0.54	4.0(3)	6.2
Ogielski[47]	Ising, $\pm J$	1.175(25)	1.3(1)	−0.22(5)	0.5	2.9(3)	6.0(5)
Bhatt–Young[46]	Ising, $\pm J$	$1.2^{+0.1}_{-0.2}$	1.3(3)	−0.3(2)	0.46*	3.2	—
Singh–Chakravarty[57]	Ising, $\pm J$	1.2(1)	—	—	—	2.9(5)	—
Bhatt–Young[58]	Ising, Gaussian	0.9	1.6(4)	−0.4(2)	—	—	—
Kawashima–Young[49]	Ising, $\pm J$	1.11(4)	1.7(3)	−0.35(5)	0.55*	4.0*	—
Iñigues *et al.*[59]	Ising, Gaussian	1.02(5)	1.5(3)	—	—	—	—
Marinari *et al.*[60]	Ising, Gaussian	0.98(5)	2.00(15)	−0.36(6)	0.64*	4.72*	—
Berg–Janke[61]	Ising, $\pm J$	0.88	—	−0.37(4)	—	—	—
Palassini–Caracciolo[50]	Ising, $\pm J$	1.156(15)	1.8(2)	−0.26(4)	0.65*	4.1(5)	—
Mari–Campbell[62]	Ising, $\pm J$	1.19(1)	1.33(5)	−0.22(2)	0.52*	2.95(15)	—
Ballesteros *et al.*[51]	Ising, $\pm J$	1.138(10)	2.15(15)	−0.337(15)	0.73*	5.0*	—
Mari–Campbell[63]	Ising, $\pm J$	1.195(15)	1.35(10)	−0.225(25)	0.55*	2.95(30)	5.65(15)
Nakamura *et al.*[64]	Ising, $\pm J$	1.17(4)	1.5(3)	−0.4(1)	0.45*	3.6(6)	6.2(2)
de Courtenary *et al.*[65]	CuMn, AgMn	—	1.4*	0.4*	1.0(1)	2.2(1)	—
Bouchiat[66]	AgMn	—	1.4*	0.4*	1.0(1)	2.2(2)	—
Levy–Ogielski[67]	AgMn	—	1.3(2)	0.4*	0.9(2)	2.1(1)	5.5
Simpson[68]	CuAlMn	—	1.3*	0.5*	1.0*	1.9*	—
Coles–Williams[69]	PdMn	—	1.3*	0.4*	0.90(15)	2.0(2)	—
Vincent–Hamman[70]	CdCrInS	—	1.25(25)	0.2*	0.75(10)	2.3(4)	5.5
Kawamura[71]	Heisenberg, Gaussian	0.157(10)^c	—	—	1.1(1)^c	—	—
Hukushima–Kawamura[72]	Heisenberg, Gaussian	0.160(5)^c	1.2^c	0.8^c	1.1(1)^c	1.5(3)^c	—
Matsubara *et al.*[73]	Heisenberg, $\pm J$	0.18	—	—	—	—	—
Nakamura–Endoh[74]	Heisenberg, $\pm J$	$0.21^{+0.01}_{-0.03}$	1.1(2)	0.3*	0.72(6)	1.9(4)	4.5
Lee–Young[75]	Heisenberg, Gaussian	0.16(2)	1.1(2)	—	—	—	—
Nakamura *et al.*[64]	Heisenberg, $\pm J$	0.20(2)	0.8(2)	−0.3(3)	—	1.9(5)	6.2(5)
Kawamura–Li[76]	XY, $\pm J$	0.39(3)^c	1.2(2)^c	0.15(20)^c	—	—	7.4(10)^c
Lee–Young[75]	XY, Gaussian	0.34(2)	1.2(2)	—	—	—	—
Nakamura *et al.*[64]	XY, $\pm J$	0.43(3)	—	−0.4(2)	—	—	6.8(5)

Notes: The entries are categorized into three groups according to the anisotropy; easy-axis (top), isotropic (middle), and easy-plane (bottom). The entries with "*" are not quoted in the original paper but estimated through the scaling relations by the present authors. The entries with "c" are for the chiral glass transition.

9.3.2. *Universality class*

Early MC simulations such as those in Ref. [46] indicated a scenario that the system stays critical in the whole low-temperature region. Namely, these data could be explained by a line of critical points terminating at $T_c = 1.2$, similar to the Berezinskii–Kosterlitz–Thouless transition in the 2D XY ferromagnet. The conclusion in Ref. [50] on this issue was similar, while it was suggested that a finite-temperature transition with a algebraic divergence is most likely. With the assumption of the algebraic divergence, the critical parameters were estimated to be consistent with previous ones mentioned above. However, the data were also consistent with an exponential divergence as in the BKT transition.

A large-scale MC simulation[51] clarified the issue of the nature of the phase transition. Using special purpose machines and parallel tempering methods,[52,53] the authors succeeded to equilibrate the $\pm J$ model of sizes up to $L = 20$ down to a temperature low enough to cover a sufficiently large region around the transition point. They obtained various results consistent with an ordinary second-order phase transition. For example, they observed a clear crossing in the effective correlation length[54] $\xi(L)$ divided by the system size L. (They defined the effective correlation length as $\xi^2(L) \equiv \tilde{k}_m^{-2}(C_q(0)/C_q(\boldsymbol{k}_m) - 1)$, where $C_q(\boldsymbol{k})$ is the static structure factor, $\boldsymbol{k}_m$ is the smallest non-zero wave number compatible to the boundary condition, and $\tilde{k}^2 \equiv 4(\sin^2(k^x/2) + \sin^2(k^y/2) + \sin^2(k^z/2))$.)

The scaling dimension of the effective correlation length is zero; the curves for various system sizes should intersect at the same point similar to the Binder parameter. Indeed, it shows a common intersection point as in Fig. 9.3(b). This is in marked contrast to what we typically see in a BKT-type transition, which is shown in Fig. 9.3(b) for comparison. From this result, one concludes that the phase transition is an ordinary second-order one. For the critical temperature and the indices, see Table 9.1.

The results of Ref. [51] also provided indications in favor of an RSB nature of the system below the critical temperature. Namely, the estimates of a quantity called the G-parameter seem to have a finite value even below the critical point. The G-parameter is defined[55] as

$$G \equiv \frac{[\langle q^2 \rangle^2] - [\langle q^2 \rangle]^2}{3[\langle q^2 \rangle]^2 - [\langle q^4 \rangle]}.$$

The numerator is finite only if the sample-to-sample dependence of $P(q)$ does not vanish, while the denominator can be finite even if there is no

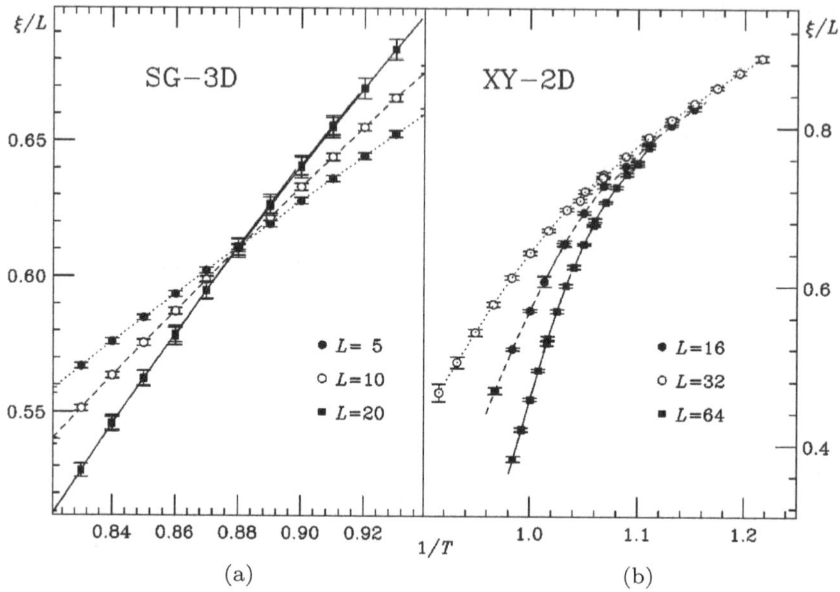

Fig. 9.3. The effective correlation length divided by the system size. (a) is for the $\pm J$ model in three dimensions, whereas (b) is for the XY model in two dimensions (from Ref. [51]).

sample dependence. Therefore, a observed non-vanishing value of G implies that $P(q)$ is not self-averaging.

However, there are a number of evidences by other groups that can be interpreted otherwise, namely, in favor of the absence of an RSB nature of the low-temperature phase, as we discuss below.

9.3.3. *Low-temperature phase of the $\pm J$ model*

There is a very active debate on the nature of the low-temperature phase of 3D spin glass models. In the following few subsections, we review various theories, arguments, and numerical calculations that were made or done with the ultimate aim to clarify the issue. But none of the pictures, RSB, droplet, or others, could be established so far. Nonetheless, the debate itself is interesting and each picture is worth being scrutinized with detailed numerical calculations. While the common belief is that the $\pm J$ model and the Gaussian-coupling model show essentially the same physics at finite temperature, we present them separately since many of the computations are done at zero temperature where the two may differ. The $\pm J$ model is discussed in the present subsection and the Gaussian-coupling model in the next.

A finite-temperature simulation on the $\pm J$ model was performed[77] using the replica-exchange MC simulation[53] for $\pm J$ model reaching down to $T = 0.2$ for the system size up to $L = 10$. The integrated overlap probability distribution

$$x(1/2) \equiv \int_0^{1/2} dqP(q)$$

was computed as a function of the temperature and the system size. Inspired by Ref. [78], the scaling — $x(1/2)$ being proportional to L^λ for small L and to L^θ for larger L — was assumed. In other words,

$$x(1/2) = TL^{-\theta}f(TL^\lambda)$$

was assumed. The fitting to the numerical results worked nicely with $\theta = 0$ yielding

$$\lambda = 0.9(1),$$

whereas a fitting with $\theta = 0.2$, the droplet prediction, turned out to be significantly worse. The estimate of λ is considerably smaller than a preceding estimate[79] while consistent with the value $\lambda = 0.72(12)$ in Ref. [80].

Zero-temperature calculations offer a powerful alternative to the finite-temperature approach with MC simulations, because one might expect that the differences between the two scenarios should be more prominent at lower temperature, and much better methods are available for systems at zero temperature than for systems at a very low but finite temperature. For three dimensional models, most methods for solving zero-temperature problems are based on heuristic optimization since no good exact method is available due to the NP-hardness of the problem. A computation of the ground states of the $\pm J$ Ising model up to $L = 14$ was done[81,82] with a heuristic algorithm called the cluster-exact approximation method.[83] Later the computation was redone,[79,84] in order to fix the problem of the biased sampling.[85] It was found that the width of the overlap distribution, $P(q)$, decreases as the system becomes larger, indicating the triviality of $P(q)$. The validity of the ultra-metric relation, $q_{12} = q_{23} < q_{31}$, was also examined, where q_{ij}'s are the overlaps among three randomly chosen ground states. The numerical results indicated that the ultra-metric relation holds for a typical triplet of ground states with relatively large mutual distances.

However, it was also found that the contribution from these triplets to $P(q)$, which constitutes the "non-trivial" (i.e. continuous) component of $P(q)$, decreases as the system becomes larger. In fact, the integrated weight of $P(q)$ systematically decreased toward zero as

$$x(q) \equiv \int_{-q}^{q} dq P(q) \propto L^{-\lambda}$$

with $\lambda = 1.25(5)$ for $q = 0.5$. This indicated a trivial structure of $P(q)$ at zero temperature. As for the exponent λ, a different estimate was obtained[77] as can be seen in the following discussion.

The authors of Ref. [78] considered $P(q)$ for the $\pm J$ model at $T = 0$ and pointed out that the ground-state space may be dominated by a single valley or a single pure state, no matter which picture is valid. They argued that even if there are multiple valleys in the energy landscape of the discrete energy model, the distribution of the overlap $P(q)$ may still be trivial, indicating that it is impossible to discriminate between the two scenarios by a computation of $P(q)$ such as the one mentioned above.

The underlying assumption of the argument is the many-valley structure in the phase space and the sponge-like structure of low-lying excitations in the real space. (The latter is further discussed in Sections 9.3.6 and 9.3.7.) A droplet typically has a finite volume and when flipped it takes a state in a valley to another state in the same valley. In contrast, a typical sponge-like cluster is supposed to occupy a finite fraction of the whole system and flipping it generally causes a transition from one valley to another.

This argument is based on an estimate of the entropy by counting zero-energy excitations from a particular ground state. In general, the continuous part of $P(q)$, if any, is caused by excitations of various scales whose excitation free energies are smaller than the temperature. In particular, at zero temperature, it is caused by zero-energy "excitations". They argued that $P(q)$ has a trivial structure even if there are multiple valleys. To see this, it suffices to consider the case of two valleys. Each valley contributes to $P(q)$ according to its weight, that is the number of distinct spin configurations in the valley. This number is roughly the same as the number of zero-energy droplets. We can choose two representative configurations, one from each valley, that can be transformed to each other by flipping a sponge-like cluster. Then, the zero-energy droplet excitations in one configuration differ from those in the other configuration only on (or near) the boundary of the sponge-like cluster. Therefore, if A is the area of the surface of the sponge-like cluster, the droplets differ only at A positions. This difference results in the entropy difference of $\mathcal{O}(A^{1/2})$. (The power 1/2 comes from the assumption that the number of the droplet-like excitations located on the surface is a random variable.) It follows that the contribution (to the continuous part of $P(q)$) from one valley differs from that from the other typically by a factor $e^{\pm(\text{const})\times A^{1/2}}$. We now see that almost certainly $P(q)$ is dominated by a contribution from a single valley in the thermodynamic limit where $A \to \infty$, leading to a trivial structure in $P(q)$. It is clear that the presence of more than two valleys does not affect the result as long as the number of valleys does not grow too fast as the system size increases. Based on this result, they argued that a trivial $P(q)$ for the $\pm J$ model in three dimensions at $T = 0$[84] does not necessarily indicate the absence of the RSB.

However, it is not known how the number of valleys depends on the system size. It is not too unrealistic to assume that the number of valleys grows faster than or proportional to the number of spins, N. If so, a typical minimum entropy difference between two valleys would be less than $A^{1/2}/N < N^{-1/2} \to 0$. Therefore, it is not clear if this argument really invalidates the numerical evidence in Ref. [84] for triviality of the energy landscape structure.

9.3.4. *Low-temperature phase of the Gaussian-coupling model*

The GS of models with the Gaussian bond distribution is unique up to a trivial degeneracy due to the Z_2 symmetry. However, it is still possible to extract useful information about low-lying excitations from zero-temperature computations. The authors of Ref. [86] attempted to use the same strategy that they used for two dimensional models with the Gaussian bond distribution[44] (see Section 9.2.3) to discriminate between the two scenarios in three dimensions. They estimated the probability of a domain wall passing through a small imaginary box placed inside the system. A clear decreasing behavior as a function of the system size was found, in favor of the droplet picture. However, the amount of the total decrease that they could observe by changing the system size was only of a factor of 1.3 or 1.4, due to a severe system size limitation. (In the case of two dimensions, the same quantity varies by almost an order of magnitude as can be seen in Fig. 9.4.)

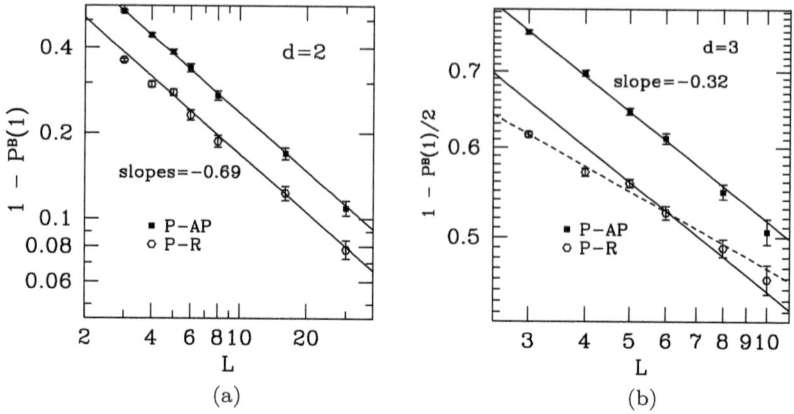

Fig. 9.4. The probability of a domain wall passing through the box. (a) For the EA model with the Gaussian bond distribution in two dimensions at $T = 0$ and $H = 0$, whereas (b) is for the same model in three dimensions (from Ref. [86]).

In addition to the response to the boundary condition, the effect of a weak perturbation applied to the whole system was also examined.[87] The perturbation was chosen in favor of the excited states against the GS. For each bond realization, they first obtained the GS, which we denote $\{S_i^{(0)}\}$. Then to each coupling constant J_{ij}, the following perturbation was added:

$$\Delta J_{ij} \equiv -\Delta J S_i^{(0)} S_j^{(0)},$$

where $\Delta J \equiv \epsilon/N_{\mathrm{bond}}$ was chosen to be of $\mathcal{O}(L^{-d})$. With this perturbation, the GS energy is increased, exactly, by ϵ, whereas excited-state energies are changed according to their similarity to the GS. Therefore, by adjusting the amplitude of the perturbation and examining the overlap between the original GS and the GS of the perturbed system, one may obtain some information about the excitations.

The RSB picture predicts the existence of many spin states that are local minima of the Hamiltonian. These states differ from the GS by only an infinitesimal amount of energy per spin but differ by a macroscopic number of spins such that the average Hamming distance between them (normalized by the system size) converges to some finite value in the large system size limit. This means for the bulk perturbation considered in Ref. [87] that the overlap between two such minima, the one with the perturbation and the other without, should be truly smaller than unity in the thermodynamic limit, no matter how small the perturbation may be. Therefore, in the RSB picture, $1 - [q]$ should be finite, where q is the overlap between the two minima. On the other hand, in the droplet picture, $1 - [q]$ should depend on the system size by a power law being characterized by the droplet energy exponent θ and an exponent related to the geometrical properties of the excitations. To be specific, since $1 - [q]$ is dominated by the contribution from droplets whose linear size is comparable to the system size L, it is roughly equal to the volume fraction of a typical droplet of size L multiplied by the probability of such a droplet being excited. Namely,

$$1 - [q] \propto \frac{L^{d_v}}{L^d} \times \frac{\Delta_L}{\epsilon_L}, \tag{9.10}$$

where Δ_L is the energy gain by the droplet excitation due to the perturbation and ϵ_L is the droplet excitation energy without the perturbation. If the droplets are compact, as assumed in Ref. [87], only the second factor matters since $d_v = d$. For Δ_L, the authors of Ref. [87] assumed that $\epsilon \sim \mathcal{O}(1)$ when integrated over the whole system, which is equivalent to setting $\Delta_L \propto L^{-d+d_s}$, where d_s is the fractal dimension of the boundary of droplets. Based on this, the following scaling form for the overlap was proposed[87]:

$$v(\epsilon, L) \equiv 1 - [q] \sim \tilde{V}(\epsilon / L^{d-d_s-\theta}). \tag{9.11}$$

For the link-overlap q_l, a similar argument yields

$$s(\epsilon, L) \equiv 1 - [q_l] \sim L^{-(d-d_s)} \tilde{S}(\epsilon / L^{d-d_s-\theta}). \tag{9.12}$$

In the system size range that could be explored (namely, $L \leq 8$), the numerical results for $[q]$ and $[q_l]$ show that $v(\epsilon, L)$ and $s(\epsilon, L)$ obey the power law and decrease to zero as the system size increases, contradicting the RSB picture. However, fitting the numerical results to the forms (9.11) and (9.12) yielded

$$d - d_s = 0.42(2) \quad \text{and} \quad d - d_s + \theta = 0.44(2).$$

This implies

$$\theta = 0.02(3).$$

The difference between the value and the stiffness exponent ($\sim$0.2) obtained from the domain wall calculation is statistically significant, i.e. the result contradicts not only to the RSB picture but also to the droplet picture unless this seeming contradiction is caused by a large correction to scaling.

Concerning the possible source of the correction, two arguments[42,88,89] were presented to explain the inconsistency observed in numerical results within the framework of the droplet argument. It was pointed out[88] that if one assumes a "clean" scaling with no correction for the droplet excitation energy, the contribution from small droplets gives rise to a correction to scaling in quantities such as the spin glass susceptibility and the magnetic susceptibility. The droplet excitation energy computed with the condition used in Refs. [40] and [41] and the link-overlap computed in Ref. [87] may have been affected by such a correction to scaling. Another possible source of a correction to scaling is the interaction between domain walls. It was argued[42,89] that the energy of a domain wall may be increased by the presence of another domain wall and that the energy shift due to this interaction may have the form $l^{-\omega'}$, where l is the distance between two domain walls. Similarly, a "self-interaction" of the domain wall may give rise to a correction to scaling in the droplet excitations as $E = Al^{\theta} + Bl^{-\omega}$. In this case, l stands for the size of the droplets. As the correction-to-scaling exponent due to this mechanism, the authors of Ref. [89] quoted the value $\omega \sim 0.13(2)$.[90]

In response to some reports (see the following paragraphs in this section) contradicting to their conclusion, the authors of Ref. [91] further pursued the ground-state nature of the three dimensional system with the Gaussian bond distribution, along the same line as their own preceding calculation.[87] This time, however, larger systems ($L = 12$) were dealt with the branch-and-cut algorithm.[92] This algorithm guarantees that the states found are the true GSs. In addition, a greater care was taken for various possibilities of fitting functions and different sources of corrections to scaling. Specifically, they considered a few different fitting functions for size-dependent quantities such as the link-overlap and the box overlap. Fitting functions consistent with the droplet, the TNT, and the RSB scenario, respectively, were considered. (For the TNT scenario, see Sections 9.3.6 and 9.3.7.) It was found that the size dependence of the surface-to-volume ratio, i.e. $1 - q_l$, could be explained by any one of three pictures and also that the size dependence of the box overlap could be explained by any one of the three pictures. The estimates of the exponent $\mu \equiv d - d_s + \theta$ turned out to depend on the boundary condition. With the free boundary condition, the estimates are $d - d_s = 0.44(3)$, $\mu = 0.63(3)$, and $\theta = 0.19(6)$, whereas for the periodic boundary condition, they are $d - d_s = 0.43(2)$, $\mu = 0.42(3)$, and $\theta = -0.01(3)$. The latter set of values are consistent with their preceding estimates,[87] whereas the former are not.

Another set of numerical results, presented in Refs. [93–95], seem to contradict with the results of Refs. [87] and [86] presented above. As for the calculation based on the small imaginary box placed at the center of the system, it was argued[93] that the probability of the state inside the box being affected by the change in the boundary condition should obey the scaling

$$P_{\text{change}}(L_{\text{box}}, L) \sim g(L_{\text{box}}/L), \tag{9.13}$$

if the droplet argument is correct. However, the numerical results in Ref. [93] for $L_{\text{box}} = 2, 3, 4$ and $L = 12$ did not fit in this scaling.

In Ref. [93], the scaling property of the domain wall induced by the anti-periodic boundary condition in the x-direction was also examined. In particular, the probability of the domain wall not intersecting a plane perpendicular to the x axis was measured. It was found that the non-intersecting probability approaches zero as the system size increases:

$$P_L(\text{“The domain wall does not intersect the plane”}) \propto L^{-\gamma},$$

with $\gamma = 1.5 - 2.0$. Based on this observation, they suggested that the domain wall is space filling, i.e. $d = d_{\text{s}}$.

In Refs. [94] and [95], zero-temperature calculations were performed with the conditions analogous to the ones in Refs. [86] and [87]. In Ref. [94], the effect of the anti-periodic boundary conditions, which were imposed in the x direction, was compared with the periodic one. The links perpendicular to the yz plane were treated separately from those parallel to the yz plane. The result of the link-overlap for the perpendicular links, q_P, and that for the transverse ones, q_T, are shown in Fig. 9.5(a). Both kinds of the link-overlap can be fitted well by second-order polynomials in $1/L$ as shown in the figure. From the zeroth order term, they concluded that

$$\lim_{L \to \infty} 1 - q_l = 0.245(15),$$

in contrast to the droplet prediction $1 - q_l \to 0$.

In Ref. [95], the effect of the bulk perturbation at zero temperature was studied. The perturbation was the same as the one considered in Ref. [87]. The link-overlap, $q_l(q)$, was computed as a function of the bulk overlap, q. In addition to the average link-overlap, $q_l(L)$, attention was paid to the quantity $A(L) \equiv 1 - q_l(q = 0)$, where $q_l(q = 0)$ is the link-overlap between two states that have zero overlap. In the droplet picture, both quantities should converge to 0 as L goes to infinity. The numerical results are shown

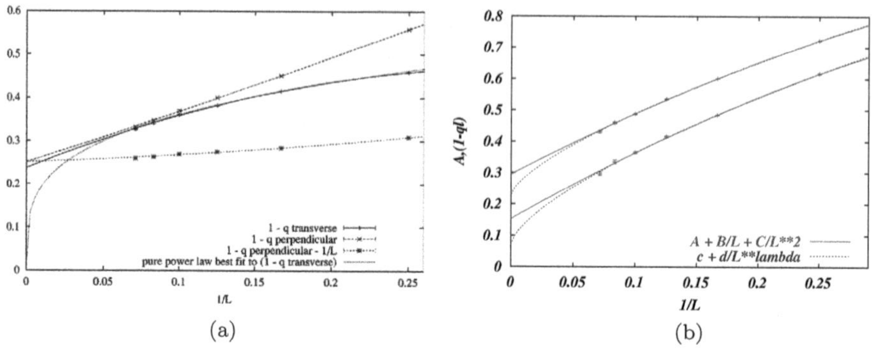

Fig. 9.5. (a) The link-overlap for the perpendicular links, q_P, and that for the transverse links, q_T, between the two ground states with and without a twist in the boundary, and (b) the link-overlap (subtracted from unity) between the two ground states with and without a bulk perturbation. (The lower set of data are for the average link-overlap, whereas the upper one is for the link-overlap restricted to those pairs of states which have zero mutual overlaps. For each set of data, the upper fitting curve is a second-order polynomial in $1/L$ and the lower one is a fractional power in $1/L$ with an additional constant term.) For both (a) and (b), the model is the three-dimensional EA model with the Gaussian bond distribution at zero temperature (from (a) Ref. [94] and (b) Ref. [95]).

in Fig. 9.5(b). It was found that in both cases of the boundary and the bulk perturbations, the size dependence of the surface-to-volume ratio of the domain wall, i.e., $1 - q_l$ can be fitted by a polynomial in $1/L$ with a finite constant term, consistent with the RSB picture.

Concerning the results presented in Ref. [93], it was pointed out[96] that the probability P_{change} is subject to a strong correction to scaling and that, taking such a correction term into account, the data presented in Ref. [93] can be fitted reasonably well with the scaling form (9.13). It was also pointed out that even the two dimensional data, which are believed to obey (9.13), could not be fitted by this scaling form without a similar correction term. As for the non-intersecting probability of the domain wall with the plane, it was simply pointed out that the observation in Ref. [93] does not necessarily imply $d = d_{\text{s}}$ because the observed behavior can be caused by domain walls that are rough but not space filling. In addition, as already mentioned above, their latest calculation[91] suggests that we have not yet accumulated numerical evidences sufficient to decide between the possible scenarios.

Before concluding this subsection, let us mention what is known about the geometrical nature of droplets. For this issue, a zero-temperature calculation was performed.[90] Using an efficient heuristic algorithm, the authors of Ref. [90] obtained GSs of 3D systems of the size $L = 6$ and $L = 10$. For each sample, they first obtained the ground state. Then, a center of the excitation is chosen randomly. They searched for the droplets of various sizes including the central spin. In their calculation, a droplet was defined as the cluster of spins of the smallest excitation energy among those which have a given volume. In spite of the limited system size, they succeeded in obtaining system-size-independent numerical results. They found that the linear scale of the droplets is described by

$$R(v) \propto v^b \quad (b \sim 0.5),$$

whereas the volume dependence of the excitation energy obeys

$$E(v) \propto v^a \quad (a \sim -0.06).$$

The first relation means that the droplets are roughly two dimensional fractal objects rather than compact ones. The second relation is rather surprising, since at first glance it might seem to contradict the existence of the finite-temperature phase transition. They claimed that this does not necessarily mean the absence of the phase transition because the basic assumption of the droplet argument may not be valid. Another possible source of the discrepancy may be the definition of the droplet with the very restrictive fixed-volume constraint.

9.3.5. *Effect of magnetic fields*

Whereas the RSB picture predicts the persistence of the spin glass phase against a small but finite magnetic field, the droplet picture predicts the contrary — the absence of the spin glass phase at any finite magnetic field. Therefore, one of the possible ways to distinguish between the two scenarios may be found in the behavior of short-range models in a finite magnetic field. However, it is rather hard to perform this task near the critical temperature (of the zero-field case). This difficulty is not only because the dynamics near the critical point is slow but also because the critical value of the magnetic field predicted by the mean-field type argument is small. To avoid at least the latter difficulty, a zero-temperature calculation was performed.[97] Using a heuristic optimization algorithm, the GS of the three dimensional EA model was computed with the Gaussian bond distribution up to $L = 12$ with various uniform magnetic fields.

In addition to the magnetic field, some constraints were also imposed at the same time to probe low-lying excitations. For example, a spin may be forced to be opposite to its natural direction in the original ground state with no constraint. Then, the lowest energy excitation among those which include the chosen spin is excited. In the RSB picture, where there are infinite number of states different from each other by a finite fraction of spins within an $\mathcal{O}(1)$ window of excitation energy, such an excitation should typically contain a finite fraction of all spins (see also Sections 9.3.6 and 9.3.7). Therefore, the volume of excited cluster tends to infinity with increasing system size. On the contrary, in the droplet scenario, the influence of the forced spin cannot reach beyond the distance determined by the magnetic field H as

$$l(H) \propto H^{-2/(d-2\theta)}.$$

Therefore, when the system size is increased beyond this length, the size of the excited clusters should stay constant independent of the system size. Some results of computation along this line of consideration were presented in Ref. [97]. They demonstrate the fragility of the ground state in a magnetic field. For example, the forced spin-flip typically drags as much as 30% of all spins at $H = 0.2$ for $L = 12$. The size of the excited cluster generally increases as the system size increases. However, neither a saturation nor a proportionality to L^d could be identified clearly.

Therefore, they tried another constraint where signs of an array of bonds along a plane are switched. This corresponds to imposing the anti-periodic boundary condition to the system. In the RSB picture, the resulting excitation again should extend to the whole system, whereas in the droplet scenario, it should not extend for systems larger than $l(H)$. They measured the probability of the excitation wrapping around the system. It was found that the wrapping probability increases for $H \leq 0.6$ as the system becomes larger, whereas it decreases for $H \geq 0.7$. Based on this result, they suggested that H_c is finite and is close to 0.65 at $T = 0$. While they did not exclude the possibility of $H_c = 0$ considering the finite-size corrections that may be present, their approach seems promising and it is probably worth putting more effort in this direction.

9.3.6. *Sponge-like excitations*

While the droplet picture presents a clear "real-space" image for the low-lying excitations, we have not discussed how low-lying excitations should look in the RSB picture. The authors of Ref. [98] discussed geometrical properties of low-lying excitations that are compatible with the RSB picture. They proposed that low-lying excitations may have a "sponge"-like shape. A sponge-like cluster has a characteristic length scale l_c. They argued that excitations smaller than this scale obey the droplet scaling, whereas beyond this scale, the droplet predictions do not apply. A sponge-like object is defined by the following properties:

(1) It is a connected object, and so is its complement.
(2) It occupies a finite fraction of the whole system's volume.
(3) It spans the whole system, and so does its complement.
(4) It has a characteristic length scale different from the system size and the lattice constant. When coarse-grained beyond this length scale, the object occupies the space uniformly, whereas below this scale, non-uniformity can be seen.
(5) Its surface contains a finite fraction of its volume, i.e. the surface-to-volume ratio is finite and independent of the system size up to a finite-size correction.

They argued that in finite dimensions, three for instance, there may be excitations of sponge-like clusters with excitation energy of $\mathcal{O}(1)$, independent of the system size. At the same time, the energy barrier that must be overcome in order to excite such a cluster diverges as the system size increases. Namely, two pure states in the RSB picture can be transformed into one another by flipping spins in one or more of the sponge-like clusters. Since they occupy a finite fraction of the whole system, as soon as thermal fluctuations are introduced, they give rise to the non-trivial structure in the overlap distribution function, $P(q)$. In addition, because of the property (9.5), such excitations yield a non-trivial functional form for the link-overlap distribution $P(q_l)$.

Obviously, the two properties (9.4) and (9.5) mentioned above are closely related to each other because the surface-to-volume ratio may be determined by the characteristic length scale as $S/V \sim l_c^{-1}$. It is important to estimate the length scale, l_c, in order to check the applicability of the picture to individual cases. An attempt was made[99] to estimate this ratio numerically by considering the EA spin glass model with a Gaussian bond distribution. For each bond realization, the ground state was obtained using a heuristic optimization technique. Then, the ground state search was performed once more, this time under the constraint that a pair of spins (chosen randomly) have the opposite relative orientation as compared to the one in the first ground state.

In the droplet picture, the difference of two ground states is supposed to be a droplet of a size that is smaller than the distance of the two chosen spins, whereas in the sponge picture, it should be identified with a sponge if one of the chosen spin is included in the sponge and the other is not. In the sponge picture, such an event should happen with a finite and system-size-independent probability because at the scale larger than l_c the sponge is uniform and the fraction of the sponge is independent of system size.

Calculations of the system up to $L = 11$ were performed.[99] In order to check the property (9.3), the authors measured the probability of the event that the excited cluster spans the whole system. It was found that the probability does not show a strong system size dependence, indicating that large excitations with $\mathcal{O}(1)$ excitation energy exists.

9.3.7. *TNT picture — introduction of a new scaling length*

However, their other findings[99] indicated that the sponge picture defined above does not exactly describe the model, either. Particularly disturbing was the result of the surface-to-volume ratio of the system-spanning clusters obtained by the procedure mentioned in Section 9.3.6; the ratio decreases as a function of the system size. A power law yielded a reasonable fit, i.e. $S/V \propto L^{-0.3}$, suggesting l_c (defined as the V/S) tends to infinity in the limit $L \to \infty$. If one compares the ground state with the (sponge) excited state and focuses on a small box located around the center of the system, the domain boundary S would never pass through this box in the infinite system size limit. The link-overlap is therefore always complete (i.e. 1) within the box. This is in agreement with the result[86,87] mentioned in Section 9.3.4. The increase of the characteristic length l_c was also confirmed in Ref. [100]. It follows that the distribution of the link-overlap should have a trivial delta-peak structure while that of the ordinary overlap may have a non-trivial structure. This new scenario with the new system-size-dependent scaling length $l_c(L)$ is called[99] the "TNT" scenario, an abbreviation of "Trivial $P(q_l)$ and Non-Trivial $P(q)$".

A closely related calculation was done by the authors of Ref. [101], who performed a MC simulation of the EA model with the Gaussian bond distribution. They reached rather low temperatures ($T \sim 0.2J$) employing the exchange MC method.[53] The overlap distribution $P(q)$ was measured. The zero-overlap probability $P(0)$ stays constant, in contrast to the prediction of the droplet picture ($P(0) \propto L^{-\theta}$ with $\theta \sim 0.2$), consistent with the RSB picture and the TNT picture. However, the temperature dependence of

$P(0)$ at a fixed system size agreed with the droplet picture, i.e., $P(0) \propto T$. They also obtained the link-overlap distribution. Its width turned out to be proportional to $L^{-\mu_l}$, where μ_l was estimated as

$$\mu_l = 0.76(3)$$

from an extrapolation to $T = 0$. The TNT picture discussed above assumes the presence of excitations with size-independent energy, i.e. $\theta = 0$, whereas the surface-to-volume ratio of excited clusters goes to zero. Based on this picture together with the above estimate of μ_l, they obtained (by setting $\theta = 0$ in (9.8))

$$d - d_{\rm S} = 0.38(2)$$

which is consistent with the zero-temperature calculation[87] $d - d_{\rm S} = 0.42(2)$.

9.3.8. *Arguments supporting the droplet picture*

In Ref. [102], the notion of pure states was re-examined for the disordered systems. The relationship between the appearance of $P_J(q)$ and the number of pure states was discussed. It was suggested that $P_J(q)$ is an erroneous indicator of the multiplicity of pure states. In the ferromagnetic 2D Ising model below the critical temperature, one has an example of $P(q)$ being non-trivial while there are only two pure states. Namely, when the anti-periodic boundary condition is imposed in both the directions, because of the arbitrariness of the position of domain boundaries, $P(q)$ includes a continuous part. On the other hand, in the 3D random field model, one can see the case where $P_J(q)$ consists of a single delta peak, although there are two pure states. This is because the free-energy difference between the two pure states, one with $\langle S_i \rangle = m > 0$ and the other with $\langle S_i \rangle = -m$, diverges being proportional to $N^{1/2}$. Therefore, $P_J(q)$ for a given sample consists of only one delta peak.

The argument was elaborated in Refs. [10] and [21], in which the authors exploited the translation ergodicity. They proved that any translationally invariant quantity measured for a particular bond realization J is equal to its bond configuration average, provided that the distribution of each bond is independent and spatially uniform. Since $P_J(q)$ is obviously translationally invariant, it follows that $P_J(q)$ is self-averaging, i.e. does not depend on J. This result is in contrast with what we know for the SK model. Furthermore, they argued that since $P_J(q)$ is self-averaging, it would be improbable that $P_J(q)$ has a continuous part. The reason for this is that there are only countably many pure states, and therefore, $P(q)$ consists of countably many delta peaks. Consequently, we would have to choose many but only countably many numbers from the interval $-1 \leq q \leq 1$ to locate these delta functions. They argued that existence of such countably many "preferred" locations, yet independent of J, is very implausible. Their argument made it clear that we must be careful about using the notion of the pure state in the disordered systems.

However, it was pointed out[103] that the very existence of the pure states that was implicitly assumed in Ref. [21] is questionable. This means that the pure states that appear in (9.3) may have only a metaphoric meaning and should not be taken too strictly for disordered systems. From this point of view, one may say that the non-existence of the unique thermodynamic limit, or, more specifically, a chaotic system-size dependence of various quantities may signify the RSB nature of the spin glass systems.

9.4. Models in Four or Higher Dimensions

Numerical simulations are easier in four or higher dimensions than in three. This is presumably because $d = 4$ is well separated from the lower critical dimension. The existence of a finite-temperature phase transition in four dimensions was established already in an early numerical study[58] where a clear crossing in the Binder parameter, defined in (9.9), could be observed. The critical point was located at $T_c = 1.75(5)$ and the critical exponents were estimated to be $\nu = 0.8 \pm 0.15$ and $\eta = -0.3 \pm 0.15$ for the Gaussian bond distribution. The scaling relation $\gamma/\nu = 2 - \eta$ yields $\gamma = 1.8(4)$, which agrees well with the result $\gamma = 2.0(4)$ of a high-temperature series expansion[104] for the $\pm J$ bond distribution. More precise estimates of the critical indices were obtained in Ref. [105]:

$$T_c = 1.80(1), \quad \nu = 0.9(1), \quad \eta = -0.35(5), \quad \text{and} \quad \gamma = 2.1(2).$$

Based on an off-equilibrium simulation, they also obtained an estimate of the EA order parameter below the critical point as a function of the temperature. A non-vanishing order parameter excludes the possibility that the transition is of the BKT type. As for the nature of the low-temperature phase, the observation made in Ref. [106] that $P(0)$ is system-size independent below the critical temperature indicates that the droplet picture is inappropriate for the 4D spin glass. For a detailed discussion on other results obtained before 1997, the readers are referred to Ref. [103]. Here, we only mention one of the latest computations for clarifying the nature of the low-temperature phase in four dimensions.

The replica exchange method was used in Ref. [101] to study the $4d$ spin glass model with a Gaussian bond distribution at rather low temperatures $T \sim 0.2J \sim 0.1T_c$. The zero-overlap probability $P(0)$ was measured in four dimensions as well as in three. In both cases, $P(0)$ was found to be independent of the system size for a fixed temperature and roughly linearly dependent on the temperature for a fixed system size. The authors of Ref. [101] summarized their results as being consistent with the TNT picture mentioned above. They also estimated the exponent characterizing the size dependence of the variance of the link-overlap distribution, namely, $\theta + 2(d - d_s)$, where d_s is the fractal dimension of droplet boundaries (see the discussion in subsection 9.3.4). The extrapolated value to the zero temperature was $\theta + 2(d - d_s) = 0.35(6)$, while they did not rule out the possibility of this value being zero.

For the GS properties in four dimensions, we refer the reader to Ref. [107], in which the ground states of the model in four dimensions up to $L = 7$ were computed. The stiffness exponent was estimated as $\theta_s = 0.64(5)$.

Although somewhat misplaced in this section on higher dimensional models, we mention here the one-dimensional spin glass model with *long-range* interactions, i.e. $J_{ij} = c_\sigma \epsilon_{ij}/r_{ij}^\sigma$, where ϵ_{ij} is a Gaussian random number with zero mean and variance one and r_{ij} the Euclidean distance between spin i and j on a finite ring (of perimeter L) embedded in the 2D space. This system, on the borderline between finite-dimensional and mean-field models, was studied in Refs. [108] and [109] for system sizes up to $L = 512$. Finite-temperature simulations as well as the study of low-energy excitations yielded results that are inconsistent with the droplet picture but (partially) consistent with the TNT picture.

9.5. Aging

The out-of-equilibrium dynamics of spin glasses has become a very rich field in the recent years and an excellent review on the intensive theoretical work that has been performed on it until 1998 can be found in Ref. [110] and an overview over the experimental situation until that date can be found in Refs. [111] and [112]. Since then, a number of interesting developments have occurred, and we will focus on them with the prerequisites necessary to understand them. We start with the theoretical concept of a length scale that evolves in time and is flexible enough to account for a number of numerical and experimental results. Then we focus on two-time quantities that are typically measured experimentally and their behavior in various temperature protocols (i.e. aging histories) — showing effects like memory and rejuvenation. Finally, we discuss the theory for violations of the fluctuation–dissipation theorem and the numerical and experimental evidences for it.

9.5.1. *A growing length scale during aging?*

Spin glasses have an order parameter which is (*cum grano salis*, i.e. disregarding the potential complications arising from a possible RSB) $q_{\mathrm{EA}} = [\langle S_i \rangle^2]$. In comparison to structural glasses,[113] this is a very lucky situation, not only as a starting point for equilibrium theory, which we have discussed in the last few sections, but also as a starting point for non-equilibrium dynamics. An example is the common picture of the dynamical evolution of a system out of equilibrium quenched into the phase with non-vanishing q_{EA}, in which larger and larger regions of space become ordered.[114] Thus, it appears natural to postulate a time-dependent length scale $L(t)$ for these equilibrated domains for spin glasses. We choose such a picture, motivated by coarsening systems,[114] as our starting point for reviewing the out-of-equilibrium dynamics of spin glasses.

Within the droplet theory,[115] the slow out-of-equilibrium dynamics of a finite-dimensional spin glass is due to thermally activated growth of locally equilibrated regions. The growth of such regions is supposed to happen through domain wall movements as in the context of pinned domain walls in random field systems[116] or in elastic manifolds in a disordered environment[117] via a thermally activated process overcoming a free energy barrier. It is assumed that the typical energy barrier $B_{L(t)}$ scales with the typical size $L(t)$ of the domains reached after a time t as

$$B_{L(t)} \sim \Delta(L(t)/L_0)^{\psi}, \tag{9.14}$$

where ψ is the barrier exponent characteristic of the particular system under consideration and Δ is some constant. Since the dynamics is activated, the typical time to overcome a barrier B grows exponentially with B/T, T being the temperature; $t_{L,\mathrm{activated}} \sim \exp(B_L/k_B T)$. Hence, one would expect that after a time t domains of size

$$L(t) \sim \left(\frac{k_B T}{\Delta} \ln(t/\tau_0) \right)^{1/\psi} \tag{9.15}$$

are equilibrated (τ_0 being a microscopic time scale), i.e. the typical domain size grows logarithmically slowly with t. Numerically, one can determine the typical domain size $L(t)$ directly via the spatial two-replica correlation function

$$G(r,t) = \frac{1}{N}\sum_{i=1}^{N}\langle S_i^a(t)S_{i+r}^a(t)S_i^b(t)S_{i+r}^b(t)\rangle, \tag{9.16}$$

where a and b denote two replicas of the same system (i.e. the same disorder realization). For the first time, this spatial correlation function was studied numerically for the 2D EA spin glass model in Ref. [118], for the 3D EA model in Refs. [119–121], and later also in Refs. [122–125]. The extraction of the typical time-dependent domain-size, $L(t)$, from the correlation function $G(r,t)$ given in (9.16) is itself a delicate issue. In the studies in Refs. [119–120], the definition $L_T(t) = \int dr\, G_T(r,t)$ was used, where the subscript T signifies the temperature dependence. This form would be justified if $G_T(r,t)$ is of a pure exponential form $G_T(r,t) \sim \exp(-r/L_T(t))$. Also a scaling form $G_T(r,t) \sim \tilde{g}(r/L_T(t))$ has been checked in these early works using the length scale obtained via the integral method, and a good data collapse was obtained for all parameter values (times and temperatures) used. A more flexible functional form was assumed in Ref. [121] that was also capable of fitting more accurate estimates of $G_T(r,t)$ obtained later[123, 126]:

$$G_T(r,t) \sim r^{-\beta(T)}\,\tilde{g}(r/L_T(t)). \tag{9.17}$$

The best-fit values of temperature exponent $\beta(T)$ were found to be constant around 0.5 in three dimensions and decreasing from 1.6 to 0.9 in the temperature range 1.0–0.5 in four dimensions (where $T_c \approx 1.8$). In Ref. [125], only the tail of $G_T(r,t)$ was used to extract $L_T(t)$ assuming a pure exponential form for it. All in all, these different forms do not cause large variations in the estimates of $L_T(t)$, the reason simply being that the typical domain sizes reached in the times t accessible to Monte-Carlo studies to date are only of a few lattice spacings. It should be noted that these functional forms imply that $G_T(r,t)$ decays to zero with the distance r even in the limit $t \to \infty$, whereas in the droplet theory, one would expect a non-vanishing large distance limit: $\lim_{r\to\infty}\lim_{t\to\infty} G_T(r,t) = q_{\mathrm{EA}}^2(T) + \mathcal{O}(r^{-\theta})$. Functional forms respecting such an asymptotic behavior can also be devised to fit the data of $G_T(r,t)$ well.[126]

Suppose that we now have estimates of $L(t)$ obtained in one way or the other from numerical simulations or from experiments. (Experimentally, there is no straightforward way by which one could possibly measure $G(r,t)$ or its space and/or time Fourier transform via scattering or similar techniques. Still there is an interesting experimental development[127] on this point which we discuss further below.) When fitted to the logarithmic growth law (9.15), the estimated value of ψ was approximately 0.7[120] and was only acceptable if a finite offset length was introduced. However, a power law form with a temperature-dependent exponent

$$L(t) \propto t^{\alpha(T)} \tag{9.18}$$

fits the data well, with $\alpha(T) \approx 0.16\, T/T_c$.[120,121] This observation may indicate that we must replace the power law length dependence of the barriers (9.14) by the logarithmic dependence

$$B_{L(t)} \propto \Lambda \ln L(t) \tag{9.19}$$

as suggested in Refs. [128] and [129], or it may serve as a motivation to modify the simple domain growth picture discussed so far in a way that has been first discussed in Ref. [124], later also in Refs. [125, 126, 130] and [131]. The essential idea in these discussions is that the initial coarsening process is still influenced by critical fluctuations — which is plausible since (i) the temperature that are usually studied are not too far from the critical temperature $T/T_c \geq 0.7$ and (ii) in the initial stage the equilibrated length scales are still very small.

Close to T_c, the equilibrium correlation length is given by $\xi(T) \sim L_0|1 - T/T_c|^{-\nu}$, and even at $T < T_c$, the dynamics on length scales $L < \xi$ might be dominated by critical fluctuations rather than activated processes. Hence, one might expect that as long as $L < \xi$, the typical size of equilibrated domains grows with time t as $L(t) \sim l_0(t/t_0)^{1/z}$ (l_0 and t_0 being of microscopic length and time scales, respectively), and only for $L(t) > \xi$, activated dynamics obeying (9.15) sets in. Thus, a crossover from critical to activated dynamics happens at a time of order $\tau_0(T)/t_0 \sim (\xi(T)/l_0)^z$, when the typical domain size reaches $\xi(T)$:

$$L(t)/\xi(T) \propto g(t/\tau_0(T)) \quad \text{with } g(x) \sim \begin{cases} x^{1/z} & \text{for } x \ll 1 \\ (\ln x)^{1/\psi} & \text{for } x \gg 1. \end{cases} \tag{9.20}$$

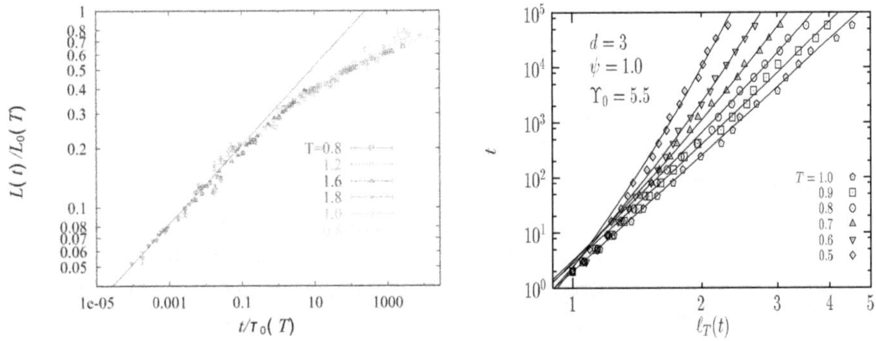

(a) (b)

Fig. 9.6. (a) Scaling plot of $L(t)$ of the $4d$ EA model according to (9.20), where the SG transition temperature $T_c = 2.0$ and the critical exponent $\nu = 0.93$ are fixed, but the dynamical exponent z is obtained to be 4.98(5) from the best scaling (from Ref. [125]). (b) Growth laws of the coherence length in the $3d$ EA model with fitting curves according to (9.21) (from Ref. [126]).

Thus, if one plots $L(t) \cdot (T_c - T)^{\nu}$ vs. $t \cdot (T_c - T)^{z\nu}$, one expects a data collapse for the time-dependent typical domain size at different temperatures. Indeed, such a data collapse has been observed in numerical studies of the 4D EA model [see Fig. 9.6(a)].[124]

One can go one step further[126, 132] and write down a more explicit form for the relation between length and time scales, similar to (9.15) but now taking into account critical fluctuations:

$$t(L) \approx \tau_0 L^z \exp\left(\frac{\Delta(T)L^{\psi}}{k_B T}\right) \quad \text{with } \Delta(T) \approx \Delta_0/\xi(T)^{\psi}, \tag{9.21}$$

where Δ_0 is an energy scale of order T_c. Here, following Ref. [115], the temperature-dependent free energy scale $\Delta(T)$ is assumed to vanish at T_c since one might expect that, in analogy to random bond ferromagnets, the surface tension of the domain walls vanishes at T_c. The pre-factor $\tau_0 L^z$ reflects the critical dynamics on short length scales (when the exponential term is still small) and might be interpreted as a renormalized microscopic time scale $\tilde{\tau}_0$ due to critical fluctuations. In this way, we obtain $t(L)/\tilde{\tau}_0 \approx f(L/\xi(T))$, with $f(x) = \exp(c\,x^{\psi})$ (and $c = \Delta_0/k_B T$) in reminiscence of (9.20). Note, however, that now the short-time critical dynamics is absorbed into $\tilde{\tau}_0$.

In Ref. [126], numerical data for the typical domain size $L(t)$ for the 3D and 4D EA model, obtained via MC simulations, were successfully fitted to the form (9.21), [see Fig. 9.6(b)]. For the barrier exponent, $\psi = 1.0$ was obtained in three dimensions using $\nu = 1.65$ and $z = 7.0$, while in four dimensions $\psi = 2.3$ was obtained using $\nu = 0.8$ and $z = 5.9$.

As we mentioned above, experimentally it appears to be impossible to have a direct access to spatial correlations in spin glass models. However, an indirect way to estimate a typical size of correlated volumes has recently been suggested in Ref. [127]. When measuring the waiting-time (t_w)-dependent thermo-remnant magnetization $M_{\rm TRM}(t, t_w)$, which is a two-time quantity discussed further below, its logarithmic derivative $S(t) = -d\,M_{\rm TRM}(t, t_w)/d(\ln t)$ at the time t after the field has been switched off has a characteristic peak at an *effective* waiting time $t_w^{\rm eff}$. One observes that the peak position depends on the strength of the field that is applied. Since this time scale is related, via the thermal activation $\tau_{\rm relax} \sim \tau_0 \exp(\Delta/k_B T)$, to the typical value of the free energy barriers that can be explored over experimental time scales: $\Delta(t_w) \sim k_B T \ln(t_w/\tau_0)$, the shift in the peak position of $S(t)$ contains information on the field dependence of the free energy barriers: $\Delta(t_w) - E_z \sim k_B T \ln(t_w^{\rm eff}/\tau_0)$, where in Ref. [127] it is assumed that E_z is the the magnetic (or Zeeman) energy associated with the change in the magnetic field.The latter is connected to the number N_s of spins that are involved in the barrier shift induced by the field change via $E_z \approx N_s \chi_{\rm fc} H^2$, where $\chi_{\rm fc}$ is the field-cooled susceptibility per spin and H is the magnetic field strength. If one assumes the N_s spins to be effectively locked together, one gets an estimate for a correlated volume $N_s \propto \xi^3$, i.e. an estimate for the coherence length ξ, as a function of the waiting time t_w.

By analyzing their data for the peak position of $S(t)$ to obtain $\xi(t_w)$ in the way just described, the authors of Ref. [127] found that it fits to (9.18) with $\alpha(T) \approx 0.169\, T/T_g$, in agreement with the numerical results reported in Refs. [120] and [121]. A fit to the logarithmic form (9.15) works equally well, but the best-fit value of ψ turns out to be rather large ($\approx$5), which does not agree with the numerical estimates cited above. These experiments probe a time window that is different from the one accessible to numerical simulations; each is roughly six decade wide but centered around times roughly 10 decades apart. Thus, only the algebraic growth law (9.18), or the form (9.21) that interpolates between critical and activated dynamics, would consistently match the data of both time windows.

9.5.2. *Two time quantities: Isothermal aging*

The out-of-equilibrium properties of glassy materials, quenched rapidly below the glass transition temperature T_g and then aged *isothermally* (i.e. at constant temperature $T < T_g$), manifest themselves most prominently in two-time quantities, typically response functions, susceptibilities, or correlation functions. The magnetic response function is defined as

$$R(t+t_w, t_w) = N^{-1} \sum_i \left. \frac{\delta \langle S_i(t+t_w) \rangle}{\delta h_i(t_w)} \right|_{h_i=0} . \tag{9.22}$$

In experiments as well as numerical simulations, one usually does not apply field pulses after a waiting time but one switches a constant magnetic field on or off after a waiting time t_w — the corresponding susceptibility is then related to the response function via a time integral

$$\chi(t+t_w, t_w) = \int_{t_w}^{t+t_w} dt' R(t+t_w, t'). \tag{9.23}$$

If one switches the field on at time t_w after the quench and measures the susceptibility, one calls it *zero-field-cooled* (ZFC), whereas if it is switched off, it is called *field-cooled* (FC). In the former case, the magnetization, which is within the *linear response regime* simply related to the susceptibility via $M(t, t_w) = h \cdot \chi(t, t_w)$, increases with the time t spent in the field, whereas in the latter case, it decreases with the time t after the field has been switched off — it is then also called the *thermo-remnant magnetization* (TRM).

Very useful for the development of theoretical concepts is the two-time spin auto-correlation function

$$C(t+t_w, t_w) = N^{-1} \sum_i \langle S_i(t+t_w) S_i(t_w) \rangle. \tag{9.24}$$

In *equilibrium*, it is related to the response function, the susceptibility, and the ZFC/FC magnetization via a *fluctuation–dissipation theorem* (FDT). (We discuss the violation of the FDT in Section 9.5.4.) However, since we are out of equilibrium, we have first to regard C as being an independent quantity.

The huge amount of experimental data that have been collected are nicely overviewed in Refs. [111] and [112]. Let us here simply state the main results. For the field-cooled magnetization, one observes that the remnant magnetization can be decomposed into a stationary part $M_{\rm eq}(t)$ that depends on t only and an aging part $M_{\rm aging}(t + t_w, t_w)$. Sometimes, this decomposition is assumed to be multiplicative, $M(t + t_w, t_w) = M_{\rm eq}(t) \cdot M_{\rm aging}(t + t_w, t_w)$, as in the universal short-time dynamics for coarsening at the critical point,[133, 134] and sometimes additive, $M(t + t_w, t_w) = M_{\rm eq}(t) + M_{\rm aging}(t + t_w, t_w)$, as in coarsening dynamics of pure systems[114] and in the out-of-equilibrium theory of mean-field spin glasses.[110] Since $M_{\rm aging}(t + t_w, t_w)$ is approximately constant for $t \ll t_w$, both forms usually fit the data well, although they are fundamentally different. For instance, the multiplicative decomposition is incompatible with the existence of a plateau in $C(t + t_w, t_w)$ that occurs in mean-field theories of spin glasses at low temperatures for long waiting times.[110] Analogously, it is also hardly compatible with a non-vanishing long time limit of $\lim_{t_w \to \infty} C(t + t_w, t_w)$, which is, within the droplet theory, identical to the order parameter $q_{\rm EA}$.

The most remarkable feature of the aging part is that it scales to a good approximation with t/t_w

$$M_{\rm aging}(t + t_w, t_w) \approx \tilde{M}_{\rm aging}(t/t_w). \qquad (9.25)$$

In ZFC experiments, instead of applying a constant field after a waiting time t_w, one usually applies a small oscillating field with frequency ω, which essentially means that one measures the t-Fourier transform $\chi(\omega, t_w)$ of $\chi(t + t_w, t_w)$. The frequency ω plays the role of an inverse observation time $1/t$, with $t \ll t_w$. The waiting time t_w is usually in the range of hours or more, while ω is in the range of 0.1–100 Hz, i.e., t is less than 10 seconds. Again, $\chi(t + t_w, t_w)$ can be decomposed into a stationary part $\chi_{\rm eq}(\omega)$ and an aging part $\chi_{\rm aging}(\omega, t_w)$ that now scale to a good approximation with ωt_w:

$$\chi_{\rm aging}(\omega, t_w) \approx \tilde{\chi}_{\rm aging}(\omega t_w) \qquad (9.26)$$

corresponding to the aforementioned t/t_w scaling for $M_{\rm FC}$. Numerically, one also has access to the spin autocorrelation function (9.24), which can also be decomposed into two parts (see Fig. 9.7). In Ref. [120], it has been shown that (i) the equilibrium part $C_{\rm eq}(t)$ decays algebraically with a very small temperature-dependent exponent $x(T)$ and (ii) that $C_{\rm age}(t + t_w, t_w)$ again adheres to a good approximation in the t/t_w scaling

$$C_{\rm eq}(t) \approx At^{-x(T)} \quad \text{and} \quad C_{\rm age}(t + t_w, t_w) \approx \tilde{c}(t/t_w), \qquad (9.27)$$

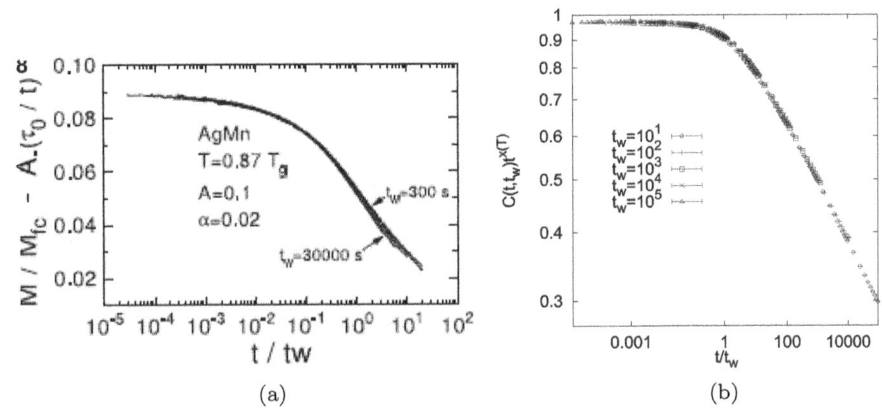

Fig. 9.7. (a) Aging part of the FC magnetization $M_{FC}(t+t_w, t_w)$ measured in an AgMn spin glass: the estimated stationary contribution $M_{\rm eq}(t)$ has been subtracted from the full measured value. The data are plotted versus t/t_w. The deviations from perfect t/t_w scaling occur for $t \sim t_w$ (sub-aging) and can be accounted for by choosing $h(t+t_w)/h(t_w)$ with $h(t)$ given in (9.30) as a scaling variable (from Ref. [111]). (b) Autocorrelation function $C(t + t_w, t_w)$, divided by $t^{-x(T)} \propto C_{\rm eq}(t)$, vs. t/t_w for $T = 0.4$ in the $3d$ EA model (from Ref. [120]).

where the scaling function $\tilde{c}$ behaves as $\tilde{c}(x) \propto x^{-\lambda(T)}$ with $\lambda(T)$ being a temperature-dependent exponent much larger than $x(T)$. This behavior was also found for the $4d$ EA model.[105] Later studies[126] focused on small but systematic deviations in the aging part from a simple t/t_w scaling behavior; the values of $C_{\rm age}(t + t_w, t_w)$ for a fixed t/t_w ratio show a slight tendency to decrease with increasing t_w, which is called *sub-aging* and is interpreted in terms of an *effective relaxation time* t_{eff} that is smaller than the actual waiting time t_w. We shall return to this point later.

Within a domain growth (or droplet) picture of the out-of-equilibrium dynamics of spin glasses,[115] only *one* characteristic length scale $L(t)$ for a given time t after the quench is assumed to exist. One would then expect that two-time quantities depend only on the ratio of the two length scales that are equilibrated within the two times t_w and $t + t_w$

$$O_{\rm aging}(t + t_w, t_w) \sim \tilde{o}(L(t + t_w)/L(t_w)), \tag{9.28}$$

where $O(t + t_w, t_w)$ is any two-time observable (such as R, χ, M or C) and $\tilde{o}$ is the corresponding scaling function. The first observation is that an algebraic growth law, (9.18) for instance, would then be completely compatible with the t/t_w scaling of the aging part of various observable reported so far. Note also that as long as $L(t)$ depends algebraically on t, the use of the scaling variable $L(t_w)/L(t)$ is equivalent to using $L(t+t_w)/L(t_w)$, both end up to be some function of t/t_w.

In the case of logarithmic domain growth (9.15), on the other hand, the scaling with $L(t_w)/L(t)$ is different from the scaling with $L(t + t_w)/L(t_w)$, and one should note that droplet scaling,[115] as it is usually applied to the experimental and numerical data, is assumed to work with $L(t_w)/L(t)$,[124, 131, 135] where $L(t_w)$ is the typical size of the domains after waiting time t_w and $L(t)$ is the typical size of the droplets being activated in the time t or being polarized by an ac field of frequency $\omega = 1/t$. The basic physical idea behind this scaling form is that the presence of a domain wall effectively reduces the excitation gap of droplets that it touches.

However, assuming the simple logarithmic growth law (9.15), $L(t) \approx ((k_BT/\Delta)\ln(t/\tau_0))^{1/\psi}$, the numerical as well as the experimental data turn out not to scale with $L(t_w)/L(t)$ as suggested by the droplet theory[115, 135] (it does not scale with $L(t+t_w)/L(t_w)$, either, which is incompatible with the experimental observation of sub-aging). For some time, this observation was taken as an indication that the droplet theory might be inappropriate to describe the out-of-equilibrium properties of spin glasses, at least for the time scales accessible to experiments and to numerical studies. However, recently it was pointed out[131] that, taking into account the crossover from critical to activated dynamics already discussed in the last subsection, the apparent inconsistency of the data with the droplet theory disappears. There experimental data for the (imaginary part of the) susceptibility $\chi(\omega, t_w)$, measured for the 3D Ising spin glass $Fe_{0.5}Mn_{0.5}TiO_3$ and the 3D Heisenberg spin glass Ag (11% at Mn), were presented. The aging part $\chi''_{\rm aging}(\omega, t_w)$ (obtained after subtracting a fitted equilibrium part $\chi''_{\rm eq}$) was shown to scale as $L(t_w)/L(1/\omega)$ with $L(t)$ following the logarithmic growth law (9.15), where the microscopic time and energy scales τ_0 and Δ are replaced by temperature-dependent characteristic time and energy scales, $\tau_c(T)$ and $\Delta(T)$. This replacement captures the features of the critical, short-time dynamics close to the critical point: $\tau_c(T) \sim \tau_m\xi^z$ and $\Delta(T) \sim \xi^{-\psi}$ ($\xi = |1 - T/T_c|^{-\nu}$ being the correlation length), which is in the same spirit as (9.20) and (9.21).

The authors of Ref. [131] found a very good data collapse when plotting their data for $(\chi''(\omega, t_w) - \chi_{\rm eq})/\chi''(\omega, t_w)$ versus the scaling variable $\ln(t/\tau_c(T))/\ln(\omega^{-1}\tau_c(T))$ (which is $L(t)/L(1/\omega)$, since $\Delta(T)$ and k_BT cancel in this ratio). Given the critical exponents ν and z from earlier studies, the fitting parameters are ψ, the whole equilibrium part $\chi_{\rm eq}$ (for which an explicit functional form predicted by the droplet theory was used), and a microscopic time scale τ_m. They found $\psi \approx 1.9$ for the Ising spin glass, which was much larger than what had been reported before in experiments[136] (see also Refs. [130] and [132]) and in the numerical simulations discussed in the previous subsection, and $\psi \approx 1.3$ for the Heisenberg spin glass. The authors of Ref. [131] showed that their experimental data are also compatible with the ωt_{w} scaling.

So far, we have discussed the compatibility of numerical and experimental data with a simple (or simplified) domain growth picture. Mean-field models, however, show a more complicated behavior.[110,137] The mean-field theory for the out-of-equilibrium dynamics predicts "ultra-metric behavior" in the time domains[138–140]:

$$O_{\text{aging}}(t+t_w, t_w) = \sum_i \mathcal{O}_i(h_i(t+t_w)/h_i(t_w)), \tag{9.29}$$

where the infinite sum over the index i refers to various large-time sectors[110,138] defined by the ratio of $h_i(t+t_w)/h_i(t_w)$ being of order 1, where the different (unknown) functions $h_i(t)$ represent different physical mechanisms at work during aging and need not necessarily be related to domain sizes. They are monotonously increasing and grow differently such that when $0 < h_i(t+t_w)/h_i(t_w) < 1$ holds for an index i, then $h_j(t+t_w)/h_j(t_w) = 1$ for all larger indices $j > i$ — implying that if two times t_1 and t_2 belong to the time sector defined by h_i and t_2 and t_3 to the one defined by h_j with $j > i$, then t_1 and t_3 also belong to the sector defined by h_i. One simple example for $h_i(t)$ is

$$h_i(t) = \exp\{(t/t_0)^{1-\mu_i}/(1-\mu_i)\}, \tag{9.30}$$

with $0 < \mu_i < 1$ ($\mu = 1$ yields a t/t_w-scaling, $\mu = 0$ indicates a time-translational invariance).

A consequence of such an Ansatz is, for instance, a hierarchy in these large-time sectors, called *dynamic ultrametricity*, which for the correlation function C means that in the limit of large times $t_1 < t_2 < t_3$,

$$C(t_3, t_1) = \min\{C(t_2, t_1), C(t_3, t_2)\} \tag{9.31}$$

holds. A number of consequences can be drawn from this ansätze.[110,137] Among them is a particular form of the violation of the fluctuation–dissipation relation, which we discuss later. Let us here just state that the mean-field theory predicts a richer scenario for glassy out-of-equilibrium dynamics than the domain growth picture with only one waiting-time-dependent length scale. However, as rich as this scenario is, all experimental data for real spin glasses and all numerical data for spin glasses with short-range interactions obtained for isothermal aging up to now can, to our knowledge, be scaled nicely with one single large-time domain, i.e. with only one term appearing in the infinite sum (9.29).

For instance, the aforementioned sub-aging property of two-time quantities that has been observed in experimental data[111, 141] (and also in analytically tractable coarsening, non-spin glass, models[142, 143]) can easily be accounted for by taking only a single term in the sum (9.29) with $h(t)$ being the function defined in (9.30) and with the exponent μ now being a fitting parameter. This form was actually used in the earliest experiments on aging in polymer glasses[144] and later also in spin glass experiments.[111] The exponent μ commonly turns out to be not very different from 1, accounting for the fact that the deviations from simple t/t_w scaling are usually very small. In numerical simulations of $3d$ models, the scaling with t/t_w, i.e. $\mu = 1$, is almost perfect. In four dimensions, the numerical data for $C(t + t_w, t_w)$ tend to show *super-aging*,[126] i.e. the values of $C_{\text{age}}(t + t_w, t_w)$ for fixed ratio t/t_w show a slight tendency to increase with increasing t_w. This can be interpreted by an effective relaxation time that grows faster than t_w, and it can be shown that the aging part $C_{\text{age}}(t + t_w, t_w)$ scales nicely with $h(t+t_w)/h(t_w)$, when $\mu > 1$ for $h(t)$ in (9.30) is chosen (which bears unfortunately mathematical inconsistencies[145]). However, the data can also be scaled with a domain growth scaling form (9.21) and a non-vanishing long time limit for the equilibrium part.[126]

9.5.3. *More complicated temperature protocols*

From what we learned from isothermal aging experiments described in the last subsection, it follows that the dynamics of a spin glass depends crucially on the age of the system, i.e. the time t_w that it spent in the glass phase, i.e. at $T < T_g$. Moreover, as experiments have impressively demonstrated,[111, 112] it depends sensitively on changes of temperature (and also field) *during* this waiting time t_w. A major goal of their systematic study described in this subsection is to understand the different phenomena that are measurable by applying different protocols (i.e. temperature variations during aging) and relating them to a theoretical picture, if not a theory, of the out-of-equilibrium dynamics of spin glasses.

The simplest experiments in this direction are small-temperature-shift experiments.[146] The protocol is as follows. First the system is rapidly quenched below T_c to a temperature T_1 where it is aged for a waiting time t_w. Then the temperature is shifted to a new temperature $T_2 = T_1 \pm \Delta T$ and the measurement is started. The first systematic numerical study of such a protocol was performed for the $3d$ EA model in Ref. [147] and a little later in Ref. [126]. It was observed that the decay of the correlation function is slower for negative shifts ($T_2 < T_1$) when compared with the same function aged isothermally at T_1. For small shifts ΔT, the functional form of $C(t + t_w, t_w)$ is the same as that for isothermal aging, and it can be matched with it using an effective waiting time $t_w^{\rm eff} < t_w$ for $T_2 < T_1$. If one assumes that the barriers that the system can surmount during a time t_w at temperature T_1 are of the same size as those that it can surmount during a time $t_w^{\rm eff}$ at temperature T_2, one obtains a good agreement with the numerical data, only if one replaces the microscopic time τ_0 by the typical time scale for critical dynamics as in (9.21).[126]

Obviously, such an interpretation of the numerical results implies that successive aging at two different temperatures will add to each other in a full accumulative way. The effective age $t_w^{\rm eff}$ is a monotonically increasing function of t_w and also depends on the pair (T_1, T_2): $t_w^{\rm eff} = f_{T_1,T_2}(t_w)$, and t_w can be related to $t_w^{\rm eff}$ by the inverse: $t_w = f^{-1}_{T_1,T_2}(t_w^{\rm eff})$. In [148], twin experiments — negative shift from T_1 to T_2 and positive shift from T_2 to T_1 — were considered. A criterion for purely accumulative aging would be $f^{-1}_{T_1,T_2}(t) = f_{T_2,T_1}(t)$, and in contrast to the aforementioned numerical results, deviations from it were observed in these experiments for $\Delta T/T_c \geq 0.01$, i.e. still very small shifts. The authors then concluded that positive and negative temperature shifts both cause a restart of aging and these findings were then interpreted as a symmetrical *temperature-chaos* effect.[148–150]

The temperature chaos in spin glasses is one of the basic ingredients of the droplet theory[115] and implies that the equilibrium configurations at two different temperatures T_1 and $T_2 = T_1 \pm \Delta T$ below T_g are uncorrelated beyond a length scale called the *overlap length*, $l_{\Delta T} \sim (\Delta T)^{-1/\zeta}$, where ζ is the chaos exponent. A number of experimental results, in particular obtained from temperature-shift or temperature-cycling experiments,[112] have been interpreted within this scenario. Numerical estimates of the value of ζ are based on small variations of the *couplings* J_{ij} (via $J_{ij} \to J_{ij} + \delta_{ij}$ with δ_{ij} small and random) at $T = 0$[151,152] rather than variations of the temperature, simply because ground states of spin glasses, in particular in

$d = 2$ can be obtained more easily than equilibrated configurations at small T. The reported estimates are $\zeta \approx 1$ both in two and three dimensions. While the same exponent was reported for temperature chaos (via MC simulations) in the 2D case, recent large-scale numerical studies of the $3d$ EA model did not show any evidence for temperature chaos in spin glasses on the length and time scales that could be probed.[153] Even if temperature chaos exists in 3D spin glasses, the overlap length might be much larger than the length scales that have been equilibrated during a particular waiting time, which then makes the chaotic rearrangements due to the temperature shift invisible. However, there still might be some effect due to dangerously irrelevant droplets whose free energy gaps are quite small, as discussed in Refs. [125,150] and [154].

The next temperature protocols that we discuss here are larger shifts and cycles $T_1 \to T_2 \to T_1$.[111] The first shift experiment was reported in Ref. [155], whereas cycle experiments can be found in Ref. [156]. The first qualitative numerical study of this kind was reported in Ref. [157]. In the theory of the asymptotic out-of-equilibrium dynamics of mean-field spin glass models, one can understand these effects on the basis of infinitely many time scales organized in a hierarchical — ultra-metric — way (which we will discuss in Section 9.5.4).

A systematic numerical study of large-shift and cycle experiments was recently performed for the $3d$ and $4d$ EA model in Ref. [126]. The basic message of large temperature shift experiments is that, independent of the sign of T_1–T_2, aging is "restarted" at the new temperature. This *rejuvenation effect* can nicely be observed in experiments as well as in simulations through the measurement of the susceptibility $\chi(\omega, t_w)$. Rejuvenation after a negative temperature shift comes from fast modes,[126, 158, 159] which were equilibrated at T_1, but fall out of equilibrium and are slow at T_2. Therefore, one should expect to see this phenomenon if the spin configurations in the equilibrated regions (on length scales $\leq L_{T_1}(t_w)$) are sufficiently different at the two temperatures. This mechanism is obviously qualitatively different from the interpretation involving the notion of temperature chaos (see above), which implies that length scale smaller than the overlap length are essentially unaffected by the temperature shift, while larger length scales are completely reshuffled by the shift. In this picture, rejuvenation is thus attributed to *large* length scales, and strong rejuvenation effects therefore require a very small overlap length.

No clear rejuvenation effects have ever been observed in simulations of the 3d EA model.[157,160–162] This was first attributed to the fact that the overlap length was perhaps numerically large, so that no large-scale reorganization could be observed on the time scale of the simulation. However, in the fast-mode mechanism mentioned above, the crucial ingredient is the small-scale reorganization due to a temperature shift. When comparing the spatial correlation function $G_T(r, t_w)$ at two different temperatures and waiting times such that the length scales $L_{T_1}(t_{w_1}) = L_{T_2}(t_{w_2})$, there is nearly no difference even for quite large $\Delta T = T_1 - T_2$ in three dimensions but significant differences in four dimensions,[126] which is compatible with the observation that the exponent $\beta(T)$ in (9.17) remains roughly constant for a large temperature interval in three dimensions, whereas it varies significantly in four dimensions. This observation suggests that the 4d EA model should be more favorable in studying temperature shift/cycle protocols. The numerical simulations of the 4d EA model show indeed a clear rejuvenation effect in $\chi(\omega, t_w)$, increasing smoothly with ΔT, which one would not expect within the temperature chaos picture (see Fig. 9.8(b)).

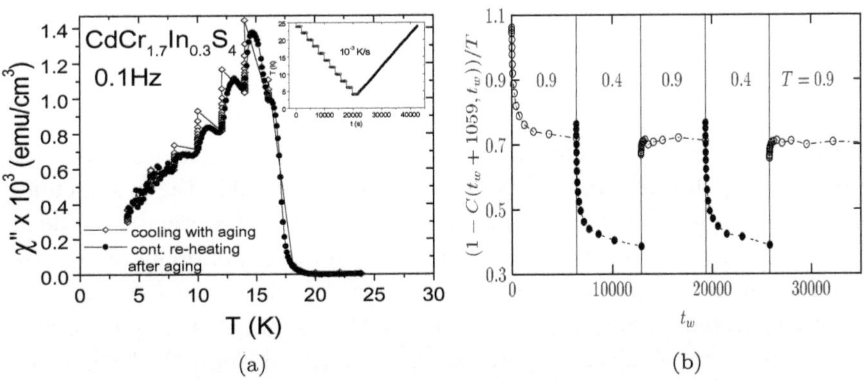

Fig. 9.8. (a) Series of "dip" imprinted on the a.c. susceptibility by successive stops at different temperatures while the system is cooled. Further cooling "rejuvenates" the system (i.e. the susceptibility goes up). However, the dips are one by one remembered by the system when it is heated again. For more details, see Ref. [163]. (The figure is adopted from Ref. [132]). (b) Evolution of the "a.c." correlation function taken from MC simulations of the 4d EA model in the procedure $T = \infty \to T_1 = 0.9 \to T_2 = 0.4 \to T_1 \to T_2$, showing, as in experiments the coexistence of rejuvenation and memory effects (from Ref. [126]).

The experimental procedure for cycles is $T = \infty \to T_1 \to T_2 < T_1 \to T_1$. The time spent at T_1 is t_s and the time spent at T_2 is t'_s. The spectacular *memory effect* arises when the temperature is shifted back to T_1. It is observed that although aging was fully restarted at T_2, the system has a strong memory of the previous aging at T_1. The dynamics at T_1 proceeds almost as if no cycle to T_2 has been performed.[155] The coexistence of rejuvenation and memory was made even more impressive in a multiple step, or *dip* experiment,[163] in which the temperature was decreased to T_1, then maintained for some time, then further decreased to $T_2 < T_1$, again maintained, then further decreased to $T_3 < T_2$, etc. At some lowest temperature, the sample was then re-heated with the same rate as it was cooled before, but now without the interrupts at $T_1, T_2, T_3, \ldots$. As shown in Fig. 9.8(a), on the way down to the final temperature, the susceptibility $\chi(\omega)$ as a function of the temperature showed clear dips at $T_1, T_2, T_3, \ldots$, which is the rejuvenation effect. Most remarkably, the same dips reappear in $\chi(\omega)$ on the way up, which is the memory effect. The spin glass has memorized the individual temperatures at which the constant cooling was interrupted for some time.

According to Ref. [132], the memory effect is a simple consequence of the separation of time and length scales, also observable in simpler slowly coarsening systems with activated dynamics.[159] When the system is at $T_2 < T_1$, rejuvenation involves very small length scales as compared to the length scales involved in the aging at T_1. Thus, when the temperature is shifted back to T_1, the correlation of length scale $L_{T_2}(t'_s)$ nearly instantaneously re-equilibrates at T_1. The memory is just stored in the intermediate length scales, between $L_{T_2}(t'_s)$ and $L_{T_1}(t_s)$. How such a storage mechanism actually works in a microscopic model for a finite-dimensional spin glass, like the EA model, is a challenging question. Progress has been made for simpler microscopic models, such as the $2d$ Mattis model, a simple coarsening model,[164] or the directed polymer in a random medium.[165] Also, experiments on random *ferromagnetic* systems,[166–170] on random *ferroelectric* systems,[171,172] and on frustrated systems without a spin glass transition at finite temperatures[173] show strong rejuvenation effects and a weak memory effect. This indicates that the existence of "many pure states" as predicted by Parisi's equilibrium solution of the Sherrington–Kirkpatrick model and the interpretation of these non-equilibrium effects based on it (in terms of a diffusion in a hierarchical space[111,174]) are not necessary prerequisites to observe rejuvenation and memory in the out-of-equilibrium dynamics.

Nevertheless, the hierarchical picture might be relevant in more complex situations and a concrete implementation was proposed in Ref. [175] in terms of a thermally activated random energy model (for an overview, see Ref. [110]). Here, to each level of a hierarchical tree, a transition temperature is associated, such that for each level the dynamics is stationary for higher temperatures and aging for lower temperatures. A small decrease of temperature induces some rejuvenation by driving out of equilibrium a new level of the tree, while freezing out the dynamics at the upper levels, thereby allowing the memory to be conserved. This model was recently studied further,[176] where it was shown numerically that the rejuvenation/memory effect is indeed already reproduced with two levels. A real space interpretation of this hierarchical tree was proposed in Refs. [175] and [177] and further developed in Ref. [159] in terms of a multi-scale dynamics. Low levels of the tree correspond to short-wavelength modes, which are only frozen at low temperatures, while large-wavelength modes are frozen at a higher temperature and constitute a *backbone*, where memory is imprinted. In the context of domain walls, this picture is particularly clear[159] and should apply directly to disordered ferromagnets where the slow dynamics comes from the motion of these pinned domain walls.[166–170]

9.5.4. *Violation of the fluctuation–dissipation theorem*

In a system in equilibrium, the response R to an external magnetic field (Eq. (9.22)) and the autocorrelation function C (Eq. (9.24)) depend only on the time difference t, i.e. $R(t+t_w,t_w) = R_{\rm eq}(t)$ and $C(t+t_w,t_w) = C_{\rm eq}(t)$, and are related to each other through the fluctuation–dissipation theorem (FDT)

$$R_{\rm eq}(t) = -\frac{1}{T}\frac{\partial C_{\rm eq}(t)}{\partial t}. \tag{9.32}$$

In a system that is out-of-equilibrium, this relation is generally not valid and the violation of FDT can be parameterized through a violation factor $X(t+t_w,t_w)$:

$$R(t+t_w,t_w) = -\frac{X(t+t_w,t_w)}{T}\frac{\partial C(t+t_w,t_w)}{\partial t_w}, \tag{9.33}$$

where the differentiation is done with respect to t_w, not the time difference t as in (9.32). In analytic studies of mean-field systems, it can be shown[138] that for large times ($t_w \to \infty$), X depends on t and t_w only through the value of the correlation function. Hence, $X(t+t_w, t_w) = X[C(t+t_w, t_w)]$. In particular, when $C > q_{\text{EA}}$, it is $X = 1$ and the FDT is recovered. In essence, the asymptotic dependence of X on the values of C alone underlies the decomposition into asymptotic time domains discussed in Section 9.5.2 under (9.29). It turns out[178] that the *effective temperature* $T_{\text{eff}}(t+t_w, t_w) = T/X(t+t_w, t_w)$ is precisely the temperature that would be read on a thermometer with response time t (or frequency $\omega \sim 1/t$) when connected to the magnetization at time t_w. A "fast" thermometer of response time $t \ll t_w$ will then probe the stationary regime for which $X = 1$ and thus measure the heat bath temperature.

Different scenarios of the FDT violation can occur. **(A)** $X(C) = 1$ for all C implies the system is in equilibrium, e.g. in the paramagnetic phase of a spin glass. **(B)** Coarsening systems with $q_{\text{EA}} = m^2$ have $X(C) = 1$ for $C > q_{\text{EA}}$ and $X(C) = 0$ for $C < q_{\text{EA}}$. **(C)** The so-called *discontinuous* spin glasses, such as the spherical mean-field spin glass model with p-spin interactions ($p \geq 3$),[138] have $X(C) = 1$ for $C > q_{\text{EA}}$ and $X(C) = x_1 < 1$ for $C < q_{\text{EA}}$, where x_1 is a constant. This corresponds to the existence of only *one* asymptotic time domain (see (9.29)), and the two-time quantities are expected to scale with only one (possibly unknown) function $h(t)$[179] **(D)** The so-called *continuous* spin glasses, like the SK model, have $X(C) = 1$ for $C > q_{\text{EA}}$ and $X(C)$ continuously varying, non-constant function of C for $C < q_{\text{EA}}$.[138]

A convenient way to extract the function $X(C)$ is to measure the autocorrelation function $C(t+t_w, t_w)$ and the susceptibility $\chi(t+t_w, t_w)$ (9.23) for a situation in which a (infinitesimal) small homogeneous external field h is switched on after a time t_w (in which case the induced magnetization is zero-field-cooled and $\chi(t+t_w, t_w) = \lim_{h\to 0} M_{ZFC}(t+t_w, t_w)/h$). In this case, we have

$$\begin{aligned}\chi(t+t_w, t_w) &= \int_{t_w}^{t+t_w} dt' R(t+t_w, t') \\ &= \frac{1}{T}\int_{t_w}^{t+t_w} dt'\, X[C(t+t_w, t')] \frac{\partial C(t+t_w, t')}{\partial t'}. \qquad (9.34)\end{aligned}$$

As long as the function $C(t+t_w, t')$ is monotonously increasing with t' for fixed time $t+t_w$, the substitution of $X(t+t_w, t')$ by $X[C(t+t_w, t')]$ is legitimate. One should keep in mind then that $X[C]$ still has a t_w dependence (that we could indicate by $X_{t_w}[C]$), which however vanishes in the limit

$t_w \to \infty$. Hence, one gets (with $C(t+t_w, t+t_w) = 1$):

$$\chi(t+t_w, t_w) = \int_{C(t+t_w,t_w)}^{1} dC\, X[C] \quad \text{or for fixed } t_w: \quad \frac{d\chi}{dC} = -\frac{X[C]}{T}, \tag{9.35}$$

which implies that for fixed waiting time the slope of a parametric plot of $\chi(t+t_w, t_w)$ versus $C(t+t_w, t_w)$ yields $X_{t_w}[C]$. In the limit $t_w \to \infty$, one should obtain the desired FDT violation $X[C] = \lim_{t_w\to\infty} X_{t_w}(C)$.

The first study of the FDT violation function $X(C)$ in the $3d$ EA model was performed in Ref. [180] (the first report of an FDT violation was actually already presented earlier in Ref. [181]). The non-constant part of $X(C)$ still showed a small but systematic t_w dependence for the small-C branch of the curve $X(C)$. Since no tendency to approach a straight line (scenario B or C) could be observed, the data were interpreted as an indication for scenario D, which implies that the 3D EA model has an out-of-equilibrium dynamics that is similar to the SK model. Later studies confirmed this result, in three dimensions[123] as well as in four dimensions.[182] However, in Ref. [125] it was proposed that the numerical data for the FDT violation in the $3d$ EA model (as well as those for the spatial correlations and the two-time quantities) can be interpreted in view of an extended droplet theory and that for the $3d$ EA model, scenario B (essentially slow coarsening in the spirit of the droplet model) is appropriate (see Fig. 9.9(a)). Since numerical simulations can cover only a few decades of the waiting time t_w and there is, as mentioned above, still a significant t_w dependence in $X(C, t_w)$ in the numerical data,[183, 184] a definite statement is hard to make.

For a clear experimental demonstration of the FDT violation, one has to measure both response and autocorrelation function (in Ref. [186] possible indications of the FDT violation based only on the response have been discussed). Such an experimental evidence of the violation of the FDT in spin glasses has been presented for the first time in Ref. [185] (for a structural glass, glycerol, it has been measured in Ref. [187], for a colloidal glass, laponite, in Ref. [188]), where autocorrelation function $C(t+t_w, t_w)$ has been measured via the magnetization fluctuations in a spin glass sample and the susceptibility $\chi(t+t_w, t_w)$ in the usual way by applying a magnetic field (see Fig. 9.9(b)). The χ vs. C plots show still a strong waiting time dependence and give some room for extrapolations to the long time limit — one that is suggested by the authors has similarities to the one

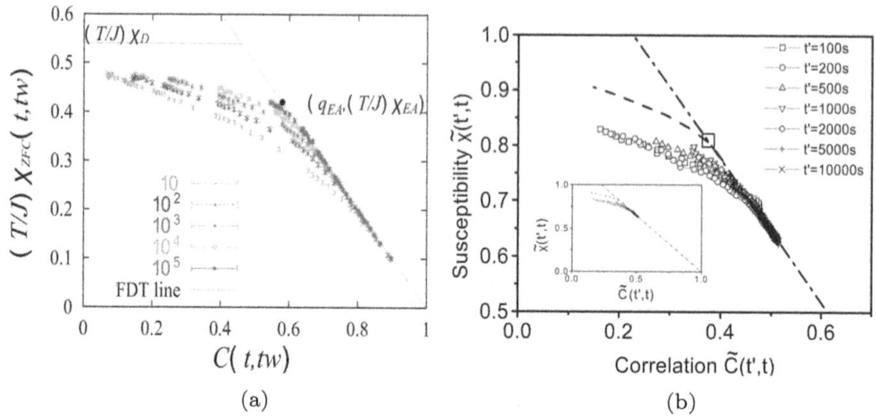

Fig. 9.9. (a) Parametric plot $(T/J)\chi_{\rm ZFC}(t+t_w, t_w)$ vs $C(t+t_w, t_w)$ at $T/J = 1.2$ for the 4*d* EA model. The straight tangent lines represent the FDT. The convergence points of the break points of time translational invariance $(q_{\rm EA}, (T/J)\chi_{\rm EA})$ and the break points of FDT $(q_{\rm D}, (T/J)\chi_{\rm D})$ (from Ref. [125]) (b) FDT plot obtained from the experiments performed in Ref. [185]. Relaxation measurements are plotted versus correlation functions for each t'. The dot–dashed line (FDT line) is calculated for $T = 0.8Tg = 13.3\,{\rm K}$, from a calibration obtained with a copper sample. The dashed line represents the scaling extrapolation for $t' \to \infty$. The branching point with the FDT line corresponds to $\widetilde{C} = q_{\rm EA}$ (square symbol, with size giving the error range). In the Inset, the same data are shown in the whole range (from Ref. [185]).

for the SK model. On the other hand, the aging part of their data for C and χ scale perfectly with $h(t_w)/h(t)$, where $h(u) = \exp(u^{1-\mu}/(1-\mu))$ with $\mu = 0.85$, which would not correspond to what one expects for the SK model.

The action for the long-time asymptotic behavior of the generating functional for the correlation and response functions is invariant under a time re-parameterization $t \to h(t)$, simply because in the long-time regime a re-parameterization of time with a monotonous function does not change the ultra-metric relation in the time ordering. The global invariance under the re-parameterization group (RpG) for mean-field models was recently extended[189,190] to its local variant for short-range models to treat fluctuations. In analogy to Heisenberg magnets, where the global spin rotation invariance lead to the existence of spin waves in finite-dimensional models via the Goldstone modes, the authors proposed that local time re-parameterizations play a similar role for the asymptotic dynamics of short-range spin glasses. One consequence of this observation is that one expects fluctuations in the local correlation functions and local susceptibilities

$$C_r(t+t_w,t_w) = \overline{S}_r(t+t_w)\overline{S}_r(t_w),$$
$$\chi_r(t+t_w,t_w) = \overline{S}_r(t+t_w)|_{h_r(t)=h\theta(t)}/h, \tag{9.36}$$

where $\overline{S}_r(t)$ means a slightly coarse-grained spin value at position r and time t, i.e. it involves a spatial average of the spin values over a small volume centered around r and a time average over a small time window centered around t. If the volume is extended to the system size, the usual global quantities discussed above are recovered. These spatially fluctuating quantities then have a joint probability distribution $\rho(C_r, \chi_r)$ that stretches along the FDT relation $\chi(C)$ for the global quantities χ and C, as has been shown numerically in Ref. [190].

9.5.5. *Hysteresis in spin glasses*

Due to the complex energy landscape of spin glasses, one expects strong hysteresis effects, which were studied recently via MC simulations[191–194] also using new techniques such as the recently introduced first order reversal curve (FORC) method.[195]

In Ref. [193], the zero-temperature dynamics of the $2d$ EA spin glass model was simulated with a varying external field H. The procedure used there was as follows. The magnetic field is changed in small steps, first downward from positive saturation and then upward from a reversal-field H_R. After each field step, the effective local field h_i of each spin S_i is calculated: $h_i = \sum_j J_{ij}S_j - H$. A spin is unstable if $h_iS_i < 0$. A randomly chosen unstable spin is flipped and then the local fields at neighboring sites are updated. This procedure is repeated until all spins become stable.

The first important observation was of a memory effect in the hysteresis of the $2d$ EA spin glass model that emerged when the magnetic field was first decreased from its saturation value and then increased again from some reversal-field H_R. It was found that the EA spin glass exhibited a singularity at the negative of the reversal field, $-H_R$, in the form of a kink in the magnetization of the reversal curve (see Fig. 9.10).

One can describe this effect within a phenomenological approach to hysteretic systems, the Preisach model.[196] In the Preisach model, a magnetic system is described as a collection of independent two-state (± 1) switching units, or "hysterons". Unlike Ising spins, which always align with their local field, the hysteron's state changes from -1 to $+1$ at a field $H_b + H_c$,

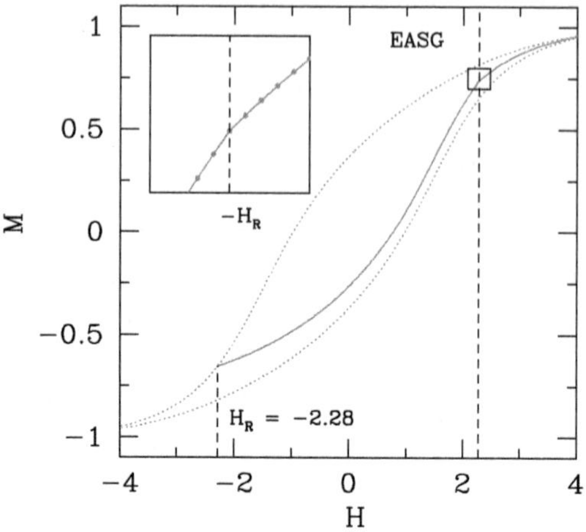

Fig. 9.10. Reversal curve (solid line) and major hysteresis loop (dotted line) for a two-dimensional (2D) EA spin glass with 10^4 spins and $H_R = -2.28$. In the inset, a kink is seen around $-H_R$ (from Ref. [193]).

different from the field $H_b - H_c$, required to switch the hysteron from $+1$ to -1. Different systems are distinguished by their different distributions $\rho(H_b, H_c)$ of hysterons of a given bias H_b and coercivity H_c. Here, $\rho(H_b, H_c)$ is the so-called Preisach function.

This function was extracted by a tool developed for analyzing experimental data of hysteretic systems.[195] A family of FORCs with different H_R was generated, with $M(H, H_R)$ denoting the resulting magnetization as a function of the applied and reversal fields. Computing the mixed second-order derivative $\rho(H, H_R) = -(1/2)[\partial^2 M/\partial H \partial H_R]$ and changing variables to $H_c = (H - H_R)/2$ and $H_b = (H + H_R)/2$, the local coercivity and bias, respectively, yielded the "FORC distribution" $\rho(H_b, H_c)$. For phenomenological Preisach models, the FORC distribution is equal to the Preisach function. However, the definition of the FORC distributions is more general, because it is extracted from numerical or experimental data and is thus model independent.

Figure 9.11 shows the FORC diagram of the EA spin glass. The ridge along the H_c axis in the range $1.5 < H_c < 4.0$ corresponds to the kinks of Fig. 9.10. Thus, FORC diagrams capture the reversal-field memory effect in the form of a ridge along the H_c axis. In Fig. 9.11 also the experimentally determined FORC diagram of thin films of well-dispersed single-domain magnetic Co-γ-Fe_2O_3 particles provided by Kodak Inc is shown.

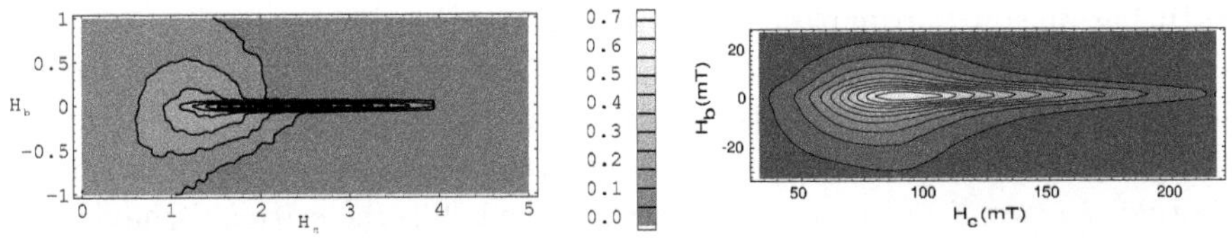

(a) (b)

Fig. 9.11. (a) FORC Diagram of the EASG. Note the ridge along the H_c axis. (b) Experimental FORC diagram of a Kodak sample. Note the similarity to the FORC diagram of the EASG shown in the left diagram (from Ref. [193]).

It clearly exhibits the horizontal ridge associated with the reversal-field memory effect. This striking similarity between the experimentally determined FORC diagram of the Co-γ-Fe_2O_3 films and the numerically determined FORC diagram of the EASG indicates not only that Co-γ-Fe_2O_3 films exhibit reversal-field memory but also that frustration may be a component of the physics of the Co-γ-Fe_2O_3 films. Note that, for instance, the $2d$ random-field Ising model does NOT exhibit a reversal-field memory and that its FORC diagram has a vertical ridge rather than a horizontal one.[193]

9.6. Equilibrium Properties of Classical *XY* and Heisenberg Spin Glasses

9.6.1. *Continuous spin models in three dimensions*

Compared to Ising spin glass models, less effort has been devoted to the models with continuous spins. One of the reasons might be that the minimal model of spin glasses, the Ising spin glass model, already shows a highly non-trivial behavior and poses serious difficulties to reach a final conclusion. Nonetheless, the continuous spin models are worth studying for several reasons. An obvious reason is that, in many real spin glass materials such as AuFe and CuMn, the magnetic anisotropy is much smaller than the (isotropic) exchange interaction. For these materials, therefore, the Heisenberg spin glass models would be more realistic than the Ising spin glass models.

However, the continuous spin models in three dimensions had long been thought of as models without a spin glass transition, which is another reason why the continuous spin models have not been studied so extensively until recently. The scenario that seemed to explain this apparent disagreement between the models and the materials was that the anisotropy is a relevant perturbation and, no matter how small it may be, it changes the system to an Ising-like spin glass, giving rise to a finite-temperature phase transition. This is quite analogous to what happens when a small amount of Ising anisotropy was introduced to the isotropic Heisenberg model (with no disorder) in two dimensions. If this scenario is valid, the critical behavior of any spin glass material should be the same and fall into the Ising spin glass universality class. However, there are some qualitative differences between experimental results for materials with a small anisotropy and numerical results for the Ising spin glasses. In particular, in a series of recent torque experiments[197, 198] of Heisenberg spin glass samples with varying degrees of anisotropies, indications for a freezing transition even within an external field were found (implying the applicability of a mean-field or RSB scenario rather than a droplet scenario to these Heisenberg spin glass systems). More surprisingly, the transition in spin glass materials with lower anistropy (more Heisenberg-like) appeared to be more robust against the application of an external field than those with higher anisotropy (more Ising like), which is in contrast with the traditional view that higher anisotropy implies a stronger tendency to order. These observations call for further theoretical investigations, and here, we report on recent theoretical efforts to scrutinize the existence of a transition in pure Heisenberg spin glasses without anisotropy. Below, we present two sets of evidences that seem to contradict each other. (The conflict has not been solved yet.) While the first set of evidences supports a novel phase transition via the chirality scenario (see below), the other suggests an apparently more familiar spin glass transition.

The first one was proposed by Kawamura and coworkers (for a review, see Ref. [199]). It was suggested that a phase transition may exist in a three dimensional Heisenberg model even if there is no anisotropy and belong to a new universality class different from that of the Ising spin glass model. In a series of numerical works (e.g. Ref. [71]), the origin of the disagreement between the experimental and numerical results was scrutinized by reconsidering the role of the magnetic anisotropy in the mechanism of the spin glass transition. The resulting hypothesis was that there is a phase transition already in the isotropic system but the spins do not show long-range correlations even below this critical point. It is therefore difficult, if not impossible, to detect an anomalous behavior in spin–spin correlations, which had been focused on in most preceding numerical simulations. Only the anisotropy gives rise to the long-range correlations in spins, not the transition itself, by coupling them to other degrees of freedom that have been already ordered. It was proposed that these degrees of freedom relevant for the transition are the chiralities defined as

$$\chi_i = \boldsymbol{S}_i \cdot (\boldsymbol{S}_{i+\delta} \times \boldsymbol{S}_{i+\delta'}),$$

where δ and δ' are two distinct unit lattice vectors.

The dynamical behavior of the 3D Heisenberg model with and without a uniaxial spin anisotropy was studied,[71] and a clear aging phenomena, for both the isotropic and the anisotropic cases, was found in the autocorrelation function of the chirality at sufficiently low temperatures. For the isotropic Heisenberg spin glass model in three dimensions with the Gaussian bond distribution, the critical temperature and the exponent were estimated as

$$T_{\mathrm{ch}} = 0.157(1), \quad \beta_{\mathrm{ch}} = 1.1(1).$$

Such a clear aging phenomena, however, was observed in the bare spin degrees of freedom only when the anisotropy exists. It was therefore suggested[72] that this phase transition is not accompanied by a freezing of spins in the isotropic case. As can be seen in Fig. 9.12, the estimate of the Binder parameter for the bare spins decreases as the system size increases at any temperature and does not show any crossing. In contrast, the Binder parameter of the chirality crosses.

An evidence of a finite-temperature phase transition was also found for the plane-rotator model (i.e. the XY model with two-component spins) in three dimensions. In Ref. [76], an equilibrium MC simulation was performed and a crossing in the Binder parameter defined in terms of the chirality

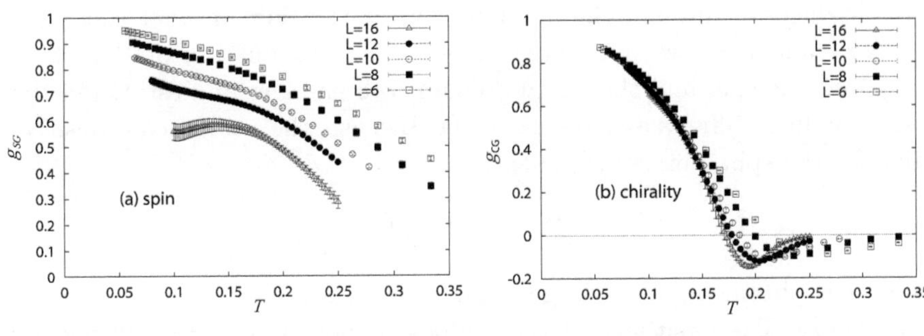

Fig. 9.12. The binder parameter for the $\pm J$ Heisenberg spin glass model in three dimensions. The left panel is for the spin overlap distribution and the right panel is for the chirality overlap distribution (from Ref. [72]).

was observed while, again, no such crossing was observed for the Binder parameter defined with the bare spins. For the $3d$ plane-rotator model with $\pm J$ bond distribution, their estimates are

$$T_{\rm ch} = 0.39(3), \quad \nu = 1.2(2), \quad \eta_{\rm ch} = 0.15(20), \quad z_{\rm ch} = 7.4(10).$$

It was further argued that, if a small but finite anisotropy exists, the spins and the chiralities are coupled, and therefore the spin sector, being dragged by anomalies in the chirality sector, also shows anomalies at the transition point. It follows that, as a function of the anisotropy, the transition temperature continuously approaches its isotropic limit. Moreover, the singular part of dragged quantities, i.e. those which diverge when the system is anisotropic and do not diverge when it is isotropic, are proportional to the quantity that drags them, namely, the chirality. The spin glass susceptibility, for instance, is given by

$$\chi_{\rm SG}(T, D) \sim D^4 (T - T_c)^{-\gamma_{\rm SG}^{\rm chiral}}.$$

The exponent $\gamma_{\rm SG}$ characterizing the spin glass susceptibility is the same as the one characterizing the chiral spin glass susceptibility. Their estimates of critical indices were compared with those of real materials, and many experimental estimates are closer to those of chirality transition than the Ising spin glass transition (see Table 9.1).

The second set of evidences contradicts these findings. In a few recent reports on the XY and the Heisenberg spin glasses in three dimensions, it was suggested that spins are ordered at the same finite temperature as the chiral degrees of freedom even in the isotropic case. The first such evidence was obtained[200] through the computation of the domain wall energy of the XY model in two and three dimensions. The authors of Ref. [200] computed the spin and the chiral domain wall energies and estimated the corresponding stiffness exponents. In the 3D case, they obtained a positive value for the spin stiffness exponent,

$$\theta_S^{\rm spin} = 0.056(11),$$

while they obtained a much larger stiffness exponent for the chiral domain-wall energy. The positivity of the stiffness exponent was reconfirmed by another computation based on the Coulomb gas representation.[201] However, since the estimated value is close to zero and the actual increase observed in the domain wall energy was only about 10% for the whole range of system sizes explored, these results on the stiffness exponent alone could not be taken as a conclusive evidence.

Additional evidences came from MC simulations at a finite temperature,[73, 202, 203] in which the Binder parameters for the magnetization and the chirality of the 3D $\pm J$ Heisenberg model were computed. A common crossing of the curves for different system sizes was found in both Binder parameters. The two crossing temperatures, one for the spin and the other for the chirality, were close to each other. Based on this result, the presence of a finite-temperature phase transition at which both the spin and the chiral degrees of freedom are ordered was suggested.

This result was confirmed in Ref. [74] where again the $\pm J$ Heisenberg spin glass in three dimensions was studied. There the relaxation of several quantities at temperatures close to the critical one was measured, starting from a completely random initial configuration. In the time regime where the time-dependent length scale discussed in Section 9.5.1 is smaller than the system size, the size dependence was absent in the auto-correlation function and the observed time dependence could be regarded as the one in the thermodynamic limit. Right at the critical temperature, this time dependence was well described by an algebraic function for diverging quantities. For example, for the spin glass susceptibility, we have

$$\chi_{\mathrm{SG}} \propto t^{\lambda}, \tag{9.37}$$

with $\lambda = \gamma_{\mathrm{SG}}/z\nu$.[204] Using this asymptotic form, the critical temperature as well as the critical exponents (divided by $z\nu$) can be determined through the observation of the time dependence of some quantity. The exponent $z\nu$ can be obtained from the temperature dependence of the characteristic time

$$\tau(\Delta T) \propto (\Delta T)^{-z\nu},$$

where $\Delta T \equiv T - T_{\mathrm{sg}}$. In addition, z was estimated from the results of relatively small system sizes for which the correlation length was larger than the system size. In such a setting, the size dependence of the correlation time near the critical point should be described by

$$\tau(\Delta T = 0) \propto L^{z}.$$

The estimated critical temperature and exponents for the Heisenberg spin glass model with $\pm J$ bond distribution in three dimensions were

$$T_{\mathrm{sg}}/J = 0.21\,{}^{+0.01}_{-0.03}, \quad \gamma_{\mathrm{sg}} = 1.9(4), \quad \nu = 1.1(2), \quad \beta = 0.72(6).$$

It should be noted that this estimate of ν agrees with some of the experiments on spin glass materials that appear not to be in the Ising universality class.

A clearer demonstration of the existence of the spin ordering was presented in Ref. [75], in which an equilibrium MC simulation was performed covering not only the critical point but also a fairly large region in the low-temperature phase. Instead of the Binder parameter, which is an indirect probe of the presence of the long-range order, they used the correlation length itself as in the estimation of the transition temperature of the Ising spin glass model[51] (see Fig. 9.3). For both the XY and the Heisenberg spin glass with the Gaussian bond distribution in three dimensions, a clear crossing in the correlation length divided by the system size as a function of the temperature was observed (Fig. 9.13). The critical indices were estimated as

$$T_{\rm sg} = 0.34(2), \quad \nu = 1.2(2) \quad (XY)$$

and

$$T_{\rm sg} = 0.16(2), \quad \nu = 1.1(2) \quad \text{(Heisenberg)}.$$

It is rather puzzling to have two conflicting sets of numerical evidences, represented, e.g. by Figs. 9.12 and 9.13, as each would pass as a convincing evidence if the other did not exist. This puzzling situation has not been resolved yet. It is at least clear that there is an unexpectedly large correction

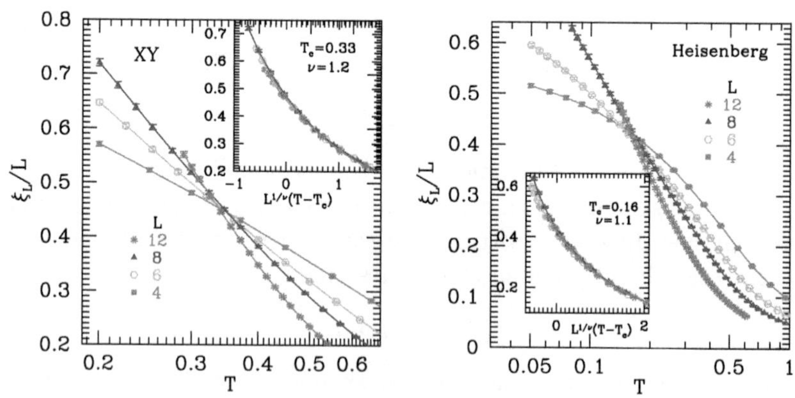

(a) (b)

Fig. 9.13. Correlation lengths of bare spins of the XY model (a) and of the Heisenberg model (b) divided by the system size (from Ref. [75]).

to scaling and that we need to answer why one of the seemingly legitimate methods for detecting the long-range order failed in the present case.

9.6.2. *Continuous spin models in higher dimensions*

The Heisenberg spin glass model with Gaussian couplings was studied in four and in five dimensions via equilibrium MC simulations in Ref. [205]. As in the study of the 3D case[72] the Binder parameter for spins and chiralities was calculated. In the 4D case, it was found that the spin Binder parameters for various system sizes do not intersect while the chirality Binder parameters do. In contrast, in the five dimensional case, both kinds of the Binder parameters intersect. Based on these results, it was suggested that the spin chirality separation occurs only in four or lower dimensions, while in five or higher dimensions, an ordinary phase transition, at which both degrees of freedom freeze, takes place. In five dimensions, the transition temperature was estimated as $T_c = 0.60(2)$ and the critical indices were estimated as

$$\alpha = -1.0(3), \ \beta_{\rm SG} = 0.7(3), \ \gamma_{\rm SG} = 1.7(2), \ \nu_{\rm SG} = 0.6(2).$$

Indications were found that the chiral order phase is characterized by one-step replica symmetry breaking unlike the mean-field model. The findings for the 4D model were not as clear as those for the five dimensional model. In four dimensions, the chiral transition temperature was estimated to be $T_c = 0.38(2)$, but no reliable estimates for the critical indices for the 4D model could be obtained. This may be because four dimensions is marginal or nearly marginal and the critical region is rather narrow.

For these numerical evidences based on the Binder parameters, however, one has to be aware that the same ambiguities might be present in four and five dimensions as in the 3D case discussed in the previous subsection. Therefore, while the existence of a phase transition has been established in four and five dimensions as well as in three, we still cannot make a conclusive statement concerning the nature of the phase transition in continuous spin models for the same reason as in the 3D case.

9.6.3. *Potts spin glasses*

Potts spin variables are discrete variables that can take on q different states $S_i \in \{1, 2, \ldots, q\}$ and that interact with one another in such a way that only two cases are discriminated: either both interacting spins are in the same state or they are in different states. The Hamiltonian of the q-state Potts glass is given by

$$H = -\sum_{(i,j)} J_{ij}(q\delta_{S_i,S_j} - 1), \tag{9.38}$$

where the coupling constants J_{ij} are quenched random variables and the sum runs over all interacting spin pairs (i.e. all possible spin pairs in a mean-field model).

The q-state infinite-range Potts glass with $q > 4$[206,207] displays a dynamical phase transition at a temperature T_d, where the dynamics freezes and, for instance, the spin auto-correlations do not decay any more (i.e. they reach asymptotically a non-vanishing plateau value). The second, static, transition takes place at a lower temperature T_c, below which the EA spin glass order parameter has a non-vanishing value and the replica symmetry is broken. On the mean-field level, there is a close connection between Potts glasses for $q \geq 4$ and Ising spin glass models with p-spin interactions for $p \geq 3$, which again display similar self-consistency equations as the mode-coupling equations describing structural glasses above the glass transition temperature.[208,209] Therefore, Potts spin glasses can also be regarded as simple prototypical models for structural glasses.

The question unanswered to date is whether this analytically well-established mean-field scenario with the two separate transitions is also valid in finite-dimensional models with short-range interactions. Recently, the q-state Potts glass was thoroughly investigated via MC simulations in three dimensions,[210,211] on the simple cubic lattice, with $q = 10$ for a discrete, bimodal bond distribution as well as a Gaussian distribution. In both cases, the first two moments of the distribution were chosen such that no ferromagnetic ordering of the Potts spins could occur. It was found that for all temperatures investigated, the spin glass susceptibility remained finite, the spin glass order parameter remained zero, and the specific heat had only a smooth Schottky-like peak. These results could be understood quantitatively by considering small but independent clusters of spins. These observations imply that there is no static phase transition at any nonzero temperature. Consistent with these findings, only very minor size effects were observed, which implied that all correlation lengths of the models remained very short.

Moreover, the auto-correlation function $C(t)$ of the Potts spins was computed. While in the Gaussian model $C(t)$ shows a smooth uniform decay, the correlation for the $\pm J$ model has several distinct steps. These steps correspond to the breaking of bonds in small clusters of ferromagnetically coupled spins (dimers, trimers, etc.). The relaxation times follow simple Arrhenius laws, with activation energies that are readily interpreted within the cluster picture, giving an evidence that the system does not have a dynamic transition at a finite temperature. Hence, one can conclude that all the transitions known for the mean-field version of the model are completely wiped out in three dimensions.

It should also be mentioned that the mean-field model of the Potts spin glass was studied numerically via MC simulations.[212–214] It turned out that it is extremely difficult to control the finite-size effects in the mean-field model and that the fourth-order cumulant $g_4(N,T)$ and the Guerra parameter $G(N,T)$ could not be used to locate the static transition temperature for the system sizes investigated. Also the spin autocorrelation function $C(t)$ showed strong finite-size effects and did not display a plateau even for temperatures around the dynamical critical temperature T_d.

9.7. Weak Disorder

While there are only very few exactly known facts about spin glass models, they serve as helpful guides to numerical studies. This is particularly the case in the study of random spin systems with a weak disorder. An example is the findings based on the gauge invariance (a brief review can be found in Ref. [215]). A line in the phase space, commonly referred to as the Nishimori line (simply N-line, hereafter), was found on which the free energy can be computed exactly. The line is associated with the crossover from the high-temperature ferromagnetic region to the low-temperature disorder-dominated region. The results derived from the gauge invariance seem to have a close connection to a number of renormalization group studies. In the following, we present a gross survey of these findings and related works.

9.7.1. *Phase diagram of the discrete spin models*

It was pointed out[216] that the gauge invariance of the partition function under the transformation $S_i \to \sigma_i S_i$ (with an arbitrary choice of $\sigma_i \in \{-1,+1\}$ for every site) can be used to derive a number of exact relations among various quantities in spin glass models. This gauge transformation relates thermal fluctuations to the geometrical properties of the bond configurations. Consequently, a *geometric temperature*, T_p, can be defined that depends only on the bond distribution. This temperature appears in various exact relationships derived from this gauge invariance. An example is the identity for correlation functions in $\pm J$ models

$$[\langle S_i S_j \rangle_T] = [\langle S_i S_j \rangle_T \langle S_i S_j \rangle_{T_p}], \tag{9.39}$$

where $\langle \cdots \rangle_T$ is the thermal average at temperature T. In particular, $\langle \cdots \rangle_{T_p}$ means the thermal average with the real temperature and the geometrical temperature are equal to each other. The symbol $[\cdots]$ denotes, as usual, the bond configuration average. The geometric temperature T_p is defined as

$$p = (1 + \tanh(J/T_p))/2 \quad \text{or} \quad T_p = \frac{2J}{\log \frac{p}{1-p}},$$

where p is the concentration of the ferromagnetic bonds. The higher the randomness ($p \to 1/2$), the higher the geometric temperature ($T_p \to \infty$). The line in the $p - T$ phase diagram defined by $T_p = T$ is the N-line. Along this line, (9.39) infers $m = q$, i.e. the spin glass order is always accompanied by the ferromagnetic order. Therefore, no part of the line lies inside the pure spin glass phase. (Here, we define the spin glass phase by the conditions $m = 0$ and $q > 0$.) Another consequence of (9.39) is that the ferromagnetic long-range order is absent at any temperature if the concentration p is such that the point (p, T_p) falls in the paramagnetic region in the p–T plane. This can be understood by an inequality directly derived from (9.39):

$$|[\langle S_i S_j \rangle_T]| \leq [|\langle S_i S_j \rangle_{T_p}|]. \tag{9.40}$$

If the point (p, T_p) belongs to the paramagnetic region, the right-hand side of the inequality is zero in the limit of $R_{ij} \to \infty$, which makes the left-hand side also zero, meaning the absence of ferromagnetic order at (p, T). Assuming that the topology of the $T - p$ phase diagram is as depicted in Fig. 9.14(a), this fact implies that any phase in the region $p < p_N$ (or $T_p > T_N$) cannot have a ferromagnetic long-range order where p_N is the ferromagnetic bond concentration at the N-point. Here the N-point (p_N, T_N) is defined as the point at which the paramagnetic phase boundary intersects with the N-line. It follows that the phase boundary separating the ferromagnetic phase from the other low-temperature phase (paramagnetic or spin glass) must be strictly vertical or bend toward the ferromagnetic side (see Fig. 9.14).

The inequality (9.40) also means that the ferromagnetic order is maximal on the N-line when p is fixed. When $T_p < T_N$, as we decrease the temperature departing from the N-line, the magnetization decreases and

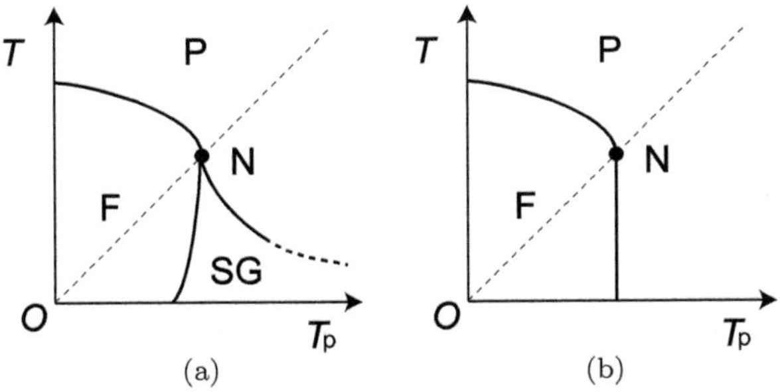

Fig. 9.14. The schematic p–T phase diagram of the $\pm J$ Ising model for (a) the generic case and (b) the 2D case. The thin dashed lines represent the N-line.

the spin glass order parameter becomes larger than the square of the magnetization squared. This hints a cross-over from the purely ferromagnetic region to the randomness-dominating region at the temperature $T = T_p$.

It was further argued[217] that the phase boundary below the N-line is strictly vertical. The argument is based on the fact that the free energy on the N-line is the same as the geometric entropy, i.e. the entropy of the frustration distribution at p

$$S(p) \equiv -\sum_F P(F) \log P(F),$$

where F is a configuration of frustrated plaquettes and $P(F)$ is the probability of F being realized. Since the free energy (or its derivative with respect to some external field) must have some singularity at the N-point when one moves along the N-line, so does the geometric entropy. The latter singularity is, however, solely due to the geometric properties of the bond configuration, in which the temperature does not play any role. Therefore, it is reasonable to suspect that the same singularity with the geometric origin may affect the spin system defined on it at $p = p_{\rm N}$ regardless of the temperature. This leads to a temperature-independent transition point, i.e., the strictly vertical phase boundary (see Fig. 9.14). Note that the argument also applies to models with continuous degrees of freedom, such as the XY model and the Heisenberg model, if the ferromagnetic phase exists in the pure model. It was claimed[217] that the vertical phase boundary between the ferromagnetic phase and the other is universal.

In order to substantiate this argument, the authors of Ref. [218] considered the random Z_q model with gauge symmetry, $\mathcal{H} = -\sum_{(ij)} V(S_i - S_j + J_{ij})$, where S_i, S_j, J_{ij} take on a value $0, 1, 2, \ldots$ or $q-1$, and $V(\cdots)$

is a periodic function of the period q. This model includes the $\pm J$ Ising spin glass model as a special case. The sample-to-sample fluctuations of the energy along the N-line was computed via the exact relationship between the sample-to-sample fluctuations and the thermal fluctuations

$$[(\Delta E)^2] \equiv [E^2] - [E]^2 = N_{\rm B}(\{V^2\} - \{V\}^2) - [\langle H^2\rangle - \langle H\rangle^2].$$

Here, $E \equiv \langle \mathcal{H} \rangle$, $N_{\rm B}$ the number of nearest-neighbor pairs, and $\{\cdots\}$ denotes a single-bond average defined by $\{Q(l)\} \equiv \sum_{l=0}^{q-1} \exp(-\beta V(l))Q(l)/\sum_{l=0}^{q-1} \exp(-\beta V(l))$. The sample-to-sample fluctuations are proportional to the "geometric specific heat", i.e. the energy differentiated by $K_p \equiv 1/T_p$. The results showed that $[(\Delta E)^2]$ takes its maximum at the N-point. The authors of Ref. [218] argued that this indicates the existence of a singularity in the geometric nature of the system at the N-point, though the nature of the singularity is not known possibly because of the size limitation in their calculation.

It is interesting to note that the correlation length exponent ν at the zero-temperature phase transition from the paramagnetic phase to the ferromagnetic phase in two dimension was found[31,219] to be very close to (or perhaps exactly equal to) the one for 2D percolation. This fact, together with the geometric mechanism of the transition, makes it very attractive to speculate that the transition is due to a percolation of something that is defined geometrically. However, this "something" has not yet been identified.

9.7.2. *Dynamical properties*

Based on the gauge invariance of the model, several interesting exact relations were derived[220] for the dynamics. Particularly important is the one that relates the auto-correlation function with the totally aligned initial state and that with the initialization at $T = T_p$. Specifically,

$$[\langle S_i(t+t_w)S_i(t_w)\rangle_T^{(\mathrm{F})}] = [\langle S_i(t+t_w)S_i(T_w)\rangle_T^{T_p}]. \tag{9.41}$$

The superscripts specify the initial spin configurations: "(F)" indicates that the initial state is the totally aligned ferromagnetic state, and "T_p" an equilibrium spin configuration at T_p. The left-hand side is the autocorrelation function of a spin S_i at the t_wth Monte Carlo step (MCS) after the initialization (at $t = 0$) and itself at the $(t + t_w)$th step. At $t > 0$, the system evolves with the dynamics of a finite temperature T. The right-hand side is the auto-correlation function after a quenching from the temperature T_p to T (or a sudden heating up, depending upon whether $T > T_p$ or $T < T_p$). Of particular interest is the relation obtained by setting $t_w = 0$ and $T = T_p$ in the above

$$[\langle S_i(t)\rangle_{T_p}^{(\mathrm{F})}] = [\langle S_i(t)S_i(0)\rangle_{T_p}^{T_p}].$$

This indicates the equivalence of the equilibrium auto-correlation function on the N-line to the non-equilibrium relaxation of the single-spin expectation value starting from the all-aligned condition. By using this relation, we can simply measure the value of a spin without equilibrating at all to obtain the equilibrium autocorrelation function. This is a considerable advantage from the computational point of view.

9.7.3. *The renormalization group approach for the discrete models*

The Harris criterion[221] is well known as the criterion by which one can decide whether introduction of a weak disorder to a pure system is relevant or irrelevant: If the specific heat diverges at the critical point of the pure system, i.e. the specific heat exponent α is positive, the universality class of the transition in the disordered system will be different from that of the pure system. In particular, it was argued that if the disorder is relevant, the specific heat divergence is smeared out by the effect of the quenched disorder.

The argument was elaborated further with the help of the renormalization group theory.[222] The analysis of the randomly diluted m-vector model showed that the quenched disorder is relevant around the "pure" fixed point (denoted as "P" hereafter), if the specific exponent α is positive at P. As a result, the existence of another fixed point, called the random fixed point (denoted as "R"), was suggested, and the renormalization group flow leads the system to R when starting in the vicinity of P. However, because of the small cross-over exponent, i.e. α in this case, the critical region where the true critical behavior of R can be observed may be very narrow in many cases, such as the random Ising model in three dimensions ($\alpha \sim 0.11$). For instance, results of MC simulations of the site-disordered Ising system[223] were interpreted in terms of a non-universal behavior, i.e. continuously varying critical exponent depending on the strength of the disorder. It was argued that this apparent non-universal behavior can be interpreted as a cross-over from P to R.

As for the critical indices at R, the following values for 3D systems were reported[224] based on the four-loop order field-theoretic renormalization group calculation

$$\beta = 0.348, \ \gamma = 1.321, \ \nu = 0.671, \ \eta = 0.032. \tag{9.42}$$

In Ref. [225] the role of the gauge invariance in the renormalization group method was scrutinized. It was found that the defining condition of the N-line, $T_p = T$ is invariant under a certain renormalization group transformation. More specifically, in an extended phase space an invariant manifold exists that intersects with the p–T plane with the intersection being the N-line. A fixed point is located on this manifold and can be identified with the one corresponding to the multi-critical point that is shared by three phases: paramagnetic, ferromagnetic, and spin glass. One of the directions of the two scaling axes is also found to be parallel to the temperature axis while the other is tangent to the N-line. The fixed point is unstable in both directions.

The authors of Ref. [226] studied the site-diluted Ising model with MC simulation. Their estimates for the critical indices were

$$\beta = 0.3546(18), \ \gamma = 1.342(5), \ \nu = 0.6837(24), \ \eta = 0.0374(36).$$

These are very close to the corresponding estimates of the indices for the pure model such as[227]:

$$\beta = 0.3250(15), \ \gamma = 1.241(2), \ \nu = 0.6300(15).$$

Therefore, it is technically rather difficult to discriminate a critical point corresponding to the random fixed point from the pure fixed point.

The numerical renormalization group approach based on MC simulation[228] suggested the simplest flow diagram consistent with all of these predictions. The variance and the mean of the domain wall free energy were computed and analyzed with the method of the domain wall renormalization group.[28] For the randomly diluted spin systems in three dimensions, the flow diagram turned out to be the simplest one consistent with preceding theoretical predictions (see Fig. 9.15). The RG flow diagram for the bond-diluted model consists of two finite-temperature fixed points (the pure fixed point (P) and the random fixed point (R)) together with two zero-temperature fixed points (the ferromagnetic fixed point (F) and the zero-temperature percolation fixed point (Z)). The two eigen values of the

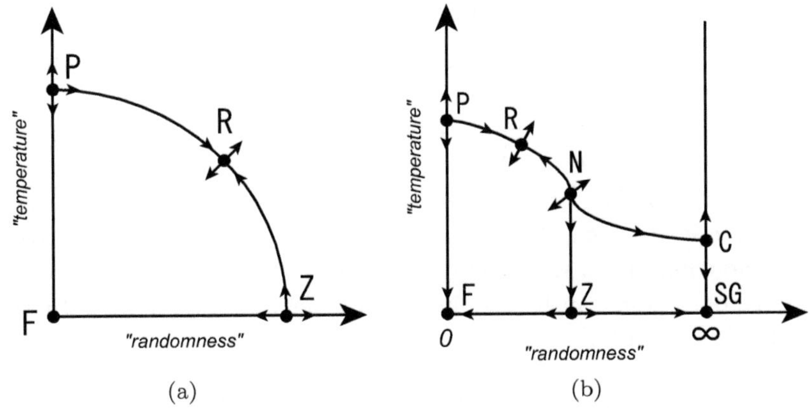

Fig. 9.15. A schematic RG flow diagram for (a) non-frustrated models (e.g. the randomly diluted ferromagnet) and (b) frustrated models (e.g. the $\pm J$ model) in three dimensions.

linearized RG transformation were evaluated at R. The positive one turned out to be

$$y_1 = 1.47(4),$$

which is consistent with the previous estimates of $\nu = 1/y \sim 0.68$ mentioned above. The other negative one was

$$y_2 = -1.3(4).$$

A similar calculation for the site-diluted system yielded a result consistent with these.

When in addition the system is randomly frustrated, the simplest flow diagram must contain at least three more fixed points: the multi-critical or Nishimori fixed point (N), the spin glass critical fixed point (C), and the spin glass zero-temperature fixed point (SG). The findings in Ref. [228] were indeed consistent with this simplest flow diagram (Fig. 9.15). In this case, the exponent at R was estimated as

$$y_1 = 1.52(2) \quad \text{and} \quad y_2 = -0.42(13).$$

While the positive one agrees with the estimate for the non-frustrated system, the other does not agree. The reason for this has not been clarified yet.

9.7.4. *The location of the multi-critical point*

For the location of the multi-critical point, or the N-point, of the $\pm J$ model, there are a number of numerical estimates. In particular, the non-equilibrium simulation was employed effectively to deduce the equilibrium properties. The basic idea of the method can be traced back to the dynamical scaling hypothesis.[229, 230] Various scaling relations, such as (9.37), can be derived from it. It is assumed that a generating function, which is an extension of the equilibrium free energy, exists and that the time dependence of various quantities can be obtained as its derivatives. The hypothetical generating function has the form

$$f(\Delta T, h, L, t) = L^{-d}\tilde{f}(\Delta T L^{y}, hL^{y_h}, tL^{-z}) \tag{9.43}$$

analogous to the ordinary form of the finite-size scaling. While this form is presumably valid in the near-equilibrium time regime, it was suggested[133] that there exists an initial time regime where the dynamics is governed by another critical exponent independent of those which appear in (9.43).

Using the scaling relations derived from (9.43), the non-equilibrium relaxation of the system was studied. The authors of Ref. [231] measured the magnetization $m(t) \equiv [\langle S_i(t)\rangle_T^{(F)}]$ on the N-line, which is proved to be the same as the equilibrium autocorrelation function $q(t)$ as mentioned above (see Section 9.7.2). Then they used the following scaling form:

$$m(t) = L^{-\beta y}\tilde{q}((T - T_{\rm N})L^{y}, tL^{-z})$$

that can be derived from (9.43). In particular, the long-time asymptotic form at $T = T_{\rm N}$,

$$m(t) \sim t^{-\beta y/z}$$

was used to obtain the effective exponent at a finite time,

$$\lambda(t) \equiv -\frac{d\log m(t)}{d\log t},$$

which should converge to $\lambda \equiv \beta y/z$ as $t \to \infty$ at the critical point. A good convergence to a finite value was observed around the N-point for the 2D and 3D models, although the convergence in the latter case seemed more unstable. They quoted the values

$$p_{\rm N} = 0.8872(8), \quad \lambda = 0.021(1) \tag{9.44}$$

for the 2D model and

$$p_{\rm N} = 0.7673(3), \quad \lambda = 0.090(3) \tag{9.45}$$

for the 3D model.

In two dimensions, the exact location of the N-point was predicted by Ref. [232]. The authors studied the duality transformation formulated in Ref. [233], applying it to a random model with Z_q symmetry. The model is defined in terms of variables, ξ_i, each taking one of q values ($\xi = 0, 1, 2, \ldots, q-1$). The partition function is defined as

$$Z = \sum_{\{\xi_i\}} e^{\sum_{(ij)} V(\xi_i - \xi_j + J_{ij})},$$

where $V(l)$ is a periodic function defined on integers with the period q and J_{ij} is a quenched random variable that takes on $0, 1, 2, \ldots, q-1$. They obtained an equation that yields a (possibly exact) value of the critical concentration p_c. The resulting values of p_c agreed with numerical estimates for various models such as the $\pm J$ Ising model, and the three-state Potts gauge glass. In addition, a similar equation derived for the random Ising model with Gaussian bond distribution yielded a value very close to the numerically obtained critical value of $J/\Delta J$, where J and ΔJ are the mean and the standard deviation of the bond distribution, respectively.

The replica method was employed in the derivation of the key equation; the duality relation of the n replica system was considered. As reported in Ref. [234], a quantity, which we denote as α, defined with the physical parameters, such as p and T for the Ising model, has a very simple transformation rule under the duality transformation:

$$\alpha \to q^n/\alpha.$$

Therefore, $\alpha = q^{n/2}$ defines a manifold that is invariant under the duality transformation. The virtue of this equation is that we can explicitly take its $n \to 0$ limit, resulting in

$$\sum_{l=0}^{q-1} p_l \log\left(\sum_{\eta=0}^{q-1} e^{V(\eta+l)-V(l)}\right) = \frac{1}{2}\log q, \tag{9.46}$$

where p_l is the probability of J_{ij} being l. If a well-defined duality equation in the $n \to 0$ limit exists and if a self-dual point lies on our real p–T plane, the condition (9.46) must be satisfied on it. The problem is that we do not know what the $n \to 0$ self-dual equation is nor whether a self-dual point lies on our p–T phase diagram.

However, with the assumption that a self-dual point coincides with the N-point, it becomes possible to obtain a number, which may be an exact value of p_c. Specifically, the intersection of the N-line and the line defined by (9.46) for the $\pm J$ Ising model is located at

$$p_c = 0.889972\cdots$$

which agrees with preceding numerical estimates of the N-point such as

$$p_c = 0.8905(5)$$

in Ref. [235]. This remarkable agreement, together with similar agreements for a few other models, seems a little too much to regard as a mere coincidence; the exact transition points of these models may have been derived from the duality.

9.7.5. *Phase diagram of the random XY model in two dimensions*

The effect of weak randomness in the XY model in two dimensions is of special relevance due to its relationship to various other models, such as the Coulomb gas and Josephson junction arrays. The model discussed most frequently is the random-phase XY model

$$\mathcal{H} = -\sum_{(ij)} J_{ij}\cos(\theta_i - \theta_j - \phi_{ij}),$$

with J_{ij} being a uniform constant with $J > 0$ and ϕ_{ij} a quenched random variable. However, other models, such as the one with random J_{ij} and uniform ϕ_{ij}, most likely have essentially the same property.

In an early study,[236] the random phase model was studied and it was suggested that a sufficiently strong disorder destroys the quasi-long-range order of the pure XY model. More strikingly, the amount of the disorder sufficient to destroy the quasi-order is vanishing as we approach the zero temperature. Consequently, the system undergoes a re-entrant phase transition as the temperature is lowered with the magnitude of the disorder being fixed at a sufficiently small value.

The latter prediction concerning the re-entrance phase transition was corrected later by a number of groups.[237–241] Here, we follow the heuristic argument given by the authors of Ref. [238]. They argued that there is a disorder-induced phase transition at zero temperature in the 2D XY model. Generalizing the Kosterlitz argument for the KT transition, they considered the balance between the energy cost for creating a topological excitation (vortex) and the energy gain due to the interaction of the vertex with the disordered potential. They estimated the probability $p_{\rm defect}$ of creating a vortex on a given site being energetically favored. It depends algebraically on the area of the system A with an exponent α

$$p_{\rm defect} \sim A^{-\alpha},$$

where α depends on the magnitude of the disorder and the temperature. The ordered phase is stable against the introduction of disorder when the total number of sites on which the creation of the vortex is favored is zero in the thermodynamic limit. That is, $Ap_{\rm defect} \to 0$. Therefore, $\alpha = 1$ defines the phase boundary. This result contradicts preceding studies such as the one mentioned above,[236, 242] which suggest that the ground state of the pure system is unstable at any finite disorder.

They argued that the preceding renormalization group theories failed to capture the correct physics because they neglected the fluctuation in the local energy gain due to disorder potential, which can be very large with a small but finite probability. In particular, we cannot neglect such a fluctuation when it exceeds the thermal fluctuation. This means that the previous argument may fail near and below the N-line because the N-line can be regarded as the cross-over line below which the geometrical fluctuation dominates (see Section 9.7.1).

They also discussed a finite-temperature phase diagram introducing the thermal fluctuations in their argument. This yields another interesting feature of their results; the straight phase boundary between the ordered phase and the disordered phase in the T–T_p phase diagram, in agreement with the Nishimori's claim (see Section 9.7.1). Generalizing the simple argument based on the energy balance, they computed the exponent α for finite temperature:

$$\alpha = \begin{cases} \dfrac{T^*}{T_p} & (T_p > \lambda T) \\[2ex] \dfrac{T^*}{T_p}\left(\dfrac{T_p}{\lambda T}\left(2 - \dfrac{T_p}{\lambda T}\right)\right) & (T_p < \lambda T), \end{cases} \tag{9.47}$$

where T^* and λ are model-dependent parameters. The phase boundary is again given by $\alpha = 1$. Therefore, for the region where the disorder fluctuation dominates ($T_p > \lambda T$), the phase boundary is $T_p = T^*$, independent of the temperature, whereas it depends on the temperature in the region where the thermal fluctuation dominates ($T_p < \lambda T$). If one identifies, quite naturally, the line $T_p = \lambda T$ with the N-line, the result is perfectly in parallel with Nishimori's picture based on the gauge invariance.

In fact, the gauge invariance can be used for deriving exact results for continuous spin models, such as the XY models and the Heisenberg models. In Ref. [215], the author derived a number of exact equations for the Villain model on the N-line. For example, the exact solution for the energy along the N-line was obtained. As we have seen in Section 9.7.1 for the discrete models, one can argue, based on the exact results, that the N-line ($T = T_p$) is the cross-over line separating the purely ferromagnetic region and the disorder dominant region. Interestingly, the renormalization group theory on the XY model[237] indicates that there is a freezing transition or a cross-over at $T = 2T_p$ with the difference of a factor 2 from the N-line. It is plausible that this difference is only due to the approximation involved in the renormalization group theory, and they both reflect the same physics.

9.8. Quantum Spin Glasses

A quantum spin glass is a magnetic system that can be described by a quantum mechanical Hamiltonian with spin glass like features (randomness and frustration). In such a system, a spin glass phase may exist while at the same time quantum fluctuations play an important role, possibly a dominant role, in particular, in the absence of thermal fluctuations at zero temperature. Such a Hamiltonian is, for instance, the spin-1/2 Heisenberg spin glass

$$H = \sum_{(ij)} J_{ij}(\sigma_i^x\sigma_j^x + \sigma_i^y\sigma_j^y + \sigma_i^z\sigma_j^z), \tag{9.48}$$

where $\sigma^{x,y,z}$ are Pauli spin-1/2 operators, J_{ij} are random exchange interactions (e.g. Gaussian), and the sum runs over all nearest neighbors on some d-dimensional lattice. Another example is the Ising spin glass in a transverse field

$$H = -\sum_{(ij)} J_{ij}\sigma_i^z\sigma_j^z + \Gamma\sum \sigma_i^x, \tag{9.49}$$

where Γ denotes the transverse field strength. This Hamiltonian becomes diagonal if Γ is zero, in which case it reduces simply to the classical Ising spin glass that we have discussed in the previous sections. Thus, the role of the parameter Γ is to tune the strength of quantum fluctuations, and they do not play a role in the equilibrium statistical physics of a diagonal Hamiltonian. An important experimental realization of this model Hamiltonian is the system $\mathrm{LiHo}_x\mathrm{Y}_{1-x}\mathrm{F}_4$,[243] an insulating magnetic material in which the magnetic ions (Ho) are in a doublet state due to crystal field splitting. The interactions between Ho ions can be described by an Ising model with dipolar couplings. For $x = 1$, the system is a ferromagnet with a critical temperature of $T_c = 1.53$ K at $\Gamma = 0$, and as x is reduced, the critical temperature decreases. For concentrations below 25% Ho and above 10% Ho, a thermal phase transition to a spin glass phase occurs indicated by a diverging nonlinear susceptibility (for instance, at $x = 0.167$ the spin glass transition temperature is $T_g = 0.13$ K at $\Gamma = 0$). If a transverse field is applied ($\Gamma > 0$), the spin glass transition temperature decreases monotonically to zero (see Fig. 9.16). This particular point, at zero temperature and at a critical field strength, is what we denote as a quantum phase transition point.[244]

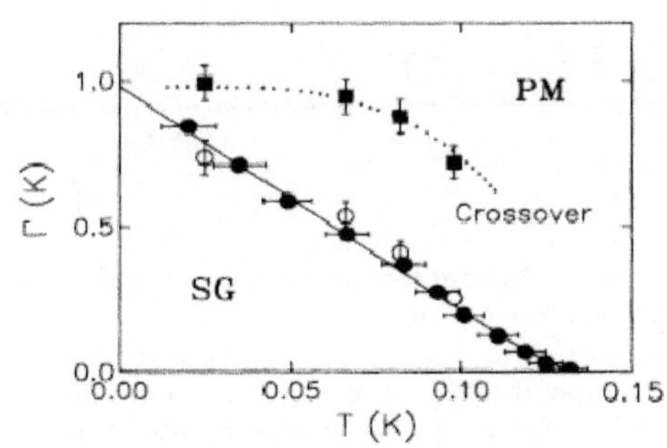

Fig. 9.16. Phase diagram of $\mathrm{LiHo}_{0.167}\mathrm{Y}_{0.833}\mathrm{F}_4$ according to the measurement of the nonlinear susceptibility (from Ref. [243]).

Earlier reviews on quantum spin glasses and in particular the Ising spin glass in a transverse field can be found in Refs. [245–247]. Here, we try to focus on a number of new developments that have been made since then.

9.8.1. *Random transverse Ising models*

The generic phase diagram for the EA Ising spin glass model in a transverse field Γ is shown in Fig. 9.17 for two dimensions and for three dimensions. In the 3D case, starting from the classical spin glass transition temperature T_c for $\Gamma = 0$ the critical temperature decreases monotonically with increasing transverse field strength Γ until it reaches $T = 0$. One expects that the universality class of the transition at any non-vanishing temperature is the same as the one for the classical Ising spin glass transition at T_c. The zero-temperature quantum phase transition, however, establishes a new universality class. This transition exists in any dimension, including one and two dimensions. A critical value Γ_c for the transverse field strength separates a disordered or paramagnetic phase for $\Gamma > \Gamma_c$ from an ordered phase for $\Gamma < \Gamma_c$. This transition is characterized by a diverging length scale $\xi \sim |\Gamma - \Gamma_c|^{-\nu}$ and a vanishing characteristic frequency $\omega \sim \Delta E \sim \xi^{-z}$. The latter is the quantum analog of "critical slowing-down" in the critical dynamics of classical, thermally driven transitions. The new and most

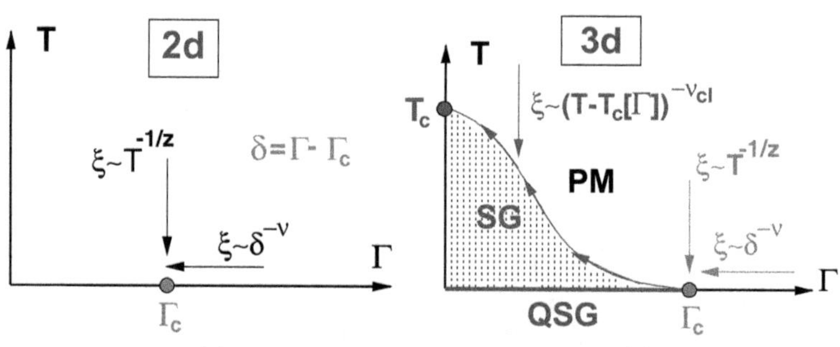

(a) (b)

Fig. 9.17. (a) Generic phase diagram for the 2D Ising spin glass in a transverse field Γ. Since no spin glass phase is present in $d = 2$ for $T > 0$, only a quantum spin glass phase and a quantum phase transition at $T = 0$ exist. Approaching the quantum critical point at Γ_c by decreasing the temperature T, the correlation length diverges like $T^{-1/z}$, where z is the dynamical critical exponent (if z is formally infinite, it increases logarithmically). (b) Generic phase diagram of a 3D Ising spin glass in a transverse filed. The classical transition temperature (at $\Gamma = 0$) is T_c and the corresponding classical correlation length exponent is ν_{cl}.

important property occurring at zero temperature in the random transverse Ising model is the *infinite randomness fixed point* (IRFP) that governs the quantum critical behavior at the critical value Γ_c of the transverse field.[248] One feature of the IRFP is that the dynamical exponent z is formally infinite, the relation between length and energy scales is not algebraic but exponential: $\Delta E \sim \exp(-A\xi^{\psi})$.

To describe this scenario, we generalize the discussion of the transverse Ising spin glass by including random *ferromagnetic* interactions $J_{ij} > 0$ as well, because many more analytical and numerical results are available for the ferromagnetic rather than the spin glass case and the same main features are expected to hold in both cases.

Let us start with a review of the 1D case, in which the sign (if it can be negative) of the nearest-neighbor couplings can be gauged so that all interactions are ferromagnetic and the resulting model is the random Ising chain in a transverse field or a random transverse-field Ising model (RTIM) in one dimension:

$$\mathcal{H} = -\sum_i J_i \sigma_i^z \sigma_{i+1}^z + \sum_i h_i \sigma_i^x. \quad (J_i > 0) \tag{9.50}$$

A uniform transverse field is represented by $h_i = \Gamma$ for all sites. Since this case and the case of random transverse fields turn out to belong to the same universality class, we also consider a random transverse field here. The couplings J_i and the transverse fields h_i are random variables with distributions $\pi(J)$ and $\rho(h)$, respectively. The Hamiltonian in (9.50) is closely related to the transfer matrix of a classical 2D layered Ising model, which was first introduced and studied by McCoy and Wu.[249] Extensive researches on this model were initiated by Fisher[250] with an application of the Ma–Dasgupta–Hu renormalization group scheme,[251] followed by numerical and analytical work.[252–259] We briefly summarize the results. The quantum control parameter of the model is given by

$$\delta = \frac{[\ln h]_{av} - [\ln J]_{av}}{\mathrm{var}[\ln \mathrm{h}] + \mathrm{var}[\ln \mathrm{J}]}. \tag{9.51}$$

For $\delta < 0$, the system is in the ordered phase with a non-vanishing average magnetization, whereas the region $\delta > 0$ corresponds to the disordered phase. There is a phase transition in the system at $\delta = 0$ with rather special properties, which differs in several respects from the usual second-order phase transitions of pure systems. One of the most striking phenomena is that some physical quantities are not self-averaging, which is due to very broad, logarithmic probability distributions. As a consequence, the *typical value* (which is the value in an frequent event) and the *average value* of such quantities can be drastically different. Thus, the critical behavior of the system is primarily determined by rare events that give dominating contributions to the averaged values of various observables.

The average bulk magnetization is characterized by an exponent β, which is $\beta = 2 - \tau$, where $\tau = (1+\sqrt{5})/2$ is the golden mean. The average spin–spin correlation function $C(r) = [\langle \sigma_i^z \sigma_{i+r}^z \rangle]_{av}$ involves the average correlation length ξ, which diverges at the critical point as $\xi \sim |\delta|^{-\nu_{\rm av}}$, with $\nu_{\rm av} = 2$. On the contrary, the typical correlations have a faster decay, since $\xi_{\rm typ} \sim |\delta|^{-\nu_{\rm typ}}$ with $\nu_{\rm typ} = 1$.

Close to the critical point, the relaxation time t_r is related to the correlation length as $t_r \sim \xi^z$, where z is the dynamical exponent. The random transverse-field Ising spin chain is strongly anisotropic at the critical point, since according to the RG picture[250] and numerical results,[260]

$$\ln t_r \sim \xi^{1/2}, \tag{9.52}$$

which corresponds to $z = \infty$. On the contrary, the relaxation time is related to the inverse of the energy-level spacing at the bottom of the spectrum $t_r \sim (\Delta E)^{-1}$. Then, as a consequence of (9.52), some quantities (such as specific heat, bulk and surface susceptibilities, etc.) have an essential singularity at the critical point, and the correlation function of the critical energy density has a stretched exponential decay, in contrast to the usual power law behavior.

Away from the critical point in the disordered phase, the rare events with strong correlations still play an important role, up to the point $\delta = \delta_G$. Above this point, all transverse fields are bigger than the interactions. In the region $0 < \delta < \delta_G$, which is called the Griffiths–McCoy phase, the magnetization is a singular function of the uniform longitudinal field H_z as $m_{\rm sing} \sim |H_z|^{1/z}$, where the dynamical exponent z varies with δ. At the two borders of the Griffiths–McCoy phase, it behaves as $z \approx 1/2\delta \times (1+\mathcal{O}(\delta))$[250] as $\delta \searrow 0$ and $z = 1$ as $\delta \nearrow \delta_G$, respectively.

All these results could be obtained and understood by the application of a Ma–Dasgupta–Hu renormalization group scheme,[250] in which strong bonds or fields are successively decimated either by the elimination of spins (in case of large transverse fields) or by the formation of strongly coupled clusters (in case of large ferromagnetic bonds). With a decreasing energy scale Δ of the bonds and fields to be decimated, the typical size L of these strongly coupled clusters increases as

$$L \sim |\ln \Delta|^{1/\psi} \tag{9.53}$$

defining an exponent ψ that is $1/2$ in the random transverse field Ising chain. Such a cluster typically contains

$$\mu \sim L^{\phi\psi} (= |\ln \Delta|^{\phi}) \tag{9.54}$$

spins that essentially behave collectively (for instance, in response to the application of a longitudinal magnetic field H — and thus generating a huge contribution to the spin susceptibility). This defines another exponent ϕ, which is $(1 + \sqrt{5})/2$ in the RTIM. Finally, there is the correlation length exponent ν that defines the characteristic length scale of spin–spin correlations away from the critical point.

The RG runs into a fixed point that is fully determined by the geometrical features of the clusters that are generated asymptotically — very much in reminiscence of the percolation fixed point in conventional percolation. This picture is expected to also hold for higher dimensional RTIMs and even for the spin glass case. Therefore, we summarize its essence here. The distribution of the random bonds and fields not yet decimated during the RG procedure becomes broader and broader. Hence, the name *infinite randomness fixed point* (IRFP). It is characterized by the three exponents ψ, ϕ, and ν and the critical behavior of the physical observables is determined by them. For instance, the correlation function (at criticality) for two spins at site i and j with a distance r from each other is simply given by their probability to belong to the same cluster of size r: $[C_{ij}]_{\text{av}} \sim |\mathbf{r}_i - \mathbf{r}_j|^{-2(d-\phi\psi)}$. Other relations follow straightforwardly from this scheme[248]:

lowest energy scale:	$-\ln \Delta \sim L^{\psi}$
magnetic moment:	$\mu \sim (-\ln \Delta)^{\phi}$
average correlations:	$[C_{ij}]_{\text{av}} \sim |\mathbf{r}_i - \mathbf{r}_j|^{-2(d-\phi\psi)}$
typical correlations:	$-[\ln C_{ij}]_{\text{av}} \sim \kappa_{ij} |\mathbf{r}_i - \mathbf{r}_j|^{\psi}$
finite T-susceptibility:	$\chi \sim T^{-1}(-\ln T)^{2\phi - d/\psi}$
finite H-magnetization:	$M \sim (-\ln H)^{-\phi + d/\psi}$.

Away from the critical point ($\delta \neq 0$), the correlation length is finite and its average and typical value scale differently:

$$\begin{aligned} &\text{average correlation length:} && \xi_{\text{av}} \sim \delta^{-\nu} \\ &\text{typical correlation length:} && \xi_{\text{typ}} \sim \xi_{\text{av}}^{1-\psi} \\ &\text{spontaneous magnetization:} && M_0 \sim (-\delta)^{\nu(d-\phi\psi)}. \end{aligned}$$

In spite of the finiteness of the average correlation length away from the critical point, still arbitrarily large strongly coupled clusters exist — though with an exponentially small probability — leading to algebraically decaying correlations in imaginary time. Phenomenologically, one can see that as follows.[261,262] Let L be the size of a region of *strongly coupled* spins. In a random system in the paramagnetic phase they occur with an exponentially small probability $P(L) \propto \exp(-\lambda L^d)$. For instance, in the diluted ferromagnet, strongly coupled regions are connected clusters and their probability is p^V, where V is the region's volume and p is the site occupation probability ($0 < p < 1$). Then, λ is given by $\lambda = |\ln p| > 0$. The special feature of transverse field Ising systems is that in first-order perturbation theory, the gap of a finite system containing L^d spins is exponentially small: $\Delta_0 \sim \exp(-sL^d)$. An exponentially small gap means an exponentially large tunneling time, and combining the two observations on cluster probability and relaxation time, one obtains an algebraical decay for the spin–spin correlation function: $C(\tau) = [\langle\sigma_i(\tau)\sigma_i(0)\rangle]_{\text{av}} \sim \tau^{-\lambda/s} = \tau^{-d/z(\delta)}$. The parameter $z(\delta) = s/d\lambda$ is called the *dynamical exponent* in the Griffiths phase and it varies continuously with the distance from the critical point. The consequences, e.g. for the susceptibility are dramatic: $\chi(\omega = 0) = \int_0^{1/T} d\tau\, C(\tau) \propto T^{-1+d/z(\delta)}$, which implies that for $z > d$ the susceptibility diverges for $T \to 0$ even away from the critical point. Since in random transverse field Ising system $z(\delta)$ grows without bounds for $\delta \to 0$ (and thus merging with the critical dynamical exponent at $\delta = 0$, which is infinite), there is always a region around the critical point, where the susceptibility diverges.

In general, the dynamical exponent $z(\delta)$ introduced above is expected to determine all singularities occurring in the Griffiths–McCoy phase close to an IRFP[248]:

$$\begin{aligned} &\text{dynamical exponent:} && z(\delta) \propto \delta^{-\psi\nu} \\ &\text{lowest energy scale:} && \Delta \sim L^{-z(\delta)} \end{aligned}$$

finite H-magnetization: $M \sim H^{1/z(\delta)}$

susceptibility: $\chi(\omega = 0) \sim T^{-1+d/z(\delta)}$

nonlinear susceptibility: $\chi_{nl}(\omega = 0) \sim T^{-1+d/3z(\delta)}$

specific heat: $c \sim T^{d/z(\delta)}$.

The above relations summarize the scaling predictions at and close to an IRFP, and in 1D, they have been confirmed many times, both analytically and numerically.[252–259] In higher dimensions $d \geq 2$, the randomly diluted Ising ferromagnet in a transverse field is a show case for a quantum phase transition governed by an IRFP. The site-diluted model is defined by the Hamiltonian

$$H = -J\sum_{(ij)} \varepsilon_i \varepsilon_j \sigma_i^z \sigma_j^z - \Gamma \sum_i \varepsilon_i \sigma_i^x \tag{9.55}$$

and the bond-diluted model by

$$H = -J\sum_{(ij)} \varepsilon_{ij} \sigma_i^z \sigma_j^z - \Gamma \sum_i \sigma_i^x, \tag{9.56}$$

where ε_i and ε_{ij} are random variables that take on the values 1 with probability p and 0 with probability $1 - p$. Its phase diagram is depicted in Fig. 9.18.

Along the vertical line starting from the point $(p, \Gamma) = (p_c, 0)$ up to the multi-critical point, the transition from the paramagnetic to the ferromagnetic phase is a *classical percolation transition.*[263, 264] Denoting the distance

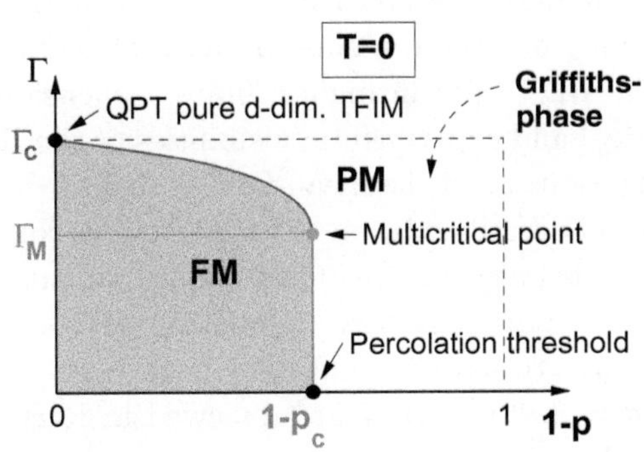

Fig. 9.18. Phase diagram of the diluted Ising ferromagnet in an transverse field Γ at zero temperature $T = 0$.

from the critical point with $\delta = p_c - p$, the connectivity correlation length diverges upon approaching the percolation point $1 - p_c$ as $\xi \sim |\delta|^{-\nu_{\rm perc}}$. The number of spins M in the percolating cluster at $p = p_c$ scales with the linear system size L as $M \sim L^{D_{\rm perc}}$, where $D_{\rm perc}$ is the fractal dimension of the percolating clusters. For small values of the transverse field Γ, one expects the percolating cluster to be fully magnetized, which implies that the gap scales as $\Delta \sim \exp(-L^{D_{\rm perc}})$. This means that $\psi = D_{\rm perc}$ in the IRFP scenario described above. Moreover, the connectivity correlation function at the percolation threshold p_c decays as $C(r) \sim r^{-(d-2+\eta_{\rm perc})}$, which means that the exponent ϕ is given by the relation $2(d - \psi\phi) = (d - 2 + \eta_{\rm perc})$. To summarize, the exponents characterizing the IRFP in the randomly diluted ferromagnet in a transverse field are related to the classical percolation exponents (which are exactly known in dimensions $d = 2$ and in $d > 6$) via

$$\nu = \nu_{\rm perc}, \quad \psi = D_{\rm perc}, \quad \phi = (d + 2 - \eta_{perc})/D_{\rm perc}. \tag{9.57}$$

For the random bond ferromagnet in a transverse field in dimensions $d \geq 2$, one has to rely only on numerical calculations: In the 2D ferromagnetic case, quantum MC simulations[265, 266] provided an evidence for an infinite randomness fixed point with $\psi \approx 0.42$ and $\phi \approx 2.1$. Later, a numerical implementation of the Ma–Dasgupta–Hu RG scheme indeed provided another evidence for an infinite randomness fixed point[267] with $\psi = 0.42 \pm 0.06$, $\phi = 2.5 \pm 0.4$, which agrees with the QMC estimate within the error bars, and $\nu = 1.07 \pm 0.15$. For random Ising ferromagnets in a transverse field, the existence of the IRFP dominating the quantum critical behavior thus appears to be confirmed for finite dimensions. Strictly speaking, detailed numerical studies have only be performed for $d = 1$ and $d = 2$ up to now, but there seems to be no strong argument against the existence of the IRFP also in higher, finite, dimensions, although one expects the numerical and experimental visibilities of the IRFP to diminish for increasing dimensions d. In the mean-field limit ($d \to \infty$), the quantum phase transition is *not* described by an IRFP and obeys conventional scaling laws. In particular, z is finite and Griffiths–McCoy singularities are absent.

What about the spin glass case? Quantum MC simulations on the Ising spin glass with a transverse field have been performed for the cases $d = 2$[262, 268] and $d = 3$,[269, 270] they are reviewed in Refs. [246] and [247]. The main result is that the numerical data appeared to be compatible with a finite value for the dynamical exponent in $d = 2$ and 3 and that the critical behavior can be described by conventional scaling laws. However, the existence of a Griffiths–McCoy phase away from the critical point has been uncovered, with a continuously varying dynamical exponent describing the singularities of the susceptibility and nonlinear susceptibility. In contrast to the quantum MC simulations of the random bond ferromagnets, no cluster algorithm could be used in the quantum spin glass case, which restricted the system sizes and in particular the temperatures to rather small values (note that anisotropic finite-size scaling demands that the temperature has to decrease exponentially with the system size at a quantum critical point described by an IRFP). In addition, a homogeneous rather than a random transverse field has been used, which causes strong cross-over effects and the true asymptotic scaling behavior might be more difficult to extract. Therefore, it might very well be that the indications found for the absence of an IRFP in the $2d$ and $3d$ quantum spin glass are still pre-asymptotic and that studies using larger system sizes and more sophisticated simulation methods could detect evidence for the IRFP also here.

Finally, a word about the consequences of the aforementioned theoretical developments for the experiments. There it was observed that upon approaching the quantum critical point, the divergence of the nonlinear susceptibility was drastically suppressed indicating even the absence of a divergence at zero temperature. The numerical results, on the other hand, hint at a strong divergence of the nonlinear susceptibility at the quantum critical point — even more than the IRFP scenario. Up to now, no clear reason for the discrepancy has been pinned down. The possibility of a second-order transition turning a first-order one at low temperatures has been raised,[271] but this possibility can definitely be ruled out for a system that can be described by the Hamiltonian (9.50) that we discussed here. We do not think that dipolar interactions of a magnetically diluted system cause substantial modifications of the picture that emerged for short-range interactions. At this point, one cannot rule out the possibility that the transverse field Ising Hamiltonian with quenched disorder is simply not a sufficiently detailed description of $LiHo_{0.167}Y_{0.833}F_4$.

9.8.2. *Mean-field theory*

As a mean-field model of quantum Ising spin glass, we consider the Sherrington–Kirkpatrick model in a transverse field

$$H = -\sum_{(i,j)} J_{ij}\sigma_i^z\sigma_j^z - \Gamma\sum_i \sigma_i^x. \tag{9.58}$$

The first sum is over all pairs of spins and the couplings J_{ij} are quenched random variables that obey the Gaussian distribution with zero mean and variance J^2/N, where N is the number of spins. Γ is the strength of the transverse field. Although no exact solution has been found for finite Γ, the phase diagram of this model has been well delineated. At zero transverse field, the transition is the well-known classical transition of the SK model at $T_c(\Gamma = 0) = J$. For sufficiently high temperature and/or sufficiently large Γ, thermal and/or quantum fluctuations destroy the spin glass order, yielding a paramagnet.[272] For low T and small Γ, one finds an SG ordered phase, apparently with broken replica symmetry.[273] MC calculation, numerical spin summation,[274] and perturbation expansion[275] in $1/\Gamma$ have determined the phase boundary to some precision. As in the classical model, the infinite-range interactions apparently wipe out the Griffiths singularities discussed in Section 9.8.1. The critical behavior along the line $T_c(\Gamma)$ is expected to be the same as the classical critical behavior, i.e. the non-linear susceptibility diverges as $\chi_{nl} \sim (T - T_c(\Gamma))^{-\gamma}$ with $\gamma = 1$, the specific heat exponent is $\alpha = 0$, etc.

The zero-temperature quantum critical point $\Gamma_c(T = 0)$ is in a different universality class and has been studied in Refs. [276–278]. The static approximation — the approximation usually applied to small field values in which the imaginary time correlation function $C(\tau) = \langle\sigma_i(\tau)\sigma_i(0)\rangle$ is assumed to be time independent — is not valid at $T = 0$ (large fields) and the full functional form of $C(\tau)$ has to be explored. The dynamical self-consistency equations obtained via standard manipulations[272,279] were analyzed at $T = 0$ at the quantum critical point in Refs. [276–278], and it turned out that the quantum critical point is located at $\Gamma_c \approx 0.7J$. At $\Gamma > \Gamma_c$ (and zero temperature), $C(\tau)$ decays exponentially with τ as $\tau \to \infty$, indicating a gap Δ in the corresponding spectral density; at $\Gamma = \Gamma_c$, $C(\tau)$ decays as $1/\tau^2$, and in the ordered phase, $C(\tau) \to q_{\rm EA}$. The Fourier transform of $C(\tau)$ has the form $C(\omega) \sim \text{const.} - \sqrt{\omega^2 - \Delta^2}$ for $\Gamma \geq \Gamma_c$, which is responsible for the $1/\tau^2$ behavior at Γ_c and it turned out that the correlation time diverges as $\xi_\tau \sim 1/\Delta \sim [(\Gamma - \Gamma_c)^{-1}\ln(\Gamma - \Gamma_c)]^{1/2}$. Thus, we can define an exponent $z\nu$, anticipating anisotropic scaling in space and time in the short-range model, which takes the value $z\nu = 1/2$ in the infinite-range model. Since $C(\tau \to \infty) = q_{\rm EA}$ is the Edwards–Anderson order parameter, we may also define $q_{\rm EA} = (\Gamma_c - \Gamma)^\beta$, and it was found that $\beta = 1$. At $\Gamma = \Gamma_c$, one expects $C(\tau) \sim \tau^{-\beta/z\nu}$, which is satisfied with the values obtained. The nonlinear susceptibility diverges as $1/\Delta$, which implies with $\chi_{nl} \sim (\Gamma - \Gamma_c)^{-\gamma}$ that $\gamma = 1/2$. Studying Gaussian fluctuations around the saddle-point solution valid for infinite range, one finds[278] for the correlation length exponent above the upper critical dimension (i.e. $d \geq 8$) that $\nu = 1/4$ and therefore $z = 2$. Moreover, $\eta = 0$ in mean-field theory. The complete collection of critical exponents obtained so far in comparison with the classical model ($T > 0$, where we assume to cross the phase boundary under a non-vanishing angle) are as follows:

	β	γ	ν	z
quantum ($T = 0$)	1/2	1/2	1/4	2
classical ($T = 0$)	1	1	1/2	–

(9.59)

Note that as a consequence of the absence of Griffiths singularities in mean-field models the dynamical exponent z is finite in contrast to the IRFP scenario that is supposedly valid for the finite-dimensional models. In a longitudinal field, one obtains, in analogy to the classical case, an AT manifold in the T, Γ, h phase diagram below where the replica symmetry is broken and the system is in the SG phase.

The dynamics of the model (9.58) in the paramagnetic phase has been studied in Ref. [280], where the dynamical single-site self-consistency equations have been iteratively solved using a quantum MC scheme developed in Ref. [281]. They mapped the spin glass transition line in the Γ–T plane using the stability criterion $1 = J\chi_{\text{loc}}$, where $\chi_{\text{loc}} = \int_0^\beta d\tau\, C(\tau)$ is the local susceptibility. They found a second-order transition line ending at a quantum critical point at $T = 0$ in agreement with the argument presented above. Going down in temperature to $T \sim 0.01J$ and extrapolating the results to $T = 0$, they determined a precise value for the critical field $\Gamma_c = 0.76 \pm 0.01$, which lies between previous estimates.[273, 276] The asymptotic form of $C(\tau) \sim \tau^{-2}$ found in Ref. [276] was also confirmed. A comparison of the results for the low-frequency susceptibility with the experimental curves obtained for $LiHo_{0.167}Y_{0.833}F_4$ in Ref. [243] yields a good agreement.

A different class of mean-field spin glass models have been studied in Ref. [271] — simplified insofar as spherical spins rather than Ising spins were considered and more general insofar as p-spin interactions were considered. The quantum fluctuations are introduced via a kinetic energy rather than the transverse field. The corresponding quantum spherical p-spin glass

Hamiltonian is defined by

$$H = \frac{1}{2M}\sum_{i=1}^{N}\hat{p}_i^2 - \sum_{i_1,\ldots,i_p} J_{i_1,\ldots,i_p} s_{i_1}\cdots s_{i_p}, \tag{9.60}$$

where s_i are "soft-spins" fulfilling the spherical constraint $\sum_{i=1}^{N} s_i(t)^2 = N$ for all times t. Quantum mechanics is introduced into the classical p-spin glass via the canonical momenta $\hat{p}_i$ that fulfill the commutation relation $[\hat{p}_i, s_j] = -i\hbar\delta_{ij}$. The multi-spin coupling constants are taken from a Gaussian distribution with zero mean and variance $\tilde{J}p!/(2N^{p-1})$, with $\tilde{J}$ being a constant of $\mathcal{O}(1)$.

Before we discuss this model, we want to clarify the connection to the SK model in a transverse field discussed above. The replacement of Ising spins $S_i = \pm 1$ by continuous spins $s_i \in [-\infty, +\infty]$ is often performed in the theory of critical phenomena — the discrete symmetry is then enforced by a quartic term $\sum_i s_i^4$ in the Hamiltonian (this is just the usual Φ^4 Ginzburg–Landau theory for critical phenomena with a discrete symmetry), which also restricts automatically the spin length. Analytically, the quartic term causes extra complications in all computations, saddle point evaluations, RG calculations, dynamical formalism, etc. — for which reason one often skips it and replaces it by a spherical constraint (either strictly or via a Lagrangian parameter having the same effect as a chemical potential). Unfortunately, the classical spherical mean-field spin glass model with the usual two-spin interactions does not have a non-trivial spin glass phase. Therefore, generalizations to p-spin interactions are sometimes considered.[138] At this point, a clear connection to the original magnetic system of interest is already lost. Nevertheless, one might expect that one can learn something about possible scenarios.

Finally, spherical spins cannot be quantized in the same way as Ising spins via the application of a transverse field. Therefore, they are usually quantized via the introduction of a kinetic energy term as in (9.60). In addition, various analytical techniques available for interacting soft spins with kinetic energy, such as the Schwinger–Keldysh formalism,[282] are not available for spin operators. The microscopic details of the quantum dynamics described by either a transverse field or a kinetic energy term might be very different; on large timescales, however, one expects a similar behavior for the following reason. To see this, let us consider a model that consists of two terms — an arbitrary classical Hamiltonian, H_{cl}, that is diagonal in the z-representation of the spins, and the transverse-field term. Performing a Trotter decomposition of the partition function of this model, one obtains

$$\mathrm{Tr}e^{-\beta(\Gamma\sigma^x+H_{\mathrm{cl}}(\sigma^z))} = \lim_{\Delta\tau\to 0}\prod_{\tau=1}^{L_\tau}\langle S_\tau|e^{-\Delta\tau[\Gamma\sigma^x+H_{\mathrm{cl}}(\sigma^z)]}|S_{\tau+1}\rangle$$

$$\propto \lim_{\Delta\tau\to 0}\sum_{S_1,\dots,S_{L_\tau}}\exp\left(-\Delta\tau\left[\sum_{\tau=1}^{L_\tau}K(S_\tau - S_{\tau+1})^2 + H_{\mathrm{cl}}(S_\tau)\right]\right), \qquad (9.61)$$

where L_τ is the number of Trotter slices in the imaginary time direction, $\Delta\tau = \beta/L_\tau$ and K is given by $e^{-2K} = \tanh(\Delta\Gamma)$. For $\Delta\tau \ll 1$, it is $K = |\ln(\Delta\tau\Gamma)|/2$. In the last step, we neglected a constant factor $\cosh(\Delta\tau\Gamma)^{L_\tau}$. If we choose $\Delta\tau$ as a small time cut-off (representing the typical spin flip time), we can approximate the last Trotter sum as the imaginary time path integral

$$Z \approx \int \mathcal{D}S(\tau)\exp\left(\int_0^\beta d\tau\left[\frac{M}{2}\left(\frac{\partial S}{\partial\tau}\right)^2 + H_{\mathrm{cl}}(S(\tau))\right]\right), \qquad (9.62)$$

where $M = 2K\Delta\tau = \Delta\tau|\ln(\Gamma\Delta\tau)|$. The first term in the integral of the action is identical to what one would obtain for the kinetic energy if one writes down the imaginary time path integral for the partition sum of the Hamiltonian (9.60). In this way, the transverse-field term and the kinetic energy term are related.

In Ref. [271], the equilibrium properties of the model were obtained using a replicated imaginary-time path integral formalism[279] and analyzing the dynamical self-consistency equations for the spin auto-correlation function $C(\tau)$ arising in the limit $N \to \infty$ from a saddle point integration. The result for the phase diagram, EA order parameter and linear susceptibility in the case $p = 3$ are depicted in Fig. 9.19, where the parameter $\Gamma = \hbar^2/(JM)$ has been used — resembling the transverse field strength (since for $\Gamma \to 0$ one recovers the classical case). Above a temperature T^*, one has a continuous transition at a critical point $\Gamma = \Gamma_c(T)$ from a paramagnetic phase with vanishing EA order parameter to a spin glass phase with $q_{\mathrm{EA}} \neq 0$ and one-step replica-symmetry-breaking (1RSB). Although the EA order parameter jumps discontinuously, the transition is of second order: There is no latent heat (as in the classical case $\Gamma = 0$) and the susceptibility has only a cusp. This is due to the fact that the parameter m characterizing the Parisi order parameter function $q(x)$ (which is a step function with a single step at $x = m$) is unity at the transition. However, for temperatures

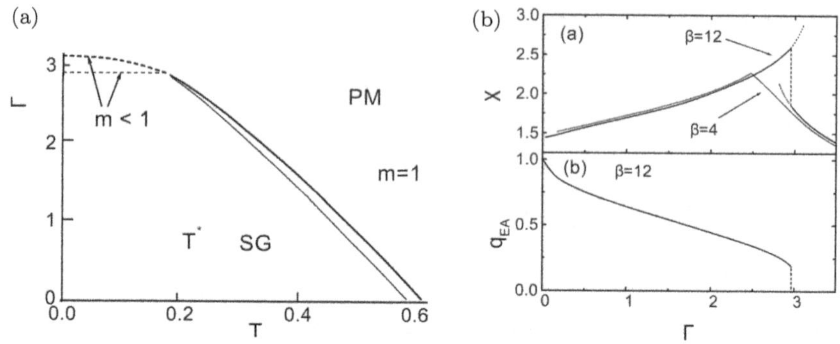

Fig. 9.19. (a) Static (thin lines) and dynamic (thick lines) phase diagrams of the p-spin model for $p = 3$. Solid and dashed lines represent second- and first-order transitions, respectively. (b) Magnetic susceptibility (a) and Edwards–Anderson order parameter (b) of the p=3 model (from Ref. [271]).

below T^* this parameter jumps at the transition, too, and the transition becomes discontinuous; for $T < T^*$, the transition is of the first order with latent heat and a discontinuous susceptibility (see Fig. 9.19).

9.8.3. *Mean-field theory — Dissipative effects*

An important question that arises for interacting quantum spins at low temperatures are the effects of a dissipative environment.[283,284] This is usually described in terms of its collective excitations, lattice vibrations, spin or charge fluctuations, etc., which may be thought of as an ensemble of independent quantum harmonic oscillators.[285–289] A concrete example of a single quantum degree of freedom, a spin-1/2 or a so-called two-level system (TLS), coupled to a bath of bosons is the well-known spin-boson model[283,284]:

$$H = H_S + H_B + H_{SB}, \tag{9.63}$$

where H_S, H_B, and H_{SB} denote the Hamiltonian of the system, the bath and their coupling, respectively. These are given by

$$\begin{aligned} H_S &= -\Gamma\sigma^x, \\ H_B &= \frac{1}{2}\sum_n (p_n^2/m_n + m_n\omega_n^2 x_n^2), \\ H_{SB} &= -\sum_n c_n x_n \sigma^z, \end{aligned} \tag{9.64}$$

where Γ is the transverse field (or tunneling matrix element in the context of TLSs) and n is the index enumerating an infinite number of harmonic oscillators with coordinates and momenta, x_n and p_n, and mass and frequency, m_n and ω_n, respectively. The constant c_n is the coupling between oscillator n and the spin. The spectral density of the environment, $I(\omega) = \pi \sum_n (|c_n|^2/(m_n\omega_n)\delta(\omega - \omega_n))$, is commonly assumed to take the standard form[284]

$$I(\omega) = 2\alpha\hbar(\omega/\omega_{\rm ph})^{s-1}\omega e^{-\omega/\omega_c}, \qquad (9.65)$$

where α is a dimensionless coupling constant, ω_c a high frequency cut-off (which can be set to $\omega_c = \infty$ if $0 < s < 2$), and $\omega_{\rm ph}$ a phonon frequency necessary in the non-ohmic ($s \neq 1$) case to keep α dimensionless.

With standard techniques,[283, 290] one can integrate out the oscillator degrees of freedom to express the partition function of the system solely in terms of the spin variables

$$Z = {\rm Tr}\, e^{-\beta H} = \int \mathcal{D}\sigma(\tau)\mathcal{T}\exp(-S/\hbar), \qquad (9.66)$$

where $\int \mathcal{D}\sigma(\tau)$ denotes a path integral over all spin configurations (in time), $\mathcal{T}$ is the imaginary time ordering operator, and the action is

$$S = -\int_0^{\beta\hbar} d\tau\Gamma\sigma^x(\tau) - \frac{1}{2}\int_0^{\beta\hbar} d\tau \int_0^{\beta\hbar} d\tau' K(\tau - \tau')\sigma^z(\tau)\sigma^z(\tau'). \qquad (9.67)$$

The kernel $K(\tau)$ is related to the spectral density $I(\omega)$ and is for the ohmic case ($s = 1$) essentially an inverse square $K(\tau - \tau') \propto \alpha/(\tau - \tau')^2$. The effect of the dissipative environment is therefore a long-range interaction of the quantum spin in imaginary time. In analogy to the Ising model with inverse square interactions[291] depending on the strength of the coupling constant α, the system is ferromagnetically ordered or paramagnetic in the imaginary time direction; for large α, the spin is frozen, and for small α, the spin will tunnel.

Indeed, for the ohmic case, at zero temperature, there is a phase transition at $\alpha = 1$.[286, 287] For $\alpha < 1$, there is tunneling and two distinct regimes develop. If $\alpha < 1/2$, the system relaxes with damped coherent oscillations; in the intermediate region $1/2 < \alpha < 1$, the system relaxes incoherently. For $\alpha > 1$, quantum tunneling is suppressed and $\langle \sigma^z \rangle \neq 0$, signaling that the system remains localized in the state in which it was prepared. These results also hold for sub-Ohmic baths while weakly damped oscillations persist for super-Ohmic baths.[283] At finite temperatures (but low enough such that thermal activation can be neglected), there is no localization, but the probability of finding the system in the state it was prepared decreases slowly with time for $\alpha > \alpha_c$.

These conclusions hold for a single spin interacting with a bath. The question then arises as to which are the effects of the interplay between the spin–spin interactions and the spin–noise coupling in the physics of the interacting system. In Ref. [282], the effect of a dissipative bath on a mean-field spin glass model with p-spin interactions has been investigated. They studied the dissipative spin–boson system (9.64) for N interacting spins $H = H_S + H_B + H_{SB}$, where the bath Hamiltonian is the same, the coupling Hamiltonian gets an additional sum over the spin index i, and H_S is now the p-spin Hamiltonian with transverse field

$$H_S = -\Gamma \sum_{i=1}^{N} \sigma_i^x - \sum_{i_1,\ldots,i_p} J_{i_1,\ldots,i_p} \sigma_{i_1}^z \cdots \sigma_{i_p}^z . \tag{9.68}$$

The second term, namely, the multi-spin interaction term is the same as the one in (9.60). For the reason explained in the last section, it is analytically easier to study spherical spins instead of quantum spin-1/2 degrees of freedom and the quantization of the spherical spins is done via the introduction of a kinetic energy term. The partition function then reads

$$Z = \int \mathcal{D}\sigma(\tau) \exp(-S/\hbar) , \tag{9.69}$$

with the action

$$\begin{aligned} S = \int_0^{\hbar\beta} d\tau \Bigg[\frac{M}{2} \sum_i \left(\frac{\partial s_i(\tau)}{\partial \tau} \right)^2 - \sum_{i_1 < \ldots < i_p} J_{i_1,\ldots,i_p} s_{i_1}(\tau) \cdots s_{i_p}(\tau) \\ + z \sum_i [s_i^2(\tau) - 1] \Bigg] - \int_0^{\hbar\beta} d\tau \int_0^{\hbar\beta} d\tau' K(\tau - \tau') s_i(\tau) s_i(\tau'), \end{aligned} \tag{9.70}$$

where the first term is the kinetic energy term already motivated in (9.61) and (9.62) replacing the transverse field term, the second is the p-spin interaction term, the third a term with the Lagrangian multiplier z enforcing the spherical constraint and the last term is the long-range interaction imaginary time (9.67) that is generated by the integration over the bath variables.

Starting from (9.70), the saddle point equations for the self-consistent single-spin dynamics were derived[282] and the phase diagram was computed. Analogous to the non-dissipative case discussed in Section 9.8.2, a critical line with a second-order section (close to the classical critical point $(T_d, \Gamma = 0)$) and a first-order section (close to the quantum critical point $(T = 0, \Gamma_d)$) was obtained in the presence of a dissipative environment. The second-order critical line is determined by the condition $m = 1$, the first-order critical line is defined as the locus of the points where a marginally stable solution first appears with decreasing Γ for T fixed. For each Γ and α, this defines a *dynamic* transition temperature $T_d(\Gamma, \alpha)$. The qualitative features of the phase diagram, similar to those found for the isolated system, see the discussion in the previous section. Note that the line $T_d(\Gamma, \alpha)$ lies always *above* $T_s(\Gamma, \alpha)$, the static critical line that we shall discuss below.

In Fig. 9.20(b), the dynamic phase diagrams obtained for $p = 3$ and three values of the coupling to an Ohmic bath, $\alpha = 0, 0.25, 0.5$ are shown. The full line and the line points represent second- and first-order transition, respectively.

The first observation that can be made is that in the limit $\Gamma \to 0$, the transition temperature is independent of the strength of the coupling to the bath. This is a consequence of the fact that in the limit $\Gamma \to 0$, the partition function is essentially determined by the zero-frequency components of the pseudo-spin which are decoupled from the bath. However, this result is

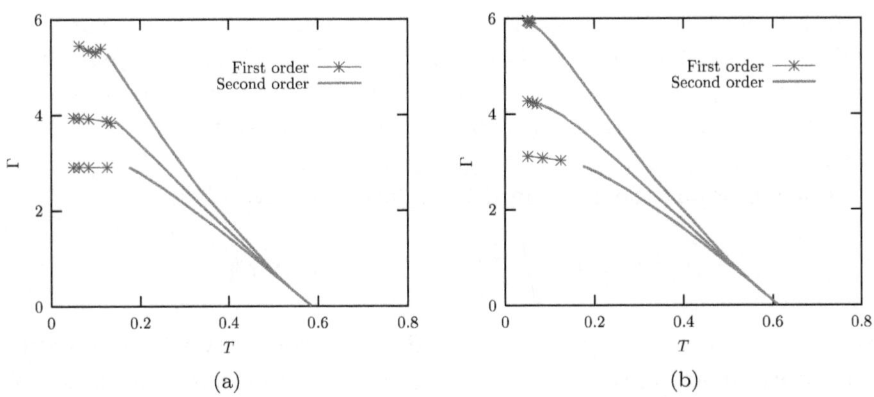

Fig. 9.20. Static (a) and dynamic (b) phase diagrams for the $p = 3$ spin model coupled to an Ohmic bath ($s = 1$). The couplings to the bath are $\alpha = 0$, 0.25, and 0.5 from bottom to top. The solid line and line points represent second- and first-order transitions, respectively (from Ref. [282]).

non-trivial from a dynamical point of view, since it implies that the dynamic transition of a classical system coupled to a colored classical bath is not modified by the latter.

The second observation is that the size of the region in the phase space where the system is in the ordered state increases with α. Coupling to the dissipative environment thus stabilizes this state. This follows from simple physical considerations. The interaction term in the action favors spin glass order. Coupling to the bath favors localization and its effect is to reduce the effective tunneling frequency. Therefore, in the presence of the bath, the value of the bare tunneling frequency needed to destroy the ordered state must increase with α. Even if the localized state and the glassy state may seem superficially similar, they are indeed very different. In the former, the correlation function $C(t+t_w,t_w)$ approaches a plateau as a function of t and never decays toward zero, while in the latter, the relaxation first approaches a plateau but it eventually leaves it to reach zero for $t \gg t_w$. The fact that the coupling to the environment favors the ordered state also reflects itself in the value taken by the order parameters $C(\tau)$ and $q_{\rm EA}$. As α increases, $q_d(\tau)$ reaches a higher plateau level at long imaginary times.

9.8.4. *Mean-field theory — Dynamics*

The out-of-equilibrium dynamics in real time of the quantum spherical p-spin glass model coupled to a dissipative environment, which was discussed in the last subsection, was actually studied earlier[292] than the equilibrium properties. The response and correlation function are defined in analogy to the classical case; $C(t+t_w,t_w) = N^{-1}\sum_i(s_i(t+t_w)s_i(t_w) + s_i(t_w)s_i(t+t_w))$ (note that the time evolution is now governed by the quantum dynamics and C has to be symmetrized in the operators $s_i(t+t_w)$ and $s_i(t_w)$) and $R(t+t_w,t_w) = N^{-1}\sum_i \delta s_i(t+t_w)/\delta h_i(t_w)$.

In equilibrium, the quantum FDT relates $R(t)$ and $C(t)$:

$$R(t) = \frac{2i}{\hbar}\theta(t)\int \frac{d\omega}{2\pi} e^{-i\omega t}\tanh(\beta\hbar\omega/2)C(\omega). \tag{9.71}$$

Away from the critical line, C and R decay to zero very fast with oscillations. Approaching the critical line $T_d(\alpha)$, the decay slows down, and if $T_d > 0$, a plateau develops in C. At the critical line, the length of the plateau tends to infinity.

In the glassy phase (below the transition), the system does not reach equilibrium. For small time differences, the dynamics is stationary and time translational invariance as well as the QFDT hold: $\lim_{t_w\to\infty} C(t+t_w,t_w) = q + C_{\rm eq}(t)$. For large time differences the dynamics is non-stationary, time translational invariance or the QFDT does not hold, and the correlations decay from q to 0. The decay of C becomes monotonic in the aging regime, which implies $C_{\rm aging}(t+t_w,t_w) = c(h(t_w)/h(t+t_w))$ (see Section 9.5.2). One can generalize the QFDT in the same spirit as the classical FDT was generalized (see Section 9.5.4)

$$R(t+t_w,t_w) = \frac{2i}{\hbar}\theta(t)\int \frac{d\omega}{2\pi} e^{-i\omega t} \tanh(X(t+t_w,t_w)\beta\hbar\omega/2) C(t+t_w,\omega), \tag{9.72}$$

with $C(t,\omega) = 2{\rm Re}\int_0^t ds \exp[i\omega(t-s)]C(t,s)$. Again, as in the classical case (see Section 9.5.4), $T_{\rm eff} \equiv T/X(t+t_w,t_w)$ acts as an effective temperature in the system. For a model with two time sectors, it is proposed

$$X(t+t_w,t_w) = \begin{cases} X_{\rm st} = 1 & \text{if } t \le \mathcal{T}(t_w) \\ X_{\rm age}(\hbar,T) & \text{if } t > \mathcal{T}(t_w). \end{cases}$$

where $X_{\rm age}$ is a non-trivial function of $\hbar$ and T and $\mathcal{T}(t_w)$ is a certain time scale that separates the stationary and aging time regimes. When t and t_w are widely separated, the integration over ω in (9.72) is dominated by $\omega \sim 0$. Therefore, the factor $\tanh(X_{\rm age}(t+t_w,t_w)\beta\hbar\omega/2)$ can be substituted by $X_{\rm age}\beta\hbar\omega/2$ (even at $T=0$ if $X_{\rm age}(\hbar,T) = x(\hbar)T$ when $T\sim 0$). Hence,

$$R_{\rm age}(t+t_w,t_w) \sim \theta(t) X_{\rm age}\beta\partial_{t_w} C_{\rm age}(t+t_w,t_w) \tag{9.73}$$

and one recovers, in the aging regime, the *classical* modified FDT.[138, 293]

The self-consistency equations for $C(t+t_w,t_w)$ and $R(t+t_w,t_w)$ were evaluated numerically in Ref. [292]. An example of the solution is shown in Fig. 9.21 for $p=3$. In all figures the following parameters have been chosen: zero temperature $T=0$, the width of the coupling distribution $J=1$, the frequency cut-off for the oscillator bath is set to $\omega_c = 5$, the mass in the kinetic energy term is $M=1$, and the strength of the quantum fluctuations $\tilde{\hbar} = \alpha\hbar$ (where α is the spin bath coupling strength) is $\tilde{\hbar} = 0.1$.

These plots demonstrate the existence of the stationary and aging regimes. For $t < \mathcal{T}(t_w)$ (e.g. $\mathcal{T}(40) \sim 5$), time translational invariance and fluctuation dissipation theorem are established, while beyond $\mathcal{T}(t_w)$, they break down. For $\tilde{\hbar} = 0.1$, the plateau in C is at $q \sim 0.97$. C oscillates around q but is monotonous when it goes below it. In the inset, the dependence of $q_{\rm EA}$ on $\tilde{\hbar}$ for $T=0$ is presented. Quantum fluctuations generate a $q_{\rm EA} < 1$

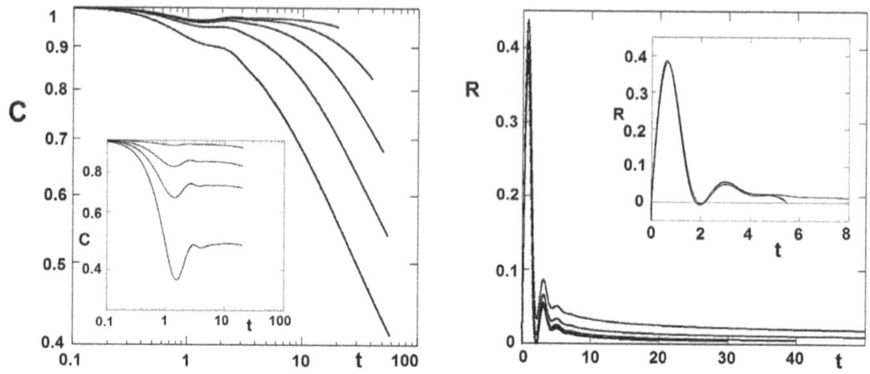

(a) (b)

Fig. 9.21. (a) The correlation function $C(\tau + t_w, t_w)$ vs τ for the $p = 3$ quantum spherical p-spin SG model. The waiting times are, from bottom to top, $t_w = 2.5, 5, 10, 20, 40$. $q_{\mathrm{EA}} \sim 0.97$. In the inset, the same curves for $t_w = 40$ and, from top to bottom, $\tilde{\hbar} = 0.1, 0.5, 1, 2$. (b) The response function for the same model as in the left part. The waiting times increase from top to bottom. In the inset, check of FDT in the stationary regime. The full line is $R(t + t_w, t_w)$ for $t + t_w = 40$ fixed and $t_w \in [0, 40]$. The dots are obtained from Eq. (9.72) with $X_{\mathrm{st}} = 1$, using the numerical data for $C_{\mathrm{stat}}(t) = C(t + t_w, t_w) - q$ ($q_{\mathrm{EA}} \sim 0.97$, see left part). In both cases, the response is plotted against t (from Ref. [292]).

such that the larger is $\tilde{\hbar}$, the smaller is q_{EA}. The addition of thermal fluctuations has a similar effect, the larger is T, the smaller is q_{EA}. In order to check the FDT in the stationary regime, in the inset of the right part of Fig. 9.21 a comparison is shown of $R(t + t_w, t_w)$ from the numerical algorithm for $t + t_w = 40$ fixed and $t_w \in [0, 40]$ (full line) with $R(t + t_w, t_w)$ from Eq. (9.72) with $X = 1$ using $C_{\mathrm{stat}}(t) = C(t + t_w, t_w) - q$, $q \sim 0.97$ obtained from the algorithm (dots). The accord is very good if $t \leq \mathcal{T}(t_w) \sim 5$. Finally, when one plots parametrically the integrated response χ vs. C, one finds that for $C < q \sim 0.97$, the χ vs C curve approaches a straight line of *finite* slope $1/T_{\mathrm{eff}} = X_{\mathrm{age}}/T \sim 0.60$.

9.8.5. *Heisenberg quantum spin glasses*

The spin-1/2 Heisenberg quantum spin glass is defined by the Hamiltonian (9.48) where the random exchange interactions J_{ij} can be ferromagnetic and anti-ferromagnetic. The system cannot be studied efficiently with quantum Monte-Carlo methods, due to the sign problem arising from the frustration. Therefore, not much is known about these models in finite dimensions, and the mean-field theory also becomes tractable only in certain limits and approximations.

9.8.5.1. *Finite dimensions*

In Ref. [294] and later in Ref. [295], small clusters of the 2D Heisenberg quantum spin glass were studied using exact diagonalization. The average total spin in the ground state turned out to scale as $S \propto \sqrt{N}$, where N is the number of sites. The spin glass order parameter in the ground state extrapolates to a small but non-vanishing value in the thermodynamic limit and the spin stiffness does not scale to zero in the thermodynamic limit. Ma–Dasgupta–Hu renormalization group studies[251, 296] were performed for randomly frustrated spin ladders[297] and in $d = 2$ and 3[298] for various lattices and spin values. The general idea of this RG procedure was already described in Section 9.8.1; large energies (in the form of exchange interactions) are successively decimated and ferromagnetic bonds lead to large spin formation and anti-ferromagnetic bonds are to a spin reduction or even elimination in case of equal effective spins connected by the bond to be decimated. In two and three dimensions, it was also observed that the final magnetic moment to be eliminated increased with system size as $\sqrt{N}$, which corresponds to the aforementioned observation for the ground state spin in small clusters. In addition, the Ma–Dasgupta–Hu RG calculations in Ref. [298] showed that the probability distribution of the low-energy excitation scales as $P(\Delta) \sim \Delta^{\omega}$, with $\omega = 0$ for $2d$ and $3d$, and that the dynamical critical exponent is $z = 2$ in $d = 2$ and $z = 3/2$ in $d = 3$.

9.8.5.2. *Mean-field model*

The first analytical treatment of the mean-field model of the Heisenberg quantum spin glass was performed in Ref. [279] applying the replica theory. Although the solution was confined to the paramagnetic state, the arguments for the existence of a low-temperature spin glass phase were given and the critical temperature was estimated.

Later a Landau theory for quantum rotors on a regular d-dimensional lattice was studied in Ref. [278], which is defined by the Hamiltonian

$$H = \frac{g}{2}\sum_{i}\hat{\mathbf{L}}_i^2 - \sum_{\langle ij\rangle} J_{ij}\hat{\mathbf{n}}_i\hat{\mathbf{n}}_j, \tag{9.74}$$

where $\hat{\mathbf{n}}_i$ are M-component vectors of unit length ($\hat{\mathbf{n}}_i^2 = 1$) and represent the orientation of the rotors on the surface of a sphere in M-dimensional rotor space. The operators $\hat{\mathbf{L}}_{i\mu\nu}$ ($\mu < \nu$, $\mu, \nu = 1, \ldots, M$) are the $M(M-1)/2$ components of the angular momentum $\hat{\mathbf{L}}_i$ of the rotor: the first term in H is the kinetic energy of the rotor with $1/g$ being the moment of inertia. The different components of $\hat{\mathbf{n}}_i$ constitute a complete set of commuting observables and the state of the system can be described by a wave function $\Psi(n_i)$. The action of $\hat{\mathbf{L}}_i$ on Ψ is given by the usual differential form of the angular momentum $\hat{\mathbf{L}}_{i\mu\nu} = -i(n_{i\mu}\partial/\partial_{i\nu} - n_{i\nu}\partial/\partial_{i\mu})$. The difference between rotors and Heisenberg–Dirac quantum spins is that the components of the latter at the same site do not commute, whereas the components of the $\hat{\mathbf{n}}_i$ do.

In Ref. [278], a Landau theory for this model is derived and it is shown that for a suitable distribution of exchange constants J_{ij} this model displays spin-glass and quantum paramagnetic phases and a zero-temperature quantum phase transition between them. The mean-field phase diagram near the $T = 0$ critical point is mapped out as a function of T, the strength of the quantum coupling g and applied fields. The spin glass phase has replica symmetry breaking. Moreover, the consequences of fluctuations in finite dimensions are considered, and above $d = 8$, the transition turned out to be controlled by a Gaussian fixed point with mean-field exponents. Below $d = 8$, a runaway RG flow to strong coupling was found.

Recently, the mean-field Heisenberg quantum spin glass model was generalized from the $SU(2)$ spin algebra to an $SU(N)$ symmetry and solved in the limit $N \to \infty$.[299] Certain universal critical properties are shown to hold to all orders in $1/N$. A spin glass transition was found for all values of the spin S and the phase diagram as a function of the spin S and temperature T was described. The quantum critical regime associated with the quantum transition at spin value $S = 0$ and the various regimes in the spin glass phase at high spin are analyzed. The specific heat is shown to vanish linearly with temperature.

The out-of-equilibrium dynamics of the the same model in the same limit $N \to \infty$, but coupled to a thermal bath, was studied in Ref. [300]. It was found that the model displays a dynamical phase transition between a paramagnetic and a glassy phase. In the latter, the system remains out-of-equilibrium and displays an aging phenomenon, which we characterize using both analytical and numerical methods. In the aging regime, the quantum fluctuation–dissipation relation is violated and replaced over a very long time range by its classical generalization, as in models involving simple spin algebras studied previously.

In the context of Heisenberg spin glasses also, the work on metallic spin glasses should be mentioned, which were first considered in Ref. [301] and later more extensively in Refs. [302–304]. The main ingredient of a metallic spin glass is an itinerant electron system with random (and frustrated) exchange interactions between the electron spins. Thus, in contrast to the spin glass systems discussed so far, the spins are not fixed to particular sites but can diffuse (quantum mechanically) from site to site. These systems are motivated by experiments on heavy-fermion compounds such as $Y_{1-x}\mathrm{U}_x\mathrm{Pd}$ $_3$,[305] which appear to show a paramagnetic to spin glass transition with increasing doping, x, in a metallic regime. To be concrete, the Hamiltonian studied in Ref. [303] is

$$H = -\sum_{i<j,\alpha} t_{ij} c_{i\alpha} c_{j\alpha} - \sum_{i<j,\mu} J_{ij} S_{i\mu} S_{j\mu} + H_{\text{int}}, \tag{9.75}$$

where $c_{i\alpha}$ annihilates an electron on site i with spin $\alpha =\uparrow, \downarrow$, and the spin operator is given by $S_{i\mu} = \sum_{\alpha\beta} c^+_{i\alpha} \sigma^\mu_{\alpha\beta} c_{i\beta}/2$, with σ^μ indicating the Pauli spin matrices. The sites i, j lie on a d-dimensional lattice, the hopping matrix elements t_{ij} are short-ranged and possibly random, and the J^μ_{ij} are Gaussian random exchange interactions, possibly with spin anisotropies. The remainder H_{int} includes other possible short-range interactions between the electrons, and the resulting total Hamiltonian H has a metallic ground state.

Starting from this Hamiltonian, in Ref. [303], an effective field theory for the vicinity of a zero-temperature quantum transition between a metallic spin glass ("spin density glass") and a metallic quantum paramagnet was introduced. Following a mean-field analysis, a perturbative renormalization-group study was performed, and it was found that critical properties are dominated by static disorder-induced fluctuations and that dynamic quantum mechanical effects are dangerously irrelevant. A Gaussian fixed point was found to be stable for a finite range of couplings for spatial dimensionality $d > 8$, but disorder effects always lead to runaway flows to strong coupling for $d \leq 8$. Moreover, scaling hypotheses for a *static* strong-coupling critical field theory were proposed. The nonlinear susceptibility has an anomalously weak singularity at such a critical point.

In Ref. [304], the competition between the Kondo effect and RKKY interactions near the zero-temperature quantum critical point of an Ising-like metallic spin glass was studied. In the "quantum critical regime," non-analytic corrections to the Fermi liquid behavior were found for the specific heat and uniform static susceptibility, while the resistivity and NMR relaxation rate have a non-Fermi liquid dependence on temperature.

9.9. Conclusions and Remaining Problems

In this review, we have tried to provide an overview on the recent developments in spin glasses. We concentrated on the topics to which substantial efforts have been devoted in recent years and for which, in our view, the most significant progress has been achieved: (1) The numerical investigation of the equilibrium thermodynamics of finite-dimensional spin glass models with short-range interactions using new and powerful methods, such as combinatorial optimization for ground-state calculations (through which excited states can also be studied) and extended-ensemble MC methods (e.g. the replica exchange method) for the location and characterization of the phase transition. The most challenging task remaining here is still the unification of the two paradigmatic pictures: the droplet picture and the mean-field scenario. Although being unsolved for several decades now, we see promising steps toward the unification of these two approaches. (2) The experimental and theoretical investigation of non-equilibrium dynamics and aging phenomena in spin glasses, especially the study of fluctuation–dissipation theorem violations in the glassy phase and the concept of an effective (non-equilibrium) temperature. Still a lot of work has to be done, experimentally in particular, in order to put the fascinating theoretical ideas on a firm and consistent experimental ground. (3) The theoretical exploration of quantum effects in spin glasses, statically and dynamically, for mean-field models as well as finite-dimensional models. Here the most demanding challenge appears to be handling the *real* time quantum dynamics at low temperatures for realistic (i.e. finite-dimensional short-range interacting) models.

Unfortunately, we could not cover in this review many "glassy" topics related to and inspired by the spin glass world. First and most actively pursued in recent years is the theory of the structural glass transition and the glass "phase". A lot of progress has been made to make contact between spin glasses and structural glasses, the main difference being that the disorder is self-induced in the latter during the freezing process, whereas spin glasses live with frozen, time-independent disorder. Recently, it was discussed extensively that this might not be a major obstacle in relating the two systems. From a theoretical point of view, a more substantial difference between them with concrete experimental consequences is that spin glasses have a well-defined order parameter. A diverging correlation length and a divergent susceptibility defined in terms of this order parameter signify the transition. In the theory of structural glasses, apparently one cannot depend on these helpful vehicles.

Other related issues are vortex, gauge, and Bragg glasses. All three appear in the context of disordered superconductors. The gauge glass model is essentially an XY spin glass model with a random (quenched) vector potential, originally devoted for describing amorphous granular superconductors but later also taken as a paradigmatic model for amorphous high-T_c superconductors. It can be analyzed in the same way as XY spin glasses and also has the same order parameter. The major questions concern the existence of a finite-temperature phase transition with or without screening effects. The vortex glass model is a model of interacting, elastic magnetic flux lines in a random potential, which freezes at low temperatures into a glassy phase (which however escapes a clean theoretical description, since an order parameter is hard to define). If the disorder is weak, topological defects can be neglected and an elastic description is possible, starting from the Abrikosov flux line lattice and taking into account its small elastic deformation via thermal fluctuations and disorder. For such weakly disordered elastic systems, a Bragg glass phase is predicted in which the true long-range order of the Abrikosov flux line lattice is transformed into a quasi long-range order. This glassy phase manifests itself also via an extremely sluggish dynamics. Upon increasing the strength of the disorder or the density of the lines (via an increased magnetic field), topological defects will proliferate, the quasi-long-range order of the Bragg glass phase vanishes, and the system becomes a vortex glass.

Finally, one should mention Bose, Fermi, and Coulomb glass models, which occur in the context of the low-temperature physics of quantum-mechanical, disordered, electronic, or bosonic systems. The origin of the interesting physics is the competition between the quenched random potential and the interactions between particles, usually long ranged as in the Coulomb case. The major issues concern the phase transition between conducting (metallic or superconducting) and insulating phases, in which the particles are localized. Again, the name "glass" for the low-temperature phases is justified by the anomalously slow dynamics present here.

To conclude this tour through the glass zoo, we hope to have demonstrated that a review on spin glasses should not only be useful for people working in the field of frustrated magnets but also for those encountering a strong disorder and strong interactions at low temperatures in other fields of condensed matter physics.

Acknowledgments

N.K.'s work was supported by the grant-in-aid (Program No. 14540361) from Monka-sho, Japan. H.R. acknowledges financial support from the Deutsche Forschungsgemeinschaft (DFG) and from the European Community's Human Potential Programme under contract HPRN-CT-2002-00307, DYGLAGEMEM, and thanks the KITP at UC Santa Barbara for its hospitality, where he composed large parts of this review. The authors are indebted to H. E. Castillo, C. Chamon, L. Cugliandolo, K. Hukushima, H. Kawamura, J. Kurchan, C. M. Newman, D. L. Stein, and A. P. Young for useful comments.

References

1. J. L. van Hemmen and I. Morgenstern, *Heidelberg Colloquium on Spin Glasses*, Lecture Notes in Physics 192 (Springer, 1983).
2. J. L. van Hemmen and I. Morgenstern, *Heidelberg Colloquium on Spin Glasses*, Lecture Notes in Physics 275 (Springer, 1986).
3. K. H. Fischer and J. A. Hertz, *Spin Glasses* (Cambridge, 1991).
4. J. A. Mydosh, *Spin Glasses: An Experimental Introduction* (Taylor and Francis, London, Washington, DC, 1993).
5. A. P. Young (editor), *Spin Glasses and Random Fields* (World Scientific, 1998).
6. P. E. Jönsson, H. Yoshino, and P. Nordblad, *Phys. Rev. Lett.* **89**, 097201 (2002).
7. K. Binder and A. P. Young, *Rev. Mod. Phys.* **58**, 801 (1986).
8. M. Mézard, G. Parisi, and M. A. Virasoro, *Spin Glass Theory and Beyond* (World Scientific, Singapore, 1987).
9. E. Marinari *et al.*, *J. Stat. Phys.* **98**, 973 (2000).
10. C. M. Newman and D. L. Stein, *J. Phys.: Condens. Matter* **15**, R1319 (2003).
11. G. Parisi, *Phys. Rev. Lett.* **43**, 1754 (1979).
12. G. Parisi, *J. Phys. A: Math. Gen.* **13**, 1101 (1980).
13. G. Parisi, *J. Phys. A: Math. Gen.* **13**, 1887 (1980).
14. G. Parisi, *Phys. Rev. Lett.* **50**, 1946 (1983).
15. D. Sherrington and S. Kirkpatrick, *Phys. Rev. Lett.* **35**, 1792 (1975).
16. M. Mezard, G. Parisi, N. Sourlas, G. Toulouse, and M. Virasoro, *Phys. Rev. Lett.* **52**, 1156 (1984).
17. M. Mezard, G. Parisi, N. Sourlas, G. Toulouse, and M. Virasoro, *J. Phys. (Paris)* **45**, 843 (1984).

18. D. J. Thouless, P. W. Anderson, and R. G. Palmer, *Phil. Mag.* **35**, 593 (1977).
19. J. R. L. de Almeida and D. J. Thouless, *J. Phys. A: Math. Gen.* **11**, 983 (1978).
20. F. Guerra and F. L. Toninelli, *J. Math. Phys.* **43**, 3704 (2002); *J. Math. Phys.* **43**, 6224 (2002); *Commun. Math. Phys.* **230**, 71 (2002); cond-mat/0205123, cond-mat/0208579.
21. C. M. Newman and D. L. Stein, *Phys. Rev. Lett.* **76**, 515 (1996).
22. M. A. Moore and A. J. Bray, *J. Phys. C* **17**, (1984) L613.
23. A. J. Bray and M. A. Moore, in J. L. van Hemmen and I. Morgenstern (eds.), *Heidelberg Colloquium on Spin Glasses* (Springer, 1986), p. 121.
24. D. S. Fisher and D. A. Huse, *Phys. Rev. Lett.* **56**, 1601 (1986).
25. D. S. Fisher and D. A. Huse, *Phys. Rev. B* **38**, 373 (1988).
26. D. S. Fisher and D. A. Huse, *Phys. Rev. B* **38**, 386 (1988).
27. B. Drossel, H. Bokil, M. A. Moore, and A. J. Bray, *Eur. Phys. J. B* **13**, 369 (2000).
28. W. L. McMillan, *Phys. Rev. B* **29**, 4026 (1984).
29. W. L. McMillan, *Phys. Rev. B* **30**, 476 (1984).
30. A. K. Hartmann and A. P. Young, *Phys. Rev. B* **64**, 180404 (2001).
31. N. Kawashima and H. Rieger, *Europhys. Lett.* **39**, 85 (1997).
32. C. Amoruso, E. Marinari, O. C. Martin, and A. Pagnani, *Phys. Rev. Lett.* **91**, 087201 (2003).
33. T. Shirakura and F. Matsubara, *J. Phys. Soc. Jpn.* **65**, 3138 (1996).
34. N. Kawashima and M. Suzuki, *J. Phys. A* **25**, 4985 (1992).
35. N. Kawashima, H. Hatano, and M. Suzuki, *J. Phys. A: Math. Gen.* **25**, 4985 (1992).
36. S. Liang, *Phys. Rev. Lett.* **69**, 2145 (1992).
37. Y.-C. Lin, N. Kawashima, F. Iglói, and H. Rieger, *Prog. Theor. Phys. Suppl.* **138**, 479 (2000).
38. H. Rieger *et al.*, *J. Phys. A* **30**, 3939 (1997).
39. M. Ney-Nifle and A. P. Young, *J. Phys. A* **30**, 5311 (1997).
40. N. Kawashima and T. Aoki, *Int. J. Mod. Phys. C* **10**, 1453 (1999).
41. N. Kawashima, *J. Phys. Soc. Jpn.* **69**, 987 (2000).
42. A. K. Hartmann and M. A. Moore, *Phys. Rev. Lett.* **90**, 127201 (2003).
43. A. A. Middleton, *Phys. Rev. Lett.* **83**, 1672 (1999).
44. M. Palassini and A. P. Young, *Phys. Rev. B* **60**, R9919 (1999).
45. T. Shirakura and F. Matsubara, *Phys. Rev. Lett.* **79**, 2887 (1997).
46. R. N. Bhatt and A. P. Young, *Phys. Rev. Lett.* **54**, 924 (1985).
47. A. T. Ogielski, *Phys. Rev. B* **32**, 7384 (1985).
48. E. Marinari, G. Parisi, and F. Ritort, *J. Phys. A: Math. Gen.* **27**, 2687 (1994).
49. N. Kawashima and A. P. Young, *Phys. Rev. B* **53**, R484 (1996).
50. M. Palassini and X. Caracciolo, *Phys. Rev. Lett.* **82**, 5128 (1999).
51. H. G. Ballesteros *et al.*, *Phys. Rev. B* **62**, 14237 (2000).
52. M. Tesi, E. J. van Resburg, E. Orlandini, and S. G. Whillington, *J. Stat. Phys.* **82**, 155 (1996).

53. K. Hukushima and K. Nemoto, *J. Phys. Soc. Jpn.* **65**, 1604 (1996).
54. F. Cooper, B. Freedman, and D. Preston, *Nucl. Phys. B* **210**, 210 (1989).
55. E. Marinari *et al.*, *Phys. Rev. Lett.* **81**, 1698 (1998).
56. K. Gunnarsson, P. Svedlindh, P. Nordblad, L. Lundgren, H. Aruga, and A. Ito, *Phys. Rev. B* **43**, 8199 (1991).
57. R. R. P. Singh and S. Chakravarty, *Phys. Rev. B* **36**, 559 (1987).
58. R. N. Bhatt and A. P. Young, *Phys. Rev. B* **37**, 5606 (1988).
59. D. Iñigues, G. Parisi, and J. J. Ruiz-Lorenzo, *J. Phys. A* **29**, 4337 (1996).
60. E. Marinari, G. Parisi, and J. J. Ruiz-Lorenzo, *Phys. Rev. B* **58**, 14852 (1998).
61. B. A. Berg and W. Janke, *Phys. Rev. Lett.* **80**, 4771 (1998).
62. P. O. Mari and I. A. Campbell, *Phys. Rev. E* **59**, 2653 (1999).
63. P. O. Mari and I. A. Campbell, *Phys. Rev. B* **65**, 184409 (2002).
64. T. Nakamura, S. Endoh, and T. Yamamoto, *J. Phys. A: Math. Gen.* **36**, 10895 (2003).
65. N. de Courtenary, H. Bouchiat, H. Hurdequint, and A. Fert, *J. Phys.* (*Paris*) **47**, 71 (1986).
66. H. Bouchiat, *J. Physique* **47**, 71 (1986).
67. L. P. Levy and A. T. Ogielsky, *Phys. Rev. Lett.* **57**, 3288 (1986).
68. M. Simpson, *J. Phys. F* **9**, 1377 (1979).
69. B. R. Coles and G. Williams, *J. Phys. F* **18**, 1279 (1988).
70. E. Vincent and J. Hammann, *J. Phys. C:Solid State Phys.* **20**, 2659 (1987).
71. H. Kawamura, *Phys. Rev. Lett.* **24**, 5421 (1998).
72. K. Hukushima and H. Kawamura, *Phys. Rev. E* **61**, 1008(R) (2000).
73. F. Matsubara, T. Shirakura, and S. Endoh, *Phys. Rev. B* **64**, 092412 (2001).
74. T. Nakamura and S. Endoh, *J. Phys. Soc. Jpn.* **71**, 2113 (2002).
75. L. W. Lee and A. P. Young, *Phys. Rev. Lett.* **90**, 227203 (2003).
76. H. Kawamura and M. S. Li, *Phys. Rev. Lett.* **29**, 187204 (2001).
77. M. Palassini and A. P. Young, *Phys. Rev. B* **63**, 140408(R) (2001).
78. F. Krzakala and O. C. Martin, *Europhys. Lett.* **53**, 749 (2001).
79. A. K. Hartmann, *Europhys. Lett.* **45**, 619 (1999).
80. B. A. Berg, U. E. Hansmann, and T. Celik, *Phys. Rev. B* **50**, 16444 (1994).
81. A. K. Hartmann, *Europhys. Lett.* **40**, 429 (1997).
82. A. K. Hartmann, *Europhys. Lett.* **44**, 249 (1998).
83. A. K. Hartmann and H. Rieger, *Optimization Algorithms in Physics* (Wiley, Berlin, 2002).
84. A. K. Hartmann, *Euro. Phys. B* **13**, 539 (2000).
85. A. Sandvik, *Europhys. Lett.* **45**, 745 (1999).
86. M. Palassini and A. P. Young, *Phys. Rev. Lett.* **83**, 5126 (1999).
87. M. Palassini and A. P. Young, *Phys. Rev. Lett.* **85**, 3017 (2000).
88. A. A. Middleton, *Phys. Rev. B* **63**, 060202(R) (2001).
89. M. A. Moore, cond-mat/0203469.
90. J. Lamarcq, J.-P. Bouchaud, O. C. Martin, and M. Mézard, *Europhys. Lett.* **58**, 321 (2002).

91. M. Palassini, F. Liers, M. Jünger, and A. P. Young, *Phys. Rev. B* **68**, 064413 (2003).
92. M. Jünger, G. Reinelt, and S. Thienel, in W. Cook, L. Lovasz, and P. Seymour (eds.) *DIMACS Series in Discrete Mathematics and Theoretical Computer Science*, Volume 20 (American Mathematical Society, 1995).
93. E. Marinari and G. Parisi, *Phys. Rev. Lett.* **85**, 3332 (2000).
94. E. Marinari and G. Parisi, *Phys. Rev. B* **62**, 11677 (2000).
95. E. Marinari and G. Parisi, *Phys. Rev. Lett.* **86**, 3887 (2001).
96. M. Palassini and A. P. Young, *Phys. Rev. Lett.* **85**, 3333 (2000).
97. F. Krzakala, J. Houdayer, E. Marinari, O. C. Martin, and G. Parisi, *Phys. Rev. Lett.* **87**, 197204 (2001).
98. J. Houdayer and O. C. Martin, *Europhys. Lett.* **49**, 794 (2000).
99. F. Krzakala and O. C. Martin, *Phys. Rev. Lett.* **85**, 03013 (2000).
100. J. Houdayer, F. Krzakala, and O. C. Martin, *Eur. Phys. J. B* **18**, 467 (2000).
101. H. G. Katzgraber, M. Palassini, and A. P. Young, *Phys. Rev. B* **63**, 184422 (2001).
102. D. Huse and D. Fisher, *J. Phys. A* **20**, L997 (1987).
103. E. Marinari, G. Parisi, and J. J. Ruiz-Lorenzo, in A. P. Young (ed.), *Spin Glasses and Random Fields* (World Scientific, 1998).
104. R. R. P. Singh and S. Chakravarty, *Phys. Rev. Lett.* **57**, 245 (1986).
105. G. Parisi, F. Ricci-Tersenghi, and J. J. Ruiz-Lorenzo, *J. Phys. A* **29**, 7943 (1996).
106. J. D. Reger, R. N. Bhatt, and A. P. Young, *Phys. Rev. Lett.* **64**, 1859 (1990).
107. A. K. Hartmann, *Phys. Rev. E* **60**, 05135 (1999).
108. H. G. Katzgraber and A. P. Young, *Phys. Rev. B* **67**, 134410 (2003).
109. H. G. Katzgraber and A. P. Young, *Phys. Rev. B* **68**, 224408 (2003).
110. J.-P. Bouchaud, L. F. Cugliandolo, J. Kurchan, and M. Mézard, in A. P. Young (ed.), *Spin glasses and Random Fields* (World Scientific Singapore, 1998), pp. 161–224.
111. E. Vincent, J. Hamman, M. Ocio, J.-P. Bouchaud, and L. F. Cugliandolo, in M. Rubi (ed.), *Complex Behaviour of Glassy Systems*, Lecture Notes in Physics **492** (Springer-Verlag, Berlin, 1997), pp. 184–219.
112. P. Nordblad and P. Svedlindh, in A. P. Young (ed.), *Spin Glasses and Random Fields* (World Scientific Singapore, 1998), pp. 1–28.
113. C. A. Angell, *Science* **267**, 1924 (1995).
114. A. J. Bray, *Adv. Phys.* **43**, 357 (1994).
115. D. S. Fisher and D. A. Huse, *Phys. Rev. B* **38**, 373 (1988); **38**, 386 (1988).
116. For a review see T. Natterman, in A. P. Young (ed.), *Spin Glasses and Random Fields* (World Scientific Singapore, 1998).
117. D. S. Fisher and D. A. Huse, *Phys. Rev. B* **43**, 10728 (1991).
118. H. Rieger, B. Steckemetz, and M. Schreckenberg, *Europhys. Lett.* **27**, 485 (1994).
119. H. Rieger, in D. Stauffer (ed.), *Annual Review of Computers and Physics II* (World Scientific, Singapore, 1995).

120. J. Kisker, L. Santen, M. Schreckenberg, and H. Rieger, *Phys. Rev. B* **53**, 6418 (1996).
121. E. Marinari, G. Parisi, F. Ritort, and J. J. Ruiz-Lorenzo, *Phys. Rev. Lett.* **76**, 843 (1996).
122. T. Komori, H. Yoshino, and H. Takayama, *J. Phys. Soc. Jpn.* **68**, 3387 (1999).
123. E. Marinari, G. Parisi, F. Ricci-Tersenghi, and J. J. Ruiz-Lorenzo, *J. Phys. A* **33**, 2373 (2000).
124. K. Hukushima, H. Yoshino, and H. Takayama, *Prog. Theor. Phys. Suppl.* **138**, 568 (2000).
125. H. Yoshino, K. Hukushima, and H. Takayama, *Phys. Rev. B* **66**, 064431 (2002).
126. L. Berthier and J.-P. Bouchaud, *Phys. Rev. B* **66**, 054404 (2002).
127. Y. G. Joh, R. Orbach, G. G. Wood, J. Hamman, and E. Vincent, *Phys. Rev. Lett.* **82**, 438 (1999).
128. G. J. Koper and H. J. Hilhorst, *J. Phys. (France)* **49**, 429 (1988).
129. H. Rieger, *J. Phys. A* **26**, L615 (1993).
130. V. Dupis, E. Vincent, J.-P. Bouchaud, J. Hammann, A. Ito, and H.A. Katori, *Phys. Rev. B* **64**, 174204 (2001).
131. P. E. Jönsson, H. Yoshino, P. Nordblad, H. Aruga Katori, and A. Ito, *Phys. Rev. Lett.* **88**, 257204 (2002).
132. J.-P. Bouchaud, V. Dupuis, J. Hamman, and E. Vincent, *Phys. Rev. B* **65**, 024439 (2001).
133. H. K. Janssen, B. Schaub, and B. Schmittmann, *Z. Phys. B.* **73**, 539 (1989).
134. D. Huse, *Phys, Rev. B* **40**, 304 (1989).
135. A. G. Schins, A. F. M. Arts, and H. W. de Wijn, *Phys. Rev. Lett.* **70**, 2340 (1993).
136. J. Mattson, T. Jönsson, P. Nordblad, and H. Aruga Katori, *Phys. Rev. Lett.* **74**, 4305 (1995).
137. For a comprehensive review over the various analytical predictions of mean field theory for glassy dynamics see: L. F. Cugliandolo, eprint cond-mat/0210312.
138. L. F. Cugliandolo and J. Kurchan, *J. Phys. A* **27**, 5749 (1994).
139. L. Berthier, J.-L. Barrat, and J. Kurchan, *Phys. Rev. E* **63**, 016105 (2001).
140. E. Bertin and J. P. Bouchaud, *J. Phys. A* **35**, 3039 (2002).
141. M. Picco, F. Ricci-Tersenghi, and F. Ritort, *Eur. Phys. J. B* **21**, 211 (2001).
142. L. Berthier, *Eur. Phys. J. B* **17**, 689 (2000).
143. D. S. Fisher, P. Le Doussal, and C. Monthus, *Phys. Rev. E* **64**, 066107 (2001).
144. L. C. E. Struik, *Physical Aging in Amorphous Polymers and Other Materials* (Elsevier, Amsterdam, 1978).
145. J. Kurchan, *Phys. Rev. E* **66**, 017101 (2002).
146. J. Hamman, M. Lederman, M. Ocio, R. Orbach, and E. Vincent, *Physica A* **185**, 278 (1992).

147. T. Komori, H. Yoshino, and H. Takayama, *J. Phys. Soc. Jpn.* **69**(Suppl. A), 228 (2000).
148. P. E. Jönsson, H. Yoshino, and P. Nordblad, *Phys. Rev. Lett.* **89**, 097201 (2002).
149. Ludovic Berthier *et al.*, *Phys. Rev. Lett.* **90**, 059701 (2003).
150. P. E. Jönsson, H. Yoshino, and P. Nordblad, *Phys. Rev. Lett.* **90**, 059702 (2003).
151. A. J. Bray and M. A. Moore, *Phys. Rev. Lett.* **58**, 57 (1987).
152. H. Rieger, L. Santen, U. Blasum, M. Diehl, and M. Jünger, *J. Phys. A* **29**, 3939 (1996).
153. A. Billoire and E. Marinari, *J. Phys. A* **33**, L265 (2000).
154. H. Yoshino, K. Hukushima, and H. Takayama, cond-mat/0202110.
155. P. Réfrégier, E. Vincent, J. Hamman, and M. Ocio, *J. Phys.* (*France*) **48**, 1533 (1987) .
156. L. Lundgren, P. Svedlindh, and O. Beckman, *J. Magn. Magn. Mat.* **31–34**, 1349 (1983).
157. H. Rieger, *J. Physique I* **4**, 883 (1994).
158. L. F. Cugliandolo and J. Kurchan, *Phys. Rev. B* **60**, 922 (1999).
159. J. P. Bouchaud, in M. E. Cates and M. R. Evans (eds.), *Soft and Fragile Matter* (Institute of Physics, Bristol, 2000), e-print cond-mat/9910387.
160. H. Takayama, H. Yoshino, and T. Komori, cond-mat/9909228.
161. M. Picco, F. Ricci-Tersenghi, and F. Ritort, *Phys. Rev. B* **63**, 174412 (2000).
162. L. W. Bernardi, H. Yoshino, K. Hukushima, H. Takayama, A, Tobo, and A. Ito, *Phys. Rev. Lett.* **86**, 720 (2001).
163. K. Jonason, E. Vincent, J. Hamman, J.-P. Bouchaud, and P. Nordblad, *Phys. Rev. Lett.* **81**, 3243 (1998).
164. H. Yoshino, A. Lemaitre, and J. P. Bouchaud, *Eur. Phys. J. B* **20**, 367 (2001).
165. M. Sales and H. Yoshino, *Phys. Rev. E* **65**, 066131 (2002).
166. K. Jonason, J. Mattsson, and P. Nordblad, *Phys. Rev. Lett.* **77**, 2562 (1996).
167. K. Jonason and P. Nordblad, *Europ. Phys. J. B* **10**, 23 (1999).
168. E. Vincent, V. Dupuis, M. Alba, J. Hammann, and J.-P. Bouchaud, *Europhys. Lett.* **50**, 674 (2000).
169. J. Hamman, E. Vincent, V. Dupuis, M. Alba, M. Ocio, and J.-P. Bouchaud, *J. Phys. Soc. Jp.* **69**(Suppl. A), 296 (2000).
170. E. Vincent, F. Alet, J. Hamman, M. Ocio, and J. P. Bouchaud, *Europhys. Lett.* **50**, 674 (2000).
171. P. Doussineau, T. de Lacerda-Aroso, and A. Levelut, *Europhys. Lett.* **46**, 401 (1999).
172. J.-P. Bouchaud, P. Doussineau, T. de Lacerda-Aroso, and A. Levelut, *Eur. Phys. J B* **21**, 335 (2001).
173. A. V. Kityk, M. C. Rheinstädter, K. Knorr, and H. Rieger, *Phys. Rev. B* **65**, 144415 (2002).
174. E. Vincent, J. Hamman, and M. Ocio, *Recent Progress in Random Magnets* (World Scientific, Singapore, 1992).

175. J.-P. Bouchaud and D. S. Dean, *J. Phys. I (France)* **5**, 265 (1995).
176. M. Sasaki and K. Nemoto, *J. Phys. Soc. Jpn.* **69**, 2283 (2000); **69**, 2642 (2000).
177. L. Balents, J.-P. Bouchaud, and M. Mézard, *J. Phys. I (France)* **6**, 1007 (1996).
178. L. F. Cugliandolo, J. Kurchan, and L. Peliti *Phys. Rev. E* **55**, 3898 (1997).
179. L. F. Cugliandolo and J. Kurchan, *Phys. Rev. Lett.* **71**, 173 (1993).
180. S. Franz and H. Rieger, *J. Stat. Phys.* **79**, 749 (1995).
181. J. O. Andersson, J. Mattsson, and P. Svedlindh, *Phys. Rev. B* **46**, 8297 (1992).
182. E. Marinari, G. Parisi, F. Ricci-Tersenghi, and Juan J. Ruiz-Lorenzo, *J. Phys. A* **31**, 2611 (1998) .
183. L. Berthier, J.-L. Barrat, and J. Kurchan, *Eur. Phys. J. B* **11**, 635 (1999).
184. A. Barrat and L. Berthier, *Phys. Rev. Lett.* **87**, 087204 (2001).
185. D. Herisson and M. Ocio, *Phys. Rev. Lett.* **88**, 257202 (2002).
186. L. F. Cugliandolo, D. R. Grempel, J. Kurchan, and E. Vincent, *Europhys. Lett.* **48**, 699 (1997).
187. T. S. Grigera and N. E. Israeloff, *Phys. Rev. Lett.* **83**, 5038 (1999).
188. L. Bellon and S. Ciliberto, preprint cond-mat/0201224.
189. C. Chamon, M. P. Kennett, H. E. Castillo, and L. F. Cugliandolo, *Phys. Rev. Lett.* **89**, 217201 (2002).
190. H. E. Castillo, C. Chamon, L. F. Cugliandolo, and M. P. Kennett, *Phys. Rev. Lett.* **88**, 237201 (2002).
191. F. Pazmandi, G. Zarand, and G. T. Zimanyi, *Phys. Rev. Lett.* **83**, 1034 (1999).
192. G. Zaránd, F. Pázmándi, K. F. Pál, and G. T. Zimányi, *Phys. Rev. Lett.* **89**, 150201 (2002).
193. H. G. Katzgraber, F. Pazmandi, C. R. Pike, Kai Liu, R. T. Scalettar, K. L. Verosub, and G. T. Zimanyi, *Phys. Rev. Lett.* **89**, 257202 (2002).
194. H. G. Katzgraber, Gary Friedman, and G. T. Zimanyi, *Physica B* **343**, 10 (2004); cond-mat/0307178.
195. C. R. Pike, A. P. Roberts, and K. Verosub, *J. Appl. Phys.* **85**, 6660 (1999).
196. F. Preisach, *Z. Phys.* **94**, 277 (1935).
197. D. Petit, L. Fruchter, and I. A. Campbell, *Phys. Rev. Lett.* **83**, 5130 (1999).
198. D. Petit, L. Fruchter, and I. A. Campbell, *Phys. Rev. Lett.* **88**, 207206 (2002).
199. H. Kawamura, *Can. J. Phys.* **79**, 1447 (2001).
200. J. Maucourt and D. R. Grempel, *Phys. Rev. Lett.* **80**, 770 (1998).
201. N. Akino and J. M. Kosterlitz, *Phys. Rev. B* **66**, 054536 (2002).
202. F. Matsubara, S. Endoh, and T. Shirakura, *J. Phys. Soc. Jpn.* **69**, 1927 (2000).
203. S. Endoh, F. Matsubara, and T. Shirakura, *J. Phys. Soc. Jpn.* **70**, 1543 (2001).
204. A. Sadic and K. Binder, *J. Stat. Phys.* **35**, 517 (1984).
205. D. Imagawa and H. Kawamura, *Phys. Rev. B* **67**, 224412 (2003).
206. T. R. Kirkpatrick and D. Thirumalai, *Phys. Rev. B* **36**, 8552 (1987).

207. T. R. Kirkpatrick and D. Thirumalai, *Phys. Rev. B* **37**, 5342 (1988).
208. U. Bengtzelius, W. Götze, and A. Sjölander, *J. Phys. C* **17**, 5915 (1984).
209. W. Götze, J. P. Hansen, D. Levesque, and J. Zinn-Justin (eds.), *Liquids, Freezing and the Glass Transition* (Amsterdam, North Holland, 1990), p. 287.
210. C. Brangian, W. Kob, and K. Binder, *Europhys. Lett.* **59**, 546 (2002); *J. Phys. A*; cond-mat/0211195.
211. C. Brangian, W. Kob, and K. Binder, *J. Phys. A: Math. Gen.* **36**, 10847 (2003).
212. C. Brangian, W. Kob, and K. Binder, *J. Phys. A : Math. Gen.* **35**, 191 (2002).
213. C. Brangian, W. Kob, and K. Binder, *Phil. Mag. B* **82**, 663 (2002).
214. C. Brangian, W. Kob, and K. Binder, *Europhys. Lett.* **53**, 756 (2001).
215. H. Nishimori, *Physica A* **306**, 68 (2002).
216. H. Nishimori, *Prog. Theor. Phys.* **66**, 1169 (1981).
217. H. Nishimori, *J. Phys. Soc. Jpn.* **55**, 3305 (1986).
218. H. Nishimori, C. Falvo, and Y. Ozeki, *J. Phys. A: Math. Gen.* **39**, 8171 (2002).
219. R. R. P. Singh and J. Adler, *Phys. Rev. B* **54**, 364 (1996).
220. Y. Ozeki, *J. Phys.: Condens. Matter* **9**, 11171 (1997).
221. A. B. Harris, *J. Phys. C* **7**, 1671 (1974).
222. G. Grinstein and A. Luther, *Phys. Rev. B* **13**, 1329 (1976).
223. H.-O. Heuer, *J. Phys. A: Math. Gen.* **26**, L333 (1993).
224. I. O. Mayer, *J. Phys. A: Math. Gen.* **22**, 2815 (1989).
225. P. Le Doussal and A. B. Harris, *Phys. Rev. Lett.* **61**, 625 (1988).
226. H. G. Ballesteros *et al.*, *Phys. Rev. B* **58**, 2740 (1998).
227. J. C. LeGuillow and J. Zinn-Justin, *Phys. Rev. B* **21**, 3976 (1980).
228. K. Hukushima, *J. Phys. Soc. Jpn.* **69**, 631 (2000).
229. M. Suzuki, *Prog. Theor. Phys.* **58**, 1142 (1977).
230. M. Suzuki, *Phys. Lett. A* **58**, 435 (1978).
231. Y. Ozeki and N. Ito, *J. Phys. A: Math. Gen.* **31**, 5451 (1998).
232. H. Nishimori and K. Nemoto, *Physica A* **321**, 108 (2003).
233. F. Y. Wu and Y. K. Wang, *J. Math. Phys.* **17**, 439 (1976).
234. H. Nishimori, *J. Phys. C* **12**, L905 (1979).
235. F. D. A. Aarao Reis, S. L. A. de Queiroz, and R. R. dos Santos, *Phys. Rev. B* **60**, 6740 (1999).
236. M. Rubinstein, B. Shraiman, and D. R. Nelson, *Phys. Rev. B* **27**, 1800 (1983).
237. T. Nattermann, S. Scheidl, S. E. Korshunov, and M. S. Li, *J. Phys.* (*France*) I **5**, 565 (1995).
238. M.-C. Cha and H. A. Fertig, *Phys. Rev. Lett.* **74**, 4867 (1995).
239. L.-H. Tang, *Phys. Rev. B* **54**, 3350 (1996).
240. S. Scheidl, *Phys. Rev. B* **55**, 457 (1997).
241. D. Carpentier and P. Le Doussal, *Nuc. Phys. B* **588**, 565 (2000).
242. D. R. Nelson, *Phys. Rev. B* **27**, 2902 (1983).

243. W. Wu, B. Ellmann, T. F. Rosenbaum, G. Aeppli, and D. H. Reich, *Phys. Rev. Lett.* **67**, 2076 (1991); W. Wu, D. Bitko, T. F. Rosenbaum, and G. Aeppli, *Phys. Rev. Lett.* **71**, 1919 (1993).
244. S. Sachdev, *Quantum Phase Transitions* (Cambridge University Press, 1999).
245. B. K. Chakrabarti, A. Dutta, and P. Sen, *Quantum Ising Phases and Transitions in Transverse Ising Models* (Springer, Heidelberg, 1996).
246. H. Rieger and A. P. Young, in J. M. Rubi and C. Perez-Vicente (eds.), *Quantum Spin Glasses, Lecture Notes in Physics* **492**, Complex Behaviour of Glassy Systems (Springer Verlag, Berlin-Heidelberg-New York, 1997), p. 254.
247. R. N. Bhatt, in A. P. Young (ed.), *Spin glasses and Random Fields* (World Scientific, Singapore, 1998).
248. D. S. Fisher, *Physica A* **263**, 222 (1999).
249. B.M. McCoy and T.T. Wu, *Phys. Rev.* **176**, 631 (1968); **188**, 982 (1969); *B.M. McCoy, Phys. Rev.* **188**, 1014 (1969).
250. D. S. Fisher, *Phys. Rev. Lett.* **69**, 534 (1992); *Phys. Rev. B* **51**, 6411 (1995).
251. S. K. Ma, C. Dasgupta, and C.-K. Hu, *Phys. Rev. Lett.* **43**, 1434 (1979); C. Dasgupta and S. K. Ma, *Phys. Rev. B* **22**, 1305 (1980).
252. F. Iglói and H. Rieger, *Phys. Rev. Lett.* **78**, 2473 (1997).
253. F. Iglói and H. Rieger, *Phys. Rev. B* **57**, 11404 (1998).
254. H. Rieger and F. Iglói, *Europhys. Lett.* **39**, 135 (1997).
255. F. Iglói, D. Karevski and H. Rieger, *Europ. Phys. J. B* **1**, 513 (1998); **5**, 613 (1998).
256. F. Iglói and H. Rieger, *Phys. Rev. E* **58**, 4238 (1998).
257. F. Iglói, R. Juhász and H. Rieger, *Phys. Rev. B* **59**, 11308 (1999).
258. H. Rieger and F. Iglói, *Phys. Rev. Lett.* **83**, 3741 (1999).
259. F. Iglói, R. Juhász, and H. Rieger, *Phys. Rev. B* **61**, 11552 (2000).
260. A. P. Young and H. Rieger, *Phys. Rev. B* **53**, 8486 (1996).
261. M. J. Thill and D. A. Huse, *Physica A* **214**, 321 (1995).
262. H. Rieger and A. P. Young, *Phys. Rev. B* **54**, 3328 (1996).
263. T. Senthil and S. Sachdev, *Phys. Rev. Lett.* **77**, 5292 (1996).
264. T. Ikegami, S. Miyashita and H. Rieger, *J. Phys. Soc. Jap.* **67**, 2761 (1998).
265. C. Pich, A. P. Young, H. Rieger, and N. Kawashima, *Phys. Rev. Lett.* **81**, 5916 (1998).
266. H. Rieger and N. Kawashima, *Europ. Phys. J. B* **9**, 233 (1999).
267. O. Motrunich, S.-C. Mau, D. A. Huse, and D. S. Fisher, *Phys. Rev. B* **61**, 1160 (2000).
268. H. Rieger and A. P. Young, *Phys. Rev. Lett.* **72**, 4141 (1994).
269. M. Guo, R. N. Bhatt, and D. A. Huse, *Phys. Rev. Lett.* **72**, 4137 (1994).
270. M. Guo, R. N. Bhatt, and D. A. Huse, *Phys. Rev. B* **54**, 3336 (1996).
271. L. F. Cugliandolo, D. R. Grempel, and C. A. da Silva Santos, *Phys. Rev. Lett.* **85**, 2589 (2000); *Phys. Rev. B* **64**, 014403 (2001).
272. Y. V. Fedorov and E. F. Shender, *Zh. Eksp. Teor. Fiz.* **43**, 526 (1986) [*JETP Lett.* **43**, 681 (1986)].
273. Y. Y. Goldschmidt and P. Y. Lai, *Phys. Rev. Lett.* **64**, 2467 (1990).

274. K. D. Usadel and B. Schmidt, *Solid State Comm.* **64**, 975 (1987).
275. T. Yamamoto and H. Ishii, *J. Phys. C.* **20**, 6053 (1987).
276. J. Miller and D. A. Huse, *Phys. Rev. Lett.* **70**, 3147 (1993).
277. J. Ye, S. Sachdev, and N. Read, *Phys. Rev. Lett.* **70**, 4011 (1993).
278. N. Read, S. Sachdev, and J. Ye, *Phys. Rev. B* **52**, 384 (1995).
279. A. J. Bray and M. A. Moore, *J. Phys. C* **13**, L655 (1980).
280. M. J. Rozenberg and D. R. Grempel, *Phys. Rev. Lett.* **81**, 2550 (1998).
281. D. R. Grempel and M. J. Rozenberg, *Phys. Rev. Lett.* **80**, 389 (1998).
282. L. F. Cugliandolo, D. R. Grempel, G. Lozano, H. Lozza, and C. A. da Silva Santos, *Phys. Rev. B* **66**, 014444 (2002).
283. A. J. Leggett *et al.*, *Rev. Mod. Phys.* **59**, 1 (1987); **67**, 725 (1995).
284. U. Weiss, *Series Modern Condensed Matter Physics*, Vol. 2 (World Scientific, Singapore, 1993).
285. A. O. Caldeira and A. J. Leggett, *Phys. Rev. Lett.* **46**, 211 (1981); *Ann. Phys. N.Y.* **149**, 374 (1983).
286. A. J. Bray and M. A. Moore, *Phys. Rev. Lett.* **49**, 1545 (1982).
287. S. Chakravarty, *Phys. Rev. Lett.* **49**, 681 (1982); S. Chakravarty and A. J. Leggett, *ibid.* **52**, 5 (1984).
288. R. P. Feynman and F. L. Vernon, Jr., *Ann. Phys. N.Y.* **24**, 118 (1963).
289. N. V. Prokof ev and P. C. E. Stamp, *Phys. Rev. Lett.* **80**, 5794 (1998).
290. H. Grabert, P. Schramm, and G.-L. Ingold, *Phys. Rep.* **168**, 115 (1988).
291. P. W. Anderson and G. Yuval, *Phys. Rev. Lett.* **32**, 89 (1969); *J. Phys. C* **4**, 607 (1971); P. W. Anderson, G. Yuval, and D. R. Hammann, *Phys. Rev. B* **1**, 4464 (1970).
292. L. F. Cugliandolo and G. Lozano, *Phys. Rev. Lett.* **80**, 4979 (1998).
293. T. Giamarchi and P. Le Doussal, *Phys. Rev. B* **53**, 15206 (1996).
294. Y. Nonomura and Y. Ozeki, *J. Phys. Soc. Jpn.* **64**, 2710 (1995).
295. J. Oitmaa and O. P. Sushkov, *Phys. Rev. Lett.* **87**, 167206 (2001).
296. D. S. Fisher, *Phys. Rev. B* **50**, 3799 (1994).
297. R. Mélin, Y.-C. Lin, P. Lajkó, H. Rieger, and F. Iglói, *Phys. Rev. B* **65**, 104415 (2002).
298. Y.-C. Lin, R. Mélin, H. Rieger, and F. Iglói, *Phys. Rev. B* **68**, 024424 (2003).
299. A. Georges, O. Parcollet, and S. Sachdev, *Phys. Rev. Lett.* **85**, 840 (2000); *Phys. Rev. B* **63**, 134406 (2001).
300. G. Biroli and O. Parcollet, *Phys. Rev. B* **65**, 094414 (2002).
301. J. A. Hertz, *Phys. Rev. B* **19**, 4796 (1979).
302. R. Oppermann and M. Binderberger, *Ann. Phys. (N.Y.)* **3**, 494 (1994).
303. S. Sachdev, N. Read, and R. Oppermann, *Phys. Rev. B* **52**, 10286 (1995).
304. A. Sengupta and A. Georges, *Phys. Rev. B* **52**, 10295 (1995).
305. W. D. Wu *et al.*, *Phys. Rev. Lett.* **72**, 3722 (1994).

CHAPTER 10

FRUSTRATED MAGNETIC THIN FILMS: SPIN WAVES AND SKYRMIONS

Hung T. Diep[*,‡] and Hiroshi Koibuchi[†,§]

[*] *Laboratoire de Physique Théorique et Modélisation*
Université de Cergy-Pontoise, CNRS UMR 8089
2 Avenue Alphonse Chauvin, 95302 Cergy-Pontoise Cedex, France
[†] *Department of General Engineering, National Institute of Technology (KOSEN), Sendai College*
8 Nodayama, Medeshima-Shiote, Natori-shi, Miyagi 981-1239, Japan
[‡] *diep@u-cergy.fr*
[§] *koi-hiro@sendai-nct.ac.jp*

We present in this chapter a review of our recent works on frustrated magnetic thin films. We show theoretically and numerically that we can obtain the desired macroscopic effects by modifying the surface parameters. Surface effects when combined with frustration effects give rise to striking phenomena such as surface ordering and surface phase transition distinct from the bulk ones. We describe the methods employed to study a number of frustrated magnetic thin films. Examples are shown and discussed with focus on the frustration effect. In particular, the well-studied case of helimagnetic films and the effect of the Dzyaloshinskii–Moriya interaction (DMI) in thin layers are shown. Results on the spin waves are presented. Effects of an applied magnetic field on the spin configuration are shown. In particular, we show that a crystal of skyrmions is formed in the phase space and is stable at finite temperatures. Effects of elasticity and uniaxial strains on the skyrmion structure and the phase transition are shown.

10.1. Introduction

Surface effects in thin films have been the subject of extensive experimental, theoretical and numerical investigations since the 1970s. The rapid development of the surface physics is due to technological advances in the fabrication and the characterization of films very thin down to a single atomic layer.[1–3] This is also due to numerous industrial applications.[4]

The aim of this chapter is to introduce some fundamental theories and simulations to study the surface effects on the properties of frustrated films. The surface effect when combined with the frustration effect gives rise to unexpected phenomena. The chapter focuses thus on theoretical and numerical methods and results. The reader can find results on a particular domain of experimental surface physics he/she wishes to learn in numerous books and reviews.

Most of the examples shown in this chapter are of frustrated thin films that have been recently investigated by our group. Of course, 2D cases, namely monolayers, belong to the family of thin films. However, there are no surface effects as seen in thin films with a thickness where surface atomic layers do not have the same number of neighbors as the interior layers. Because of this and as a result of the other symmetry rupture mechanisms, surface parameters are often different from the bulk ones, giving rise to numerous surface effects.

The case of frustrated 2D Ising systems has been treated in Chapter 1 using exact methods. Most of the frustration effects are found in these exactly solved models, such as the partial disorder, the re-entrance, and the disorder line. The reader is referred to Chapter 1 for further details.

Sections 10.2–10.4 are mainly devoted to frustrated thin films, where spins are localized on rigid lattices. Surface effects are outlined. In particular, we show in detail the existence of surface spin waves and the surface phase transition, which occurs at a temperature distinct from that of the bulk transition in a helimagnetic film. In the case where a Dzyaloshinskii–Moriya (DM) interaction competes with the ferromagnetic interaction, we show that a skyrmion crystal is generated when an external magnetic field is applied. The stability of this crystal at finite temperatures and its transition to the disordered phase are studied. In Section 10.5, we show the effects of elasticity on properties of skyrmions. We study a 2D skyrmion system on fluctuating surfaces with periodic boundary conditions using a Monte Carlo (MC) simulation technique. In this model, not only spins but also lattice vertices are integrated into the partition function of the model.

We find that skyrmions are stable even on fluctuating lattices, contrary to initial expectations. Effects of stress and uniaxial strains are also shown and discussed.

10.2. Surface Effects in Helimagnets

In this section we study the properties of a helimagnetic thin film in zero field with quantum Heisenberg spins by using the Green's function method. The surface spin configuration is calculated by minimizing the spin interaction energy. It is shown that the angles between spins near the surface are strongly modified with respect to the bulk configuration. Taking into account this surface spin reconstruction, we calculate self-consistently the spin wave spectrum and the layer magnetizations as functions of temperature up to the disordered phase. The spin wave spectrum shows the existence of a surface-localized branch which causes a low surface magnetization. We show that quantum fluctuations give rise to a cross-over between the surface magnetization and interior-layer magnetizations at low temperatures. We calculate the transition temperature and show that it depends strongly on the surface conditions. The results are in agreement with the existing experimental observations on the stability of helical structure in thin films and on the insensitivity of the transition temperature with the film thickness. MC simulations for the classical spin model are also carried out for comparison with the case of quantum spins.

10.2.1. *Introduction*

Helimagnetic structures were discovered a long time ago by Yoshimori[5] and Villain.[6] In the simplest model, the helimagnetic ordering is non-collinear due to a competition between the nearest-neighbor (NN) and next-nearest-neighbor (NNN) interactions: for example, a spin in a chain turns an angle θ with respect to its previous neighbor. Low-temperature properties in bulk helimagnets, such as spin waves[7–10] and heat capacity,[11] have been extensively investigated. Helimagnets belong to a class of frustrated vector-spin systems. In spite of their long history, the nature of the phase transition in bulk helimagnets and other non-collinear magnets, such as stacked triangular XY and Heisenberg antiferromagnets, has been elucidated only recently[12–14] (see also reviews in Chapters 3 and 4).

Hereafter, we study a helimagnetic thin film with the quantum Heisenberg spin model. Surface effects in thin films have been widely studied theoretically, experimentally, and numerically, in the last three decades.[2,3] Nevertheless, surface effects in helimagnets have only been recently studied: surface spin structures,[15] MC simulations,[16] and magnetic field effects on the phase diagram in Ho[17] have been studied in a few experiments.[18,19]

Helical magnets present potential applications in spintronics with predictions of spin-dependent electron transport in these magnetic materials.[20–22] We shall use the Green's function (GF) method to study a quantum spin model on a helimagnetic thin film of body-centered cubic (BCC) lattice. The GF method has been initiated by Zubarev[23] for collinear bulk magnets (ferromagnets and antiferromagnets) and by Diep-The-Hung *et al.* for thin films with collinear spin configurations.[24] For non-collinear magnets, the GF method has also been developed for bulk helimagnets[10] and for frustrated thin films.[25,26] Note that the surface effects in thin films of stacked triangular antiferromagnets with non-collinear 120° spin configuration have been investigated by the equation of motion method.[27] However, the model in these works did not possess a surface spin reconstruction and does not belong to the family of helical structures as described below in our model.

In helimagnets, the crystal rupture at the surface modifies the competing forces acting on the surface spins. As a consequence, as will be shown below, the angles between the neighboring spins become non-uniform across the film, making the calculation of physical quantities harder.

Note that for illustration, we use below the BCC lattice structure, but the results obtained here are valid for different lattices, and not restricted to the BCC crystal, provided modifications are given for the coordination number and therefore on the value of the critical value $(J_2/J_1)_c$ (J_1: NN interaction, J_2: NNN interaction). For example, the BCC case has $(J_2/J_1)_c = 1$ while the simple cubic (SC) lattice has $(J_2/J_1)_c = 1/4$ (see Chapter 1, Section 1.1.2 and Fig. 10.3).

10.2.2. *Model and classical ground state*

We recall that bulk helical structures are due to the competition of various kinds of interaction.[5,6,28–30] To generate a bulk helimagnetic structure, the simplest way is to take a ferromagnetic interaction between NNs, say J_1 (>0), and an antiferromagnetic interaction between NNNs, $J_2 < 0$. It is shown in Chapter 1 that if $|J_2|$ is larger than a critical value $|J_2^c|$, the classical GS spin configuration is helimagnetic.[7–9] Since our purpose is to investigate the helimagnetic structure near the surface and surface effects, let us consider the case of a helimagnetic structure only in the c-direction perpendicular to the film surface. In such a case, we assume a non-zero J_2 only on the c-axis. This assumption simplifies formulas but does not change the physics of the problem since including the uniform helical angles in two other directions parallel to the surface will not introduce additional surface effects. Note that the bulk case of the above quantum spin model has been studied by the Green's function method.[10] Let us recall that the helical structure in the bulk is planar: spins lie in planes perpendicular to the c-axis: the angle between two NNs in the adjacent planes is a constant and is given by $\cos\alpha = -J_1/J_2$ for a BCC lattice. The helical structure exists therefore if $|J_2| \geq J_1$, namely $|J_2^c|(\text{bulk}) = J_1$.

For films, we consider the helical ordering along one direction, namely along the c-axis perpendicular to the film surface. We suppose a film of BCC lattice with N_z layers having two symmetrical surfaces. The Hamiltonian is written as

$$\mathcal{H}_e = -\sum_{\langle i,j \rangle} J_{i,j} \mathbf{S}_i \cdot \mathbf{S}_j, \tag{10.1}$$

where $J_{i,j}$ denotes the interaction between two spins $\mathbf{S}_i$ and $\mathbf{S}_j$ occupying the lattice sites i and j.

To calculate the classical GS surface spin configuration, we write down the expression of the energy of spins along the c-axis, starting from the surface

$$\begin{aligned} E = & -Z_1 J_1 \cos(\theta_1 - \theta_2) - Z_1 J_1 [\cos(\theta_2 - \theta_1) \\ & + \cos(\theta_2 - \theta_3)] + \cdots \\ & - J_2 \cos(\theta_1 - \theta_3) - J_2 \cos(\theta_2 - \theta_4) \\ & - J_2 [\cos(\theta_3 - \theta_1) + \cos(\theta_3 - \theta_5)] + \cdots , \end{aligned} \tag{10.2}$$

where $Z_1 = 4$ is the number of NNs in a neighboring layer and θ_i denotes the angle of a spin in the ith layer made with the Cartesian x-axis of the layer. The interaction energy between two NN spins in the two adjacent layers i and j depends only on the difference $a_i \equiv \theta_i - \theta_{i+1}$. The GS configuration corresponds to the minimum value of E. The results calculated for various J_2/J_1 are shown in Fig. 10.1 for a film of $N_z = 8$ layers.

The values obtained are shown in Table 10.1 where we see strong deviations of the bulk angle while moving toward the surface.

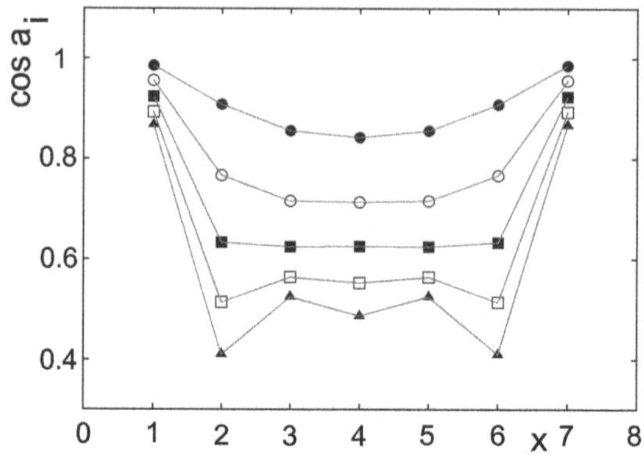

Fig. 10.1. Cosinus of $a_1 = \theta_1 - \theta_2, \ldots, a_7 = \theta_7 - \theta_8$ across the film for $J_2/J_1 = -1.2, -1.4, -1.6, -1.8, -2$ (from top) with $N_z = 8$. x indicates the film layer i where the angle a_i with the layer $(i+1)$ is shown. The values of the angles are given in Table 10.1: a strong rearrangement of spins near the surface is observed.

Table 10.1. Values of $\theta_{n,n+1} = \alpha_n$ between two adjacent layers are shown for various values of J_2/J_1.

J_2/J_1	$\theta_{1,2}$	$\theta_{2,3}$	$\theta_{3,4}$	$\theta_{4,5}$	α(bulk)
−1.2	9.79°	24.73°	31.15°	32.54°	33.56°
−1.4	17.07°	39.92°	44.28°	44.41°	44.42°
−1.6	22.52°	50.73°	51.38°	51.30°	51.32°
−1.8	26.66°	59.04°	55.66°	56.48°	56.25°
−2.0	29.84°	65.76°	58.31°	60.85°	60°

Notes: Only angles of the first half of the 8-layer film are shown: other angles are, by symmetry, $\theta_{7,8} = \theta_{1,2}$, $\theta_{6,7} = \theta_{2,3}$, $\theta_{5,6} = \theta_{3,4}$. The angles are in degrees. The last column shows the value of the angle in the bulk case (infinite thickness). For presentation, angles are shown with two digits.

Note that (i) the results shown above are obtained by iteration with errors less than 10^{-4} degrees, (ii) strong angle variations are observed near the surface with oscillations for strong J_2, (iii) the angles at the film center are close to the bulk value α (last column), meaning that the surface reconstruction affects just a few atomic layers, and (iv) the bulk helical order is stable at just a few atomic layers away from the surface even for films thicker than $N_z = 8$. This helical stability has been experimentally observed in holmium films.[31]

Using the numerical steepest descent method described in Refs. [25] and [26], we obtain the same result as shown above.

Let us discuss now the effect of the anisotropy on the GS configuration. We know that according to the theorem of Mermin and Wagner,[32] continuous isotropic spin models such as XY and Heisenberg spins do not have long-range ordering at finite temperatures in 2D. Since we are dealing with the Heisenberg model in a thin film, it is useful to add an anisotropic interaction to create a long-range ordering and a phase transition at finite temperatures. The full Hamiltonian is given as

$$\mathcal{H} = \mathcal{H}_e + \mathcal{H}_a, \tag{10.3}$$

where $\mathcal{H}_e$ is given by Eq. (10.1) and the anisotropic part $\mathcal{H}_a$ of the Hamiltonian is chosen as

$$\mathcal{H}_a = -\sum_{\langle i,j \rangle} I_{i,j} S_i^z S_j^z \cos\theta_{ij}, \tag{10.4}$$

where $I_{i,j}$ is the interaction between the z spin components lying in the plane of the spins, i.e. the planes perpendicular to the c-axis as determined above. $I_{i,j}$ is supposed to be positive, small compared to J_1, and limited to NN on the c-axis.

The form of Eq. (10.4) simplifies a lot: since the interaction is limited to NN, it suffices to replace in Eq. (10.2) the parameter J_1 by $J_1' = J_1 + I_1$, where $I_1 = I_{i,j}$ for any NN pairs (i, j). The GS calculation is done exactly in the same manner. That is the reason why we choose the form of Eq. (10.4). The GS configuration is slightly modified but the method and the general aspects of the results described above remain valid. Of course, the calculation of the spin wave spectrum and layer magnetizations presented below take into account the GS modification at each value of I_1. Choosing another form of anisotropy, for example taking a standard single-ion anisotropy $-I(S_i^z)^2$ along the spin local axis, will add just a constant in Eq. (10.2) [because $(S_i^z)^2 = 1$ in the GS]: it will not therefore affect the GS configuration. Therefore, this form is not our interest.

In the following, using the GS spin configuration we calculate the spin wave excitation and properties of the film such as the zero-point spin contraction, the layer magnetizations, and the critical temperature.

10.2.3. *Green's function method*

Let us define the local spin coordinates as follows: the quantization axis of spin $\mathbf{S}_i$ is on its ζ_i axis which lies in the plane, the η_i axis of $\mathbf{S}_i$ is along the c-axis, and the ξ_i axis forms with η_i and ζ_i axes a direct trihedron. Since the spin configuration is planar, all spins have the same η axis. Furthermore, all spins in a given layer are parallel. Let $\hat{\xi}_i$, $\hat{\eta}_i$, and $\hat{\zeta}_i$ be the unit vectors on the local (ξ_i, η_i, ζ_i) axes. We use the following local transformation which has been described in Ref. [1]:

$$\mathbf{S}_i = S_i^x \hat{\xi}_i + S_i^y \hat{\eta}_i + S_i^z \hat{\zeta}_i, \tag{10.5}$$

$$\mathbf{S}_j = S_j^x \hat{\xi}_j + S_j^y \hat{\eta}_j + S_j^z \hat{\zeta}_j. \tag{10.6}$$

We have (see Fig. 10.2)

$$\begin{aligned}
\hat{\xi}_j &= \cos\theta_{ij}\hat{\zeta}_i + \sin\theta_{ij}\hat{\xi}_i,\\
\hat{\zeta}_j &= -\sin\theta_{ij}\hat{\zeta}_i + \cos\theta_{ij}\hat{\xi}_i,\\
\hat{\eta}_j &= \hat{\eta}_i,
\end{aligned}$$

where $\cos\theta_{ij} = \cos(\theta_i - \theta_j)$ is the angle between two spins i and j.

Note that in the laboratory coordinate system, namely in the film coordinates, the z-direction coincides with the c-direction or the $\hat{\eta}$ axis perpendicular to the film surface, while the x and y directions are taken to be the BCC crystal axes in the film plane.

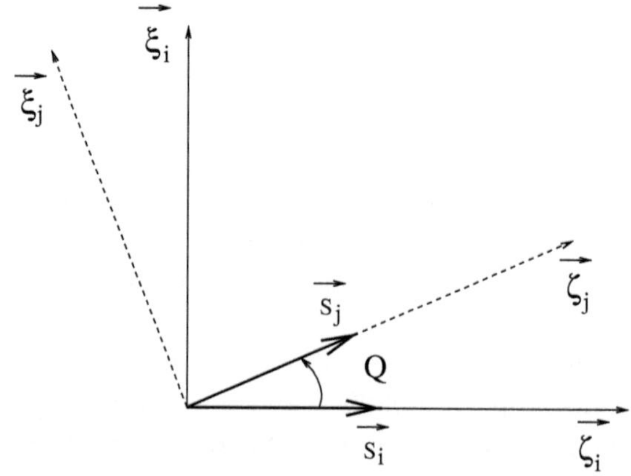

Fig. 10.2. Local coordinates in a xy-plane perpendicular to the c-axis. Q denotes $\theta_j - \theta_i$.

Replacing these into Eqs. (10.5) and (10.6) to express $\mathbf{S}_j$ in the $(\hat{\xi}_i, \hat{\eta}_i, \hat{\zeta}_i)$ coordinates, then by calculating $\mathbf{S}_i \cdot \mathbf{S}_j$, we obtain the following exchange Hamiltonian from (10.1):

$$\begin{aligned}\mathcal{H}_e = -\sum_{\langle i,j\rangle} J_{i,j} \bigg\{ &\frac{1}{4}(\cos\theta_{ij} - 1)(S_i^+ S_j^+ + S_i^- S_j^-) \\ &+ \frac{1}{4}(\cos\theta_{ij} + 1)(S_i^+ S_j^- + S_i^- S_j^+) \\ &+ \frac{1}{2}\sin\theta_{ij}(S_i^+ + S_i^-)S_j^z - \frac{1}{2}\sin\theta_{ij} S_i^z(S_j^+ + S_j^-) \\ &+ \cos\theta_{ij} S_i^z S_j^z \bigg\}. \end{aligned} \tag{10.7}$$

10.2.4. *General formulation for non-collinear magnets*

We define the following two double-time Green's functions in the real space:

$$\begin{aligned} G_{i,j}(t,t') &= \langle\langle S_i^+(t); S_j^-(t')\rangle\rangle \\ &= -i\theta(t-t')\langle [S_i^+(t), S_j^-(t')]\rangle, \end{aligned} \tag{10.8}$$

$$\begin{aligned} F_{i,j}(t,t') &= \langle\langle S_i^-(t); S_j^-(t')\rangle\rangle \\ &= -i\theta(t-t')\langle [S_i^-(t), S_j^-(t')]\rangle. \end{aligned} \tag{10.9}$$

Note that the equation of motion of the first function generates functions of the second type, and vice versa. The next steps in the general Green's function formalism have been described elsewhere.[1,33] We show here only some details and the results obtained from that method.

10.2.5. *BCC helimagnetic films*

In the case of a BCC thin film with a (001) surface, writing the equations of motion of the above Green's functions and using the Tyablikov decoupling scheme,[34] we obtain a closed system of coupled equations. For simplicity, all NN interactions $(J_{k,k'}, I_{k,k'})$ are taken to be equal to (J_1, I_1) and all NNN interactions are taken to be equal to J_2 in the following. Furthermore, the Cartesian components of the spin position $\mathbf{R}_i$ will be denoted by (ℓ_i, m_i, n_i).

We make the in-plane Fourier transforms

$$\begin{aligned} G_{i,j}(t,t') = \frac{1}{\Delta}\int\!\!\int_{BZ} d\mathbf{k}_{xy} \frac{1}{2\pi}\int_{-\infty}^{+\infty} d\omega e^{-i\omega(t-t')} \\ \times g_{n_i,n_j}(\omega, \mathbf{k}_{xy}) e^{i\mathbf{k}_{xy}\cdot(\mathbf{R}_i - \mathbf{R}_j)}, \end{aligned} \tag{10.10}$$

$$F_{k,j}(t,t') = \frac{1}{\Delta}\int\int_{BZ} d\mathbf{k}_{xy}\frac{1}{2\pi}\int_{-\infty}^{+\infty} d\omega e^{-i\omega(t-t')} \times f_{n_k,n_j}(\omega,\mathbf{k}_{xy})e^{i\mathbf{k}_{xy}\cdot(\mathbf{R}_k-\mathbf{R}_j)}, \tag{10.11}$$

where ω is the spin wave frequency, $\mathbf{k}_{xy}$ denotes the wave vector parallel to xy planes, and $\mathbf{R}_i$ is the position of the spin at the site i. n_i, n_j, and n_k are, respectively, the z-component indices of the layers where the sites $\mathbf{R}_i$, $\mathbf{R}_j$, and $\mathbf{R}_k$ belong to. The integral over $\mathbf{k}_{xy}$ is performed in the first Brillouin zone (BZ) whose surface is Δ in the xy reciprocal plane. For convenience, we denote $n_i = 1$ for all sites on the surface layer, $n_i = 2$ for all sites of the second layer, and so on.

We finally obtain the following matrix equation (see details in Ref. [33]):

$$\mathbf{M}(\omega)\mathbf{h} = \mathbf{u}, \tag{10.12}$$

where $\mathbf{M}(\omega)$ is a square matrix of dimension $(2N_z \times 2N_z)$ and $\mathbf{h}$ and $\mathbf{u}$ are the column matrices which are defined as follows:

$$\mathbf{h} = \begin{pmatrix} g_{1,n'} \\ f_{1,n'} \\ \vdots \\ g_{n,n'} \\ f_{n,n'} \\ \vdots \\ g_{N_z,n'} \\ f_{N_z,n'} \end{pmatrix}, \quad \mathbf{u} = \begin{pmatrix} 2\langle S_1^z\rangle\delta_{1,n'} \\ 0 \\ \vdots \\ 2\langle S_{N_z}^z\rangle\delta_{N_z,n'} \\ 0 \end{pmatrix}, \tag{10.13}$$

where, taking $\hbar = 1$ hereafter, the matrix $\mathbf{M}(\omega)$ is

$$\begin{pmatrix}
\omega + A_1 & 0 & B_1^+ & C_1^+ & D_1^+ & E_1^+ & 0 & 0 & 0 & 0 & 0 & 0 \\
0 & \omega - A_1 & -C_1^+ & -B_1^+ & -E_1^+ & -D_1^+ & 0 & 0 & 0 & 0 & 0 & 0 \\
\cdots & \cdots & \cdots & \cdots & \cdots & \cdots & \cdots & \cdots & \cdots & \cdots & \cdots & \cdots \\
\cdots & D_n^- & E_n^- & B_n^- & C_n^- & \omega + A_n & 0 & B_n^+ & C_n^+ & D_n^+ & E_n^+ & \cdots \\
\cdots & -E_n^- & -D_n^- & -C_n^- & -B_n^- & 0 & \omega - A_n & -C_n^+ & -B_n^+ & -E_n^+ & -D_n^+ & \cdots \\
\cdots & \cdots & \cdots & \cdots & \cdots & \cdots & \cdots & \cdots & \cdots & \cdots & \cdots & \cdots \\
0 & 0 & 0 & 0 & 0 & 0 & D_{N_z}^- & E_{N_z}^- & B_{N_z}^- & C_{N_z}^- & \omega + A_{N_z} & 0 \\
0 & 0 & 0 & 0 & 0 & 0 & -E_{N_z}^- & -D_{N_z}^- & -C_{N_z}^- & -B_{N_z}^- & 0 & \omega - A_{N_z}
\end{pmatrix}, \tag{10.14}$$

where

$$\begin{aligned} A_n &= -8J_1(1+d)[\langle S^z_{n+1}\rangle \cos\theta_{n,n+1} \\ &\quad + \langle S^z_{n-1}\rangle \cos\theta_{n,n-1}] \\ &\quad - 2J_2[\langle S^z_{n+2}\rangle \cos\theta_{n,n+2} \\ &\quad + \langle S^z_{n-2}\rangle \cos\theta_{n,n-2}], \end{aligned}$$

where $n = 1, 2, \ldots, N_z$, $d = I_1/J_1$, and

$$\begin{aligned} B^{\pm}_n &= 4J_1\langle S^z_n\rangle(\cos\theta_{n,n\pm1} + 1)\gamma, \\ C^{\pm}_n &= 4J_1\langle S^z_n\rangle(\cos\theta_{n,n\pm1} - 1)\gamma, \\ E^{\pm}_n &= J_2\langle S^z_n\rangle(\cos\theta_{n,n\pm2} - 1), \\ D^{\pm}_n &= J_2\langle S^z_n\rangle(\cos\theta_{n,n\pm2} + 1). \end{aligned}$$

In the above expressions, $\theta_{n,n\pm1}$ is the angle between a spin in the layer n and its NN spins in layers $n \pm 1$ etc. and $\gamma = \cos\left(\frac{k_x a}{2}\right)\cos\left(\frac{k_y a}{2}\right)$.

The spin wave spectrum ω of the present system is obtained by solving $\det|\mathbf{M}| = 0$. For each value (k_x, k_y), there are $2N_z$ eigenvalues of ω corresponding to the two opposite-spin precessions as in antiferromagnets (the dimension of $\det|\mathbf{M}|$ is $2N_z \times 2N_z$). Note that the above equation depends on the values of $\langle S^z_n\rangle$ $(n = 1, \ldots, N_z)$. Even at temperature $T = 0$, these z-components are not equal to 1/2 because of the so-called zero-point spin contraction[1] which occurs in systems with antiferromagnetic interactions. This contraction is not uniform in our system because of the existence of the film surfaces as seen below. So the solution of $\det|\mathbf{M}| = 0$ should be found by iteration.

Once ω is obtained, the solution for $g_{n,n}$ is given by

$$g_{n,n}(\omega) = \frac{|\mathbf{M}|_{2n-1}}{|\mathbf{M}|}, \tag{10.15}$$

where $|\mathbf{M}|_{2n-1}$ is the determinant made by replacing the $(2n-1)$th column of $|\mathbf{M}|$ by $\mathbf{u}$ given by Eq. (10.13) (note that $g_{n,n}$ occupies the $(2n-1)$th line of the matrix $\mathbf{h}$). Writing now

$$|\mathbf{M}| = \prod_i [\omega - \omega_i(\mathbf{k}_{xy})], \tag{10.16}$$

we see that $\omega_i(\mathbf{k}_{xy})$, $i = 1, \ldots, 2N_z$, are poles of $g_{n,n}$. Using the above form, $g_{n,n}$ can be expressed as

$$g_{n,n}(\omega) = \sum_i \frac{D_{2n-1}(\omega_i(\mathbf{k}_{xy}))}{[\omega - \omega_i(\mathbf{k}_{xy})]}, \tag{10.17}$$

where $D_{2n-1}(\omega_i(\mathbf{k}_{xy}))$ is

$$D_{2n-1}(\omega_i(\mathbf{k}_{xy})) = \frac{|\mathbf{M}|_{2n-1}(\omega_i(\mathbf{k}_{xy}))}{\prod_{j\neq i}[\omega_j(\mathbf{k}_{xy}) - \omega_i(\mathbf{k}_{xy})]}. \tag{10.18}$$

Next, using the spectral theorem which relates the correlation function $\langle S_i^- S_j^+\rangle$ to the Green's function,[23] we have

$$\begin{aligned} \langle S_i^- S_j^+\rangle = \lim_{\varepsilon\to 0} \frac{1}{\Delta} \int\int d\mathbf{k}_{xy} \int_{-\infty}^{+\infty} \frac{i}{2\pi}(g_{n,n'}(\omega + i\varepsilon) \\ - g_{n,n'}(\omega - i\varepsilon)) \frac{d\omega}{e^{\beta\omega} - 1} e^{i\mathbf{k}_{xy}\cdot(\mathbf{R}_i - \mathbf{R}_j)}, \end{aligned} \tag{10.19}$$

where ϵ is an infinitesimal positive constant and $\beta = (k_B T)^{-1}$, k_B being the Boltzmann constant.

Using the Green's function presented above, we can calculate by self-consistently iterating various physical quantities as functions of temperature T. The magnetization $\langle S_n^z\rangle$ of the nth layer is given by

$$\begin{aligned} \langle S_n^z\rangle &= \frac{1}{2} - \langle S_n^- S_n^+\rangle \\ &= \frac{1}{2} - \lim_{\epsilon\to 0} \frac{1}{\Delta} \int\int d\mathbf{k_{xy}} \int_{-\infty}^{+\infty} \frac{i}{2\pi}[g_{n,n}(\omega + i\epsilon) \\ &\quad - g_{n,n}(\omega - i\epsilon)] \frac{d\omega}{\mathrm{e}^{\beta\omega} - 1} \end{aligned} \tag{10.20}$$

Replacing Eq. (10.17) in Eq. (10.20) and making use of the following identity

$$\frac{1}{x - i\eta} - \frac{1}{x + i\eta} = 2\pi i \delta(x), \tag{10.21}$$

we obtain

$$\langle S_n^z \rangle = \frac{1}{2} - \frac{1}{\Delta} \int\int dk_x dk_y \sum_{i=1}^{2N_z} \frac{D_{2n-1}(\omega_i)}{e^{\beta\omega_i} - 1}, \tag{10.22}$$

where $n = 1, \ldots, N_z$. As $\langle S_n^z \rangle$ depends on the magnetizations of the neighboring layers via $\omega_i (i = 1, \ldots, 2N_z)$, we should solve by iteration the equation (10.22) written for $n = 1, \ldots, N_z$, to obtain the magnetizations of layers $1, 2, 3, \ldots, N_z$ at a given temperature T. Note that by symmetry, $\langle S_1^z \rangle = \langle S_{N_z}^z \rangle$, $\langle S_2^z \rangle = \langle S_{N_z-1}^z \rangle$, $\langle S_3^z \rangle = \langle S_{N_z-2}^z \rangle$, and so on. Thus, only $N_z/2$ self-consistent layer magnetizations are to be calculated.

The value of the spin in the layer n at $T = 0$ is calculated by

$$\langle S_n^z \rangle (T = 0) = \frac{1}{2} + \frac{1}{\Delta} \int\int dk_x dk_y \sum_{i=1}^{N_z} D_{2n-1}(\omega_i), \tag{10.23}$$

where the sum is performed over N_z negative values of ω_i (for positive values, the Bose–Einstein factor is equal to 0 at $T = 0$).

The transition temperature T_c can be calculated in a self-consistent manner by iteration, letting all $\langle S_n^z \rangle$ tend to zero, namely $\omega_i \to 0$. Expanding $e^{\beta\omega_i} - 1 \to \beta_c \omega_i$ on the right-hand side of Eq. (10.22), where $\beta_c = (k_B T_c)^{-1}$, we have by putting $\langle S_n^z \rangle = 0$ on the left-hand side

$$\beta_c = 2\frac{1}{\Delta} \int\int dk_x dk_y \sum_{i=1}^{2N_z} \frac{D_{2n-1}(\omega_i)}{\omega_i}. \tag{10.24}$$

There are N_z such equations using Eq. (10.22) with $n = 1, \ldots, N_z$. Since the layer magnetizations tend to zero at the transition temperature from different values, it is obvious that we have to look for a convergence of the solutions of the equations Eq. (10.24) to a single value of T_c.[33]

10.2.6. *Results from the Green's function method*

Let us take $J_1 = 1$ and negative J_2 with $|J_2| > J_1$. The non-uniform GS spin configuration across the film has been determined in Section 10.2.2 for each value of J_2/J_1. Using the values of $\theta_{n,n\pm1}$ and $\theta_{n,n\pm2}$ to calculate the matrix elements of $|\mathbf{M}|$ and then solving $\det|\mathbf{M}| = 0$ we find the eigenvalues ω_i $(i = 1, \ldots, 2N_z)$ for each $\mathbf{k}_{xy}$ with a input set of $\langle S_n^z \rangle$ $(n = 1, \ldots, N_z)$ at a given T. Using Eq. (10.22) for $n = 1, \ldots, N_z$, we calculate the output $\langle S_n^z \rangle$ $(n = 1, \ldots, N_z)$. Using this output set as input, we calculate again $\langle S_n^z \rangle$ $(n = 1, \ldots, N_z)$ until the input and output are identical within a desired precision P. Numerically, we use a Brillouin zone of 100^2 wave vector values, and we use the obtained values $\langle S_n^z \rangle$ at a given T as input for a neighboring T. At low T and up to $\sim\frac{4}{5}T_c$, only a few iterations suffice to get $P \leq 1\%$. Near T_c, several dozens of iteration are needed to get convergence. The results are as follows.

10.2.7. *Spectrum*

The spin wave spectrum is calculated as described above for each value of J_2/J_1. It depends on the temperature via the temperature dependence of layer magnetizations. Figure 10.3 shows the spin wave frequency ω vs. $k_x = k_y$ for an 8-layer film with $J_2/J_1 = -1.4$ at a low temperature $T = 0.1$ (in unit of $J_1/k_B = 1$). We observe that

(i) There are eight positive and eight negative modes corresponding to two opposite spin precessions as in antiferromagnets.[1] The negative sign does not mean spin wave negative energy, but it indicates just the precession contrary to the trigonometric sense.
(ii) There are two degenerate acoustic surface branches lying at a low energy on each side. This degeneracy comes from the two symmetrical surfaces of the film. These surface modes propagate parallel to the film surface but are damped from the surface inward.

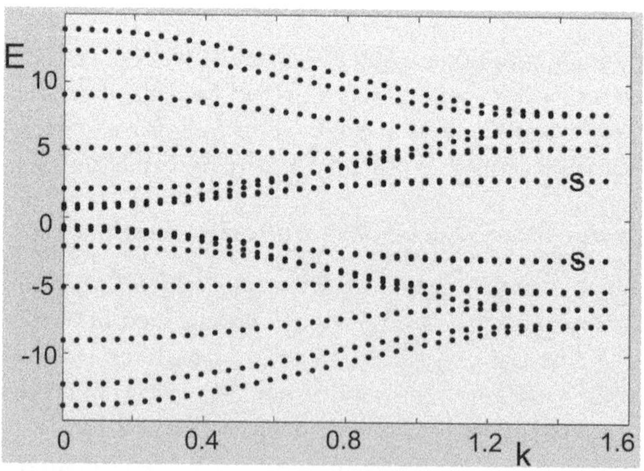

Fig. 10.3. Spectrum $E = \hbar\omega$ vs. $k \equiv k_x = k_y$ for $J_2/J_1 = -1.4$ at $T = 0.1$ for $N_z = 8$ and $d = 0.1$. The surface branches are indicated by s.

(iii) For spin magnitude $S \neq 1/2$, the spectrum is shifted toward higher frequency since it is proportional to S.
(iv) Surface spin wave spectrum (and bulk spin waves) can be experimentally observed by inelastic neutron scattering in ferromagnetic and antiferromagnetic films.[2,3]
(v) As T increases, layer magnetizations decrease, reducing therefore the spin wave energy (not shown).

10.2.8. *Spin length at $T = 0$ — Transition temperature*

It is known that in antiferromagnets, quantum fluctuations give rise to a contraction of the spin length at zero temperature.[1] In helimagnets, the NNN antiferromagnetic interaction yields similar quantum fluctuations. We will see here that a spin under a stronger antiferromagnetic interaction has a stronger zero-point spin contraction. The spins near the surface serve for such a test. In the case of the film considered above, spins in the first and in the second layers have only one antiferromagnetic NNN while interior spins have two NNN, so the contraction at a given J_2/J_1 is expected to be stronger for interior spins. This is indeed verified with the results shown in Fig. 10.4. Furthermore, when $|J_2|/J_1$ increases, namely the antiferromagnetic interaction becomes stronger, we observe stronger contractions. Note that the contraction tends to zero when the spin configuration becomes ferromagnetic, namely when J_2 tends to -1.

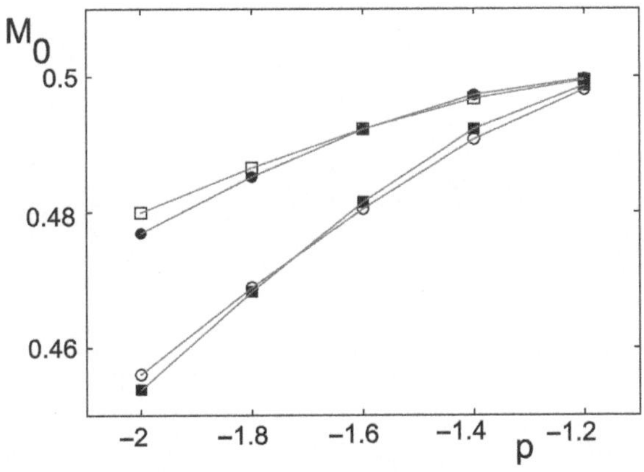

Fig. 10.4. Spin lengths of the first four layers at $T = 0$ for several values of $p = J_2/J_1$ with $d = 0.1$, $N_z = 8$. Black circles, void circles, black squares, and void squares are for first, second, third, and fourth layers, respectively. See text for comments.

10.2.9. *Layer magnetizations*

Let us show an example of the magnetization, layer by layer, from the film surface to the interior, in Fig. 10.5 for the case where $J_2/J_1 = -1.4$ with $N_z = 8$. We see that the surface magnetization, which has a large value at $T = 0$ as seen in Fig. 10.4, crosses the interior layer magnetizations at $T \simeq 0.42$ to become much smaller than interior magnetizations at higher temperatures. This crossover phenomenon is due to the competition

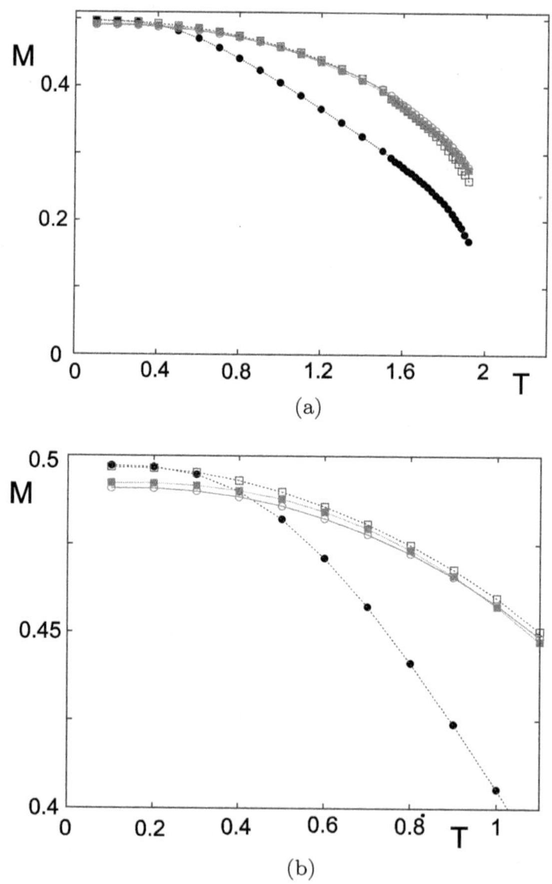

Fig. 10.5. Layer magnetizations as functions of T for $J_2/J_1 = -1.4$ with $d = 0.1$, $N_z = 8$ (a). Zoom of the region at low T to show crossover (b). Black circles, blue void squares, magenta squares, and red void circles are for first, second, third, and fourth layers, respectively. See text for comments.

between quantum fluctuations, which dominate low-T behavior, and the low-lying surface spin wave modes which strongly diminish the surface magnetization at higher T. Note that the second-layer magnetization makes also a crossover at $T \simeq 1.3$. Similar crossovers have been observed in quantum antiferromagnetic films[35] and quantum super-lattices.[36]

Note that though the layer magnetizations are different at low temperatures, they will tend to zero at a unique transition temperature as seen below. The reason is that as long as an interior layer magnetization is not zero, it will act on the surface spins as an external field, preventing them to become zero.

The temperature where layer magnetizations tend to zero is calculated by Eq. (10.24). Since all layer magnetizations tend to zero from different values, we have to solve self-consistently N_z Eq. (10.24) to obtain the transition temperature T_c. The results for several $p = J_2/J_1$ are shown in Fig. 10.6.

10.2.10. *Effect of anisotropy and surface parameters*

The results shown above have been calculated with an in-plane anisotropy interaction $d = 0.1$. Let us show now the effect of d. Stronger d will enhance all the layer magnetizations and increase T_c. Figure 10.7 shows the surface magnetization vs. T for several values of d (other layer magnetizations are not shown to preserve the figure clarity). The transition temperatures are 2.091 ± 0.010, 2.313 ± 0.010, 2.752 ± 0.010, 3.195 ± 0.010, and 3.630 ± 0.010

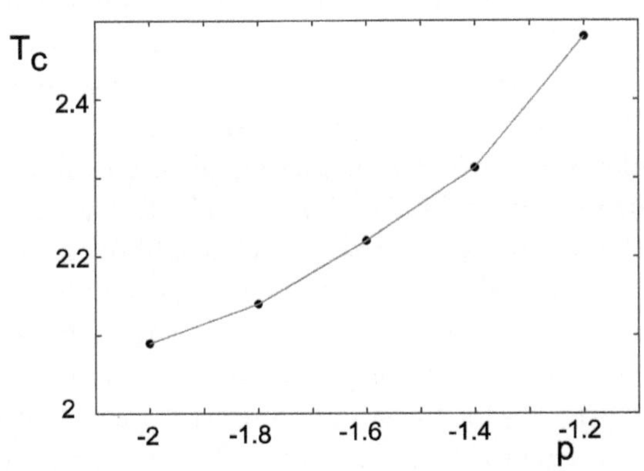

Fig. 10.6. Transition temperature vs. $p = J_2/J_1$ for an 8-layer film with $d = 0.1$. See text for comments.

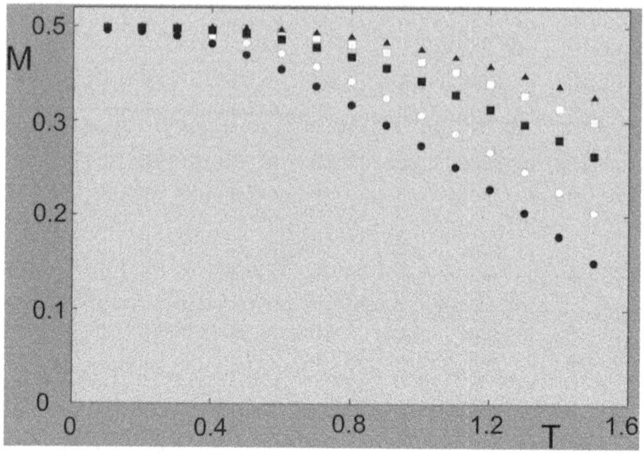

Fig. 10.7. Surface magnetization versus T for $d = 0.05$ (circles), 0.1 (void circles), 0.2 (squares), 0.3 (void squares), and 0.4 (triangles), with $J_2/J_1 = -1.4$, $N_z = 16$.

for $d = 0.05$, 0.1, 0.2, 0.3, and 0.4, respectively. These values vs. d lie on a remarkable straight line (not shown).

Let us examine the effects of the surface anisotropy and exchange parameters d_s and J_1^s. As seen above, even in the case where the surface interaction parameters are the same as those in the bulk, the surface spin wave modes exist in the spectrum. These localized modes cause a low surface magnetization observed in Fig. 10.5. Here, we show that with a weaker NN exchange interaction between surface spins and the second-layer spin, namely $J_1^s < J_1$, the surface magnetization becomes much smaller. This is shown in Fig. 10.8. We observe again here the crossover of layer magnetizations at low T due to quantum fluctuations as discussed earlier. Note that the value $J_1^s = 0.5$ is a very particular value: the GS configuration is a uniform configuration with all angles equal to 60°, namely there is no surface spin rearrangement. This can be explained if we look at the local field acting on the surface spins: the lack of neighbors is compensated by this weak positive value of J_1^s so that their local field is equal to that of a bulk spin. There is thus no surface reconstruction. Nevertheless, as T increases, thermal effects will strongly diminish the surface magnetization as seen in Fig. 10.8(b). As for the surface anisotropy parameter d_s, it affects strongly the layer magnetizations and the transition temperature: we show

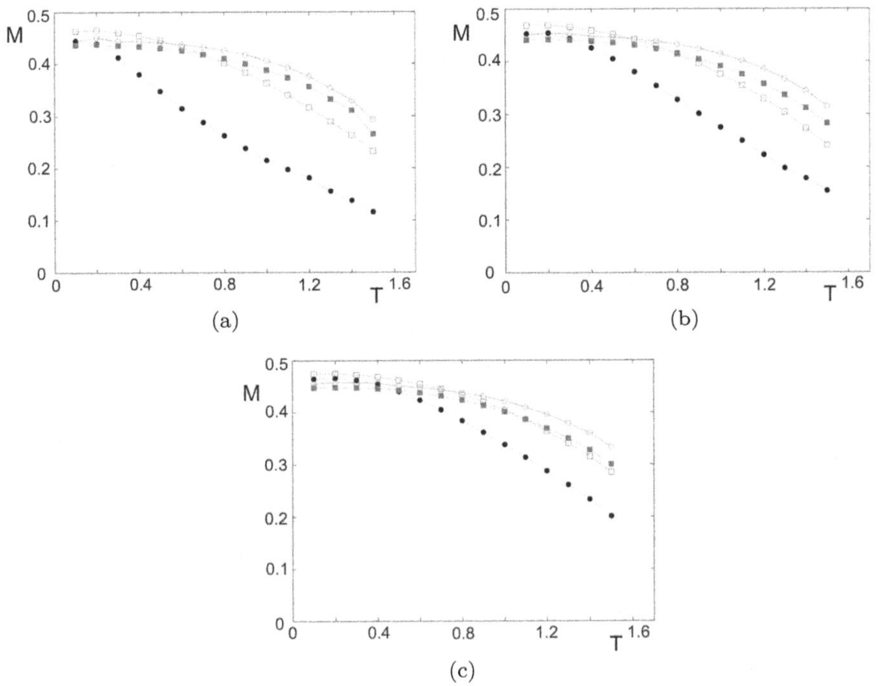

Fig. 10.8. Layer magnetizations as functions of T for the surface interaction (a) $J_1^s = 0.3$, (b) 0.5, and (c) 0.7, with $J_2/J_1 = -2$, $d = 0.1$, and $N_z = 16$. Black circles, blue void squares, magenta squares and red void circles are for first, second, third and fourth layers, respectively.

in Fig. 10.9 the surface magnetizations and the transition temperature for several values of d_s.

10.2.11. *Effect of the film thickness*

We have performed calculations for $N_z = 8, 12$, and 16. The results show that the effect of the thickness at these values is not significant: the difference lies within convergence errors. Note that the classical ground state of the first four layers is almost the same: for example, here are the values of cosinus of the angles of the film first half for $N_z = 16$ which are to be compared with the values for $N_z = 8$ given in Table 10.1, for $p = J_2/J_1 = -2$

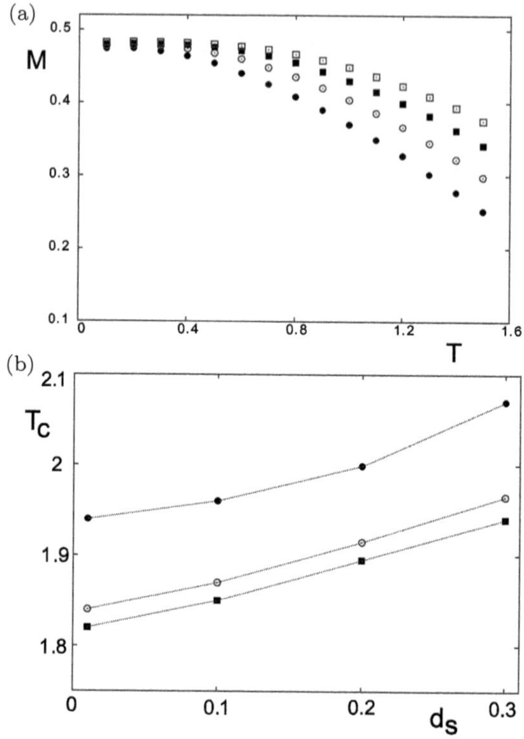

Fig. 10.9. (a) Surface magnetization versus T for $d_s = 0.01$ (circles), 0.1 (void circles), 0.2 (squares), and 0.3 (void squares), with $J_1^s = 1$, $J_2/J_1 = -2$, and $N_z = 16$. (b) Transition temperature vs. d_s for $J_1^s = 0.7$, 0.5, and 0.3 (curves from up to down), with $J_2/J_1 = -2$, $d = 0.1$, and $N_z = 16$.

(in the parentheses angles are provided in degrees):

0.86737967 (29.844446), 0.41125694 (65.716179), 0.52374715 (58.416061), 0.49363765 (60.420044), 0.50170541 (59.887100), 0.49954081 (60.030373), 0.50013113 (59.991325), 0.49993441 (60.004330).

From the fourth layer, the angle is almost equal to the bulk value (60°).

At $p = J_2/J_1 = -2$, the transition temperature is 2.090 ± 0.010 for $N_z = 8$, 2.093 ± 0.010 for $N_z = 12$, and 2.103 ± 0.010 for $N_z = 16$. These are the same within errors. The difference can be seen at smaller thicknesses. However, for helimagnets in the z direction, thicknesses smaller than 8 do not allow to see fully the surface helical reconstruction which affects the first four layers: to study surface helical effects in such a situation would not make sense.

At this stage, it is interesting to note that our result is in agreement with experiments: it has been experimentally observed that the transition temperature does not vary significantly in MnSi films in the thickness range 11–40 nm.[18] One possible explanation is that the helical structure is very stable as seen above: the surface perturbs the bulk helical configuration only at the first four layers, so the bulk dominates the transition. This has been experimentally seen in holmium films.[31]

10.2.12. *Classical helimagnetic films: Monte Carlo simulation*

To appreciate quantum effects causing cross-overs of layer magnetizations presented above at low temperatures, we show here some results of the classical counterpart model: spins are classical XY spins of amplitude $S = 1$. We take the XY spins rather than the Heisenberg spins for comparison with the quantum case because in the latter case we have used an in-plane Ising-like anisotropy interaction d. MC simulations have been carried out over film samples of $100 \times 100 \times 16$. Periodic boundary conditions are applied in the xy plane. One million MC steps are discarded to equilibrate the system and another million MC steps are used for averaging. The layer magnetizations versus T are shown in Fig. 10.10 for the case where surface interaction $J_1^s = 1$ (Fig. 10.10(a)) and 0.3 (Fig. 10.10(b)) with $J_2/J_1 = -2$ and $N_z = 16$. One sees that (i) by extrapolation there is no spin contraction at $T = 0$ and there is no crossover of layer magnetizations at low temperatures, (ii) from the intermediate temperature region up to the transition, the relative values of layer magnetizations are not always the same as in the quantum case: for example at $T = 1.2$, one has $M_1 < M_3 < M_4 < M_2$ in Fig. 10.10(a) and $M_1 < M_2 < M_4 < M_3$ in Fig. 10.10(b) which are not the same as in the quantum case shown in Fig. 10.8(a). Our conclusion is that even at temperatures close to the transition, helimagnets may have slightly different behaviors according to their quantum or classical nature. Extensive MC simulations with size effects and detection of the order of the phase transition are beyond the scope of this book.

10.2.13. *Summary*

We have studied in this section surface effects in a helimagnet of BCC lattice with quantum Heisenberg spins. Note that the method presented

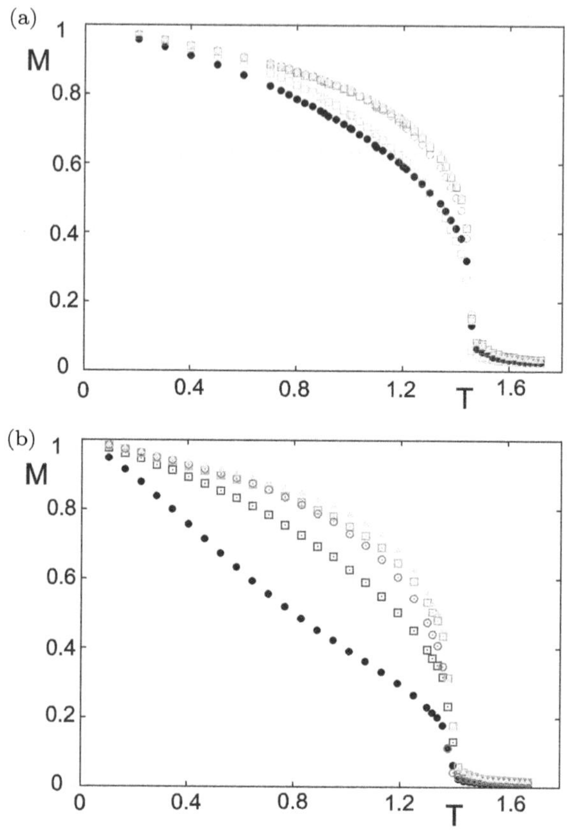

Fig. 10.10. Monte Carlo results: Layer magnetizations as functions of T for the surface interaction $J_1^s = 1$ (a) and 0.3 (b) with $J_2/J_1 = -2$ and $N_z = 16$. Black circles, blue void squares, cyan squares, and red void circles are for first, second, third, and fourth layers, respectively.

above can be applied to any lattice structure, and the results found above are valid for general helimagnetic structures, not limited to BCC crystals. Note also that the results have been shown for the case of spin $S = 1/2$ where quantum fluctuations are strong at low temperatures. Rare earth elements Ho and Dy with helical structures along the c-axis and ferromagnetic in the basal planes which are very similar to the present model are expected to bear the same features as what has been shown above. However, at low temperatures, due to their larger spin amplitudes, $S = 7/2$ and $5/2$ for Ho and Dy, quantum fluctuations are certainly weaker and the crossover may occur with a smaller difference. Numerical applications of our formalism should be performed to get precise values for these cases. The following features are among the striking features found in the present section: (i) the crossover of layer magnetizations at low temperatures due to the competition between quantum fluctuations and thermal effects, (ii) the existence of low-lying surface spin wave modes which cause a low surface magnetization, (iii) a strong effect of the surface exchange interaction (J_1^s) which drastically modifies the surface spin configuration and gives rise to a very low surface magnetization, (iv) the transition temperature varies strongly with the helical angle but it is insensitive to the film thickness in agreement with experiments performed on MnSi films[18] and holmium,[31] and (v) the classical spin model counterpart gives features slightly different from those of the quantum model, both at low and high temperatures.

Let us make some comments on works of other similar models. Mello *et al.*[15] have treated almost the same model as ours using a hexagonal anisotropy which corresponds to the case of Dy in which the helical angle is $\simeq 60°$. However, the authors of this work studied only the classical spin configuration at $T = 0$. Unlike our work, no treatment at finite T has been considered by these authors. Rodrigues *et al.*[17] have used exactly the same model as Mello *et al.* but with application to the Ho case. They have used the mean-field estimation to establish the phase diagram in the space (T, H) (H: magnetic field) and shown the surface effects. There has been however no detail given on surface spin configuration, surface magnetization, and quantum effects, unlike what we have done here. The MC work of Cinti *et al.*[16] was based on a classical spin model with a Hamiltonian, very different from ours, including a six-constant interaction (a kind of dipolar interaction) in the c-direction. The comparison with our work is impossible.

We emphasize that the general Green's function method[33] allows us to study at a microscopic level surface spin waves and their physical consequences at finite temperatures in systems with non-collinear spin configurations such as helimagnetic films. It can be used in more complicated situations such as helimagnets with Dzyaloshinskii–Moriya (DM) interactions as in Ref. [19] and in Section 10.3.

We have also studied the above model of helimagnets with the simple cubic lattice, without[37] and with a magnetic field applied perpendicularly to the film surface.[38] For the latter case, we found that under the applied magnetic field, the spins in the GS make different angles across the film. When the temperature increases, the layers with large xy spin components undergo a phase transition where the transverse (in-plane) xy ordering is destroyed. This transverse transition is possible because the xy spin components do not depend on the field. Other layers with small xy spin components, namely large z components, do not make a transition because the ordering in S_z is maintained by the applied field. The transition of a number of layers with large xy spin components, not all layers, is a new phenomenon discovered in this system.

To conclude this section, we mention some other studies of frustrated thin films which are not helimagnetic but there is a surface spin rearrangement or a surface spin configuration different from the bulk one: the case of a frustrated triangular antiferromagnetic layer on a ferromagnetic film[25] and the case of an FCC antiferromagnetic film where the frustration is fully realized.[26] In these cases, the surface effects are very strong and there is a surface phase transition separate from the bulk one. We have also calculated the critical exponents in these cases and found that critical exponents in thin films have values between 2D and 3D corresponding to spin models.[39]

10.3. Dzyaloshinskii–Moriya (DM) Interaction in Thin Films: Spin Waves

We study in this section magnetic properties of a system of quantum Heisenberg spins interacting with each other via a ferromagnetic exchange interaction J and an in-plane DM interaction D. The non-collinear ground state due to the competition between J and D is determined. We employ the self-consistent Green's function theory described in the previous section to calculate the spin wave spectrum and the layer magnetizations at finite T in two dimensions as well as in a thin film with surface effects. Analytical details and the validity of the method are shown and discussed. Numerical solutions are shown for realistic physical interaction parameters. Discussion on possible experimental verifications is given.

10.3.1. *Introduction*

The DM interaction was proposed to explain the weak ferromagnetism which was observed in antiferromagnetic Mn compounds. The phenomenological Landau–Ginzburg model introduced by Dzyaloshinskii[40] was microscopically derived by Moriya.[41] The interaction between two spins $\mathbf{S}_i$ and $\mathbf{S}_j$ is written as

$$\mathbf{D}_{i,j} \cdot \mathbf{S}_i \wedge \mathbf{S}_j, \tag{10.25}$$

where $\mathbf{D}_{i,j}$ is a vector which results from the displacement of non-magnetic ions located between $\mathbf{S}_i$ and $\mathbf{S}_j$, for example in Mn–O–Mn bonds. The direction of $\mathbf{D}_{i,j}$ depends on the symmetry of the displacement.[41]

There has been a large number of investigations on the effect of the DM interaction in various materials, both experimentally and theoretically for weak ferromagnetism in perovskite compounds (see references cited in Refs. [42] and [43], for example). However, the interest in the DM interaction goes beyond the weak ferromagnetism: for example, it has been recently shown in various works that the DM interaction is at the origin of topological skyrmions[20–22,30,44–51] and new kinds of magnetic domain walls.[52,53] The increasing interest in skyrmions results from the fact that skyrmions may play an important role in the electronic transport which is at the heart of technological application devices.[4]

In the following, we are interested in the SW properties of a system of spins interacting with each other via a DM interaction in addition to the symmetric isotropic Heisenberg exchange interaction. The competition between these interactions gives rise to a non-collinear spin configuration in the GS. The DM interaction yields a non-trivial SW behavior as will be seen below. Note that there has been just a small number of works dealing with the SW properties in DM systems.[54–58]

10.3.2. *Model and ground state*

We consider a thin film of simple cubic (SC) lattice of N layers stacked in the y direction perpendicular to the film surface. For the reason which is shown below, we choose the film surface as a xz plane.

The Hamiltonian is given by

$$\mathcal{H} = \mathcal{H}_e + \mathcal{H}_{\mathrm{DM}}, \tag{10.26}$$

$$\mathcal{H}_e = -\sum_{\langle i,j\rangle} J_{i,j}\mathbf{S}_i \cdot \mathbf{S}_j, \tag{10.27}$$

$$\mathcal{H}_{\mathrm{DM}} = \sum_{\langle i,j\rangle} \mathbf{D}_{i,j} \cdot \mathbf{S}_i \wedge \mathbf{S}_j, \tag{10.28}$$

where $J_{i,j}$ and $\mathbf{D}_{i,j}$ are the exchange and DM interactions, respectively, between two Heisenberg spins $\mathbf{S}_i$ and $\mathbf{S}_j$ of magnitude $S = 1/2$ occupying the lattice sites i and j.

The SC lattice can support the DM interaction in the absence of the inversion symmetry[30,41,47] as in MnSi. The absence of inversion symmetry can also be achieved by the positions of non-magnetic ions between magnetic ions. The vector $\mathbf{D}$ between two magnetic ions is defined as $\mathbf{D}_{i,j} = A\mathbf{r}_i \wedge \mathbf{r}_j$, where $\mathbf{r}_i$ is the vector connecting the non-magnetic ion to the spin $\mathbf{S}_i$ and $\mathbf{r}_j$ is that to the spin $\mathbf{S}_j$, A being a constant. One sees that $\mathbf{D}_{i,j}$ is perpendicular to the plane formed by $\mathbf{r}_i$ and $\mathbf{r}_j$.

For simplicity, let us consider the case where the in-plane and inter-plane exchange interactions between NN are both ferromagnetic and denoted by $J_{\parallel}$ and $J_{\perp}$, respectively. The vector $\mathbf{D}_{i,j}$ is chosen by supposing the situation where non-magnetic ions are positioned in the plane xz. With this choice, $\mathbf{D}_{i,j}$ is perpendicular to the xz plane and our model gives rise to a planar spin configuration, namely the spins lie in xz planes. This simplifies our calculation. The DM interaction is supposed to be between NN in the plane with a constant D.

Due to the competition between the exchange J term which favors the collinear configuration and the DM term which favors the perpendicular one, we expect that the spin $\mathbf{S}_i$ makes an angle $\theta_{i,j}$ with its neighbor $\mathbf{S}_j$. Therefore, the quantization axis of $\mathbf{S}_i$ is not the same as that of $\mathbf{S}_j$. Let us call $\hat{\zeta}_i$ the quantization axis of $\mathbf{S}_i$ and $\hat{\xi}_i$ its perpendicular axis in the xz plane. The third axis $\hat{\eta}_i$, perpendicular to the film surface, is chosen in such a way to make $(\hat{\xi}_i, \hat{\eta}_i, \hat{\zeta}_i)$ an orthogonal direct frame. Writing $\mathbf{S}_i$ and $\mathbf{S}_j$ in their respective local coordinates, one has

$$\mathbf{S}_i = S_i^x \hat{\xi}_i + S_i^y \hat{\eta}_i + S_i^z \hat{\zeta}_i, \tag{10.29}$$

$$\mathbf{S}_j = S_j^x \hat{\xi}_j + S_j^y \hat{\eta}_j + S_j^z \hat{\zeta}_j. \tag{10.30}$$

We choose the vector $\mathbf{D}_{i,j}$ perpendicular to the xz plane, namely

$$\mathbf{D}_{i,j} = D e_{i,j} \hat{\eta}_i, \tag{10.31}$$

where $e_{i,j} = +1(-1)$ if $j > i$ $(j < i)$ for NN on the xz plane. Because $\mathbf{D}_{i,j} = A\mathbf{r}_i \wedge \mathbf{r}_j$, the permutation $j \rightleftarrows i$ should yield $\mathbf{D}_{i,j} = -\mathbf{D}_{j,i}$. Equation (10.31) satisfies this condition. Note that $e_{j,i} = -e_{i,j}$.

To determine the GS, the easiest way is to use the steepest descent method: we calculate the local field acting on each spin from its neighbors and we align the spin in its local-field direction to minimize its energy. Repeating this for all spins and iterating many times until the convergence is reached with a desired precision (usually at the 6th digit, namely at $\simeq 10^{-6}$ per cents), we obtain the lowest energy state of the system (see Ref. [25]). Note that we have used several thousands of different initial conditions to check the convergence to a single GS for each set of parameters. Choosing

$\mathbf{D}_{i,j}$ lying perpendicular to the spin plane (i.e. xz plane) as indicated in Eq. (10.31), we determine the GS as a function of D. An example is shown in Fig. 10.11 for $\theta = \pi/6$ ($D = -0.577$) with $J_{\parallel} = J_{\perp} = 1$. We see that each spin has the same angle with its four NN in the plane (angle between NN in adjacent planes is zero). Let us show the relation between θ and $J_{\parallel}$: the energy of the spin $\mathbf{S}_i$ is written as

$$E_i = -4J_{\parallel}S^2 \cos\theta - 2J_{\perp}S^2 + 4DS^2 \sin\theta, \tag{10.32}$$

where $\theta = |\theta_{i,j}|$ and care has been taken on the signs of $\sin\theta_{i,j}$ and $e_{i,j}$ when counting NN, namely two opposite NN have opposite signs. The minimization of E_i yields

$$\frac{dE_i}{d\theta} = 0 \Rightarrow -\frac{D}{J_{\parallel}} = \tan\theta \Rightarrow \theta = \arctan\left(-\frac{D}{J_{\parallel}}\right). \tag{10.33}$$

The value of θ for a given $\frac{D}{J_{\parallel}}$ is precisely what is obtained by the steepest descent method.

In the present model, the DM interaction is supposed between the NN in the plane xz, so in the GS the angle between in-plane NN is not zero. We show in Fig. 10.11 the relative orientation of the two NN spins in the plane xz.

The spin configuration is planar, meaning that the turn angle is in the plane: each spin turns an angle θ with respect to its NN, and it is in both direction in the plane (see Fig. 10.11(b)). There is no helicoidal configuration. Note that the spin structure is different from that of MnSi where the turn angle is in a plane perpendicular to the screw axis. This case has been studied in the previous section (see details in Refs. [33] and [37]), while in the present DM interaction case, it lies along each of the two axes x and z of the xz plane (see Fig. 10.11(b)).

In the film plane xz, the spin $\mathbf{S}_i$ makes an angle $\theta_{i,j}$ with the spin $\mathbf{S}_j$. Define the local coordinates $(\hat{\zeta}_i, \hat{\xi}_i, \hat{\eta}_i)$ and $(\hat{\zeta}_j, \hat{\xi}_j, \hat{\eta}_j)$ for $\mathbf{S}_i$ and $\mathbf{S}_j$ where the quantization axes are, respectively, $\hat{\zeta}_i$ and $\hat{\zeta}_j$, as indicated in Fig. 10.2. Note that the perpendicular axes $\hat{\eta}_i$ and $\hat{\eta}_j$ coincide.

Now, expressing the local frame of $\mathbf{S}_j$ in the local frame of $\mathbf{S}_i$, we have

$$\hat{\zeta}_j = \cos\theta_{i,j}\hat{\zeta}_i + \sin\theta_{i,j}\hat{\xi}_i, \tag{10.34}$$

$$\hat{\xi}_j = -\sin\theta_{i,j}\hat{\zeta}_i + \cos\theta_{i,j}\hat{\xi}_i, \tag{10.35}$$

$$\hat{\eta}_j = \hat{\eta}_i, \tag{10.36}$$

so that

$$\mathbf{S}_j = S_j^x(\cos\theta_{i,j}\hat{\zeta}_i - \sin\theta_{i,j}\hat{\xi}_i),$$
$$+S_j^y\hat{\eta}_i + S_j^z(\cos\theta_{i,j}\hat{\zeta}_i + \sin\theta_{i,j}\hat{\xi}_i). \tag{10.37}$$

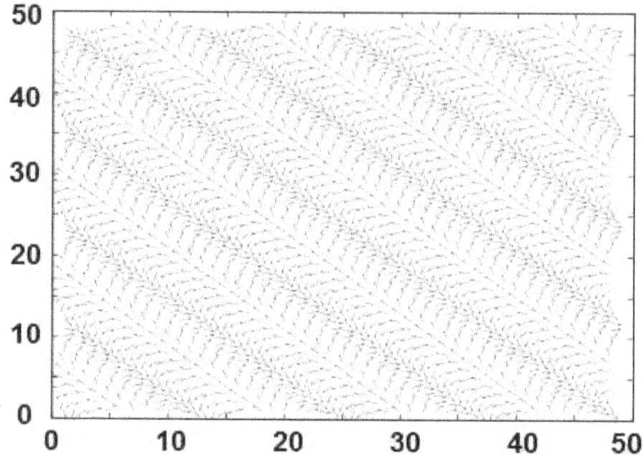

(a)

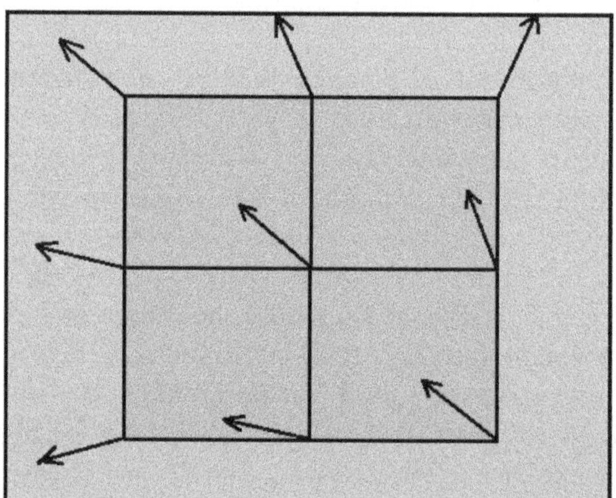

(b)

Fig. 10.11. (a) The GS is a planar configuration on the xz plane. The figure shows the case where $\theta = \pi/6$ ($D = -0.577$) along the x and z axes, with $J_{\parallel} = J_{\perp} = 1$, obtained by using the steepest descent method; (b) a zoom is shown around a spin with its nearest neighbors in the xz plane.

We have

$$\begin{aligned}\mathbf{S}_i \wedge \mathbf{S}_j &= (-S_i^z S_j^y - S_i^y S_j^x \sin\theta_{i,j} + S_i^y S_j^z \cos\theta_{i,j})\hat{\xi}_i \\ &\quad +(S_i^x S_j^x \sin\theta_{i,j} + S_i^z S_j^z \sin\theta_{i,j})\hat{\eta}_i \\ &\quad +(S_i^x S_j^y - S_i^y S_j^z \sin\theta_{i,j} - S_i^y S_j^x \cos\theta_{i,j})\hat{\zeta}_i. \end{aligned} \tag{10.38}$$

From Eq. (10.31), we obtain

$$\begin{aligned}\mathcal{H}_{\mathrm{DM}} &= \sum_{\langle i,j\rangle} \mathbf{D}_{i,j} \cdot \mathbf{S}_i \wedge \mathbf{S}_j \\ &= D\sum_{\langle i,j\rangle}(S_i^x S_j^x e_{i,j} \sin\theta_{i,j} + S_i^z S_j^z e_{i,j} \sin\theta_{i,j}) \\ &= \frac{D}{4}\sum_{\langle i,j\rangle}[(S_i^+ + S_i^-)(S_j^+ + S_j^-)e_{i,j}\sin\theta_{i,j} \\ &\quad +4S_i^z S_j^z e_{i,j}\sin\theta_{i,j}], \end{aligned} \tag{10.39}$$

where we have replaced $S^x = (S^+ + S^-)/2$. Note that $e_{i,j}\sin\theta_{i,j}$ is always positive since for an NN on the positive axis direction, $e_{i,j} = 1$ and $\sin\theta_{i,j} = \sin\theta$, where θ is positively defined, while for a NN on the negative axis direction, $e_{i,j} = -1$ and $\sin\theta_{i,j} = \sin(-\theta) = -\sin\theta$.

We emphasize that for non-collinear spin configurations, the local spin coordinates allow one to use the commutation relations between spin operators of a spin which are valid only when the z spin component is defined on its quantification axis. This method has been applied for helimagnets shown in the previous section and in Refs. [7, 9] and [10].

10.3.3. *Results from Green's function method*

Expressing the Hamiltonian in the local coordinates, we obtain

$$\begin{aligned}\mathcal{H} = -\sum_{\langle i,j\rangle} J_{i,j}\bigg\{&\frac{1}{4}\left(\cos\theta_{i,j} - 1\right)\left(S_i^+ S_j^+ + S_i^- S_j^-\right) \\ &+\frac{1}{4}\left(\cos\theta_{i,j} + 1\right)\left(S_i^+ S_j^- + S_i^- S_j^+\right)\end{aligned}$$

$$
\begin{aligned}
&+\frac{1}{2}\sin\theta_{i,j}\left(S_i^+ + S_i^-\right)S_j^z - \frac{1}{2}\sin\theta_{i,j}S_i^z\left(S_j^+ + S_j^-\right) \\
&+\cos\theta_{i,j}S_i^z S_j^z\Big\} \\
&+\frac{D}{4}\sum_{\langle i,j\rangle}[(S_i^+ + S_i^-)(S_j^+ + S_j^-)e_{i,j}\sin\theta_{i,j} \\
&+4S_i^z S_j^z e_{i,j}\sin\theta_{i,j}]. \qquad (10.40)
\end{aligned}
$$

Note that unlike the sinus term of the DM Hamiltonian, Eq. (10.39), the sinus terms of $\mathcal{H}_e$, namely the third line of Eq. (10.40), are zero when summed up on opposite NN (no $e_{i,j}$ to compensate). The third line disappears therefore in the following.

It is known that in two dimensions (2D) there is no long-range order at finite temperature for isotropic spin models with short-range interaction.[32] Thin films having small thickness need an anisotropic interaction to stabilize the ordering at finite T. We use the following anisotropy between $\mathbf{S}_i$ and $\mathbf{S}_j$ which stabilizes the angle determined above between their local quantization axes S_i^z and S_j^z:

$$\mathcal{H}_a = -\sum_{\langle i,j\rangle} I_{i,j} S_i^z S_j^z \cos\theta_{i,j}, \qquad (10.41)$$

where $I_{i,j}$ is supposed to be positive, small compared to $J_{\parallel}$, and limited to NN. Hereafter, for simplicity we take $I_{i,j} = I_1$ for NN pair in the xz plane. As it turns out, this anisotropy helps stabilize the ordering at finite T in 2D as discussed. It helps also stabilize the SW spectrum at $T = 0$ in the case of thin films but it is not necessary for 2D and 3D at $T = 0$. The total Hamiltonian is finally given by

$$\mathcal{H} = \mathcal{H}_e + \mathcal{H}_{DM} + \mathcal{H}_a \qquad (10.42)$$

Using the Green's function method described in the previous section with the details given in Refs. [25, 26, 33, 37] and [59], we finally obtain the following matrix equation:

$$\mathbf{M}(E)\,\mathbf{h} = \mathbf{u}, \qquad (10.43)$$

where $\mathbf{M}(E)$ is a square matrix of dimension $(2N \times 2N)$ and $\mathbf{h}$ and $\mathbf{u}$ are the column matrices which are defined as follows:

$$\mathbf{h} = \begin{pmatrix} g_{1,n'} \\ f_{1,n'} \\ \vdots \\ g_{n,n'} \\ f_{n,n'} \\ \vdots \\ g_{N,n'} \\ f_{N,n'} \end{pmatrix}, \quad \mathbf{u} = \begin{pmatrix} 2\langle S_1^z \rangle \delta_{1,n'} \\ 0 \\ \vdots \\ 2\langle S_N^z \rangle \delta_{N,n'} \\ 0 \end{pmatrix}, \tag{10.44}$$

where $E = \hbar\omega$ and $\mathbf{M}\,(E)$ is given by

$$\begin{pmatrix} E+A_1 & B_1 & C_1 & 0 & 0 & 0 & 0 & 0 & 0 \\ -B_1 & E-A_1 & 0 & -C_1 & 0 & 0 & 0 & 0 & 0 \\ \cdots & \cdots & \cdots & \cdots & \cdots & \cdots & \cdots & \cdots & \cdots \\ \cdots & 0 & C_n & 0 & E+A_n & B_n & C_n & 0 & 0 \\ \cdots & 0 & 0 & -C_n & -B_n & E-A_n & 0 & -C_n & 0 \\ \cdots & \cdots & \cdots & \cdots & \cdots & \cdots & \cdots & \cdots & \cdots \\ 0 & 0 & 0 & 0 & 0 & C_N & 0 & E+A_N & B_N \\ 0 & 0 & 0 & 0 & 0 & 0 & -C_N & -B_N & E-A_N \end{pmatrix}, \tag{10.45}$$

with

$$\begin{aligned} A_n = & -J_{\parallel}[8\langle S_n^z \rangle \cos\theta(1+d_n) \\ & -4\langle S_n^z \rangle \gamma(\cos\theta + 1)] \\ & -2J_{\perp}(\langle S_{n-1}^z \rangle + \langle S_{n+1}^z \rangle) \\ & -4D\sin\theta\langle S_n^z \rangle \gamma \\ & +8D\sin\theta\langle S_n^z \rangle, \end{aligned} \tag{10.46}$$

$$\begin{aligned} B_n = & \, 4J_{\parallel}\langle S_n^z \rangle \gamma(\cos\theta - 1) \\ & -4D\sin\theta\langle S_n^z \rangle \gamma, \end{aligned} \tag{10.47}$$

$$C_n = 2J_{\perp}\langle S_n^z \rangle, \tag{10.48}$$

where $n = 1, 2, \ldots, N$, $d_n = I_1/J_{\parallel}$, $\gamma = (\cos k_x a + \cos k_z a)/2$, k_x and k_z denote the wave vector components in the xz planes, and a is the lattice constant. Note that (i) if $n = 1$ (surface layer) then there are no $n - 1$ terms in the matrix coefficients, (ii) if $n = N$ then there are no $n + 1$ terms. Besides, we have distinguished the in-plane NN interaction $J_{\parallel}$ from the inter-plane NN one $J_{\perp}$.

In the case of a thin film, the SW eigenvalues at a given wave vector $\mathbf{k} = (k_x, k_z)$ are calculated by diagonalizing the matrix (10.45).

10.3.4. *Two dimensions: Spin wave spectrum and magnetization*

Consider just one single xz plane. The above matrix is reduced to two coupled equations

$$
\begin{aligned}
(E + A_n)g_{n,n'} + B_n f_{n,n'} &= 2\langle S_n^z \rangle \delta(n, n') \\
-B_n g_{n,n'} + (E - A_n) f_{n,n'} &= 0,
\end{aligned} \tag{10.49}
$$

where A_n is given by (10.46) but without the $J_{\perp}$ term for the 2D case considered here. Coefficients B_n and C_n are given by (10.47) and (10.48) with $C_n = 0$. The poles of the Green's functions are the eigenvalues of the SW spectrum which are given by the secular equation

$$
\begin{aligned}
&(E + A_n)(E - A_n) + B_n^2 = 0 \\
&[E + A_n][E - A_n] + B_n^2 = 0 \\
&E^2 - A_n^2 + B_n^2 = 0 \\
&E = \pm\sqrt{(A_n + B_n)(A_n - B_n)},
\end{aligned} \tag{10.50}
$$

where $\pm$ indicates the left and right SW precessions. Several remarks are in order:

(i) if $\theta = 0$, we have $B_n = 0$ and the last three terms of A_n are zero. We recover then the ferromagnetic SW dispersion relation

$$
E = 2ZJ_{\parallel}\langle S_n^z \rangle (1 - \gamma), \tag{10.51}
$$

where $Z = 4$ is the coordination number of the square lattice (taking $d_n = 0$),

(ii) if $\theta = \pi$, we have $A_n = 8J_{\parallel}\langle S_n^z\rangle$, $B_n = -8J_{\parallel}\langle S_n^z\rangle\gamma$. We recover then the antiferromagnetic SW dispersion relation

$$E = 2ZJ_{\parallel}\langle S_n^z\rangle\sqrt{1-\gamma^2} \tag{10.52}$$

(iii) in the presence of a DM interaction, we have $0 < \cos\theta < 1$ $(0 < \theta < \pi/2)$. If $d_n = 0$, the quantity in the square root of Eq. (10.50) is always ≥ 0 for any θ. It is zero at $\gamma = 1$. The SW spectrum is therefore stable at the long-wavelength limit. The anisotropy d_n gives a gap at $\gamma = 1$.

As said earlier, the necessity to include an anisotropy has a double purpose: it permits a gap and stabilizes a long-range ordering at finite T in 2D systems.

Figure 10.12 shows the SW spectrum calculated from Eq. (10.50) for $\theta = 30°$ ($\pi/6$ radian) and $80°$ (1.396 radian). The spectrum is symmetric for positive and negative wave vectors and for left and right precessions. Note that for small θ (i.e. small D) E is proportional to k^2 at low k (cf. Fig. 10.12(a)), as in ferromagnets. However, as θ increases, we observe that E becomes linear in k as seen in Fig. 10.12(b). This is similar to antiferromagnets. The change of behavior is progressive with increasing θ, and we do not observe a sudden transition from k^2 to k behavior. This feature is also observed in thin films as seen below.

It is noted that with the anisotropy d, we avoid the logarithmic divergence at $k = 0$ so that we can observe a long-range ordering at finite T in 2D. We show in Fig. 10.13 the magnetization M ($\equiv \langle S^z\rangle$) calculated by Eq. (10.22) for one layer using $d = 0.001$. It is interesting to observe that M depends strongly on θ: at high T, larger θ yields stronger M. However, at $T = 0$ the spin length is smaller for larger θ due to the so-called zero-point spin contraction[1] calculated by Eq. (10.23). As a consequence, there is a crossover of magnetizations with different θ at low T as shown in Fig. 10.13.

10.3.5. *The case of thin films: Spin wave spectrum, layer magnetizations*

In the 2D case shown above, there is no need to use a small anisotropy d at $T = 0$. However, in the case of thin films, due to the lack of neighbors at the surface, the introduction of a DM interaction destabilizes the spectrum at long wavelength $\mathbf{k} = 0$. Depending on θ, we have to use a value for d_n larger or equal to a "critical value" d_c to avoid imaginary SW energies at

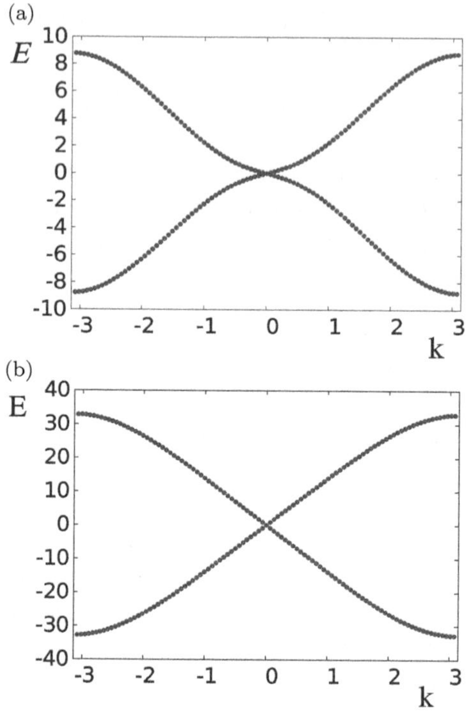

Fig. 10.12. SW spectrum $E(k)$ vs. $k \equiv k_x = k_z$ for (a) $\theta = 0.524$ radian and (b) $\theta = 1.393$ in two dimensions at $T = 0.1$. Positive and negative branches correspond to right and left precessions. A small d ($= 0.001$) has been used to stabilize the ordering at finite T in 2D. See text for comments.

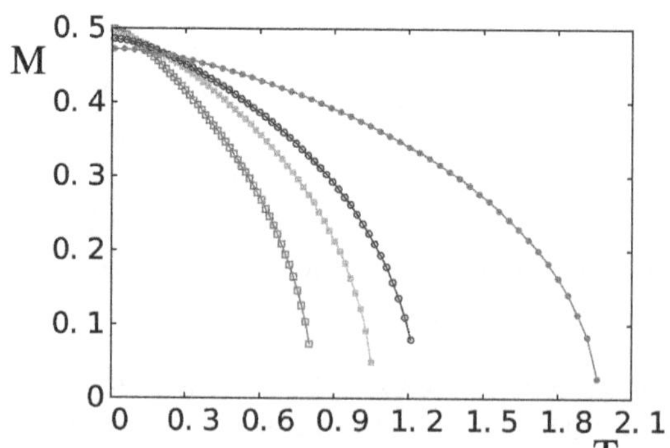

Fig. 10.13. Magnetization M vs. temperature T for a monolayer (2D) $\theta = 0.175$ (radian), $\theta = 0.524$, $\theta = 0.698$, $\theta = 1.047$ (void magenta squares, green filled squares, blue void circles and filled red circles, respectively), with $d = 0.001$. See text for comments.

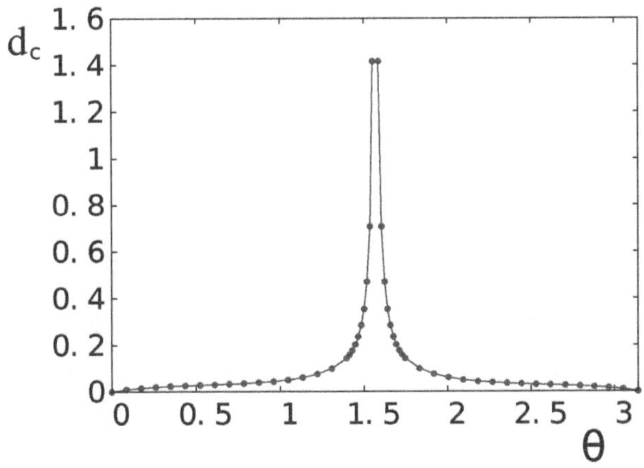

Fig. 10.14. Value d_c above which the SW energy $E(\mathbf{k} = 0)$ is real as a function of θ (in radian), for a four-layer film. No spin wave excitations are possible close to $\theta = \pi/2$.

$\mathbf{k} = 0$. The critical value d_c is shown in Fig. 10.14 for a four-layer film. Note that at the perpendicular configuration $\theta = \pi/2$, no SW excitation is possible: SW cannot propagate in a perpendicular spin configuration since the wave vectors cannot be defined.

We show now an SW spectrum at a given thickness N. There are $2N$ energy values and half of them are positive and the other half negative (left and right precessions): E_i ($i = 1, \ldots, 2N$). Figure 10.15 shows the case of a film of eight layers with $J_{\parallel} = J_{\perp} = 1$ for a weak and a strong value of D (small and large θ). As in the 2D case, for strong D, E is proportional to k at small k for the first mode (cf. Fig. 10.15(b)). The upper modes remain in the k^2 behavior.

Figure 10.16 shows the layer magnetizations of the first four layers in a eight-layer film (the other half is symmetric) for several values of θ. In each case, we see that the surface layer magnetization is smallest. This is a general effect of the lack of neighbors for surface spins even when there is no surface-localized SW.[1]

The spin length at $T = 0$ for an eight-layer film is shown in Fig. 10.17 as a function of θ. One observes that the spins are strongly contracted at large θ.

Let us discuss about the surface effect in the present model. We know that for the simple cubic lattice, if the interactions are the same everywhere in the film, then there is no surface-localized modes, and this is true with

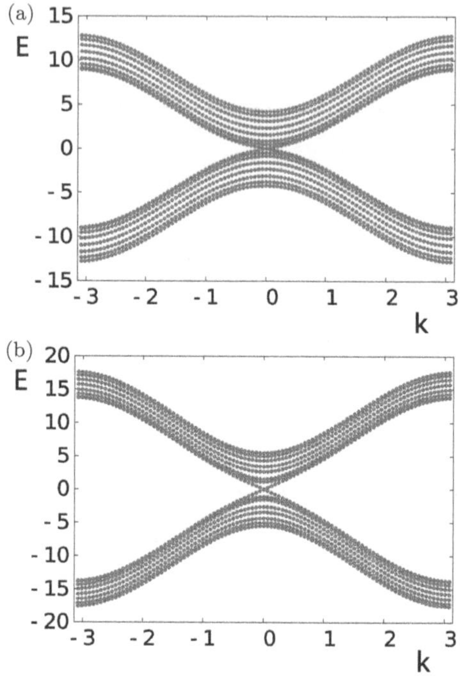

Fig. 10.15. SW spectrum $E(k)$ vs. $k \equiv k_x = k_z$ for a thin film of eight layers: (a) $\theta = \pi/6$ (in radian) and (b) $\theta = \pi/3$, using $d = d_c$ for each case ($d_c = 0.012$ and 0.021, respectively). Positive and negative branches correspond to right and left precessions. Note the linear-k behavior at low k for the first mode at the large θ case.

the DM interaction (see spectrum in Fig. 10.15) and without DM interaction (see Ref. [24]). In order to create surface modes, we have to take the surface exchange interactions different from the bulk ones. Low-lying branches of surface modes are "detached" from the bulk spectrum as seen in the SW spectrum shown in Fig. 10.18(a) with $J^s_{\parallel} = 0.5$, $J^s_{\perp} = 0.5$. These surface modes strongly affect the surface magnetization as observed in Fig. 10.18(b): the surface magnetization is strongly diminished with increasing T. The role of surface-localized modes on the strong decrease of the surface magnetization as T increases has been analyzed in Ref. [24].

We show now the effect of the film thickness in the present model. The case of thickness $N = 12$ is shown in Fig. 10.19(a) with $\theta = \pi/6$ where the layer magnetizations vs. T are shown in detail. The gap at $k = 0$ due to d is shown in Fig. 10.19(b) as a function of the film thickness N for $d = 0.1$ and $\theta = \pi/6$, at $T = 0$. We see that the gap depends not only on d but also on

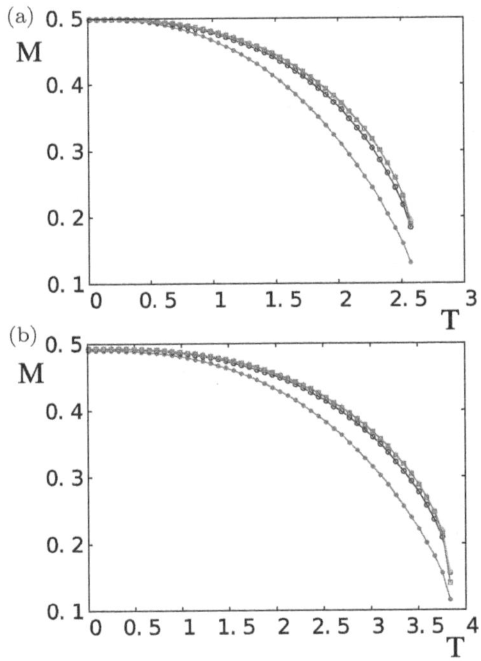

Fig. 10.16. Eight-layer film: layer magnetizations M vs. temperature T for (a) $\theta = \pi/6$ (radian) and (b) $\theta = \pi/3$, with $d = 0.1$. Circles, void circles, void triangles, and squares correspond, respectively, to the first, second, third, and fourth layer.

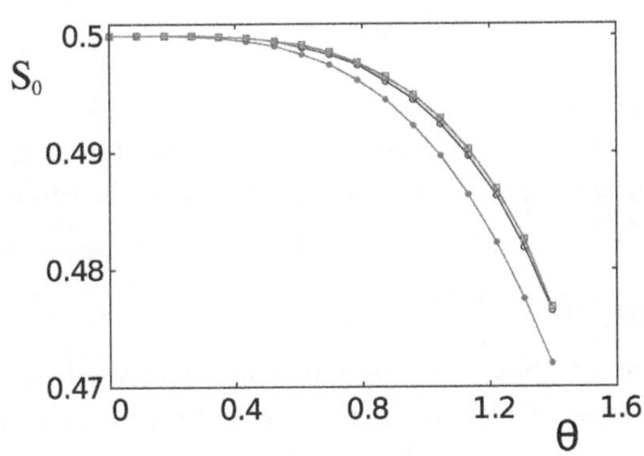

Fig. 10.17. Spin length S_0 at $T = 0$ of the first four layers as a function of θ, for $N = 8$, $d = 0.1$. Circles, void circles, void triangles, and squares correspond, respectively, to the first, second, third, and fourth layer.

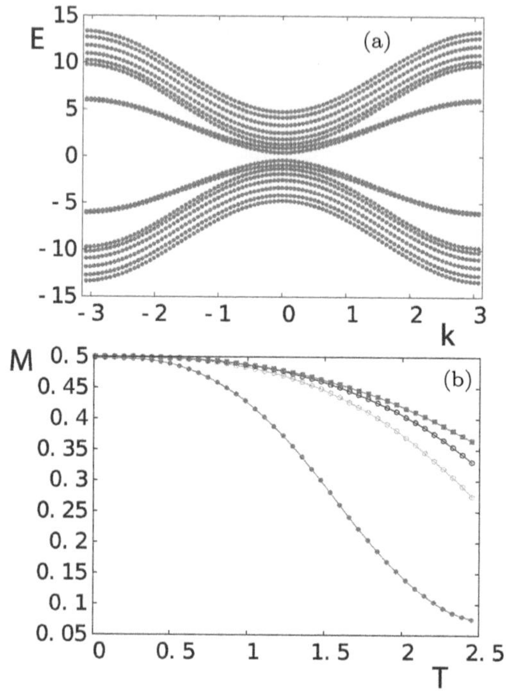

Fig. 10.18. Surface effect: (a) SW spectrum $E(k)$ vs. $k = k_x = k_z$ for a thin film of 8 layers: $\theta = \pi/6$, $d = 0.2$, $J^s_{\parallel} = 0.5$, $J^s_{\perp} = 0.5$, the gap at $k = 0$ is due to d. The surface-mode branches are detached from the bulk spectrum. (b) Layer magnetizations versus T for the first, second, third, and fourth layer (circles, void circles, void circles, and filled squares, respectively). See text for comments.

the value of the surface magnetization which is larger for thicker films. The transition temperature T_c vs. the thickness N is shown in Fig. 10.19(c), where one observes that T_c tends rapidly to the bulk value (3D) which is $\simeq 2.82$ for $d = 0.1$.

10.3.6. *Discussion*

Several results found above can be experimentally verified. Let us confine ourselves in the case of thin films in this section, although the following discussion applies also in the 2D and 3D crystals.

The first striking aspect is the very particular effect of the DM interaction in the SW spectrum: we have seen in Fig. 10.15(b) that the first mode is linear in k for small k, but not the upper modes. This is not the case of ferromagnetic and antiferromagnetic interactions in thin films: for

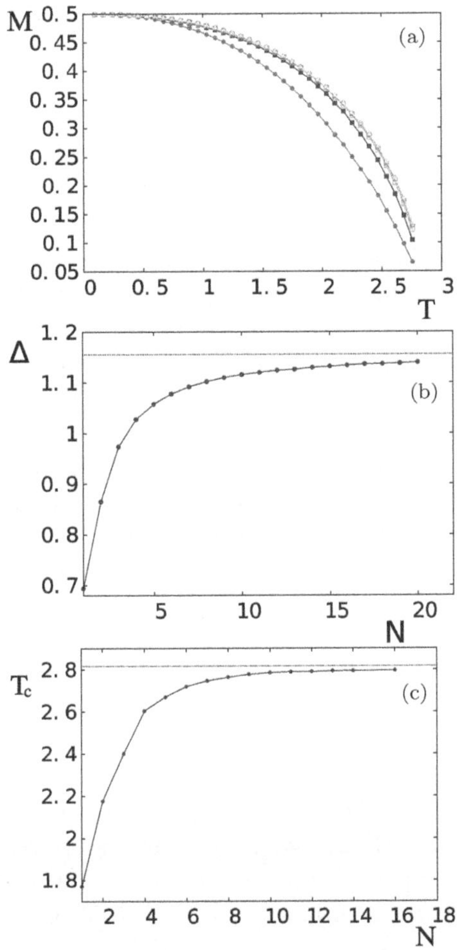

Fig. 10.19. Twelve-layer film: (a) layer magnetizations vs. T for $\theta = \pi/6$ and anisotropy $d = 0.1$. The first two curves from below (circles and squares) correspond, respectively, to first and second layers. The magnetizations of the other (interior) layers are not distinguishable at the figure scale, (b) gap at $k = 0$ as a function of film thickness N for $\theta = \pi/6$, $d = 0.1$, at $T = 0$, (c) critical temperature T_c vs. the film thickness N calculated with $\theta = \pi/6$ and $d = 0.1$. Note that for infinite thickness (namely 3D), $T_c \simeq 2.8$ for $d = 0.1$ (horizontal line).

a ferromagnetic interaction, all modes have the k^2 behavior at small k,[1] and for antiferromagnetic interactions, all modes have the linear-k behavior. Thus, the DM interaction affects only the first mode, and the J-term in the Hamiltonian, Eq. (10.4), maintains the ferromagnetic behavior for higher energy modes in the SW spectrum. Experiments which measure SW spectra such as neutron scattering[60] and SW resonance can verify if the first mode in the SW spectrum shown in Fig. 10.15(b) is linear in k for large DM interactions. For SW higher energy modes, there has been an experimental breakthrough with the spin-polarized electron energy loss spectroscopy (SPEELS): this technique allows us to detect very high energy surface magnons, up to 240 meV. It has been proved to be very efficient to probe the dispersion of magnons in the ultrathin ferromagnetic films.[61] We believe that using such high-efficient experimental means, the effect of the DM interaction shown in this section can be verified.

The second striking feature is the surface effect on the SW spectrum and the surface magnetization shown, respectively, in Figs. 10.18(a) and 10.18(b). These can be experimentally verified with the above-mentioned techniques. Note that the slope of the first SW mode allows us to deduce the interaction parameter, and therefore, the DM interaction strength D. Note that neutron scattering can be used for a complete determination of the magnetic properties of a system at finite temperatures.[60]

The third interesting result is the dependence of the critical temperature T_c on the film thickness shown in Fig. 10.19(c). This can be completely determined by neutron scattering and SPEELS, which measure the magnetization as a function of temperature of films as thin as a few atomic layers.[60,61]

We believe that, in spite of the cumbersome mathematical details shown in Section 10.3.3, numerical results coming out from the formulation shown in Sections 10.3.4 and 10.3.5 are very clear and physically meaningful. We expect therefore experimental verifications on materials showing a DM interaction such as those mentioned in Refs. [42, 43, 52] and [4].

10.3.7. *Summary*

By a self-consistent Green's function theory, we obtain the expression of the SW dispersion relation in 2D and in a thin film. Due to the competition between ferromagnetic interaction J and the perpendicular DM interaction D, the GS is non-collinear with an angle θ which is shown to explicitly depend on the ratio D/J. The spectrum is shown to depend on θ and the layer magnetization is calculated self-consistently as a function of temperature up to the critical temperature T_c.

We have obtained new and interesting results. In particular, we have showed that (i) the SW excitation in 2D is stable at $T = 0$ with the non-collinear spin configuration induced by the DM interaction D without the need of an anisotropy; (ii) in the case of thin films, we need a small anisotropy d to stabilize the SW excitations because of the lack of neighbors at the surface; (iii) the SW energy E depends on D, namely on θ: at the long wave length limit, E is proportional to k^2 for small D but E is linear in k for strong D, in 2D, and in a thin film; (iv) quantum fluctuations are inhomogeneous for layer magnetizations near the surface.

10.4. Stability and Phase Transition of Skyrmion Crystals Generated by DM Interaction

In this section, we show that we can generate a crystal of skyrmions in two dimensions using a Heisenberg Hamiltonian including the ferromagnetic interaction J, the DM interaction D, and an applied magnetic field $\mathbf{H}$. This skyrmion crystal exists in a region of (D, H). The stability of this skyrmion crystalline phase at finite temperatures is shown by a study of the time dependence of the order parameter using MC simulations. We observe that the relaxation is very slow and follows a stretched exponential law.

10.4.1. *Introduction*

The word skyrmion has been named after Skyrme[62] who used a topological soliton to model a particle in nuclear physics. Skyrmions in condensed matter physics have been extensively investigated during the last 20 years (see Refs. [4, 63] and [64] in references list).

There are several mechanisms and interactions leading to the appearance of skyrmions in various kinds of matter. The most popular one is certainly the DM interaction.[40,41] The DM interaction comes from the second-order perturbation of the exchange interaction between two spins which is not zero only under some geometrical conditions of non-magnetic atoms found between them.[41] The order of magnitude of DM interaction D is therefore, where perturbation theory obliges, small. The explicit form of the DM interaction is given in the previous section. However, we can think that the demonstration of Moriya[41] is for a special case and the general Hamiltonian may have the same form but with different microscopic origins which do not necessarily imply a small value for D. We take therefore the liberty to use D as a parameter.

The DM interaction has been shown to generate skyrmions in various kinds of crystals. For example, it can generate a crystal of skyrmions in which skyrmions arrange themselves in a periodic structure.[65–68] Skyrmions have been shown to exist in crystal liquids.[63,64,69] A single skyrmion has also been observed.[48] Skyrmion crystals have been found to exist in thin films.[49,70] Direct observation of the skyrmion Hall effect has been realized.[71] Artificial skyrmion lattices have been devised for room temperatures.[72] Experimental observations of skyrmion lattices have been realized in MnSi in 2009[47,73] and in doped semiconductors in 2010.[48] At this stage, it should be noted that skyrmion crystals can also be created by competing exchange interactions without DM interactions.[74,75] So, mechanisms for creating skyrmions are multiple.

We note that SW excitations in systems with a DM interaction in the helical phase without skyrmions have been shown in the previous section. This problem has been investigated by several authors.[54–59]

Applications of skyrmions in spintronics have been largely discussed[4] and their advantages compared to early magnetic devices such as magnetic bubbles have been pointed out in a recent review by Kang *et al.*[76] Among the most important applications of skyrmions, let us mention skyrmion-based racetrack memory,[77] skyrmion-based logic gates,[78,79] skyrmion-based transistor,[80–82] and skyrmion-based artificial synapse and neuron devices.[83,84]

In the following, we show results on a skyrmion crystal created by the competition between the nearest-neighbor ferromagnetic interaction J and the DM interaction of magnitude D under an applied magnetic field $\mathbf{H}$. These results, recently obtained by MC simulation,[85] show that the skyrmion crystal is stable at finite temperatures up to a transition temperature T_c where the topological structure of each skyrmion and the periodic structure of skyrmions are destroyed.

10.4.2. *Model and ground state*

The DM interaction energy between two spins $\mathbf{S}_i$ and $\mathbf{S}_j$ is written as

$$\mathbf{D}_{i,j} \cdot \mathbf{S}_i \wedge \mathbf{S}_j, \tag{10.53}$$

where $\mathbf{D}_{i,j}$ is a vector which results from the displacement of non-magnetic ions located between $\mathbf{S}_i$ and $\mathbf{S}_j$, for example in Mn–O–Mn bonds in the historical papers.[40,41] The direction of $\mathbf{D}_{i,j}$ depends on the symmetry of the displacement.[41] $\mathbf{D}_{i,j}$ is proportional to the vector product $\mathbf{R} \wedge \mathbf{r_{i,j}}$ of the vector $\mathbf{R}$ which specifies the displacement of the ligand (for example, oxygen) and the unit vector $\mathbf{r}_{i,j}$ along the axis connecting the magnetic ions (spins) i and j (see Fig. 10.20). We write

$$\mathbf{D}_{i,j} \propto \mathbf{R} \wedge \mathbf{r}_{i,j}. \quad (10.54)$$

Note that the spin product is antisymmetric with respect to the spin inversion and the DM vector is antisymmetric with the exchange i and j. Therefore, Eq. (10.53), is invariant with the permutation $j \leftrightarrows i$. The energy sum on opposite neighbors does not cancel out.

We consider here for simplicity the 2D case where the spins are on a square lattice. We are interested in the stability of the skyrmion crystal generated in a system of spins interacting with each other via a DM interaction and a symmetric isotropic Heisenberg exchange interaction in an applied field perpendicular to the xy plane. All interactions are limited to NN in the plane. We choose $\mathbf{R}$ and $\mathbf{r}_{i,j}$ as shown in Fig. 10.20. $\mathbf{D}_{i,j}$ is thus perpendicular to the xy plane. We write

$$\mathbf{D}_{i,j} = D\hat{z}, \quad (10.55)$$

where all constants are absorbed in D and $\hat{z}$ is the unit vector along the z direction. This choice implies that the spins lie in the xy plane in the

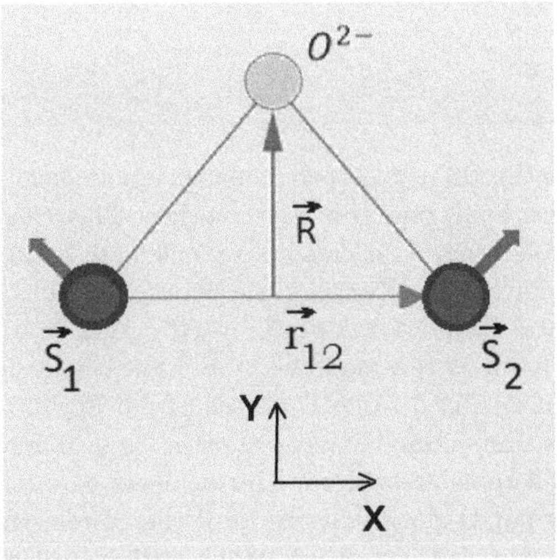

Fig. 10.20. Schema of DM interaction, where the spins are in the XY plane. $\mathbf{D}_{i,j}$ is perpendicular to the XY plane.

zero-field GS as shown in the previous section. The full Hamiltonian is

$$\begin{aligned}\mathcal{H} &= -J\sum_{\langle ij\rangle}\mathbf{S_i}.\mathbf{S_j} + D\sum_i [S_i^y S_{i+x}^z - S_i^z S_{i+x}^y - S_i^x S_{i+y}^z \\ &\quad + S_i^z S_{i+y}^x] - H\sum_i S_i^z \\ &= -J\sum_{\langle ij\rangle}\mathbf{S_i}.\mathbf{S_j} + D\sum_i [S_i^y(S_{i+x}^z - S_{i-x}^z) \\ &\quad - S_i^z(S_{i+x}^y - S_{i-x}^y) - S_i^x(S_{i+y}^z - S_{i-y}^z) \\ &\quad + S_i^z(S_{i+y}^x - S_{i-y}^x)] - H\sum_i S_i^z. \end{aligned} \tag{10.56}$$

The local field acting on the i-th spin is

$$\mathcal{H}_i = -S_i^x H_i^x - S_i^y H_i^y - S_i^z H_i^z y, \tag{10.57}$$

where the components are given by

$$H_i^x = J\sum_{NN} S_j^x + D(S_{i+y}^z - S_{i-y}^z), \tag{10.58}$$

$$H_i^y = J\sum_{NN} S_j^y - D(S_{i+x}^z - S_{i-x}^z), \tag{10.59}$$

$$H_i^z = J\sum_{NN} S_j^z + D(S_{i+x}^y - S_{i-x}^y) - D(S_{i+y}^x - S_{i-y}^x) + H. \tag{10.60}$$

To determine the GS in a perpendicular magnetic field $\mathbf{H}$, we minimize the energy of each spin, one after the another. This can be numerically achieved as the following. At each spin, we calculate its local-field components acting on it from its NN using the above equations. Next, we align the spin in its local field, i.e. taking $S_i^x = H_i^x/\sqrt{(H_i^x)^2 + (H_i^y)^2 + (H_i^z)^2}$, etc. The denominator is the modulus of the local field. In doing so, the spin modulus is normalized to be 1. As seen from Eq. (10.57), the energy of the spin $\mathbf{S}_i$ is minimum. We take another spin and repeat the same procedure until all spins are visited. This achieves one iteration. We have to do a sufficient number of iterations until the system energy converges. For the skyrmion case, it takes about 1000 iterations to have the fifth-digit convergence. We have used random initial configurations, and we observed that a number of them lead to metastable states. This is seen by comparing the energies of several thousands of initial configurations and examining the snapshots. A good GS always has a perfect hexagonal structure as shown below. Note that the GS configuration does not depend on the system size $N \times N$ if it is large enough. This size is from $N = 50$ for the value of $D = 1$ and $H = 0.5$ used in the example shown below.

An example of GS is displayed in Fig. 10.21 for the crystal size 50×50: a crystal of skyrmions is seen using $D = 1$ and $H = 0.5$ (in unit of $J = 1$). Note that we use here an example of large D for clarity, but we will show later in the phase diagram that the skyrmion crystal exists in a much smaller region for small D.

Let us give some comments on Fig. 10.21:

(i) The size of each skyrmion depends on the value of D for a given H. In the case presented in this figure the diameter of each skyrmion is 10 lattice spacings. For larger D, the diameter is reduced, there are more skyrmions for a given lattice size. For small D, the system tends to a single large skyrmion centered at the middle of the lattice.
(ii) The skyrmions form a triangular lattice (see Fig. 10.21). Note that the underlined lattice is a square lattice where each site is occupied by a spin. There is thus a six-fold degeneracy of the skyrmion lattice due to the global rotation of the spin by $2\pi/6$,
(iii) All the skyrmions have the same chirality. There is thus, in addition to the six-fold degeneracy mentioned above, a two-fold chirality degeneracy of the configuration shown in Fig. 10.21. Note that two neighboring skyrmions are separated by spins perfectly aligned along the field (the spin at the center of each skyrmion is perfectly aligned antiparallel to the field).

In Fig. 10.22(a), we show a GS at very small H where domains of long and round islands of up-spins separated by labyrinths of down-spins are mixed. When H is increased, vortices begin to appear. The GS is a mixing of long islands of up-spins and vortices as seen in Fig. 10.22(b) obtained with $D = 1$ and $H = 0.25$. This phase can be called "labyrinth phase" or "stripe phase".

It is interesting to note that skyrmion crystals with texture similar to those shown in Figs. 10.21 and 10.22 have been experimentally observed in various materials,[47,49,72,86] but the most similar skyrmion crystal was observed in 2D Fe(0.5)Co(0.5)Si by Yu *et al.* using Lorentz transmission electron microscopy.[48]

(a)

(b)

(c)

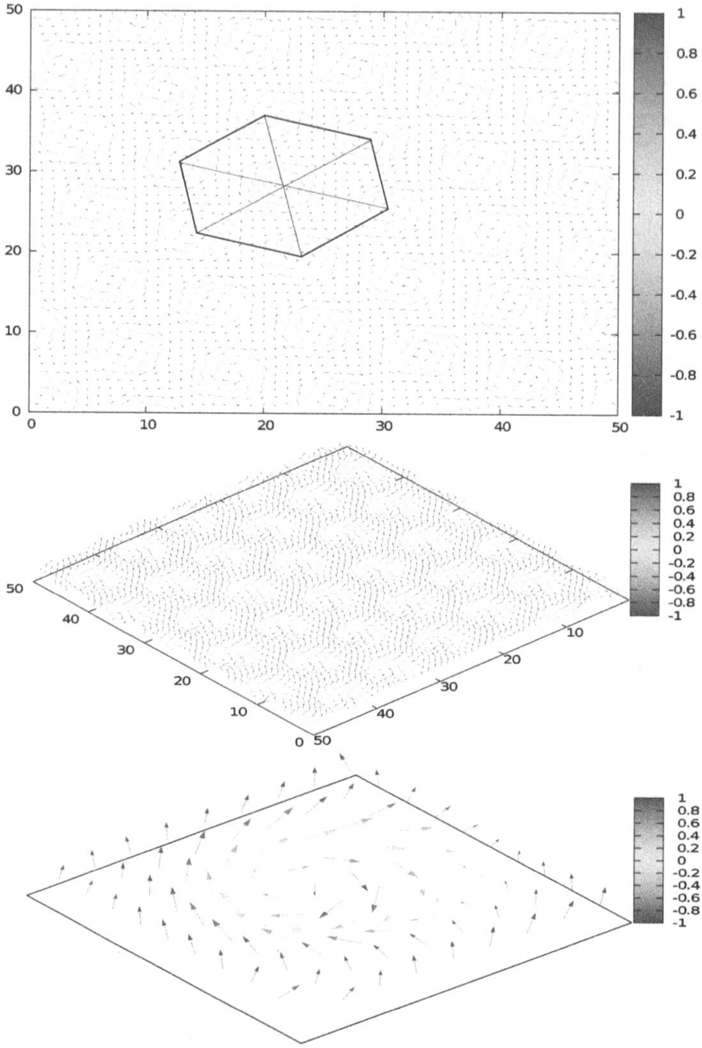

Fig. 10.21. With GS for $D/J = 1$ and $H/J = 0.5$, a crystal of skyrmions is observed. (a) Skyrmion crystal viewed in the xy plane, (b) a 3D view, (c) zoom of the structure of a single vortex. The value of S_z is indicated on the color scale. See text for comments.

We have performed the GS calculation taking many values in the plane (D, H). The phase diagram is established in Fig. 10.23. Above the blue line is the field-induced ferromagnetic phase. Below the red line is the labyrinth phase with a mixing of skyrmions and rectangular domains. The skyrmion

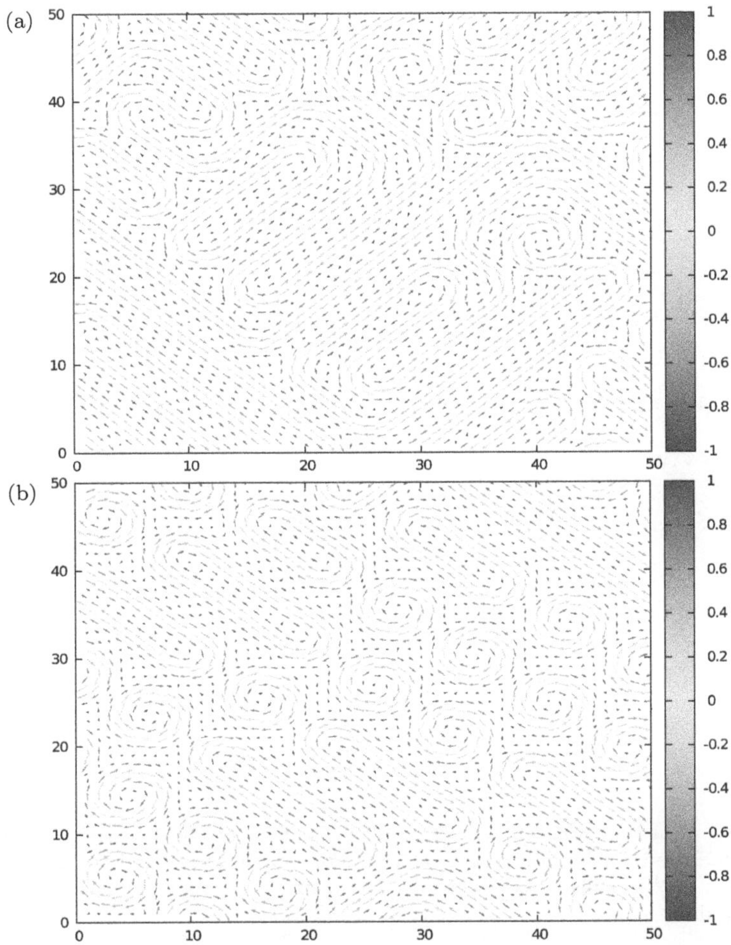

Fig. 10.22. (a) GS for $D/J = 1$ and $H/J = 0.05$, a mixing of domains of long and round islands, (b) GS for $D/J = 1$ and $H/J = 0.25$, a mixing of domains of long islands and vortices. We call these structure the "labyrinth phase".

crystal phase is found in a narrow region between these two lines, down to infinitesimal D. In order to enlarge the stability region of skyrmions, we may need to include other kinds of interaction such as dipole–dipole interaction, anisotropies, and/or other competing interactions.

Although effects of the temperature (T) are to be shown in the next section, we anticipate here the phase diagram shown in Fig. 10.24 in the space (T, H) for an overview before the detailed data presentation below.

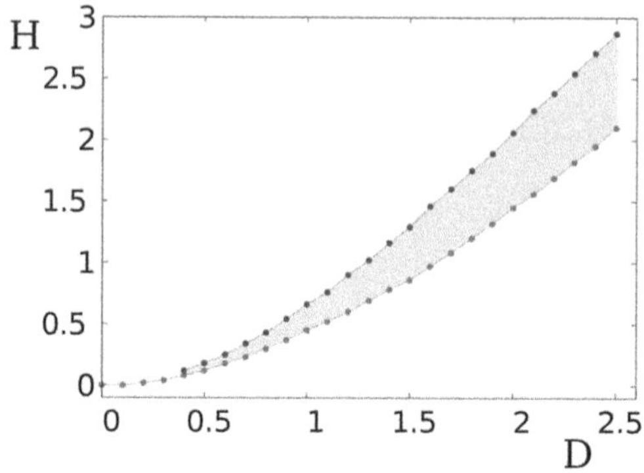

Fig. 10.23. Phase diagram in the (D, H) plane for size $N = 100$.

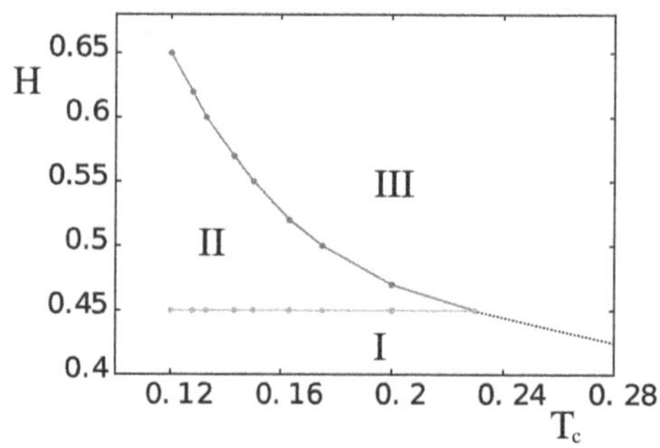

Fig. 10.24. Phase diagram in the (T, H) plane for $D = 1$ with $N = 100$. Phases I, II, and III indicate the labyrinth phase, the skyrmion crystal phase, and the field-induced ferromagnetic phase. The discontinued line separating phase I and III at high T is not a transition line.

The definitions of phases I, II, and III are given in the caption. Note that the discontinued line is not a transition line as it schematically indicates a region where spins turned progressively to the field at high T.

In Section 10.4.3, we are interested in the stability of the skyrmion crystal phase as the temperature T is increased from zero.

10.4.3. *Stability of Skyrmion crystal at finite temperatures*

Using the above GS configuration, we heat the system from $T = 0$ to a temperature T during an equilibrating time t_0 before averaging physical quantities over the next 10^6 MC steps per spin. The time t_0 is the "waiting time" during which the system relaxes before we perform averaging during the next t_a.

The definition of an order parameter for a skyrmion crystal is not obvious. Taking advantage of the fact that we know the GS, we define the order parameter as the projection of an actual spin configuration at a given T on its GS and we take the time average. This order parameter is thus defined as

$$M(T) = \frac{1}{N^2(t_a - t_0)} \sum_i \left| \sum_{t=t_0}^{t_a} \mathbf{S}_i(T,t) \cdot \mathbf{S}_i^0(T=0) \right|, \qquad (10.61)$$

where $\mathbf{S}_i(T,t)$ is the ith spin at the time t, at temperature T, and $\mathbf{S}_i(T=0)$ is its state in the GS. The order parameter $M(T)$ is close to 1 at very low T where each spin is only weakly deviated from its state in the GS. $M(T)$ is zero when every spin strongly fluctuates in the paramagnetic state. The above definition of $M(T)$ is similar to the Edward–Anderson order parameter used to measure the degree of freezing in spin glasses[87]: we follow each spin with time evolving and take the spatial average at the end.

We show in Fig. 10.25 the order parameter M vs. T (red data points) as well as the average z spin component (blue data points) calculated by the projection procedure for the total time $t = 10^5 + 10^6$ MC steps per spin. As seen, both two curves indicate a phase transition at $T_c \simeq 0.26J/k_B$. The fact that M does not vanish above T_c is due to the effect of the applied field. It should be said that each skyrmion has a center with spins of negative z

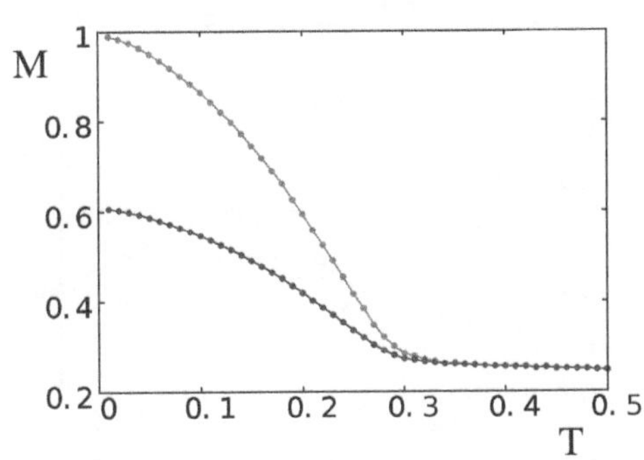

Fig. 10.25. Red circles: order parameter defined in Eq. (10.61) vs. T, for $H = 0.5$ and $N = 1800$, averaged during $t_a = 10^5$ MC steps per spin after an equilibrating time $t_0 = 10^5$ MC steps. Blue crosses: the projection of the S_z on S_z^0 of the ground state as defined in Eq. (10.61) but for the z components only. See text for comments.

components (the most negative at the center), the spins turn progressively to positive z components while moving away from the center.

We can also define another order parameter: since the field acts on the z direction, in the GS and in the skyrmion crystalline phase we have both positive and negative S_z. In the paramagnetic state, the negative S_z will turn to the field direction. We define thus the following parameters using the z spin components

$$Q_+(T) = \frac{1}{N^2(t_a - t_0)} \sum_{S_i^z>0} \sum_{t_0}^{t_a} S_i^z(T,t), \tag{10.62}$$

$$Q_-(T) = \frac{1}{N^2(t_a - t_0)} \sum_{S_i^z<0} \sum_{t_0}^{t_a} S_i^z(T,t). \tag{10.63}$$

Figure 10.26 shows Q_+ and Q_- vs. T. As seen, at the transition Q_+ undergoes a change of curvature and Q_- becomes zero. All spins have positive S_z after the transition due to spin reversal by the field.

The results obtained at the end of the simulation may depend on the overall time $t = t_0 + t_a$. In simple systems, the choices of t_0 and t_a can be guided by testing the time dependence of physical quantities, and the values of t_0 and t_a are chosen when physical quantities do not depend on these run times. However, in disordered systems such as spin glasses and in complicated systems such as frustrated systems, the relaxation time is very long and out of the reach of simulation time. In such cases, we have to

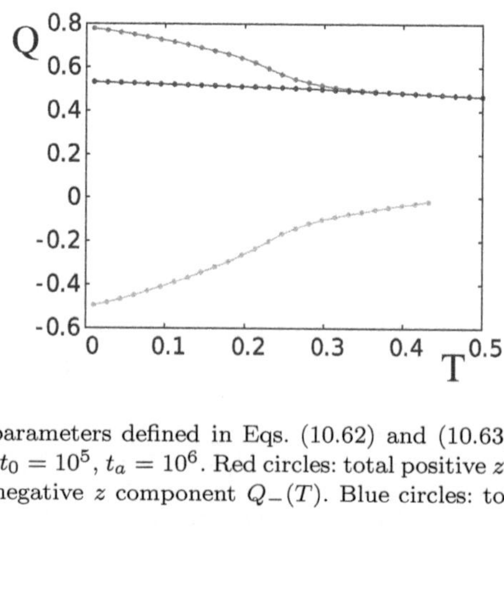

Fig. 10.26. Order parameters defined in Eqs. (10.62) and (10.63) vs. T, for $H = 0.5$, $D = 1$ and $N = 800$, $t_0 = 10^5$, $t_a = 10^6$. Red circles: total positive z component ($Q_+(T)$). Green circles: total negative z component $Q_-(T)$. Blue circles: total z component. See text for comments.

recourse to some scaling relations in order to deduce the values of physical quantities at equilibrium.[88,89] We show below how to obtain the value of an order parameter at the infinite time.

In order to detect the dependence of $M(T)$ on the total MC time $t = t_0 + t_a$, we calculate the average of $M(T)$ over 10^6 MC steps per spin, after a waiting time t_0 as said above. We record the values of $M(T)$ in different runs with t_0 varying from 10^4 to 10^6 MC steps per spin. We plot these results as a function of different total time t in Fig. 10.27 for three temperatures. As said above, to find the value extrapolated at the infinite time, we use the stretched exponential relaxation defined by

$$M(T,t) = A \exp\left[-(t/\tau)^{\alpha}\right] + c, \tag{10.64}$$

where t is the total simulation time, α is the stretched exponent, A a temperature-dependent constant, and τ the relaxation time. Note that this definition, without the constant c, has been used by many previous authors in the context of spin glasses.[90–94] We have introduced c which is the infinite time limit of $M(T)$. We have taken t from 10^4 to 10^6 MC steps per spin in the simulation. At the infinite time limit, c is zero for $T \gg T_c$, and $c \neq 0$ for $T < T_c$. Figure 10.27 shows $M(T,t)$ as a function of time t in unit of 10^3 MC steps per spin, for three temperatures $T = 0.01$, 0.094, and 0.17. As seen, the fit with Eq. (10.64) presented by the continuous line is very good for the whole range of t.

Note that

(i) The precision of all parameters is between 1% and 5% depending on the parameter.
(ii) The value of α can vary a little bit according to the choice and the precision of the other parameters in the fitting, but this variation is within a very small window of values around the value given above. For example, at $T = 0.17$, α can only be in the interval $[0.8 \pm 0.02]$. The value of α varies with temperature: at low T, $\alpha = 0.6$, and at a higher T, we have $\alpha = 0.8$. This variation has been seen in other systems, in particular in spin glasses.[95]
(iii) The stretched exponential behavior is believed to be due to the non-uniform spin configuration since collinear spin states (ferromagnetic an antiferromagnetic orderings) follow a non-stretched exponential law.[88] The stretched exponential law was found in disordered phases such as spin glasses.[90–94] Of course, if D is small, the configuration tends to a ferromagnetic one and the relaxation tends to a non-stretched law.

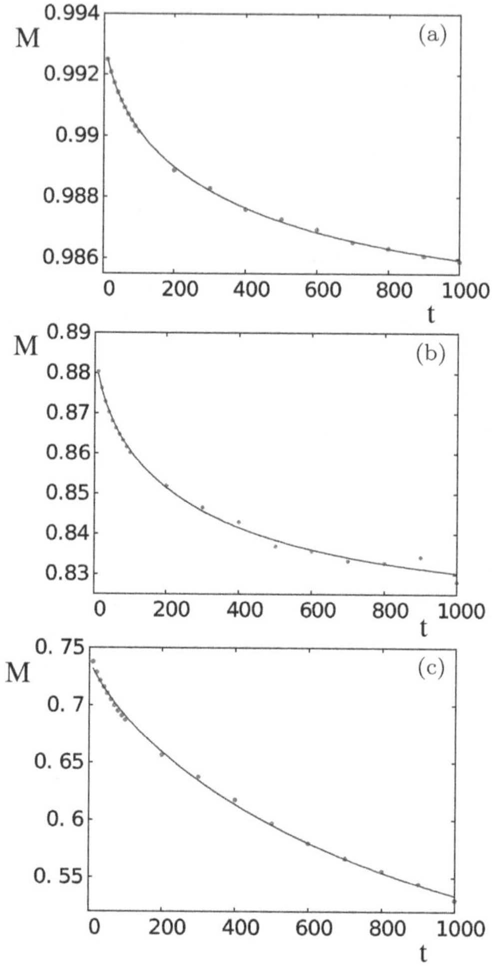

Fig. 10.27. The order parameter M defined by Eq. (10.61) vs. MC time t in unit of 1000 MC steps per spin, for $H = 0.5$, $D = 1$, and $N = 800$ (a) $T = 0.01$, values of fitting parameters $\alpha = 0.6$, $A = 0.008 \pm 0.00001$, $\tau = (364 \pm 19)10^3$, $c = 0.984505 \pm 0.00012$; (b) $T = 0.094$, $\alpha = 0.6$, $A = 0.0653 \pm 0.0013$, $\tau = (277 \pm 36)10^3$, $c = 0.822 \pm 0.018$); (c) $T = 0.17$, $\alpha = 0.8$, $A = 0.31 \pm 0.01$, $\tau = 891 \pm 100)10^3$, $c = 0.43 \pm 0.017$.

(iv) The relaxation time, within statistical errors, is approximately constant at low T, but it increases rapidly when T tends to T_c as seen in the value of τ at $T = 0.17$. This increase is a consequence of the so-called critical slowing down when the system enters the critical region.

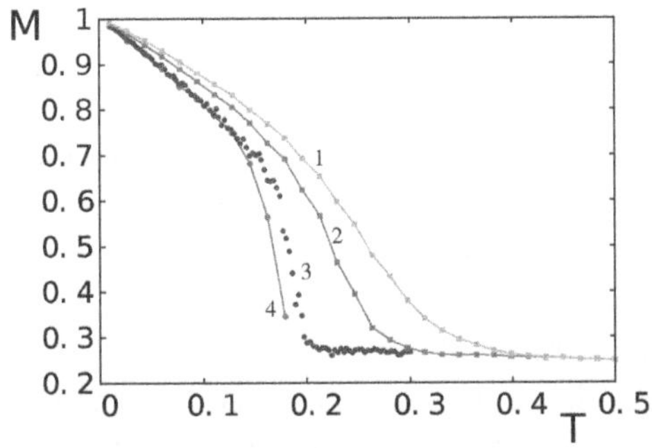

Fig. 10.28. The order parameter $M(T)$ vs. T for several waiting times t, for $H = 0.5$, $D = 1$, and $N = 800$: curves 1, 2, 3, and 4 are, respectively, for $t = 10^5$, 2×10^5, 10^6, ∞ (by fitting with Eq. (10.64)).

Let us show $M(T)$ as a function of T in Fig. 10.28 using the results of different run times from $t = 10^4 + 10^6$ MC steps per spin to $t = 10^6 + 10^6$. The values of $M(T)$ at infinite time for each T deduced from Eq. (10.64) is also shown. We see that while the total time $10^5 + 10^6$ MC steps per spin is sufficient at low T, it is not enough at higher T. That was the reason why we should use Eq. (10.64) to find the value of $M(T)$ at infinite time to be sure that the skyrmion crystal is stable at finite temperatures. Curves of $M(T)$ at infinite time for various H with $D = 1$ (for example curve 4 in Fig. 10.28 for $H = 0.5$) are used to establish the phase diagram shown in Fig. 10.24.

We have studied finite-size effects on the phase transition at T_c and we have seen that from $N = 800$, all curves coincide: there is no observable finite-size effects for $N \geq 800$. The present 2D skyrmion lattice phase therefore remains stable for the infinite dimension, unlike the ferromagnetic Heisenberg model in 2D.[32] Note that the nature of the ordering at low T and the phase transition observed here, in spite of the vortex nature of skyrmions, are not those of the Kosterlitz–Thouless vortex mechanism observed in ferromagnetic XY spin systems[96] since our vortices are stable without changing their position up to the transition temperature.

10.4.4. *Summary*

With a simple model including a ferromagnetic interaction J and a DM interaction D under an applied magnetic field H in two dimensions, we obtained a skyrmion crystal in a region of the phase space (D, H). The spin model is the classical Heisenberg model. We have numerically determined the ground state by minimizing the local energy, spin by spin, using an iteration procedure. The skyrmion lattice so obtained is then heated to a finite temperature by the use of MC simulations. We have shown that the skyrmion lattice is stable up to a finite temperature T_c beyond which the system becomes disordered. We have also shown that the relaxation follows a stretched exponential law. We believe that such a stability can be experimentally observed in real systems.

Perhaps, our model is the simplest model which is able to generate a skyrmion crystal without the need to include more complicated interactions such as long-range dipolar interactions, easy and uniaxial anisotropies.[44,68,86] Note that other authors have shown that skyrmion crystals can also be generated only with frustrated short-range interactions without even the DM interaction.[74,75] So, we have to keep in mind that there are many interaction mechanisms of different nature which can generate skyrmion crystals. The origin of a skyrmion crystal determines its structure, its stability, its dynamics, and in short its properties (see a discussion on this point in Ref. [86]). Experiments can be used therefore to determine the interaction mechanism which is at the origin of the formation of a given skyrmion crystal.

10.5. Elastic Effects in Membranes with DM Interaction

In this section, we would like to see whether the skyrmion phase is stable under large lattice deformations.[97] For this purpose, it is interesting to assume the lattice degrees of freedom as a dynamical variable. Here, we should note that "dynamical" variable refers to the variable that is integrated in the partition function in the context of statistical mechanics. Therefore, a fluctuating lattice is considered to be a direct tool for this purpose, where the lattice fluctuation is allowed not only into the in-plane direction but also into the out-of-plane direction.[98–105] Skyrmions on fluctuating lattice are interesting also from the view point of liquid crystal skyrmions.[63,64]

It is widely accepted that skyrmion configurations, which are observed on materials such as FeGe, MnSi, or Cu_2OSeO_3, are promising candidates for future computer memories because of their dramatic energy-saving property in electric transportation. Due to the topological nature of the spin configuration, its stability against external stimuli such as thermal fluctuations is one of its most interesting properties.[40,41,45,46,62,106] For these reasons, many experimental and theoretical studies have been conducted by assuming competing interactions, such as ferromagnetic interaction, DM, and Zeeman energy under constant magnetic fields.[47,66,79,107–109] In particular, skyrmions in 2D systems have attracted a great deal of attention due to their potential applications in memory devices.[110–113] However, the creation/annihilation of skyrmions and the stability of the skyrmion phase are yet to be studied in the context of such applications.

Along this research direction, several investigations have been conducted on the responses to mechanical stresses, which have a non-trivial effect on the spin configuration.[114,115] In Ref. [116], Shi and Wang numerically find that skyrmions are stabilized due to the magnetoelastic coupling even without an external magnetic field by introducing an explicit magnetoelastic interaction energy. This confirms that skyrmions may be generated by various mechanisms. We come back to this point later in the next section. Nii *et al.* recently reported that MnSi is an anisotropic elastic object,[117] and they also reported the experimental results of a stress-induced skyrmion-to-conical phase transition in MnSi, where the applied uniaxial stresses create/annihilate skyrmions under suitable external magnetic fields.[118] It is also reported by Seki *et al.* that skyrmions in Cu_2OSeO_3 can be stabilized by a uniaxial tensile strain,[50,119] and Levatic *et al.* reported that mechanical pressure significantly increases the skyrmion pocket in Cu_2OSeO_3 (see Ref. [120]) and in MnSi.[121] From these experiments, it was clarified that uniaxial mechanical stimuli stabilize/destabilize the skyrmion phase, or, in other words, a small lattice deformation enhances the stability/instability of the skyrmion phase. This indicates that the interplay between two different degrees of freedom spin and lattice, which can be rephrased by spin and orbit, plays an important role in mechanical responses, which has been extensively studied via DM interaction energy. Chen *et al.* numerically studied skyrmions in MnSi by using a 3D model and suggested that anisotropies in DM and FM interactions play an important role in their stability under uniaxial stress in a wide range of temperatures including the low-temperature region.[122] These anisotropies are naturally expected when crystalline lattices are considerably deformed. Thus, it is interesting to study the skyrmion stability under lattice deformations at finite temperatures.

The DM interaction on the continuous 2D plane depends on both spins and coordinates, and hence, we are able to calculate mechanical properties such as surface tension under the presence of skyrmions if the coordinate, or equivalently the surface, is treated as a dynamical variable.[98–105] Moreover, DM interaction is invariant under an arbitrary rotation around the axis perpendicular to the surface. However, its discrete version, which will be described in the next section, only has a discrete symmetry on 2D rigid lattices because the rigid lattices have only discrete symmetries. In contrast, the fluctuating lattices, where the edge direction of the lattice becomes almost random, are not regarded as a rigid lattice by any rotation. For this reason, the discrete DM interaction is expected to be influenced by the surface fluctuations.

Therefore, it is worth studying the skyrmions to see the interplay between spins and surface elasticity on fluctuating surfaces. In this section, we calculate the surface tension τ by MC simulations to study the influences of skyrmion configurations on τ. Three different spin configurations are expected on the surface: stripe, skyrmion, and FM phases, as in the case of rigid lattices.[48,68,85,86] Phase diagrams are also obtained in the space temperature vs. external magnetic field. In this calculation, mechanical stress can be applied to the elastic lattices only in the form of strain, which is given by fixing the boundary to a suitable size or shape. Such an effective stress becomes isotropic in general, because the lattice structure changes relatively freely due to the fact that the vertex position is a dynamical variable. Nevertheless, the uniaxial stress condition is simulated by modifying the boundary lengths of the rectangular lattice, for example from its isotropic lengths (L_1, L_2) to $(L_1/\xi, \xi L_2), (\xi \neq 1)$. As a result of this geometry modification, it will be checked whether the stripe domain axis aligns along the uniaxial stress direction. We also examine whether this uniaxial stress condition influences the stability of skyrmion phase under an external magnetic field.

10.5.1. *Triangulated lattices and skyrmion model*

10.5.1.1. *Triangulated lattices*

A triangulated lattice with a square boundary is used for the simulations. Triangular lattices have already been used to study skyrmions using non-DM[75] and DM interaction energies.[65] The total number of vertices N is given by $N = L^2$, where L is the number of vertices on one boundary edge. In Fig. 10.29(a), an initial configuration of the lattice of $N = 100 (\Leftrightarrow L = 10)$ is shown, where the boundary vertices (BVs) are denoted by small spheres.

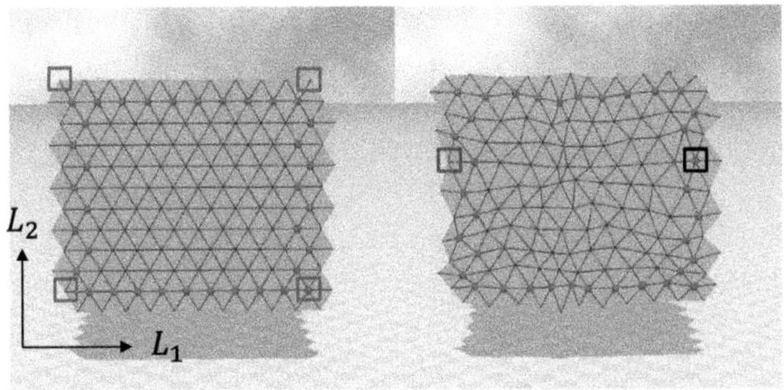

(a) (b)

Fig. 10.29. (a) The initial configuration of triangulated lattices of size $(L_1, L_2) = (10, 10)$, where L_1 and L_2 are the total number of vertices along the horizontal (L_1) and vertical (L_2) directions, respectively, and (b) a snapshot of the lattice after a sufficiently large number of MC updates. PBC is assumed in both directions. The BVs are denoted by small spheres, and the virtual vertices connected by the PBC are represented outside. The vertices enclosed by squares without the small sphere are virtual points corresponding to the real vertex with a small sphere in (a) and (b).

The lattice size $N = 100$ is small enough to visualize the lattice structure including the BVs, and it is 10^2 times smaller than the size of lattices used in the simulations. Let a be the edge length of the initial regular triangle, which forms the initial lattice in Fig. 10.29(a), then the side lengths along the horizontal (L_1) and vertical (L_2) directions are

$$((L-1)a, (L-1)a\sqrt{3}/2), \quad a = 1. \tag{10.65}$$

We call this a the lattice spacing[123] of the undeformed lattice shown in Fig. 10.29(a). Note that each bond length changes due to the surface fluctuations when elasticity is turned on.

If the simulation results such as τ are compared with experimental data, this a can be fixed to a suitable value.[124] In this paper, we are interested in the response of τ to isotropic and uniaxial strains under the variation of external magnetic field and temperature, and therefore we fix a to $a = 1$, $a = 0.75$, and $\xi^{\pm 1}a$ ($\xi^2 = 0.9, 1.1$ for $a = 1$), which are direction dependent, in the simulations.

The vertices inside the boundary are allowed to move three-dimensionally, while the BVs can move only in the 2D plane parallel to the boundary square with the PBC, or in other words, the BVs are moving into the in-plane directions under the condition that the vertices are connected to the opposite BVs, which are placed at the positions $\pm L_{1,2}$ from the original positions. To describe this periodic boundary condition in more detail, we plot in Fig. 10.29(b) a snapshot of a typical configuration after a sufficiently large number of MC updates, which will be described below. In Figs. 10.29(a) and 10.29(b), the vertices outside the BVs are the opposite BVs. Thus, the lattices used in the simulations are summarized as follows: (i) the BVs move only in the plane defined by the boundary square, (ii) the PBCs are imposed on the BVs.

We should comment on the reason why the above-mentioned PBC is imposed on the BVs. First of all, the boundary condition imposed on the surfaces is necessary for the calculation of the surface tension τ, which will be described below in detail. For this reason, the PBC assumed here is slightly different from the ordinary PBC, which is simply introduced to lower the boundary effects. Moreover, the PBC assumed here on the lattice should be connected or close to experimental ones, where the z direction is mechanically fixed at the boundary. Therefore, it is reasonable to assume that the BVs are prohibited at least from moving in the z direction. It is also possible to assume that the BVs are completely fixed for the evaluation of τ. Hence, the problem is whether the results depend on the boundary conditions or not. Although the case that the BVs are completely fixed is not examined, we expect that the results are independent of these boundary conditions, because the BVs move only slightly as we will see in the snapshots of lattices below.

Let us define the area A_p of the boundary frame without deformation by elastic effects, by using L_1, L_2, and a, as follows:

$$A_p = (L_1 - 1)(L_2 - 1)\sqrt{3}a^2/2, \quad a = 1. \tag{10.66}$$

Under the elastic effect, the membrane fluctuates out of the xy plane. However, we see that A_p is identical with the projected membrane area if the BVs are completely fixed on the boundary frame. A_p is not identical to the area of the projected membrane if the BVs are allowed to move in the xy plane at $z=0$ (see Fig. 10.29(b)). This boundary condition is mechanically imposed on materials to pin the membrane boundary in the $z = 0$ plane so that fluctuations in the z direction do not drift the whole system in the space during the simulation (see Fig. 10.30(a) for illustration). Such a mechanically pinning BC does not restrain microscopic in-plane movements of particles.

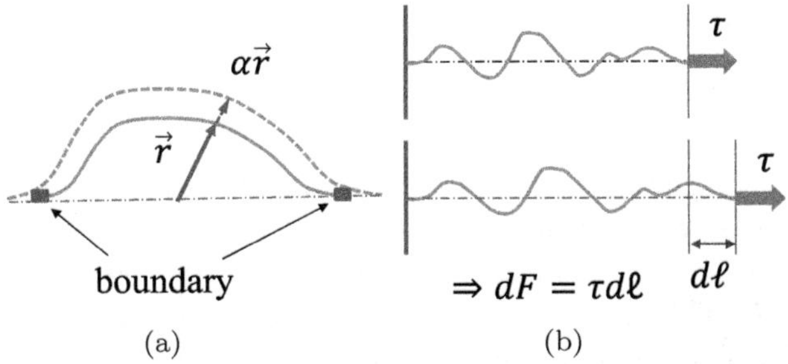

Fig. 10.30. (a) An illustration of a surface (denoted by $\mathbf{r}$) spanning a fixed boundary and its scaled surface (denoted by $\alpha\mathbf{r}$), where the boundary frame remains fixed, (b) illustrations for the energy increment ($dF = \tau d\ell$) accumulated in the surface by external tensile force τ. In this surface extension, the increment of projected area is given by $dA_p = d\ell$, where the side length is fixed to 1 and remains unchanged for simplicity.

10.5.1.2. *Skyrmion model*

We start with a continuous version of DM interaction.[40,41,46] On curved surfaces, the continuous Hamiltonian is given by

$$S_{\rm DM} = \int \sqrt{g} d^2 x g^{ab} \frac{\partial \mathbf{r}}{\partial x^a} \cdot \left(\sigma \times \frac{\partial \sigma}{\partial x^b} \right), \tag{10.67}$$

where $x^a (a = 1, 2)$ denotes a parameter or a local coordinate of the 2D surface, g is the determinant of the metric g_{ab}, and g^{ab} is its inverse. The symbol $\mathbf{r}(\in \mathbf{R}^3)$ denotes the surface position given by a three-component vector such as $\mathbf{r} = (X, Y, Z)$, where X, Y, Z are functions of local coordinates (x^1, x^2) such that $X(x^1, x^2), Y(x^1, x^2), Z(x^1, x^2)$. Therefore, the position vector $\mathbf{r}$ is mathematically considered to be a mapping from a two-dimensional parameter space M to $\mathbf{R}^3$ such that $\mathbf{r} : M \ni (x^1, x^2) \mapsto (X, Y, Z) \in \mathbf{R}^3$. This parameter space M is a manifold in general, and the components of the metric g_{ab} are functions on M. This metric tensor g_{ab} allows us to calculate the length of tangential vectors on M. For example, let $\partial\mathbf{r} = (\partial X, \partial Y, \partial Z)$, where each component is a tangential vector of M such as $\partial X = g^{ab}\partial_a X \partial_b$, which is the gradient vector of X, and $\partial_b = \partial/\partial x^b$ are the basis vectors of the tangential plane of M. From this, we have $|\partial\mathbf{r}|^2 = \partial\mathbf{r} \cdot \partial\mathbf{r} = (\partial X)^2 + (\partial Y)^2 + (\partial Z)^2 = (g^{ab}\partial_a X \partial_b)^2 + (g^{ab}\partial_a Y \partial_b)^2 + (g^{ab}\partial_a Z \partial_b)^2 = g^{ab}\partial_a \mathbf{r} \cdot \partial_b \mathbf{r}$, where the relations $\partial_a \cdot \partial_b = g_{ab}$ and $g^{ab}g^{cd}g_{bd} = g^{ac}$ are used. Thus, we find that the Gaussian bond potential S_1 is given by the length squares of $\partial\mathbf{r}$:

$$S_1 = \int dA|\partial\mathbf{r}|^2 = \int \sqrt{g}d^2xg^{ab}\frac{\partial\mathbf{r}}{\partial x^a}\cdot\frac{\partial\mathbf{r}}{\partial x^b}, \quad (10.68)$$

where $\int dA = \int \sqrt{g}d^2x$ denotes the area integral on M. The important point to note is that the combination of $\partial_a\mathbf{r}\cdot\partial_b\mathbf{r}$ with g^{ab} becomes a scalar function, which is invariant under arbitrary coordinate transformation on M. This is the reason why S_1 is meaningful as an energy functional for polymeric materials such as membranes. In the case of DM interaction, the spin components are also functions on M such that $\sigma(x^1,x^2)=(\sigma_X(x^1,x^2),\sigma_Y(x^1,x^2),\sigma_Z(x^1,x^2))$. From this, it is easy to see that $\partial\sigma_X,\partial\sigma_Y,\partial\sigma_Z$, which are components of $\partial\sigma$, are tangential vectors on M such that $\partial\sigma_X = g^{ab}\partial_a\sigma_X\partial_b$, for example. Using this symbol $\partial\sigma$, we find that the DM interaction can be written as $S_{\rm DM} = \int dA\partial\mathbf{r}\cdot(\sigma\times\partial\sigma) = \int \sqrt{g}d^2xg^{ab}\partial_a\mathbf{r}\cdot\left(\sigma\times\frac{\partial\sigma}{\partial x^b}\right)$, which is the same as the expression in Eq. (10.67). Thus, the combination of $\partial_a\mathbf{r}\cdot\left(\sigma\times\frac{\partial\sigma}{\partial x^b}\right)$ with g^{ab} is also a scalar function on M. Therefore, we find that $S_{\rm DM}$ is meaningful for an energy functional as well as S_1.

It is clear from the expression of $S_{\rm DM}$ in Eq. (10.67) that the surface shape and the spin are interacting with each other. Here, we consider the case of Euclidean metric $g_{ab}=\delta_{ab}$ for simplicity. Thus, we have

$$S_{\rm DM} = \int d^2x\frac{\partial\mathbf{r}}{\partial x^a}\cdot\left(\sigma\times\frac{\partial\sigma}{\partial x^a}\right), \quad (10.69)$$

Note also that the expression of $S_{\rm DM}$ in Eq. (10.69) is identical with $S'_{\rm DM} = \int d^2x\sigma\cdot(\nabla\times\sigma)$ if $\partial\mathbf{r}/\partial x^a$ in $S_{\rm DM}$ is regarded as the unit tangential vector along x^a direction. We should note that $\partial_a\mathbf{r}$ and $\sigma\times\partial_a\sigma$ in $S_{\rm DM}$ are vectors in $\mathbf{R}^3$, and the latter one is a pseudo-vector, which remains unchanged under the inversion $\mathbf{r}\to-\mathbf{r}$ in $\mathbf{R}^3$. Hence, $\sum_a\partial_a\mathbf{r}\cdot(\sigma\times\partial_a\sigma)$ or the energy $S_{\rm DM}$ is a pseudo-scalar in $\mathbf{R}^3$. For this reason, $S_{\rm DM}$ is invariant under an arbitrary rotation around the axis perpendicular to the lattice at least, because a pseudo-scalar shares the same property with a scalar under space rotations. This allows us to choose a rigid lattice to define the discrete Hamiltonian, which will be introduced below. Real materials also have a crystalline structure. In contrast, fluctuating lattices are not always connected to the crystalline lattice by any rotation. For this reason, it is interesting to study the skyrmion stability on fluctuating lattices. Moreover, an interaction of spins and lattices can be reflected in the frame tension if the lattice is treated as a dynamical variable as mentioned in Section 10.1.

To study skyrmions on the fluctuating lattices, we have to combine the skyrmion model and the elastic surface model.[100–105] If the two Hamiltonians are merged, the total Hamiltonian becomes slightly lengthy, but the definition itself is straightforward. Indeed, the Hamiltonian S is given by a linear combination of several terms such as

$$\begin{aligned} S(\mathbf{r},\sigma) &= S_1 + \kappa S_2 + \lambda S_F + \delta S_{\rm DM} + S_B, \\ S_1 &= \sum_{ij} \ell_{ij}^2, \quad S_2 = \sum_{ij} (1 - \mathbf{n}_i \cdot \mathbf{n}_j), \\ S_F &= \sum_{ij} (1 - \sigma_i \cdot \sigma_j), \quad S_{\rm DM} = \sum_{ij} \mathbf{e}_{ij} \cdot \sigma_i \times \sigma_j, \\ S_B &= \sum_i \sigma_i \cdot \mathbf{B}, \quad \mathbf{B} = (0, 0, B). \end{aligned} \tag{10.70}$$

The unit of energy is given by $k_B = 1$ (in k_BT), which we call simulation unit, where k_B and T are the Boltzmann constant and the temperature, respectively. The symbols $\mathbf{r}, \sigma$ in S denote the vertices and spins such that $\mathbf{r} = \{\mathbf{r}_1, \mathbf{r}_2, \ldots, \mathbf{r}_N\}$, $\sigma = \{\sigma_1, \sigma_2, \ldots, \sigma_N\}$, where $\mathbf{r}_i \in \mathbf{R}^3$ ($\mathbf{r}_i \in \mathbf{R}^2$ for BVs) and $\sigma_i \in S^2$ (unit sphere). The first and second terms are the discrete Hamiltonians of the surface model of Helfrich and Polyakov.[98,99] S_1 and S_2 are called Gaussian bond potential and bending energy, respectively, and $\ell_{ij} (= |\mathbf{r}_j - \mathbf{r}_i|)$ is the length of bond ij, and $\mathbf{n}_i$ and $\mathbf{n}_j$ in S_2 are the unit normal vectors of the triangles that share the bond ij. The parameter κ is the bending rigidity of the surface.[100–104] We should note that the first term S_1 depends on the size of projected area of lattice because S_1 is the sum of bond length squares of surface which spans the boundary of projected area A_p, and for this reason the model becomes dependent on the lattice spacing a. Indeed, for sufficiently large (small) a, fluctuations of the surface spanning the boundary are expected to be suppressed (enhanced). Here we should also emphasize that since the boundary is fixed during the simulations, A_p is not changed with the surface fluctuations because it depends only on a as seen in Eq. (10.66).

The third term λS_F is the energy for describing the FM interaction between two neighboring spins with the interaction strength λ. The fourth term $\delta S_{\rm DM}$ describes the DM interaction with the interaction coefficient δ, where $\mathbf{e}_{ij} (= (\mathbf{r}_j - \mathbf{r}_i)/|\mathbf{r}_j - \mathbf{r}_i|)$ is the unit tangential vector from $\mathbf{r}_i$ to $\mathbf{r}_j$. This vector $\mathbf{e}_{ij}$ corresponds to $\partial \mathbf{r}$ in the continuous $S_{\rm DM}$ in Eq. (10.67) and varies on the fluctuating lattices. It is also possible to use $\ell_{ij} (= \mathbf{r}_j - \mathbf{r}_i)$ for $\partial \mathbf{r}$; however, here we choose to use the unit length vector $\mathbf{e}_{ij}$ for $\partial \mathbf{r}$. The continuous expression $\sigma \times \partial_a \sigma$ can also be reduced to $\sigma_j \times \sigma_i$ in the discrete

$S_{\rm DM}$, because the differential $\partial_a\sigma$ is replaced by the difference $\sigma_j-\sigma_i$. Thus, the term $S_{\rm DM}$ plays a role in the interaction between σ and $\mathbf{r}$, the spins and the surface shape. The final term S_B is the energy for the external magnetic field $\mathbf{B}=(0,0,B)$, which is perpendicular to the plane of the BVs.

We note that effects of surface fluctuations on spins are taken into consideration only via the DM interaction energy $S_{\rm DM}$. The tangential vector $\mathbf{e}_{ij}$ in $S_{\rm DM}$ is deformed by the surface fluctuations as mentioned above. On the contrary, the FM interaction can also reflect surface fluctuations, because the magnetoelastic coupling is not negligible on the fluctuating surfaces due to the spin orbit coupling.[125] However, in the case of skyrmion deformations in a strained crystal, it is expected that the deformation of DM interaction plays an essential role, which is precisely checked in Ref. [126]. So, in this section, we simply use the ordinary undeformed S_F in Eq. (10.70). In other words, we neglect the effect of magnetoelastic coupling in the FM interaction and we study only effects of surface elasticity on spins via the DM interaction deformation.

The partition function is given by

$$Z(\beta,\kappa,\lambda,\delta,B)=\sum_\sigma\int\prod_{i=1}^{N-N_{\rm BD}}d\mathbf{r}_i\int\prod_{i=1}^{N_{\rm BD}}d\mathbf{r}_iC(A_p)\exp[-\beta S(\mathbf{r},\sigma)], \tag{10.71}$$

$$\beta=1/T,\quad(\Leftrightarrow k_B=1),$$

where $\sum_\sigma$ denotes the sum of all possible values of σ. The symbol β is the inverse temperature, and the simulation unit is defined by the relation $k_B=1$, and, hence, the temperature T is the reduced temperature. The symbols $\int\prod_i d\mathbf{r}_{i=1}^{N-N_{\rm BD}}$, $\int\prod_i d\mathbf{r}_{i=1}^{N_{\rm BD}}$ denote the 3D and 2D multiple integrations of the vertices inside and on the boundary, respectively, where $N_{\rm BD}$ is the total number of vertices on the boundary. The symbol $C(A_p)$ denotes a constraint on the integration of the boundary vertices so that the projected area remains fixed to A_p. This constraint is a function of $\mathbf{r}_1,\ldots,\mathbf{r}_{N_{\rm BD}}$. However this function $C(\mathbf{r}_1,\ldots,\mathbf{r}_{N_{\rm BD}};A_p)$ is too complex to write down with A_p, and for this reason, we use $C(A_p)$ for the short notation.[127]

10.5.1.3. *Frame tension*

As mentioned in Section 10.1, only strains can be imposed on the elastic lattices by fixing the edge lengths of the rectangular boundary frame.[127,128] The imposed strains uniquely determine the corresponding stresses due to the stress–strain correspondence such as the stress–strain diagram, at least in the elastic deformation region specific to the materials. From this correspondence, stresses are effectively imposed on the elastic lattices via strains.

The integration variable $\mathbf{r}$ in Z of Eq. (10.71) can be changed to $\mathbf{r} \to \alpha\mathbf{r}$, where α is called the scale parameter, and we have the scaled partition function $Z[\alpha; A_p(\alpha)]$ such that[127]

$$Z[\alpha; A_p(\alpha)] = \alpha^{3(N-N_B)+2N_B} \sum_\sigma \int \prod_{i=1}^{N-N_B} d\mathbf{r}_i \int \prod_{i=1}^{N_B} d\mathbf{r}_i C(A_p(\alpha)) \exp[-\beta S(\alpha\mathbf{r}, \sigma)],$$
$$S(\alpha\mathbf{r}, \sigma) = \alpha^2 S_1 + \kappa S_2 + \lambda S_F + \delta S_{\rm DM} + S_B. \tag{10.72}$$

The second argument $A_p(\alpha)$ in $Z[\alpha; A_p(\alpha)]$ is given by $A_p(\alpha) = \alpha^{-2}A_p$, which comes from the assumption that the projected area A_p remains fixed while the scale of the lattice is changed by α (Fig. 10.30(a)). The factor $\alpha^{3(N-N_B)+2N_B}$ on the right-hand side of Eq. (10.72) comes from the integration measures, and only S_1 changes to αS_1 in S because all other terms in S are scale independent. Since Z is independent of α, we have $Z[\alpha; A_p(\alpha)] = Z[1; A_p(1)]$. From this, we obtain $dZ[\alpha; A_p(\alpha)]/d\alpha|_{\alpha=1} = [\partial Z/\partial\alpha + (\partial Z/\partial A_p)(dA_p/d\alpha)]_{\alpha=1} = 0$. Dividing both sides of this equation by Z, we have

$$3N - N_B - 2\left[\sum_\sigma \int \prod_i d\mathbf{r}_i C(A_p)\beta S_1 \exp(-\beta S) + A_p \frac{\partial Z}{\partial A_p}\right] Z^{-1} = 0. \tag{10.73}$$

The expression $\partial Z/\partial A_p$ corresponds to the derivative of the constraint $C(A_p)$. To calculate $\partial Z/\partial A_p$, we assume for the surface with the projected area A_P that the free energy F is given by

$$F(A_p) = \tau \int_{A_0}^{A_p} dA, \tag{10.74}$$

where τ is called frame tension because the surface spans the fixed boundary frame with projected area A_p. We should note that the expression in Eq. (10.74) for the free energy can be called macroscopic energy. The other macroscopic energies such as bending energy corresponding to S_2 in Eq. (10.70) are not included in this free energy, because such bending energy is not explicitly dependent on A_p. On the contrary, the Hamiltonian S in Eq. (10.70) is a microscopic energy, which is defined by the sum of local interaction energy between the neighboring vertices or triangles i and j. In this microscopic formulation, the function $C(A_p)$ is a constraint on the multiple integration and is too complex to be expressed with A_p as mentioned above. However, this constraint on A_p turns to be very simple in the macroscopic energy in Eq. (10.74). For this reason, the free energy in Eq. (10.74) is used to calculate the frame tension τ as a macroscopic physical quantity from the microscopic perspective.[127] Thus, by using $Z = \exp(-\beta F)$, we have $\partial Z/\partial A_p = -\beta\tau Z$. Using the symbol $\langle S_1\rangle$ for the first term of $[\cdots]$ on the left-hand side of Eq. (10.73), we obtain $\tau = [2\beta\langle S_1\rangle - (3N - N_B)]/(2\beta A_p)$, and hence we have the formula for the frame tension τ such that[127]

$$\tau = \frac{\langle S_1\rangle - (3N - N_B)T}{2A_p}. \tag{10.75}$$

Note that $\langle S_1\rangle$ implicitly depends on T. Note also that this τ can be written as $\tau = [\langle S_1\rangle - \langle S_1^0\rangle]/A_p$, where $\langle S_1^0\rangle$ is $\langle S_1\rangle$ for $\tau = 0$. In the calculation of τ, we use the formula in Eq. (10.75), in which $\langle S_1^0\rangle$ is not necessary.

Note that the projected area A_p used in τ is different from the real projected area of the lattice, though the difference of A_p and the real projected area of the lattice is expected to be very small. The projected area A_p of the boundary frame is the only area that is connected with experimental measurements of τ. This comes from the fact that the elastic energy by tensile force accumulated in the surface is given by "force $\times$ increment of the projected area" (Fig. 10.30(b)). The important point to note is that the A_p in Eq. (10.74) is simply the area of boundary frame.

10.5.1.4. *Monte Carlo technique*

The spins σ and the vertex positions $\mathbf{r}$ are updated using the standard Metropolis MC technique.[129,130] A new σ_i' at the vertex i, generated independently of the old σ_i by using three different uniform random numbers, is accepted with the probability $\mathrm{Min}[1, \exp -\Delta S]$, where $\Delta S = S(\mathrm{new}) - S(\mathrm{old})$. We should note that the three different uniform random numbers $\sigma_{x,y,z} \in (-0.5, 0.5]$ are generated with the constraint $\sigma_x^2 + \sigma_y^2 + \sigma_z^2 \leq 0.25$. This constraint makes the point $(\sigma_x, \sigma_y, \sigma_z)$ uniform in the ball of radius 0.5. Then, the vector $(\sigma_x, \sigma_y, \sigma_z)$ is normalized such that $\sigma_x^2 + \sigma_y^2 + \sigma_z^2 = 1$. Thus, the distribution of this unit vector $(\sigma_x, \sigma_y, \sigma_z)$ is expected to be uniform on the unit sphere. Because of this independence of σ_i' on the old σ_i, the rate of acceptance is not controllable. A new $\mathbf{r}_i' = \mathbf{r}_i + \Delta\mathbf{r}$ is also updated in the probability $\mathrm{Min}[1, \exp -\Delta S]$. In this $\mathbf{r}_i'$, the symbol $\Delta\mathbf{r}$ denotes a random three dimensional vector inside a small sphere of radius R_0, which is determined such that the rate of acceptance for $\mathbf{r}_i'$ is approximately 50%. One Monte Carlo sweep (MCS) consists of simultaneously by updating $\mathbf{r}$ and σ at a given spin and repeating this for all spins.

These MC updates for σ and $\mathbf{r}$ are started after a generation of skyrmion configuration, which is the GS obtained by the technique described in Ref. [85]. This GS is generated by the iteration for σ, and during this iteration, the variables $\mathbf{r}$ are fixed to the initial configuration, like in the snapshot in Fig. 10.29(a). Therefore, the GS is influenced by both thermal fluctuations of spins and surface fluctuations after the start of MC updates for σ and $\mathbf{r}$. However, since the GS is determined depending on the parameters λ, δ, and B, it is close to the equilibrium state for sufficiently low T and sufficiently large κ. The MC simulation starting with the GS configuration is called "slow heating of the system from the GS".

This procedure is used in strongly degenerate systems, strongly disordered systems, systems with many meta-stable states and systems with competing interactions if we know the GS. Starting with initial random configurations, namely cooling the system from high T, except for simple systems with long-range ordering and small unit cell such as ferromagnets, it is well known that MC simulations encounter problems to equilibrate the system at low temperatures. This is not a problem of simulation time. This comes from the fact that the system is often stuck in meta-stable states at low T. One way to get rid of this is to determine the GS by other methods such as the steepest descent method used in Ref. [85] and slowly heat the system from its GS. Even in this case, the relaxation time is very long. This happens in strongly non-uniform systems such as spin glasses or systems with huge unit cells such as skyrmion crystals.[85]

In general, when the GS has several degenerate configurations, only one of them is chosen by the system if we are able to cool the system down to $T = 0$. But this is impossible for complex systems as said above. What we do is to choose a ground state and make a slow heating. To see if the system is equilibrated or not, we have to make a finite-time scaling to deduce properties at the infinite time. This is very similar in spirit with the finite-size scaling used to deduce properties at the infinite crystal size. We have previously performed a finite-time scaling for the 2D skyrmion crystal of the rigid lattice.[85] We have seen that skyrmions need much more than 10^6 MC steps per spin to relax to equilibrium. The order parameter follows a stretched exponential law as in spin glasses and is stabilized at non-zero values for $T < T_c$ at the infinite time. If there is no skyrmion stability, we would not have non-zero values of the order parameter for $T < T_c$ at the infinite time. Based on the work of Ref. [85] on the same model without elasticity, we have taken sufficient run times as indicated below. We note that in the present work, as soon as T is not zero, the non-uniform deformation of the lattice excludes invariance by global rotation.

The lattice size for the simulations is $N = 10^4 (\Leftrightarrow L = 100)$. This lattice size is relatively small compared to those (lattices up to $L = 800$) in Ref. [85], where the phase transition between the skyrmion and paramagnetic phases is studied. However, we do not go into details in the phase transitions in this paper, and therefore, the lattice size $N = 10^4$ is considered to be sufficient. The total number of MCSs is $3 \times 10^8 \times 2.9 \times 10^9$ including $1 \times 10^7 \times 2.3 \times 10^9$ thermalization MCSs, and physical quantities are calculated for every 1000 MCSs. This total number of MCSs depends on the fluctuations of lattice, and it is small (large) for simulations on the lattices with small (large) fluctuations. The reason for this dependence is that the skyrmion configuration is almost uninfluenced (considerably influenced) if the lattice fluctuations are relatively small (large).

10.5.2. *Results under isotropic strain*

10.5.2.1. *Phase diagram and snapshots*

Our focus is on the frame tension, which is expected to be influenced by spin configurations because of the DM interaction as mentioned in Section 10.5.1.2. To study this, we first show a phase diagram on the B–T plane in Fig. 10.31, where snapshots obtained at several points are also shown. The parameters a, λ, and κ are fixed to $a = 1$, $\lambda = 0.5$, and $\kappa = 25$. In the figure, the letters ferro, skx, and str denote the FM, skyrmion crystal, and stripe phases, and sk-fe and st-sk denote the intermediate phases between these three different phases. The stripe pattern is identical with that observed in Refs. [48, 85, 86] and [68]. The typical configuration of skx is shown in Fig. 10.31(a) at $T = 0.05$, which is sufficiently low. The criteria for distinguishing one from another is as follows: along the vertical dashed line at $T = 0.3$, we start from the skx phase at $B = -1$ (Fig. 10.31(b)) and go downward, then skyrmions start to connect and form oblong islands.

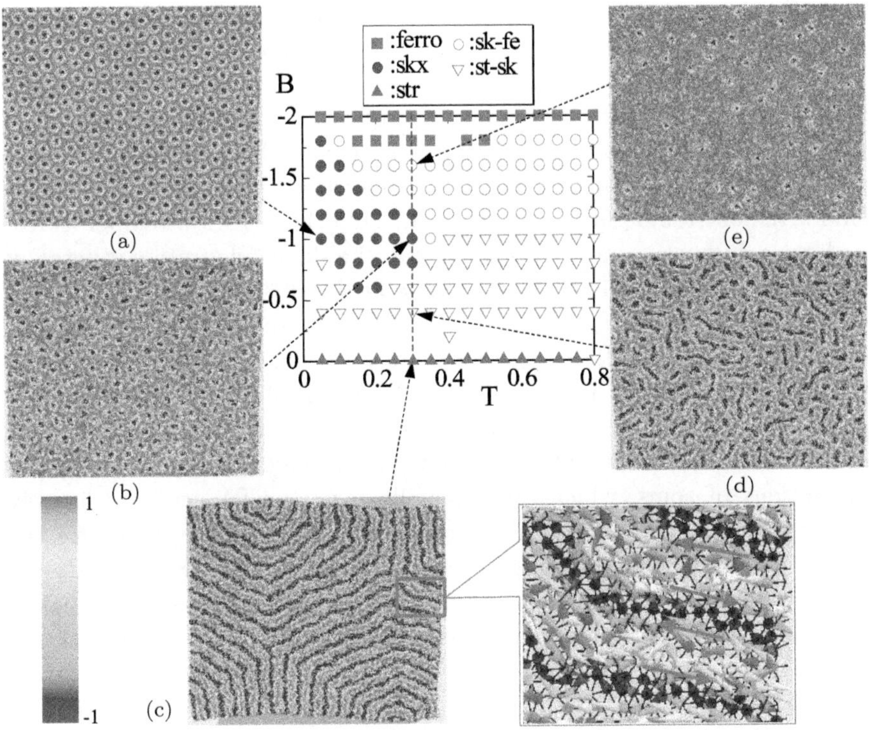

Fig. 10.31. A phase diagram on the surface B vs. T, with snapshots at (a) skx, (b) skx, (c) str, (d) st-sk, and (e) sk-fe phases, and the color variation corresponds to σ_z. The results are obtained on the lattice of $a=1$ and with the parameters $\kappa=25$, $\lambda=0.5$, and $\delta=1$. The vertical dashed line is drawn at $T=0.3$, where the skx phase terminates when T is increased. The variable σ is drawn by a cone (cylinder) in the enlarged (original) snapshot. The arrows in the enlarged snapshot denote the direction of σ. The lattice size is $N=10^4$.

This phase is the st-sk phase denoted by $\bigtriangledown$. The length of islands increases when $|B|$ decreases, and for sufficiently small $|B|$, the islands are connected to each other like in Fig. 10.31(c), which we call the stripe phase. In the enlarged snapshot of Fig. 10.31(c), we see that the direction of spins along the stripes changes alternatively.[48,68,86] The configuration typical to the st-sk is shown in Fig. 10.31(d) at $B=-0.4$. On the contrary, if we start from the skx phase at $B=-1$ and go upward along the dashed line, we find that the total number of skyrmions decreases, and this is the sk-fe phase like in the snapshot of Fig. 10.31(e), where $B=-1.6$. For sufficiently large $|B|$, the skyrmion configuration completely disappears, and the phase changes from sk-fe to ferro denoted by the solid square symbol.

The main purpose of this phase diagram is to find where the skyrmion phase appears, and for this reason, the diagram includes empty regions without symbols where no simulation is performed because skyrmions are not expected. The skyrmion phase appears for the intermediate region of B and for the low-temperature region as expected there. At low temperatures, the lattice fluctuations are suppressed, and the skx configuration is actually smooth at $T = 0.05$ (Fig.10.31(a)), while at $T = 0.3$, which is the highest temperature at which the skx can be seen, and the skx configuration is relatively fluctuated (Fig. 10.31(b)). For $T = 0.4$, no skx configuration appears. In the regions of $T(> 0.3)$ close to $T = 0.3$, the initial ground states are still skx; however, these skx states disappear after a sufficiently large number of MC sweeps for the updates of both σ and $\mathbf{r}$. This is due to the thermal fluctuations of spins, which are originally expected for such a relatively large T region. The surface fluctuations are not always the reason for the destruction of skx configurations, because, at the same temperature $T = 0.3$ on the rigid lattice, the skx phase is separated from the sk-fe phase (using the terminology in this paper).[85] We should note that the phase diagram also indicates the stability of skyrmions at a sufficiently low temperature; however, this is independent of whether the lattice is fluctuated or not, because skyrmions in the same 2D model are stable in the region $T \leq 0.3$ on the rigid lattice.[85] As mentioned above, different kinds of interactions can give rise to the birth of skyrmions. Skyrmions in our model come from the competition of only two interactions, $S_{\rm DM}$ and S_F, in the ground state. That is the reason why our skyrmions appear in such a low-temperature region, and there is only the skyrmion phase below the paramagnetic phase. In other models and in experiments, various kinds of interactions (more than two) may compete and can give rise to various phases with increasing T, the skyrmion phase is one of these phases and can be found at high temperature.

The frame tension τ is obtained by fixing the projected area A_p of the boundary as in Eq. (10.75), and A_p is determined by the lattice spacing a when the total number of vertices is fixed. Therefore, it is interesting to see the dependence of τ on a, and, for this purpose, we change the lattice spacing from $a = 1$ to $a = 0.75$, and, as a result, the side length of the lattice is also reduced by 3/4 from the lattice used in Fig. 10.31. It is also interesting to see whether the transition temperature $T = 0.3$, between the

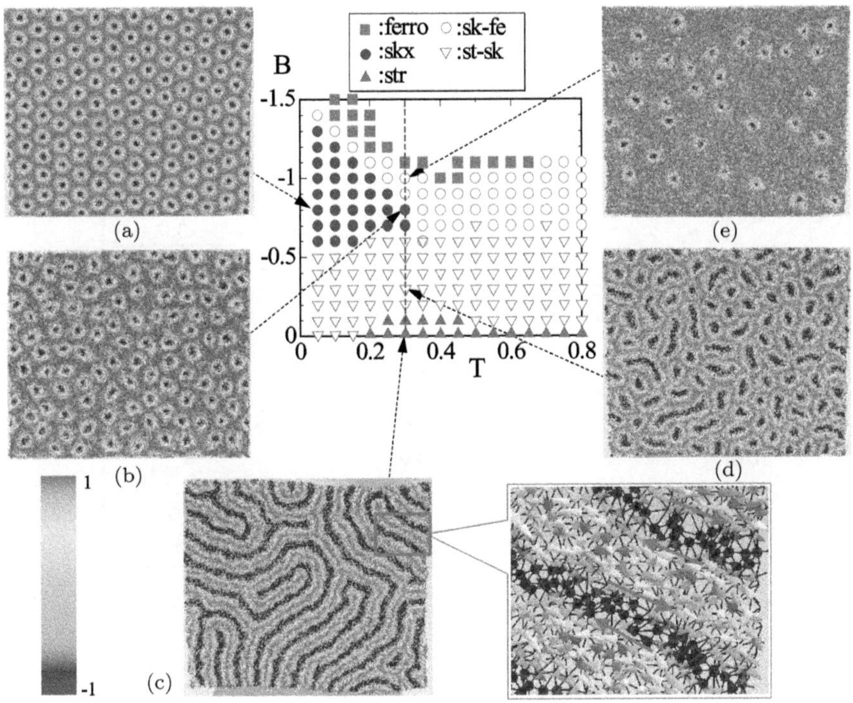

Fig. 10.32. A phase diagram on the surface B vs. T, with snapshots at (a) skx, (b) skx, (c) str, (d) st-sk, and (e) sk-fe phases, and the color variation corresponds to σ_z. The results are obtained on the lattice of $a=0.75$ and with the parameters $\kappa=15$, $\lambda=1$, and $\delta=1$. The vertical dashed line is drawn at $T=0.3$, where the skx phase terminates when T is increased. The lattice size is $N=10^4$.

skx and sk-fe phases in Fig. 10.31, changes depending on the lattice spacing a. To further increase the surface fluctuations, we also reduce the bending rigidity to $\kappa=15$ from $\kappa=25$ assumed for Fig. 10.31.

We find from Fig. 10.32 that the phase structure is almost the same as that in Fig. 10.31. The temperature separating the skx and sk-fe phases remains the same at $T=0.3$. This implies that the surface fluctuation, if it is small enough, does not strongly influence the phase structure of skyrmions.

10.5.2.2. *Frame tension and interaction energies*

Now we plot the frame tension τ vs. the magnetic field B in Figs. 10.33(a) and 10.33(b). The data are obtained along the dashed lines on the

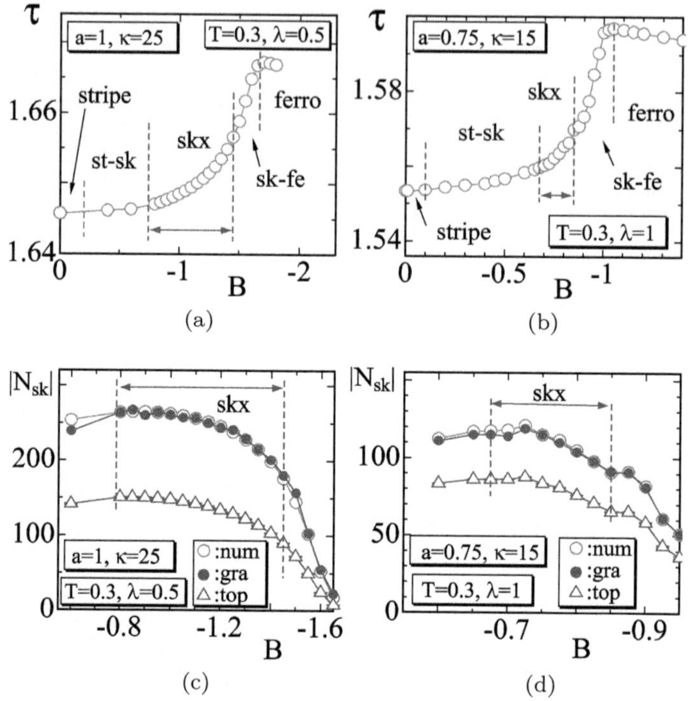

Fig. 10.33. (a) and (b) The frame tension τ vs. B obtained along the dashed line on the phase diagram in Figs. 10.31 and 10.32, (c) and (d) the total number of skyrmions $|N_{\text{sk}}|$ in the skx phase obtained along the same dashed lines, where $|N_{\text{sk}}|$ is obtained by the three different techniques "num", "gra", and "top". The numerical technique "num" denoted by the symbol (◯) is not always correct outside the skx phase. Inside the skx phase in both (c) and (d), $|N_{\text{sk}}|$ increases with decreasing τ ($\Leftrightarrow$ increasing p) except the boundary region close to the st-sk in (d).

phase diagrams in Figs. 10.31 and 10.32. The vertical dashed lines on Figs. 10.33(a) and 10.33(b) represent the phase boundaries. As we find from the figures, the value of τ in Fig. 10.33(a) is larger than that in Fig. 10.33(b). This is because, as explained in the previous subsection, the projected area of the lattice in Fig. 10.33(a) is larger than that in Fig. 10.33(b), and consequently, the surface in Fig. 10.33(a) is relatively expanded and smooth compared with that in Fig. 10.33(b). We also find in both cases that τ increases with increasing $|B|$ or equivalently τ decreases with decreasing $|B|$ in the skx phase and also in the sk-fe phase.

The total number $|N_{\text{sk}}|$ of skyrmions plotted in Figs. 10.33(c) and 10.33(d) is calculated/measured by three different techniques.

The first one is to calculate the topological skyrmion number

$$N_{\rm sk} = \frac{1}{4\pi}\int d^2x\sigma\cdot\frac{\partial\sigma}{\partial x_1}\times\frac{\partial\sigma}{\partial x_2} \tag{10.76}$$

numerically on triangulated surfaces. The sign (plus or minus) of $N_{\rm sk}$ depends on that of B_z, and hence, we use the absolute symbol $|\ |$ for the total number of skyrmions. The discrete form of $N_{\rm sk}$ in Eq. (10.76) is expressed by

$$N_{\rm sk} = \frac{1}{4\pi}\sum_{\Delta}(\sigma_2-\sigma_1)\times(\sigma_3-\sigma_1), \tag{10.77}$$

where σ_i denotes σ at the vertex i of the triangle Δ (Fig. 10.34(a)). In this discretization, the integration $\int d^2x$ and derivatives $\partial_i\sigma$ are simply replaced by $\int d^2x \to \sum_\Delta$ and $\partial_1\sigma \to \sigma_2-\sigma_1$, $\partial_2\sigma \to \sigma_3-\sigma_1$. We should note that there are three possible local-coordinate origins in one triangle, and hence, summing over the contributions to $N_{\rm sk}$ from these three coordinates with the numerical factor 1/3, we calculate $N_{\rm sk}$ in Eq. (10.77). The obtained value is denoted by "top" on Figs. 10.33(c) and 10.33(d). No numerical cutoff is necessary for this calculation of $N_{\rm sk}$ by Eq. (10.77).

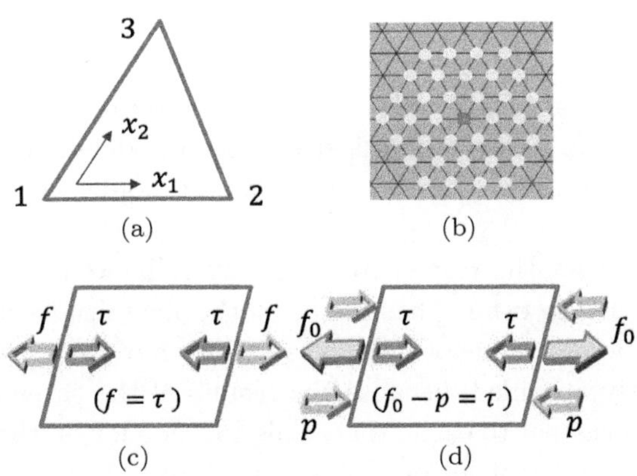

Fig. 10.34. (a) A triangle 123 with local coordinate axes x_1 and x_2; (b) A vertex (solid square) and its third-nearest neighbor vertices (solid circles) for counting the total number of skyrmions. (c) External force f and material response τ are balanced: if f is increased (decreased), then τ is increased (decreased) and vice versa. (d) f can be replaced by $f_0 - p$ under constant f_0 using a pressure p: if p is increased (decreased), then τ is decreased (increased) and vice versa. Units of these forces τ, f, and p are assumed to be the same.

The second technique to calculate $N_{\rm sk}$ is as follows: in the case $B_z < 0$, we assume that there uniquely exists a vertex i such that $\sigma_i^z < \sigma_j^z$ for all vertices $j(\neq i)$ inside the range up to three-nearest neighbors in the skx phase (Fig. 10.34(b)). Here, the condition $\sigma_i^z < 0$ is imposed on the vertex i. This assumption for the uniqueness of i for identifying the minimum σ_z is reasonable, because the spins in the skyrmion configuration have values uniformly on the unit sphere. Thus, we expect that the vertices determined by this technique are in one-to-one correspondence with skyrmions. Therefore, $|N_{\rm sk}|$ is obtained by counting these uniquely determined vertices. The value of $|N_{sk}|$ obtained by this numerical technique is denoted by "num" on the figures. The third technique to obtain $|N_{\rm sk}|$ is a graphical technique, which uses a graphics software. The detailed information of this graphical technique is written in the appendix of Ref. [97]. The corresponding data plotted on Figs. 10.33(c) and 10.33(d) are denoted by "gra".

The results obtained by the techniques "num" and "gra" are almost the same at least in the skx and sk-fe phases. Moreover, from Fig. 10.33(c), we find that the $|N_{sk}|$ obtained by "num" and "gra" decrease to $|N_{sk}|\to 0$ for sufficiently large $|B_z|$ close to the ferro phase. This implies that "num" and "gra" correctly count the skyrmions except the stripe phase, where $|N_{sk}|$ of "num" is slightly different from that of "gra" (see Fig. 10.33(c)). Indeed, $|N_{sk}|$ obtained by both "gra" and "num" are unreliable in the st-sk phase because a long island is counted as skyrmions by these techniques. On the contrary, the results obtained by the technique "top" are smaller than those obtained by the other two techniques. This error observed in the technique "top" is simply due to the numerical errors caused by the discretization technique described above, though this discretization is simple enough and straightforward.

We consider that the results by "top" are reliable not only in the skx phase but also in the other phases except the magnitude of $|N_{sk}|$. Thus, apart from the precise value of $|N_{sk}|$, the variation of $|N_{sk}|$ vs. B in the skx phase is correctly obtained, because the results of the three techniques on this point are consistent to each other. This consistency on the the variation of $|N_{sk}|$ vs. B is sufficient for our present purpose.

To compare the results of τ with the experimental data in Ref. [118], we should replace τ by the corresponding external pressure p, because the experimental data are expressed by using pressure. We should remind ourselves of the fact that τ is a response of the material to the applied external pressure p. Therefore, the replacement of τ with p is possible using the action–reaction principle or the equilibrium principle of forces. Recalling that τ is balanced with an external mechanical force f applied to the boundary in this case, we understand that the positive τ implies that the corresponding external force f is a tensile force (Fig. 10.34(b)). Units of f and τ are assumed to be the same (this means that τL is written as τ for simplicity). Moreover, a decrease of the tensile force f can be understood as a result of the fact that a small pressure $p(>0)$ is applied in addition to the constant tensile force $f_0(>0)$ such that $f = f_0 - p$, where the unit of p is also assumed to be same as that of f (Fig. 10.34(c)). Thus, the increment (decrement) of τ is equivalent with the decrement (increment) of p. To summarize, the frame tension τ is identified with the external tensile force f, and this f can be replaced by $-p$ using the constant f_0 such that $f = f_0 - p$.

Using this p, the observed fact that the decrease of τ in the skx phase as $|B|$ decreases (from the right to left on the B axis) in Figs. 10.33(a) and (b) should be understood such that an increase of p is caused by a decrease of $|B|$. At the same time, we find from Figs. 10.33(c) and 10.33(d) that $N_{\rm sk}$ increases with decreasing $|B|$ in the relatively large $|B|$ region in the skx phase indicated by the arrows ($\leftrightarrow$). In this sense, it is also understood that the sk-fe phase is changed to the skx phase by an increment of p, which is caused by a decrease of $|B|$. To summarize, our results show that $N_{\rm sk}$ increases with increasing p ($\leftrightarrow$ decreasing τ) and a topological phase transition from the sk-fe to skx also occurs when p is increased. Thus, we find that this numerically observed phenomenon is qualitatively consistent with the experimentally observed fact that skyrmions are created by a lateral pressure to a 3D material.[118]

The FM energy S_F/N_B vs. B and the DM interaction $S_{\rm DM}$ vs. B are plotted in Figs. 10.35(a) and 10.35(b), and 10.35(c) and 10.35(d), where N_B is the total number of bonds. We find that both energies rapidly change in the sk-fe phase or at the boundary between the skx and sk-fe phases.

Next, the convergence speed of simulations is discussed. In Fig. 10.36(a), we plot $|N_{sk}|$ vs. MCS obtained at $a=1$ and $\kappa=25$, where $|N_{sk}|$ denoted by "top" on the figures indicates that $|N_{sk}|$ is calculated by the discrete topological charge in Eq. (10.77). Note that the $|N_{sk}|$ calculated by "num" shows the same behavior in their variations as the one plotted in Figs. 10.36(a) and 10.36(b), though the values $|N_{sk}|$ themselves are different from each other as demonstrated in Figs. 10.33(c) and 10.33(d). Therefore, we use $|N_{sk}|$ by "top" vs. MCS to find the thermalization MCS.

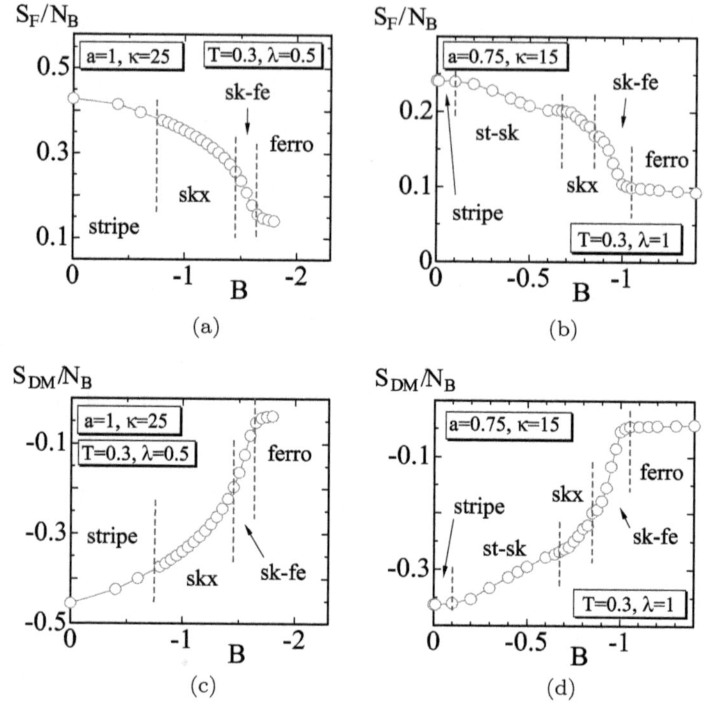

Fig. 10.35. (a) and (b) The FM energy S_F/N_B vs. B obtained along the dashed line on the phase diagram in Figs. 10.31 and 10.32, (c) and (d) the DM interaction energy S_{DM} along the same dashed lines.

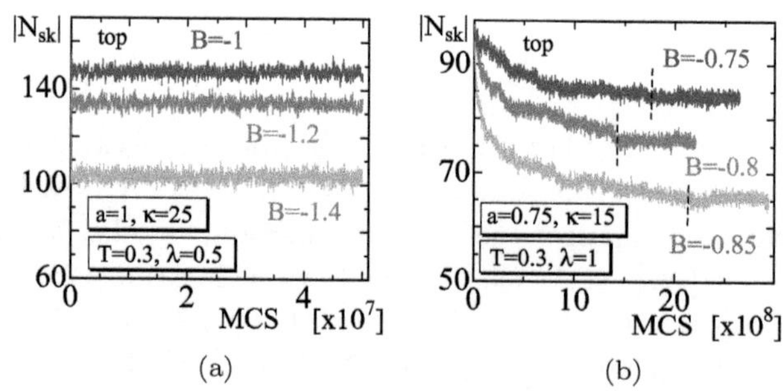

Fig. 10.36. The total skyrmion number $|N_{sk}|$ vs. MCS for (a) $a = 1$ and $\kappa = 25$ and (b) $a = 0.75$ and $\kappa = 15$, where $|N_{sk}|$ is the discrete topological charge calculated by Eq.(10.77). The convergence speed of $|N_{sk}|$ in (b) is relatively low compared to that in (a) and depends on the magnetic field B, where $B = -0.85$ is close to the phase boundaries between the skx and sk-fe phases.

From these plots in Figs. 10.36(a) and 10.36(b), we find that the skyrmion configurations remain almost unchanged during MC updates at least in the case of $a=1$ and $\kappa=25$, where the initial configurations immediately change to those equilibrium configurations in the early stage of MC iterations. In contrast, we find that the convergence is very slow for $a=0.75$ and $\kappa=15$ from the plots in Fig. 10.36(b). In this case, the surface fluctuation is relatively large and destroys the skyrmion configurations; however, the skx phase becomes stable after a sufficiently large number of MCSs. This implies that the skx phase is stable even on surfaces where the surface fluctuation is considerably large compared to the case $a = 1$ and $\kappa = 25$. The total number of MCS for thermalization is 1×10^7, which is fixed independent of B_z, for $a = 1$ and $\kappa = 25$, while it is determined depending on B_z for $a = 0.75$ and $\kappa = 15$. The vertical dashed lines denote the thermalization MCSs in Fig. 10.36(b). Since the configuration for $B=-0.85$ is close to the phase boundary between the skx and sk-fe phases, the convergent speed is relatively slow compared to the other two in Fig. 10.36(b).

10.5.3. *Results under uniaxial strains*

It is also interesting to see that the results are obtained under uniaxial stresses, although the 2D simulation results cannot always be compared to those of experiments where 3D materials are always targeted. To realize the uniaxial stress condition, we deform the edge length of the boundary by changing the lattice spacing a to be direction dependent such that $(a_1, a_2)=(a/\xi, a\xi)$ as mentioned in Section 10.1. We should note that this constraint can be called "uniaxial strains". Indeed, by this modification of lattice spacing from a to a_1 (a_2) in L_1 (L_2) direction in Eq. (10.65), we effectively have the modified edge length in the unit of a such that

$$(\xi^{-1}L_1, \xi L_2). \tag{10.78}$$

We should note that the projected area remains unchanged under this modification. It is more straightforward to use these effective edge lengths $(\xi^{-1}L_1, \xi L_2)$ for the direction-dependent lattice spacing (a_1, a_2), because the lattice deformation by uniaxial strains is simply expressed by ξ in this expression. Note also that both L_1 and L_2 originally denote the total number of lattice points, and the expression in Eq. (10.78) has the meaning of edge lengths only when a is multiplied.

The condition $\xi<1$ corresponds to the case that the edge length $\xi^{-1}L_1$ (ξL_2) is longer (shorter) than its original L_1 (L_2), and the corresponding

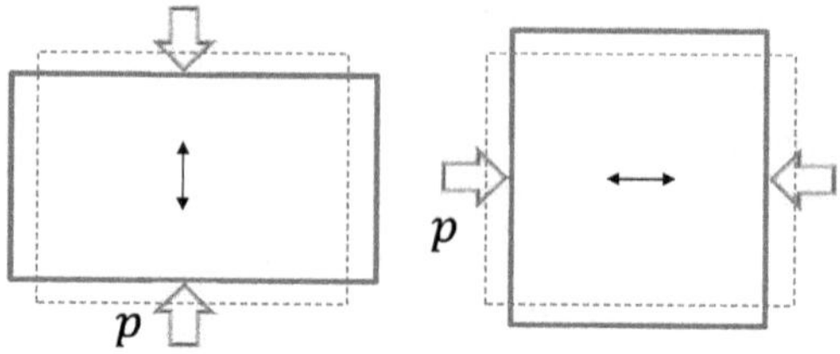

Fig. 10.37. A boundary shape corresponding to the constraints for (a) $\xi < 1$ and (b) $\xi > 1$ in Eq. (10.78). The shapes of boundary in (a), (b) correspond to the conditions that additional stresses p are applied along L_2 (↕) and L_1 (↔) directions, respectively. The "strain relief" direction is same as the p direction in both (a) and (b).

tensile force along L_1 (L_2) direction is expected to be smaller (larger) than that along L_2 (L_1) direction. Therefore, by using the pressure p in Fig. 10.34(d), we understand that $\xi<1$ ($\xi>1$) corresponds to the condition that a uniaxial stress p is applied along the L_2 (L_1) direction as shown in Figs. 10.37(a) and 10.37(b).

Note also that the "strain relief" direction is identical with the p direction in both Figs. 10.37(a) and 10.37(b). The reason is as follows: The surface is expanded by a strong external force f at the boundary frame in both L_1 and L_2 directions as described in Section 10.5.1.3 (see also Fig. 10.34(c)). Therefore, if an additional pressure p is applied to the L_2 direction, the side length along the L_2 direction is slightly reduced like in Fig. 10.37(a), and, as a consequence, the corresponding stress τ is also "relieved" such that $\tau = f - p$.

In Fig. 10.38, we show snapshots obtained under $\kappa = 25$, $T = 0.3$, $\lambda = 0.5$, which are the same as those for Fig. 10.31. The parameter ξ is fixed to $\xi^2 = 0.9 (\Leftrightarrow \xi \simeq 0.9487)$ for Figs. 10.38(a)–10.38(d) (upper row) and $\xi^2 = 1.1 (\Leftrightarrow \xi \simeq 1.049)$ for Figs. 10.38(e)–10.38(h) (lower row). The four snapshots in the upper (lower) row are obtained under $B = -1$, $B = -1$, $B = 0$, $B = -0.4$ ($B = -0.8$, $B = -0.8$, $B = 0$, $B = -0.3$), which correspond to Figs. 10.31(a)–10.31(d) (Figs.10.32(a)–10.32(d)). As mentioned above, $\xi^2 = 0.9$ corresponds to an additional uniaxial stress being applied to L_2 ($\Leftrightarrow$ vertical (↕)) direction, and we find from Figs. 10.38(a) and 10.38(b) that the skyrmion phase is not influenced by this uniaxial stress, while the stripe domains including islands align along this direction. In contrast, $\xi^2 = 1.1$ corresponds to an additional uniaxial stress being applied in the L_1 ($\Leftrightarrow$ horizontal (↔)) direction, and we find that the results are consistent with those

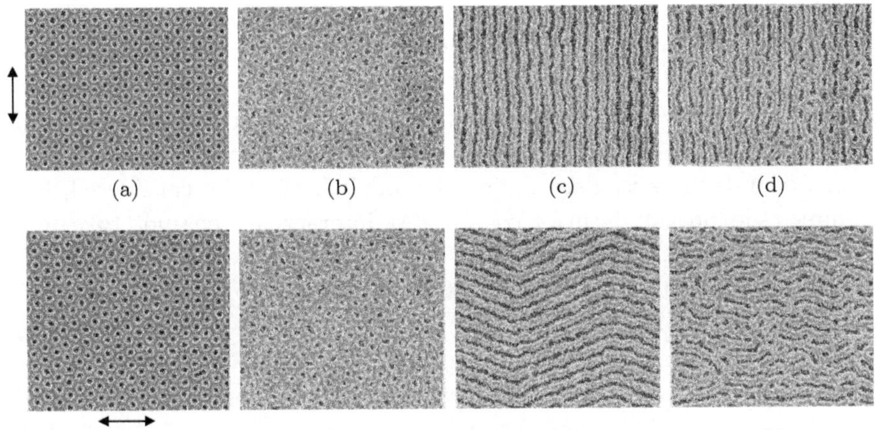

Fig. 10.38. Snapshots of surfaces, where an effective stress is applied to the directions (↕) for (a)–(d) and (↔) for (e)–(h). The parameter ξ in Eq. (10.78) is $\xi^2=0.9(\Leftrightarrow \xi\simeq 0.9487)$ for (a)–(d) and $\xi^2=1.1(\Leftrightarrow \xi\simeq 1.049)$ for (e)–(h). Both (a)–(d) and (e)–(h) correspond to Figs. 10.31(a)–10.31(d), where the lattice spacing is $a=1$ and the parameters are $\kappa=25$, $\lambda=0.5$, and $\delta=1$.

in the case for L_2 direction. Indeed, we see no influence on the skyrmion phase (Figs. 10.38(a) and 10.38(e)) except one pair of skyrmions merged in the down-right corner of Fig. 10.38(e), the configuration of which is only slightly influenced by the strain because it is obtained at the boundary between skx and st-sk phases. The stripe domains almost align along the stress direction (Figs. 10.38(c), 10.38(d) and 10.38(g), 10.38(h)). Almost the same results are obtained on rigid lattices with uniaxial strain conditions. Therefore, we find no remarkable influence of lattice fluctuations on the skyrmion configurations.

The stripe directions in Figs. 10.38(c), 10.38(d) and 10.38(g), 10.38(h) are different from the experimental results in Ref. [131]. One possible explanation is that skyrmions in the experimental material may have another physical origin than the DM interaction. We know that skyrmions can be generated from a DM interaction as in the present model, by the frustration in a triangular antiferromagnet, by a long-range dipole–dipole interaction, by different kinds of anisotropies. The response of the system to an external parameter depends on the physical origin generating skyrmions. Different materials have different origins of skyrmions; our model does not certainly correspond to materials used in Ref. [131]. The magnetoelastic coupling included in $S_{\rm DM}$ is isotropic and not always reflected in the responses of stripe domains under the uniaxial strains, contrary to the experimental result in Ref. [126], where it is reported that the anisotropic deformation of skyrmions under uniaxial strains is caused by a direction-dependent DM interaction coefficient. Again, we emphasize that different materials have different kinds of skyrmions. Different models have to be considered. For example, the present form of S_F for FM interaction does not take into account the lattice deformation.

We should note also that the stripe pattern in Fig. 10.38(g) is different from the zigzag spin spiral observed in the Fe double layer by Hsu *et al.*[132,133] In their experiment, the influence of uniaxial strain relief on the spin spiral was studied by STM measurements, and the authors reported that the $\mathbf{q}$ vector of the zigzag spin spiral is perpendicular to the "strain relief" direction, which is parallel to the zigzag pattern. The "strain relief" direction in our 2D model is also parallel to the stripe direction as mentioned above, as was the case of the zigzag pattern in their analysis. However, in their analysis, the direction of spins is vertical to the stripe direction, while in our 2D calculation the spin direction is parallel to the stripe direction or vertical to the 2D plane.

It is also interesting to see the lattice structures, although these do not always correspond to those of real materials. The snapshots of lattices are shown by reducing the size of the symbol (cone) for σ in Figs. 10.39(a)–10.39(h), which correspond to the top-right corner of the spin configurations in Figs. 10.38(a)–10.38(h). We find that at low temperature $T=0.05$, the lattices (Figs. 10.39(a) and 10.39(e)) are relatively close to the regular lattice as expected, while the other lattices considerably fluctuate because the temperature $T = 0.3$ is relatively high. These observations indicate that the fluctuation of lattices mainly depends on the temperature. More precisely, the three different straight (=horizontal and two oblique) lines are almost straight and the spaces or separations between them are almost uniform in Figs. 10.39(a) and 10.39(e), while in the other snapshots these lines deviate from the straight line and the separations are not uniform. We can also see that the separation of the horizontal line (oblique lines) in some part of the snapshots in the upper (lower) row becomes slightly narrow. These lattice deformations are due to the imposed uniaxial strains. In addition, combining the observed fact that the lattices are considerably fluctuated at $T=0.3$ with another fact that the skx phase is stable observed from the variation of $|N_{sk}|$ vs. MCS in Fig. 10.36(a), we confirm that the

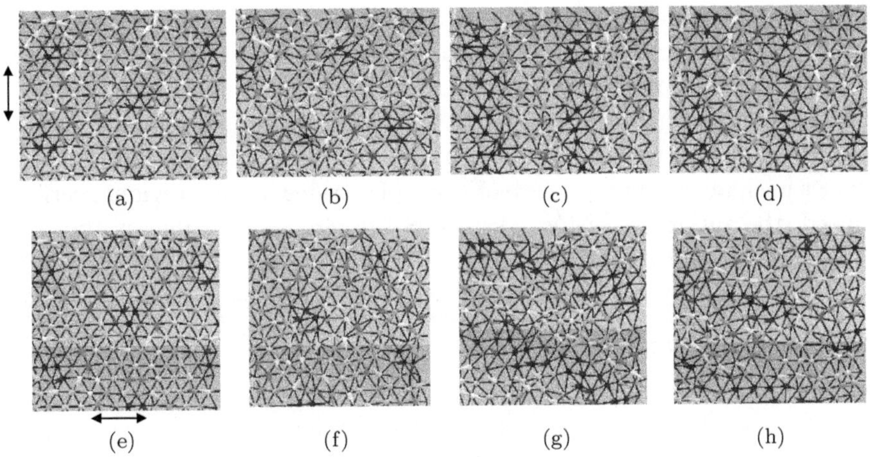

Fig. 10.39. Snapshots of lattices with a small symbol (cone) for σ corresponding to the top-right corner of those of spin configurations in Figs. 10.38(a)–10.38(h). An effective stress is applied to the directions (↕) for (a),(b),(c),(d) and (↔) for (e),(f),(g),(h).

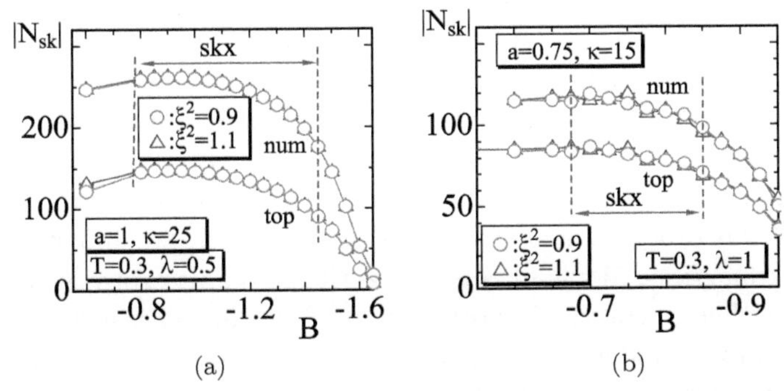

Fig. 10.40. The total number of skyrmions $|N_{\rm sk}|$ vs. B under uniaxial strain conditions $\xi^2 = 0.9$ and $\xi^2 = 1.1$ with (a) $a = 1$, $\kappa = 25$, $T = 0.3$, $\lambda = 0.5$, and (b) $a = 0.75$, $\kappa = 15$, $T = 0.3$, $\lambda = 1$. These parameters correspond to those assumed in Fig. 10.31 and Fig. 10.32, respectively. The data plotted in the upper (lower) part of the figures are obtained by the "num" ("top") techniques.

skx phase is stable on fluctuating surfaces. Indeed, we can see from the snapshots in Figs. 10.39(b) and 10.39(f) in the skx phase that the lattice structure is considerably fluctuated as described above.

To see the influence of uniaxial strains on the total number $|N_{\rm sk}|$ of skyrmions, we plot $|N_{\rm sk}|$ in Figs. 10.40(a) and 10.40(b) under the conditions corresponding to those of Figs. 10.33(c) and 10.33(d). It is almost clear that there is no influence of uniaxial strains on $|N_{\rm sk}|$, because the results remain unchanged from those of the isotropic case in Figs. 10.33(c) and 10.33(d). Other physical quantities such as τ, S_0, and $S_{\rm DM}$ also remain almost unchanged. We should also note that the convergence speed of simulations is almost unchanged from the case of isotropic strains in the previous section. Indeed, the variations of $|N_{\rm sk}|$ vs. MCS are almost the same as those in Figs. 10.36(a) and 10.36(b).

10.5.4. *Conclusions*

We study a 2D skyrmion system on fluctuating surfaces with periodic boundary conditions using a Monte Carlo simulation technique. In this model, not only spins but also lattice vertices are integrated into the partition function of the model. From the obtained results we conclude that skyrmions are stable even on fluctuating surfaces, contrary to initial expectations.

The DM interaction is assumed together with the ferromagnetic interaction and Zeeman energy under an external magnetic field. We assume that the lattice fluctuations are only reflected in the DM interaction, and the magnetoelastic coupling in the ferromagnetic interaction is not taken into consideration. Note that in our model the competition of S_F and $S_{\rm DM}$ suffices to create a skyrmion crystal. To keep the model as simple as possible in order to analyze the elasticity effect, we did not include long-range magnetic dipole–dipole interaction and anisotropy in the model. These ingredients will yield another kind of skyrmions which give rise to another behavior under elasticity and uniaxial strains. In addition to those skyrmion Hamiltonians, the Helfrich–Polyakov Hamiltonian for membranes is assumed. A surface governed by these Hamiltonians spans a rectangular frame of a fixed projected area, and the surface (or frame) tension is calculated. In this lattice calculation, the effect of stress is introduced by strains, which are imposed on the lattices by fixing the edge lengths of the boundary constant with periodic boundary conditions. In this sense, this effective stress is automatically introduced by the boundary frame. No mechanical stress can be imposed on the lattice as an input, but it is obtained as an output of the simulations.

We obtain numerical results that are consistent with the experimentally observed fact that a lateral pressure imposed on a 3D material creates skyrmions under a longitudinal magnetic field. Our result indicates that skyrmions are stable on 2D fluctuating surfaces, where there is no crystalline structure in contrast to the rigid lattices including real crystalline materials.

We further examine how uniaxial strains influence spin configurations. To impose a uniaxial strain on the lattice, we assume an anisotropic condition for the edge length of the rectangular boundary. As a result of this calculation, we find that the effective pressure or stress increment makes the stripe domain align to the stress direction under the same magnetic field and temperature as in the case without the uniaxial strain. As mentioned in the preceding section, the responses of stripe domains under uniaxial strains depend on the physical mechanism which generate skyrmions. Our model is not consistent with the reported experimental results of Ref. [131], but the fact that our stripes are parallel to uniaxial strains may be found in other materials and may have interesting transport applications. Of course, a more general magnetoelastic coupling included not only in $S_{\rm DM}$ but also in S_F should be considered in our future work.[126,134] In contrast, almost no difference is observed in the skyrmion phase, and this also confirms the stability of the skyrmion phase on fluctuating surfaces.

We should like to comment on the studies of skyrmion stability on fluctuating lattices from the view point of Thiele equation.[135] In Ref. [136], an interaction between skyrmion pairs on two separated surfaces was studied by Thiele equation. The authors in Ref. [136] assumed that the bilayer lattice is flat, and hence, effects of surface fluctuations were not taken into consideration. Therefore, to clarify the influence of the surface fluctuations on the skyrmion stability, it is interesting to study their model on a bilayer which consists of fluctuating and triangulated lattices. One possible technique is simply to assume that the fluctuating lattices are frozen and only spin variables are time dependent.

Finally, we also comment on a possibility of experimental studies. Skyrmions can only be observed on the materials such as FeGe, MnSi, or Cu_2OSeO_3 as mentioned in the Introduction, and we have no possibility at present to observe and study skyrmions on elastic surfaces such as ferroelectric polymers under the condition similar to those described in this paper. However, it seems possible to measure the effects of local stresses even on the above-mentioned materials such as FeGe by using the technique developed for individual manipulation of superconducting magnetic vortex.[137] It is interesting to study the stability of skyrmions under local stress or strains. This may be the first step toward the understanding of the role of surface fluctuations in the skyrmion stability.

Acknowledgments

The authors acknowledge S. El Hog and F. Kato for discussions and the data analyses.

References

1. See for example H. T. Diep, *Theory of Magnetism — Application to Surface Physics* (World Scientific, Singapore, 2014).
2. J. A. C. Bland and B. Heinrich (eds.), *Ultrathin Magnetic Structures*, Vol. I and II (Springer-Verlag, 1994).
3. A. Zangwill, *Physics at Surfaces* (Cambridge University Press, 1988).
4. A. Fert, V. Cros, and J. Sampaio, *Nat. Nanotechnol.* **8**, 152 (2013).
5. A. Yoshimori, *J. Phys. Soc. Jpn* **14**, 807 (1959).
6. J. Villain, *Phys. Chem. Solids* **11**, 303 (1959).
7. I. Harada and K. Motizuki, *J. Phys. Soc. Jpn* **32**, 927 (1972).
8. E. Rastelli, L. Reatto, and A. Tassi, Quantum fluctuations in helimagnets, *J. Phys. C* **18**, 353 (1985).
9. H. T. Diep, Low-temperature properties of quantum Heisenberg helimagnets, *Phys. Rev. B* **40**, 741 (1989).
10. R. Quartu and H. T. Diep, Phase diagram of body-centered tetragonal Helimagnets, *J. Magn. Magn. Mater.* **182**, 38 (1998).
11. S. M. Stishov, A. E. Petrova, S. Khasanov, G. Kh. Panova, A. A. Shikov, J. C. Lashley, D. Wu, and T. A. Lograsso, Magnetic phase transition in the itinerant helimagnet MnSi: Thermodynamic and transport properties, *Phys. Rev. B* **76**, 052405 (2007).
12. H. T. Diep, Magnetic transitions in helimagnets, *Phys. Rev. B* **39**, 397 (1989).
13. V. Thanh Ngo and H. T. Diep, Stacked triangular XY antiferromagnets: End of a controversial issue on the phase transition, *J. Appl. Phys.* **103**, 07C712 (2007).
14. V. Thanh Ngo and H. T. Diep, Phase transition in Heisenberg stacked triangular antiferromagnets: End of a controversy, *Phys. Rev. E* **78**, 031119 (2008).
15. V. D. Mello, C. V. Chianca, Ana L. Danta, and A. S. Carriç, Magnetic surface phase of thin helimagnetic films, *Phys. Rev. B* **67**, 012401 (2003).
16. F. Cinti, A. Cuccoli, and A. Rettori, Exotic magnetic structures in ultrathin helimagnetic holmium films, *Phys. Rev. B* **78**, 020402(R) (2008).
17. L. J. Rodrigues, V. D. Mello, D. H. A. L. Anselmo, and M. S. Vasconcelos, *J. Mag. Mag. Mater.* **377**, 24 (2014).

18. E. A. Karhu, S. Kahwaji, M. D. Robertson, H. Fritzsche, B. J. Kirby, C. F. Majkrzak, and T. L. Monchesky, Helical magnetic order in MnSi thin films, *Phys. Rev. B* **84**, 060404(R) (2011).
19. E. A. Karhu, U. K. Rößler, A. N. Bogdanov, S. Kahwaji, B. J. Kirby, H. Fritzsche, M. D. Robertson, C. F. Majkrzak, and T. L. Monchesky, Chiral modulation and reorientation effects in MnSi thin films, *Phys. Rev. B* **85**, 094429 (2012).
20. J. Heurich, J. König, and A. H. MacDonald, *Phys. Rev. B* **68**, 064406 (2003).
21. O. Wessely, B. Skubic, and L. Nordstrom, *Phys. Rev. B* **79**, 104433 (2009).
22. F. Jonietz, S. Mühlbauer, C. Pfleiderer, A. Neubauer, W. Munzer, A. Bauer, T. Adams, R. Georgii, P. Böni, R. A. Duine, K. Everschor, M. Garst, and A. Risch, *Science* **330**, 1648 (2010).
23. D. N. Zubarev, *Usp. Fiz. Nauk* **187**, 71 (1960) [translation: Soviet Phys.-Uspekhi **3**, 320 (1960)].
24. Diep-The-Hung, J. C. S. Levy and O. Nagai, Effect of surface spin-waves and surface anisotropy in magnetic thin films at finite temperatures, *Phys. Stat. Solidi (b)* **93**, 351 (1979).
25. V. Thanh Ngo and H. T. Diep, Effects of frustrated surface in Heisenberg thin films, *Phys. Rev. B* **75**, 035412 (2007), Selected for the Vir. *J. Nan. Sci. Tech.* **15**, 126 (2007).
26. V. Thanh Ngo and H. T. Diep, Frustration effects in antiferromagnetic face-centered cubic Heisenberg films, *J. Phys: Condens. Matter.* **19**, 386202 (2007).
27. E. Meloche, C. M. Pinciuc, and M. L. Plumer, *Phys. Rev. B* **74**, 094424 (2006); E. Meloche, M. L. Plumer and C. M. Pinciuc, *Phys. Rev. B* **76**, 214402 (2007).
28. P. Bak and M. H. Jensen, *J. Phys. C* **13**, L881 (1980).
29. M. L. Plumer and M. B. Walker, *J. Phys. C* **14**, 4689 (1981).
30. S. V. Maleyev, Cubic magnets with Dzyaloshinskii–Moriya interaction at low temperature, *Phys. Rev. B* **73**, 174402 (2006).
31. V. Leiner, D. Labergerie, R. Siebrecht, Ch. Sutter, and H. Zabel, *Physica B* **283**, 167 (2000).
32. N. D. Mermin and H. Wagner, *Phys. Rev. Lett.* **17**, 1133 (1966).
33. H. T. Diep, Quantum Theory of Helimagnetic Thin Films, *Phys. Rev. B* **91**, 014436 (2015).
34. S. V. Tyablikov, *Ukr. Mat. Zh.* **11**, 289 (1959); S. V. Tyablikov, *Methods in the Quantum Theory of Magnetism* (Plenum Press, New York, 1967).
35. H. T. Diep, Quantum effects in antiferromagnetic thin films, *Phys. Rev. B* **43**, 8509 (1991).
36. H. T. Diep, Theory of antiferromagnetic superlattices at finite temperatures, *Phys. Rev. B* **40**, 4818 (1989).
37. S. El Hog and H. T. Diep, Helimagnetic thin films: Surface reconstruction, surface spin-waves, and magnetization, *J. Mag. Mag. Mater.* **400**, 276–281 (2016).

38. S. El Hog and H. T. Diep, Partial phase transition and quantum effects in helimagnetic films under an applied magnetic field, *J. Mag. Mag. Mater.* **429**, 102–109 (2017).
39. Pham Phu, X. T.; Ngo, V. Thanh; Diep, H. T. Critical behavior of magnetic thin films. *Surf. Sci.* **603**, pp 109–116 (2009).
40. I. E. Dzyaloshinskii, Thermodynamical theory of "Weak" ferromagnetism in antiferromagnetic substances, *Sov. Phys. JETP* **5**, 1259 (1957).
41. T. Moriya, Anisotropic superexchange interaction and weak ferromagnetism, *Phys. Rev.* **120**, 91 (1960).
42. I. A. Sergienko and E. Dagotto, Role of the Dzyaloshinskii–Moriya interaction in multiferroic perovskites, *Phys. Rev. B* **73**, 094434 (2006).
43. Claude Ederer. and Nicola A. Spaldin, Weak ferromagnetism and magnetoelectric coupling in bismuth ferrite, *Phys. Rev. B* **71**, 060401(R) (2005).
44. Shi-Zeng Lin, Avadh Saxena, and Cristian D. Batista, *Phys. Rev. B* **91**, 224407 (2015).
45. A. N. Bogdanov and D. A. Yablonskii, Thermodynamically stable vortics in magnetically ordered crystals: The mixed state of magnets, *Sov. Phys. JETP* **68**, 101 (1989).
46. U. K. Rößler, A. N. Bogdanov, and C. Pfleiderer, Spontaneous skyrmion ground states in magnetic metals, *Nature (London)* **442**, 797 (2006).
47. S. Mühlbauer, B. Binz, F. Jonietz, C. Pfleiderer, A. Rosch, A. Neubauer, R. Georgii, and P. Böni, *Science* **323**, 915 (2009).
48. X. Z. Yu, Y. Onose, N. Kanazawa, J. H. Park, J. H. Han, Y. Matsui, N. Nagaosa, and Y. Tokura, Real-space observation of a two-dimensional skyrmion crystal, *Nature (London)* **465**, 901 (2010).
49. X. Z. Yu, N. Kanazawa, Y. Onose, K. Kimoto, W. Z. Zhang, S. Ishiwata, Y. Matsui, and Y. Tokura, Near room-temperature formation of a skyrmion crystal in thin-films of the helimagnet FeGe, *Nat. Mater.* **10**, 106 (2011).
50. S. Seki, X. Z. Yu, S. Ishiwata, and Y. Tokura, Observation of skyrmions in a multiferroic material, *Science* **336**, 198 (2012).
51. T. Adams, A. Chacon, M. Wagner, A. Bauer, G. Brandl, B. Pedersen, H. Berger, P. Lemmens, and C. Pfleiderer, *Phys. Rev. Lett.* **108**, 237204 (2012).
52. M. Heide, G. Bihlmayer, and S. Blügel, Dzyaloshinskii–Moriya interaction accounting for the orientation of magnetic domains in ultrathin films: Fe/W(110), *Phys. Rev. B* **78**, 140403(R) (2008).
53. S. Rohart and A. Thiaville, Skyrmion confinement in ultrathin film nanostructures in the presence of Dzyaloshinskii–Moriya interaction, *Phys. Rev. B* **88**, 184422 (2013).
54. H. Puszkarski and P. E. Wigen, Effect of Dzialoshinsky-Moriya Interactions on Propagation of Spin Waves in Ferromagnets: Dynamical Canting, *Phys. Rev. Lett.* **35**, 1017 (1975).
55. Kh. Zakeri, Y. Zhang, J. Prokop, T.-H. Chuang, N. Sakr, W. X. Tang, and J. Kirschner, Asymmetric spin-wave dispersion on Fe(110): Direct evidence of the Dzyaloshinskii–Moriya interaction, *Phys. Rev. Lett.* **104**, 137203 (2010).

56. W. Wang, M. Albert, M. Beg, M.-A. Bisotti, D. Chernyshenko, D. Cortés-Ortuño, I. Hawke, and H. Fangohr, Magnon-driven domain-wall motion with the Dzyaloshinskii–Moriya Interaction, *Phys. Rev. Lett.* **114**, 087203 (2015).
57. A. A. Stashkevich, M. Belmeguenai, Y. Roussigné, S. M. Cherif, M. Kostylev, M. Gabor, D. Lacour, C. Tiusan, and M. Hehn, Experimental study of spin-wave dispersion in Py/Pt film structures in the presence of an interface Dzyaloshinskii–Moriya interaction, *Phys. Rev. B* **91**, 214409 (2015).
58. Jung-Hwan Moon, Soo-Man Seo, Kyung-Jin Lee, Kyoung-Whan Kim, Jisu Ryu, Hyun-Woo Lee, R. D. McMichael, and M. D. Stiles, Spin-wave propagation in the presence of interfacial Dzyaloshinskii–Moriya interaction, *Phys. Rev. B* **88**, 184404 (2013).
59. Sahbi El Hog, H. T. Diep, and H. Puszkarski, Theory of magnons in spin systems with Dzyaloshinskii–Moriya interaction, *J. Phys.: Condens. Matter* **29**, 305001 (2017). arXiv:1612.04147, hal-01415320 (Dec. 2016).
60. J. W. Lynn, Magnetic neutron scattering, *J. Appl. Phys.* **75**, 6806 (1994). See also the basics of neutron scattering and the flexibility of BT-7 triple-axis spectrometer in the SPINS Experiment Handout given at http://www.ncnr.nist.gov/instruments/bt7_-new/index.html.
61. M. Plihal, D. L. Mills, and J. Kirschner, Spin wave signature in the spin polarized electron energy loss spectrum of ultrathin Fe films: Theory and experiment, *Phys. Rev. Lett.* **82**, 2579 (1999).
62. T. H. R. Skyrme, A unified field theory of mesons and baryons, *Proc. Roy. Soc. A* **260** 127 (1961); *Nucl. Phys.* **31**, 556 (1962).
63. A. N. Bogdanov, U. K. Rößler, and A. A. Shestakov, Skyrmions in liquid crystals, *Phys. Rev. E* **67**, 016602 (2003).
64. A. O. Leonov, I. E. Dragumov, U. K. Rößler, and A. N. Bogdanov, Theory of skyrmion states in liquid crystals, *Phys. Rev. E* **90**, 042502 (2014).
65. H. D. Rosales, D. C. Cabra, and P. Pujol, Three-sublattice Skyrmions crystal in the antiferromagnetic triangular lattice, *Phys. Rev. B* **92**, 214439 (2015); arXiv:1507.05109v1 (2015).
66. S. Buhrandt and L. Fritz, Skyrmion lattice phase in three-dimensional chiral magnets from Monte Carlo simulations, *Phys. Rev. B* **88**, 195137 (2013).
67. H. Y. Kwon, S. P. Kang, Y. Z. Wu, and C. Won, Magnetic generated by Dzyaloshinskii–Moriya interaction, *J. Appl. Phys.* **113**, 133911 (2013).
68. H. Y. Kwon, K. M. Bu, Y. Z. Wu, and C. Won, Effect of anisotropy and dipole interaction on long-range order magnetic structures generated by Dzyaloshinskii–Moriya interaction, *J. Mag. Mag. Mater.* **324**, 2171 (2012).
69. P. J. Ackerman, R. P. Trivedi, B. Senyuk, J. V. D. Lagemaat, and I. I. Smalyukh, *Phys. Rev. E* **90**, 012505 (2014).
70. M. Ezawa, Giant Skyrmions Stabilized by Dipole–Dipole interactions in thin ferromagnetic films, *Phys. Rev. Lett.* **105**, 197202 (2010).
71. Wanjun Jiang, Xichao Zhang, Guoqiang Yu, Wei Zhang, Xiao Wang, M. Benjamin Jungfleisch, John E. Pearson, Xuemei Cheng, Olle Heinonen, Kang L. Wang, Yan Zhou, Axel Hoffmann, and Suzanne G. E. te Velthuis, Direct observation of the skyrmion Hall effect, *Nat. Phys.* **13**, 162–169 (2017).

72. D. A. Gilbert, B. B. Maranville, A. L. Balk, B. J. Kirby, P. Fischer, D. T. Piercen J. Unguris, J. A. Borchers, and K. Liu, Realization of ground-state artificial skyrmion lattices at room temperature, *Nat. Comm.*, published 8 Oct. 2015, DOI:10.1038/ncomms9462.
73. A. Bauer and C. Pfleiderer, Magnetic phase diagram of MnSi inferred from magnetization and ac susceptibility, *Phys. Rev. B* **85**, 214418 (2012).
74. S. Hayami, S.-Z. Lin, and C. D. Batista, Bubble and skyrmion crystals in frustrated magnets with easy-axis anisotropy, *Phys. Rev. B* **93**, 184413 (2016).
75. T. Okubo, S. Chung, and H. Kawamura, Multiple-states and the skyrmion lattice of the triangular-lattice heisenberg antiferromagnet under magnetic fields, *Phys. Rev. Lett.* **108**, 017206 (2012).
76. W. Kang, Y. Huang, X. Zhang, Y. Zhou, and W. Zhao, Skyrmion-electronics: An overview and outlook, *Proceedings of the IEEE*, August 2016, DOI: 10.1109/JPROC.2016.2591578.
77. S. S. P. Parkin, M. Hayashi, and L. Thomas, Magnetic domain-wall racetrack memory, *Science* **320**(5873), 190 (2008).
78. Xichao Zhang, Motohiko Ezawa, and Yan Zhou, Magnetic skyrmion logic gates: conversion, duplication and merging of skyrmions, *Scientific Reports* **5**, 9400 (2015).
79. Y. Zhou and M. Ezawa, A reversible conversion between a skyrmion and a domain-wall pair in junction geometry, *Nat. Commun.* **5**, Art. no. 4652 (2014).
80. X. Zhang, Y. Zhou, M. Ezawa, G. P. Zhao, and W. Zhao, Magnetic skyrmion transistor: skyrmion motion in a voltage-gated nanotrack, *Sci. Rep.* **5**, Art. no. 11369 (2015).
81. J. Kim *et al.*, Voltage controlled propagating spin waves on a perpendicularly magnetized nanowire, arXiv: 1401.6910 (2014).
82. Y. Shiota *et al.*, Quantitative evaluation of voltage-induced magnetic anisotropy change by magnetoresistance measurement, *Appl. Phys. Expr.* **4**(4), Art. no. 43005 (2011).
83. Yangqi Huang, Wang Kang, Xichao Zhang, Yan Zhou, and Weisheng Zhao, Magnetic skyrmion-based synaptic devices, *Nanotechnology* **28**, 08LT02 (2017).
84. Sai Li, Wang Kang, Yangqi Huang, Xichao Zhang, Yan Zhou, and Weisheng Zhao, Magnetic skyrmion-based artificial neuron device, *Nanotechnology* **28**, 31LT01 (2017).
85. Sahbi El Hog, Aurélien Bailly-Reyre, H.T. Diep, Stability and phase transition of skyrmion crystals generated by Dzyaloshinskii–Moriya interaction, *J. Magnetism and Magnetic Materials* **455**, 32–38 (2018).
86. X. Z. Yu, M. Mostovoy, Y. Tokunaga, W. Z. Zhang, K. Kimoto, Y. Matsui, Y. Kaneko, N. Nagaosa, and Y. Tokura, Magnetic stripes and skyrmions with helicity reversals, *PNAS* **109**, 8856 (2012).

87. M. Mézard, M. Parisi, and M. Virasoro, *Spin Glass Theory and Beyond An Introduction to the Replica Method and Its Applications* (World Scientific, Singapore, 1986).
88. J. C. Phillips, *Rep. Prog. Phys.* **96**, 1193 (1996) and references therein.
89. A. T. Ogielski, *Phys. Rev. B* **32**, 7384 (1985).
90. C. de Dominicis, H. Orland, and F. Lainée, *J. de Physique Lett.* **46**, 463 (1985).
91. R. M. C. de Almeida, N. Lemke, and I. A. Campbell, Brazilian *J. Phys.* **30**, 701 (2000); *ibid. J. Mag. Mag. Mater.* **226–230**, 1296 (2001).
92. A. Billoire and I. A. Campbell, *Phys. Rev. B* **84**, 054442 (2011).
93. I. S. Suzuki and M. Suzuki, *Phys. Rev. B* **78**, 214404 (2008).
94. A. Malinowski, V. L. Bezusyy, R. Minikayev, P. Dziawa, Y. Syryanyy, and M. Sawicki, *Phys. Rev. B* **84**, 024409 (2011).
95. V. Thanh Ngo, D. Tien Hoang, H. T. Diep, and I. A. Campbell, Effect of disorder in the frustrated ising FCC antiferromagnet: Phase diagram and stretched exponential relaxation, *Mod. Phys. Lett. B* **28**, 1450067 (2014).
96. J. M. Kosterlitz and D. J. Thouless, Long range order and metastability in two dimensional solids and superfluids, *J. Phys. C: Solid State Phys.* **5**, L124-126 (1972); Ordering,metastability and phase transitions in two-dimensional systems, *J. Phys. C: Solid State Phys.* **6**, 1181–203 (1973).
97. S. El Hog, F. Kato, H. Koibuchi, and H.T. Diep, *J. Mag. Mag. Mat.* (submitted); arXiv:1910.01760.
98. W. Helfrich, *Z. Naturforsch* **28c**, 693 (1973).
99. A. M. Polyakov, *Nucl. Phys. B* **268**, 406 (1986).
100. K. J. Wiese, Polymerized membranes, a review, in C. Domb and J. L. Lebowitz (eds.), *Phase Transitions and Critical Phenomena 19* (Academic Press, 2000), pp. 253–498.
101. D. Nelson, The statistical mechanics of membranes and interfaces, in D. Nelson, T. Piran, and S. Weinberg, (eds.), *Statistical Mechanics of Membranes and Surfaces*, 2nd edn. (World Scientific, 2004), pp. 1–17.
102. G. Gompper and D. M. Kroll, Triangulated-surface models of fluctuating membranes, in D. Nelson, T. Piran, and S. Weinberg, (eds.), *Statistical Mechanics of Membranes and Surfaces*, 2nd ed. (World Scientific, 2004), pp. 359–426.
103. Y. Kantor and D. R. Nelson, *Phys. Rev. A* **36**, 4020 (1987).
104. K. Essafi, J. P. Kownacki, and D. Mouhanna, *Phys. Rev. E* **89**, 042101(1-5) (2014).
105. H. Noguchi, *J. Phys. Soc. Jpn.* **78**, 041007 (2009).
106. A. N. Bogdanov, U. K. Rössler, and C. Pfleiderer, Modulated and localized structures in cubic helimagnets, *Physica B* **359**, 1162–1164 (2005).
107. W. Munzer, A. Neubauer, T. Adams, S. Muhlbauer, C. Franz, F. Jonietz, R. Georgii, P. Boni, B. Pedersen, M. Schmidt, A. Rosch, and C. Pfleiderer, *Phys. Rev. B* **81**, 041203(R) (2010).

108. S. Dong, S. Yunoki, X. Zhang, C. Sen, J.-M. Liu, and E. Dagotto, *Phys. Rev. B* **82**, 035118 (2010).
109. J. Iwasaki, M. Mochizuki, and N. Nagaosa, *Nature Comm.* **4**, 1463 (2013).
110. A. N. Bogdanov and D. A. Yablonskii, *Zh. Eksp. Teor. Fiz.* **95**, 178–182 (1989).
111. J. Fukuda and S. Žumer, *Nature Comm.* **2**, 246 (2011).
112. S. Banerjee, J. Rowland, O. Erten, and M. Randeria, *Phys. Rev. X* **4**, 031045 (2014).
113. U. Güngördü, R. Nepal, O. A. Tretiakov, K. Belashchenko, and A. A. Kovalev, *Phys. Rev. B* **93**, 064428 (2016).
114. T. Nakajima, S. Mitsuda, T. Nakamura, H. Ishii, T. Haku, Y. Honma, M. Kosaka, N. Aso, and Y. Uwatoko, *Phys. Rev. B* **83**, 220101(R) (2011).
115. R. Ritz, M. Halder, C. Franz, A. Bauer, M. Wagner, R. Bamler, A. Rosch, and C. Pfleiderer, *Phys. Rev. B* **87**, 134424 (2013).
116. Y. Shi and J. Wang, *Phys.Rev. B* **97**, 224428 (2018).
117. Y. Nii, A. Kikkawa, Y. Taguchi, Y. Tokura, and Y. Iwasa, *Phys. Rev. Lett.* **113**, 267203 (2014).
118. Y. Nii, T. Nakajima, A. Kikkawa1, Y. Yamasaki, K. Ohishi, J. Suzuki, Y. Taguchi, T. Arima, Y. Tokura, and Y. Iwasa, *Nat. Comm.* **6**, 8539 (2015).
119. S. Seki, Y. Okamura, K. Shibata, R. Takagi, N. D. Khanh, F. Kagawa, T. Arima, and Y. Tokura, *Phys. Rev. B* **96** 220404(R) (2017).
120. AI. Levatić, P. Popčević, V. Šurija, A. Kruchkov, H. Berger, A. Magrez, J. S. White, H. M. Ronnow, and I. Živković, *Sci. Rep.* **6**, 21347 (2016).
121. A. Chacon, A. Bauer, T. Adams, F. Rucker, G. Brandl, R. Georgii, M. Garst, and C. Pfleiderer, *Phys. Rev. Lett.* **115** 267202 (2015).
122. J. Chen, W. P. Cai, M. H. Qin, S. Dong, X. B. Lu, X. S. Gao, and J.-M. Liu, *Sci. Rep.* **7**, 7392 (2017).
123. M. Creutz, *Quarks, Gluons and Lattices* (Cambridge University Press, Cambridge, 1983).
124. Y. Takano and H. Koibuchi, *Phys. Rev. E* **95**, 042411(1–11) (2017).
125. D. Sander, *Rep. Prog. Phys.* **62**, 809 (1999).
126. K. Shibata, J. Iwasaki, N. Kanazawa, S. Aizawa, T. Tanigaki, M. Shirai, T. Nakajima, M. Kubota, M. Kawasaki, H. S. Park, D. Shindo, N. Nagaosa, and Y. Tokura, *Nat. Nanotech.* **10**, 589 (2015).
127. J. F. Wheater, *J. Phys. A Math. Gen.* **27**, 3323 (1994).
128. W. Cai, T. C. Lubensky, P. Nelson, and T. Powers, *J. Phys. II France* **4**, 931 (1994).
129. N. Metropolis, A. W. Rosenbluth, M. N. Rosenbluth, and A. H. Teller, *J. Chem. Phys.* **21**, 1087 (1953).
130. D. P. Landau, *Phys. Rev. B* **13**, 2997 (1976).
131. J. Dho, Y. N. Kim, Y. S. Hwang, J. C. Kim, and N. H. Hur, *Appl. Phys. Lett.* **82**, 1434–1436 (2003).
132. P.-J. Hsu, A. Finco, L. Schmidt, A. Kubetzka, K. von Bergmann, and R. Wiesendanger, *Phys. Rev. Lett.* **116**, 017201 (2016).

133. A. Finco, P.-J. Hsu, A. Kubetzka, K. von Bergmann, and R. Wiesendanger, *Phys. Rev. B* **94**, 214402 (2016).
134. J. Wang, Y. Shi, and M. Kamlah, *Phys. Rev. B.* **97**, 024429(1–7) (2018).
135. A. A. Thiele, *Phys. Rev. Lett.* **30**, 230–233 (1973).
136. W. Koshibae and N. Nagaosa, *Sci. Rep.* **7**, 42645(1–10) (2017).
137. A. Kremen, S. Wissberg, N. Haham, E. Persky, Y. Frenkel, and B. Kalisky, *Nano Lett.* **16**, 1626–1630 (2016).

Lightning Source UK Ltd.
Milton Keynes UK
UKHW022102210620
365284UK00004B/155

9 789811 214134